Ear, Nose and Throat Diseases of the Dog and Cat

Ear, Nose and Throat Diseases of the Dog and Cat

Richard G. Harvey BVSc DVD Dip ECVD FSB
The Veterinary Centre, Cheylesmore
Coventry, UK

Gert ter Haar BVSc DVM, PhD, MRCVS, DECVS
President of the International Veterinary Ear Nose and Throat Association
Senior Lecturer Soft Tissue Surgery, Head of the ENT, Brachycephaly and
Audiology Clinics
Royal Veterinary College
Hatfield, UK

CRC Press
Taylor & Francis Group
Boca Raton London New York

CRC Press is an imprint of the
Taylor & Francis Group, an **informa** business

CRC Press
Taylor & Francis Group
6000 Broken Sound Parkway NW, Suite 300
Boca Raton, FL 33487-2742

First issued in paperback 2018

© 2017 by Taylor & Francis Group, LLC
CRC Press is an imprint of Taylor & Francis Group, an Informa business

No claim to original U.S. Government works

ISBN-13: 978-1-4822-3649-1 (hbk)
ISBN-13: 978-0-367-13317-7 (pbk)

Visit the Taylor & Francis Web site at
http://www.taylorandfrancis.com

and the CRC Press Web site at
http://www.crcpress.com

CONTENTS

The authors are indebted to colleagues who have provided illustrations, in particular Dr David Duclos, DVM, who put his entire picture bank at our disposal. Professors Sue Dawson, BVMS PhD MRCVS and Lynette Cole, DVM, MS, DACVD, and Drs Louis Gotthelf DVM and Marge Scherk DVM, DABVP (Feline Practice) provided most valuable advice. Dr Richard Lam, DECVI, provided excellent quality diagnostic imaging figures. We also owe thanks to Dr. Alex Stoll for his microscopic artwork and Joop Fama for his fantastic intraoperative photographs. The authors would also like to thank Drs Harari and DeLauche for their contributions. Other colleagues provided clinical pictures for us and these are credited in situ. We are most grateful to all of these colleagues, who have all helped us to create this book.

It goes without saying that our families suffered as we searched databases, begged illustrations and wrote text. To them, a special thanks.

Richard Harvey
Gert Ter Haar

PREFACE

This book builds on its predecessor, which concentrated on ear disease. However, ear, nose and throat medicine is not only a speciality within human medicine, it is also the subject of text books. It seemed appropriate therefore to produce a canine and feline ENT text.

Ear disease in dogs has been reported to account for between 10.2 and 16.6% cases in small animal practice (O'Neill, Inoue, Świecicka). The reorts by O'Neill and Inoue reported respiratory disease at around 2.8 to 5.7%. Similar data is not available for cats. Consideration of these data suggest that clinicians in busy practices can expect to see an ENT case almost on a daily basis.

The breadth of this book encompasses both medicine and surgery of the ear, nose and throat. The book is written as a teaching aid for students and as a resource for practitioners.

The authors have attempted to make the text relevant to readers around the English speaking world and we hope that you find that it becomes an essential part of your day to day reference material.

O'Neill DG, Church DB, McGreevy PD, Thomson PC, Brodbelt. Prevalence of disorders recorded in dogs attending primary-care veterinary practices in England. *PLOS One* 2014, **9**: 1-16.

Inoue M, Hasegawa A, Hosoi Y, Sugiura K. Breed, gender and age pattern of diagnosis for veterinary care in insured dogs in Japan during fiscal year 2010. *Preventive Veterinary Medicine* 2015 **119**: 54-60.

Świecicka N, Bernacka H, Fac E, Zawiślak J. Prevalence and commonest causes for otitis externa in dogs from two Polish veterinary clinics. Bulagarian Journal of Veterinary Medicine 2015, **18: 65-73**

5-ALA	5-aminolaevulinic acid	FCV	feline calicivirus
ACTH	adrenocorticotrophic hormone	FDM	fluid-displacement method
AMP	adenosine monophosphate	FeLV	feline leukaemia virus
ANA	antinuclear antibody	FHV	feline herpesvirus
AR	acoustic rhinometry	FIV	feline immunodeficiency virus
ARHL	age-related hearing loss	FNAB	fine-needle aspiration biopsy
BAER	brainstem auditory evoked response	GCPS	Glasgow Composite Pain Scale
BAL	bronchoalveolar lavage	HN	head and neck
BERA	brainstem-evoked response audiometry	IDA	inner dynein arm
BOAS	brachycephalic obstructive airway syndrome	Ig	immunoglobulin
		IL-4	interleukin-4
CAV	canine adenovirus	IHC	inner hair cell
CIRD	canine infectious respiratory disease	KCS	keratoconjunctivitis
CITB	canine infectious tracheobronchitis	LBO	lateral bulla osteotomy
CKCS	Cavalier King Charles Spaniel	LECR	lateral ear canal resection
CM	cochlear microphonics	LPR	lymphoplasmacytic rhinitis
CN	cranial nerve	MIC	minimum inhibitory concentration
CNS	central nervous system	MGCS	modified Glasgow coma scale
COX	cyclo-oxygenase	MOE	main olfactory epithelium
CPG	central pattern generator	MPC	mutant prevention concentration
CRD	complex repetitive discharge	MRI	magnetic resonance imaging
CRI	continuous rate infusion	mRNA	messenger ribonucleic acid
C&S	culture and sensitivity testing	MRSA	methicillin-resistant *Staphylococcus aureus*
CSF	cerebrospinal fluid		
CT	computed tomography	MRSP	methicillin-resistant *Staphylococcus pseudintermedius*
DFI	disease-free interval		
DLE	discoid lupus erythematosus	MST	median survival time
DPOAE	distortion product otoacoustic emission	MSW	mutant selection window
DSS	dioctyl sodium sulfosuccinate	NIHL	noise-induced hearing loss
ECG	electrocardiogram	NO	nitric oxide
EDTA	ethylenediaminotetraacetic acid	NSAID	non-steroidal anti-inflammatory drug
EFA	essential fatty acid	OAE	otoacoustic emission
ELISA	enzyme linked immunosorbent assay	ODA	outer dynein arm
EMG	electromyography	OHC	outer hair cell
ENT	ear, nose and throat	OME	otitis media with effusion
EOAE	evoked otoacoustic emission	OMI	otitis media and interna
ER	epiglottic retroversion	PCD	primary ciliary dyskinesia
ERO	evoked response olfactometry	PCMX	parachlorometaxylenol

PCR	polymerase chain reaction	SVCA	stria vascularis cross-sectional area
PE	pemphigus erythematosus	TECA	total ear canal ablation
PF	pemphigus foliaceus	TECA–LBO	total ear canal ablation with lateral bulla osteotomy
PSOM	primary secretory otitis media		
RAR	rapidly adapting receptor	TEOAE	transient evoked otoacoustic emission
SAR	slowly adapting receptor	Th	T helper cell
SCC	squamous cell carcinoma	Tris	tromethamine
SGC	spiral ganglion cell	UOS	upper oesophageal sphincter
SNHL	sensorineural hearing loss	URT	upper respiratory tract
SO	septal organ	VBO	ventral bulla osteotomy
SOAE	spontaneous otoacoustic emission	VECA	vertical ear canal ablation
SPF	specific pathogen-free	VNO	vomeronasal organ
SPL	sound pressure level	VOC	volatile organic compound
SSD	silver sulfadiazine	VSB	Vibrant Soundbridge

PART 1

THE EAR

CHAPTER 1

ANATOMY AND PHYSIOLOGY OF THE EAR

1.1 INTRODUCTION

A knowledge of the normal anatomy, physiology and the resident flora of the ear is an essential prerequisite to recognising disease and instituting treatment or surgery. For many years students and clinicians have had to rely on successive anatomical studies and bacteriological culture techniques to provide this information. However, recent advances in imaging technology have allowed visualisation of the components of the middle and inner ear without the need for dissection. Furthermore, it is possible that our understanding of the otic microflora will be transformed by the increased availability and use of 16s messenger ribonucleic acid (mRNA) gene assays.

1.2 GROSS AND MICROSCOPIC ANATOMY OF THE EAR

The ear of the dog and cat is composed of three parts, reviewed by Cole[1]: the external ear, the middle ear and the inner ear[2–4]. Together, these components allow the animal to:

- Locate a sound and the direction from which it emanates.
- Orientate the head in relation to gravity.
- Measure acceleration and rotation of the head.

Selective breeding, of dogs in particular, has resulted in a wide variation in relative size and shape of the components of the external ear. Consider, for example, the French Bulldog, the Cocker Spaniel, the German Shepherd Dog, the St Bernard and the Persian cat. The pinnal shape and carriage, the diameter of the external ear canal, the degree of hair and amount of soft tissue within the external ear canal and the shape of the skull within which the middle and inner ear lie, vary from one breed or species, to another.

Despite this anatomic variation, the essential relationship between the various components of the external, middle and inner ear is preserved.

Pinna

The evolutionary role of the pinna has been as an aid to sound collection and point-of-origin location (Figs 1.1A, B). However, selective breeding of dogs has resulted in pinnae that often appear to have been designed more as lids to prevent access by foreign bodies or as vehicles to carry ornate displays of exuberant growths of hair. Despite these changes the functionality of the ear appears to have been maintained.

In most breeds of cats the pinna is held erect, with the exception of the Scottish Fold Cat where the distal portion of the scapha is folded rostroventrally[5].

The pinna is composed of a sheet of cartilage covered on both sides by skin, which is more firmly adherent on the concave aspect than on the convex aspect[3,4,6]. The cartilage sheet that supports the pinna is a flared extension of the auricular cartilage. Proximally this becomes rolled to form the vertical ear canal and part of the horizontal ear canal. The major part of the external auditory canal is contained within the auricular cartilage.

The portion of the flared auricular cartilage that forms the body of the pinna is called the scapha, although the free edges of the pinna are termed the rostral border of the helix and the caudal border of the helix, respectively[2,4]. The anthelix is the medial ridge with the prominent tubercle that is situated on the medial aspect of the entrance to the vertical ear canal (Fig. 1.2).

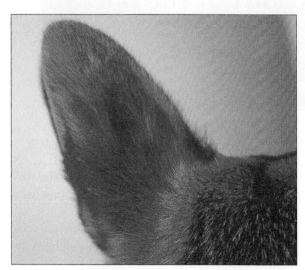

Fig. 1.1A Archetypal pinna, convex side, in this case of a German Shepherd Dog. Note the even distribution of short hairs on the convex aspect.

The anthelix is a good landmark to illustrate to owners where to apply topical otic medication. Thus, for treatment of an atopic ear, where inflammation is as much on the proximal concave aspects of the pinna as in the vertical ear canal, the owners are instructed to squeeze the bottle gently and apply, in a circular motion, on, and around, the anthelix for three seconds (counting, slowly, one, two, three).

Opposite the anthelix is an irregularly shaped, dense plate of cartilage, called the tragus (Fig. 1.3)[2,4]. This is extended caudally and medially to the antitragus and thus creates the caudal boundary of the opening into the external ear canal. The rostral border of the opening is demarcated by the medial and lateral crus of the helix[2].

The tragus is an essential surgical landmark in aural surgery.

The auricular cartilage becomes rolled proximal to the scapha and is termed the concha[3]. The scutiform

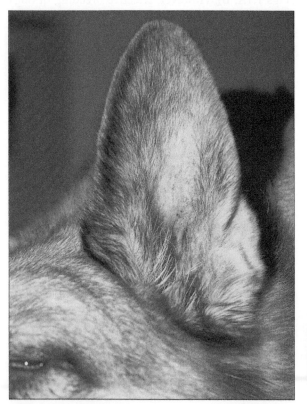

Fig. 1.1B Concave side of pinna of the animal depicted in 1.1A. There is a variable amount of hair around the margins. The glabrous central region is confluent with the epithelial lining of the external ear canal.

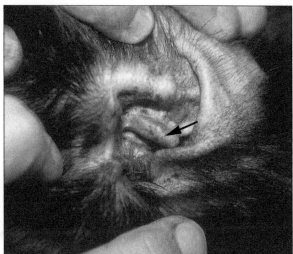

Fig. 1.2 The prominent tubercle of the anthelix (arrow) is clearly visible on the medial aspect of the entrance to the external ear canal. Note areas of overlap at the pinnal margins.

cartilage is rostromedial to the horizontal canal, closely associated with muscular tissue[3], and it forms no part of the external ear, although the associated pad of fat, the corpus adiposum auricula, may help to provide support to the horizontal portions of the external ear canal[2].

Generally the pinna is haired on the convex surface and in some breeds, such as the Cocker Spaniel and Papillon, for example, markedly so. The concave aspect may be lightly haired on the free edges and towards the tip, but towards the base it becomes essentially glabrous and is tightly adherent to the underlying cartilage. A few fine hairs are usually present around the entrance to the external ear canal. In breeds with hirsute ear canals, such as the Cocker Spaniel, there may be profuse hair growth along the entire length of the ear canal.

The blood supply to the pinna arises from the great auricular artery, a branch of the maxillary artery[7]. The great auricular artery ascends dorsally towards the pinna just deep to the caudomedial aspect of the vertical ear canal[6]. Avascular necrosis of the pinna may follow damage to this blood vessel during resection of the vertical ear canal. The great auricular artery divides at the base of the convex aspect of the pinna. Its branches ascend the convex aspect of the pinna (Fig. 1.4), wrapping around the helicine margins and penetrating the plate of the scapha to supply the concave surface[5]. The majority of the foramina through which the vessels pass to the concave aspect of the pinna are located about one-third of the way along the longitudinal aspect of the scapha[2]. The auricular veins drain via the internal maxillary vein into the jugular vein.

> When incising to drain an aural haematoma, the incision is made longitudinally along the centre of the convex aspect of the pinna to avoid these vessels.

The sensory and motor innervation of the pinna is extremely complex[8]. The convex surface of the pinna is innervated by:

- Branches of the second cervical nerve (Fig. 1.5A):
  - The dorsal cutaneous branch of the second cervical nerve;
  - The great auricular nerve (a branch of the ventral cutaneous branch of C2).
- Branches of the trigeminal nerves:
  - The third cervical nerve;
  - The mandibular branch of the trigeminal nerve (Fig. 1.5B).

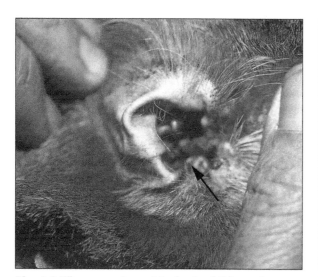

Fig. 1.3 The tragus (arrow) is clearly visible on the lateral aspect of the entrance to the vertical ear canal, in this case of a cat.

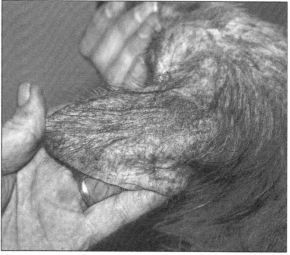

Fig. 1.4 The convex aspect of the pinna of an elderly dog with an endocrinopathy. Note the clearly visible blood vessels.

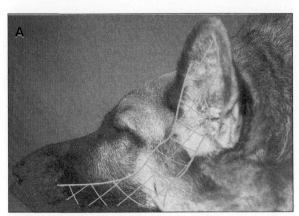
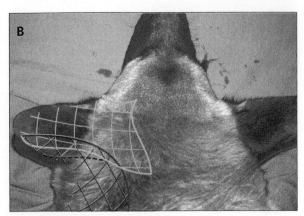

Figs 1.5A, B Lateral (A) and dorsal (B) views of a German Shepherd Dog marked up to illustrate the principal areas of innervation of the convex aspect of the pinna. Note that the central portion is innervated by branches of the cervical nerve and that overlap zones occur at the pinnal margins; A: area innervated by the mandibular branch of the trigeminal nerve; B: branches of the second cervical nerve innervate the majority of the dorsal aspect of the pinna; the greater occipital nerve innervated the bulk of the area, the greater auricular nerve innervates the caudal part.

The concave surface of the pinna is innervated by:

- Branches of the facial nerve:
  - Middle internal auricular branch of the facial nerve (Figs 1.6A–C);
  - Lateral internal auricular branch of the facial nerve;
  - Caudal internal auricular branch of the facial nerve (Fig. 1.6D).
- The second cervical nerve.

Movement of the pinna is provided by facial nerve innervation[3].

External ear canal (external auditory meatus)

The external auditory meatus serves to conduct sound waves to the tympanum. It is contained within the vertical and horizontal portions of the external ear canal. Proximally, the external auditory meatus abuts the tympanum and distally it is defined within, and bounded by, the medial faces of the various cartilaginous components at the base of the pinna. The vertical ear canal and the distal part of the horizontal canal are contained within the rolled plate (the concha) of the auricular cartilage. The free edges of the rolled auricular cartilage overlap on the medial aspect of the vertical canal, and the lumen contained within becomes progressively narrower proximally.

> This has great practical implications. When the epithelial lining becomes inflamed, oedematous and hyperplastic, it cannot expand outwards, only inwards, because of the confines of the auricular cartilage. The result is a progressive loss of luminal volume, with all the attendant problems of increased humidity and surface maceration.

The vertical canal deviates medially, just dorsal to the level of the tympanum, towards the external acoustic process. The annular cartilage is interposed between the proximal end of the auricular cartilage and the distal end of the external ear canal[2]. The average length of canal within the annular cartilage is 1.2 cm (0.8–1.9 cm)[9]. The ligamentous unions between these cartilaginous and bony tubes allow great freedom of motion to the pinna.

The physical parameters of the vertical canal (length and volume) correlate with body weight[9,10]. The average length of the external ear canal within the auricular cartilage is 4.1 cm (2.2–5.7 cm) and its average diameter, at the level of the tragus, is 5.8 cm (2.1–7.9 cm)[9].

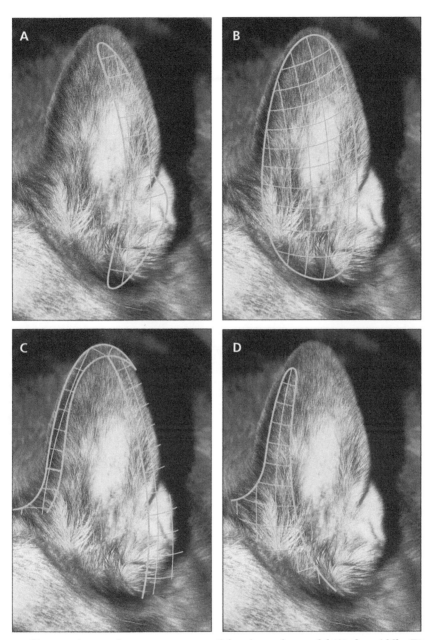

Figs 1.6A–D **Areas of innervation on the concave aspect of the pinna: the caudal (A) the middle (B), and the lateral (C) internal auricular branches of the facial nerve; D: the mandibular branch of the trigeminal nerve.**

The volume and the surface area of the external ear canal also correlate with body weight[11]. Thus, a 5.4 kg Shi-Tzu has an external ear canal with approximately 20 cm^2 area and volume 4.5 cm^3, whereas in a 20 kg dog the surface area is 38.5 cm^2 and the volume 9.6 cm^3.

This has implications for treatment. The volume of medication required to deliver the appropriate amount of drug adequately to cover the surface area of the external ear canal will be approximately twice as much in the 20 kg dog compared to that required for the 5 kg dog[11]. Manufacturers' data sheets rarely reflect this.

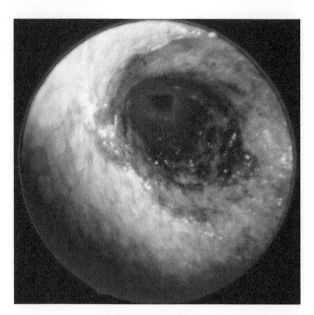

The epithelium and dermal tissues that line the bony and cartilaginous components of the external ear canal result in a smooth inner surface to the canal (Fig. 1.7). The epithelium is sparsely haired in most, but not all, breeds and it is rich in adnexal glands (see below, 1.3 Microscopic structure of the external ear canal).

The fine detail of the relationship between the external acoustic meatus and the surrounding bony tissues varies with breed[12]. Thus, in the American Pit Bull Terrier the external acoustic meatus sits deeper, and is better protected, than, for example, that of a German Shepherd cross (Figs 1.8A, B). It is also rather more difficult to examine otoscopically[11].

Fig. 1.7 Otoscopic picture of a normal external ear canal. Note the smooth epithelial lining and the occasional small clump of cerumen.

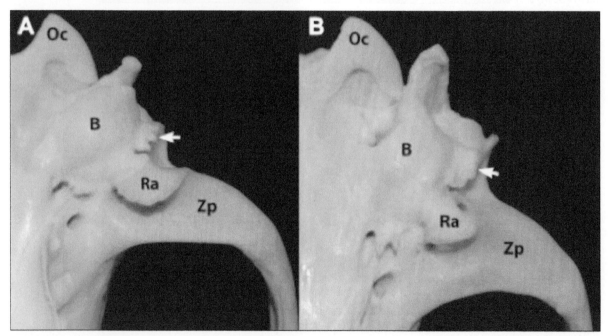

Figs 1.8A, B External ear, dog. A: Macerated skull, Pit Bull. The external acoustic meatus (arrow) opens into an area that is made shallow by a broad retroarticular process (Ra) and a longer zygomatic process of the temporal bone (Zp). The angle that is formed by the plateau of bone approximates 90°, or slightly less; B: macerated skull, shepherd cross dog. The external acoustic meatus (arrow) opens into an area that is more open with a small, shallower, retroacoustic process (Ra) and a shorter zygomatic process of the temporal bone (Zp). In this dog, the angle formed is more obtuse providing greater access to this ear by otoscopic examination. B, tympanic bulla; Oc occipital condyles. (Courtesy of Dr BL Njaa, Centre for Veterinary Health Sciences, Oklahoma State University, Stillwater, OK). (By permission, Saunders, Philadelphia. Njaa BL, Cole LK, Tabacca N. Practical otic anatomy and physiology of the dog and cat. *Veterinary Clinics of North America Small Animal Practice* 2012;42:1109–26.)

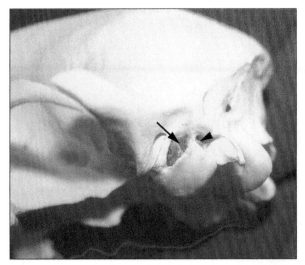

Fig. 1.9 **The caudolateral aspect of the canine skull. Note the close proximity of the stylomastoid foramen (arrowhead) and the external acoustic meatus (arrow).**

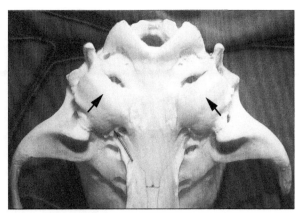

Fig. 1.10 **The caudoventral aspect of the canine skull. The paired bullae are clearly visible (arrows).**

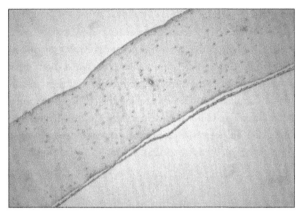

Fig. 1.11 **Photomicrograph of a section of normal bulla. Note the thin bone and the secretory epithelium. (Courtesy David Shearer.)**

Medial to the vertical portion of the external ear canal are branches of the auricular and superficial temporal arteries. The parotid salivary gland overlies the lateral and proximal portions of the vertical ear canal[7]. Deep to the parotid gland are the facial nerve, the internal maxillary vein and branches of the external carotid artery[7]. The facial nerve emerges from the skull (via the stylomastoid foramen) and passes beneath the rostroventral aspect of the horizontal canal (Fig. 1.9). Branches of the facial nerve and the auriculotemporal branch of the mandibular portion of the trigeminal nerve pass rostral to the vertical ear canal[7].

Middle ear

The middle ear and auditory (Eustachian) tube comprise a functional physiologic unit with protective, drainage and ventilatory capabilities[13,14]. The middle ear is composed of the tympanum, the ossicles, the auditory tube and the tympanic cavities (Fig. 1.10)[1,2,4]. The middle ear cavities are lined with secretory epithelium (Fig. 1.11). Epithelia such as this not only secrete liquid, they also absorb gas[13]. This tends to result in a slight negative pressure within the normal middle ear cavity[13]. The composition of the gas in the normal middle ear cavity of both dogs and cats has been described[15]. It appears to correlate closely to the com-

position of the capillary blood, rather than reflecting gaseous exchange along the auditory tube.

The three ossicles transmit sound waves impacting upon the tympanic membrane to the oval window. At this point the mechanical energy of the ossicles is transduced to pressure waves within the inner ear, subsequently to be interpreted as sound. Pressure and internal homeostasis within the inner ear is equilibrated across the round window membrane.

Tympanum

The gross appearance of the canine and feline tympanic membrane is similar (Fig. 1.12A)[4,5,12]. The canine tympanum is a thin, semi-transparent membrane[16] with a

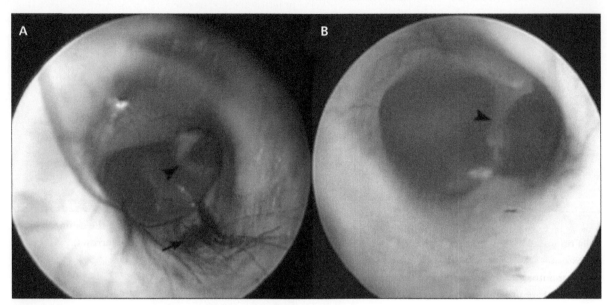

Figs 1.12A, B Video otoscopic image of the tympanic membrane. A: Tympanic membrane, right ear, dog. A prominent tuft of hair is immediately distal to the tympanic membrane (arrow), the stria mallearis is distinctly 'C' shaped (arrowhead); B: tympanic membrane, right ear, cat. The stria mallearis in cats is much more straightened and perpendicular (arrowhead), lacking the 'C' shape observed in dogs. (Courtesy Dr LK Cole, College of Veterinary Medicine, The Ohio State University, Columbus, OH. Used by permission Saunders, Philadelphia, PA. Njaa BL, Cole LK, Tabacca N. Practical otic anatomy and physiology of the dog and cat. *Veterinary Clinics of North America Small Animal Practice* 2012;42:1109–26.)

rounded, elliptical outline; its mean size is 15 × 10 mm. The shorter dimension is nearly vertical, the long axis is directed ventral, medial and cranial, and it has a surface area of approximately 63.3 sq mm[2,3,17]. The feline tympanum is more circular in shape (8.7 × 6 mm) and has a surface area of approximately 41 sq mm[5,14,18].

The tympanum is divided into two sections[12]. The smaller, dorsal, portion of the tympanic membrane is the *pars flaccida*. The pars flaccida is usually opaque, pink or white in colour. It is confined to the upper quadrant of the tympanic membrane and bound ventrally by the lateral process of the malleus[4,12,17]. The larger part of the external aspect of the tympanum is thin, tough and glistening, the *pars tensa*. The outline of the manubrium of the malleus (*stria mallearis*) is clearly visible (Fig. 1.12B). The manubrium inserts under the epithelium on the medial aspect of the tympanum and exerts tension onto it, resulting in a concave shape to the intact membrane, rather similar to the speaker cone in a loudspeaker[4,19].

- Tympanometry gives an indirect measurement of the compliance of the tympanic membrane, and the pressure within the middle ear[1]. An abnormal tympanometric test may indicate a ruptured tympanic membrane or otitis media.
- In a study of 100 cases of canine otitis externa[20] rupture of the tympanic membrane was negatively associated with underlying allergic disease and positively associated with grass awns, particularly in Cocker Spaniels.

A branch of the mandibular branch of the trigeminal nerve innervates the tensor tympani muscle. It is this muscle that is responsible for maintaining the tension of the tympanic membrane as it vibrates in response to sound waves. The stapedial branch of the facial nerve innervates the stapedius muscle, which is responsible for reducing movement of the stapes ossicle[3].

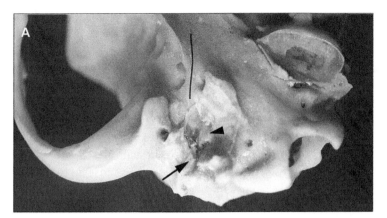
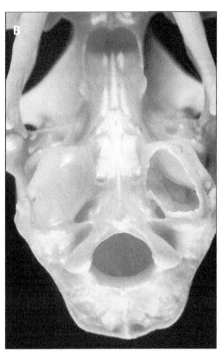

Figs 1.13A, B A: The caudolateral aspect of the canine skull with the bulla removed. Three of the four ports of communication are visible: the external acoustic meatus, arrow, the round window, situated within the promontory, arrowhead, the entrance to the auditory tube, whose direction of travel is delineated by the piece of dark nylon.
B: The caudolateral aspect of the feline skull. The bulla has been opened and the septum is now clearly visible (arrow). The promontory can be seen medial and deep to the septum.

Tympanic cavities

The tympanic cavity proper, at least in anatomical terms, is one of three intercommunicating, air-filled cavities that lie directly behind the tympanum. The smallest cavity, the epitympanic cavity, lies dorsal to the tympanum and is almost entirely occupied by the head of the malleus and its articulation with the incus[2,4]. The tympanic cavity proper is bounded laterally by the tympanic membrane. Its greatest dimension is less than 1 cm. Opposite the tympanum is a bony protuberance, the promontory, within which lies the cochlea. The mean volume of the canine middle ear cavity is about 2.5 cm[3] and this increases with body weight[9]. The volume of the feline middle ear cavity is about 0.9 cm[3,21].

The largest of the three tympanic cavities is the ventral (or fundic) cavity[2]. Getty *et al.*[2] compared the canine tympanic bulla to the shell of an egg, the long axis of which is some 15 mm in length, set at an angle of 45° to a sagittal plane. The width and depth of the chamber are 8–10 mm[3]. There is an elliptical opening in the dorsal wall that communicates with the tympanic cavity proper[2].

There are four ports of communication into the middle ear, all of which enter the tympanic cavity proper (Fig. 1.13A). Three of the ports, the tympa-
num, the round window and the vestibular window, into which the stapes connects, have membranes across their lumen. The auditory tube does not have such a membrane. The tympanum, as outlined above, is located on the lateral aspect of the tympanic cavity. The vestibular window is on the dorsolateral aspect of the promontory, on the caudomedial wall of the tympanic cavity. The round window, across whose membrane the cochlea communicates with the middle ear[22], is located in the caudal portion of the tympanic cavity proper. The ostium of the auditory tube is at the rostral extremity of the tympanic cavity[2,23].

The ventral chamber of the feline tympanic cavity (Fig. 1.13B) is characterised by a near complete bony septum[5,24–26]. It is this septum that is visible upon opening the ventral wall of the tympanic bulla, and it divides the ventral cavity into two. The larger, ventromedial chamber is entered via a bulla osteotomy surgical approach and the smaller, dorsolateral chamber, in effect the tympanic cavity proper, lies beyond the septum. The two chambers communicate via the space between the septum and the caudomedial wall of the tympanic cavity[5,24–26]. The round window of the cochlea, the promontory and the postganglionic fibres of the cervical sympathetic trunk (to the eye and orbit) are in this region of the dorsomedial wall and are thus vulnerable to damage, particularly if the septum is removed during surgery[5]. Damage to these nerves leads to ipsilateral pupillary dilation and retraction of the third eyelid[5]. Sectioning of the nerves results in Horner's syndrome[5]. After Garosi[27]:

- Facial nerve paralysis: facial paralysis with ipsilateral drooping and inability to move the ear and lip, a widened palebral fissure, loss of ability to blink. As long as the parasympathetic nerve supply is unaffected there is no effect on the tear film. Vestibulocochlear nerve damage: vestibular syndrome: head tilt, ataxia, circling, nystagmus.
- Loss of sympathetic innervation to the eye – Horner's syndrome: miosis, ptosis (drooping of the upper eyelid), protrusion of the third eyelid, enopthlamos (retraction of eyeball as orbital smooth muscle loses tone).

Ossicles

The ossicular chain comprises three bones: the malleus, the incus and the stapes[2,4]. The manubrium of the malleus is embedded in the central fibrous portion of the tympanum and is visible via otoscopy as a pale line across the tympanic membrane. In the dog it is about 1 cm in length[2]. The malleus articulates with the incus, about 4 mm in length, in the epitympanic cavity dorsal to the tympanum[4]. The incus subsequently articulates with the head of the stapes, the shortest bone in the body, at about 2 mm in length[2,4]. The ossicular chain does not cross the epitympanic and tympanic cavities unsupported; ligaments attach to the bony walls of the cavities. In addition to the various ligaments there are two muscles attaching to the ossicles: the tensor tympani and the stapedius. Contraction of these muscles moves the relevant ossicle in relation to the membrane with which it is associated, significantly reducing the efficiency of sound transmission. These actions may serve a protective function, particularly from the harmful effect of low frequency sound vibrations[2].

The ratio of the length of the manubrium of the malleus to the length of the long process of the incus is known as the malleus:incus ratio. In the dog it is $2.7 \pm 0.75:1$ and in the cat it is 3.1 ± 0.6:[15,28]. These ratios are almost two to three times that of humans (ratio of 1:3) and probably explains why dogs and cats are able to hear very faint sounds that are inaudible to humans[27]. Furthermore, the ratio of the weight of the stapes to the sum of the weights of the malleus and incus is constant, as is the ratio of the cross-sectional areas of the malleus and incus to the area of the tympanum[27]. Virtual computed tomography (CT) otoscopy has made visualisation of the ossicular chain and middle ear *in vivo* a possibility[29] (Figs 1.14A, B, 1.15A, B).

Auditory tube

The auditory tube arises in the dorsolateral wall of the nasopharynx and passes dorsocaudolaterally to enter the rostral aspect of the tympanic cavity proper (Figs 1.16, 1.17)[2,30]. In both the dog and cat it is in the order of 1.5–2 cm long[4,5,23]. The entrance to the auditory tubes is obscured behind the soft palate. It is roughly midline, midway between the posterior nares and the caudal border of the soft palate[23,30]. The distal end of the auditory tube is patent at all times[14,15,23], whereas the proximal, pharyngeal entrance is normally closed. It is opened by contraction of two muscles: the levator muscle and the tensor palatini muscle[14,23].

The epithelial lining of the auditory tube is continuous with that of the pharynx and it is mucociliary in nature[4,31,32]. The patency of the auditory tube is ensured by the presence of a phospholipid-based surfactant, which helps to prevent closely-apposed, protein-rich mucus layers from 'sticking together'[33]. It has been shown that application of surfactant may increase the efficacy of auditory tube function in some cases of secretory otitis media[31].

The normal functionality of the auditory tube and oropharynx has a direct impact on normal middle ear function[32,34,35], and otitis media may reflect abnormal function of the auditory tube and oropharynx[15,17,32–36].

The acoustic reflex test assesses the functionality of the stapedius to contract in the presence of a loud noise[1]. The acoustic reflex is usually absent in cases of otitis media.

Pharyngeal and auditory tube dysfunction likely plays a role in primary secretory otitis media (see Chapter 6 Diseases of the Middle Ear, Section 6.4) that has been described in the Cavalier King Charles Spaniel[37,38].

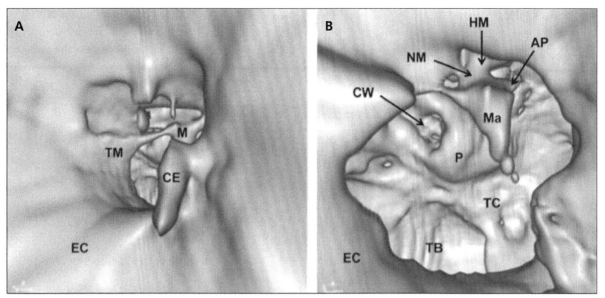

Figs 1.14A, B Virtual CT otoscopic view from the ear canal towards the middle ear (CW, cochlear window; CE, cerumen; EC, ear canal; HM, head of malleus; Ma, manubrium; NM, neck of malleus; P, promontory, TB, tympanic bulla; TC, tympanic cavity; TM, tympanic membrane. (Courtesy of American College of Veterinary Radiology, Raleigh, NC. Eom K, Kwak H, Kang H, *et al*. Virtual CT ososcopy of the middle ear and ossicles in dogs. *Veterinary Radiology and Ultrasound* 2008;49:545–50.)

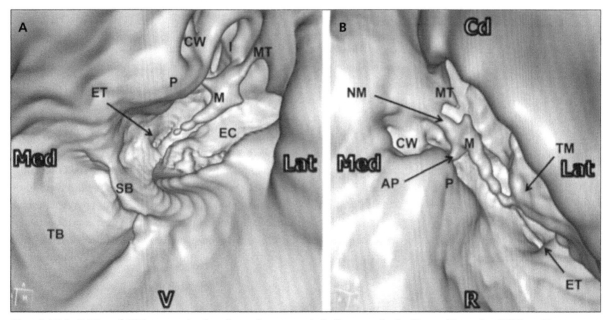

Figs 1.15A, B Virtual CT otoscopic view from the tympanic bulla towards the Eustachian tube (CW, cochlear window; EC, ear canal; ET, Eustachian tube; I, incus; M, malleus; MT tensor tympani muscle; NM, neck of malleus; P, promontory; SB, septum bulla; TM, tympanic membrane). (Courtesy of American College of Veterinary Radiology, Raleigh, NC. Eom K, Kwak H, Kang H *et al*. Virtual CT ososcopy of the middle ear and ossicles in dogs. *Veterinary Radiology and Ultrasound* 2008;49:545–50.)

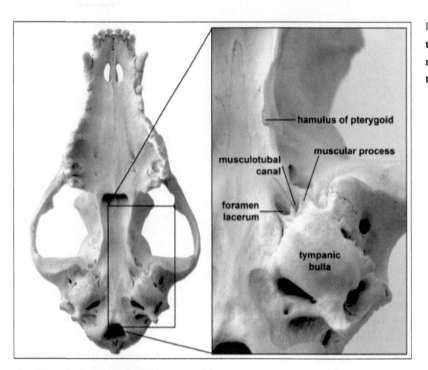

Fig. 1.16 Ventral aspect of the canine skull in which the musculotubal canal of the auditory tube can be seen.

hamulus of pterygoid

muscular process

musculotubal canal

foramen lacerum

tympanic bulla

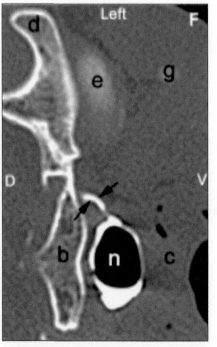

Fig. 1.17 One of a number of transverse CT images of the naso-pharynx (n) from just caudal to the hamulus process of the left pterygoid to the mid portion of the left tympanic. Contrast opacifica-tion of the auditory tube is clearly visible (small black arrows). Basiphenoid bone (b), soft palate (c), zygomatic process of the left temporal bone (d), head of left mandible (e), left foramen ovale (F), angular process of the left mandible (g), dorsal (D), ventral (V). (Courtesy of American College of Veterinary Radiology, Raleigh Nc. Cole LK, Samii VF. Contrast-enhanced computed tomographic imaging of the auditory tube in mesaticephalic dogs. *Veterinary Radiology and Ultrasound* 2007;48:125–8.)

Inner ear

The various tubular and spiral cavities within the petrous temporal bone that contain the components of the inner ear are called the labyrinth[1,3,4]. The three organs contained within the bony labyrinth are the cochlea, the semicircular canals and the vestibulum (Figs 1.18 A, B). Each labyrinthine tube and dila-tion contains a membranous sleeve supported and anchored by connective tissue trabeculae. The space between bone and membrane contains perilymph, whereas the membranous sleeve contains endolymph and the specialised sensory cells[1]. The membranous sleeves of the cochlea and the semicircular canals sim-ulate the shape of their bony counterparts, whereas that of the vestibulum is composed of the sac-like utricle and saccule, which communicate with each other, the cochlea and the semicircular canals[2,4].

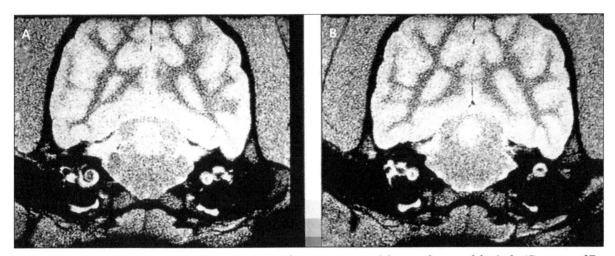

Figs 1.18A, B Two dimensional MRS scan depicting the components of the membranous labyrinth. (Courtesy of B Dayrell-Hart.)

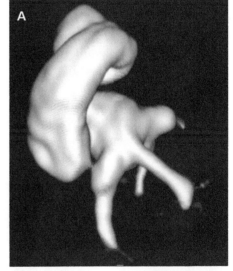

Imaging these structures is difficult[39], for they are deep within bone. Advanced imaging techniques such as 3D magnetic resonance imaging (MRI) are indicated[39], and these are only available at specialised institutions. Consequently, description relies heavily on anatomical texts. Figure 1.19 shows a 3D manipulated MRI scan depicting the components of the membranous labyrinth.

Vestibulum

The vestibulum is an irregularly shaped cavity, roughly 3 mm in diameter in dogs[1,3]. Rostrally it communicates with the *scala vestibuli*, caudally with the semicircular canals[3]. The medial wall contains two depressions: the elliptical recess of the utricle caudodorsally and the spherical recess of the saccule rostroventrally. The lateral wall contains two fenestrations: the oval window closed by the base of the stapes dorsally and the membrane-covered round window ventrally. The semicircular canals open into the vestibulum caudally. The smallest foramen on the vestibular wall is that of the minute vestibular aqueduct, which transports endolymph to the endolymphatic sac adjacent to the dura.

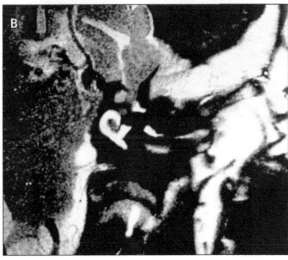

Figs. 1.19 3D reconstruction (performed in OsiriX) of the cochlear apparatus and proximal semicircular canals (A). Baseline MRI sequence used was high-resolution 3D T2W DRIVE (Driven Equilibrium) (B).

Cochlea

The spiral of the canine cochlea takes 3.25 turns around a central bony core[1]; that of the feline cochlea takes 2.75 turns[4]. The basal turn is about 4 mm in diameter and the height of the bony spiral in the dog is about 7 mm[1]. The whole is contained within the petrous temporal bone, although it bulges laterally into the tympanic cavity, the promontory. The cochlea contains the cochlear duct between two perilymph-filled chambers: the *scala vestibuli* dorsally and the *scala tympani* ventrally. These two chambers communicate at the apex of the cochlear duct[25]. The scala vestibuli communicates proximally with the vestibule, whereas the scala tympani terminates proximally at the level of the cochlear (round) window, which is covered by a membrane separating it from the middle ear.

The cochlear duct is triangular in shape with its base, the *stria vascularis*, situated at the outer lateral wall of the cochlea. A thin layer of tissue, the vestibular membrane, forms the dorsal border of the cochlear duct and this borders the scala vestibuli. The basilar membrane separates the endolymph of this duct from the perilymph of the scala tympani. Within the cochlear duct lies the organ of Corti (spiral organ), which is a sensory epithelium, composed of several types of

supporting cells and hair cells, resting on a basement membrane, the basilar membrane. The hair cells have modified microvilli on their luminal surface. The tips of these hairs are embedded in a proteinaceous membrane, the tectorial membrane, which covers the hair cells and is attached medially along the cochlear duct. These structures are involved in the transduction and transmission of sound impulses, via the cochlear portion of the eighth cranial nerve, to the brain[3].

Sound waves are transmitted from the air medium of the external ear to the solid medium of the tympanum and, via a chain of three ear ossicles that extend to the vestibular window, to the fluid medium of the perilymph in the scala vestibuli. Wave flow through the perilymph in the scala vestibuli is reflected through the basilar membrane by way of the movement triggered in the endolymph of the cochlear duct. Movement of the highly organised basilar membrane causes the hair cells of the overlying organ of Corti to move and their stereo-cilia, which are embedded in the tectorial membrane, to bend. This action causes an impulse to be generated in the cochlear neurons that synapse with the base of the hair cells. Low frequencies cause maximal vibration of the basilar membrane at the apex of the cochlear duct, whereas high frequencies affect the proximal portion of the basilar membrane maximally[2].

Peripheral vestibular system

The vestibular system is the primary sensory system that maintains an animal's balance or normal orientation relative to the gravitational field of the Earth. It comprises the peripheral vestibular system, located in the inner ear, and the central vestibular system, located in the brainstem. Details of the central vestibular system are beyond the scope of this book. The vestibular system is responsible for maintaining the

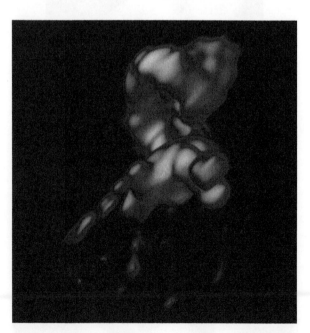

Fig. 1.20 3D volume rendering of a DESS sequence. The winding cochleas of the vestibule, and of the semi-circular canals, are clearly seen. (DESS, double echo steady state). (Courtesy of Dr Matthias Luepke, University of Veterinary Medicine Hannover.)

position of the eyes, trunk and limbs with reference to the position or movement of the head at any time. This orientation is maintained in the face of linear or rotatory acceleration and tilting of the animal. This receptor for a special type of proprioception develops in conjunction with the receptor for the auditory system, the membranous labyrinth[2].

The bony labyrinth in the petrosal part of the temporal bone consists of three communicating, perilymph-filled portions: the large vestibule, the three semicircular canals and the cochlea. There are two openings in the bony labyrinth, the vestibular and cochlear (round) windows, which are named according to the component of the bony labyrinth in which they are located. Each is covered by a membrane and the stapes ossicle is inserted in the membrane covering the vestibular window[2].

The membranous labyrinth consists of four compartments: the saccule and utricle within the bony vestibule; three semicircular ducts within the bony semicircular canals; and a cochlear duct within the bony cochlea (Fig. 1.20). The three semicircular ducts are the anterior (vertical) duct, the posterior (vertical) duct and the lateral (horizontal) duct. Each semicircular duct is oriented at right angles to the others; thus rotation of the head around any plane causes endolymph to flow within one or more of the ducts. Each of the semicircular ducts connects at both ends with the utriculus, which in turn connects to the saccule by way of the intervening endolymphatic duct and sac. The saccule communicates with the cochlear duct by the small ductus reuniens[2].

Each 0.5 mm diameter semicircular canal describes approximately two-thirds of a circle and is between 3.5 mm and 6 mm across, roughly at 90° to each other. They lie caudal and slightly dorsal to the vestibule[1,3,4]. At one end of each membranous semicircular duct is a dilatation called the ampulla. On one side of the membranous ampulla, a proliferation of connective tissue forms a transverse ridge called the crista ampullaris. This is lined on its medial surface by columnar epithelial cells, the neuroepithelium. On the surface of the crest is a gelatinous structure composed of a protein–saccharide material called the cupula, which extends across the lumen of the ampulla. The neuroepithelium is composed of two basic cell types: supporting cells and hair cells. The latter are in synaptic relationship to the dendritic zone of the vestibular portion of the vestibulocochlear nerve. The hair cells have 40–80 stereocilia and a single kinocilium on their luminar surface. These project into the overlying cupula. Movement of fluid in the semicircular duct causes deflection of the cupula that is orientated transversely to the direction of flow of endolymph. This bends the stereocilia and is the source of stimulus by way of the hair cell to the dendritic zone of the vestibular neuron. Each semicircular duct on one side can be paired with a semicircular duct on the opposite side by their common position in a parallel plane. Deviation of the stereocilia towards the kinocilium increases vestibular neuronal activity. These receptors function in dynamic equilibrium. They are not affected by a constant velocity of movement but respond to acceleration or deceleration, especially when there is rotation of the head[25].

Receptors (maculae) are present in the utriculus and saccule. They comprise thickened connective tissue on the surface of the membranous labyrinth, which is covered by a neuroepithelium composed of hair cells and supporting cells. The neuroepithelium is covered by the otolithic membrane, a gelatinous material on the surface of which are found calcareous crystalline bodies known as statoconia (otoliths). The hair cells are similar to those found in the crista ampullaris and their stereocilia and kinocilia project into the otolithic membrane. Movement of the otoliths away from these cell processes initiates an impulse in the vestibular neuron. The macula sacculi is oriented in a vertical direction (sagittal plane), while the macula utriculi is orientated in a horizontal direction (dorsal plane). Thus, gravitational forces continually affect the position of the otoliths in relation to the hair cells. These are responsible for the sensation of the static position of the head and linear acceleration or deceleration. They function in static equilibrium. The macula utriculi might be more important as a receptor for sensing changes in posture of the head whilst the macula sacculi might be more sensitive to vibrations and loud sounds[2].

1.3 MICROSCOPIC STRUCTURE OF THE EXTERNAL EAR CANAL

Normal external ear canal

The epidermis lining the external ear canal (reviewed by Cole[1]) is similar in structure to that of the interfollicular epidermis of the skin, i.e. a stratified, cornifying epithelium with adnexal organs such as hair follicles and their associated sebaceous and ceruminous (apocrine) glands (Figs 1.21A, B)[2,3,9,40]. The underlying dermis is heavily invested with elastic and collagenous fibres (Figs 1.22, 1.23).

Beneath the dermis and subcutis lie the rolled cartilaginous sheets of the auricular and annular cartilages that contain and support the external ear canal. The syndesmoses between these cartilage tubes, and between the annular cartilage and the osseous external acoustic process, allow great freedom of movement for the pinna[2]. The elastic and collagen fibres of the dermis allow a degree of freedom of movement for the external ear canal as well, and this can be exploited during otoscopic examination.

Epidermis

The epidermis is stratified and rather thin[9,40,41], being only a few layers thick (Fig. 1.24). The superficial topography is smooth and, on a microscopic level, composed of layers of flattened squames closely apposed and overlapping at the edges (Fig. 1.25). Although no laboratory

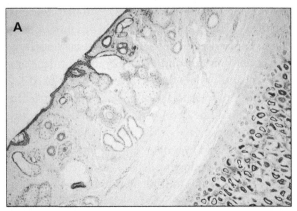
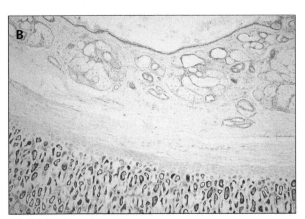

Figs 1.21A, B A: Photomicrograph of a section of normal canine external ear canal. Note the thin epidermis, the sparse hair follicles, sebaceous and ceruminous glands and the underlying cartilage; B: photomicrograph of a section of feline external ear canal. The component structures are similar to that of the canine section.

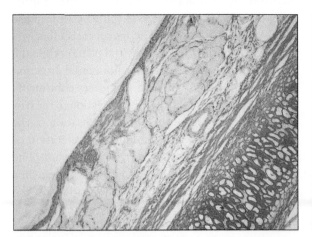
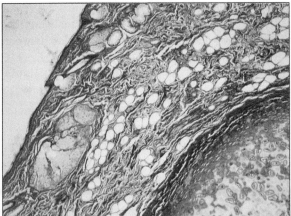

Fig. 1.22 Photomicrograph of a section of normal canine external ear canal stained with Gomorri's stain to highlight collagen and fibrous tissue within the dermis.

Fig. 1.23 Photomicrograph of a section of normal canine external ear canal stained with Masson's trichrome to highlight collagen and fibrous tissue within the dermis.

data are available, there is no reason to believe that the epidermis of the external ear canal turns over, or reacts to insult, any differently to the epidermis of the skin. The epidermis is punctuated by the hair follicles. In man, and presumably also in the dog and cat, the superficial epidermis and keratinised stratum corneum migrate, *en masse*, laterally from the tympanum[42,43]. This process is an extension of the epidermal migration on the surface of the tympanic membrane and serves to keep the proximal ear canal and tympanum free from cerumen and debris.

Hair follicles

All breeds of dog have hair follicles throughout the length of the external ear canal, although in most breeds these follicles are simple and sparsely distributed (Fig. 1.26)[40]. It has been suggested[2, 3] that the density of hair follicles decreases as one progresses toward the external acoustic meatus, but recent studies did not describe such a distribution[8,39]. The mean proportion of integument occupied by hair follicles was found to be 1.5–3.6%, with no significant spatial distribution along the canal. There was a large interdog variation[5].

However, some breeds that are predisposed to otitis externa differ from the basic pattern[40]. Thus, Cocker Spaniels exhibit a much higher concentration of hair follicles than other breeds and, furthermore, the follicles are typically compound in pattern (Fig. 1.27)[40].

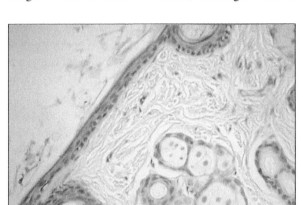

Fig. 1.24 Photomicrograph of a section of normal canine external ear canal illustrating the thin epidermis, which is only a few cells thick.

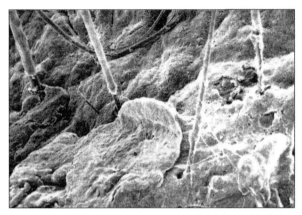

Fig. 1.25 Scanning electron micrograph of the epithelial surface of the external ear canal. (Produced by the Department of Anatomy, Royal Veterinary College, London.)

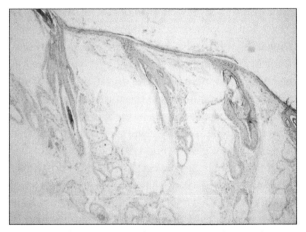

Fig. 1.26 Photomicrograph of a section of normal canine external ear canal showing simple hair follicles.

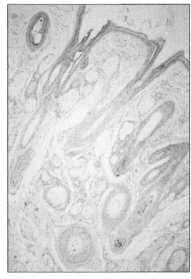

Fig. 1.27 Photomicrograph of a section of normal external ear canal from a Cocker Spaniel. Note the density of the hair follicles (compared to those in 1.26), and that they are compound, not simple.

There is no correlation between the percentage of hair follicles within the otic integument and a predisposition to otitis externa[9].

Hair is sparse or absent in the feline external ear canal[44].

Adnexal glands

Sebaceous glands are present in the upper dermis[2,3,9,40]. They are numerous and prominent (Fig. 1.28) and have a similar structure to the sebaceous glands of the skin. The mean proportion of integument occupied by sebaceous glands is 4.1–10.5%, gradually increasing from proximal to distal and peaking at the level of the anthelix[9]. There is a large interdog variation[8]. The sebaceous glands secrete

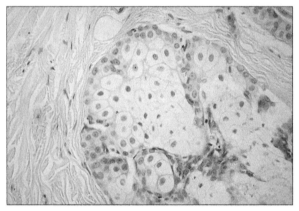

Fig. 1.28 Photomicrograph of a section of canine external ear canal showing a high power view of a sebaceous gland.

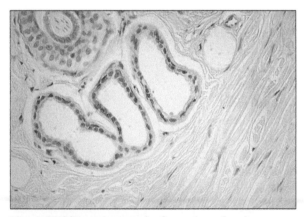

Fig. 1.29 Photomicrograph of a section of canine external ear canal showing a high power view of a ceruminous (apocrine) gland.

principally neutral lipids[45]. In the normal dog this lipid accounts for the majority of the cerumen, along with sloughed epidermal debris[45]. This high lipid content of normal cerumen helps maintain normal keratinisation of the epidermis, aids in the capture and excretion of debris that is produced within and enters the external ear canal, and results in a relatively low humidity within the lumen of the ear canal. In the cat the sebaceous glands become more prevalent and crowded proximally[41].

Ceruminous (apocrine) glands are located in the deeper dermis (Fig. 1.29)[2,3,9,40,41]. They are characterised by a simple tubular pattern and a lumen lined by a simple cuboidal-pattern epithelium. In the normal dog and cat the ducts of the apocrine glands are virtually non-apparent. The mean proportion of integument occupied by apocrine glands is 1.4–4.5%, gradually decreasing from proximal to distal and peaking at the level of the tympanic membrane[9]. There is a large interdog variation[9,40]. The apocrine glands contain acid mucopolysaccharides and phospholipids[41].

Overall, these data suggest that the ratio of apocrine to sebaceous gland decreases from proximal to distal, tending to produce a more aqueous cerumen in the deeper ear canal, possibly more conducive to epidermal migration[9]. The more lipid nature of cerumen at the distal end may facilitate water repulsion.

Breed variations

Fernando observed that the external ear canals of long-haired breeds of dogs and those with fine hair contained more sebaceous and apocrine glandular tissue, which was also better developed, than dogs with short hair[41]. Breeds predisposed to otitis externa, particularly Cocker Spaniels, also have abnormal morphometric ratios compared to normal dogs[40]. Specifically, they exhibit an increase in the overall amount of soft tissue within the confines of the auricular cartilage, an increase in the area occupied by the apocrine glands and an increase in the apocrine gland area, compared to that of the sebaceous glands (Fig. 1.30).

Overall, it may be that certain breeds of dog, again particularly the Cocker Spaniel, which are predisposed to otitis externa have increased apocrine tissue within the external ear canal. If this increased volume of apocrine tissue is actively secreting, the concentration of lipid within the cerumen will fall[9], humidity within the ear canal will rise, and maceration, followed by infection and otitis externa, will follow. Increased moisture and surface maceration creates an environment particularly favourable to gram-negative bacteria. Theoretically, the increased apocrine secretions in the ear canals of these dogs should result in a cerumen with a lower pH than normal and an environment not conducive to gram-negative colonisation. It may be that in these breeds that are predisposed the acidifying effect of increased ceruminous gland secretion is not sufficient to overcome the effects of humidity, inflammation and surface maceration.

Response to insult and injury

The epidermis of the external ear canal reacts to inflammation by increasing its rate of turnover and increasing in thickness; it becomes hyperplastic (Fig. 1.31)[46,47]. There may be surface erosions and ulceration, particularly with gram-negative infections. The dermis becomes infiltrated with inflammatory cells (Fig. 1.32) and fibrosis will follow (Figs 1.33, 1.34). The mean proportion of integument occupied by connective tissue falls from 85.9–91.5% to a mean of 66.5–75.2% in cases of chronic otitis, a reflection of the relative and absolute increase in glandular tissue[9].

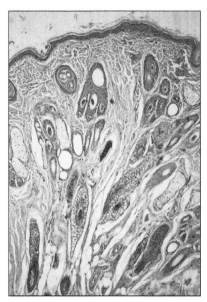

Fig. 1.30 Photomicrograph of a section of normal canine external ear canal from a Cocker Spaniel. Note the closely packed dermis compared to 1.26.

The progression of these changes is inevitable unless the otitis is treated. Identify the particular combination of primary, predisposing and perpetuating causes and institute appropriate treatment.

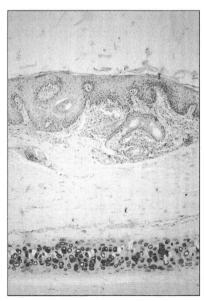

Fig. 1.31 Photomicrograph of inflamed feline external ear canal demonstrating epidermal hyperplasia.

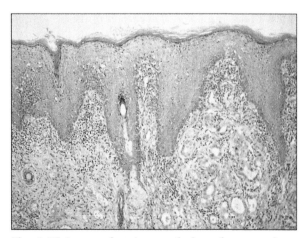

Fig. 1.32 A more chronic example from the external ear canal of a dog showing a thickened stratum corneum, epidermal hyperplasia and an inflammatory infiltrate into the dermis.

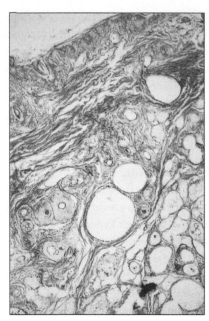

Fig. 1.33 Photomicrograph of a section from a chronically inflamed external ear canal of a dog, which has been stained with Gomorri's stain to highlight fibrosis. Compare with the normal shown in 1.22.

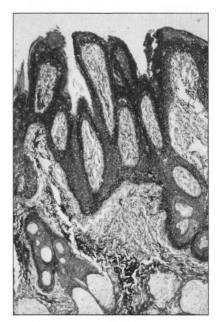

Fig. 1.34 Photomicrograph of a section from a chronically inflamed external ear canal of a dog, which has been stained with Masson's trichrome stain to highlight fibrosis. Compare with the normal shown in 1.23.

In the early stages of otitis externa there is some hyperplasia of the sebaceous glands, and their ducts may become dilated[47]. If chronic otitis persists, the apocrine glands become increasingly hyperplastic, with cystic dilatation of the glands and ducts (Figs 1.35, 1.36). Thus, in ears from dogs with chronic otitis:

- The proportion of integument occupied by sebaceous glands increases from a mean of 5.2% to a mean of 19.2%, significantly more than in normal ears[9].
- The proportion occupied by apocrine glands increases from a mean of 10.1% to 17.1%, significantly more than in normal ears[9].

This tendency for the ceruminous glands to hyperproliferate is exemplified in the Cocker Spaniel[48]. Whether this represents a breed-specific aetiopathogenic pathway is not known. The progression of clinical and histological change is particularly rapid and severe in the Cocker Spaniel[48]. In this breed there may be a predisposition to develop a markedly proliferative severe ceruminous gland hyperplasia and ectasia, resulting in end-stage otitis externa (Fig. 1.37).

> Given this breed predisposition, Cocker Spaniels with otitis externa warrant early and aggressive investigation into the cause of the otitis.

Papillary proliferation of ceruminous glands and ducts may obliterate the lumen of the external ear canal in some cases (Figs 1.38–1.41)[47]. In very chronic cases ossification of the tissues may take place.

Similar changes take place in the feline ear canal, although the papillary changes in the ceruminous glands may be sufficiently florid that discrete polypous or cerumous hyperplastic masses occur (Fig. 1.42)[47].

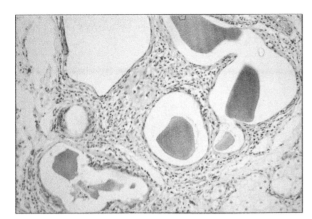

Fig. 1.35 Photomicrograph of a section from a chronically inflamed external ear canal of a dog, showing moderate dilation of the ceruminous glands.

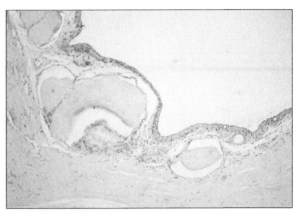

Fig, 1.36 Photomicrograph of a section from a chronically inflamed external ear canal of a dog, showing massively dilated apocrine glands. Note that they are sufficiently enlarged to distort the overlying epidermis, sufficient to reduce the diameter of the external ear canal.

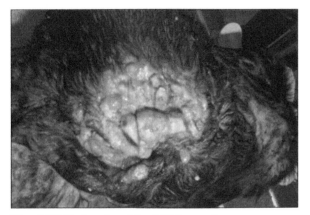

Fig. 1.37 Cocker Spaniel exhibiting end-stage hyperproliferative otitis externa. (Courtesy Dr Mona Boord.)

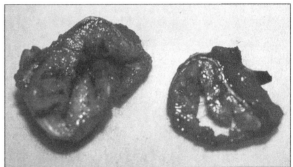

Fig. 1.38 Cross-section of a hyperplastic external ear canal of a dog, taken at surgery, demonstrating the almost complete obstruction of the external ear canal.

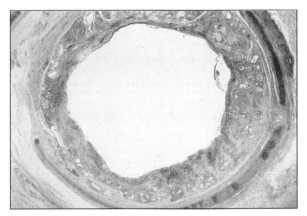

Fig. 1.39 Photomicrograph of a cross-section of a similarly affected ear canal to 1.38. The increase in soft tissue within the confines of the auricular cartilage is readily apparent.

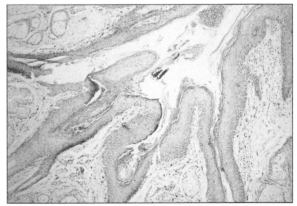

Fig. 1.40 Photomicrograph exhibiting how hyperproliferative folds can obstruct the lumen of the external ear canal.

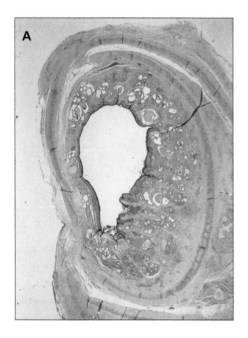

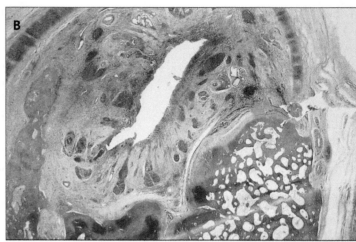

Figs. 1.41A, B Photomicrographs of chronically inflamed canine external ear canals exhibiting marked stenosis of the lumen.

Fig. 1.42 Feline ceruminous hyperplasia sufficient to obstruct the external ear canal of this cat. (Courtesy Dr Lorraine A Corriveau.)

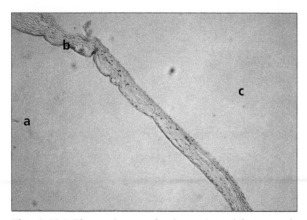

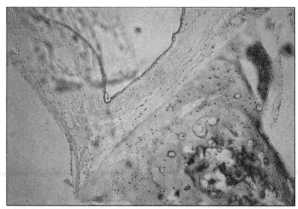

Fig. 1.43A Photomicrograph of a section of the normal canine tympanic membrane wherein the three layers (outer keratinising epithelium [a], central lamina propria [b] and inner mucosa[c]) are clearly visible.

Fig. 1.43B Photomicrograph of a section of normal canine tympanic membrane at the level of the insertion of the malleus. (Sample preparation by Dr David Shearer.)

1.4 MICROSCOPIC STRUCTURE OF THE TYMPANUM AND MIDDLE EAR

Tympanum
Microscopic structure

The components of the tympanic membrane are arranged in three layers (Figs 1.43A, B): an outer keratinising epithelium, a central lamina propria and an inner mucosa of pharyngeal origin[4,49]. There are no hair follicles, or sebaceous or apocrine glands on the tympanum[49].

The *pars flaccida* tends to be thicker than the *pars tensa* and it contains irregular, loosely packed collagen bundles[50–52], whereas the pars tensa is more dense and contains tightly packed collagen fibres[53,54]. The outer collagen bundles are radial in arrangement, whereas the inner bundles are arranged in a circular pattern, helping to maintain the acoustic properties of the membrane under varying conditions[49]. Elastin fibres are found in both the pars tensa and the pars flaccida of the dog and in neither area of the cat[53]. The reason for this disparity is not known. The lamina propria of both dogs and cats contains abundant mast cells[53] although their function is unclear[1].

Epithelial migration

Superficial epithelial cells on the human tympanum migrate centripetally from the umbo, and it appears that this is a mechanism for clearing epithelial and ceruminous debris from the tympanum, which, if it accumulated, would impair hearing[55]. The epithelial migration may also have a role in repair of tympanic perforation[56]. The cleavage line between stationary and migrating cells is at the level of the nucleated squames, rather than the stratum corneum, although the mechanism of the migration is not clear[57]. In some experimental animals, such as the guinea pig, the movement is not centripetal but rather follows the line of the underlying collagen fibres[49].

The mechanics of tympanic epithelial migration in the dog have been studied[58]. By placing minute drops of waterproof ink on the various areas of the tympanic membrane it was observed that the mean rates of migration (movement) were between 96.4 (± 43.1 SD) and 225.4 (± 128.1 SD) μm/day for the pars tensa and pars flaccida, respectively. All ink drops moved outwards, the majority in a radial direction.

Reaction to insult and puncture
Reaction to irritants

Experimental studies have demonstrated that the epithelium of the tympanum responds similarly to that of the external ear canal and other squamous epithelia when exposed to irritants. There is thickening of the tympanic epithelium, particularly of the stratum corneum, and an increase in epithelial turnover time, as measured by an increased rate of basal cell mitoses[59]. The tympanic epithelium also responds to chronic inflammation (Figs 1.44A, B) within the middle ear, even if there is no tympanic rupture. There is hyperplasia of the epithelium on the lateral side and loss of the mucosa on the medial side[60].

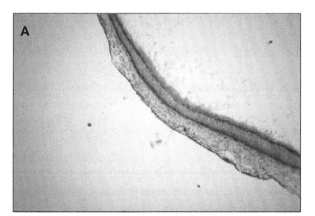
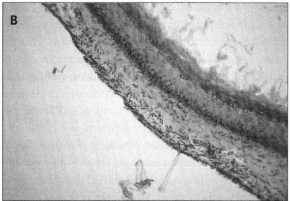

Figs 1.44A, B Low- (A) and high- power (B) photomicrographs of inflamed tympanic membrane; there is marked hyperkeratosis of the outer epithelium and there is an inflammatory infiltrate within the central lamina propria. (Sample courtesy of Dr Dominique Heripret, preparation by Dr David Shearer.)

Puncture of the tympanum

Puncture of the tympanum most commonly arises as a result of infection or trauma[61]. The well-documented sequence of haemostasis, inflammation, fibroblastic and collagenous proliferation, and epithelialisation, which characterises healing of skin wounds[62], is not followed in the tympanum[62]. In particular, the epithelium migrates over and bridges the defect before a granulation bed is established beneath[60–64]. This adaptation presumably results in a more rapid restoration of tympanic function than might be achieved if the cutaneous pattern was followed. Thus, simple, elective myringotomy wounds heal rapidly and even excision myringectomy wounds heal within 3–8 weeks[65–67]. Determinants that affect rate of healing of perforation of the tympanum include the size of the puncture and, in particular the presence, or absence, of infection[68,69].

Middle ear

The epithelial lining of the normal middle ear is a modified respiratory pattern with a squamous or cuboidal appearance (Fig. 1.45)[70]. A few ciliated and secretory cells are scattered among the epithelial cells. There is a thin lamina propria between the epithelium and the periosteum overlying the bone of the tympanic bulla[70].

When exposed to irritant chemicals, which may enter via a punctured tympanum for example, the epithelium responds with a range of inflammatory reactions proportional to the toxicity of the chemical and, presumably, its concentration and the period of exposure. There may be a mild, and potentially reversible, perivascular dermatitis with hypersecretion of mucus into the middle ear cavity[60,62,71]. If the irritation continues, the low cuboidal or squamous epithelium becomes columnar with a papillary appearance[69]. Secretory cells are visible between some of the columnar epithelial cells. Glandular or even cystic structures may be found within the granulation tissue; they may become so large that they fill the middle ear (Figs 1.46A, B). The lamina propria underlying the epidermis thickens and may take on the appearance of loose, oedematous granulation tissue[70–72] with collagen, and even bone, being laid down as chronic changes take place. New bone deposition occurs on both the luminal and extraluminal bone of the tympanic bulla. At worst, a necrotising inflammation may occur, with a thick, cellular neutrophil-rich exudate filling the middle ear cavity and an underlying osteomyelitis[60,70–72].

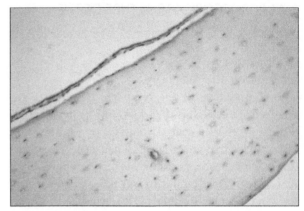

Figs 1.45 Photomicrograph of a section of normal canine bulla. Note the thin mucosa overlying the underlying bone. (Sample preparation by Dr David Shearer.)

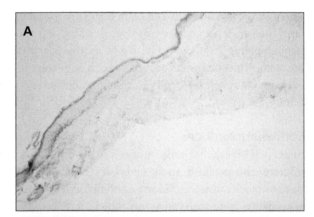

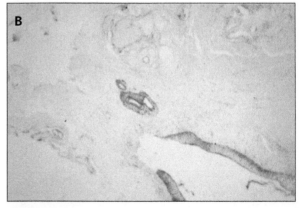

Figs 1.46A, B Low (A) and high (B) power photomicrographs of inflamed bulla associated with otitis media. Note the hyperplastic mucosa and early glandular hyperplasia, both readily apparent in (B).

Chronic inflammation within the middle ear cavity may lead to loss of the mucosal surface of the tympanum[63] and, ultimately, to cholesteatoma formation[61]. Cholesteatomas are slowly enlarging, cystic lesions within the middle ear cavity. They are lined by stratified squamous epithelium and keratin squames are shed into them[72,73]. The aetiology of cholesteatoma is poorly understood[73] (see Chapter 6 Diseases of the Middle Ear).

Auditory tube

The canine auditory tube is lined with pseudostratifed ciliated and non-ciliated columnar epithelium interspersed with goblet cells[74]. Cilia predominate at the proximal end of the auditory tube. This is presumably an adaptation to facilitate mucus clearance into the nasopharynx, while at the same time providing a mechanism to limit bacteria ascending from the nasopharynx[74,75]. Goblet cells are more numerous at the distal, tympanic end. The goblet cells may be the source of the surface tension-lowering substance detected in the canine auditory tube[76-78]. This surfactant is a complex mixture of lecithin, lipids and polysaccharides, which helps to keep the auditory tube patent. Infection or allergic reactions, for example, may compromise epithelial function, resulting in decreased clearance of middle ear secretions and decreased patency of the auditory tube. These changes have been postulated as possible causes of otitis media in man[78].

Otitis media may reflect abnormal function of the auditory tube[58,60]. For example, reflux or aspiration of pharyngeal organisms may result in middle ear infection. Malfunction of the normal homeostatic controls will also follow oedematous swelling of the epithelial lining of the auditory tube.

Round window membrane

The round window membrane is located in the medial wall of the middle ear. It separates the scala tympani of the cochlea from the cavity of the middle ear and provides a means of communication between the two[79]. The structure of the round window membrane is essentially similar to that of the tympanum. There are three layers, the middle layer being composed of connective tissue[79]. However, unlike the tympanum, both outer layers are thin epithelium. Morphological evidence suggests that the round window membrane has an important role in the movement of substances between the inner and middle ear and, as such, it plays a role in both homeostasis and ototoxicity[79].

1.5 MICROCLIMATE OF THE EXTERNAL EAR CANAL

Epithelial lining

The external ear canals are lined such that the underlying cartilaginous architecture and the intercartilaginous joints are covered by a smooth, clean epithelial surface (Fig. 1.47). The epithelial surface is composed of closely apposed squames that are covered by a variable, but usually thin, layer of cerumen and adherent debris (Figs 1.48A, B). There is a constant, outward movement of cerumen[55,57]. Squames detach (Figs 1.49A, B) and move distally in the cerumen[57,80], thus keeping the tympanum clear of debris and providing a mechanism for removing sloughed epithelial and glandular secretions from the external ear canal.

Temperature

In a series of studies the temperature within the external ear canal of dogs was 38.2–38.4°C (100.7–101.1°F)[81-83]. These studies were performed over 25 years apart, with very different technologies, and for such close results to be achieved is remarkable.

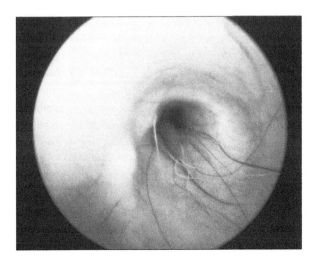

Fig. 1.47 Otoscopic view of the normal external ear canal. Note the smooth, clean surface and that the presence of hair is not abnormal.

Figs 1.48A, B Scanning electron micrographs of the normal epithelial surface of the external ear canal of a dog. Note the squames and hair emerging from hair follicles. Some cerumen is apparent. (Samples prepared by the Department of Anatomy, Royal Veterinary College.)

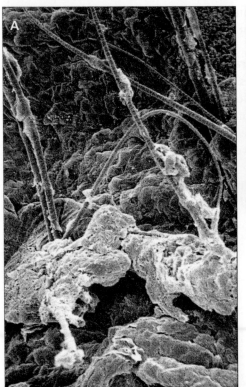

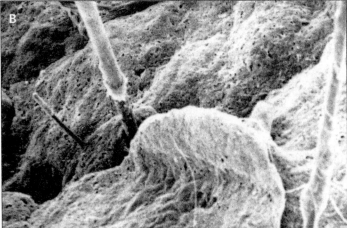

Figs 1.49A, B Scanning electron micrographs of the normal epithelial surface of the external ear canal of a dog (A) and a cat (B). Squames in varying degrees of detachment are apparent.

There was no significant difference between breeds of dog or whether there was a pendulous pinna or not[82,84]. One study found slightly lower temperatures within the external ear canal (range 28.1–34.0°C) and that the temperature within the external ear canal of German Shepherd Dogs was significantly lower (28.2 ±0.6°C) compared to other breeds[7].

Yoshida *et al.*[84], found no difference in the temperature within the external ear canal of normal dogs and that within the external ear canal of dogs with otitis externa. In contrast, Hui-Pi Huang and Hui-Mei Shih[83] reported that the temperature within the external ear canal rises significantly if otitis externa is present: mean 38.9°C (102°F). This might reflect methodology, as the latter study used infrared technology to measure temperature. The temperature within the external ear canal is a mean of 0.6°C (33.1°F) lower than rectal temperature[83].

One study was performed in Australia where the environmental temperature tends to be high[81]. Nevertheless, as the day grew progressively hotter the temperature within the external ear canal only rose 0.3°C (32.5°F) compared to a rise of 6.4°C (43.5°F) in the environment. This illustrates very well how the environment within the ear canal is effectively buffered from the external environment.

> Overall, it can be concluded that the temperature within the external ear canal does not predispose to otitis externa.

Relative humidity
The relative humidity within the external ear canal of 43 dogs, from various breeds, including those with pendulous pinnae, 84.1% to 98.6% and did not vary between breeds[85].

pH
The range of pH within the external ear canal in normal dogs is 4.6–7.2[86]. The mean pH is slightly lower in males than in females (6.1 compared to 6.2). The pH rises in otitis externa. Grono measured the pH in cases of otitis externa and found the mean to be 5.9 (range 5.9–7.2) in acute cases and 6.8 (range 6.0–7.4) in chronic cases[86]. Grono also measured the pH of the

external ear canals of dogs and recorded the bacteria that were isolated from some of these cases. Non-parametric (Mann Whitney) analysis of Grono's data by the authors showed that in cases of otitis externa associated with *Pseudomonas* spp. the pH is significantly higher (mean 6.85, $p > 0.05$) than in cases of otitis externa in which no *Pseudomonas* spp. are isolated (mean 5.7).

Cerumen in normal and otitic ears
Cerumen coats the lining of the external ear canal (Figs 1.50A, B). It is composed of lipid secretions from the sebaceous glands, ceruminous gland secretion[41], and sloughed epithelial cells. The lipid component of dogs' cerumen can vary widely, as does the type of lipid within the cerumen. Margaric (17:0), stearic (18:0), oleic (18:1) and linoleic (18:2) fatty acids are the most common[45,87]. A range of 18.2–92.6% (by weight) of lipid content was found in the external ear canals of normal dogs and in

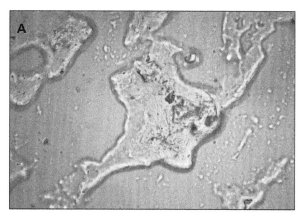

Figs 1.50A, B Scanning electron micrographs of the normal external ear canal from a dog (**A**) and cat (**B**) showing amorphous globules and clumps of cerumen.

some cases there was wide disparity between the left and right ears. This variation presumably reflects individual variation in concentration and activity of ceruminous glands. In man, cerumen type ('wet' or 'dry') is inherited as a simple mendelian trait[88]. Whether there is a simple genetic control of canine or feline cerumen type is not known. Oleic and linoleic acid have antibacterial activity[89,90], although the effects of these fatty acids, and others, if any, against bacteria and *Malassezia pachydermatis* within the ear canal is less clear[7].

Masuda *et al.* reported that total lipid (μM per ear canal) was higher in dogs with pendulous pinnae compared to those with erect pinnae[91]. Interestingly, the Pug (pendulous pinnae) and Siberian Husky (erect pinnae) were outwith these findings. The authors found no evidence to suggest that pinnal type was associated with increased or decreased carriage of *M. pachydermatis*.

In cases of otitis externa the lipid content of the cerumen falls significantly to a mean of 24.4%, compared to a mean of 49.7% from normal ears[45]. How much is an absolute decrease and how much is a relative decrease, reflecting the hypertrophy of apocrine glands and an increase in their secretion, is not known. However, given that the ceruminous secretion is generally acidic[86], this change might be considered to produce a more antibacterial environment. Balanced against this, the decreased lipid component of cerumen may account for the increase in relative humidity reported in the external ear canals of dogs with otitis externa[85]. This, plus the decrease in antibacterial activity, may allow increased bacterial multiplication.

> It may be that individual, and perhaps breed, related variations in the balance between pH and relative humidity within the external ear canal dictate the progression of otitis externa.

1.6 EFFECT OF UNDERLYING DISEASE AND THE EFFECT OF SURGERY ON THE OTIC ENVIRONMENT AND MICROFLORA

With the above changes in otic environment in mind it is apposite to consider the effect, if any, of disease on the external ear canal. No studies have been performed to looked at the changes in environment within the external ear canal in dogs and cats with disease, but studies have look at bacterial carriage, an indirect marker perhaps.

Zur *et al.* looked at the relationship between signalment, underlying disease and the carriage of common otic pathogens[92]. The study failed to find a significant relationship between primary causes (atopy, endocrinopathy) and pathogen. Pressanti *et al.* looked at microflora carriage in three groups of cats: 20 healthy cats, 15 with systemic disease and 15 with allergic dermatitis[93]. Higher numbers of malassezial yeast and bacteria were found in the ear canals of allergic cats compared to normal, and malassezial yeast was more likely to be recovered from the ears of cats with systemic disease compared to normal.

Lateral wall resection results in a fall in both temperature and humidity within the external ear canal, as might be expected. In one study of 12 dogs the temperature in the resected ears was reduced by a mean of 0.6°C (33.1°F) compared to the contralateral, normal ears[81]. Furthermore, the relative humidity in resected ear canals falls by a mean of 10%[6]. Whether this change in temperature and humidity is sufficient to explain the clinical improvement that sometimes follows lateral wall resection is debatable.

1.7 MICROBIOLOGY OF THE EXTERNAL EAR CANAL

Normal and abnormal normal flora of the canine external ear canal

Two controlled studies, using samples taken immediately postmortem, have examined carriage of bacteria in clinically normal ears. In one, samples were taken from the centre of the concave aspect of the pinna, the areas around the anthelix and the middle of the vertical ear canal[94], from both ears of 52 dogs. Positive samples were recovered from 91.8% of the pinnae, 70.5% of the area around the anthelix and 48.8% from the centre of the vertical ear canal. Around 10% of the dogs carried no detectable bacteria. Gram-positive staphylococci (coagulase-negative staphylococci and coagulase-positive *Staphylococcus intermedius*) predominated, with very little difference in recovery rates from each ear. *Pseudomonas* spp. was only recovered from around 10% of the dogs, and never from the vertical ear canal.

Yeasts (not speciated) were found in 48.8%, 81.4% and 83.7% of central pinnal, anthelical and vertical ear canal samples respectively. These recovery rates of yeast are high and might reflect the breed (Beagles have pendulous ears, with high cerumen content) and the location (a humid area of Japan).

The second study was also a postmortem study, and 30 dogs of various breeds were sampled[95]. Two areas were sampled, the horizontal ear canal, around 2 cm from the tympanic membrane, and the epithelial lining of the bulla. Bacteria were recovered from 48% of middle ear samples and 47% of horizontal canals. Species of bacteria were similar to the above with gram-positive staphylococci predominating and no *Pseudomonas* spp. being isolated. Yeasts (not speciated) were found in only 8% of middle ears.

A semi-quantitative cytological study reporting the microscopic examination of stained samples from normal ears demonstrated that very few bacteria (2–10 per high power field) could be found[96] – you need to look hard for them. Inflammatory cells are not seen in normal canine ear canals.

Only one study specifically investigated the bacterial flora of the horizontal ear canal[97]. Fourteen of the 51 ears sampled yielded no growth, and when bacteria were recovered they were reported to be in very low numbers.

The studies described above validate other, earlier, studies[98–104] that reported on samples taken by swabbing external ear canals. In summary:

- A proportion of dogs appear to carry no viable bacteria within their external ear canals.
- When bacteria are recovered, the gram-positive flora predominate, with both coagulase-negative and coagulase-positive staphylococci (*Staphylococcus pseudintermedius*) being recovered from a large percentage of normal ear canals,
- Although absolute numbers are variable, they are usually low.
- Bacteria are scarce to absent in the normal horizontal canal.
- *Pseudomonas* spp. and gram-negative organisms are rarely recovered from the external ear canal of normal dogs.

Very recently, early data about the skin microbiome has been published[105]. Using the 16s mRNA gene as a genus and speciation tool, the technique can reveal the presence of bacterial species that standard culture methods cannot. For example, *vis-à-vis* the normal dog 'one sample from the ear had 866 observed species'. As we learn more about this technique we might learn how changes in the otic biome reflect disease and, indeed, influence disease.

Consideration of the above suggests that the relationship between normal flora and disease is not straightforward.

- Potential pathogens such as coagulase-positive staphylococci can be recovered from ear canals in the absence of evidence of disease, just as they can be found on the normal interfollicular epidermis of the skin[106].
- It seems reasonable to consider that these organisms cannot proliferate unless inflammation or maceration occurs within the ear canal, and it was for this reason that August[107] considered micro-organisms as perpetuating causes rather than primary or predisposing causes of otitis externa.
- This theory also highlights the value of otic cytology as an adjunctive aid in assessing the relevance of bacteriological results. Failure to detect changes consistent with otitis externa suggests that any organisms found on culture are part of the normal flora.

The normal flora of the feline external ear canal

Controlled studies of the bacterial flora of the normal feline external ear canal have not been reported, although there is one large study looking at the normal flora of feral cats in Grenada[108]. Over 60% of 54 cats yielded no bacterial growth. In those that did, the major isolate from the cat's ear canals was coagulase-negative staphylococci, in particular *Staphylococcus felis/simulans*, a finding supported by Cox *et al.*[109].

Fungi

Two studies have looked at malassezial carriage in the ear canals of a total of 89 normal dogs: 12–38% of samples contained malassezial yeast[110,111]. The unaffected ear, contralateral to an otitic ear, in 14 dogs with unilateral otitis contained yeast in 21.4% of samples.

Two species of malassezial yeast have been regularly recovered from the normal ears of cats, *M. pachydermatis* and *M. sympodialis*, with the former being found in 24% of ears and the latter in only 4%[110–116]. Malassezial carriage varies significantly between some feline breeds[117–120]:

- Devon Rex cats and Sphynx exhibit significantly higher carriage rates of *Malassezia* spp. in their ears than Cornish Rex cats, where carriage is similar to Domestic Short Hair cats.
- *M. pachydermatis* can be found in high numbers in the ear canals of Sphynx cats, often in association with *M. nana*.

Other yeast-like fungi

A small study from Spain reported that *M. pachydermatis* represented only 3% of the yeast isolates from cases of otitis externa, and that *Candida* spp. and *Cryptococcus* spp. were more prevalent[121]. It may be that the local environment (hot and dry climate) and the microbiological methods used influenced this result.

Candida albicans has been recovered from normal ear canals of dogs and cats[95,96]. However, the organism is only rarely implicated in otitis externa. In one incident, *C. albicans* was implicated as the prime agent in an outbreak of apparently contagious otitis externa in 76 foxhounds[122].

Microbial changes associated with otitis externa

The overall changes in bacterial flora associated with otitis externa are qualitative and quantitative; the number of bacteria increases and the proportion of various species changes.

- The incidence of recovery of staphylococci in general, and of coagulase-positive staphylococci in particular, increases.
- More particularly, the incidence of recovery of *Pseudomonas* spp. and *Proteus* spp. increases.

The most common bacteria recovered from otitis externa in cats' ears were coagulase-positive staphylococci (54.8%). Gram-negative bacteria such as *Pseudomonas* spp. and *Proteus* spp. were only rarely recovered from feline otitis externa[116].

The fungal flora of the ear canal also changes in otitis externa and almost all of the increase results from an increased incidence of *M. pachydermatis*. Thus, Fraser recovered *M. pachydermatis* from 36% of normal ear canals and from 44% of cases of otitis externa[114]. However, the incidence of fungi was unchanged; indeed

the number of isolations of *Aspergillus* spp., *Penicillium* spp. and *Rhizopus* spp. was reduced in otitic ear canals.

Otitis media and otitis externa

Otitis media is thought to be an important cause of recurrent otitis externa, as the presence of bacteria in the middle ear may act as a focus for reinfection. The presence of an intact tympanum does not rule out otitis media since it can repair in the presence of otitis media. There may be quantitative and qualitative differences in microbial isolates from either side of an intact tympanum[123], perhaps suggesting that topical otic medications fail to cross the tympanum, resulting in different populations of bacteria. The implication of this is that separate samples from both the middle and external ear must be taken for bacterial culture and sensitivity testing if otitis media and otitis externa are present.

Effects of treatment on the otic microflora

Only a few published studies have investigated the otic microflora before and after appropriate treatment[117,121,122]. These studies noted:

- Dramatic reductions in the numbers of organisms recovered after treatment.
- A marked increase in the number of ears from which no growth was recorded.
- Several cases which responded well to treatment, and were pronounced cured, but in which the original organisms could still be recovered.

Therefore, clinical cure does not always necessitate microbial elimination. It may well be that in a number of cases the other agents within polypharmaceutical preparations, such as glucocorticoids and antimycotics in particular, affect inflammation, epidermal proliferation, and bacterial adhesion to such an extent that microbial multiplication is not possible. Therefore, although viable bacteria can be recovered from ear canals post-treatment, they are unable to adhere, multiply or cause infection.

1.8 PHYSIOLOGY OF HEARING

Introduction

Sound normally reaches the cochlea via the outer and middle ear, but it may also reach the cochlea by conduction through bone. The auricle and ear canal have two roles in transmitting sound to the tympanic membrane: they aid in sound localisation and they increase the sound pressure at the tympanic membrane by resonance[124–126]. The middle ear acts as an acoustic impedance transformer, transmitting energy from low-impedance air over the middle ear ossicles to the higher-impedance cochlear fluids[124–126]. Sensory transduction occurs in the cochlea, where the mechanosensory cells in the organ of Corti transform sound into a train of nerve impulses in the auditory nerve, thus conveying information to the brain. The cochlea separates sounds according to their frequency components so that different populations of hair cells become activated by sounds within different parts of the audible frequency range[124–126]. This frequency specificity is determined by the location of maximal amplitude of the travelling wave over the basilar membrane, which depends on the membrane's characteristics[124–126]. Furthermore, it depends on the outer hair cells, which play an important and active role in increasing the frequency selectivity of the basilar membrane, especially at low sound intensities[124–126].

Hearing involves the perception of sounds, which are air pressure waves generated by vibrating air molecules. Sound waves propagate in three dimensions, creating spherical shells of alternating compression and rarefaction[126]. Like all wave phenomena, sound waves have four major features: waveform, phase, amplitude and frequency. These determine the perception of sound, especially the frequency and the amplitude of the waves. Sounds composed of a single sine wave are, however, extremely rare in nature; most sounds consist of acoustically complex waveforms[126]. The frequency of a sound, expressed in cycles per second or Hertz (Hz) roughly corresponds to its pitch, whereas the amplitude, usually expressed in decibels (dB), determines its loudness. A change in the frequency composition and/or amplitude of a sound causes a change in stimulation of the ear and it is this that results in perception[126]. Humans are generally capable of hearing sounds in the frequency range of 20 Hz to 20 kHz[127], whereas in dogs the high-frequency cut-off is much higher at 43 kHz[128]. In addition, dogs are capable of hearing sounds at lower intensities than are humans.

Outer ear

As mentioned above, the auricle, the ear canal and the head influence the sound that reaches the tympanic membrane. In a free sound field, the head causes the sound pressure at the entrance of the ear canal to be higher than it would be at the same location in the field in the absence of the body[124]. This effect depends on the frequency of the sound and on the direction of the head relative to the sound source. Lower frequency sounds pass around the head, whereas middle and higher frequencies are captured by the auricle. The difference in time of arrival of a sound at the two ears is the physical basis for directional hearing in the horizontal plane, together with the difference in intensity of the sound at the two ears[124]. Intra-aural time differences in the onsets of sounds are most important for sounds below 1,500 Hz, while it is the difference in the intensity that is most important for high-frequency sounds[124]. The raised ridges of the pinna and conchae aid in reflecting sound waves into the ear canal and are believed to provide clues on the direction of the sound source as coming from the front or behind or from above or below[125].

The outer ear collects sound waves over the large area of the pinna and concha and funnels them into the narrower ear canal. The sound pressure at the tympanic membrane not only depends on the acoustic properties of the auricle and the head, but also on those of the outer ear canal. The concha of the auricle and the outer ear canal act as resonators, increasing the sound pressure at the tympanic membrane[124]. The gain is greatest near 2.5–3 kHz (the resonance frequency) in humans, being approximately 10–20 dB[124,125]. In the cat, the pinna can produce a gain of up to 21 dB in sound pressure at high frequencies[129]. A gain as a result of resonance in the auricle and outer ear canal has not been studied in dogs, but is likely to be similar to that in cats. Even though the auricle in dogs varies greatly in size and shape (more so than in cats), studies have shown that the type and position of the pinnae (erect versus pendulous ears) at least do not appear to have a large effect on the free-field audiogram[128]. Little is known about the effects of the frequent auricle movements in dogs and cats on sound localisation. Research in cats has shown that for sound sources in front of the cat, the midfrequency components provide the clue for source localisation[130].

Middle ear

The middle ear acts as an impedance transformer that matches the high impedance of the cochlear fluids to the low impedance of the air in the outer ear canal. The middle ear transformer uses two principles. First, the pressure is increased because the area of the oval window is smaller than that of the tympanic membrane[124–126]. Second, the lever action of the ossicles increases the force acting on the oval window[125]. It is the difference between the force that acts on the two windows of the cochlea that sets the cochlear fluid in motion. Normally, the force on the oval window is much larger than that acting on the round window because of the gain of the middle ear, via the ossicular chain[124]. A broken chain results in hearing loss. The gain of the middle ear is frequency dependent and the increase in sound transmission to the cochlear fluid, due to improvement in impedance matching, is close to 30 dB in the midfrequency range in humans[124]. In dogs the gain of the middle ear is unknown, but in cats the greatest transmission is produced in the range around 1–2 kHz and is also around 30 dB[131–133]. Transmission through the middle ear is affected by the middle ear muscles, which reduce the transmission of low-frequency sounds. This may serve to protect the ear to some extent from noise damage[125].

Inner ear

The movement of the stapes footplate in the oval window sets up a fluid wave inside the scala vestibuli in the cochlea, displacing the fluid toward the round window. As the basilar membrane is displaced by this travelling fluid wave, the hair cells within the organ of Corti are also displaced. This displacement causes deflection of the stereocilia, which results in opening of the ion-selective channels by the shearing forces on the stereocilia. This allows K^+ ions to enter the cell, causing depolarisation of the hair cell[126,127,134,135]. This generates a sensory action potential which, when summed with numerous hair cell sensory action potentials, generates the all-or-none nerve compound action potential. The inner hair cell clearly forms the centre of the cochlear apparatus, as it is solely responsible for transducing the mechanical energy brought by sound and initiating the receptor potential and the auditory nerve action potential[134]. Deflection of the stereocilia by the travelling wave opens and closes ion channels in the stereocilia, thereby modulating the current being driven into the hair cells, its magnitude determined by the combined effects of the positive endocochlear potential and the negative intracellular potential[127,135]. Changes in the electrical potentials can

be measured in the hair cells with fine microelectrodes, and grossly in the cochlea with larger electrodes. The potentials that can be recorded with an electrode at or near the cochlea became known as the cochlear microphonics (CM), the source of which is mainly the outer hair cells. Synchronised activity of auditory nerve fibres and of neurons in the passageway through the brainstem toward the auditory cortex can be recorded as far-field potentials with large electrodes on the skull (brainstem-evoked response audiometry, see Chapter 2 Diagnostic procedures, Section 2.10).

The range of sound frequencies and intensities that the ear is capable of handling is impressive. The frequency selectivity of the cochlea has been the topic of many studies and in the past 10 years the knowledge concerning this subject has increased dramatically. It has been known for a long time that frequency selectivity of the cochlea is based on the basilar membrane characteristics. Only recently it became known that it is the active function of the outer hair cells that provides the fine-tuning of the frequency selectivity and that it is responsible for the amplitude compression of sounds and the creation of otoacoustic emissions.

In the 19th century, Von Helmholtz proved that the ear performs a spectral analysis of sounds. He suggested that the basilar membrane functions as a series of resonators tuned to different frequencies covering the audible range[127]. This became known as the resonance theory[127]. Georg von Békésy showed that the tone of a certain frequency caused the highest vibration amplitude at a certain point along the basilar membrane and that a frequency scale could be laid out along the cochlea, with high frequencies located at the base and low frequencies at the apex[136]. Von Békésy showed that sounds set up a travelling wave motion along the basilar membrane. Travelling waves produced by sounds of high frequency reach a peak near the base of the cochlea, whereas waves produced by low-frequency sounds reach a peak closer to the apex[126,127,134,135]. The physical make-up alone, however, cannot explain the fine-tuning properties as implied by psychoacoustic data on the frequency discrimination in the auditory system[127]. It was not until much later that the actual source of the fine tuning was discovered to be the outer hair cells.

The properties of the outer hair cells were further appreciated by the studies of Brownell and co-workers, who analysed *in vitro* solitary outer hair cells removed from the cochlea. As the outer hair cells were stimulated, they were found to actually expand, stretch and contract[137]. This property is known as electromotility or as the cochlear amplifier[126,127,135]. The sharply tuned peak of the travelling wave arises because the outer hair cells, when stimulated by the movement, make an active mechanical response that amplifies the vibration of the basilar membrane as the travelling wave passes through[135]. The travelling wave therefore increases in amplitude as it passes along the cochlear duct, until it dies away abruptly when it reaches a point where the cochlear partition can no longer sustain vibrations of that particular frequency[135].

It is now also known that the ability of the basilar membrane to separate the frequency components of sounds is intensity dependent because of the active role of the outer hair cells[127]. For low sound intensities the outer hair cells actively compensate for the energy losses in the basilar membrane. The active amplification by the outer hair cells has its largest effect at low stimulus intensities and makes a smaller contribution at higher intensities[127,135]. Therefore, the frequency selectivity of the cochlea decreases with increasing stimulus intensity and the location of the maximal response of the basilar membrane shifts towards the base of the cochlea[127]. This way the cochlea compresses the amplitude of sounds before initiating nerve impulses in the auditory nerve. This has an important functional consequence, because it allows the auditory system to discriminate stimuli over a very wide range of stimulus intensities[127].

The finding of the active outer hair cells brought the understanding of the functioning of the cochlea a large step forward and answered many, but not all, unresolved questions that had been troubling researchers for many years. The active role of the outer hair cells explains why metabolic energy is necessary to maintain the normal sensitivity and frequency selectivity of the ear. Furthermore, it explains why oxygen deprivation causes the threshold of auditory nerve fibres to increase and the tuning to become wider[127]. The fact that efferent neural activity controls the function of the outer hair cells implies that the brain can modulate the mechanical properties of the basilar membrane. This will undoubtedly be the subject of many future studies.

1.9 REFERENCES

1 Cole LK. Anatomy and physiology of the canine ear. *Veterinary Dermatology* 2009;**20**:412–21.

2 Getty R, Foust HL, Presley ET, Miller ME. Macroscopic anatomy of the ear of the dog. *American Journal of Veterinary Research* 1956;**17**:364–75.

3 Fraser G, Gregor WW, Mackenzie CP, Spreull JSA, Withers AR. Canine ear disease. *Journal of Small Animal Practice* 1970;**10**:725–54.

4 Evans HE. Miller's *Anatomy of the Dog*, 3rd edn. Evans HE (ed). WB Saunders, Philadelphia 1993, pp. 988–1008.

5 Hudson LC, Hamilton WP. *Atlas of Feline Anatomy for Veterinarians*. WB Saunders, Philadelphia 1993, pp. 228–39.

6 Henderson RA, Horne RD. The pinna. In: *Textbook of Small Animal Surgery*, 2nd edn. Slatter D (ed). WB Saunders, Philadelphia 1993, pp. 1545–59.

7 Smeak DD. Total ear canal ablation and lateral bulla osteotomy. In: *Current Techniques in Small Animal Surgery*, 4th edn. Bojrab MJ (ed). Williams and Wilkins, Baltimore 1998, pp. 102–9.

8 Whalen L, Kitchell R. Electrophysiological studies of the cutaneous nerves of the head of the dog. *American Journal of Veterinary Research* 1983;44:615-27.

9 Huang H-P. *Studies of the Microenvironment and Microflora of the Canine External Ear Canal*. PhD Thesis, Glasgow University, 1993.

10 Forsythe WB. Tympanographic volume measurements of the canine ear. *American Journal of Veterinary Research* 1985;46:1351–3.

11 Wefstaedt P, Behrens B-A, Nolte I, Bouguecha A. Fine element modelling of the canine and feline outer ear canal: benefits for local drug delivery. *Berliner und Münchener Tierärztliche Wochenschrift* 2011;124:78–82.

12 Njaa BL, Cole LK, Tabacca N. Practical otic anatomy and physiology of the dog and cat. *Veterinary Clinics of North America Small Animal Practice* 2012;42:1109–26.

13 Bluestone CD, Doyle WJ. Anatomy and physiology of the eustachian tube and middle ear related to otitis media. *Journal of Allergy and Clinical Immunology* 1998;81:997–1003.

14 Bluestone CD. Eustachian tube function: physiology, pathophysiology, and role of allergy in pathogenesis of otitis media. *Journal of Allergy and Clinical Immunology* 1983;72:242–51.

15 Ostfield E, Blonder J, Crispin M, Szeinberg A. The middle ear gas composition in air-ventilated dogs. *Acta Otolaryngology* 1980;89:105–8.

16 Chole RA, Kodama K. Comparative histology of the tympanic membrane and its relationship to cholesteatoma. *Annals of Rhinology and Laryngology* 1989;98:761–6.

17 Neer TM. Otitis media. *Compendium on Continuing Education* 1982;4:410–16.

18 Nummela S. Scaling of the mammalian middle ear. *Hearing Research* 1995;85:18–30.

19 Secondi U. Structure and function of the lamina propria of the tympanic membrane in various mammals. *Archives of Otolaryngology* 1951;53:170–81.

20 Saridomichelakis MN, Farmaki R, Leontides LS, Koutinas AF. Aetiology of canine otitis externa: a retrospective study of 100 cases. *Veterinary Dermatology* 2007;18:341–7.

21 Huang GT, Rosowski JJ, Flandermeyer DT, Lynch TJ, Peak WT. The middle ear of a lion: comparison of structure and function to domestic cat. *Journal of the Acoustic Society* 1997;101:1532–49.

22 Goycoolea MV, Lundman L. Round window membrane. Structure, function, and permeability: a review. *Microscopy Research and Technique* 1997;36:201–11.

23 Rose WR. The eustachian tube. 1: General considerations. *Veterinary Medicine/Small Animal Clinician* 1978;73:882–7.

24 Boothe HW. Ventral bulla osteotomy: dog and cat. In: *Current Techniques in Small Animal Surgery*, 4th edn. Bojrab MJ (ed). Williams and Wilkins, Baltimore 1998, pp. 109–12.

25 Seim HB III . Middle ear. In: *Textbook of Small Animal Surgery*, 2nd edn. Slatter D (ed). WB Saunders, Philadelphia 1993, pp. 1568–76.

26 Trevor PB, Martin RA. Tympanic bulla osteotomy for treatment of middle-ear disease in cats: 19 cases (1984–1991). *Journal of the American Veterinary Medical Association* 1993;202:123–8.

27 Garosi LS, Lowrie ML, Swinbourne NF. Neurological manifestations of ear disease in dogs and cats. *Veterinary Clinics of North America Small Animal Practice* 2012;42:1143–60.

28 El-Mofty A, El-Serafy S. The ossicular chain in mammals. *Annals of Otology, Rhinology and Laryngology* 1967;76:903–9.

29 Eom K, Kwak H, Kang H, *et al*. Virtual CT otoscopy of the middle ear and ossicles in dogs. *Veterinary Radiology and Ultrasound* 2008;49:545–50.

30 Cole LK, Samii, VF. Contrast-enhanced computed tomographic imaging of the auditory tube in mesaticephalic dogs. *Veterinary Radiography and Ultrasound* 2007;48:125–8.

31 Hopwood PR, Bellenger CR. Cannulation of the canine auditory tube. *Research in Veterinary Science* 1980;28:382–3.

32 Cunsolo E, Marchioni D, Leo G, Incorvaia C, Presutti L. Functional anatomy of the Eustachian tube. *International Journal of Immunopathology and Pharmacology* 2010;23:S4–S7.

33 McGuire JF. Surfactant on the middle ear and Eustachian tube: a review. *International Journal of Pediatric Otorhinolaryngology* 2002;66:1–15.

34 Chandrasekhar SS, Mautone AJ. Otitis media: treatment with intranasal aerosolized surfactant. *Laryngoscope* 2004;114:472–85.

35 Chauhan B, Chauhan K. A comparative study of Eustachian tube functions in normal and diseased ears with tympanometry and videonasopharyngoscopy. *Indian Journal of Otolaryngology and Head and Neck Surgery* 2013;65:468–76.

36 Mandel EM, Swarts JD, Casselbrant ML, *et al*. Eustachian tube function as a predictor of the recurrence of middle ear effusion in children. *Laryngoscope* 2013;123:2285–90.

37 Stern-Bertholtz W, Sjöström L, Håkanson NW. Primary secretory otitis media in the Cavalier King Charles spaniel: a review of 61 cases. *Journal of Small Animal Practice* 2003;44:253–6.

38 Cole LK. Primary secretory otitis media in Cavalier King Charles spaniels. *Veterinary Clinics of North America Small Animal Practice* 2012;42:1137–42.

39 Wolf D, Lüpke M, Wefstaedt P, Klopmann T, Nolte I, Seifert H. Optimising magnetic resonance image quality of the ear in healthy dogs. *Acta Veterinaria Hungarica* 2011;59:53–68.

40 Stout-Graham M, Kainer RA, Whalen LR, Macy DW. Morphologic measurements of the external ear canal of dogs. *American Journal of Veterinary Research* 1990;51:990–4.

41 Fernando SDA. A histological and histochemical study of the glands of the external auditory canal of the dog. *Research in Veterinary Science* 1996;7:116–19.

42 Johnson A, Hawke M, Berger G. Surface wrinkles, cell ridges and desquamation in the external auditory canal. *Journal of Otolaryngology* 1984;13:345–54.

43 Revadi G, Prepageran N, Raman R, Sharzal TA. Epithelial migration on the external ear canal wall in normal and pathologic ears. *Otology & Neurology* 2011;32:504-7.

44 Scott DW. Feline dermatology: a monograph. *Journal of the American Animal Hospital Association* 1980;16:426–33.

45 Huang HP, Fixter LM, Little CJL. Lipid content of cerumen from normal dogs and otitic canine ears. *Veterinary Record* 1994;134:380–1.

46 Fernando SDA. Certain histopathologic features of the external auditory meatus of the cat and dog with otitis externa. *American Journal of Veterinary Research* 1996;28:278–82.

47 Van der Gaag I. The pathology of the external ear canal in dogs and cats. *Veterinary Quarterly* 1986;8:307–17.

48 Angus JC, Lichtensteiger C, Campbell KL, Schaeffer DJ. Breed variations in histopathologic features of chronic severe otitis externa in dogs: 80 cases (1995–2001). *Journal of the American Veterinary Medical Association* 2002;221:1000-6.

49 Lim DJ. Structure and function of the tympanic membrane: a review. *Acta Otorhinolaryngolica Belg* 1995;49:101–15.

50 Secondi U. Structure and function of the lamina propria of the tympanic membrane in various mammals. *Archives of Otolaryngology* 1951;53:170–81.

51 Filogamo G. Recherches sur la structure de la membrane du tympan chez les différents vertébrés. *Acta Anatomica* 1949;7:248–72.

52 Lim DJ. Tympanic membrane. Electron microscopic observations. Part II: Pars flaccida. *Acta Otolaryngologica* 1968;66:515–32.

53 Chole RA, Kodama K. Comparative histology of the tympanic membrane and its relationship to cholesteatoma. *Annals of Rhinology and Laryngology* 1989;98:761–6.

54 Lim DJ. Tympanic membrane. Electron microscopic observations. Part I: Pars tensa. *Acta Otolaryngologica* 1968;66:181–98.

55 Litton W. Epithelial migration over the tympanum and external canal. *Archives of Otolaryngology* 1963;77:254–7.

56 Reeve DR. Some observations on the diurnal variation of mitosis in the stratified squamous epithelium of wounded tympanum. *Cell, Tissue Research* 1977;9:253–63.

57 Broekaert D. The migratory capacity of the external auditory canal epithelium: a critical mini review. *Acta Otorhinolaryngologica Belgica* 1990;44:385–92.

58 Tabacca NE, Cole LK, Hillier A, Rajala-Schultz PJ. Epithelial migration on the canine tympanic membrane. *Veterinary Dermatology* 2011;22:502–10.

59 Monkhouse WS, Moran P, Freedman A. The histological effect on the guinea pig external ear of several constituents of commonly used aural preparations. *Clinical Otolaryngology* 1988;13:121–31.

60 Sennaroglu L, Özkul A, Gedikoglu G, Turan E. Effect of intratympanic steroid application on the development of experimental cholesteatoma. *Laryngoscope* 1998;108:543–7.

61 Boedts D. Tympanic membrane perforations. *Acta Otorhinolaryngolica Belg* 1995;49:149–58.

62 Dyson M. Advances in wound healing physiology: the comparative perspective. *Veterinary Dermatology* 1997;8:227–33.

63 Makino K, Amatsu M, Kinishi M, Mohri M. Epithelial migration in the healing process of tympanic membrane perforations. *Archives of Otorhinolaryngology* 1990;247:352–5.

64 Wang WQ, Wang ZM, Tian J. Spontaneous healing of various types of rat tympanic membrane perforation. *Zhonghua Er Bi Yan Hou Ke Zha Zhi* 2004;39:602-5.

65 Fagan P, Patel N. A hole in the drum. An overview of tympanic membrane perforations. *Australian Family Practitioner* 2002;31:707-10.

66 Truy E, Disant F, Morgon A. Chronic tympanic membrane perforation: an animal model. *American Journal of Otology* 1995;16:222–5.

67 Steiss JE, Boosinger TR, Wright JC, Pillai SR. Healing of experimentally perforated tympanic membranes demonstrated by electrodiagnostic testing and histopathology. *Journal of the American Animal Hospital Association* 1982;28:307–10.

68 Orji FT, Agu CC. Determinants of spontaneous healing in traumatic perforations of the tympanic membrane. *Clinical Otolaryngology* 2008;33:420–6.

69 Lou ZC, Tang YM, Yang J. A prospective study evaluating spontaneous healing of aetiology, size and type-different groups of traumatic tympanic membranes. *Clinical Otolaryngology* 2011;36:450–60.

70 Little CJL, Lane JG. Inflammatory middle ear disease of the dog: the pathology of otitis media. *Veterinary Record* 1991;128:293–6.

71 Mansfield PD, Steiss JE, Boosinger TR, Marshall AE. The effects of four commercial ceruminolytic agents on the middle ear. *Journal of the American Animal Hospital Association* 1997;35:479–86.

72 Little CJL, Lane JG, Gibbs C, Pearson GR. Inflammatory middle ear disease in the dog: the clinical and pathological features of cholesteatoma, a complication of otitis media. *Veterinary Record* 1991;128:319–22.

73 Preciado DA. Biology of cholesteatoma: special considerations in pediatric patients. *International Journal of Pediatric Otorhinolaryngology* 2012;76:319–21.

74 Harada Y. Scanning electron microscopic study on the distribution of epithelial cells in the eustachian tube. *Acta Otolaryngolica* 1997;83:284–90.

75 Sucheston ME, Cannon MS. Eustachian tube of several mammalian species. *Archives of Otolaryngology* 1971;93:58–64.

76 Birken EA, Brookler KH. Surface tension lowering substance of the canine eustachian tube. *Annals of Otology* 1972;81:268–71.

77 Mendenhall RM, Mendenhall AL, Tucker JH. A study of some biological surfactants. *Annals of the New York Academy of Sciences* 1966;130:902–19.

78 Bluestone CD. Eustachian tube function: physiology, pathophysiology, and role of allergy in pathogenesis of otitis media. *Journal of Allergy and Clinical Immunology* 1983;72:242–5.

79 Goycoolea MV, Lundman L. Round window membrane. Structure, function, and permeability: a review. *Microscopy Research and Technique* 1997;36:201–11.

80 Revadi G, Prepageran N, Raman R, Sharazi TA. Epithelial migration on the external ear canal wall in normal and pathologic ears. *Otology and Neurology* 2011;32:504–7.

81 Grono LR. Studies of the microclimate of the external auditory canal in the dog. 1: Aural temperature. *Research in Veterinary Science* 1970;11:307–11.

82 Huang H-P, Shih H-M, Chen K-Y. The application of an infrared tympanic membrane thermometer in comparing the external ear canal temperature between erect and pendulous ears in dogs. In: *Advances in Veterinary Dermatology*, Volume 3. Kwochka KW, Willemse T, von Tscharner C (eds). Butterworth Heinemann, Oxford 1988, pp. 57–63.

83 Huang H-P, Shih H-M. Use of infrared thermometry and effect of otitis externa on external ear canal temperature in dogs. *Journal of the American Veterinary Medical Association* 1998;213:76–9.

84 Yoshida N, Naito F, Fukuta T. Studies of certain factors affecting the microenvironment and microflora of the external ear canal of dogs in health and disease. *Journal of Veterinary Medical Science* 2002;64:1145-7.

85 Grono LR. Studies of the microclimate of the external auditory canal in the dog. III: Relative humidity within the external auditory meatus. *Research in Veterinary Science* 1970;11:316–19.

86 Grono LR. Studies of the microclimate of the external auditory canal in the dog. II: Hydrogen ion concentration of the external auditory meatus in the dog. *Research in Veterinary Science* 1970;11:312–15.

87 Huang HP, Little CJL, Fixter LM. Effects of fatty acids on the growth and composition of *Malassezia pachydermatis* and their relevance to canine otitis externa. *Research in Veterinary Science* 1993;55:119–23.

88 Roeser RJ. Physiology, pathophysiology, and anthropology/epidemiology of human ear canal secretions. *Journal of the American Academy of Audiology* 1997;8:391–400.

89 Gutteridge JMC, Lamport P, Dormandy TL. Autoxidation as a cause of antibacterial activity in unsaturated fatty acids. *Journal of Medical Microbiology* 1974;7:387–9.

90 Knapp HR, Melly MA. Bactericidal effects of polyunsaturated fatty acids. *Journal of Infectious Diseases* 1986;154:84–94.

91 Masuda A, Sukegawa T, Mizumoto N, *et al*. Study of lipid in the ear canal in canine otitis externa with *Malassezia pachydermatis*. *Journal of Veterinary Medical Science* 2000;11:1177–82.

92 Zur G, Lifshitz B, Bdolah-Abram T. The association between the signalment, common causes of canine otitis externa and pathogens. *Journal of Small Animal Practice* 2011;52:254–8.

93 Pressanti C, Drouet C, Cadiergues M-C. Comparative study of aural microflora in healthy cats, allergic cats and cats with systemic disease. *Journal of Feline Medicine and Surgery* 2014;16:992–6.

94 Aoki-Komori S, Shimada K, Tani K, Katayama M, Saiti TR, Kataoka Y. Microbial flora in the ears of healthy experimental Beagles. *Experimental Animals* 2007;56:67–9.

95 Matsuda H, Tojo M, Fukui K, Imori T, Baba E. The aerobic bacterial flora of the middle and external ears in normal dogs. *Journal of Small Animal Practice* 1984;25:269–74.

96 Ginel PJ, Lucena R, Rodriguez JC, Ortega J. A semi-quantitive cytological evaluation of normal and pathological samples from the external ear canal of dogs and cats. *Veterinary Dermatology* 2002;13:151–6.

97 Dickson DB, Love DN. Bacteriology of the horizontal ear canal of dogs. *Journal of Small Animal Practice* 1983;24:413–21.

98 Gustafson B. Otitis externa hos hund. *Nordic Veterinærmedicin* 1954;6:434–42.

99 Fraser G. Factors predisposing to canine external otitis. *Veterinary Record* 1961;73:55–8.

100 Grono LR, Frost AJ. Otitis externa in the dog. *Australian Veterinary Journal* 1969;45:420–2.

101 Sharma VD, Rhodes HE. The occurrence and microbiology of otitis externa in the dog. *Journal of Small Animal Practice* 1975;16:241–7.

102 McCarthy G, Kelly WR. Microbial species associated with the canine ear and their antibacterial sensitivity patterns. *Irish Veterinary Journal* 1982;36:53–6.

103 Chengappa MM, Maddux R, Greer S. A microbiologic survey of clinically normal and otitic ear canals. *Pet Practice* 1983;78:343–4.

104 Marshall MJ, Harris AM, Horne JE. The bacteriological and clinical assessment of a new preparation for the treatment of otitis externa in dogs and cats. *Journal of Small Animal Practice* 1974;15:401–10.

105 Hoffman AR, Paterson AM, Diesel A, *et al*. The skin microbiome in healthy and allergic dogs. *PLOS One* 2014;9:1–12.

106 Harvey RG, Lloyd DH. The distribution of *Staphylococcus intermedius* and coagulase-negative staphylococci on the hair, skin surface, within the hair follicles and on the mucous membranes of eleven dogs. *Veterinary Dermatology* 1995;5:75–81.

107 August JR. Otitis externa: a disease of multifactorial etiology. *Veterinary Clinics of North America* 1988;18:731–42.

108 Harihan H, Matthew V, Fountain J, *et al*. Aerobic bacteria from mucous membranes, ear canals, and skin wounds of feral cats in Grenada, and the antimicrobial dug susceptibility of major isolates. *Comparative Immunology, Microbiology and Infectious Diseases* 2011;34:129–34.

109 Cox HU, Hoskins JD, Newman SS, *et al*. Distribution of staphylococcal species on clinically healthy cats. *American Journal of Veterinary Research* 1985;46:1824–8.

110 Girão MD, Prado MR, Brilhante RSN, *et al.* *Malassezia pachydermatis* isolated from normal and diseased ear canals in dogs: a comparative analysis. *The Veterinary Journal* 2006;172:544–8.

111 Cafarchia C, Gallo S, Romito D, *et al.* Frequency, body distribution and population size of *Malassezia* species in healthy dogs and in dogs with localized cutaneous lesions. *Journal of Veterinary Investigation* 2005;17:316–22.

112 Defalque VE, Rosser EJ, Peterson AD. Aerobic and anaerobic microflora of the middle ear cavity in normal dogs. Harrison BA (ed). *Proceedings of the 20th Meeting of the North American Veterinary Dermatology Forum*, Florida 2005, p. 159.

113 Cafarchia C, Gallo S, Romito D, *et al.* Frequency, body distribution and population size of *Malassezia* species in healthy dogs and in dogs with localized cutaneous lesions. *Journal of Veterinary Investigation* 2005;17:316–22.

114 Crespo MJ, Abarca ML, Cabañes FJ. Occurrence of *Malassezia* spp. in the external ear canals of dogs and cats with and without otitis externa. *Medical Mycology* 2002;40:115–21.

115 Cafarchia C, Gallo S, Capelli G, Otranto D. Occurrence and population size of *Malassezia* spp. in the external ear canal of dogs and cats both healthy and with otitis. *Mycopathologia* 2005;160:143–9.

116 Dizotti CE, Coutinho DA. Isolation of *Malassezia pachydermatis* and *M sympodialis* from the external ear canal of cats with and without otitis externa. *Acta Veterinaria Hungarica* 2007;55:471–7.

117 Shokri H, Khosravi A, Rad, MA, Jamshidi S. Occurrence of *Malassezia* in Persian and Domestic Short Hair cats with and without otitis externa. *Journal of Veterinary Medical Science* 2010;72:293–6.

118 Bond R, Stevens K, Perrins N, Ahman S. Carriage of *Malassezia* spp. Yeast in Cornish Rex, Devon Rex and Domestic Short Hair cats: a cross sectional survey. *Veterinary Dermatology* 2008;19:299–304.

119 Ahman S, Bergström KE. Cutaneous carriage of *Malassezia* species in healthy and seborrhoeic Sphynx cats and a comparison to carriage in Devon Rex cats. *Journal of Feline Medicine and Surgery* 2009;11:970–6.

120 Volk AV, Belyavin CE, Varjonen K, *et al.* *Malassezia pachydermatis* and *M. nana* predominate amongst the cutaneous mycobiota of Sphynx cats. *Journal of Feline Medicine and Surgery* 2010;12:917–22.

121 Blanco JL, Guedeja-Marron J, Hontecillas R, Suarez G, Garcia M-E. Microbiological diagnosis of chronic otitis externa in the dog. *Journal of Veterinary Medicine* 1996;43:475–82.

122 McKellar, QA, Rycroft, A, Anderson L, Love J. Otitis externa in a foxhound pack associated with Candida albicans. *Veterinary Record* 1990;127:15–16.

123 Cole LK, Kwochka KW, Kowalski JJ, Hillier A. Microbial flora and antimicrobial susceptibility patterns of isolated pathogens from the horizontal ear canal and middle ear in dogs with otitis media. *Journal of the American Veterinary Medical Association* 1998;212:534–8.

124 Moller AR. Sound conduction to the cochlea. In: *Hearing: its Physiology and Pathophysiology*. Moller AR (ed). Academic Press, San Diego, 2000, pp. 29–70.

125 Pickles JO. The outer and middle ears. In: *An Introduction to the Physiology of Hearing*. Pickles JO (ed). Emerald Group Publishing, Bingley 2008, pp. 11–24.

126 Purves D, Augustine GJ, Fitzpatrick D, *et al.* The auditory system. In: *Neuroscience*. Purves D, Augustine GJ, Fitzpatrick D, Katz LC, LaMantia AS, McNamara JO (eds). Sinauer Associates, Sunderland 2001, pp. 275–96.

127 Moller AR. Physiology of the cochlea. In: Hearing: its Physiology and Pathophysiology. Moller AR (ed). Academic Press, San Diego 2000, pp. 71–94.

128 Heffner HE. Hearing in large and small dogs: absolute thresholds and size of the tympanic membrane. *Behavioral Neuroscience* 1983;97(2):310.

129 Musicant AD, Chan JC, Hind JE. Direction-dependent spectral properties of cat external ear: new data and cross-species comparisons. *Journal of the Acoustical Society of America* 1990;87:757.

130 Rice JJ, May BJ, Spirou GA, Young ED. Pinna-based spectral cues for sound localization in cat. *Hearing Research* 1992;58:132–52.

131 Nedzelnitsky V. Sound pressures in the basal turn of the cat cochlea. *Journal of the Acoustical Society of America* 1980;68(6):1676–89.

132 Rosowski JJ. Outer and middle ears. In: *Comparative Hearing: Mammals*. Fay RR, Popper AN (eds). Springer, New York 1994, pp. 172–247.

133 Voss SE, Shera CA. Simultaneous measurement of middle-ear input impedance and forward/reverse transmission in cat. *Journal of the Acoustical Society of America* 2004;116:2187–98.

134 Dallos P. The active cochlea. *Journal of Neuroscience* 1992;12:4575–85.

135 Pickles JO. The cochlea. In: *An Introduction to the Physiology of Hearing*. Pickles JO (ed). Emerald Group Publishing Limited, Bingley 2008, pp. 25–72.

136 Békésy Von G. *Experiments in Hearing*. New York: McGraw-Hill, New York 1960, pp. 1–745.

137 Brownell WE, Bader CR, Bertrand D, de Ribaupierre Y. Evoked mechanical responses of isolated cochlear outer hair cells. *Science* 1985; 227(4683):194–6.

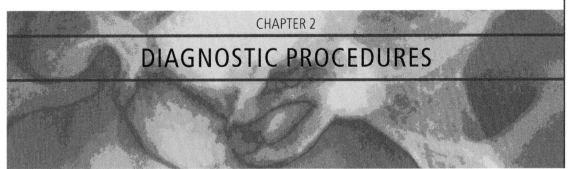

CHAPTER 2

DIAGNOSTIC PROCEDURES

2.1 INTRODUCTION

The approach to a case of otitis externa is no different to that of any disease: the signalment will allow the clinician to consider breed, age and sex predisposition to otitis externa. A case history may be sufficient to allow a working diagnosis of a foreign body to be made. In other cases it may be apparent that the dog has suffered occasional bouts of bilateral otitis externa before and, in these cases, a more detailed approach is necessary. This may include general anaesthesia, imaging and the taking and submission of samples for further investigation.

Consideration of the history and signalment will allow the clinician to make a provisional differential diagnosis, which will be further amended once the physical examination has been performed. At this point, a number of techniques can be utilised to investigate the otitis (Table 2.1), with the objective of defining, or staging, the otitis:

- Foreign body or ectoparasite (Fig. 2.1).
- Uni- or bilateral peracute otitis (Fig. 2.2).
- Uni- or bilateral chronic atopic-type otitis (Fig. 2.3).
- Uni- or bilateral chronic gram-negative-type ulcerative otitis (Fig. 2.4).
- Calcification of auricular cartilages.
- Suspicion or evidence of underlying disease (allergy, pemphigus foliaceus, seborrhoea) or known breed predisposition (Fig. 2.5).
- Otic neoplasia (Fig. 2.6).
- Presence of otitis media and interna or not (Fig. 2.7)?

Staging, or defining the type of otitis, allows for effective management. For example, atopic-type ears do not usually suffer from gram-negative complications or calcification, unlike the Cocker Spaniel-type. Understanding the likely progression, and the potential for complications, allows the clinician to put in place effective management, such as surgical intervention, before irreversible otic hyperplasia and calcification takes place, which makes surgery much more difficult.

Table 2.1 **Investigative techniques that may be employed in order to define otitis externa**
Visual examination and palpation of the auricular cartilages
Otoscopic/video-endoscopic examination of the external ear canal
Cytological examination of cerumen and exudate
Culture and sensitivity testing of exudate
Cleaning the ear canal to facilitate examination of the deeper canal and the tympanic membrane
Imaging: computed tomography (CT), radiography, including canalography
Histopathological examination of biopsy samples

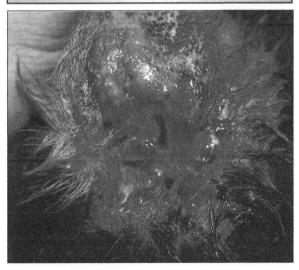

Fig. 2.1 Acute otitis externa associated with a grass awn in the horizontal portion of the external ear canal.

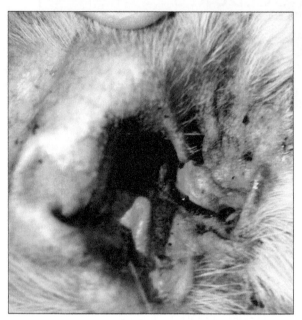

Fig. 2.2 Peracute erythematous otitis in a dog with atopy.

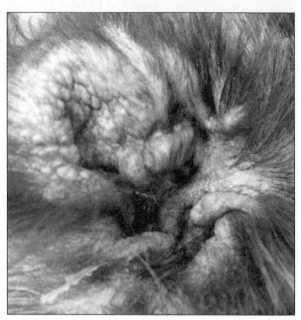

Fig. 2.3 Chronic hyperplastic atopic-type otitis externa.

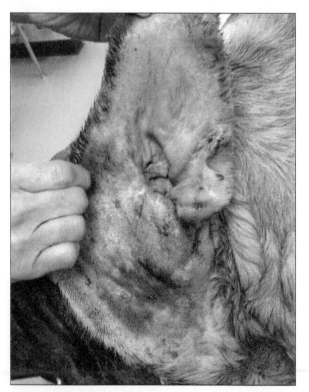

Fig. 2.4 Chronic, hyperplastic, ulcerated gram-negative otitis in a Mastiff. Note how clipping reveals extensive, periauricular and facial subacute dermatitis associated with the chronic otitis.

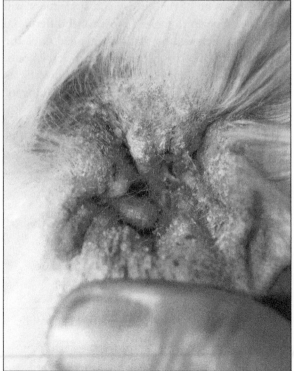

Fig. 2.5 Hyperplastic malassezial otitis externa in a West Highland White Terrier. Breed predisposition to atopy and epidermal dysplasia raise suspicion of an underlying cause to the otitis.

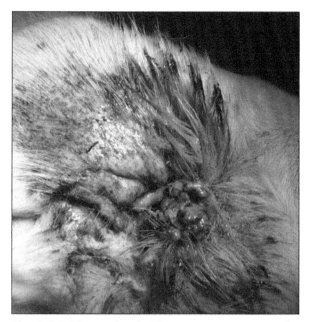

Fig. 2.6 Otitis associated with a verrucose tumour at the entrance to the external ear canal. There was associated malodour.

Fig. 2.7 Right-sided head tilt associated with otitis media and interna.

2.2 HISTORY AND SIGNALMENT

Breed

Some breeds, for example Cocker Spaniels and Persian cats, are predisposed to defects in keratinisation, which may be associated with a ceruminous otitis externa. Breeds of dog recognised as particularly prone to otitis externa include Springer Spaniels, Miniature Poodles, Shar Pei and German Shepherd Dogs. Any breed predisposed to atopy is likely to exhibit (usually) bilateral otitis externa.

- Dogs with pendulous pinnae are not necessarily predisposed to otitis externa but they may be susceptible to rapidly progressive otic disease, should otitis externa develop.
- Breeds with excessive hair within the external ear canals, such as Poodles and Maltese Terriers, and with pendulous pinnae, may be predisposed to accumulations of cerumen and malassezial overgrowth, which may provoke otitis externa[1].
- Cocker Spaniels were predisposed to grass awn-related acute otitis externa in one study of 100 cases[2].
- Cocker Spaniels appear predisposed to a rapidly developing hyperplastic otitis[3], often associated

with gram-negative infection, mandating early surgical intervention.

- Shar Pei have been reported to have narrow external ear canals[4], making otoscopic examination difficult and perhaps predisposing to stenosis.
- Yorkshire Terriers are predisposed to bilateral pinnal alopecia and hyperpigmentation.
- Longhaired breeds of cats, and show cats, are commonly affected by dermatophytosis.
- White haired cats and dogs are predisposed to actinic radiation damage to the pinnae.

Note that feline otitis is different to that seen in the dog[5]. Gram-negative complications and stenosis secondary to chronic allergic inflammation are not encountered, for example.

Age

Young animals are often affected with *Otodectes cynotis*, but this may not be associated with pruritus, particularly in kittens. In cats, the peak incidence of otitis externa is between 1 and 2 years of age, presumably reflecting exposure to, and hypersensitivity to, *O. cynotis*. In older cats bilateral otitis externa is almost always associated with *O. cynotis* infection whereas unilateral otitis externa may reflect cat bite abscess or obstructive otitis secondary to polyps or neoplasia. Young animals, particularly kittens, are predisposed to dermatophytosis.

Young dogs (and very old animals) are under-represented in studies of the incidence of otitis externa, with a peak incidence between 3 and 6 years of age. Otic foreign bodies are unusual in young animals. Underlying disease, such as a defect in keratinisation, atopy or a dietary intolerance, may cause uni- or bilateral otitis externa and may occur in young animals, particularly in predisposed breeds.

Sex

There is no sex predisposition to otitis externa in general, other than male dogs being predisposed to otic foreign bodies.

History

The key aims of history taking are:
- To identify whether or not there is any evidence of management or underlying disease that may be predisposing the animal to otitis externa, in some cases to allow a definitive diagnosis, thus allowing specific treatment.
- To obtain specific information about the otitis with regard to duration, presence of pain or head tilt and response, or otherwise to previous treatments.
- Management and lifestyle:
  - Diet: to identify deficiencies such as zinc and essential fatty acids;
  - Water intake: any polyuria/polydipsia? Endocrinopathy – do not forget to palpate the scrotum!
  - Housing: kenneled or indoors?
  - Exposure to sunlight: actinic radiation damage;
  - Exercise: swimming and hydrotherapy? Might predispose to ear disease;
  - Work: hunting dogs and those used as pointers and retrievers may be predisposed to foreign bodies? And ticks?
  - Grooming requirements: clipper burn on the pinnae, otic irritation following overzealous plucking, contagion at clipping parlour?
  - Presence of other animals (*O. cynotis*, *Sarcoptes scabiei*, dermatophytosis);
  - Hunting cat (*Spilopsyllus cuniculi*, feline poxvirus infection, ticks).

Evidence of underlying disease

Many apparently acute, and almost all recurrent or chronic cases, of otitis externa are associated with underlying disease, which may be amenable to treatment. Most, but not all will be bilateral.
- Facial, otic and pedal pruritus suggests atopy.
- Erythema in the ear, facial, neck and truncal folds, and perhaps crust, scale and erythema on the pinnae and trunk, suggest a defect in keratinisation.
- A seasonal pattern is most likely to reflect atopy or seasonal exposure to ectoparasites such as mosquitoes, flies, harvest mites (*Neotrombicula* spp.) and rabbit fleas (*S. cuniculi*).
- Dietary intolerance may be associated with otitis externa.
- Allergic contact dermatitis may affect the concave, ventral aspect of the pinnae.
- Endocrinopathies may be associated with a ceruminous otitis externa, particularly Sertoli cell tumour.
- Sudden onset of severe, ulcerative bilateral otitis, perhaps in association with other skin disease or systemic signs should raise the suspicion of drug eruption or immune-mediated disease.

Medications

Topical application of otic medication may induce an irritant or allergic contact dermatitis. The clinical sign that might suggest this is continued or increased signs of otitis in the face of repeated application of a medication. Neomycin is the most often cited agent in this regard, although propylene glycol may also be irritant.

Physical examination

Having established the immediate and past history, the dog should be given a full clinical examination. In particular, evidence of internal disease and endocrinopathies should be sought. Thus, lymph nodes and scrotum should be palpated, the oral cavity examined, the chest auscultated, the abdomen palpated and the perineum checked. Only after this general physical examination should the dermatological assessment take place.

Pinnal scratch reflex

In some pruritic canine dermatoses, rubbing the distal edge of the pinna between finger and thumb nail induces a scratch reflex from the ipsilateral hindlimb. This positive scratch reflex is most commonly associated with scabies, although it is not pathognomonic. Pediculosis, *Malassezia pachydermatis* dermatitis and atopy also may result in positive scratch reflex.

Keratoconjunctivitis, xeromycteria (see Chapter 10 Diseases of the Nasal Planum) and deafness may suggest otitis media[3].

2.3 GENERAL OTIC AND NEUROLOGICAL EXAMINATION

Gross examination of the pinnae

- Peripheral crust and scale may suggest scabies, pediculosis, a defect in keratinisation, zinc deficiency, endocrinopathy, pemphigus erythematosus, leishmaniasis or fly bite or mosquito hypersensitivity.
- Erythema on the convex aspect, particularly distal, suggests actinic radiation damage.
- Erythema on the concave aspect suggests atopy or allergic contact dermatitis.
- Alopecia may be due to pruritus (scabies, pediculosis, hypersensitivity) or dermatophytosis.
- Alopecia and hyperpigmentation may reflect an endocrinopathy.
- Curling of the pinnae in the cat is almost pathognomonic for relapsing polychondritis.
- Vesicles, pustules and crusts may be due to superficial pyoderma, pemphigus foliaceus or zinc deficiency.
- Focal ulceration and crusted erosions on the convex aspect may be due to feline cowpox.
- Punched out ulcerations on the concave aspect and pinnal margin may reflect vasculitis.

Gross examination of the external ear canal

Cleaning of the external ear canal is discussed below in section 2.4.

Note: Erythema is often associated with swelling of the soft tissues of the external ear canal and stenosis of the lumen. In some cases the stenosis is so severe that it is impossible to insert the cone of an otoscope into the canal. In these cases it may be necessary to provide symptomatic treatment with otic, and perhaps systemic, glucocorticoids and antibiotics for several days[5,6]. This should reduce swelling and facilitate proper examination.

- Malodour of the external ear canal may be associated with *M. pachydermatis* infection, gram-negative bacterial infection, devitalised tissues or neoplasia.
- The amount of hair around the entrance to and within the external ear canal should be assessed. The clinician may need to remove this hair in order to complete an otoscopic examination.
- Calcification of the otic cartilages, which can be palpated, suggests the presence of chronic otitis externa.
- The areas hidden within the cartilage folds at the entrance to the ear canal should be examined for ectoparasites, particularly ticks and trombiculid mites.
- Erythema of the vertical canal, in combination with a normal, or nearly normal, horizontal canal is very suggestive of atopy.
- Ulceration of the otic epithelium is usually associated with gram-negative bacterial infection[5], but it may be a sign of immune-mediated disease.
- The nature, colour, and odour of any discharge should be noted. However, whether any conclusions as to the causal organism, based on the physical nature of the discharge, are valid is debatable. Cytological examination of the discharge is much more reliable in this regard.
- Persian cats and some older Siamese cats may exhibit what appears to be excessive cerumen[4]. The owner should be encouraged to tolerate this rather than being overzealous with cleaning, unless infection becomes an issue.

2.4 CLEANING THE EXTERNAL EAR CANAL

- Ear cleaning and otic treatment has three phases, see Chapter 3, Section 3.2 for more detail.
- Ear cleaning to enable initial examination of the external ear canal.
- First stage management, based on examination and cytology:
  - Sampling for bacteriological susceptibility testing.
- Ear cleaning, using ear cleaners, as a means of long-term control of recurrent disease.

Given that examination of the external ear canal includes assessment of the status of the entire canal and the tympanum, it follows that cleaning of cerumen and purulent debris may be necessary.

Cytological examination of otic discharge should be taken before cleaning is initiated. The presence of erosions or ulceration of the otic epithelium, inflammatory cells and rod-shaped bacteria should prompt suspicion of *Pseudomonas aeruginosa*.

Samples should be submitted for bacteriological assessment and sensitivity testing.

- It has been suggested (Bloom[6a], citing Robson[6b]) that the application of medication into the external ear canal gives concentrations of antibacterial agents in the mg/ml range. This is many times (1,000) more than the µg/ml used in sensitivity testing. Therefore, given that the correlation with outcome (i.e. *in vitro* sensitivity testing and clinical response) is hard to predict, sensitivity testing could be reserved for cases in which there is mixed infection or a failure to respond to initial therapy.
- However, there are two caveats to consider
  - If you suspect methicillin-resistant *Staphylococcus pseudintermedius* (MRSP), it is mandatory to submit samples for susceptibility testing and so consider the implications *vis-à-vis* staff and owner safety.
  - Many dogs with gram-negative infection, in particular, require long courses of antibacterial treatment. Drugs vary in their cost and dosing requirements. Susceptibility testing gives options to allow tailoring the medication for the individual case. Factors to consider are:
    o Cost.
    o Once daily treatment is easier to comply with but carries a cost penalty.
    o Some owners can only manage liquid medications, particularly a problem with cats and small dogs.
    o Injectable only medications.

Sometimes the entire ear canal is so swollen that it is unrealistic to attempt this detailed, initial investigation without instituting some initial treatment, on a symptomatic basis[6c]:

- Topical otic glucocorticoids, perhaps within an otic polypharmaceutic, and anti-inflammatory doses of oral prednisolone (0.5–1 mg/kg daily for 3 days then every other day) for 10–14 days should allow sufficient reduction of epithelial swelling to allow examination of the deep portions of the external ear canal.
- Remember to check if the owner is administering non-steroidal anti-inflammatory agents before providing the prednisolone.

> Inflamed external ear canals are painful. Animals, not surprisingly, may resent examination, even with the relatively narrow canulae of video-otoscopes. Sedation, and most probably general anaesthesia, will be necessary.

Before examination and staging of the otitis can take place, the external ear canal must be cleaned of matted hair, ceruminous debris and any purulent discharge that have accumulated. With the dog under general anaesthesia, this may be accomplished in a variety of ways:

- Consider using a depilatory product to minimise the otic trauma associated with plucking an ulcerated and inflamed ear canal.
- Repeated flushing with a soft feeding tube and a 20 ml syringe, or a bulb syringe, using warm water.
- As above but using a commercial cleanser, or astringent cleanser, such as Epi-Otic®, or a squalene-based cleaner such as Cerumene®.
- Some clinicians prefer the foaming action of 5% dioctyl sodium sulfosuccinate (DSS) for the removal of purulent debris. Note that cleansers such as DSS, and even chlorhexidine, might have some irritant effects on the middle ear. Clinicians should flush the external ear canal copiously after cleaning.
- Using a powered delivery system, such a Otopet Earigator, Auriflush, WaterPik or the Karl Stortz VetPump®II, to facilitate removal of otic debris.
- The Entermed Enthermo ear flushing apparatus consists of a small heater in which tap water is warmed to body temperature, allowing for an unlimited supply of water at optimal pressure.

- Astringents, such as urea, and mild acids such as boric, lactic, salicylic, are considered to have a drying effect and may be used after flushing and cleaning, provided the tympanic membrane is intact. This caveat applies to ceruminolytic agents as well.

2.5 OTOSCOPIC APPEARANCE OF THE EXTERNAL EAR CANAL AND TYMPANUM

Sedation for otoscopy and aural examination

In order to examine the entire length of the ear canal properly, adequate restraint is necessary as the external ear canal must be manipulated into as straight a line as possible. This is achieved by gently grasping the pinna and pulling it, and the attached auricular cartilage, up and away from the sagittal plane.

Note that otoscope cones can carry infectious organism (*Malassezia* spp., *P. aeruginosa*) from one ear to another, and from one patient to another. This problem is particularly aggravated if otoscope cones are immersed in infrequently-changed cleansing solutions[7]. Cleansing solutions must be changed weekly, a factor that appears to be more important than the nature of the cleanser itself.

Otoscope cones are hard, often cold, and have sharp ends. It hurts when a cannula is thrust into an inflamed ear canal. Although video-otoscopes have a narrow probe, the same constraints apply. In most cases, and particularly with small dogs and cats, and animals with painful or tender ear canals, this process is resented and the clinician will require chemical restraint or general anaesthesia.

McKeever and Richardson[8] advocated a mixture that provides approximately 20 minutes sedation, sufficient to allow thorough examination and cleaning of both ears:

- Ketamine (1.36–2.2 mg/kg).
- Midazolam (0.023 mg/kg).
- Acepromazine (0.023 mg/kg).
- All mixed in the same syringe and injected slowly intravenously.

An alternative would be xylazine injection (1–2 mg/10 kg i/v), which should provide about 20 minutes of reasonable sedation.

Another alternative is detomidine (20–40 µg/kg i/v), which will produce moderate sedation and has the great advantage that it can be reversed by the intramuscular injection of its antagonist atipamezole.

Note, that there are two disadvantages with this regime:

- It is very expensive (although this can be mitigated by giving buprenorphine at the same time).
- There is a risk of inducing a cardiac arrhythmia if potentiated sulphonamides are being administered at the same time.

In many cases the discomfort associated with the otitis externa mandates a general anaesthetic.

Normal appearance of the external ear canal and tympanum

The normal external ear canal is smooth, pale in colour, and contains minimal discharge (Figs 2.8, 2.9). A small amount of pale yellow or brown cerumen (Figs 2.10, 2.11) may be seen in some cases and this is normal. Occasionally, there may be a hair shaft in the horizontal canal (Figs 2.12, 2.13). In some breeds, such as Cocker Spaniels, Miniature and Giant Schnauzers, Airedales and other terriers, for example, there are hair follicles along the whole length of the external ear canal[6,9]. The diameter of the vertical portion of the external canal varies from breed to breed but at its base, where it opposes the horizontal portion, it is 5–10 mm (0.2–0.4 in.) in diameter[9]. The horizontal canal is approximately 2 cm (0.8 in.) in length[9].

> Remember that the presence of hair in the external ear canal is normal in some dogs.

The normal tympanum is thin, pale grey in colour (described as 'rice paper'-like), and translucent (Fig. 2.14). It is visible via otoscopy in about 75% of normal ears[6]. Cerumen, debris or hair may prevent a clear view of the tympanum[10], particularly in Miniature Schnauzers and Miniature Dachshunds[11]. The shape of the tympanum is elliptical, mean 15 × 10 mm (0.6 × 0.4 in), with the short axis nearly vertical. The initial

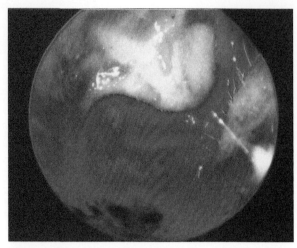

Fig. 2.8 The upper portion of the feline external ear canal.

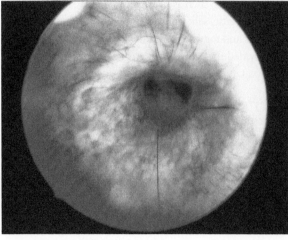

Fig. 2.9 The normal horizontal external ear canal of a dog. There is an even, pale colour with a smooth contour. A few fine hairs may be seen.

Fig. 2.10 Patchy brown cerumen adhering to the walls of a normal external ear canal.

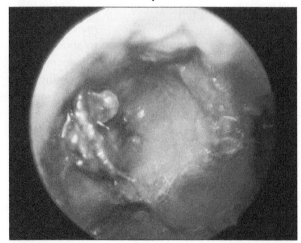

Fig. 2.11 A clump of yellowish cerumen near the tympanum. Normal.

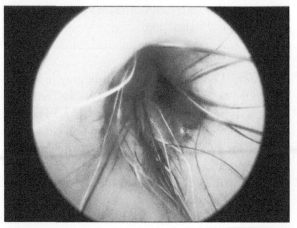

Fig. 2.12 Tufts of hair emerging from the horizontal ear canal.

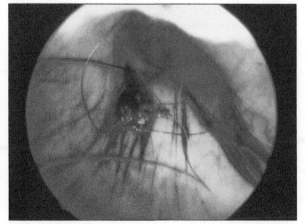

Fig. 2.13 Hair and adhering cerumen emerging from the horizontal ear canal.

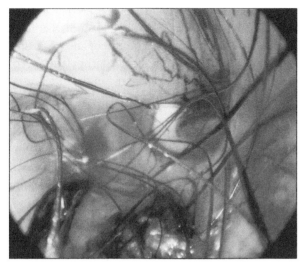

Fig. 2.14 A normal, translucent tympanic membrane, in this instance partially hidden by hair and cerumen.

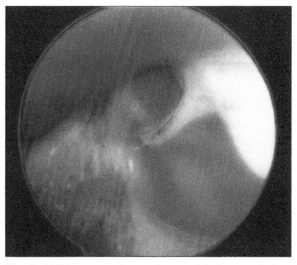

Fig. 2.15 The tympanic membrane with the manubrium of the malleus clearly visible.

otoscopic view is restricted to the posterior quadrant of the pars tensa and the pars flaccida[12,13]. Manipulation of both the external ear canal and the otoscope will usually bring the majority of the manubrium (Fig. 2.15) and the larger portion of the pars tensa into view[12,13]. The external aspect of the tympanum, as viewed with an otoscope, is divided into two unequal parts by the manubrium of the malleus. This is attached along the medial aspect of the tympanum and exerts tension onto it, resulting in a concave shape to the intact membrane.

Abnormal appearance of the external ear canal

Inflammation results in oedema, erythema, and warmth (Fig. 2.16). Given that the glandular tissues of the external ear canal are contained within a cartilaginous tube, any swelling will result in a reduction in the diameter of the lumen. In many cases the concave aspect of the pinna will also be affected (Fig. 2.17). In most cases the inflammation affects the entire ear canal but in some instances it will be localised to either the horizontal or, more usually, the vertical canal. Bilateral inflammation confined to the concave aspect of the pinna and the vertical canal, particularly if there is little discharge, is very

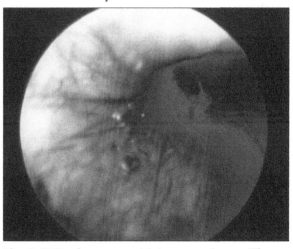

Fig. 2.16 Erythematous otitis in a case of atopy. There is erythema and some degree of swelling, resulting in loss of luminal cross-section.

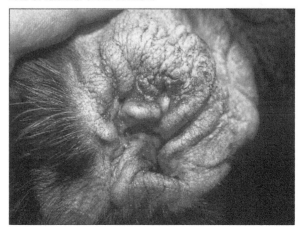

Fig. 2.17 External ear canal of an atopic dog. There is erythema, hyperplasia and lichenification.

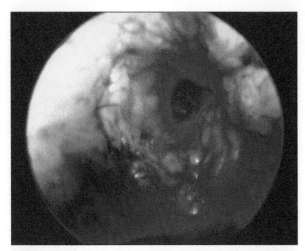

Fig. 2.18 Erythema, hyperplasia, moderate fissure formation and a tendency to ulcerate and bleed very easily, even after otoscopy, are commonly seen in ear canals of atopic dogs.

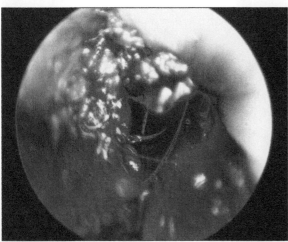

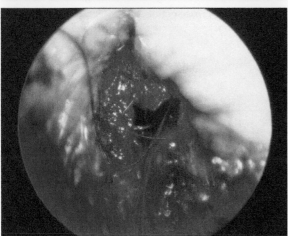

suggestive of atopy (Fig. 2.18). Indeed, erythema of the entire canal in the absence of significant discharge or other pathology is highly suggestive of allergy. Atopy, dietary intolerance and neomycin/propylene glycol sensitivity should all be considered in the differential diagnosis.

Inflammation also results in increased secretion from the glands within the epithelial lining of the canal and a shift away from a lipid to an aqueous constitution[14–16]. Continued inflammation results in maceration of the stratum corneum, loss of barrier function and the outward movement of transepidermal fluid. Discharge accumulates within the external ear canal (Figs 2.19–2.21) and microbial proliferation occurs. The colour of the discharge may vary from light yellow to dark brown and it may be aqueous, thin or pus-like in nature. Animals with severe or generalised defects in keratinisation may exhibit a greasy yellow discharge that has a purulent appearance but which may be free of pathogens and non-inflammatory in nature. Medications may result in a thin, shiny covering over the mural epithelium.

The presence of erosions and ulcers in the external ear canal should be noted (Fig. 2.22). Frank ulceration is uncommon and is usually associated with gram-negative bacterial infection[17]. Rare causes of otic ulceration are autoimmune diseases and otic neoplasms. The finding of ulcers within the external ear canal mandates

Figs 2.19–2.21 Three views showing erythema, hyperplasia, cerumen and varying degrees of luminal stenosis.

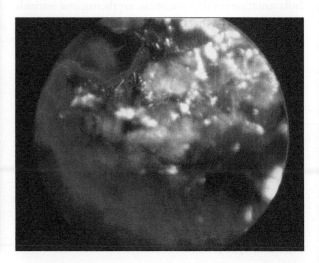

samples for both cytological evaluation and bacterial culture and sensitivity testing.

Otoscopic examination may reveal the presence of ectoparasites, such as *O. cynotis* or *Otobius megnini* (Fig. 2.23). Otodectic mites are often accompanied by the presence of a crumbly brown discharge (Fig. 2.24). Not all infestations are inflammatory and associated with pruritus.

> Do not expect to visualise *O. cynotis* unfailingly: otic cerumen mixed with liquid paraffin and examined under a microscope will allow a better chance of diagnosis if mite numbers are low.

Epidermal hyperplasia, nodules, tumours, polyps and foreign bodies within the external ear canal are easily visualised during otoscopy (Fig. 2.25), although cleaning of the external ear canal may be necessary. This is particularly the case for cats where the whole canal may fill with purulent discharge if an otic tumour or polyp is present (Fig. 2.26). Otic masses should be biopsied – pinch samples taken with biopsy forceps are adequate[17]. Elderly cats may exhibit numerous blue-black coloured papules and nodules – ceruminous cystomatosis (Fig. 2.27), sometimes numerous enough to almost occlude a portion of the external ear canal. These may be treated by surgical removal or laser ablation[5,18,19].

Usui *et al.*[11] have described a shallow depression on the ventral aspect of the external ear canal, immediately adjacent to the tympanic membrane. Further, they have described some breed-specific combinations of hair type and cerumen accumulation, which may be so closely opposed to the tympanic membrane that they engender localised microbial infection[11].

The tympanum should be examined for colour, texture and integrity; it is usually dark grey or brown in cases of otitis externa[8]. In contrast to normal dogs it is only possible to visualise the tympanum adequately in 28% of ears affected with otitis externa[10]. If tears or holes in the tympanum (Figs 2.28, 2.29) indicate that otitis media is present

Fig. 2.24 Crumbly, dry, blackish brown cerumen associated with *Otodectes cynotis* infestation.

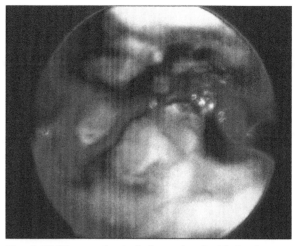

Fig. 2.22 Haemorrhagic foci associated with focal ulcerations in a case of gram-negative infection.

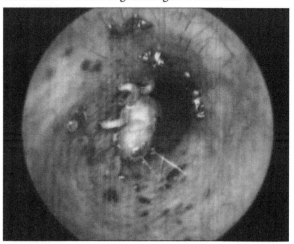

Fig. 2.23 Spinous ear tick, *Otobius megnini*, within the vertical ear canal. (Courtesy of Dr Louis Gotthelf DVM, Montgomery.)

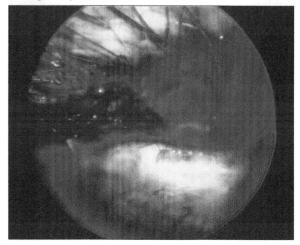

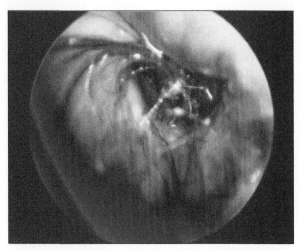

Fig. 2.25 Grass awn, cerumen and associated erythema in the external ear canal of a dog.

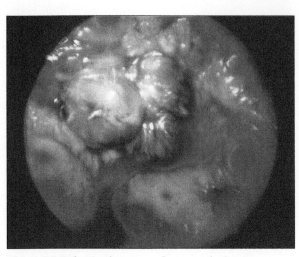

Fig. 2.26 Polyp in the external ear canal of a cat.

Fig. 2.27 Multiple blue-black papules of ceruminous cystomatosis in the external ear canal of a cat.

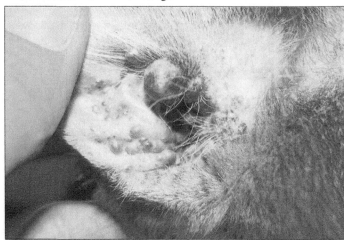

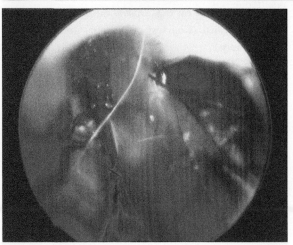

Fig. 2.28 Acute tear in the tympanic membrane of a dog associated with grass awn penetration.

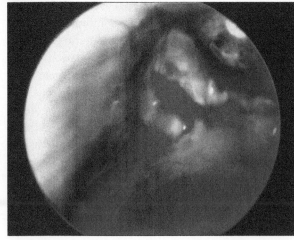

Fig. 2.29 Chronic otitis media and otitis externa have resulted in a thickened, opaque, ruptured tympanic membrane.

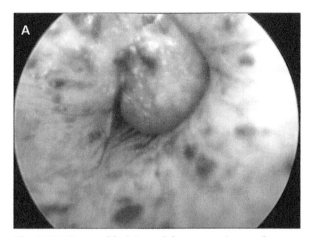 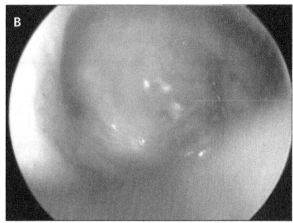

Fig. 2.30 Outward bulging of the tympanic membrane associated with otitis media in a dog (A) and a cat (B).

(although an intact tympanum does not rule out otitis media), then failure to visualise the tympanum adequately, let alone a tear, suggests that diagnosis of otitis media, exclusively via otoscopy, is not reliable. Bulging of the tympanum (Figs 2.30A, B) may indicate an accumulation of exudate within the middle ear, whereas retraction (and a concave appearance) suggests a partially filled middle ear with obstruction of the auditory tube[12].

- Otitis media is present in around 16% of dogs with acute otitis[19] but the incidence rises to 88.9% in dogs with chronic otitis media[20,21]. Do not overlook it.

- A surgical incision into a normal tympanic membrane can be expected to heal within 3–5 weeks[22]. However, defects in the tympanic membrane may heal in the presence of infection in the middle ear[23]. Thus, diagnosing otitis media on the sole basis of a ruptured tympanum is unreliable.

2.6 CYTOLOGICAL CHARACTERISTICS OF NORMAL AND ABNORMAL EARS

Otodectic otitis infestation is often associated with a crumbly, rather dryish discharge (Figs 2.31, 2.32),

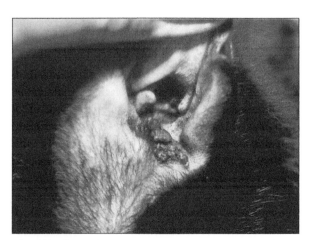

Fig. 2.31 Typical appearance of cerumen associated with *Otodectes cynotis* – dark, dry, and crumbly. As in this case, it is not always associated with inflammation in the ear canal.

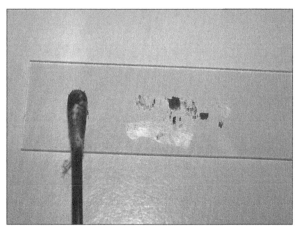

Fig. 2.32 The dry, crumbly nature of the cerumen can be appreciated when it is rolled onto a glass slide.

similar to coffee grounds[24]. However, there is no clear-cut relationship between the gross characteristics of any otic discharge and the species of micro-organism with which it is associated, e.g. staphylococcal, gram-negative or malassezial[25,26].

Cytological examination of otic exudate is a rapid, in-house test that provides diagnostically and therapeutically useful information[25–28]. Reproducibility is high with regard to detecting micro-organisms in general, and good for bacteria, but less so for yeasts[29]. In many cases, information from cytological examination of cerumen is more accurate than that from samples submitted for microbiological culture and sensitivity testing. Furthermore, the clinician can assess the significance of any

micro-organisms in the light of other ceruminal characteristics, such as the presence of nucleated squames, proteinaceous debris and inflammatory cells. This is particularly exemplified in the case illustrated (Fig. 2.33) of a Cocker Spaniel with chronic, bilateral otitis associated with a thick, greasy exudate. Gross examination of the discharge (Figs 2.34, 2.35) might suggest that a malassezial, or even a gram-negative, infection was responsible, but cytological examination (Fig. 2.36) does not support this. Although there are increased numbers of squames and microbes visible, the lack of any inflammatory cells suggests a ceruminous otitis externa, probably associated with a more generalised defect in keratinisation, rather than infection.

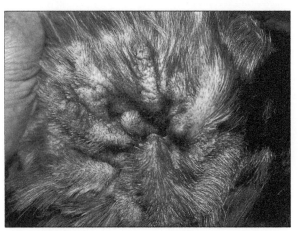

Fig. 2.33 External ear canal of a Cocker Spaniel exhibiting some hyperplasia and erythema.

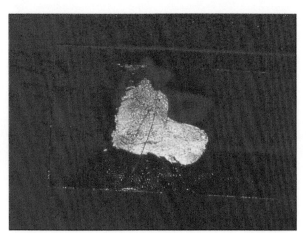

Fig. 2.34 Gross appearance of an unstained cytologic sample – thick and white.

Fig. 2.35 Gross appearance of a Diff-Quik-stained sample – thick and blue, suggestive of a high cell content.

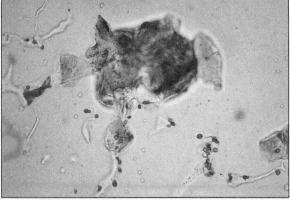

Fig. 2.36 High-power photomicrograph from the Cocker Spaniel in 2.33. Note the yeast and cocci on and around the squames, and the absence of inflammatory cells.

Samples and stains

The most useful sample for otic cytology is a swab taken from the ear canal, which is then rolled onto a clean glass slide. If feasible, samples should be taken from the horizontal portion of the external ear canal of both ears[27,28]. In large dogs it is usually possible to collect a shielded sample using, for example, an alcohol-sprayed otoscope cone. In small dogs and cats collecting a shielded sample is difficult and vertical canal samples will have to suffice[28]. If otitis media is suspected, a shielded sample should be taken from the middle ear in addition to that from the external ear canal.

Most clinicians advocate using modified Wright's stains such as Diff-Quik[24,27,28]. Alcohol-based stains are more useful than aqueous preparations (e.g. new methylene blue) because of the lipid nature of the otic discharge[27]. Griffin advocated heat fixing of obviously waxy preparations in order to prevent solvent-associated leaching of lipid[24]. Since all cerumen contains lipid it would seem appropriate to heat fix all samples, but opinion is divided on this issue[27–31]. Subtle information may be lost if heat fixing is not performed, but generally it is not necessary unless samples are to be kept for future examination. Commercial laboratories usually perform a Gram's stain because, although more time consuming, it does allow assessment of the classification of organisms by both morphology (coccus, rod, diphtheroid) and Gram's stain status. Generally, Gram staining is too cumbersome and time consuming for practitioners to consider it as a rapid, in-house stain[31].

Knowledge of morphology and Gram's stain status allows a recommendation for treatment[24,27]. In addition to allowing visualisation of the microbial populations of the external ear canal, cytology also allows the physical nature of the cerumen to be assessed, in terms of keratinaceous debris and the lipid content of the cerumen[15,32].

Stained samples should be air-dried and examined for evenness of stain and for depth of stain, which is usually deeper in intensity as the cell count increases. A coverslip should be applied prior to microscopic examination[28]. Initial low-power examination is followed by high-power examination, and this is usually sufficient for accurate classification of any micro-organisms and identification of any cells present[30,31].

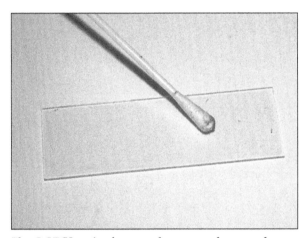

Fig. 2.37 Unstained smears from normal ear canals are all but invisible.

Gross examination of cytological preparations

Gross examination of fresh and stained smears reflects the lipid and cellular content of cerumen. Normal cerumen has a high lipid content and a low concentration of intact cells, usually squames. Unstained preparations are all but invisible in direct light (Fig. 2.37) reflecting the low cell content of cerumen. Increasing cell content, particularly if it is inflammatory in nature, is reflected in the increased opacity of the cerumen.

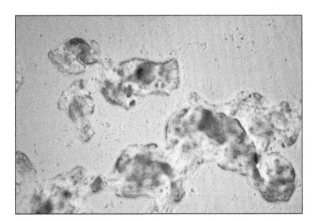

Fig. 2.38 Photomicrograph of a cytological sample from a normal ear. Note the low cell content in the cerumen.

Normal cytological characteristics

Cerumen does not take up stain because of its high lipid content (Fig. 2.38), although ghost outlines of lipid drop-

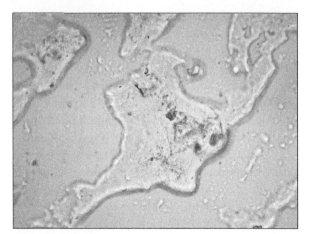

Fig. 2.39 Outlines of lipid are visible, even after fixing and staining with Diff-Quik.

lets may be seen occasionally (Fig. 2.39). Low numbers of anucleate, epithelial cells may be detected, although in normal ears they are not excessive (Fig. 2.40). When there is inflammation of the epithelial lining of the external ear canal the rate of cellular turnover increases and nucleated squames may be seen in the cerumen (Fig. 2.41). Low numbers of *M. pachydermatis* and staphylococci may be identified adhering to shed squames (Figs 2.42, 2.43). Leucocytes are usually absent from normal cerumen[4].

Abnormal cytological characteristics
Cerumen
The lipid content of cerumen from inflamed external ear canals is lower and the cell count is usually higher than that in normal ear canals[32]. This is reflected in the

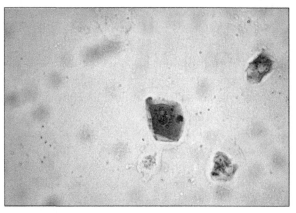

Fig. 2.40 Photomicrograph of a cytological sample from a normal ear, with only a few squames apparent.

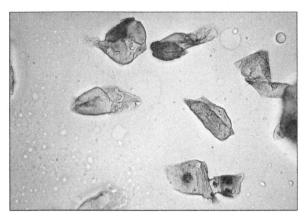

Fig. 2.41 Photomicrograph demonstrating increased numbers of squames, some of which are nucleated.

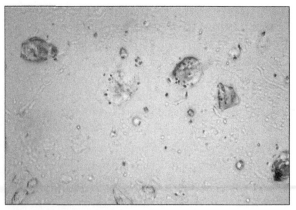

Fig. 2.42 Photomicrograph demonstrating a few anucleate squames with a few staphylococci apparent.

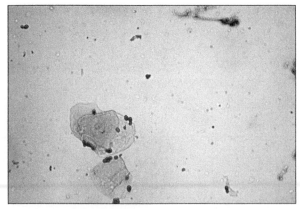

Fig. 2.43 Photomicrograph demonstrating a few anucleate squames with low numbers of yeast apparent.

gross appearance of the stained smear, which appears bluer in a sample from an otitic ear (Fig. 2.44) than in that from a normal ear (Fig. 2.45).

Keratinocytes

In acute cases of otitis externa, or cases of very short duration, there will be very little change in the epithelial shedding of squames. In chronic cases, such as those associated with defects in keratinisation or atopy, the epithelial lining of the external ear canal reacts to the inflammation and the hyperplasia may result in the appearance of both anucleate and nucleated squames and debris (Fig. 2.46).

Autoimmune disorders, in particular pemphigus foliaceus, may result in acantholysis. Single, nucleated, well-defined acanthocytes may be shed from erosions in the ear canal, often surrounded by adherent neutrophils (Fig. 2.47), i.e. a positive Tzank test.

Inflammatory cells

In samples taken from dogs and cats with acute otitis externa there may be little change in the cellular population, but in most cases there will be neutrophils and proteinaceous debris (Fig. 2.48). More chronic otitis results in the appearance of macrophages as well as neutrophils within the exudate[28]. In cases of bacterial otitis the increasing concentration of toxins may result in the appearance of toxic neutrophils (Fig. 2.49) and indicate that the external ear canal should be flushed before treatment is initiated[24].

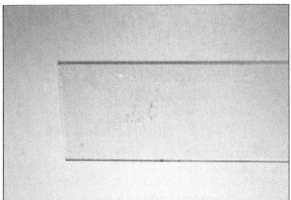

Figs 2.44, 2.45 Gross cytological samples from an otitic and a normal ear canal. The otitic ear (2.44) contains a higher cell content and appears much bluer when stained, than the smear from the normal ear (2.45).

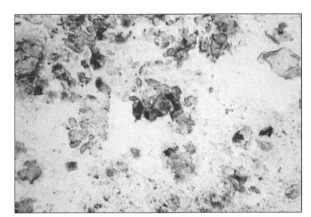

Fig. 2.46 Photomicrograph from an acute case of otitis externa. There are anucleate and nucleated squames present.

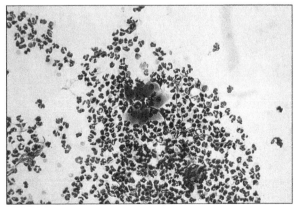

Fig. 2.47 Photomicrograph of a stained smear from a pustule in a case of pemphigus foliaceus. Compare the 'clean' neutrophils and rounded-up, nucleated squames in this Tzank test-positive smear with the cells and neutrophils in an inflammatory otitis in 2.48.

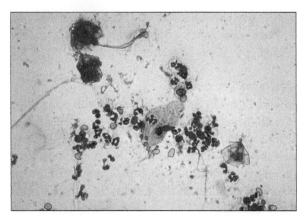

Fig. 2.48 Photomicrograph of a small group of squames with surrounding neutrophils.

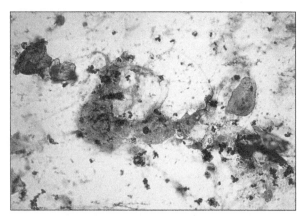

Fig. 2.49 Photomicrograph illustrating squames, proteinaceous debris and dark, pyknotic neutrophils.

Neoplastic cells

Intraluminal neoplasms may shed cells into the cerumen but it is unusual to find diagnostically useful material in this discharge[33]. Cells from adenocarcinomas tend to exfoliate in sheets or clusters, whereas squamous cell carcinomas shed large, densely-staining individual cells with prominent nucleoli (Fig. 2.50)[28].

Bacteria

The normal microbial population of the ear canal is dominated by staphylococci. In the early stages of otic inflammation the numbers of staphylococci increase, particularly the numbers of *Staphylococcus pseudintermedius* (Fig. 2.51). Care should be taken when pronouncing that staphylococci are present in cerumen or on the squames, as cocci may be confused with debris in poorly-maintained Diff-Quik (filter the stain regularly). Pigment granules within the squames (melanin granules are usually brown in colour rather than the blue-black colour of staphylococci).

Occasionally the otic flora remains staphylococcal in nature but most commonly it becomes dominated by gram-negative bacilli, in particular *Escherichia coli*, *Proteus* spp. and *Pseudomonas* spp. These changes in bacterial shape (i.e. coccoid to rod) are easily detected by microscopic examination of cytological samples (Fig. 2.52)[28,30]. Cocci cannot be reliably classified as either staphylococci or streptococci on the basis of clumping or chain formation, respectively, since this is not usually

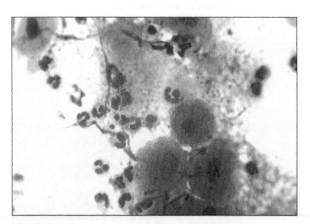

Fig. 2.50 Photomicrograph of cytological sample from an ear canal in which a squamous cell carcinoma was found. Note the typical appearance of the cells: large, densely staining with nucelioli. (Courtesy of Dr Sue Paterson.)

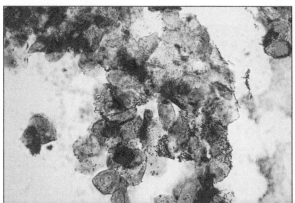

Fig. 2.51 Photomicrograph of a dense group of cells with large numbers of staphylococci apparent.

observed[27]. With experience, clinicians may be able to detect that staphylococci are larger than streptococci[27].

Generally, even in first opinion cases, the observation of bacilli should prompt sampling for bacterial culture and sensitivity testing[27]. This is particularly important if Gram's staining is not performed, since clinicians cannot differentiate *Pseudomonas* spp. from *Clostridium* spp. or *Bacillus* spp.[27]. Unless a recurrent case is involved, it is not usually necessary to submit samples from otitis externa associated with cocci for bacterial culture and sensitivity testing. Indeed, in one study testing achieved a sensitivity of 59% for gram-positive cocci and 69% for gram-negative rods, compared with 100% sensitivity with cytological examination[31].

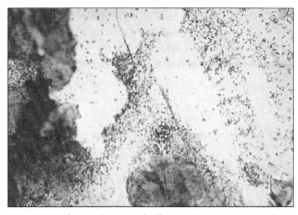

Fig. 2.52 Photomicrograph illustrating squames and vast numbers of bacilli.

Yeast

M. pachydermatis is a member of the normal canine otic microflora[32], although it has the potential for opportunistic pathogenicity. At least two species of malassezial yeast can be isolated from feline ear canals, *M. pachydermatis* and *M. sympodialis*[33]. The presence of yeast (Fig. 2.53) must, therefore, be interpreted with caution. Evidence of increased numbers of yeast (arbitrarily more than 5–10 per high-power field [Fig. 2.54][24,27,28]) and an associated inflammatory reaction (Fig. 2.55) should be sought before disease status is decided. Malassezial yeast are flask or peanut-shaped whereas candidial yeast are round in appearance, although this distinction is not easily made.

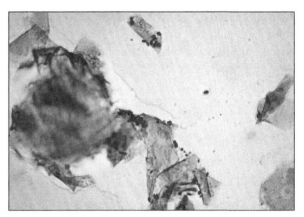

Fig. 2.53 Photomicrograph illustrating a few malassezial yeast, numbers within normal limits.

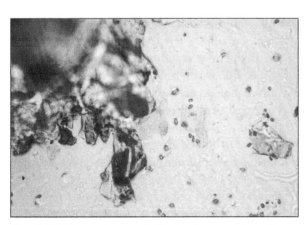

Fig. 2.54 Photomicrograph illustrating yeast in sufficient numbers to be associated with disease.

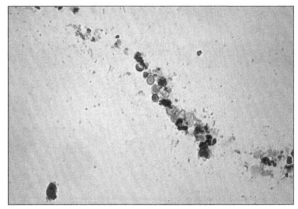

Fig. 2.55 Photomicrograph illustrating yeast and inflammatory cells.

The importance of cytological evaluation was exemplified in two studies that reported that demonstration of malassezial infection by cultural methods achieved a sensitivity of 82% and 50% respectively[24,31]. However, in another study looking at reproducibility of cytology, the agreement for yeast organisms was only moderate[29]. In addition to relative insensitivity, malassezial culture is expensive and time consuming, resulting in unnecessary cost and delay in treatment compared to cytological assessment[34,35].

Ectoparasites

It is often easy to visualise *O. cynotis* within the external ear canal simply by using an otoscope. However, since very low numbers of mites have been associated with otitis externa,[36] it is not surprising that they will be missed in some cases, particularly if there is an accumulation of debris or discharge within the canal. Therefore,

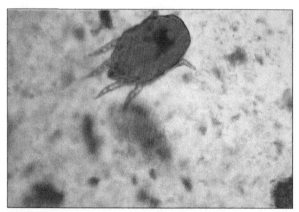

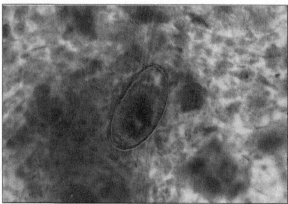

Figs 2.56, 2.57 Photomicrograph of an adult otodectic mite (2.56) and an ovum (2.57) recovered from the cerumen of an ear infected with *Otodectes cynotis*.

microscopic examination of cytological preparations is indicated if otodectic mange is suspected. Cerumen is deposited on a glass slide and mixed with mineral oil prior to microscopic examination[25]. Otodectic mites have a characteristic appearance (Figs 2.56, 2.57).

Bacterial culture and sensitivity testing

In most cases, examination of cytological samples will provide all the information necessary for effective treatment to be instituted. Microbial culture and sensitivity testing of samples from the external ear canal is, however, useful in certain cases:

- In cases of recurrent, or refractory, otitis externa.
- If ulceration of the epithelial lining of the external ear canal is present.
- If gram-negative infection is suspected.
- If otitis media is suspected (when samples from the middle ear will also be necessary).

2.7 BIOPSY OF THE EXTERNAL EAR CANAL

Taking 4 mm (0.2 in.) punch biopsy samples, under general anaesthesia, of the vertical ear canal or of lesions and masses within the external ear canal is an important method of obtaining useful information on pathological processes underway. Biopsy of the external ear canal has three main indications:

- As a means of providing information on the degree of permanence of epithelial changes. For example, some apparently permanently thickened epithelia will regress dramatically when treated with topical glucocorticoids to suppress the inflammatory reaction. However, fibrosis, in general, will not regress. Biopsy of the luminal wall can yield information on the degree of fibrosis present (see Chapter 1 Anatomy and Physiology of the Ear, Section 1.3). This information can help decide whether to opt for a surgical or medical approach.
- As a means of providing information on the aetiology of ulcerated lesions in the ear canal. Ulcers of the luminal wall may reflect gram-negative bacterial infection particularly, but also autoimmune disease and neoplasia. Biopsy of these lesions can yield information that will influence management.
- As an adjunct to surgery. Neoplastic changes within the ear canal may be fibrogranulomatous, benign or malignant. Knowing the type of

neoplasm and being able to predict its behaviour can help the surgeon plan the degree of resection necessary to remove the risk of recurrence.

2.8 IMAGING THE EXTERNAL EAR CANAL AND THE MIDDLE AND INNER EAR

Technological advances have transformed the investigation of ear disease:

- CT scanning has mostly replaced radiography for the diagnosis of middle ear disease in university and referral institutions.
- Video otoscope technology has enabled practitioners to examine and record the external ear canal in much greater detail.

However, access to CT scanning usually requires referral, which is costly and frequently entails travel. Video otoscopy requires space (often at a premium in practice premises) and a considerable cash outlay for the veterinarian. For these reasons, radiographic imaging is still the main investigative tool for many veterinarians.

CT imaging of the external ear canal, middle and inner ear

As with all imaging modalities, manipulating the animal into the desired orientation and ensuring a lack of movement, mandates general anaesthesia. It is beyond the scope of this text to detail either CT software settings or contrast materials, both of which may require optimisation to ensure the desired image and contrast detail. Readers are referred to specialist texts.

The normal computographic anatomy of the external ear canal and the middle ear of the dog and cat have been described[37,38]. The external ear canal should have an air-filled lumen and be without localised narrowing or obstruction[38] (Figs 2.58, 2.59). The tympanum is best visualised using dorsal rather than transverse imaging[37]. Interestingly, the prevalence of both otitis externa and otitis interna may be higher than previously reported, as retrospective CT studies are able to demonstrate subtle changes that owners, and indeed veterinarians, have missed[39]. Signs consistent with severe, or chronic severe, otitis externa include[38,39]:

- Narrowing of the lumen of the external ear canal (Figs 2.60, 2.61).
- Localised stenosis may suggest neoplasia, polyp or granuloma.
- Extension of soft tissue into the middle ear that suggests that the tympanum is ruptured or missing.
- Increased density of the cartilage, which indicates that calcification has occurred.

CT is particularly suitable for imaging the middle ear[37–43] in particular, because the transverse image obviates the superimposition of structures, which is a feature of radiographs[38]. The normal bulla should have a thin, crisply-defined wall and there should be broad symmetry left and right[37]. Signs consistent with otitis media include[38,39]:

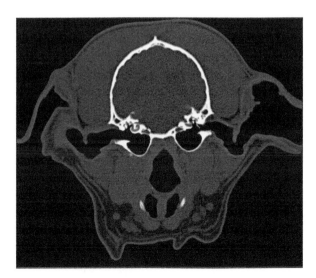

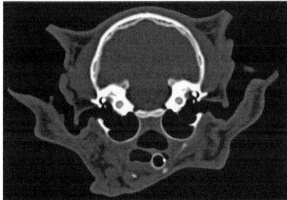

Figs 2.58, 2.59 CT scans of normal canine (2.58) and feline (2.59) external and middle ear clearly demonstrating the air-filled lumens.

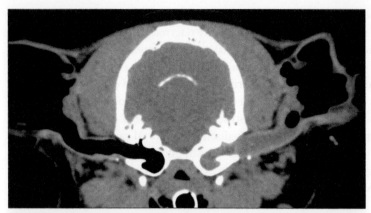

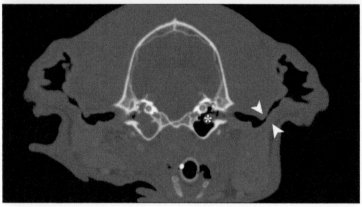

Figs 2.60, 2.61 CT scans demonstrating signs of chronic otitis externa. Fig. 2.60 shows that the left external ear canal has no effective lumen, it is filled with hyperplastic epithelium and ceruminous debris. Fig. 2.61 shows the tympanum bulging into the middle ear (asterix) and marked stenosis of the external ear canal (arrows).

- Lysis of the bulla or evidence of bony thickening.
- The presence of soft tissue or fluid within the bulla.
- Evidence of a ruptured tympanic membrane, perhaps with evidence of soft tissue protruding into the bulla.

A comparative study by Rohleder *et al.* found that CT was more accurate and reliable than radiography for the diagnosis of middle ear disease, in dogs with otitis externa[40]. However, when the middle ear disease was mild (not moderate or severe) neither diagnostic tool was superior to the other. This finding was supported in other canine and feline studies[41,42].

Ultrasound

Ultrasonography was found to be a reliable diagnostic tool, in both dogs and cats[42,43]. However, the authors of both papers agree that more studies are required to further validate ultrasound for the diagnosis of middle ear disease. Given that more veterinarians have access to ultrasound and radiography than they do to CT, it has obvious attractions, if the technique can prove reliable in non-specialist hands.

Video otoscopic examination of the ear canal

The video-otoscope (VO) is a useful and effective tool in the management of both canine and feline otitis. Its widespread use is, unfortunately, limited by expense and it is most commonly available in referral institutions. The equipment takes up considerable space in the consulting room (Fig. 2.62).

Although it has major advantages over hand-held otoscopes, animals do need to be minimally sedated if videoendoscopy is to be employed to its best advantage. More commonly, anaesthesia is induced. Its uses and indications in veterinary medicine have been reviewed by different authors[17,44a,44b,45].

It has several advantages over the hand-held otoscope:
- The VO provides a range of magnification lenses, which allows assessment of fine detail especially at the level of the tympanic membrane (Fig. 2.63).

Often the inferior magnification of the hand-held instruments does not allow the clinician to see small tears in the tympanic membrane, which can be important when deciding on topical therapy.

- The intense light source that is positioned at the tip of the endoscope, rather than at the base of the cone as is the case with hand-held devices, provides excellent illumination to allow more detailed evaluation of the structures within the ear. This also prevents the problem of shadows within the visual field created when instruments are introduced down the working channels.
- The working channels facilitate fully visualised flushing of the canal and removal of foreign bodies such as grass awns or ceruminoliths using grasping forceps.
- Minor surgical interventions can also be performed. Biopsies may be taken from the canal using grasping forceps.
- Injections can be placed into the canal using a long hollow needle inserted down the working channel or by a rigid spinal needle inserted down alongside the otoscope cone.
- When samples are to be taken from the middle ear, in the face of an intact tympanic membrane, myringotomy can be performed safely, because important structures to be avoided, such as the malleus and pars flaccida, can be visualised.
- Some VOs, especially those with wide working channels, can also be used with lasers, which can be used to ablate small lesions in the walls of the canal or to perform myringotomy.
- Modern VOs allow photographic documentation of clinical cases, which helps enhance the patient's medical record and can be used to provide colleagues and the client with a pictorial record.
- Often the ability of an owner to see the ear canal of their pet, especially when comparisons are made with a normal ear, leads to increased owner compliance with therapy.

Virtual otoscopy for evaluating the inner ear

The technique of virtual otoscopy depends on manipulating CT data with commercial software[46,47]. Although beautiful images are generated the technique has little practical application at present, see Chapter 1 Anatomy and Physiology of the Ear, Figs 1.14, 1.15).

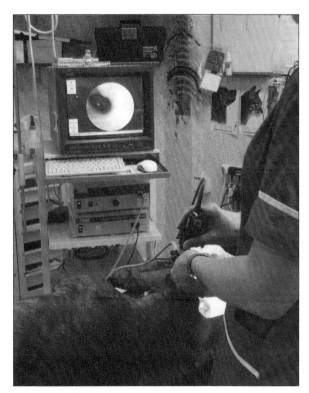

Fig. 2.62 Videoendoscope in a consulting room can occupy a great deal of space.

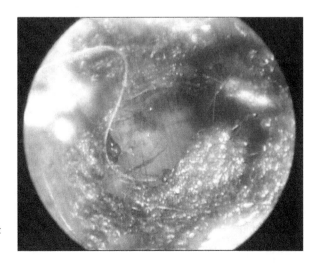

Fig. 2.63 Illustration of the quality of the image that can be obtained use a videoendoscope. In this case the accumulation of thick cerumen is readily appreciated but the tympanum can still be visualised.

Radiographic features of the normal and abnormal ear

Careful positioning is the key to radiographic interpretation of the external and middle ears. This will usually require general anaesthesia, sedation in general is not adequate. The two most useful views are the lateral oblique and rostrocaudal (open mouth). Given the individual variation between animals, comparison of one side with the other is often the only way of making a diagnosis. Therefore, perfect positioning is essential.

Radiographic examination is a useful tool in the investigation of ear disease in both the dog and the cat. Principally, it is utilised for the diagnosis of disease affecting the middle ear, although there are some indications for radiographic examination of the external ear canal.

The radiological anatomy of the petrous temporal bones, and the associated components of the middle ear, is complex and subject both to breed and individual variation, particularly in the dog[48–52]. Consequently, a thorough knowledge of the spatial relationships between the skull and the ear is essential if the radiographs are to be interpreted correctly[50]. However, pathological changes to the region are unlikely to be symmetrical. Thus, provided a good quality, symmetrical radiographic image is obtained, useful information may be acquired by comparing one side with the other[49,50].

Any one of several radiographic views will provide information on the middle ear but no one view can provide a complete picture. Therefore, at least two different radiograph views must be included in any radiographic investigation.

Lateral oblique view (Figs 2.64, 2.65)

- Advantages: good visualisation of the tympanic bulla and petrous temporal bone.
- Disadvantages:
  - Only one bulla can be visualised at a time.
  - Elevating the nose can help to increase visualisation.
  - Not easily repeatable, even in the same animal, so side-to-side comparison is difficult[53].

Positioning

The animal is placed in lateral recumbency with the head parallel to the film and the bulla of interest nearest the film. The jaw should be taped open around a syringe barrel[54]. The nose is elevated some 25–30% to achieve separation of the bullae on the plate[53,54]. The beam should be centred to the base of the ear to project the bulla clear of other structures, where it appears caudal to the left bulla. Patently the positions are reversed when the dog is in right lateral recumbency.

Interpretation

The bullae appear as thin-walled, crisply outlined bone structures with a smooth external border (Figs 2.66, 2.67)[8]. Air shadow should be visible in the external ear canal[52]. Predominantly lytic changes on the rostroventral wall of the bulla are usually associated with chronic inflammation[53]. Lytic changes within the petrous temporal bone may reflect either inflammation or neoplasia[55].

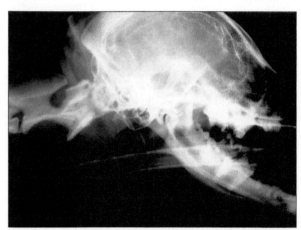

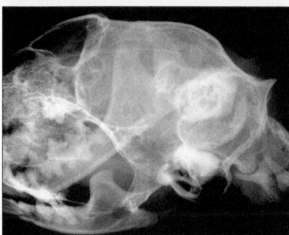

Figs 2.64, 2.65 Lateral oblique radiographs of the head of a dog (2.64) and a cat (2.65) demonstrating optimal positioning for visualising the tympanic bullae, which are clearly visible.

Rostrocaudal (open mouth) view (Figs 2.68, 2.69)

- Advantages:
  - Good visualisation of both tympanic bullae;
  - Good view for diagnosing otitis media[51].
- Disadvantages:
  - General anaesthesia is necessary and the endotracheal tube must be removed;
  - Can be difficult to obtain perfect pictures without fine-tuning, especially with brachycephalic breeds.

Positioning

The animal must be in dorsal recumbency. The head is positioned with the mandibles and hard palate vertical to the film[54]. The tongue must be brought as far forward as possible and tied to the mandible with tape[53]. The interpupillary line must be parallel to the film. In dolichocephalic breeds the primary beam is centred through the open mouth parallel to the hard palate[51]. In mesaticephalic breeds it may be necessary to angle the hard palate slightly away from the vertical (perhaps 10° or so[51]). In brachycephalic breeds the hard palate may need to be angled up to 20° away from the vertical in order to avoid superimposing the bullae on the wings

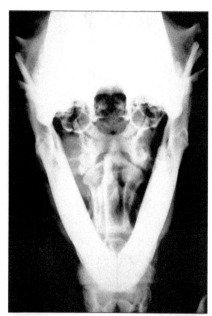

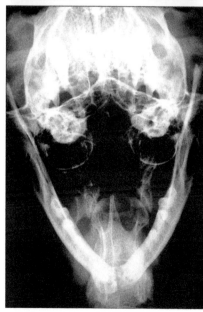

Figs 2.68, 2.69 Rostrocaudal (open mouth) radiographs of the head of a dog (2.68) and a cat (2.69) demonstrating how the tympanic bullae are skylined.

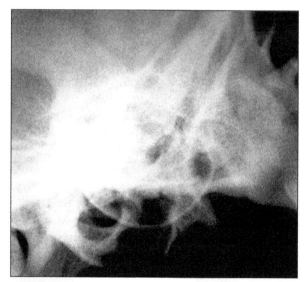

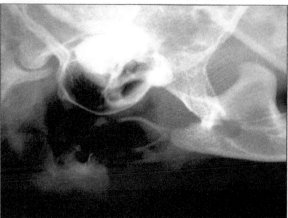

Figs 2.66, 2.67 Close-ups of the normal tympanic bulla of a dog (2.66) and a cat (2.67) in the lateral oblique view.

of the atlas[52]. Alternatively, the centre of the beam can be angled rostrocaudally, at up to 30° angling towards the hard palate. The beam can be centred on the base of the tongue.

Interpretation

The bullae appear as thin-walled, symmetrical bone opacities at the base of the skull (Fig. 2.70)[7]. Overlying

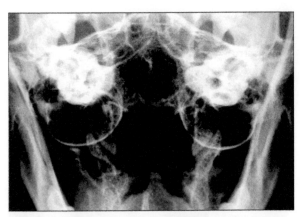

Fig. 2.70 Close-up rostrocaudal (open mouth) view of the tympanic bullae of a cat. Near perfect positioning is important for this view as one is looking for subtle changes that may only be apparent by comparing one side with the other.

soft tissues may produce the appearance of middle ear pathology. This must be interpreted with care.

Dorsoventral view (Figs 2.71, 2.72)

Better than ventrodorsal view for small breeds and cats[54].
- Advantages:
  - It is easier to achieve good bilateral symmetry with this view than with the ventrodorsal view, as the mandibles provide stability against lateral rotation[50].
  - Good view for diagnosing otitis media[5].
  - Good for visualising the external ear canal.
- Disadvantages:
  - Because the calvarium is further from the plate it is magnified and this can induce some artefactual distortion[53].
  - However, this is more than outweighed by the advantage of having the bullae close to the plate.

Positioning

The animal is placed in ventral recumbency. Care must be taken to ensure that the animal is aligned symmetrically with the interpupillary line parallel to the film[53]. The hard palate must be parallel to the table and the animal adjusted so that the base of the skull is as close to the film as possible[6]. This may require support with radiolucent blocks of foam under the

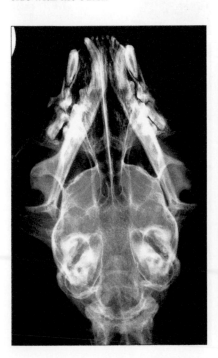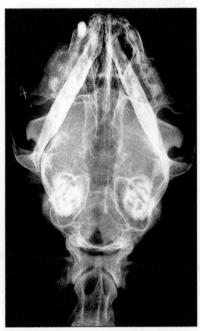

Figs 2.71, 2.72 Dorsoventral radiographs of the head of a dog (2.71) and a cat (2.72). Note the appearance of the bullae and the difficulty in visualising them using this position compared with the lateral oblique and rostrocaudal (open mouth) views.

rostral mandible or sandbags over the cervical spine or both. The beam should be centred at the intersection of two imaginary lines: a sagittal line and a lateral line at right angles to the sagittal line, drawn through the estimated position of the tympanic membranes.

Interpretation

The bullae should exhibit bilateral symmetry and appear as fine, crisp, distinct, linear bony opacities. Air shadow should be visible in the external ear canals.

Ventrodorsal view (Figs 2.73, 2.74)

Better than the dorsoventral view for deep-chested dogs[54].

- Advantages:
  - Standard view; lots of reference material.
  - Good view for diagnosing otitis media[52].
- Disadvantages:
  - This position is not suitable for brachycephalic breeds[52].
  - The sagittal crest tends to make the skull fall laterally, making exact positioning difficult[53].

Positioning

- The animal is placed in dorsal recumbency. Care must be taken to ensure that the animal is aligned symmetrically. The hard palate must be parallel

to the table. This may require support under the rostral mandible or under the cervical spine or both. Intraoral tape, positioned immediately caudal to the canine teeth and then tied to the table, may help in positioning the mandible. The beam should be centred at the intersection of two imaginary lines: a sagittal line and a lateral line, at right angles to the sagittal line, drawn through the estimated position of the tympanic membranes.

Interpretation

The superimposition of the petrous temporal bones makes the bullae walls appear thicker, and this can make evaluation of subtle changes more difficult.

Lateral view (Figs 2.75, 2.76)

- Advantages: Standard view; lots of reference material.
- Disadvantages:
  - Not ideal for visualising individual tympanic bullae as they are superimposed if a true lateral position is achieved.
  - Probably the least useful view.

Positioning

The animal is placed in lateral recumbency and the head adjusted to true lateral with the sagittal plane parallel to the film and the interpupillary line vertical[53]. This may

Figs 2.73, 2.74 Ventrodorsal radiographs of a dog (2.73) and a cat (2.74). Note the appearance of the bullae and the difficulty in visualising them using this position compared with the lateral oblique and rostrocaudal (open mouth) views.

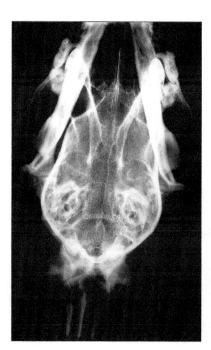
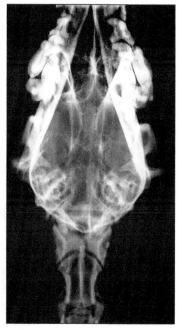

require foam padding to achieve a true lateral. The calvarium, nasal pharynx and larynx should be included in

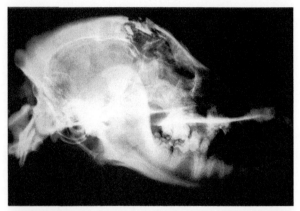

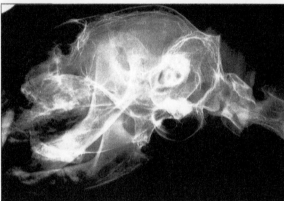

Figs 2.75, 2.76 Lateral radiographs of the head of a dog (2.75) and a cat (2.76). The bullae are visible but both left and right bullae are superimposed, making interpretation difficult.

the view. The beam should be centred between the ear and the eye.

Interpretation

The bullae appear as thin-walled, crisply outlined bony structures with a smooth external border (Figs 2.77, 2.78), but in a true lateral view they will be superimposed, making for difficult interpretation[53]. Air shadow and, if present, the thickened walls of the horizontal ear canal may be visible[53].

Visualising the external ear canal and assessing the tympanic membrane

Radiography is not commonly employed as a means of assessing pathological changes in the external ear canal. It may be possible to see air shadows (Fig. 2.79) delineating the external ear canal in some of the standard radiographic views of the ear, particularly the dorsoventral and rostrocaudal (open mouth) views. In addition, in cases of chronic otitis externa, there may be calcification of the cartilages of the external canal (Figs 2.80, 2.81) However, it is not possible to assess the integrity of the tympanic membrane or visualise the position of an obstructing luminal neoplasm without using contrast techniques.

Otoscopic and plain radiographic examinations must be performed prior to contrast studies. Significant epidermal hyperplasia or neoplastic proliferation may occlude the lumen to such an extent that adequate distribution of contrast medium is impossible. In addition,

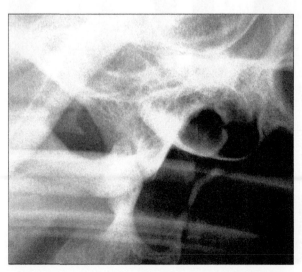

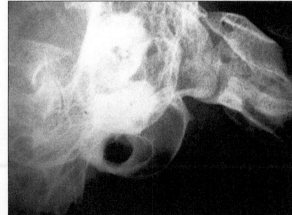

Figs 2.77, 2.78 Close-up of the normal tympanic bullae of a dog (2.77) and a cat (2.78) in the lateral view.

the presence of an exudate or mass within the bulla may prevent contrast medium from entering the middle ear. False-negative interpretation may occur if these changes are not identified prior to contrast studies[8].

A standard radiographic contrast medium is used, preferably a non-ionic, water-soluble iodine-based medium rather than an oily medium[55]. The contrast medium may be diluted 50:50 with saline, prior to instilling it into the external ear canal[55]. Care must be taken to ensure that the contrast medium is distributed evenly along the external ear canal and that none contaminates the hair on the surrounding aspects of the head[50,55]. Gentle massage of the ear canal will ensure an even distribution. Taking care to deliver all the contrast medium into the ear canal, and subsequently plugging the orifice with cotton wool, should prevent soiling of the area around the external ear[50,55].

Ventrodorsal or, preferably, rostrocaudal (open mouth) radiography will allow evaluation of the lumen of the external ear canal and permit some deductions on the state of the tympanic membrane (Fig. 2.82)[56]. If contrast medium enters the middle ear (Fig. 2.83) it is usually visualised as an opacification of the inner wall of the bulla, best seen on rostrocaudal (open mouth) views. Failure to detect contrast medium in the bulla should not be interpreted as indicating an intact tympanum.

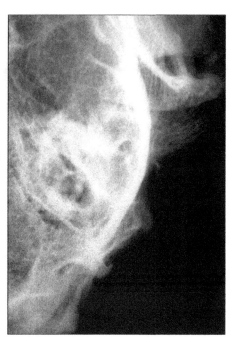

Fig. 2.79 Dorsoventral radiograph of a dog. Note the air shadow within the external ear canal.

2.9 MYRINGOTOMY

Whereas myringotomy or tympanotomy can be used to aid in the diagnosis of middle ear effusion and infection, it is more commonly used as an aid to the treatment of these conditions and is therefore described in Chapter 6 Diseases of the Middle Ear.

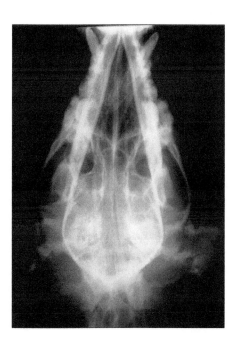
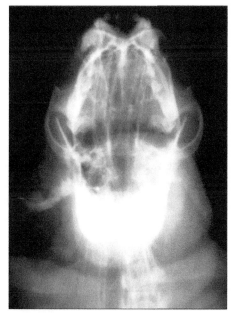

Figs 2.80, 2.81 Dorsoventral and rostrocaudal open mouth radiographs of a Cocker Spaniel with chronic otitis externa and otitis media. Note the extensive calcification of the ear canal cartilages and the changes around the left bulla.

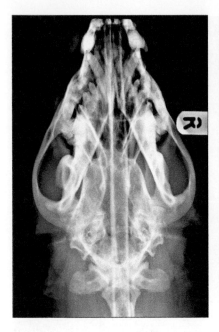 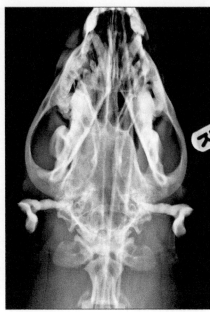

Figs 2.82, 2.83 Ventrodorsal radiographs demonstrating tympanography. In Fig. 2.83 the right tympanum is intact and the concavity is apparent. The left tympanum has been breached and contrast fills the tympanic bulla.

2.10 NEUROLOGICAL ASSESSMENT OF THE PINNA AND EXTERNAL EAR CANAL

Most neurological manifestations of ear disease reflect middle and inner ear pathology[57,58] rather than lesions in the external ear canal or on the pinna. Thus, facial nerve paralysis and/or Horner's syndrome occur on the same side as the middle ear disease (see Chapter 6 Diseases of the Middle Ear, Section 6.2). In contrast, the clinical signs of inner ear disease in the dog and cat are of deafness and/or peripheral vestibular syndrome. Evaluation of hearing loss is discussed in here, whereas evaluation of loss of balance is discussed in Chapter 7 Diseases of the Inner Ear.

The neurological assessment of the pinna is made by observing the pinnal response to physical stimulation – the pinnal reflex. Gentle pinpoint stimulation of the convex or concave surface of the pinna with a blunt instrument (such as an artery forceps) elicits a twitch of the pinna – a test of sensory innervation. Stimulation of the central portion of the pinna on the convex surface tests the 2nd cervical nerve reflex, while stimulation on the concave surface tests the facial nerve. Attempting to interpret this reflex by stimulating the cranial or the caudal margin of the pinna is inadvisable because of the overlap of different cutaneous nerves[59].

The pinnal scratch reflex is a pathological sign that must not be confused with the pinnal reflex. Rubbing the tip of the pinna between finger and thumb elicits a vigorous scratch reflex in the ipsilateral hindlimb, particularly in dogs with scabies or pediculosis. This reflex is not pathognomonic but it is very reliable.

2.11 ELECTROPHYSIOLOGICAL HEARING TESTS

Electrophysiological testing is the only objective means of testing the sense of hearing in veterinary species.

- Brainstem-evoked response audiometry (BERA) records the neurological response to sound and is the gold-standard in hearing assessment in animals.
- Otoacoustic emission testing offers great potential as a screening method for deafness in dogs and cats.
- Tympanometry measures the changes in compliance of the tympanic membrane and is useful for assessment of middle ear and auditory tube function.

Fig. 2.84 Recording of BERA with the dog in sternal recumbency under a light plane of anaesthesia. Three recording electrodes were inserted subcutaneously: a recording electrode at the base of the pinna of the ear to be stimulated, a common (ground) electrode at the base of the pinna of the other ear, and a reference electrode over the occipital protuberance on the midline. Stimulations were delivered to the ear via a flexible transmission tube, 4.5 cm long with an internal diameter of 8 mm, extending 1 cm into the vertical ear canal.

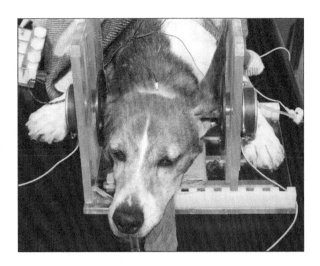

Hearing loss is a common disorder in many dog and cat breeds and auditory dysfunction and its clinical consequences can vary from mild to severe[60–62]. Taking a complete history and performing thorough physical, neurological, and otoscopic examinations usually make it possible to differentiate between conductive and sensorineural hearing loss[63]. Advanced imaging with CT or magnetic resonance imaging (MRI) is necessary, however, for proper evaluation of the middle (and inner) ear to rule out conduction deafness and for treatment planning[64–67]. Though essential for the diagnostic work-up of all patients with hearing disorders, these two techniques can only identify morphological abnormalities of the petrous bone and inner ear structures. By measuring behavioural responses to a sound, a psychoacoustic audiogram can be constructed; proper training of both the dog and the human tester is required before reliable results can be obtained, rendering the technique unsuitable for young puppies and clinical patients[62,68]. Whilst behavioural testing for clinical patients using simple whistles for instance has been used, results are often equivocal, particularly when the hearing loss is partial or unilateral. Functional abnormalities, therefore, have to be diagnosed with objective electrophysiological methods such as impedance audiometry, evoked response audiometry or cochlear microphony[69–71]. These tests are non-invasive and evaluate components of the external ear canal, the middle and inner ear, cranial nerve VIII and selected areas of the brainstem and cortex. The complete sequence of these procedures will allow differentiation between conductive and sensorineural hearing loss. The great advantage of electrodiagnostic procedures is that they do not require consciousness and they can therefore be performed even in uncooperative animals[69]. BERA is the most frequently used hearing testing modality in veterinary medicine, because it yields objective, reproducible information, and is relatively easy to perform, noninvasive, safe and cost-effective[62,70,71].

Brainstem auditory evoked response (BAER)

NB: The physiology of hearing is reviewed in Chapter 1 Anatomy and Physiology of the Ear, Section 1.8.

Auditory stimulation evokes electrical responses in the auditory pathway in a consistent manner[72,73]. Once a nerve impulse is generated in the cochlea, the signal travels along the acoustic nerve to the cochlear nuclei in the brainstem. From the cochlear nuclei, many projections lead to other nuclei in the brainstem and ultimately to the primary auditory cortex[72,73]. When amplified and averaged, these electrical changes can be recorded from scalp electrodes (far-field evoked potentials), and typically represent as a constant repeatable number of waves, normally 5–7, occurring during the first 10 ms after the presentation of a transient sound[74,75]. To record the reponse, three small needle electrodes are inserted subcutaneously on the animal's head (Fig. 2.84). A recording electrode, a reference and a ground electrode are connected to an amplifying and signal-averaging recording system, ideally coupled with an artefact rejection device. Head phones or ear inserts can be used to deliver acoustic stimulation to the ear at various rates of delivery (stimulations/second), intensities (dB), and frequencies (Hz)[76,77]. The intensity of the stimulation is most commonly calibrated in decibels (dB) and referred to as the normal hearing level (dB nHL), where 0 dB nHL is the human behavioural hearing threshold in a normal adult population or to the absolute sound pressure level of the stimulation (dB SPL)[78].

In dogs, the BAER or BERA waves have been numbered with Roman numerals analogous to the numbering in people. However, some waves are absent, bifurcated or merged in dogs, which has created confusion about correct labelling. Wave VII is usually absent in the dog and waves III and IV often merge (Fig. 2.85)[62,69,77–80]. Most authors therefore recommend and use the following criteria for labelling the BERA waves in dogs: wave I is the first recognisable wave with a positive deflection and wave V is the positive wave occurring immediately before the deep negative trough in the second half of the recording[62,69,71,79, 80].

The time between stimulation of the ear and electrophysiological response is referred to as latency. For the early latency components, the responses recorded between 0 and 10 ms after the stimulus, generators are thought to arise almost totally within the brainstem[62,69]. Therefore, this series of waves is referred to as the brainstem auditory evoked responses (BAER). Precise neuroanatomical structures were eventually associated with the different waves and they have been proven to be very similar in humans, dogs and other animal species (Figs. 2.85, 2.86)[75,80–84]. Wave I is generated by the distal part of the vestibulocochlear nerve and wave II by the intracranial portion of this nerve[72]. The other waves represent electrical activity in the cochlear nucleus, the superior olivary complex, the lateral lemniscus, the inferior colliculus, the medial geniculate nucleus and thalamocortical radiations[72].

Factors known to influence the latencies and amplitudes of the elicited waves are related to the dog being examined (e.g. sex, age and head size), body temperature and the stimulation and recording protocols[69,71,76,85,86]. The effect of anaesthesia on the BAER generally is considered to be negligible. In dogs, waves I, II and V usually

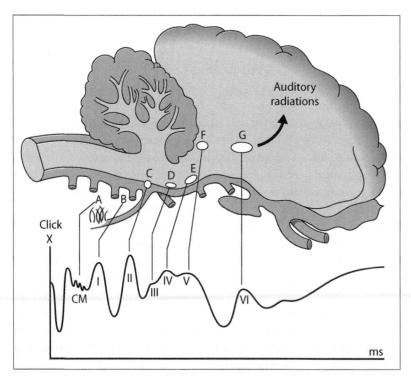

Fig. 2.85 Diagram of BAER generators. (Adapted from Mayhew IG, Washbourne JR. A method of assessing auditory function in horses. *British Veterinary Journal* **1990;146:509–18, with permission.)**

have the largest amplitudes. Wave V can be most easily used to determine auditory thresholds because it is the last to disappear with decreasing stimulus levels under normal conditions (Fig. 2.87)[87–89]. The amplitude of the waves increases with the stimulus intensity and it is highly variable. The latency of each wave decreases linearly as the acoustic stimulus intensity increases, and decreases in rectal temperature cause a lengthening of the waves' latency[77]. Inter-peak latencies are indicative of conduction time within the brainstem between generator sites and they do not change with the stimulus intensity. They can help to localise the site of deafness along the auditory pathways[74,76,81,90,91]. Normal values for latencies of the peaks and the threshold response have been established for a population of healthy dogs using click and toneburst stimulation[71,79]. Deafness due to abnormalities of the cochlea and/or cranial nerve VIII results in the total absence of recognisable recordings (isoelectric tracing, Figs 2.88, 2.89)[60,61,63,70,71].

Click stimuli are very brief (0.1 ms) and contain energy in the range of 500 to 4000 Hz[62]. Click stimulation is valuable for differentiating neurological from conduction deafness, detecting brainstem lesions and for intraoperative monitoring[71,83,86]. The click stimulus has been used most extensively and successfully in the diagnosis of congenital sensorineural hearing loss. Since deafness is complete in these cases[92], no frequency-specific thresholds are required. BERA using click stimulation is therefore useful and continues to be recommended for diagnosing the presence or total absence of auditory function for animals with congenital sensorineural hearing loss[92,93]. Testing of frequency-specific areas of the cochlea can be accomplished by use of click stimulation with high-pass masking, pure tone masking (derived response) or notch noise or white noise masking, or by direct stimulation with tone pips (tonebursts). The last procedure has been shown to be easy to perform and to yield reliable information about pure tone thresholds in humans[94–100]. Another technique for frequency-specific threshold measurement in dogs was reported in 2006, using auditory steady-state evoked potentials[101]. Frequency-specific information is essential to determine the extent of incomplete sensorineural deafness like age-related hearing loss (ARHL), noise-induced hearing loss (NIHL) and ototoxicity[60,80,97,98,102,103].

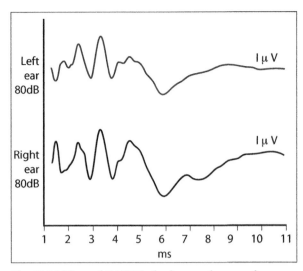

Fig. 2.86 Normal BAER in both ears; the wave forms from each ear are similar and both are similar to the stylised diagram in Fig. 2.85.

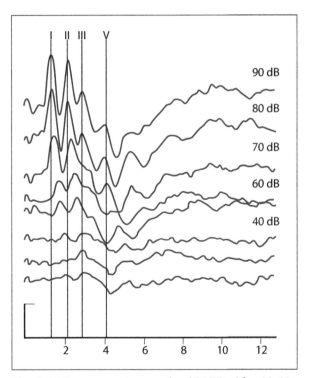

Fig. 2.87 Representative sample of BAER with a 90–20 dB SPL 8 kHz toneburst stimulation in the left ear of a healthy 5-year-old dog. The positive peaks are labelled with Roman numerals. Wave IV is missing in this tracing and wave V is identified by the waveform and latency. The vertical bar represents 1 µV.

In animals suspected of suffering from conductive hearing loss, evaluation of hearing can be performed using bone-conducted signals. In these cases the BERA is generated with a bone vibrator instead of headphones, in order to bypass the conductive hearing apparatus. However, this technique has limitations. For instance, with bone-conducted BERA the latency of wave V is about 0.5 ms longer at intensities similar to air-conducted BERA, and the maximum acoustic output from a bone vibrator is less than that of a click or a tone[104].

In a study by Wolschrijn *et al.*, the mean threshold for bone-conducted stimulus was found to vary between 50 and 60 dB, whilst the air-conducted threshold in the same population of healthy dogs was 0–10 dB[105]. Bone-conduction thresholds do not just measure inner ear function; sound applied to the skull is transmitted to the cochlea via bone, but also via the external auditory canal and the ossicles[105]. Nevertheless, wave forms and inter-peak latencies of the BERA elicited by bone-conducted stimuli and air-conducted stimuli are similar, indicating that the signals have the same origin[105].

Tympanometry and acoustic reflex testing

Tympanometry is the recording of changes in compliance (reciprocal of stiffness) of the tympanic membrane as a result of air pressure changes within the external ear canal[106,107]. Tympanometry uses impedance audiometry, which is based on the principle that the intensity of a sound wave is dependent upon the size of the cavity within which it is generated, and on the compliance or stiffness of the containing walls[108]. The reciprocal of acoustic impedance is acoustic immittance, and whilst current equipment actually measures immittance, the term 'impedance audiometry' is most commonly used. The primary purpose of impedance audiometry is to determine the status of the tympanic membrane and middle ear and auditory tube function via tympanometry. The secondary purpose of this test is to evaluate acoustic reflex pathways, which include cranial nerves

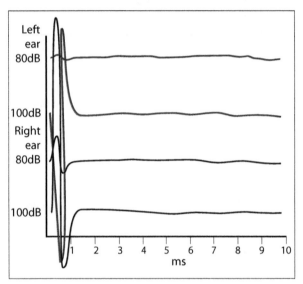

Fig. 2.88 BAER trace from a dog with bilateral deafness.

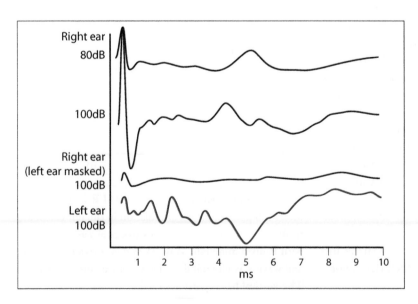

Fig. 2.89 BAER trace from a dog with unilateral (right ear) deafness.

(CN) VII and VIII and the auditory brainstem. When a sufficiently large sound is detected by the ear the stapedius muscle contracts, purportedly to protect the delicate inner ear from the damaging effects of transmitting sound waves of very high amplitude. This reflex leads to a sudden contraction of the stapedius muscle, which subsequently affects the tension within the tympanum and therefore its compliance[109].

Tympanometry records changes in middle ear immittance, while air pressure is varied in the ear canal and acoustic reflexes are recorded at a single air pressure setting (i.e. the pressure setting that provided the peak immittance reading for that particular ear on the tympanogram). Ear canal pressure is maintained at that specific setting, while tones of various intensities are presented into the ear canal and immittance is recorded. A significant change in middle ear immittance immediately after the stimulus is considered an acoustic reflex[69,108,110–112].

If the threshold for the acoustic reflex is identified, a signal some 10 dB above the threshold is given for a period of 10 seconds[109]. In normal animals and in animals with cochlear lesions, the reflex should be maintained for the entire test period. In some conditions affecting the function of CN VIII there is a characteristic decay of the reflex, within the 10-second test period[109].

There is no purpose-built veterinary equipment for impedance audiometry and the clinician must adapt those designed for humans if tympanometry is to be practised. Tympanometry involves closing the ear canal completely with a rubber plug containing three channels, one for introduction of the sound stimulus from the tympanometer (sound probe), one for recording the sound reflected after application of the stimulus (microphone) and one for the programmed variation of pressure by the tympanometer (pressure probe)[107]. In dogs and cats, the angle between the vertical ear canal and the horizontal ear canal makes aligning the equipment difficult. Establishing a good seal between the cuff and the wall of the ear canal as well as allowing the sound wave to impact squarely upon the tympanic membrane can be challenging to achieve[113]. The change in compliance of the tympanum (derived from the variation in the reflected sound in response to the programmed variation in pressure) is measured as pressure changes within the occluded external ear canal when a sound tone is generated[107]. Abnormal responses, abnormal compliance can be the result of perforation, the filling of the middle ear with fluid and/or auditory tube dysfunction, or adhesion of the tympanic membrane to surrounding structures (Figs 2.90A–C)[69,111,112,114–116]. Tympanometry has been used successfully and reliably in dogs and cats under sedation or anaesthesia. Tympanometry proved to be a reliable indicator of middle ear disease, ear canal volume and peak compliance measurements were found to be dependent on body weight in dogs[111,112,114,116].

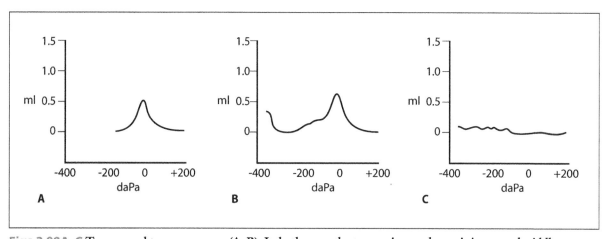

Figs 2.90A–C Two normal tympanograms (A, B). In both cases the tympanic membrane is intact and middle ear function is normal. Abnormal tympanogram (C) associated with a ruptured tympanic membrane. (Courtesy of Dr J Steiss, Scott Ritchey Research Center; Auburn University, Alabama, USA.)

Whereas in the past the use of tympanometry has been hindered by equipment costs and possible influence of anaesthetics on compliance and peak pressures values, newer lower-cost hand-held devices that can be used in conscious dogs make the methodology more accessible to veterinary practices[111,112,114,116,117]. A very recent study reported acceptable recordings from 28 of 32 (88%) ears using a hand-held device in conscious dogs without obvious otic disease[117]. Conscious dogs with otitis externa and media, perforated tympanums, middle ear effusion (PSOM) or auditory tube disorders need to be examined to validate the usefulness of these recordings in the clinical settting[117].

Cats have been used as a model in human audiology experiments and tympanometry has been reported to be a useful indicator of middle ear disease in this species[118–120]. In cats with intact tympanic cavities, measurements of acoustic input impedance were found to be effective using a test signal of 300 Hz to 4 kHz[121].

Otoacoustic emissions

Otoacoustic emissions (OAEs) are vibrations created by contractions of the outer hair cells that are located in the cochlea. It is hypothesised that outer hair cell contraction follows inner hair cell stimulation, thus enhancing the spatial resolution of the cochlea[72]. The identification and recognition of sounds of different frequency is thereby amplified. The outer hair cell contraction generates a vibration within the cochlea that is retrogradely transduced to the middle ear. There, it is transduced by the ossicles to the eardrum that is brought to vibration. These vibrations create sounds that can be recorded in the ear canal. These sounds are soft, but potentially audible, infrequently amounting to as much as 30 dB SPL[72]. For detection of these OAEs, normal middle ear function is mandatory. OAEs are generated only when the organ of Corti is in normal or near normal condition and their presence therefore, is indicative of normal middle ear and cochlear function. OAEs may emerge spontaneously, but more commonly follow acoustic stimulation. David Kemp discovered this phenomenon in 1977[122]. A clinical application of OAE is, among others, to obtain information quickly about a good or impaired ear function, such as with otitis media with effusion (OME), and with pre-existing retrocochlear lesions to demonstrate the integrity of the cochlea.

Emissions can be divided into two main classes:
- Spontaneous otoacoustic emissions (SOAEs), which can be measured without presenting a stimulus.
- Evoked otoacoustic emissions (EOAEs), which are measured during the presentation of a stimulus, or immediately after[72,123,124].

The SOAHs consist of one or more pure tones. In adults they are often very weak and therefore difficult to measure[124]. Generally speaking they will disappear with increasing hearing loss. In neonates the emissions are stronger; however, the clinical value of SOAEs is small. Spontaneous emissions have been reported in dogs as well[125].

The EOAEs on the other hand, are more easily measured. The two following tests are the most frequently performed:
- Transient evoked otoacoustic emissions (TEOAEs), in which the inner ear is activated by means of a short stimulus (usually a click), after which the emission is measured.
- Distortion product otoacoustic emissions (DPOAEs). Here two pure tones are presented which, processed in the cochlea, will generate pure tones as well.

OAEs are not only quick to obtain, but also very objective in evaluating outer hair cell integrity, and can identify ototoxicity for instance earlier than conventional pure tone audiometry[126]. In dogs evoked OAEs have been recorded and reported, but up until 2010 reports were scarce[126–128]. Recently though, Strain reported that in three litters of puppies, DPOAEs were able to distinguish reliably between deaf and hearing ears as determined by BERA[123]. Using a similar protocol, Schemera et al. reported that DPOAEs were easily obtained, robust, reliable and consistent with auditory brainstem response results in 23 puppies[129]. TEOAEs testing has been reported recently as well and appears equally reliable as DPOAEs in discriminating hearing from deaf ears in adult dogs when compared to BERA[130], and in puppies for hearing screening[131]. The clinical use of OAEs is still subject to extensive research, but OAE testing promises to become a sensitive instrument to distinguish hearing loss from normal hearing in veterinary practice. Early detection of dysfunction seems possible, as well as monitoring of cochlear alterations, and distinguishing various types of inner ear defects.

2.12 REFERENCES

1 Masuda A, Sukegawa T, Mizumoto N, *et al*. Study of lipid in the ear canal in canine otitis externa with *Malassezia pachydermatitis*. *Journal of Veterinary Medical Science* 2000;**11**:1177–82.

2 Saridomichelakis MN, Farmaki R, Leontides LS, Koutinas AF. Aetiology of canine otitis externa: a retrospective study of 100 cases. *Veterinary Dermatology* 2007;**18**:341–7.

3 Angus JC, Lichtensteiger C, Campbell KL, Schaeffer DJ. Breed variations in histologic features of chronic severe otitis externa in dogs: 80 cases (1995–2001). *Journal of the American Veterinary Medical Association* 2002;**221**:1000–6.

4 Sylvestre AM. Potential factors affecting the outcome of dogs with a resection of the lateral wall of the vertical ear canal. *Canadian Veterinary Journal* 1998;**39**:157–60.

5 Kennis RA. Feline otitis: diagnosis and treatment. *Veterinary Clinics of North America Small Animal Practice* 2012;**43**:51–6.

6 Cole LK. Otoscopic evaluation of the ear canal. *Veterinary Clinics of North America Small Animal Practice* 2004;**34**:397–410.

6a Bloom P. Diagnosis and Management of Otitis in the Real World. Proceedings Ontario Veterinary Medical Association. Westin Harbour, Toronto 2012, pp. 19–24.

6b Robson DC, Burton GG, Bassett RJ. Correlation between topical antibiotic selection, *in vitro* bacterial antibiotic sensitivity and clinical response in 17 case of canine otitis externa complicated by *Pseudomonas aeruginosa*. In: College Science Week Proceedings (Dermatology Chapter), Australian College of Veterinary Scientists.

6c Fadok VA. Otitis externa: the bane of our existence. *OVMA Annual Conference*, Westin Harbour Castle, Toronto 2014, pp. 43–7.

7 Kirby AL, Rosenkrantz WS, Ghubash RM, Neradilek B, Polissar N. Evaluation of otoscope cone disinfection techniques and contamination level in small animal private practice. *Veterinary Dermatology* 2010;**21**:175–83.

8 McKeever PJ, Richardson HW. Otitis externa. Part 2: Clinical appearance and diagnostic methods. *Companion Animal Practice* 1988;**2**:25–31.

9 Stout-Graham M, Kainer RA, Whalen LR, Macy DW. Morphologic measurements of the external ear canal of dogs. *American Journal of Veterinary Research* 1990;**51**:990–4.

10 Little CJL, Lane JG. An evaluation of tympanometry, otoscopy and palpation for assessment of the canine tympanic membrane. *Veterinary Record* 1989;**124**:5–8.

11 Usui, R, Usui R, Fufuda M, Fukui E, Hasegawa A. Treatment of canine otitis externa using video otoscopy. *Journal of Veterinary Medical Science* 2011;**73**:1249–52.

12 Fraser G, Gregor WW, Mackenzie CP, Spreull JSA, Withers AR. Canine ear disease. *Journal of Small Animal Practice* 1970;**10**:725–54.

13 Neer TM. Otitis media. *Compendium on Continuing Education* 1982;**4**:410–16.

14 Fraser G. Aetiology of otitis externa in the dog. *Journal of Small Animal Practice* 1965;**6**:445–52.

15 Huang HP, Fixter LM, Little CJL. Lipid content of cerumen from normal dogs and otitic canine ears. *Veterinary Record* 1994;**134**:380–1.

16 Huang HP, Little CJ, McNeil PE. Histological changes in the external ear canal of dogs with otitis externa. *Veterinary Dermatology* 2009;**20**:422–8.

17 Cole LK. Otoscopic evaluation of the ear canal. *Veterinary Clinics of North America Small Animal Practice* 2004;**34**:397–410.

18 Corriveau LA. Use of a carbon dioxide laser to treat ceruminous gland hyperplasia in a cat. *Journal of Feline Medicine and Surgery* 2012;**14**:413–16.

19 Boord M. Fun with lasers. In: *Advances in Veterinary Dermatology*. Torres SMF, Frank LA, Hargis AM (eds). John Wiley & Sons, New York 2013.

20 Spreull JSA. Treatment of otitis media in the dog. *Journal of Small Animal Practice* 1964;**5**:107–52.

21 Cole LK, Kwochka KW, Podell M, Hillier A, Smeak DD. Evaluation of radiography, otoscopy, pneumatoscopy, impedance audiometry and endoscopy for the diagnosis of otitis media in the dog. In: *Advances in Veterinary Dermatology*, 4th edn. Thoday KL, Foil CS, Bond R (eds). Iowa State Press, Ames 2002, pp. 49–55.

22 Steiss JE, Boosinger TR, Wright JC, Pillai SR. Healing of experimentally perforated tympanic membranes demonstrated by electrodiagnostic testing and histopathology. *Journal of the American Animal Hospital Association* 1992;**28**:307–10.

23 Cole LK, Kwochka KW, Kowalski KJ, Hillier A. Microbial flora and antibacterial sensitivity patterns of isolated pathogens from the horizontal ear canal and middle ear in dogs with otitis media. *Journal of the American Veterinary Medical Association* 1998;**212**:534–8.

24 Griffin CE. Otitis externa and otitis media. In: *Current Veterinary Dermatology*. CE Griffin CE, Kwochka KW, MacDonald JM (eds). Mosby, St Louis 1993, pp. 245–62.

25 Rosser EJ. Evaluation of the patient with otitis externa. *Veterinary Clinics of North America* 1988;**18**:765–72.

26 Huang H-P. *Studies of the Microenvironment and Microflora of the Canine External Ear Canal*. PhD Thesis, 1993. Glasgow University.

27 Kowalski JJ. The microbial environment of the ear canal in health and disease. *Veterinary Clinics of North America* 1988;**18**:743–54.

28 Chickering WR. Cytologic evaluation of otic exudates. *Veterinary Clinics of North America* 1988;**18**:773–82.

29 Lehner G, Louis CS, Mueller RS. Reproducibility of ear cytology in dogs with otitis externa. *Veterinary Record* 2010;**167**:23–6.

30 Bouassiba C, Osthold W, Mueller RS. Comparison of four different staining methods for ear cytology of dogs with otitis externa. *Tierarztliche Praxis Ausgabe Kleintiere Heimtiere* 2013;**41**:7–15.

31 Rosychuck RWA. Management of otitis externa. *Veterinary Clinics of North America* 1994;**24**:921–52.

32 Huang H-P. Canine cerumen cytology. *Chinese Society of Veterinary Science* 1995;**21**:18–23.

33 Rogers KS. Tumors of the ear canal. *Veterinary Clinics of North America Small Animal Medicine* 1988;**4**:859–68.

34 Mansfield PD, Boosinger TR, Attleburger MH. Infectivity of *Malassezia pachydermatis* in the external ear canal of dogs. *Journal of the American Animal Hospital Association* 1990;**26**:97–100.

35 Bond R, Anthony RM, Dodd M, Lloyd DH. Isolation of *Malassezia sympodialis* from feline skin. *Journal of Medical and Veterinary Mycology* 1996;**34**:145–7.

36 Frost CR. Canine otocariasis. *Journal of Small Animal Practice* 1961;**2**:253–6.

37 Russo M, Covelli EM, Meomartino L, Lamb CR, Brunetti A. Computed tomographic anatomy of the canine inner and middle ear. *Veterinary Radiology and Ultrasound* 2002;**43**:22–6.

38 Bischoff MG, Kneller SK. Diagnostic imaging of the canine and feline ear. *Veterinary Clinics of North America Small Animal Practice* 2004;**34**:437–58.

39 Foster A, Morandi F, May E. Prevalence of ear disease in dogs undergoing multidetector thin-slice computed tomography of the head. *Veterinary Radiology and Ultrasound* 2014;**58**:18–24.

40 Rohleder JJ, Jones JC, Duncan RB, Larsson MM, Waldron DL, Tromblee T. Comparative performance of radiology and computed tomography in the diagnosis of middle ear disease in 31 dogs. *Veterinary Radiology and Ultrasound* 2006;**47**:45–52.

41 Doust R, King A, Hammond G, *et al*. Assessment of middle ear disease in the dog: a comparison of diagnostic imaging modalities. *Journal of Small Animal Practice* 2007;**48**:188–92.

42 King AM, Weinrauch SA, Doust R, Hammond G, Yam PS, Sullivan M. Comparison of ulrasonography, radiology, and a single computed tomography slice for fluid identification within feline tympanic bulla. *Veterinary Journal* 2007;**173**:638–44.

43 Dickie AM, Doust R, Cromarty L, Johnson VS, Sullivan M, Boyd JS. Comparison of ultasonography, radiography and a single computed tomography slice for the identification of fluid within the canine tympanic bulla. *Research in Veterinary Science* 2003;**75**:209–16.

44a Angus JC, Campbell KL. Uses and indications for video-otoscopy in small animal practice *Veterinary Clinics of North America Small Animal Practice* 2001;**31**:809–28.

44b Radlinsky M. Advances in otoscopy. *Veterinary Clinics of North America Small Animal* 2015;**46**:171–9.

45 Usui R, Usui R, Fukuda M, Fukui E, Hasegawa A. Treatment of canine otitis externa using video otoscopy. *Journal of Veterinary Medicine and Science* 2011;**73**:1249–53.

46 Cho Y, Jeong J, Lee H, Kim, M, Kim N, Lee K. Virtual otoscopy for evaluating the inner ear with a fluid filled tympanic cavity in dogs. *Journal of Veterinary Science* 2012;**14**:419–24.

47 Eom K, Kwak H, Kang H, *et al*. Virtual CT otoscopy of the middle ear and ossicles in dogs. *Veterinary Radiology and Ultrasound* 2008;**49**:545–50.

48 Hare WCD. Radiographic anatomy of the canine skull. *Journal of the American Veterinary Medical Association* 1958;**133**:149–57.

49 Gibbs C. Radiological refresher. Part III. The head. *Journal of Small Animal Practice* 1978;**19**:539–45.

50 Rose WR. Small animal clinical otology: radiology. *Veterinary Medicine/Small Animal Clinician* 1977;**72**:1508–17.

51 Love NE, Kramer RW, Spodnick GJ, Thrall DE. Radiographic and computed tomographic evaluation of otitis media. *Veterinary Radiology and Ultrasound* 1995;**36**:375–9.

52 Douglas SW, Herrtage ME, Williamson HD. Canine radiography: skull. In: *Principles of Veterinary Radiography*. Baillière Tindall, London 1987, pp. 177–92.

53 Sullivan M. The head and neck. In: *BSAVA Manual of Small Animal Diagnostic Imaging*. R Lee (ed). British Small Animal Veterinary Association, Cheltenham 1995, pp. 16–22.

54 Wilson M, Mauragis D, Berry CR. Radiography of the small animal skull: temporomandibular joints and tympanic bulla. *Today's Veterinary Practice* 2014;**4**:53–8.

55 Trower ND, Gregory SP, Renfrew H, Lamb CR. Evaluation of canine tympanic membrane by positive contrast ear canalography. *Veterinary Record* 1998;**142**:78–81.

56 Eom K-D, Lee H-C, Yoon Y-H. Canalographic evaluation of the external ear canal in dogs. *Veterinary Radiology and Ultrasound* 2000;**41**:231–4.

57 Garosi LS, Lowrie ML, Swinbourne NF. Neurological manifestation of ear disease in dogs and cats. *Veterinary Clinics of North America Small Animal Clinician* 2012;**42**:1143–60.

58 Cook LB. Neurological evaluation of the ear. *Veterinary Clinics of North America Small Animal Clinician* 2004;**34**:425–35.

59 Whalen L, Kitchell R. Electrophysiological studies of the cutaneous nerves of the head of the dog. *American Journal of Veterinary Research* 1983;**44**:615–17.

60 Strain GM. Aetiology, prevalence and diagnosis of deafness in dogs and cats. *British Veterinary Journal* 1996;**152**:17–36.

61 Rak SG, Distl O. Congenital sensorineural deafness in dogs: a molecular genetic approach toward unravelling the responsible genes. *Veterinary Journal* 2005;**169**:188–96.

62 Wilson WJ, Mills PC. Brainstem auditory-evoked response in dogs. *American Journal of Veterinary Research* 2005;**66**:2177–87.

63 ter Haar G. Inner ear dysfunction in dogs and cats: conductive and sensorineural hearing loss and peripheral vestibular ataxia. *European Journal of Companion Animal Practice* 2007;**17**:127–35.

64 Garosi LS, Lowrie CT, Swinbourne NF. Neurological manifestations of ear disease in dogs and cats. *Veterinary Clinics of North America Small Animal Practice* 2012;**42**:1143–60.

65 Bischoff MG, Kneller SK. Diagnostic imaging of the canine and feline ear. *Veterinary Clinics of North America Small Animal Practice* 2004;**34**:437–58.

66 Eom K, Kwak H, Kang H, *et al*. Virtual CT otoscopy of the middle ear and ossicles in dogs. *Veterinary Radiology and Ultrasound* 2008;**49**:545–50.

67 Gotthelf LN. Diagnosis and treatment of otitis media in dogs and cats. *Veterinary Clinics of North America Small Animal Practice* 2004;**34**:469–87.

68 Heffner HE. Hearing in large and small dogs: absolute thresholds and size of the tympanic membrane. *Behavioral Neuroscience* 1983;**97**:310.

69 Sims MH. Electrodiagnostic evaluation of auditory function. *Veterinary Clinics of North America Small Animal Practice* 1988;**8**:913–44.

70 Knowles KE, Cash WC, Blauch BS. Auditory-evoked responses of dogs with different hearing abilities. *Canadian Journal of Veterinary Research* 1988;**52**:394–7.

71 ter Haar G, Venker-van Haagen AJ, de Groot HN, van den Brom WE. Click and low-, middle-, and high-frequency toneburst stimulation of the canine cochlea. *Journal of Veterinary Internal Medicine* 2002;**16**:274–80.

72 Moller AR. Physiology of the cochlea. In: *Hearing: its Physiology and Pathophysiology*. Moller AR (ed). Academic Press, San Diego 2000, pp. 71–94.

73 Pickles JO. The cochlea. In: *An Introduction to the Physiology of Hearing*. Pickles JO (ed). Emerald Group Publishing, Bingley 2008, pp. 25–72.

74 Jewett DL. Volume-conducted potentials in response to auditory stimuli as detected by averaging in the cat. *Electroencephalography and Clinical Neurophysiology* 1970;**28**:609–18.

75 Jewett DL, Williston JS. Auditory-evoked far fields averaged from the scalp of humans. *Brain* 1971;**94**:681–97.

76 Marshall AE. Brain stem auditory-evoked response of the nonanesthetized dog. *American Journal of Veterinary Research* 1985;**46**:966–73.

77 Bodenhamer RD, Hunter JF, Luttgen P. Brain stem auditory-evoked responses in the dog. *American Journal of Veterinary Research* 1985;**46**:1787–92.

78 Shiu JN, Munro KJ, Cox CL. Normative auditory brainstem response data for hearing threshold and neuro-otological diagnosis in the dog. *Journal of Small Animal Practice* 1997; **38**:103–7.

79 Haagen AJVV, Siemelink RJG, Smoorenburg GF. Auditory brainstem responses in the normal beagle. *Veterinary Quarterly* 1989;**11**:129–37.

80 Uzuka Y, Fukaki M, Hara Y, Matsumoto H. Brainstem auditory evoked responses elicited by tone-burst stimuli in clinically normal dogs. *Journal of Veterinary Internal Medicine* 1998;**12**:22–5.

81 Buchwald J, Huang C. Far-field acoustic response: origins in the cat. *Science* 1975;**189**:382–4.

82 Huang CM. A comparative study of the brain stem auditory response in mammals. *Brain Research* 1980;**184**:215–19.

83 Fischer A, Obermaier G. Brainstem auditory–evoked potentials and neuropathologic correlates in 26 dogs with brain tumors. *Journal of Veterinary Internal Medicine* 1994;**8**:363–9.

84 Steiss JE, Cox NR, Hathcock JT. Brain stem auditory–evoked response abnormalities in 14 dogs with confirmed central nervous system lesions. *Journal of Veterinary Internal Medicine* 1994;**8**:293–8.

85 Sims MH, Moore RE. Auditory-evoked response in the clinically normal dog: early latency components. *American Journal of Veterinary Research* 1984;**45**:2019–27.

86 Myers LJ, Redding RW, Wilson S. Abnormalities of the brainstem auditory response of the dog associated with equilibrium deficit and seizure. *Veterinary Research Communications* 1986;**10**:73–8.

87 Boettcher FA. Presbyacusis and the auditory brainstem response. *Journal of Speech Language and Hearing Research* 2002;**45**:1249–61.

88 Gates GA, Mills JH. Presbycusis. *Lancet* 2005;**366**:1111–20.

89 Poncelet L, Deltenre P, Coppens A, Michaux C, Coussart E. Brain stem auditory potentials evoked by clicks in the presence of high-pass filtered noise in dogs. *Research in Veterinary Science* 2006;**80**:167–74.

90 Achor LJ, Starr A. Auditory brain stem responses in the cat. I. Intracranial and extracranial recordings. *Electroencephalography and Clinical Neurophysiology* 1980;**48**:155–73.

91 Achor LJ, Starr A. Auditory brain stem responses in the cat. II. Effects of lesions. *Electroencephalography and Clinical Neurophysiology* 1980;**48**:174–90.

92 Strain GM. Congenital deafness and its recognition. *Veterinary Clinics of North America Small Animal Practice* 1999;**29**:895–907.

93 Strain GM. Aetiology, prevalence and diagnosis of deafness in dogs and cats. *British Veterinary Journal* 2010;**152**:17–36.

94 Kodera K, Marsh RR, Suzuki M, Suzuki J. Portions of tone pips contributing to frequency-selective auditory brain stem responses. *Audiology* 1983;**22**:209–18.

95 Davis H, Hirsh SK, Turpin LL, Peacock ME. Threshold sensitivity and frequency specificity in auditory brainstem response audiometry. *Audiology* 1985;**24**:54–70.

96 Oates P, Stapells DR. Frequency specificity of the human auditory brainstem and middle latency responses to brief tones. I. High-pass noise masking. *Journal of the Acoustical Society of America* 1997;**102**:3597–608.

97 Stapells DR, Picton TW, Durieux-Smith A, Edwards CG, Moran LM. Thresholds for short-latency auditory-evoked potentials to tones in notched noise in normal-hearing and hearing-impaired subjects. *Audiology* 1990;**29**:262–74.

98 Fausti SA, Frey RH, Henry JA, Olson DJ, Schaffer HI. Early detection of ototoxicity using high-frequency, tone-burst-evoked auditory brainstem responses. *Journal of the American Academy of Audiology* 1992;**3**:397–404.

99 Fausti SA, Olson DJ, Frey RH, Henry JA, Schaffer HI. High-frequency tone burst-evoked ABR latency-intensity functions. *Scandinavian Audiology* 1993;**22**:25–33.

100 Fausti SA, Rappaport BZ, Frey RH, *et al*. Reliability of evoked responses to high-frequency (8–14 kHz) tone bursts. *Journal of the American Academy of Audiology* 1991;**2**:105–14.

101 Markessis E, Poncelet L, Colin C, *et al*. Auditory steady-state evoked potentials (ASSEPs): a study of optimal stimulation parameters for frequency-specific threshold measurement in dogs. *Clinical Neurophysiology* 2006;**117**:1760–71.

102 ter Haar G, Venker-van Haagen AJ, van den Brom WE, van sluijs FJ, Smoorenburg GF. Effects of aging on brainstem responses to toneburst auditory stimuli: a cross-sectional and longitudinal study in dogs. *Journal of Veterinary Internal Medicine* 2008;**22**:937–45.

103 Smoorenburg GF. Risk of noise-induced hearing loss following exposure to Chinese firecrackers. *Audiology* 1993;**32**:333–43.

104 Steiss JE, Wright JC, Storrs DP. Alterations in the brain stem auditory evoked response threshold and latency-intensity curve associated with conductive hearing loss in dogs. *Progress in Veterinary Neurology* 1990;**1**:205–11.

105 Wolschrijn CF, Venker-van Haagen AJ, van den Brom WE. Comparison of air- and bone-conducted brain stem auditory evoked responses in young dogs and dogs with bilateral ear canal obstruction. *Veterinary Quarterly* 1997;**19**:158–62.

106 Penrod JP, Coulter DB. The diagnostic uses of impedance audiometry in the dog. *Journal of the American Animal Hospital Association* 1980;**6**:941–8.

107 Venker-van Haagen AJ. The ear. In: *Ear, Nose, Throat, and Tracheobronchial Diseases in Dogs and Cats*. Venker-van Haagen AJ (ed). Schlütersche Verlagsgesellschaft, Hannover 2005, pp. 1–50.

108 Hall JW, Lewis WS. Diagnostic audiology, hearing aids, and habilitation. In: *Ballenger's Otorhinolaryngology Head and Neck Surgery*. Snow JJB, Ballenger JJ (eds). BC Decker, Hamilton 2003, pp. 134–60.

109 Steiss JE, Boosinger TR, Wright JC, Storrs DP, Pillai SR. Healing of experimentally perforated tympanic membranes demonstrated by electrodiagnostic testing and histopathology. *Journal of the American Animal Hospital Association* 1992;**28**:307–10.

110 Sims MH, Weigel JP, Moore RE. Effects of tenotomy of the tensor tympani muscle on the acoustic reflex in dogs. *American Journal of Veterinary Research* 1986;**47**:1022–31.

111 Forsythe WB. Tympanographic volume measurements of the canine ear. *American Journal of Veterinary Research* 1985;**46**:1351–3.

112 Little CJ, Lane JG. An evaluation of tympanometry, otoscopy and palpation for assessment of the canine tympanic membrane. *Veterinary Record* 2011;**124**:5–8.

113 Harvey RG, Harari J, Delauche AJ. Diagnostic procedures. In: *Ear Diseases of the Dog and Cat*. Harvey RG, Harari J, Delauche AJ (eds). Manson Publishing, London 2001, pp. 43–80.

114 Cole LK, Podell M, Kwochka KW. Impedance audiometric measurements in clinically normal dogs. *American Journal of Veterinary Research* 2000;**61**:442–5.

115 Sims HS, Yamashita T, Rhew K, Ludlow CL. Assessing the clinical utility of the magnetic stimulator for measuring response latencies in the laryngeal muscles. *Otolaryngology – Head and Neck Surgery* 1996;**114**:761–7.

116 Cole LK, Kwochka KW, Podell M, Hillier A, Smeak DD. Evaluation of radiography, otoscopy, pneumotoscopy, impedance audiometry and endoscopy for the diagnosis of otitis media in the dog. In: *Advances in Veterinary Dermatology 4*. Thoday KL, Foil CS, Bond R (eds). Wiley Blackwell, Hoboken 2002, pp. 49–99.

117 Strain GM, Fernandez AJ. Handheld tympanometer measurements in conscious dogs for the evaluation of the middle ear and auditory tube. *Veterinary Dermatology* 2015;**26**:193–7.

118 Osguthorpe JD. Effects of tympanic membrane scars on tympanometry: a study in cats. *Laryngoscope* 1986;**96**:1366–77.

119 Osguthorpe JD, Lam C. Methodologic aspects of tympanometry in cats. *Otolaryngology – Head and Neck Surgery* 1981;**89**:1037–40.

120 Margolis RH, Osguthorpe JD, Popelka GR. The effects of experimentally-produced middle ear lesions on tympanometry in cats. *Acta Larynologica* 1978;**85**:428–36.

121 Lynch TJ III, Peake WT, Rosowski JJ. Measurements of the acoustic input impedance of cat ears: 10 Hz to 20 kHz. *Journal of the Acoustical Society of America* 1994;**96**:2184–209.

122 Kemp DT. Stimulated acoustic emissions from within the human auditory system. *Journal of the Acoustical Society of America* 1978;**64**:1386–91.

123 Strain GM. *Deafness in Dogs and Cats*. Strain GM (ed). CABI, Cambridge 2011.

124 Oostenbrink P, Verhaagen-Warnaar N. Otoacoustic emissions. *American Journal of Electroneurodiagnostic Technology* 2004;**44**:189.

125 Ruggero MA, Kramek B, Rich NC. Spontaneous otoacoustic emissions in a dog. *Hearing Research* 1984;**13**:293–6.

126 Sockalingam R, Filippich L, Charles B, Murdoch B. Cisplatin-induced ototoxicity and pharmacokinetics: preliminary findings in a dog model. *Annals of Otology Rhinology and Laryngology* 2002;**111**:745–50.

127 Sims MH, Rogers RK, Thelin JW. Transiently evoked otoacoustic emissions in dogs. *Progress in Veterinary Neurology* 1994;**5**:49–56.

128 Sockalingam R, Filippich L, Sommerlad S, Murdoch B, Charles B. Transient-evoked and 2F1-F2 distortion product oto-acoustic emissions in dogs: preliminary findings. *Audiology Neurotology* 1998;**3**:373–85.

129 Schemera B, Blumsack JT, Cellino AF, Quiller TD, Hess BA, Rynders PE. Evaluation of otoacoustic emissions in clinically normal alert puppies. *American Journal of Veterinary Research* 2011;**72**:295–301.

130 Gonçalves R, McBrearty A, Pratola L, Calvo G, Anderson TJ, Penderis J. Clinical evaluation of cochlear hearing status in dogs using evoked otoacoustic emissions. *Journal of Small Animal Practice* 2012;**53**:344–51.

131 McBrearty A, Penderis J. Transient evoked otoacoustic emissions testing for screening of sensorineural deafness in puppies. *Journal of Veterinary Internal Medicine* 2011;**25**:1366–71.

CHAPTER 3

EAR CLEANERS, EAR CLEANING, DEPILATORIES, CERUMINOLYTICS, ANTIBIOTICS, TOPICAL STEROIDS AND OTIC ANTIPARASITICS

3.1 INTRODUCTION

There are, literally, dozens of otic products on the veterinary market for cleaning and treating dog's ears[1,2]. Until recently, it was difficult to make a rational choice as many had similar ingredients. Furthermore, these common ingredients were often compounded with differing co-compounds, making comparison and choice difficult, particularly with cleaners and ceruminolytic preparations[1,2]. Clinicians had to depend on personal experience. In some cases one could tap into the knowledge base of expert opinion, although this in itself was not objective. The recent publication of some *in vivo* and *in vitro* studies has helped clinicians to begin to make a more rational choice when selecting a product.

3.2 OTIC CLEANSERS, CERUMINOLYTICS AND DEPILATORIES

Otic cleansers are used in-hospital to aid removal of debris and exudates from affected external ear canals (Figs 3.1, 3.2). This allows visualisation of the deeper

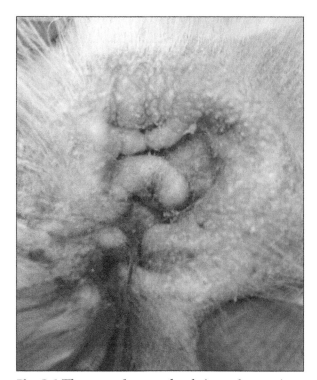

Fig. 3.1 The external ear canal and pinna of an atopic West Highland White showing erythema and adherent cerumen and crusted scale.

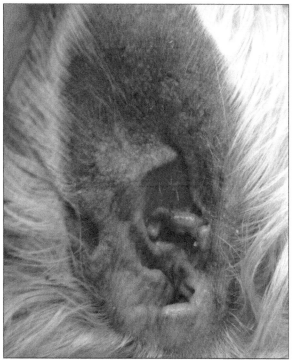

Fig. 3.2 Same ear canal as in 3.1 after plucking away hair and gentle cleaning with very dilute chlorhexidine, followed by careful removal of residual liquid. These photographs graphically demonstrate the clinical benefit of admitting dogs with otitis to facilitate treatment.

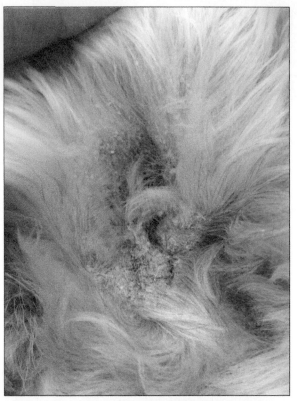

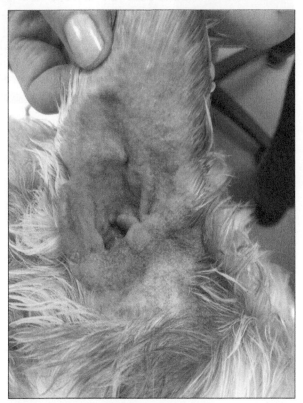

Fig. 3.3 The left ear of an atopic West Highland White prior to application of thiocycanate-based depilatory cream, Veet®. This ear was minimally affected with otitis.

Fig. 3.4 Same ear as in Fig. 3.3, 10 minutes after application and subsequent removal of cream (and hair) by gentle wiping with cotton wool balls.

portions of the canal and of the tympanum. Cleaners are also used by owners to control mild ceruminal discharge or even in an attempt to prevent recurrent otitis externa.

There are three main classes of otic cleansers and cleaners[3]:
- Ceruminolytic surfactant cleansers for in-hospital use.
- Mild leave-in cleansers for at-home use.
- Antiseptic/drying rinses.

Ceruminolytic surfactants are applied for 5–15 minutes, perhaps with mild palpation of the ear canal to aid mixing and penetration, before flushing. These agents may have irritant potential and must be thoroughly removed from the ear canal after cleaning is completed. This is particularly important if the tympanum is seen to be ruptured. Calcium or sodium dioctylsulfosuccinate (syn. docusate calcium or sodium) are good

examples, as is triethanolamine polypeptide oleate condensate. Carbamide peroxide is slightly less potent but it has humectant activity and a foaming effect helpful with purulent exudates[1,3].

Milder cleansers for home use are applied once or twice daily and excess is often shaken out by the dog. Owners may swab away further excess. These products are especially useful for waxy ears where there is no overt otitis externa. Examples of typical components include light oils, propylene glycol and phytosphingosine.

Depilatories

Thiogycolate disrupts intercysteine disulfide bonds that stabilise the keratin in hair shafts[4]. The weakened hair shaft may then be removed by gentle rubbing. Thioglycolate-based agents also appear to be general penetration enhancers[4,5], for, when applied to normal skin, they facilitate percutaneous passage of drugs.

Thioglycolate preparations have been evaluated for use as depilatory agents within the canine external ear canal[6]. Subsequent histopathological examination of biopsy samples from treated skin from within the external ear canal failed to reveal evidence of inflammation. A Cochrane review[7] of the use of commercial depilatory creams as an alternative to shaving the skin before surgery found no evidence that they had any adverse effects.

There are no veterinary depilatory creams. Fadok (Communication on VetDerm Listserv) has suggested that each dog is first tested for sensitivity by applying a small sample of the product to the concave aspect of the pinna, before applying larger quantities into the external ear canal. If the product is tolerated well it is applied to the external ear canal and left in place for 5–10 minutes, before being wiped and flushed away, with any hair (Figs 3.3, 3.4). While not suitable for all instances, the use of a depilatory was useful in the case of an atopic Miniature Schnauzer (Fig. 3.5) with hirsute ears and a profuse ceruminous discharge. The removal of the hair allowed for easy cleaning.

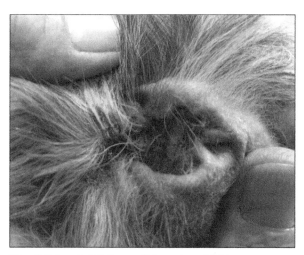

Fig. 3.5 Atopic Miniature Schnauzer with hirsute ears and a thick, clumping, ceruminous discharge. The owner found plucking the hair difficult and could not effectively clean the ear. The dog could tolerate the depilatory cream and showed no signs of post-application discomfort. Cleaning was made much less arduous for the owner.

Antiseptic and drying agents

These agents are used to dry the ear canals and prevent maceration after cleaning and swimming. Alcohols and mild acids are most commonly used. These products are often broad spectrum, they do not induce resistance and they are frequently less expensive than antibiotic-containing otic polypharmaceuticals. Examples include chlorhexidine, isopropyl alcohol, iodine, parachloro-metaxylenol (PCMX) and lactic, boric, acetic, benzoic, malic and salicylic acids. These mild acids may have direct antibacterial action and they lower the pH.

Ceruminolysis and topical antibacterial efficacy

The veterinary literature has recently benefited from publication of some peer-reviewed studies, and this has made it possible to make objective recommendations with regard to ceruminolysis, antimicrobial activity and clinical ability to resolve otitis externa. Readers should note, however:

- These studies are in no way comprehensive and, unfortunately, some products that have been tested are not widely available.
- Similarly, there are many well-recognised products that have not yet been rigorously tested.

- Experimental design might favour one product and disadvantage another and it is therefore difficult for the authors to make sweeping recommendations.

However, several products have been endorsed by the authors as they have performed well in one, or even several, tests and stand out.

All comparative studies of ceruminolysis have, to date, been performed *in vitro*. All three studies used a mix of lipids (myristic acid 33.6%, palmitic acid 33.6%, oleic acid 9.4%, cholesterol 10.9% and squalene 12.5%) designed to mimic natural canine cerumen[8]. However, normal cerumen differs from that found in otitis externa. For example, in otitis externa, cerumen might well be more aqueous and contain cellular debris and inflammatory exudate.

One product in particular was outstanding in this initial study, a mix of salicylic acid, lactic acid, oleic acid, propylene glycol, sodium lauryl sulphate, glycerin, plant extracts and water (Otoclean®, Laboratories Dr Esteve SA, Spain). The two subsequent studies[1,9] were similar in design in that a plug of the cerumen-like oil mix was placed in the test tubes and the ability of various commercial ceruminolytics to break down the oil plug was assessed. The study by Stahl *et al.*[9], found that only one-third of the products had any significant cerumi-nolytic effect *in vitro* and these were:

- Cerumene® (Schering Plough, Spain).
- Otoclean® (Laboratories Dr Esteve, Spain).
- Specicare® (Lovens Animal Health Division, Spain).

Robson *et al.*[1] found that three products gave over 80% cerumen clearing:

- Otoclean® (Laboratories Dr Esteve, Spain).
- Cerulytic® (Virbac Animal Health, USA).
- Leo Ear Cleaner (Boehringer Ingelheim, UK).

Four other products also performed well (Oto-clean®, Compounded product, Compoundia, Australia; CleanAural®; Vetxx® A/S, Denmark; and USP grade propylene).

Patently, the interior of the canine external ear canal is not the same environment as a test tube but at least these studies give an objective basis for comparison.

3.3 ANTIMICROBIAL MEDICATIONS

The three main microbes associated with otitis externa are *Staphylococcus pseudintermedius* (*S. pseudinter-medius*), *Pseudomonas aeruginosa* (*P. aeruginosa*) and *Malassezia pachydermatis* (*M. pachydermatis*). *In vitro* studies have demonstrated that Epi-Otic® (Virbac Animal Health), which contains lactic and salicylic acid, is a potent antimicrobial agent, killing both strains of these microbes within 1 minute, even when diluted five-fold[10].

The antimicrobial effect of nine commercial ear cleaners was assessed against cultures of the three most common otic pathogens[11]:

- Cleanaural Dog® (Dechra Veterinary Products, UK) proved most effective overall, killing both species of bacteria and the yeast, at dilutions up to 1/32.
- At 1/8 dilution, Sancerum® (MSD Animal Health, UK) was effective against all three organisms and at 1/4 dilution Otoclean® (Janssen Animal Health, UK) was also very effective.
- Epi-Otic® Advanced (Virbac Animal Health, UK) scored very well against *P. aeruginosa* and *M. pachydermatitis* at 1/8 dilution, but was surprisingly poor against *S. pseudintermedius*.

The authors speculated that Epi-Otic Advanced performs well *in vivo* against all three microbes in part because of the antiadhesive saccharides that are included. These would not be expected to be effective *in vitro* and demonstrate the problems in drawing too many conclusions from *in vitro* studies, notwithstanding the objectivity.

The authors of this study tried to identify which ingredient or property might be the key to their efficacy. Isopropyl alcohol is the major distinguishing feature of Cleanaural® and it is recognised as a potent antimicrobial agent. It is also a drying agent and this action may potentiate its antimicrobial efficacy *in vivo*[11].

Two studies reported the effectiveness of tromethamine ethylenediaminotetraacetic acid (Tris-EDTA) against different strains of *M. pachydermatis* and the different clinical isolates of the three main otic pathogens[12,13]. Cole *et al.* reported that Tris-EDTA and benzyl alcohol (T8 Solution™ Ear Rinse, IVX Animal Health Inc) was not effective against *M. pachydermatitis in vitro*, whereas the addition of 0.1% ketoconazole resulted in a dramatic reduction of yeast numbers[12].

The second study used Otodine® (ICF, Cremona, Italy), which contains 1.5 mg/ml chlorhexidine and 0.048 mg/ml Tris-EDTA[13], used against 150 isolates from cases of otitis externa (*Corynebacterium auriscanis*, *Escherichia coli* (*E. coli*), *M. pachydermatis* and other yeast, *Proteus mirabilis*, *P. aeruginosa*, *Streptococcus canis*, *S. pseudintermedius*, *S. aureus*). The product showed excellent activity against all organisms.

Finally, Mason *et al.* tested nine ear cleaners against 50 clinical strains of *M. pachydermatis*[14]. Five products showed excellent activity:

- CleanAural Dog® (Dechra Veterinary Products, UK).
- Epi-Otic® Ear Cleaner Advanced formula (Virbac Animal Health, UK).
- MalAcetic Aural ® (Dechra Veterinary Products, UK).
- Sancerum® (MSD Animal Health, UK).
- TrizUltra™ + Keto Otic Flush (Dechra Veterinary Products, UK).

Two products showed moderate activity:

- Otodine® (ICF, Italy).
- Surosolve® (Elanco, UK).

Criticism of these *in vitro* tests is justified. For example, in the external ear canal there is a different environment (warmth, movement, exudate), an ingredient might only be active *in vivo* (antiadhesive saccharides), the pH is different, the volume of product applied (and therefore amount of an individual medication within the external ear canal) might be different. Notwithstanding, at least objective data are facilitating building of a knowledge base.

Clinical trials

Epi-Otic® (Virbac Animal Health) exhibited potent antibacterial and antimalassezial activity in Basset Hounds[10]. In a second, open, clinical, study Epi-Otic was provided to owners of 16 dogs (31 ears in total) for twice daily application[15]. After 2 weeks there was resolution of infection in 67.7% ears and the clinical signs of otitis externa were significantly reduced.

Epi-otic® Advanced Formula was studied in a trial of 45 dogs[16]. Epi-Otic Advanced contains specific monosaccharides in a sodium docusate and non-ionic surfactant. It has properties over and above those of Epi-Otic, which prevent bacteria from adhering.

Otodine® (ICF, Italy) was tested in a placebo-controlled double-blinded study against Aurizon® (Vetoquinol, France) with good results *vis-à-vis* eliminating bacteria and yeast[17]. This was assessed by bacteriological and yeast culture, and by cytological assessment. Otodine contains Tris-EDTA and chlorhexidine, whereas Aurizon contains marbofloxacin, dexamethasone and clotrimazole.

One factor to be borne in mind when providing ear cleansers for medium-term application, is owner compliance and dog acceptance. For example, in the study by Rème *et al.* some 37% dogs showed resentment and 'marked avoidance reaction'[16].

3.4 TOPICAL ANTIBACTERIAL THERAPY

The influence of excess cerumen, purulent discharge and biofilm on antibacterial efficacy

The influence of otic and purulent discharge is twofold. It physically inhibits topically applied otic medications entering the external ear canal and reaching the otic epithelium. Furthermore, it provides a variably anaerobic environment in which free deoxyribonucleic acid (DNA), a low pH and bacterial enzymes exert variable antibacterial activity[18,19].

The implications of biofilm

A biofilm is defined as a differentiated mass of microorganism community tightly attached on a biotic or abiotic surface[20]. Bacteria or yeast, such as *P. aeruginosa*, *S. pseudintermedius* or *M. pachydermatis*, have been shown capable of forming biofilm. Two conditions must be fulfilled:

- The bacteria or yeast must be adhered to a surface, in this case the outermost plasma membrane of the epithelial cells that line the external ear canal
- The organism must form an inert organic polymer matrix[21].

The practical implication of a bacterial biofilm is that it protects the organism from effective treatment. For example, antibiotics such as aminoglycosides and enrofloxacin, must enter the cell to enable binding to intracellular targets. If biofilm hinders diffusion of an antibiotic, then the minimal inhibitory concentration (MIC) will be affected. MIC values are significantly higher for bacteria within biofilm than for the planktonic ('free living') bacteria[22]. In addition, in the case of bacteria at least, exposure to subinhibitory concentrations of antibiotic may favour the development of resistance[22]. Unfortunately, a biofilm cannot be visualised within the external ear canal; its presence may be anticipated but it cannot be demonstrated. Furthermore, there is doubt as to whether *in vitro* susceptibility testing can approximate an *in vivo* biofilm, making antibiotic selection problematic.

Clinicians must be aware of the physical, chemical and biological changes within the external ear canal and how they might hinder the effectiveness of antibacterial agents. How to address these changes?

- Consider frequent and thorough cleaning of the external ear canal to remove pus, epithelial debris and cerumen in order to facilitate medication reaching the epithelial surfaces of the external ear canal. Bear in mind that this may require repeated heavy sedation, or even general anaesthesia, with all the attendant expense and logistical implications.
- Tris-EDTA (50 mM) has been shown to disperse *P. aeruginosa* biofilm *in vitro*[22]. In combination with 0.15% chlorhexidine, Tris-EDTA has been

demonstrated to treat effectively otitis externa due to *P. aeruginosa*[23] and is the synergist treatment most frequently used in veterinary otitis cases[24–26].

- Remember that the volume of the external ear canal increases with the weight of the dog[27] and may be sufficiently large that much larger volumes of drug need to be applied. Thus a 5.4 kg Shi-Tzu has an external ear canal with approximately 20 cm² area and volume 4.5 cm³, whereas in a 20 kg dog the surface area is 38.5 cm² and the volume 9.6 cm³.
- Consider twice a day treatment, even if the data sheet suggests once daily treatment is sufficient.
- Consider concurrent systemic antibacterial treatment in an attempt to increase the concentration of antibiotic at the epithelial surface. This may require scrutiny of the sensitivity panel in an attempt to identify drugs that are available both for systemic and topical use in the dog.

Tris-EDTA

Tromethamine (Tris) buffer enhances the effects of ethylenediaminotetraacetic acid (EDTA), which is optimal around pH 8[28]. Tris-EDTA appears to exert its effect on *P. aeruginosa* in two ways[29]:

- Removing essential metal ions from the immediate environment of the bacterium by chelation, thus limiting the organism's ability to grow and multiply.
- Destabilising the outer cell membrane, leading to increased entry of antibacterial agents and perhaps lysis.

These effects can be exploited, using pretreatment with Tris-EDTA to enhance the effect of antibiotics, such as amikacin or neomycin[30] and gentamicin and marbofloxacin[31]. In effect, the MIC is reduced. This synergistic effect is so potent that that even multiresistant strains of *P. aeruginosa* can be rendered susceptible, *in vitro*[31] and *in vivo*[32,33].

Tris-EDTA may be used in two ways. There are no published data as to which method is superior to the other *vis-à-vis* clinical effectiveness, although pretreatment would seem more logical, given that the ion chelation and membranolytic processes cannot be expected to be instantaneous.

- Tris-EDTA solution is applied to the cleaned ear canal as a pretreatment wash, 10–15 minutes before the chosen antibiotic is instilled.

- The chosen antibiotic may be added to the Tris-EDTA before the mixture is instilled in the affected ear canal[28,33]. For example:
  - 10 ml of 100 mg/ml Baytril added to 90 ml Tris-EDTA gives 100 ml of 10% Baytril. Or, 12 ml Baytril is added to a 4 oz bottle of Tris-EDTA, having taken 12 ml out.
  - One 2 ml vial of 250 mg/ml Amikacin solution is added to 98 ml Tris-EDTA to make 100 ml of 5% Amikacin. Or, 2.25 ml Amikacin is added to a 4 oz Tris-EDTA bottle, having taken 2.25 ml out, giving a 5.2% solution.
  - If a glucocorticoid is required, dexamethasone can be added (24 mg per 4 oz bottle) to yield a 2% solution.
- European users will have to ensure that they stay within the prescribing cascade and must use proprietary preparations in the vast majority of clinical situations. Patently, dealing with a refractory, multiresistant pseudomonas otitis might take the clinician off the cascade, and in this case, best practice, as described above is desirable.

Antibiotic selection and antibiotic resistance

The mutant selection window (MSW) and mutant prevention concentration (MPC) are concepts based on the hypothesis that if a bacterial infection comprises sufficient numbers of bacteria (10^7–10^{10}) a small number of mutants may exist that are potentially resistant to a given antibacterial agent[34]. If an antibacterial agent is administered at the MIC, these potentially resistant bacteria might be selected, with the potential to develop a resistant infection[35].

- MIC: an *in vitro* measurement of the minimum concentration of a given antibacterial agent that will inhibit the growth of the bacterium under investigation.
- MSW: the concentration of antibacterial agent between the MIC and MPC that might facilitate selection of resistant bacteria.
- MPC: the concentration of a given antibacterial agent that will kill all bacteria, including that small percentage of potentially resistant bacteria.

The concept has been proven with respect to fluoro-quinolones[36] and other antibacterial agents, to the use of mixed antibacterial treatment[37] and to both gram-negative and gram-positive bacteria[37–40].

Within the confines of an infected external ear canal it is possible that the numbers of bacteria could be sufficiently high for the MSW to apply. From a practical point of view, it is unlikely that many commercial laboratories will offer testing of MSW/MPC concentrations. The clinician faced with a gram-negative infection, in particular, must ensure an adequate concentration of antibacterial agents reaches the otic epidermis. (See discussion above, Section 3.4 Topical antibacterial therapy.)

Systemic or topical administration?
Otic antibacterial agents
There are very few peer-reviewed studies relating to otic therapeutics. Almost all relate to a field trial of a product destined for the veterinary market. The most commonly-encountered antibacterial agents in propriety otic medications are the aminoglycosides (neomycin, framycetin, gentamicin), the fluoroquinolones (marbofloxacin, enrofloxacin, pradofloxacin, orbifloxacin) and polymyxin. Fusidic acid and silver sulfadiazine may also be used.

Aminoglycosides
The aminoglycosides are the most commonly used class of topical otic products[41], with neomycin and gentamicin predominating (Table 3.1). The drugs inhibit protein synthesis within the bacterial ribosomes and are considered bactericidal.

> Aminoglycosides are inhibited by an acid environment and therefore should not be used after applying acid ceruminolytics or cleansers.

Table 3.1 A selection of the polypharmeutical otic medications that are available to clinicians (Readers are advised to check local availability)

BRAND	ANTIBACTERIAL AGENT	ANTIFUNGAL AGENT	GLUCOCORTICOID	AVAILABILITY
Otic polypharmaceutics				
Animax,	Neomycin	Nystatin	0.1% Triamcinolone	USA
Auroto	Neomycin	Thiabendazole		Europe
Neo-Predef	Neomycin	–	0.1% Isoflupredone	USA
Oribiotic	Neomycin/Bacitracin	Nystatin	0.1% Triamcinolone	Europe, Canada
Oridermyl	Neomycin	Nystatin	0.1% Triamcinolone	Europe, Canada
Panalog	Neomycin	Nystatin	0.1% Triamcinolone	USA, Canada, Europe, Australia
Tresaderm	Neomycin	Thiabendazole	0.1% Dexamthasone	USA, Canada
Canaural	Framycetin/Fucidin	Nystatin	0.25% Prednisolone	Europe, Canada
Dexoryl	Gentamicin	Thiabendazole	Dexamethasone	Europe, USA
Easotic	Gentamicin	Miconazole	0.11% Hydrocortisone aceponate	USA, Europe, Australia
Gentacin otic	Gentamicin	–	0.1% Betamethasone	USA, Canada
Mometamax	Gentamicin	Clotrimazole	0.1% Mometasone furoate	USA, Australia
Otomax	Gentamicin	Clotrimazole	0.1% Betamethasone	Europe, Australia, Canada
Tri-Otic	Gentamicin	Clotrimazole	0.1% Betamethasone	USA
Aurizon	Marbofloxacin	Clotrimazole	0.1% Dexamethason	Europe, Canada, Australia
Baytril Otic	Enrofloxacin	Silver sulfadiazine		USA, Canada, Australia
Posatex	Orbifloxacin	Posaconazole	0.1% Mometasone furoate	USA, Europe
Surolan	Polymyxin	Miconazole	0.5% Prednisolone	USA, Canada, Europe, Australia
Conofite		Miconazole		USA

Neomycin has been the antibacterial component of otic preparations since the 1970s[42]. It is considered useful in gram-positive infections, and may be considered a first-line treatment for acute, non-specific otitis externa. Neomycin has often been implicated in allergic contact dermatitis, but clinical cases have very rarely been reported[43].

Gentamicin is also considered a first-line treatment, although its spectrum of activity against otic pathogens is likely to be limited to gram-positive bacteria[44]. Gentamicin has shown great efficacy in the clinical setting, albeit against a predominantly gram-positive infection[44]. Experimental studies into the ototoxic potential of gentamicin, particularly in the presence of a ruptured tympanic membrane, failed to find evidence of adverse effects[44]. Gentamicin in combination with ketoconazole and mazipredone proved very effective, within 7–10 days of treatment, in a total of 210 dogs with clinical otitis externa[45].

Tobramycin and **amikacin** are not available in proprietary otic preparations. They are regarded as third-line products[41] and their value is in treating gentamicin-resistant *P. aeruginosa* infection. Injectable amikacin solution may be diluted to 30–50 mg/ml (either in sterile saline or Tris-EDTA)[41]. Similarly, tobramycin injectable solution is diluted to 8 mg/ml. Synergism between amikacin (or neomycin) with Tris-EDTA was reported when the combinations were used against *S. intermedius*, *P. aeruginosa*, *P. mirabis* and *E. coli*[46]. Amikacin is synergistic with marbofloxacin against *Pseudomonas* spp.[47].

Fluoroquinolones

These drugs are bactericidal and they exert their effect by inhibiting DNA-gyrase. Enrofloxacin, marbofloxacin and orbifloxacin are concentration-dependent antibiotics, with a broad spectrum of activity. They are fast becoming first-line treatments as the proprietary products are now widely marketed in the USA and Europe. In one study 95% of dogs treated with marbofloxacin (Aurizon, Vetoquinol) responded very well, although more than 10–14 days of treatment was required in many cases[48]. Orbifloxacin was tested *in vitro* against 171 isolates of *S. intermedius* from canine otitis[49]. All strains were sensitive at concentrations equal to or twice the MIC.

Polymyxins

Polymyxins target and disrupt the outer and cytoplasmic cell membranes of gram-negative bacteria, leading to cell death[50]. In addition, the damage to the integrity of the cell membranes enables increased access for other antibacterial agents, and synergism may be demonstrated. This synergistic effect has been exploited with a combination of polymyxin B, prednisolone and miconazole (Surolan®, Vetoquinol), which was highly effective against clinical isolates of *P. aeruginosa*[51] in two randomised, blinded clinical trials[52,53]. In a three way study, comparing Surolan® with Panalog® (neomycin, nystatin, thiostrepton, triamcinolone acetonide) and Oterna® (neomycin, monosulfiram, betamethasone), the time to resolution was significantly less for Surolan, and the relapse rate was lower[53].

3.5 SYSTEMIC ANTIBACTERIAL THERAPY

Carlotti *et al.* described a study in which oral marbofloxacin at a dose of 5 mg/kg once daily (2.5 times recommended dose) was administered to dogs with *P. aeruginosa* otitis externa[54]. The ears were cleaned with saline only and a good response was reported. Note, however, the study by Hosseini *et al.* where topical Tris-EDTA/chlorhexidine gave an equally good response[55].

Cole *et al.* reported a study in which the dose of marbofloxacin was titrated against tissue concentration within samples taken at surgery for external and middle ear disease, in cases of chronic end-stage otitis externa[56]. They concluded that to achieve effective concentrations of antibiotic sufficiently high to exceed typical MIC:

- For staphylococci, where *in vitro* sensitivity testing suggested marbofloxacin sensitivity at 0.125–0.25 µg/ml, 5 mg/kg once daily would be adequate.
- For *P. aeruginosa*, with higher MIC of around 0.5 µg/ml, a dose of 20 mg/kg once daily was necessary.

Hillier has suggested that systemic treatment of *P. aeruginosa* be limited to those cases in which there is concurrent otitis media or in which there is tissue swelling and ulceration[57]. Systemic treatment might also be indicated in those cases in which the owner was not able to apply topical otic preparations to the external ear canal. Bacterial culture and sensitivity testing is mandatory.

3.6 SYSTEMIC ABSORPTION OF TOPICAL ANTIBACTERIAL AGENTS AND TOPICAL OTIC GLUCOCORTICOIDS

The plasma concentrations of gentamicin, marbofloxacin, orbiflaxacin and polymyxin were evaluated after 14 days of treatment with the parent product (Otomax®, Aurizon®, Posatex® and Surolan®, respectively)[58]. Concentrations in excess of the limit of detection were present for all except polymyxin. The authors considered that the presence of these antibacterial agents in such low concentrations might influence the development of bacterial resistance. Clinicians should bear this potential in mind in their patients receiving long-term otic antibacterial agents. Systemic distribution of low concentrations of antibacterial agents might engender resistant bacteria in any body system.

There have been four studies that investigated the adrenosuppressive potential of the glucocorticoid component of otic medications[59–62]. The tested products containing dexamethasone and triamcinolone were associated with demonstrable adrenocortical effects in 9–100% of dogs assessed. Aniya et al. also looked at the effect of vehicle on the systemic effect of otic dexamethasone[62] and found that both the vehicle and the concentration of steroid affected the results.

Interestingly, a product containing mometasone furoate was not found to cause adrenal suppression[63]. This lack of systemic effect is surprising as mometasome furoate is extremely potent when applied to dog's skin[63]. Hydrocortisone aceponate (in Easotic®, Virbac) would also be expected not to cause systemic effects[64]. Although the metabolites of hydrocortisone aceponate are systemically active, they did not adversely affect adrenocorticotrophic hormone (ACTH) testing (unpublished study[65] cited by Bizikova et al.[66]). Note that both mometasone furoate and hydrocortisone aceponate adversely affect intradermal skin testing, suggesting that there is some, albeit weak, systemic effect[66,67].

Clinicians should have regard to the potential effects of a glucocorticoid-containing otic product in cases of acute otitis externa, proper work-up notwithstanding. However, one might wish to make a more considered judgement when managing an atopic dog with chronic inflammatory otitis externa:

- Hydrocortisone aceponate spray for the pinnal lesions, particularly considering the ease of applying a spray formulation to the pinna.
- Hydrocortisone aceponate- or mometasome furoate-containing otic products for the vertical canal.

3.7 ANTIMALASSEZIAL TREATMENT

Cafarchia et al. examined the antifungal susceptibility of malassezial yeast from dogs with and without skin disease[68]. Most of the 30 strains from normal dogs and all 32 from dogs with skin disease were susceptible to itraconazole, ketoconazole and posaconazole, whereas terbinafine, miconazole and fluconazole showed the lowest activity. The authors noted that strains isolated from dogs with skin disease tended to have higher MIC values and also tended to show increased cross-reactivity. This cross-reactivity was noted by Jesus et al.[69].

Malassezial yeast can produce biofilm[70], although whether this is clinically relevant is not yet known. In vitro studies demonstrated that biofilm protects an organism from therapeutic agents, making the MIC very much higher.

Notwithstanding the above, most clinical trials have documented a good or excellent response to one or several antifungal agents, whether given topically or systemically, daily or pulsed[3,71,72].

In the light of this, the observations by Peano et al.[73] and Chiavassa et al.[74] are very pertinent. Chiavassa et al. looked at the in vitro sensitivity of malassezia yeast from 17 dogs with acute otitis externa and from 25 dogs with chronic otitis externa. Although the MIC for both miconazole and clotrimazole were increased in the samples from chronic otitis it would not have had an impact on the clinical response. Both authors point out that in clinical practice, the concentration of any antifungal agent within the external ear canal was likely to be several thousand times greater than even the highest MIC measures in vitro and makes any discussion of in vitro resistance almost meaningless.

Peano et al.[73] concluded that 'apparently resistant malassezial otitis' was more likely a manifestation of two things:

- Failure to recognise underlying disease.
- Poor management, particularly a failure to clean the external ear canal properly of all ceruminous debris before apply antimalassezial medication.

3.8 METHICILLIN-RESISTANT STAPHYLOCOCCI

Methicillin-resistant staphylococci carry the mecA gene, which in *S. aureus* confers resistance to methicillin and all other β-lactam antibiotics, including cephalosporins and carbapenems[75]. However, at least currently, methicillin-resistant *S. pseudintermedius* (MRSP) is usually sensitive to cephalosporins, fluoroquinolones, clindamycin and sulphonamides (reviewed by Frank and Loeffler[76]). Note, however, that in the USA, some strains of MRSP are now resistant not only to the penicillin derivatives but also to the other agents listed[77]. *S. pseudintermedius* from the external ear canal accounted for 38% of the methicillin-resistant staphylococci in one study[78].

Most probably, the first indication that a clinician is dealing with methicillin-resistant *S. pseudintermedius* will be the laboratory report from a sample submitted because of a failure to respond. Because of the widespread resistance shown by MRSP, the choice of treatments may be very limited. Off-label use of veterinary drugs and using drugs primarily used in human medicine may be the only option[77]. Chloramphenicol, tetracyclines, aminoglycosides and rifampin should be considered if susceptibility tests are supportive[77]. If there is widespread resistance to these then other options include[77]:

- Vancomycin: 15 mg/kg by slow i/v injection every 8 hours.
- Linezolid: first in class human drug where it is used for methicillin-resistant *S. aureus* (MRSA) and vancomycin-resistant staphylococcal infections; dose is 10 mg/kg twice a day, by mouth. It is very expensive.
- Remember that there are several appropriate topical antibacterial preparations available; preparations containing mupirocin, silver sulfadiazine or fusidic acid have been shown to have good activity against multiresistant staphylococci[78,79].
- Consider chlorhexidine–Tris-EDTA combinations.

The zoonotic implications, both in the veterinary practice and at the pet's home, of MRSP infection in domestic pets has been reviewed[76]. There is little evidence, to date, that human carriage of, or infection with, MRSP occurs, even when they are in prolonged contact with, for example, a dog with an infected ear canal. Notwithstanding, it is essential to ensure that the practice infection control policy is compliant with best-practice, particularly with regard to effective hand washing, which is known to limit the spread of staphylococci[76].

3.9 OTIC ANTIPARASITIC TREATMENTS

Although the *Otodectes cynotis* mite is confined to the external ear canal it does have the capacity to survive in the environment for up to 2 weeks at 10°C[80]. Thus, in refractory or recurrent otoacariasis, attention should be paid to treatment of the local environment. For the most part, the parasite is easy to eliminate with otic antiparasitics:

- Selamectin: two applications at 30 day intervals[81–83].
- Moxidectin: 2.5% + 10% imidacloprid, two applications at 28 day intervals[84].
- Thiabendazole daily for 7 days[83,85].
- Ivermectin:
  - Topical (i.e. applied onto the skin between the scapulae) twice at 14 day intervals[86].
  - As an otic preparation (1% injectable ivermectin diluted 1:9 with propylene glycol), daily for 21 days[87].
  - Systemic ivermectin, typically 200 µg/kg[83].
- Monosulfiram[88].
- Permethrin: 1% otic preparation (Oridermyl®), once daily for 10 days[89].
- Non-acaricidal treatment shown to be effective, both topically administered[90]:
  - Miconazole nitrate, polymyxin B sulphate and prednisolone acetate (Surolan®) twice daily for 14 days.
  - Diethanolamine fusidate, framycetin sulphate, nystatin and prednisolone (Canaural®) twice daily for 14 days[90].

Ticks, such as *Otobius megnini* within the external ear canal and *Ixodes* spp. on the pinna, may be easily removed, with gentle traction. Many veterinary practices sell small plastic tools that make the removal of the tick easy. Note that removal is more easily accomplished if the ticks are killed first, for example by local application of vegetable or mineral oil, or a specific agent such as amitraz or fipronil[91–93]. Recent improvements in tick control products, and the availability of sustained release agents such as fluralanar or afoxolaner in collars and tablets[94–97] in particular, makes prevention a real possibility for the dog and cat owner.

3.10 REFERENCES

1 Robson D, Morton D, Burton G, Bassett R. *In vitro* ceruminolytic activity of 23 ear cleaners against standardized synthetic canine cerumen: preliminary results. *Proceedings of the Australian and New Zealand College of Veterinary Scientists* Gold Coast, 2008, pp. 98–103.

2 Nuttall T, Cole LK. Ear cleaning: the UK and US perspective. *Veterinary Dermatology* 2004;**15**:127–36.

3 Miller WH, Griffin CE, Campbell KL (eds). Disease of the eyelids, claws, anal sacs and ears. In: *Muller and Kirk's Small Animal Dermatology*, 7th edn. Elsevier, St Louis 2013, pp. 724–73.

4 Lee J-N, Jee S-H, Chan C-C, *et al.* The effects of depilatory agents as penetration enhancers on human stratum corneum structures. *Journal of Investigative Dermatology* 2008;**128**:2240–7.

5 Weiland L, Croubels S, Baert K, *et al.* Pharmacokinetics of a lidocaine patch 5% in dogs. *Journal of Veterinary Medicine. A, Physiology, Pathology, Clinical Medicine* 2006;**53**:34–9.

6 Hammond DL, Conroy JD, Woody BJ. The histological effects of a chemical depilatory on the auditory canal of dogs. *Journal of the American Animal Hospital Association* 1990;**26**:551–4.

7 Tanner J, Norrie P, Melen K. Preoperative hair removal to reduce surgical site infection. *Cochrane Database Systemic Reviews* 2011, issue 11. http://onlinelibrary.wiley.com/doi/10.1002/14651858.CD004122.pub4/pdf.

8 Sánchez-Leal, Mayós I, Homedas J, Ferrer L. *In vitro* investigation of ceruminolytic activity of various otic cleansers for veterinary use. *Veterinary Dermatology* 2006;**17**:121–7.

9 Stahl J, Mielke S, Pankow W-R, Kietzman M. Ceruminal diffusion activities and ceruminolytic characteristics of otic preparations. *BMC Veterinary Research* 2013;**9**:70–8.

10 Lloyd DH, Bond R, Lamport I. Antimicrobial activity *in vitro* and *in vivo* of a canine ear cleanser. *Veterinary Record* 1998;**143**:111–12.

11 Swinney A, Fazakerley J, McEwan N, Nuttall T. Comparative *in vitro* antimicrobial efficacy of commercial ear cleaners. *Veterinary Dermatology* 2008;**19**:373–9.

12 Cole LK, Dao HL, Paivi JR-S, Cheyney M, Torres AH. *In vitro* activity of an ear rinse containing tromethamine EDTA, benzyl alcohol and 0.1% ketoconazole on *Malassezia* organisms from dogs with otitis externa. *Veterinary Dermatology* 2007;**18**:115–19.

13 Guardabassi L, Ghibaudo G, Damburg P. *In vitro* antimicrobial activity of a commercial ear antiseptic containing chlorhexidine and Tris-EDTA. *Veterinary Dermatology* 2009;**21**:282–6.

14 Mason CL, Steen S, Paterson S, Cripps PJ. Study to assess *in vitro* antimicrobial activity of nine ear cleaners against 50 *Malassezia pachydermatis* isolates. *Veterinary Dermatology* 2013;**24**:362–5.

15 Cole LK, Kwochka KW, Kowalski JJ, Hillier A, Hoshaw-Woodward SL. Evaluation of an ear cleanser for the treatment of infectious otitis externa in dogs. *Veterinary Therapeutics* 2003;**4**:12–23.

16 Rème CA, Pin D, Collinot C, Cadiergues MC, Joyce JA, Fontaine J. The efficacy of an antiseptic and microbial anti-adhesive ear cleanser in dogs with otitis externa. *Veterinary Therapeutics* 2006;**7**:15–26.

17 Bouassiba C, Osthold W, Mueller RS. *In vivo* efficacy of a commercial ear antiseptic containing chlorhexidine and Tris-EDTA. A randomized, placebo-controlled, double blinded comparative trial. *Tierärztliche Praxis. Ausgabe K, Kleintiere/Heimtiere* 2012;**40**:161–70.

18 Bryant RE, Mazza JA. Effect of the abscess environment on the antimicrobial activity of ciprofloxacin. *American Journal of Medicine* 1989;**87**:23S–27S.

19 Wagner C, Sauermann R, Joukhadar C. Principles of antibacterial penetration into abscess fluid. *Pharmacology* 2006;**78**:1–10.

20 Blakenship JR, Mitchell AP. How to build a biofilm: a fungal perspective. *Current Opinion in Microbiology* 2006;**9**:588–94.

21 Dunne WM Jr. Bacterial adhesion: seen any good biofilms lately? *Clinical Microbiology Reviews* 2002;**15**:155–66.

22 Pye CC, Yu A, Weese JS. Evaluation of biofilm production by *Pseudomonas aeruginosa* from canine ears and the impact of biofilm on antimicrobial susceptibility *in vitro*. *Veterinary Dermatology* 2013;**24**:446–9.

23 Banin E, Brady KM, Greenberg EP. Chelator-induced dispersal and killing of *Pseudomonas aeruginosa* cells in biofilm. *Applied and Environmental Microbiology* 2006;**72**:2064–9.

24 Hosseini J, Zdovc I, Golob M, *et al*. Effect of treatment with TRIS-EDTA/Chlorhexidine topical solution on canine *Pseudomonas aeruginosa* otitis externa with or without concomitant treatment with oral fluoroquinolones. *Slovenian Veterinary Research* 2012;**49**:133–40.

25 Pye CC, Singh A, Weese JS. Evaluation of tromethamine edetate disodium dehydrate on antimicrobial susceptibility of *Pseudomonas aeruginosa* in biofilm *in vitro*. *Veterinary Dermatology* 2014;**25**:120–3.

26 Griffin CE. Otitis: key points and tips. *Proceedings 17th Annual North Carolina Veterinary Conference*, Raleigh, 2012, pp. 79–83.

27 Wefstaedt P, Behrens B-A, Nolte I, Bouguecha A. Fine element modelling of the canine and feline outer ear canal: benefits for local drug delivery. *Berliner und Münchener Tierärztliche Wochenschrift* 2011;**124**:78–82.

28 Bloom P. Diagnosis and management of otitis in the real world. *Proceedings Ontario Veterinary Medical Association*, Westin Harbor, Toronto, 2012, pp. 19–24.

29 Lambert RJW, Hanlon GW, Denyer SP. The synergistic effect of EDTA/antimicrobial combinations on *Pseudomonas aeruginosa*. *Journal of Applied Microbiology* 2004;**96**:244–53.

30 Sparks TA, Kemp DT, Wooley RE, Gibbs PS. Antimicrobial effect of combinations of EDTA-TRIS and amikacin or neomycin on the microorganisms associated with otitis externa in dogs. *Veterinary Research Communications* 1994;**18**:241–9.

31 Buckley LM, McEwan NA, Nuttall T. Tris-EDTA significantly enhances antibiotic efficacy against multidrug-resistant *Pseudomonas aeruginosa in vitro*. *Veterinary Dermatology* 2013;**24**:519–25.

32 Farca AM, Piromali G, Maffei F, *et al*. Potentiating effect of EDTA-Tris on the activity of antibiotics against resistant bacteria associated with otitis, dermatitis and cystitis. *Journal of Small Animal Practice* 1997;**38**:243–5.

33 Fadok VA. Otitis externa: the bane of our existence. *OVMA Annual Conference*, Westin Harbour Castle, Toronto 2014, pp. 43–7.

34 Blondeau JM. New concepts in antimicrobial susceptibility testing: the mutant prevention concentration and the mutant selection window. *Veterinary Dermatology* 2009;**20**:383–96.

35 Davidson RJ, Cavalcanti R, Brunton JL, *et al*. Resistance to levofloxacin and failure of treatment of pneumococcal pneumonia. *New England Journal of Medicine* 2002;**346**:747–50.

36 Firsov A A, Vostrov SN, Lubenko IY, Drlica K, Portnoy YA, Zinner SH. *In vitro* pharmacodynamic evaluation of the mutant selection window hypothesis using four fluoroquinolones against *Staphylococcus aureus*. *Antimicrobial Agents and Chemotherapy* 2003;**47**:1604–13.

37 Firsov AA, Smirnova MV, Lubenko IY, Vostrov SN, Portnoy YA, Zinner SH. Testing the mutant selection window hypothesis with *Staphylococcus aureus* exposed to daptomycin and vancomycin in an *in vitro* dynamic model. *Journal of Antimicrobial Chemotherapy* 2006;**58**:1185–92.

38 Shimizu T, Harada K, Kataoka Y. Mutant prevention concentration of orbifloxacin: comparison between *Escherichia coli*, *Pseudomonas aeruginosa*, and *Staphylococcus pseudintermedius* of canine origin. *Acta Veterinaria Scandinavica* 2012;**55**:1–7.

39 Awji EG, Tassew DD, Lee J-S, *et al*. Comparative mutant prevention concentration and mechanism of resistance to veterinary fluoroquinolones in *Staphylococcus pseudintermedius*. *Veterinary Dermatology* 2012;**23**:376–82.

40 Liu MT, Sheng MY, Znang Y, Li Y. Combined application of ciprofloxacin and tobramycin on mutant selective windows of ciprofloxacin against *Pseudomonas aeruginosa*. *Zhonghua Yi Xue Za Zhi* 2011;**91**:1427–31. (Article in Chinese, abstract in English.)

41 Morris DO. Medical therapy of otitis externa and otitis media. *Veterinary Clinics of North America Small Animal Practice* 2004;**34**:541–55.

42 Marshal MJ, Harris AN, Horne JE. The bacteriological and clinical assessment of a new preparation for the treatment of otitis externa in dogs and cats. *Journal of Small Animal Practice* 1974;**15**:401–10.

43 White SD. Contact dermatitis in the dog and cat. *Seminars in Veterinary Medical Surgery* 1991;**6**:303.

44 Engelen M, De Bock M, Hare J, Goossens L. Effectiveness of an otic product containing miconazole, polymixin B and prednisolone in the treatment of canine otitis externa: multi field trial in the US and Canada. *International Journal of Applied Research in Veterinary Medicine* 2010;**8**:21–30.

45 Strain GM, Merchant SR, Neer TM, Bedford BL. Ototoxicity assessment of gentamicin sulfate preparation in dogs. *American Journal of Veterinary Research* 1995;**56**:971–9.

46 Kiss G, Radványi Sz, Szigeti G, Lukáts B, Nagy G. New combination for the therapy of canine otitis externa II. Efficacy *in vitro* and *in vivo*. *Journal of Small Animal Practice* 1997;**38**:57–60.

47 Sparks TA, Kemp DT, Wooley RE, Gibbs PS. Antimicrobial effect of combinations of EDTA-Tris and amikacin or neomycin on the microorganisms associated with otitis externa in dogs. *Veterinary Research Communications* 1994;**18**:241–9.

48 Prescott JF. Drug interactions. In: *Antimicrobial Therapy in Veterinary Medicine*. Giguére S, Prescott JF, Dowling PM (eds). John Wiley & Sons, Somerset 2013, p. 307.

49 Rougier S, BorellD, Pheulpin S, Woehrlé F, Boisramé B. A comparative study of two antimicrobial/anti-inflammatory formulations in the treatment of canine otitis externa. *Veterinary Dermatology* 2005;**16**:299–307.

50 Ganiére JP, Médaille C, Etoire F. *In vitro* antimicrobial activity of orbifloxacin against *Staphylococcus intermedius* isolates from canine skin and ear infections. *Research in Veterinary Science* 2004;**77**:67–71.

51 Vaara M. Novel derivatives of polymyxins. *Journal of Antimicrobial Chemotherapeutics* 2013;**68**:1213–19.

52 Pietsmann S, Meyer M, Voget M, Cieslicki M. The joint *in vitro* action of polymyxin B and miconazole against pathogens associated with canine otitis externa from three different European countries. *Veterinary Dermatology* 2013;**24**:439–45.

53 Studdert VP, Hughes KL. A clinical trial of a topical preparation of miconazole, polymixin and prednisolone in the treatment of otitis externa in dogs. *Australian Veterinary Journal* 1991;**68**:193–5.

54 Carlotti DN, Guaguére E, Koch HJ, Guiral V, Thomas E. Marbofloxacin for the systemic treatment of *Pseudomonas* spp. suppurative otitis externa in the dog. In: *Advances in Veterinary Dermatology Volume 3*. Kwochka KW, Willemse T, von Tscharner C (eds). Butterworth Heinemann, Oxford 1996, pp. 463–4.

55 Hosseini J, Zdovc I, Golob M, *et al*. Effect of treatment with TRIS-EDTA/Chlorhexidine topical solution on canine *Pseudomonas aeruginosa* otitis externa with or without concomitant treatment with oral fluoroquinolones. *Slovenian Veterinary Research* 2012;**49**:133–40.

56 Cole LK, Papich MG, Kwochka KW, *et al*. Plasma and ear tissue concentrations of enrofloxacin and its metabolite ciprofloxacin in dogs with chronic end-stage otitis externa after intravenous administration of enrofloxacin. *Veterinary Dermatology* 2009;**1**:51–9.

57 Hillier A. Treatment of *Pseudomonas* otitis in the dog. http://veterinarymedicine.dvm360.com/vetmed/article/articleDetail.jsp?id=179408.

58 Voget M, Armbruster M, Meyer M. Antibiotic resistance levels in dogs with otitis externa treated routinely with various topical preparations. *Berliner und Münchener tierärztliche Wochenschrift* 2012;**125**:441–8.

59 Moriello KA, Fehrer-Sawyer SL, Meyer DJ, *et al.* Adrenocortortical suppression associated with topical otic administration of glucocorticoid in dogs. *Journal of the American Veterinary Medical Association* 1988;**193**:329–31.

60 Reeder CJ, Griffin CE, Polissar NL, *et al.* Comparative adrenocortical suppression in dogs with otitis externa following topical administration of four different glucocorticoid-containing medications. *Veterinary Therapeutics* 2008;**9**:111–12.

61 Ghubash R, Marsella R, Kunkle G. Evaluation of adrenal function in small breed dogs receiving otic glucocorticoids. *Veterinary Dermatology* 2004;**15**:363–8.

62 Aniya J, Griffin CE. The effect of otic vehicle and concentration of dexamethasone on liver enzyme activities and adrenal function in small breed healthy dogs. *Veterinary Dermatology* 2008;**9**:226–31.

63 Kimura T, Doi K. Dorsal skin reactions of hairless dogs to topical treatment with corticosteroids. *Toxicologic Pathology* 1999;**27**:528–35.

64 Nuttall T, Mueller R, Bensignor E, *et al.* Efficacy of 0.0584% hydrocortisone aceponate spray in the management of canine atopic dermatitis: a randomized, double blind, placebo-controlled study. *Veterinary Dermatology* 2009;**20**:191–8.

65 Reme C. Introduction to cortavance: a topical diester glucocorticoid developed for veterinary dermatology. In: *Virbac International Derm Symposium: Advances in Topical Glucocorticoid Therapy*, Nice, France, Virbac SA, 2007, pp. 15–18.

66 Bizikova P, Linder KE, Paps, J, Olivry T. Effect of a novel topical diester glucocorticoid spray on immediate- and late-phase cutaneous allergic reactions in Maltese-beagle atopic dogs: a placebo-controlled study. *Veterinary Dermatology* 2010;**21**:71–80.

67 Marcia Murphy K, OLivry T. The influence of mometasone furoate ear solution on intradermal test immediate reactions in dogs with atopic dermatitis. *Veterinary Dermatology* 2014;**26**:32–4.

68 Cafarchia C, Figueredo LA, Iatta R, Montagna MT, Otranto D. *In vitro* antifungal susceptibility of *Malassezia pachydermatis* from dogs with and without skin disease. *Veterinary Microbiology* 2012;**155**:395–8.

69 Jesus FO, Lautert C, Zanatte RA, *et al. In vitro* susceptibility of fluconazole-susceptible and -resistant isolates of *Malassezia pachydermatis* against azoles. *Veterinary Microbiology* 2011;**152**:161–4.

70 Figueredo LA, Cafarchia C, Otranto D. Antifungal susceptibility of *Malassezia pachydermatis* biofilm. *Medical Mycology* 2013;**51**:863–7.

71 Kiss G, Radványi S, Szigeti G. New combination for the therapy of canine otitis externa. 1. Microbiology of otitis externa. *Journal of Small Animal Practice* 1997;**38**:51–6.

72 Pinchbeck LR, Hillier A, Kowalski JJ, Kwochka KW. Comparison of pulse treatment versus once daily administration of itraconazole for the treatment of *Malassezia pachydermatitis* dermatitis and otitis in dogs. *Journal of the American Veterinary Medical Association* 2002;**220**:1807–12.

73 Peano A, Beccati M, Chiavassa E, Pasquetti M. Evaluation of the antifungal susceptibility of *Malassezia pachydermatis* to clotrimatozle, miconazole and thiabendazole using a modified CLSI M27-A3 microdilution method. *Veterinary Dermatology* 2012;**23**:131–5.

74 Chiavasa E, Tizzani P, Peano A. *In vitro* sensitivity of *Malassezia pachydermatis* strains isolated from dogs with chronic and acute otitis. *Mycopathologia* 2014;**178**:315–9.

75 Chambers HF. Methicillin resistance in staphylococci: molecular and biochemical basis and clinical implications. *Clinical Microbiology Reviews* 1997;**10**:781–91.

76 Frank LA, Loeffler A. Methicillin-resistant *Staphylococcus pseudintermedius*: clinical challenge and treatment options. *Veterinary Dermatology* 2012;**23**:283–93.

77 Papich MG. Antibiotic treatment of resistant infections in small animals. *Veterinary Clinics of North America Small Animal Practice* 2013;**43**:1091–107.

78 Ruscher C, Lübke-Becker A, Wieklinski CG, *et al.* Prevalence of methicillin-resistant *Staphylococcus pseudintermedius* isolated from clinical samples of companion animals and equidaes. *Veterinary Microbiology* 2009;**136**:197–201.

79 Acikel C, Oncul O, Ulkur E, *et al*. Comparison of silver sulfadiazine 1%, mupirocin 2%, and fusidic acid 2% for topical antibacterial effect in methicillin-resistant staphylococci-infected, full-skin thickness rat burn wounds. *Journal of Burn Care and Rehabilitation* 2003;**24**:37–41.

80 Otranto D, Milillo P, Mest P, De Caprariis D, Perrucci S, Capelli G. *Otodectes cynotis* (Acari: Psoroptidae): examination of survival off-the-host under natural and laboratory conditions. *Experimental and Applied Acarology* 2004;**32**:171–9.

81 Six RH, Clemence RG, Thomas CA, *et al*. Efficacy and safety of selamectin against *Sarcoptes scabiei* on dogs *and Otodectes cynotis* on dogs and cats presented as veterinary patients. *Veterinary Parasitology* 2000;**91**:291–309.

82 Fisher MA, Shanks DJ. A review of the off-label use of selamectin (Stronghold/Revolution) in dogs and cats. *Acta Veterinaria Scandinavica* 2008;**50**:46–51.

83 Curtis CF. Current trends in the treatment of *Sarcoptes*, *Cheyletiella* and *Otodectes* mites infestations in dogs and cats. *Veterinary Dermatology* 2004;**15**:108–14.

84 Arther RG, Davies WL, Jacobson JA, Lewis VA, Settje TL. Clinical evaluation of the safety and efficacy of 10% imidacloprid +2.5% moxidectin topical solution for the treatment of ear mites (*Otodectes cynotis*) infestation in dogs. *Veterinary Parasitology* 2015;**210**:64–8.

85 De Sousa CP, Correia TR, Melo RM, *et al*. Miticidal efficacy of thiabendazole against *Otodectes cynotis* (Hering. 1838) in dogs. *Revista Brasileira de Parasitologia Veterinária* 2006;**15**:143–6.

86 Pagé N, de Jaham C, Paradis M. Observations on topical ivermectin in the treatment of otoacariosis, cheyletiellosis, and toxocariosis in cats. *Canadian Veterinary Journal* 2000;**41**:773–6.

87 Huang HP, Lein YH. Otic ivermectin in the treatment of feline *Otodectes* infestation. *Veterinary Dermatology* 2000;**11**:46S.

88 Evans JM, Jemmett JE. Otitis externa: the case for polypharmacy. *New Zealand Veterinary Journal* 1978;**26**:280–3.

89 Josée R, Bédard C, Moreau M. Treatment of feline otitis externa due to *Otodectes cynotis* and complicated by secondary bacterial and fungal infection with Oridermyl auricular ointment. *Canadian Veterinary Journal* 2011;**52**:277–82.

90 Engelen MACM, Anthonissens E. Efficacy of non-acaridal containing otic preparations in the treatment of otoacariasis in dogs and cats. *Veterinary Record* 2000;**147**:567–9.

91 Searle A, Jensen CJ, Atwell RB. Results of a trial of fipronil as an adulticide on ticks (*Ixodes holocyclus*) naturally attached to animals in the Brisbane area. *Australian Veterinary Practice* 1995;**25**:157–8.

92 Folz SD, Ash KA, Conder GA, Rector DL. Amitraz: a tick and flea repellent and tick detachment drug. *Journal of Veterinary Pharmacology and Therapeutics* 1986;**9**:150–6.

93 Hunter JS, Keister DM, Jeannin P. A comparison of the tick control efficacy of Frontline Spray against the American dog tick and brown dog tick. *Proceedings of the 41st Annual Meeting of the American Association of Veterinary Parasitologists*, Louisville 1996, p. 51.

94 Stanneck D, Rass J, Radeloff I, *et al*. Evaluation of the long-term efficacy and safety of an imidacloprid 10% flumethrin 4.5% polymer matrix collar (Seresto®) in dogs and cats naturally infected with fleas and/or ticks in multicentre clinical field studies in Europe. *Parasites and Vectors* 2012;**5**:66–71.

95 Stanneck D, Ebbinghaus-Kintscher U, Schoenhense ET, *et al*. The synergistic action of imidacloprid and flumethrin and their release kinetics from collars applied for ectoparasite control in dogs and cats. *Parasites and Vectors* 2012;**5**:73–85.

96 Rohdich N, Roepke RK, Zschieche D. A randomized, blinded, controlled and multi-centred field study comparing the efficacy and safety of Bravecto (fluraner) against Frontline (fipronil) in flea- and tick-infested dogs. *Parasites and Vectors* 2014;**7**:83.

97 Beugnet F, Liebenberg J, Halso L. Comparative efficacy of two oral treatments for dogs containing either afoxolaner or fluraner against *Rhipicephalus sanguinus* sensu lato and *Dermacentor reticulatus*. *Veterinary Parasitology* 2015;**209**:142–5.

4.1 INTRODUCTION

Dermatoses, other than neoplasia, are rarely confined to the pinna, the exceptions being aural haematoma, which is discussed elsewhere (Chapter 18 Surgery of the Ear, Section 18.2), some of the vascular diseases and symmetrical pinnal alopecia/hyperpigmentation. Some diseases, such as scabies and dermatophytosis, may initially present with lesions on the pinna before spreading elsewhere. Other diseases, such as atopy, initially may affect the pinna but the upper aspect of the external ear canal is usually involved as well. Finally, the pinna may be involved as part of a generalised condition such as an endocrinopathy, a defect in keratinisation, or pemphigus foliaceus.

In common with skin elsewhere on the body, the pinnal skin responds with a range of primary and secondary lesions. However, primary lesions, other than nodules, are rare, as is chronic lichenification.

> Bacterial skin infection, such as superficial pyoderma, very rarely affects the pinna. Thus, the presence of papules and pustules is most commonly associated with pemphigus foliaceus, itself a rare condition.

Scale, crust, alopecia and self-excoriation are the most common secondary lesions.

4.2 PINNAL TRAUMA

Trauma to the pinna, particularly in cats, is relatively common. Fight wounds result in torn pinnae (Fig. 4.1) and, often, prolific bleeding. Surprisingly, abscess formation after a bite on the pinna appears to be uncommon. In dogs, pinnal trauma more often results from head shaking rather than fighting. Otodectic mange, otic foreign bodies, otitis media and facial pruritus may all result in damage to the pinna, typically trauma to the periphery (Fig. 4.2).

Overt, or subclinical, bleeding disorders (e.g. von Willebrand factor deficiency in Doberman Pinchers) may present as marginal wounds on the pinna that continually bleed and may take a long time to heal (Figs 4.3A, B).

Fig. 4.1 Lacerated pinnal margin in an entire tom cat due to repeated fight wounds.

Fig. 4.2 Traumatic wounds on the pinnal margin of a Hungarian Viszla with otitis media. Head shaking has resulted in repeated damage.

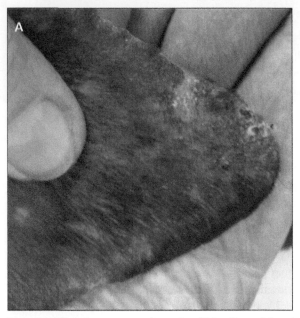
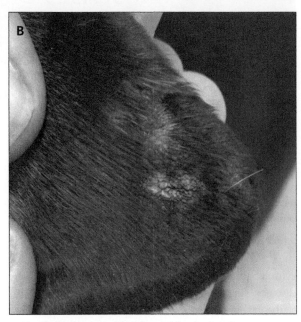

Fig. 4.3 Chronic trauma and continual bleeding of the pinna (A) of a Doberman Pinscher with previously unrecognised von Willebrand factor deficiency. This wound had still not entirely healed 3 months later (B), although it had stopped bleeding.

Aural haematoma

Many texts, clinicians and owners ascribe aural hematoma (Figs 4.4A, B) to pinnal trauma[1,2]. However, the exact relationship between pinnal trauma and the development of the haematoma is unclear[3]. Currently, the aetiology of aural haematoma is unknown. Lesions that are left untreated result in scarring and distortion of the pinna (Fig. 4.4B). Management of aural haematoma is discussed in Chapter 18 Surgery of the Ear, Section 18.2.

4.3 PINNAL DISEASE CHARACTERISED BY CRUST AND SCALE

Pinnal crust and scale, particularly peripheral crusting, is quite commonly seen. The differential diagnosis of focal peripheral pinnal crusting includes:

- Trauma.
- Scabies.
- *Spilopsyllus cuniculi* infestation.
- Stable fly bites.
- Vasculitis.
- Cold agglutinin disease.
- Ischaemic dermatopathy.
- Leishmaniasis.

Ectoparasites

Sarcoptes scabiei and *Notoedres cati*

Scabies is an intensely pruritic disease of dogs that results from infestation with the mite *Sarcoptes scabiei* (*S. scabiei*). The mite exhibits a typical distribution pattern with early lesions usually appearing on the distal pinna, the elbows and the hocks – 'launch points'[4].

An erythematous papular dermatitis is usually present on the elbows and hocks, whereas the pinnal lesions are typically crust and scale (Fig. 4.5). With time these lesions spread along the periphery of the pinna, but accompanying pruritus may result in crust, scale, alopecia and self-excoriation on the dorsum of the pinna (Fig. 4.6)[5]. Rubbing the tip of the pinna between fingertip and thumbnail usually evokes an intense scratching action from the ipsilateral hindlimb – a pinnal scratch reflex.

> Scabies is one of the few dermatoses that can induce steroid-resistant pruritus (Table 4.1).

Localised sarcoptic mange has been reported in dogs, particularly if they are being routinely treated with a

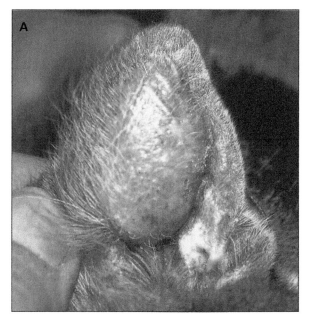

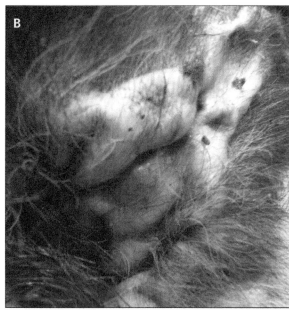

Fig. 4.4 Aural hematoma on the pinna of a dog (A). Distorted pinna following an untreated aural haematoma (B).

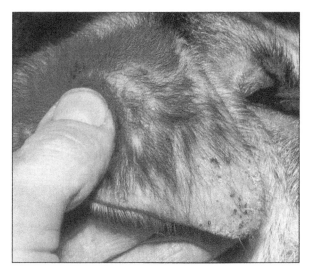

Fig. 4.5 Classic lesions of scabies. Peripheral pinnal alopecia in association with fine crust and scale.

flea treatment that has little acaricidal activity. The mite is not killed but the typical pattern of pruritus and lesions does not develop. The pinna was noted to be a predilection point for this variant[6].

Microscopic examination of multiple skin scrapes may reveal the typical shape of a sarcoptic mite (Fig. 4.7). Clinicians can maximise recovery of mites by

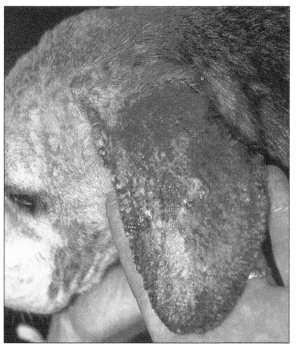

Fig. 4.6 Severe self-trauma has resulted in alopecia, scale and self-excoriation in this Beagle with scabies.

taking care to collect skin scrapes from lesions that are not heavily traumatised, or from the periphery of

Table 4.1 **Steroid refractory pruritus**
Pruritus which fails to respond to prednisolone at 0.5–1 mg/kg twice a day is described as steroid refractory. The differential diagnosis is limited:
Scabies
Malassezial dermatitis
Occasional cases of food allergy
Mycosis fungoides
Calcinosis cutis
Necrolytic migratory erythema
Allergic contact dermatitis
Pelodera dermatitis

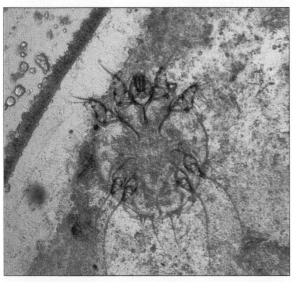

Fig. 4.7 **Photomicrograph of an adult** *Sarcoptes scabiei* **mite.**

the pinna. Many veterinary laboratories now offer an enzyme linked immunosorbent assay (ELISA) test for scabies, based on the test developed by Bornstein *et al.*[7].

Feline scabies is caused by *Notoedres cati*[5]. This mite also exhibits a preference for the pinnae, although in contrast to *S. scabiei* infestation, the early lesions are typically on the rostral periphery of the pinna, rather than the tip. Notoedric mange is often restricted to the pinnae and head (Fig. 4.8) and generalised infections are unusual[5]. Mites are usually plentiful and easily found on

microscopic examination of skin scrapes. The detailed treatment of scabies and other ectoparasitic diseases is discussed in Chapter 3 Ear Cleaners, Section 3.9.

Feline scabies due to *S. scabiei* had been reported, albeit rarely[8]. All five cats exhibited progressive and non-responsive dermatitis with crusting on both the convex and the concave aspects of the pinna. Interestingly, all five also exhibited crusting on the nose, although not on the nasal planum. All responded to topical antiparasitic therapy.

Response to appropriate antiparasitic treatment is fairly prompt, perhaps 7–10 days before some improvement is noticed. After 2 weeks pruritus should be waning. Although 'environmental tidy up' is recommended, there is no requirement for potent insecticidal/acaricidal environmental treatments as scabies mites cannot survive off-host.

Pediculosis

Pediculosis in dogs and cats causes variable pruritus and variable secondary lesions such as scale, knotted hair, alopecia and serohaemorrhagic crust, particularly on the pinnae (Fig. 4.9) but also on the trunk. The lice are quite large (Fig. 4.10) and they may be seen with the naked eye. Egg cases on the hair shafts (nits) may also be seen on gross examination or microscopic examination of hair. However, diagnosis of pediculosis may be difficult as *Trichodectes canis* (the biting louse of dogs) can move surprisingly quickly. Furthermore, because of the supposition that the parasite is easily killed, a diagnosis of pediculosis is often not considered. Treatment of pediculosis is not difficult in individual animals, since the insect is confined to the host and is easily killed by most insecticides. Infection in groups of animals may be more difficult to control.

Demodicosis

Demodicosis is an uncommon cause of pinnal alopecia in dogs. Although the mite is considered to be a member of the normal flora of haired areas of the skin of healthy dogs[9], it is very difficult to find on normal dogs[10]. Pinnal lesions are usually seen in association with more extensive lesions of generalised demodicosis (Fig. 4.11) rather than with the localised form[5].

Demodicosis in cats is associated with two species of demodecid mites, *Demodex cati* and *D. gatoi*[5,11,12], and

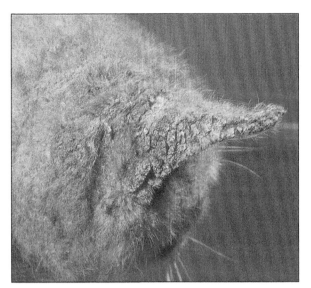

Fig. 4.8 *Notoedres cati* infestation causing spectacular crust on the pinna and adjacent areas of the head of a cat.

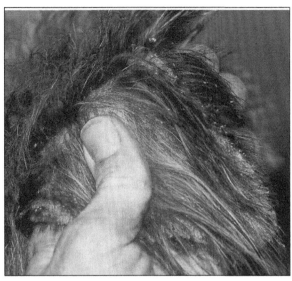

Fig. 4.9 Pediculosis resulting in traumatic alopecia to the pinna.

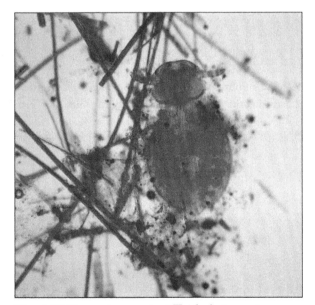

Fig. 4.10 Photomicrograph of *Trichodectes canis*.

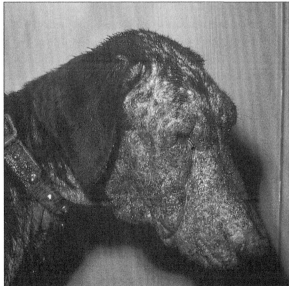

Fig. 4.11 Adult-onset generalised demodicosis in a Doberman Pinscher resulting in scale, crust and alopecia.

perhaps a third[13]. *D. cati*, like *D. canis*, inhabits the follicular crypts of haired skin. Lesions consist of patchy, multifocal or generalised erythema, papules, scale, crusts and alopecia[5]. The pinnae and head are predisposed sites[5]. *D. gatoi*, in contrast to *D. cati*, is shorter and broader and inhabits the stratum corneum. Lesions include pruritus, scale and crust. *D. gatoi*, like the putative third species[13], appears to be contagious[14].

Diagnosis of both canine and feline demodicosis is made by identifying the mites on microscopic examination of skin scrapings (Fig. 4.12).

The treatment of canine demodicosis involves suppressing infection and eliminating the mite[15-17]:

- Systemic antibiotics are indicated for all but the most local cases of demodicosis.
- Consider clipping to facilitate shampoo and dipping.
- Shampoo with a cleansing shampoo, such as Paxcutol®, immediately prior to dipping.
- Dip once weekly with amitraz at 0.025–0.05 mg/kg thoroughly, ensuring that the dog is entirely wet with dip. Allow to air-dry and ensure that the dog does not become wet between treatments.
- Or ivermectin (0.3–0.6 mg/kg) q12 h p/o, DO NOT USE IN COLLIES or COLLIE CROSSES, see Chapter 3 Ear Cleaners.
- Or milbemycin oxime 1–2 mg/kg q24 h p/o, REMEMBER, start with a low dose (0.5 mg/kg) and increase slowly – there are rare cases of toxicity.
- Consider moxidectin, usually in combination with 10% imidacloprid, as a weekly spot-on. It may be used for localised demodicosis but NOT for generalised disease.
- Fluralaner 25-56 mg/kg every 12 weeks, PO (personal communication)

Feline demodicosis similarly responds to antiparasitic treatment:

- There is good evidence for efficacy of 2% lime sulphur dips weekly[15].
- Fair evidence for amitraz rinses at 0.0125–0.025% weekly.
- Fair evidence for the use of doramectin at 600 µg/kg once weekly by subcutaneous injection.

Harvest mite infestation

Trombiculidiasis results from infestation with the parasitic larval stage of free living mites[18]. The most common infestation results from contact with larvae of *Neotrombicula autumnalis* and *Eutrombicula alfredugesi*, typically in the late summer; hence the name 'harvest mite'. Clusters of mites gather on the tip of the pinna, or within the deep folds of skin covering the cartilage at the entrance to the vertical portion of the external ear canal. Moderate to severe pruritus results and this, together with dried serous exudate, may result in crust formation. Diagnosis is straightforward as the six-legged larva is unmistakable (Fig. 4.13).

Insect bite dermatitis

The rabbit flea, *Spilopsyllus cuniculi*, is a seasonal cause of pinnal dermatitis that has been reported in hunting cats in the UK and Australia[19,20]. The flea coordinates its maximum activity to coincide with the rabbit's parturition, thus the seasonality can be predicted. Small,

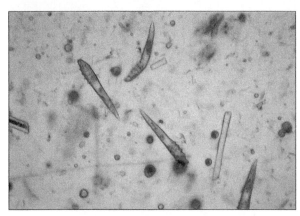

Fig. 4.12 Photomicrograph of canine demodectic mites illustrating the unmistakable shape of the mite.

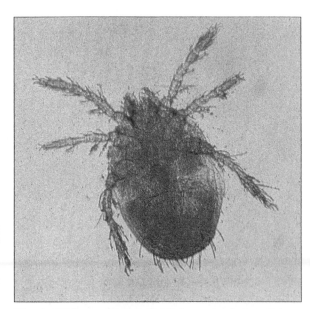

Fig. 4.13 Photomicrograph of a harvest mite (*Neotrombicula autumnalis*).

dark brown fleas may be seen attached to the tips of the pinnae. Focal alopecia (Fig. 4.14A) may be associated with these fleas. In some cases there is marked alopecia and crusting (Fig. 4.14B).

Stable flies (*Stomoxys calcitrans*) may cause a serosanguinous, crusted dermatitis, which may progress to a granulomatous dermatitis. Lesions are principally confined to the tips of the pinnae (Fig. 4.15) and they are usually confined to dogs with erect pinnae and access to stables[5,21]. Black flies (*Simulium* spp.) and other small biting flies may cause a papular dermatitis on the dorsal aspect of the pinna, which may be accompanied by focal alopecia (Fig. 4.16)[5,21]. These lesions may look very similar to those associated with *S. cuniculi* but no parasites will be found.

Mosquito bites may cause a seasonal dermatitis in cats[22]. The lesions are a result of hypersensitivity[23] and they consist of papules, erythematous erosions, alopecia

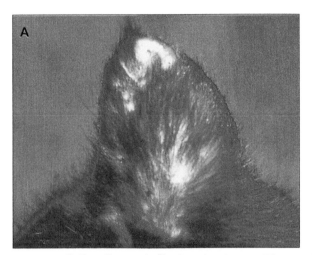
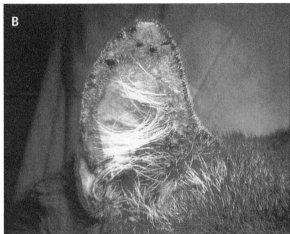

Fig. 4.14 *Spilopsyllus cuniculi* **infestation in cats. The tiny fleas can be seen at the tip of the pinna (A). The associated crust and alopecia follows prolonged infestation (B).**

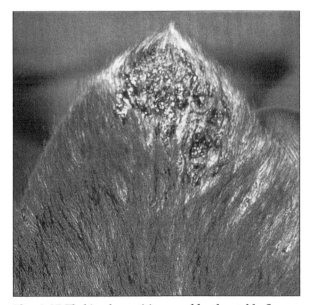

Fig. 4.15 **Fly bite dermatitis caused by the stable fly** *Stomoxys calcitrans*. **Relatively focal, tightly adherent crust at the tip of the pinna.**

Fig. 4.16 **More extensive alopecia with a papulocrustous dermatitis on the pinna of a cat due to biting flies.**

and occasionally hypopigmentation on the dorsum of the face and pinnae. Affected animals may also exhibit moderate polylymphadenopathy, mild pyrexia and symmetrical erythema, fissuring and hyperkeratosis of the footpads. Lesions resolve with the onset of winter or with effective screening to prevent exposure to mosquitoes.

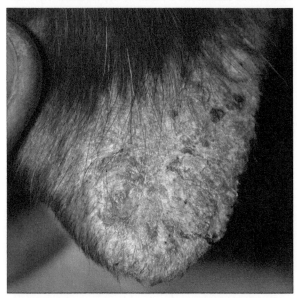

Fig. 4.17 Severe pinnal crusting overlying a papular dermatitis in a dog with leishmaniasis. (Courtesy of Dr Manolis Saridomichelakis, University of Thessaly, Greece.)

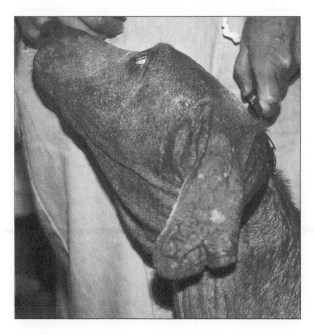

Infectious diseases
Leishmaniasis

Canine leishmaniasis is a severe, often fatal disease of dogs caused by parasites of the genus Leishmania[24]. The disease has zoonotic potential[25]. Classically, leishmaniasis is characterised by exercise inability, chronic wasting and systemic signs such as intermittent pyrexia, anaemia, and polylymphadenopathy[26]. Cutaneous lesions often accompany the systemic signs. Alopecia is often first noted on the pinna and face and is symmetrical and often accompanied by a fine dry scale[27,28]. Pinnal lesions might include:

- Exfoliative dermatitis.
- Peripheral, well-circumscribed deep dermal ulcers.
- Solitary to multiple papular or nodular dermatitis (Fig. 4.17).
- Sterile pustular dermatitis.

The diagnosis and management of leishmaniasis is beyond the scope of this book and readers are referred to current texts and reviews for details of up-to-date treatment recommendations[24].

Malassezia pachydermatis

Dermatitis caused by the yeast *Malassezia pachydermatis* is frequently found as a secondary complication in many chronic dermatoses, such as atopy and idiopathic defects in keratinisation, for example[29]. Malassezial dermatitis may also occur in the absence of discernible underlying disease. Some breeds of dog appear to be predisposed to malassezial dermatitis and these include Basset Hounds, Cocker Spaniels, West Highland White Terriers and Miniature Poodles[30].

The clinical signs consist of erythematous dermatitis, occasionally papular, associated with a variable, typically yellowish grey, greasy scale (Fig. 4.18) and significant malodour in some cases[29,30]. In some dogs the pinnal dermatitis may affect the concave aspect of the pinna only as an extension of malassezial otitis externa, but in some breeds, notably the Basset Hound and West Highland White Terrier, both sides of the pinna, as well

Fig. 4.18 Severe *Malassezia pachydermatis* dermatitis in a Weimaraner. Note the erythema, alopecia and scale. The dog exhibited severe steroid-refractory pruritus.

as adjacent areas of the head and neck, may be affected. Pruritus is often intense and it may be steroid resistant.

Diagnosis is based on demonstrating the yeast in tape strips, skin scrapes, or on contact plate cultures[31]. The approach to treatment is based on identifying and treating the underlying disease and removing the yeast[30,32]:

- Topical shampoos (miconazole, ketoconazole, chlorhexidine) twice weekly. Note that some animals, particularly Basset Hounds, require regular shampoo therapy to maintain remission.
- Systemic ketoconazole 5–10 mg/kg twice a day or itraconazole 5mg/kg once daily, for 3 weeks.

Immune-mediated and autoimmune disease

Pemphigus foliaceus

Pemphigus foliaceus (PF) is a superficial pustular dermatosis[33]. The pustules in PF are a result of a loss of cohesion between keratinocytes due to immunoglobulin G (IgG) targeting the protein desmocollin-1, a component of the desmosomes[34].

In the case of PF the pustules are the primary lesion, but they may be very transient and most commonly cases are presented with crusting, erosion and alopecia on the nasal planum, face and pinnae (Fig. 4.19). Mueller *et al.* reviewed 91 cases of PF and found pustules in 36 cases, crusts in 79 cases and footpad lesions in 32 cases[35]. Facial lesions were present in 46 cases and lesions were found on the concave aspects of the pinnae in 46 cases.

In cats pustules are extremely rare. The most common presenting sign in the cat is symmetrical facial crusting that may affect the pinna (Figs 4.20A, B). Pruritus is variable[36].

Lesions in the dog are usually symmetrical and they are often generalised. Pinnal lesions may appear first and they may be very severe with pruritus, crusting and alopecia. Careful examination of these cases may reveal primary lesions, tiny papules and pustules (Fig. 4.21).

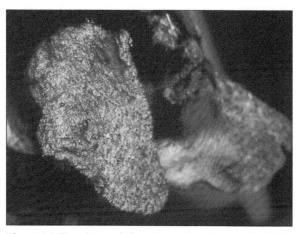

Fig. 4.19 Crusting and alopecia on the pinna, periorbital region and face of a dog with pemphigus foliaceus.

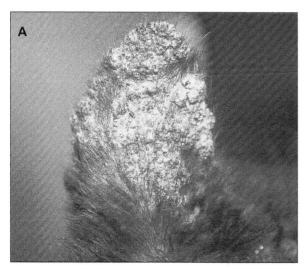

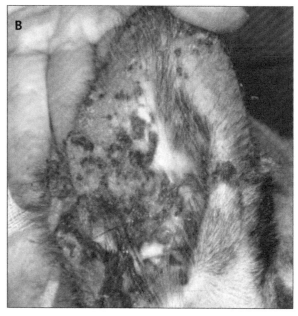

Fig. 4.20 Crusting on the pinna of a cat with pemphigus foliaceus, almost confluent in (A) and appearing as multitudes of crusted papules in (B).

Other, much rarer immune-mediated dermatoses in this group that may affect the pinnae include pemphigus erythematosus (Fig. 4.22) and systemic lupus erythematosus.

The definitive diagnosis of these conditions is based on clinical examination, cytology, histopathological examination of biopsy samples and immunological investigations[37]. Although most of these dermatoses require systemic therapy (see extended discussion in

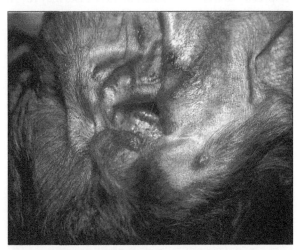

Fig. 4.21 Erythematous papules and pustules on the concave aspect of a pinna. These are primary lesions of pemphigus foliaceus.

Chapter 10 Diseases of the Nasal Planum) it is possible that those dogs with mainly pinnal lesions may respond to topical treatment with tacrolimus ointment[38,39].

Vasculitis

Vasculitis is usually an inflammatory process in which cell-mediated (neutrophils, lymphocytes or macrophages) damage occurs within the walls of blood vessels[40,41]. The cellular infiltrate may occur in response to antigen–antibody complex deposition in the vascular endothelium. This immune complex deposition may be associated with an underlying disease such as systemic lupus erythematosus or drug eruption. More commonly the underlying cause remains unidentified[40]. The damaged endothelium results in extravasation of erythrocytes and underperfusion of the tissue served by the blood vessel. There is erythema, oedema and sloughing of necrosed tissue producing 'punched out' ulcers, which may occur in the centre of the pinna or on the periphery (Fig. 4.23).

Initial investigations should try to identify an underlying disease. In idiopathic cases, immunosuppressive doses of prednisolone (1.1–2.2 mg/kg once or twice daily) may be necessary, although some cases have been reported to respond to pentoxyphylline at doses of 15 mg/kg three times a day, or up to 30 mg/kg twice a day[42].

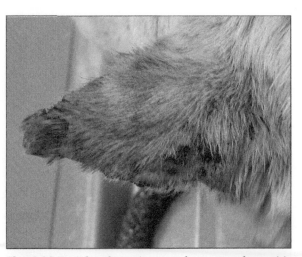

Fig. 4.22 Peripheral crusting, papulocrustous dermatitis and patchy alopecia on the pinna of a dog with pemphigus erythematosus. (Courtesy of Dr DH Bhang, College of Veterinary Medicine, Seoul National University, Seoul, South Korea.)

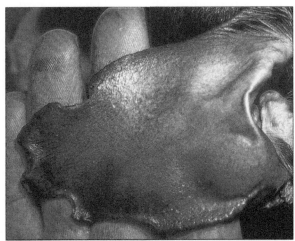

Fig. 4.23 'Punched out' ulcers on the periphery of the pinna of a dog with vasculitis.

Proliferative thrombovascular necrosis of the pinna

This is also known as pinnal margin vasculopathy. Morris cites Dachshunds and Rhodesian Ridgebacks as perhaps being predisposed[42]. Peripheral pinnal erythema and crusting (Fig. 4.24A) are early signs and cyanosis might be noted[42]. Local necrosis results in fissures and notches (Fig. 4.24B). Tetracycline and niacinamide (250 mg of each, three times daily, for dogs less than 10 kg in weight and 500 mg of each, three times daily, for dogs over 10 kg in weight) can be tried[41,42], or pentoxifylline at 15 mg/kg three times a day or 30 mg/kg twice a day. Morris also recommends oral vitamin E, in conjunction with pentoxifylline, twice daily at a dose of 200 IU (small dogs), 400 IU (medium dogs) or 600 IU (large dogs)[42].

Fenbendazole-related drug eruption was reported to be the cause of bilateral necrosis of the distal third of both pinnae in one dog[43]. Patently, the degree of tissue damage and the extent of the lesions is much worse than in proliferative thrombovascular pinnal necrosis, described above. However, drug eruption remains an important differential, albeit a very rare one.

Environmental dermatoses

Actinic dermatitis and pinnal squamous cell carcinoma

Actinic (solar) dermatitis results from exposure to ultraviolet B radiation and it is most commonly seen on the tips of the pinnae of white haired dogs and cats[5]. Lesions on the nasal planum may also be seen. In dogs a predilection site is the glabrous skin of the ventral abdomen, especially in dogs who lie in strong sunshine[44].

Early lesions consist of erythema, fine scale and alopecia, which may be bilateral. With continued exposure there is preneoplastic and finally neoplastic transformation to squamous cell carcinoma[45–47]. These lesions are erosive and crusted and may be haemorrhagic (Figs 4.25A, B). If exposure cannot be prevented, tattooing may be helpful, although amputation of the pinnal tip to below the hair line is the most effective prophylactic option for at-risk cats. It is also the most effective treatment once neoplastic transformation has occurred.

Of interest is the recent research into the role of cyclo-oxygenase 2 (COX-2) in the generation of the inflammatory pathways that result from solar irradia-

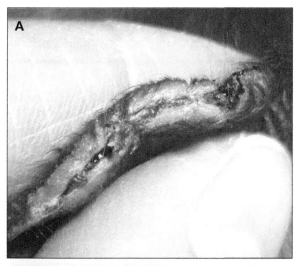

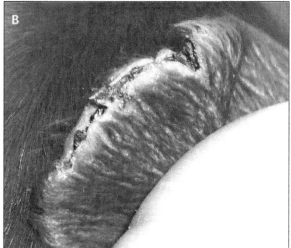

Fig. 4.24 Proliferative thrombovascular necrosis of the pinna of a Miniature Dachshund. Note the ulceration and crusting (A) and early fissure formation (B). (Courtesy of Juliet Owens, Hook Norton, UK.)

tion. Demonstration of COX-2 in lesional skin has been demonstrated[48]. Furthermore, there is clinical evidence that administration of a selective COX-2 blocker (e.g. firocoxib) resulted in histopathological and clinical improvement in five dogs[44].

Dermatoses associated with cold temperatures

Frost bite, cold agglutinin disease, and cryoglobulinaemia/cryofibrinogenaemia may affect the tips of the pinnae, in addition to the tip of the tail and the digits.

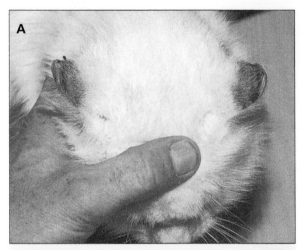

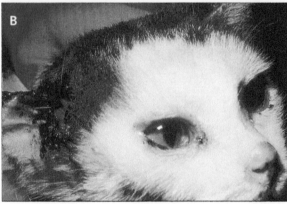

Fig. 4.25 Advanced actinic dermatitis on the pinnae of a cat (A) and advanced neoplastic transformation on another cat (B).

Frost-bitten pinnae

Frost-bitten pinnae may appear pale or cyanosed and they are cold to the touch. Lesions of frostbite are best treated with gentle warming.

Cold agglutinin disease

Cold agglutinin disease is an immune-mediated disease in which autoagglutination of erythrocytes occurs in distal extremities when the temperature within the tissue falls to a critical level[49]. In humans it is usually seen in conjunction with autoimmune haemolytic anaemia[49]. Microthromboses occur in small end-arterioles and capillaries, with ischaemic necrosis ensuing. Early lesions consist of oedema and erythema and this may be followed by the appearance of crust and even ulceration. Cold agglutinin disease is managed by immunosuppressive doses of prednisolone (1.1–2.2 mg/kg q12–24 h). In humans rituximab has been reported to be very effective[49].

Cryoglobulins and cryofibrinogins

Cryoglobulins and cryofibrinogins are proteins that precipitate out of cold serum, only to redissolve when the serum is warmed[50,51]. The precipitating proteins induce microthrombi and vasculitis. The clinical signs in one case were limited to well-demarcated areas of necrosis at the distal end of the pinnae (Fig. 4.26)[51]. The only treatment required was to keep the dog indoors during the winter months.

Fig. 4.26 Distal pinnal necrosis due to precipitation of cryoglobulins and cryofibrinogins in small blood vessels at the tip of the pinna, induced by exposure to cold weather. (Illustration from Nagata M, Nanko H, Hashimoto K, Ogawa M, Sakashita E. Cryoglobulinaemia and cryofibronogenaemia: a comparison of canine and human cases. *Veterinary Dermatology* 1998;9:277–81, with permission.)

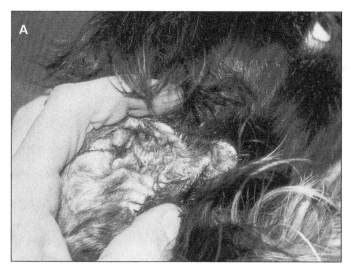

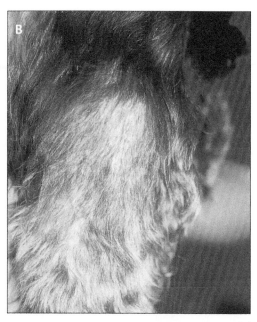

Fig. 4.27 Otitis externa (A), note the scale and crust formation. Pinnal alopecia (B) in a Cocker Spaniel with an idiopathic defect in keratinisation.

Hyperkeratoses

Idiopathic defects in keratinisation

A defect in keratinisation may be primary or secondary. Primary defects reflect inherent errors in the control of the various processes associated with epidermal turnover, sebaceous gland function and the production of hair. The disease is best documented in American Cocker Spaniels[52,53] but other breeds such as Basset Hounds, West Highland White Terriers, Labrador Retrievers, Doberman Pinschers, Irish Setters and Springer Spaniels may also be affected. A heritable defect in keratinisation may spontaneously occur in any breed. For example, a keratinisation defect was reported in Norfolk Terriers[54].

An hereditary seborrhoea oleosa has been described in Persian cats[55].

Affected animals show varying degrees of papules, scale, crust and alopecia (Figs 4.27A, B). The scale may be very greasy or dry. Affected animals frequently have chronic otitis externa and a greasy, ceruminous exudate often coats the hairs on the concave aspect of the pinna (Fig. 4.27B).

Diagnosis is based on ruling out the differential diagnoses and on the histopathological findings. Management is based on the individual and involves control of secondary infection (*Staphylococcus pseudintermedius* and *M. pachyder-*

Fig. 4.28 Peripheral hyperkeratosis resulting in crust and scale in a Miniature Dachshund.

matis in particular), suppressing exuberant keratinisation (retinoids, vitamin A, prednisolone, for example) and degreasing the skin with shampoos. Pruritus in some individuals may be so severe that systemic glucocorticoids are indicated, notwithstanding the potential for side-effects. Clinicians are referred to specialist texts for detailed discussion of these conditions[53,56,57].

Canine ear margin dermatosis

This presents as a peripheral hyperkeratosis and is regarded as a primary keratinisation defect. It affects principally the Miniature Dachshund[58]. Mild lesions are confined to the pinnal periphery and consist of nonpruritic, greasy or scaly plugs attached to the pinnal margin (Fig. 4.28). Severe cases may affect the body of

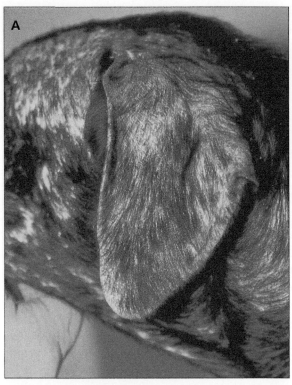
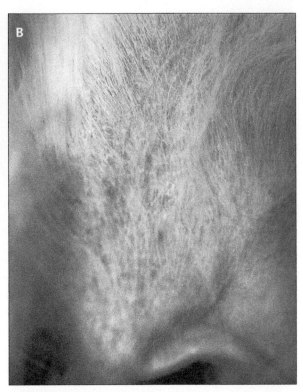

Fig. 4.29 **A: Labrador Retriever with advanced sebaceous adenitis. Note the surface scale and alopecia; B: Standard Poodle with sebaceous adenitis. Note the tightly adherent surface scale and alopecia on this pinna.**

the pinna. Treatment consists of keratolytic and kerato-plastic shampoos such as tar, sulphur and salicylic acid, benzoyl peroxide or selenium sulphide.

Sebaceous adenitis

Sebaceous adenitis is a disease of unknown aetiology characterised by an inflammatory response directed against the sebaceous glands[59]. The damage results in a lack of sebum, which in turn results in follicular disruption, alopecia, surface scale and, frequently, sec-ondary superficial pyoderma (Figs 4.29A, B). There are predisposed breeds (Standard Poodle, Akita, Hungar-ian Vizsla and Chow Chow) but any breed, even cross breeds, may be affected[59].

There is considerable variation in presentation, both on a breed and an individual basis.

• In some shorthaired breeds, classically the Hungarian Vizsla, there may be a granulomatous response. This may present as a 'moth-eaten' coat particularly on the trunk. The head and pinna are

often affected early in the course of the disease[59] and in some individuals there may be a very severe ulcerative dermatitis[60].

• In individuals and breeds with dense coats (Akitas[61]) or long hair (Havanese[62], for example) or other types (Standard Poodle[63]), the early signs are more subtle, with a loss of undercoat, perhaps, a slight thinning of the coat, fine scale and follicular casts. The pinnae and dorsal trunk are usually the first places to be affected. Astute owners may notice that the crimp and nature of the coat changes before alopecia is apparent. The alopecia gradually progresses to affect the trunk, head and limbs[59,61–63].

Diagnosis is based on clinical suspicion but it must be confirmed by histopathological examination of multi-ple (>3) biopsy samples. Cyclosporine has been proven effective in controlling the progression of the disease[64]. For best results it seems that a combination of topical and systemic therapy is required, although not all cases respond:

- Oral cyclosporine, 5 mg/kg once daily.
- Three times a week aggressive, labour intensive, topical therapy as follows:
  - Either sulphur and salicylic acid (Sebolytic®) or ethyl lactate Etiderm®) shampoo, left on for 10 minutes and then rinsed off.
  - Generic baby oil applied to the entire body and left on for 2 hours. Rinse off and clean the coat with shampoo again.
  - Spritz a mixture of 70% propylene glycol in water onto the affected areas, as necessary to keep the scale under control.

The treatment of sebaceous adenitis with cyclosporine (ciclosporin) has been shown to result in suppression of lesions, both clinically and histologically and, furthermore, there is evidence of some regeneration of sebaceous gland tissue[63,65].

The administration of cyclosporine can be associated with side-effects, the most common, and most troubling for the owner, of which are gastrointestinal in nature, and usually transient[66]. Vomiting, anorexia, variable stool and diarrhoea can occur in up to 30% dogs and in a small proportion of dogs are so severe that owners will elect to cease therapy. In these animals immunosuppressive doses of prednisolone (1.1–2 mg/kg bid) may be tried, although the response is less predictable than that seen with cyclosporine.

Idiopathic benign lichenoid keratosis

Idiopathic benign lichenoid keratosis affecting the pinnae has been reported in four dogs[66]. Affected animals present with multiple, wart-like papules, or hyperkeratotic plaques. Histologically the lesions are distinct. Complete surgical excision is curative.

Lichenoid psoriasiform dermatosis

Lichenoid psoriasiform dermatosis is an extremely rare dermatosis that has only been reported in young Springer Spaniels in the USA[67], raising the possibility that it is a genodermatosis. Affected animals exhibit erythematous papules and lichenoid plaques on the concave aspects of the pinnae, external ear canals and other, predominantly ventral, parts of the head and trunk. Diagnosis is based on the almost pathognomonic signalment and clinical signs and is confirmed by histopathological examination of biopsy samples. The condition is frustrating to treat, requiring combinations of topical antimicrobial shampoo and systemic antibacterial agents, usually with systemic glucocorticoids, to effect even a modicum of control. Interestingly, the exact same lesions were seen in three dogs treated with cyclosporine and the authors speculated that the lesions might reflect atypical staphylococcal infections[68].

Lupoid dermatosis

Lupoid dermatosis was reported as a rare, heritable disease of German Short Haired Pointers, although it has been reported in other breeds[69]. The disease is characterised by progressive, non-pruritic scale and alopecia of the face, pinnae (Fig. 4.30) and trunk. There is no effective treatment.

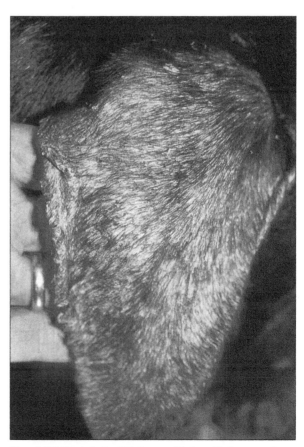

Fig. 4.30 Shorthaired Pointer with lupoid dermatosis. Note the surface scale and alopecia.

Nutritional dermatoses

Zinc deficiency

Zinc deficiency occurs as two clinical syndromes in the dog[70,71] but has not been reported in the cat. Occasional clusters of cases, affecting any breed of dog, may occur when a deficient diet is fed[72]. Some breeds, for example the Alaskan type (Siberian Husky, Alaskan Malamute, Samoyed), may manifest a physiological inability to absorb zinc, Type 1. Affected animals require zinc supplementation for life. The other presentation, Type 2, is primarily seen in young dogs of any breed, which are fed a diet that is absolutely deficient, or relatively deficient, in zinc. Once the deficiency is identified, and the diet is corrected, these animals usually require only short-term zinc supplementation[70,71].

The major clinical sign in such cases is crusting, localised, lesional alopecia and erythema of the pinnae and perioral, periorbital, perianal and perivulval sites, as well as crusting over pressure points[71]. Pinnal crusting occurs in around 30% cases and seems to be particularly pruritic.

Zinc methionine tablets are the recommended route of delivering the supplement, at manufacturer's advised dosages, typically around 1.7 mg/kg a day of zinc methionate[70,71].

Adverse food reaction

Dietary intolerance and food allergy are rare conditions that may be associated with any combination of clinical signs: pruritus, lesions or both[73]. The prevalence varies. For example, in one study of 130 pruritic dogs, some 26% were diagnosed with adverse food reaction[74]. The authors admitted that this prevalence was higher than in most reported studies[75], in the order of 5–7% of pruritic dogs.

The unpredictability of the clinical signs implies that food allergy should always be included in differential diagnoses for animals with chronic pinnal dermatosis. Approximately one-half of affected animals will have a history of steroid-resistant pruritus[76,77]. There are no primary or pathognomic lesions. The nature of the clinical signs will almost always reflect pruritus, although the degree and duration of the pruritus will dictate the changes: erythema, alopecia, crust and scale and, ultimately, hyperpigmentation and lichenification.

Diagnosis of dietary intolerance and food allergy depends on eliminating alternative diagnoses from the working differential and then instituting dietary testing. There is no evidence that intradermal skin testing for food allergy or the serological assessment of food allergens by ELISA, for example, is of any diagnostic or predictive value[73,78].

The diagnosis is based on feeding an elimination diet, probably a hydrolysed protein diet[79,80] for a minimum of 6 weeks:

- Demonstrate that the clinical signs resolve. This may take at least 6 weeks and in some cases the diet may need to be fed for 12 weeks[76]. This procedure should not be undertaken lightly.
  - It may be necessary to administer low-dose prednisolone (0.5 mg/kg every other day) to suppress the pruritus, if it is severe, until the diet kicks in.
- Demonstrate that the signs recur when normal food is reinstituted. Most animals react within 2–10 days.
- Demonstrate that the signs abate upon refeeding the hypoallergenic diet.

At this point there are two choices:

- Continue feeding the hypoallergenic diet and gradually add back the previous components of the diet, one at a time, in order to identify what can, and cannot, be tolerated. It should then be possible to identify an appropriate commercial diet.
- Continue feeding the hypoallergenic diet *ad infinitum* and try to find, by trial and error, what commercial diet the dog can tolerate.

4.4 FOCAL LESIONS PRINCIPALLY AFFECTING THE DORSAL PINNA

Dermatophytosis

Dermatophytosis is one of the most common conditions affecting the pinnae of cats, particularly kittens, whereas it is much less common in dogs[81]. Dermatophytosis is more common in young cats than in mature individuals, unless their immune system is compromised by ill health[81]. The most common lesions consist of broken, dull-looking hairs and a thin, often silvery scale (Fig. 4.31)[81]. However, dermatophytosis may also cause a papular dermatitis, erythematous dermatitis

(Fig. 4.32) and, in some cases, a granulomatous reaction, although these are much less common on the pinnae than on the trunk or limbs.

Diagnosis is by examination of skin scrapes, examination of the affected area under Wood's light, and ultimately, by culture[81,82], the latter being the gold standard[83].

Current treatment guidelines do not recommend 'spot' treatment, even of localised lesions, and instead advise a combination of systemic medication and topical therapy is used[82]:

- Itraconazole, 0.5 mg/kg once daily, 'week on and week off' for three cycles, as an initial treatment phase.
- Terbinafine, 34–45.7 mg/kg once daily[84]. For ease of dosing, Moriello *et al.*[85] suggest:
  - Cats <2.8 kg receive one-quarter of a tablet (62.5 mg) per dose;
  - Cats 2.8–5.5 kg receive one-half a tablet (125 mg) per dose;
  - Cats >5.5 kg receive one tablet.
- Topical treatment[83,85,86]:
  - Enilconazole solution, twice a week.
  - Miconazole, with or without chlorhexidine shampoo twice a week.
  - Lime sulphur rinses, twice weekly, applied with a commercial plant sprayer to ensure thorough wetting of the coat and skin.

Cure is defined as two sequential negative cultures, taken at weekly or biweekly sampling intervals.

Feline poxvirus

Feline poxvirus infection (synonym: cowpox) is caused by an Orthopoxvirus, presumably transmitted by bites from small rodents[87]. There is a seasonal incidence that peaks in the autumn, and, understandably, there is an increased incidence in rural, outdoor cats[88]. Primary lesions appear as transient erythematous papules or erosions, usually on the anterior part of the body[87,88]. Some 10–14 days later secondary lesions appear as papules, which rapidly crust and become alopecic erosions (Fig. 4.33). Contagion to cats in the same household appears uncommon[88]. The disease has zoonotic potential[88–90] and suitable precautions against infection should be taken.

Fig. 4.31 Patchy alopecia caused by dermatophytosis on the head and pinna of a cat.

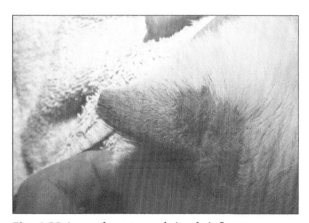

Fig. 4.32 An erythematous, obviously inflammatory, patch of alopecia on the pinna of a cat due to dermatophytosis.

Fig. 4.33 Multiple erythematous erosions on the head and pinna of a cat with feline pox virus infection. These are secondary lesions resulting from haematogenous spread from the primary lesion.

Supportive treatment with systemic antibacterial agents is usually sufficient treatment[91]. Glucocorticoids should be avoided as they may predispose to systemic infection and severe complications, even death, may ensue.

4.5 PINNAL DERMATOSES CAUSING PAPULES AND NODULES

Nodules on the pinnae may reflect many aetiologies. It is almost impossible to distinguish between them with confidence on the basis of clinical appearance. Cytological examination may be helpful but punch or excision biopsy is indicated in all cases in which the aetiology is unknown.

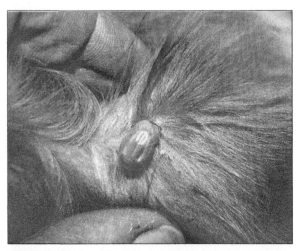

Fig. 4.34 *Ixodes ricinus* on a cat.

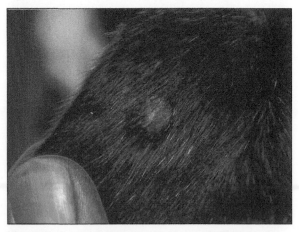

Fig. 4.35 An erythematous papule, caused by a focal granulomatous reaction to a tick bite on the pinna of a dog.

Tick infestation

Owners often mistakenly identify ticks as tumours. The most common ticks found on the pinnae of dogs are the American dog tick (*Dermacentor variablis*), the lone star tick (*Amblyomma americanum*), the brown dog tick (*Rhipicephalus sanguinus*) and *Ixodes* spp. (Fig. 4.34). The latter are the species least likely to be found on the pinnae and head[92,93]. It has been speculated that the prevalence of ticks on the head and pinnae simply reflects the inability of the host to remove them easily from this area of the body[93].

In addition to causing local irritation and occasionally postbite granulomas (Fig. 4.35), ticks are frequently associated with vectoring infectious disease. Different species of ticks in different countries carry different organisms[94,95], and often carry mixtures of organisms. Common examples include:

- *Borrelia burgdorferi*.
- *Babesia* spp.
- *Rickettsia* spp.

Ticks may be removed with gentle traction, and many veterinary practices sell plastic tools which make the removal easy. Removal is more easily accomplished if the ticks are killed first, for example by application of vegetable or mineral oil, or a specific agent such as amitraz or fipronil[96–98]. Recent improvement in tick control products and availability of sustained release collars[99,100], and tablets containing fluralanar or afoxolaner[101,102] in particular, makes prevention a real possibility for the dog and cat owner.

Neoplastic lesions
Papilloma

Cocker Spaniels, Kerry Blue Terriers, and male dogs in general are predisposed[103,104]. Papillomas are usually well demarcated (Fig. 4.36), superficial, occasionally pedunculated papules (i.e. <1 cm in diameter)[105], which tend to bleed if traumatised[7]. Many papillomas exhibit a frond-like appearance.

Canine cutaneous histiocytoma

Canine cutaneous histiocytomas are most likely of Langerhans cell origin[106,107]. These are the most common neoplasms of the dog and 50% of cases occur in dogs under 2 years of age[107]. Pedigree dogs are predisposed, particularly the Boxer, English Bulldog, and Scottish Terrier[107,108]. The most common site for the tumour is the dorsal pinna[107,108]. These are solitary, elevated, domed, erythematous, well-demarcated, minimally pruritic lesions (Fig. 4.37) that are often reported to show rapid growth. Occasionally they are ulcerated and very rarely they have been reported to be multiple. Canine cutaneous histiocytomas are benign lesions and response to surgery is close to 100%[107,108]. The principal differential diagnoses are mast cell tumours and granulomas.

Mast cell tumour

Mast cell tumours are common in the dog[109]. Although mast cell tumours may occur in young dogs (hence the inclusion of canine cutaneous histiocytoma as a differential diagnosis) the incidence increases with age, peaking at 8 or 9 years old[109,110]. Boxers and other breeds with Bulldog ancestry are predisposed, but mast cell tumours may occur in any dog[109,110]. The clinical appearance of the tumour is very variable but single cutaneous lesions less than 3 cm in diameter are the most common presentation (Fig. 4.38)[109,110]. The diagnosis of mast cell tumour is made by a combination of cytology and histopathology.

Clinical management of all but single nodules has been transformed with the advent of masitinib[111,112]. However, masitinib would not normally be indicated for solitary pinnal mast cell tumours, which are best treated by simple surgical excision. The selection of various treatment modalities for extensive canine mast cell neoplasia is beyond the scope of this book and readers are referred to specialist oncology texts.

Granulomatous lesions

Granulomas may be solitary or multiple (Fig. 4.39). Clinically, they are usually non-pruritic and are usually nodular in appearance, slowly growing, and domed in appearance[113]. Occasionally they may be exudative, or ulcerated and crusted (Fig. 4.40), particularly when associated with eosinophilic furunculosis, a peracute nasal dermatitis[114]. Occasionally, clinicians may encounter what appears to be a more chronic eosinophilic granulomatous dermatitis (Fig. 4.41). Whether, or how, these relate to the peracute furunculotic variant is not known.

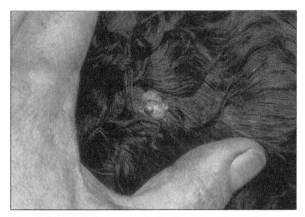

Fig. 4.36 Well-demarcated, flesh-coloured papilloma on the pinna of a Cocker Spaniel.

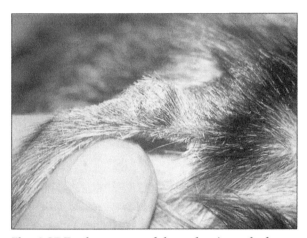

Fig. 4.37 Erythematous nodule on the pinna of a dog – histiocytoma.

Fig. 4.38 Large, erythematous, eroded mast cell tumour at the base of the pinna of a Newfoundland dog.

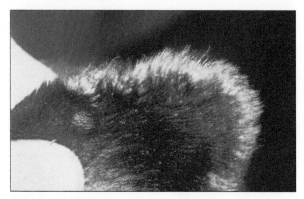

Fig. 4.39 Solitary domed mass on the pinna of a dog, a tick bite granuloma.

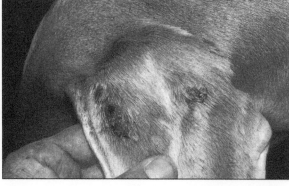

Fig. 4.40 Acute eosinophilic furunculosis/granulomas on the dorsal pinna of a Weimaraner, which presented with acute nasal eosinophilic furunculosis. These lesions were intensely pruritic.

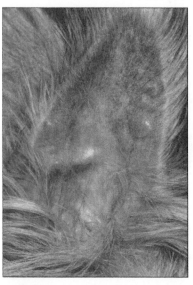

Fig. 4.41 Terrier exhibiting chronic, minimally pruritic eosinophilic granulomae on the ventral aspect of the pinna.

The management of these nodules is primarily aimed at defining the aetiology so that specific treatment may be provided[113]. The clinical approach may therefore include aspiration cytology, serology, histology, microbiology and fungal culture (Figs 4.42A, B). Staphylococci, *Nocardia* spp., *Actinomycetes* spp., mycobacteria, atypical mycobacteria, dermatophytes, subcutaneous mycoses (Fig. 4.43), deep mycoses, algae, ticks, *Leishmania* spp. (Fig. 4.44) and foreign bodies are all potential causes of granulomatous dermatitis[115].

Sterile, idiopathic, granulomatous dermatitis may also be encountered[116–120]. Lesions typically present as firm, non-painful and non-pruritic, often multiple,

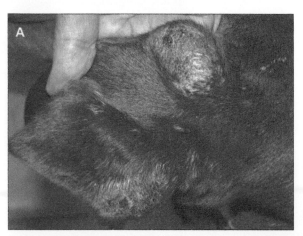
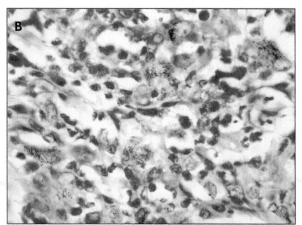

Fig. 4.42 A: Large leproid granulomas on the pinna of a dog; B: photomicrograph of the acid-fast stained histopathological preparation from the mass shown in (A). Intracellular bacteria are visible. (Courtesy of Dr Lissandro Conceição, Departamento de Veterinária, Universidade Federal de Viçosa, Viçosa, Minas Gerais, Brazil.)

papules, nodules or plaques, typically on the head and pinna (Fig. 4.45). Treatment options for idiopathic sterile pyogranulomae include:

- Oxytetracycline and niacinamide:
  - 250 mg of each, three times a day for dogs weighing <10 kg.
  - 500 mg of each, three times a day for dogs weighing >10 kg.
- Prednisolone: 1–2 mg/kg twice daily, tapering to alternate days over 2–3 weeks.
- Cyclosporine: 5 mg/kg once daily.
- Azathioprine: 1–2 mg/kg once daily tapering to twice a week.

4.6 DERMATOSES PRINCIPALLY AFFECTING THE CONCAVE ASPECT OF THE PINNA

Atopy

Dogs with atopy frequently have otitis externa[121–124]. Lesions within the upper portion of the external ear canal are frequently associated with erythema, and even lichenification, on the concave aspect of the pinnae (Figs 4.46A, B).

Otitis externa and any accompanying pinnal lesions do not respond well to immunotherapy[125], often requiring long-term additional management measures over and above immunotherapy, for example. The author's experience with oclacitinib (Apoquel® Zoetis), albeit on a few cases, suggest that it might have a much better effect on the otic and pinnal lesions than immunotherapy.

Pinnal lesions cannot be managed in isolation from the accompanying otitis externa, which is discussed elsewhere (Chapter 5 Aetiology and Pathogenesis of Otitis Externa, Section 5.5). However, in early cases, application of hydrocortisone aceponate (Cortavance®, Virbac) may be very effective. Some dogs appear to resent the application of Cortavance, in which case a steroid-containing otic preparation may be applied.

Allergic contact dermatitis

Allergic contact dermatitis is rare in dogs, and published case reports even more so[124]. Allergic contact dermatitis appears to be extremely rare in the cat[125].

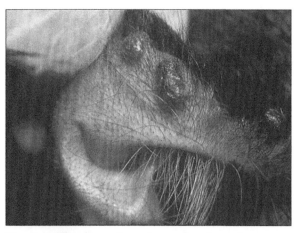

Fig. 4.43 Two crusted papules on the pinna of a cat, *Cryptococcus* **spp.**

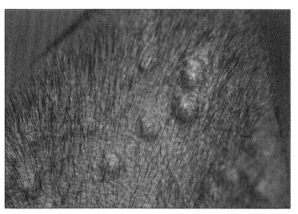

Fig. 4.44 Multiple, small, granulomatous papules on the pinna of a dog with leishmaniasis. (Courtesy of Dr Manolis Saridomichelakis, University of Thessaly, Greece.)

Fig. 4.45 Multiple papules and nodules on the pinna: sterile pyogranuloma.

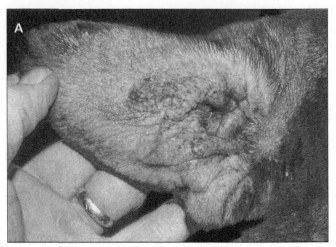

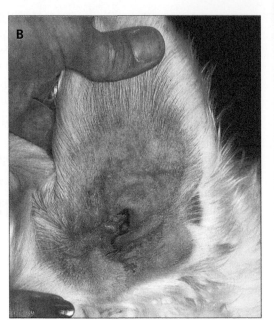

Fig. 4.46 Hyperplasia and lichenification on the concave aspect of the pinnae of an atopic dog (A) often accompanies erythema of the vertical ear canal (B).

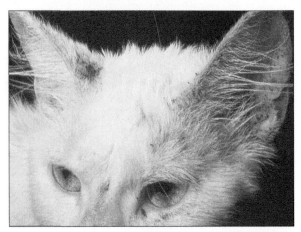

Fig. 4.47 Erythematous pinna as a result of allergic contact dermatitis to topical application of Vaseline.

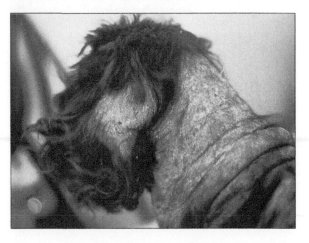

Most commonly it affects middle-aged animals since the induction period is long[126]. If the allergen is contained within something on which the animal lies, the ventral aspects of the pinnae may be involved, because the convex surface of the pinna is haired.

The glabrous ventral aspect of the pinna also may be involved in allergic contact dermatitis to topical otic medications, such as neomycin and propylene glycol (Fig. 4.47). Drug eruption may also result in widespread alopecia that may include the pinnae (Fig. 4.48).

The management of allergic contact dermatitis is based upon identifying the allergen with exclusion/provocative or closed patch testing, neither of which is easy. Failure to identify the allergen, and avoiding contact, usually condemns the animal to systemic glucocorticoid therapy, with all the attendant side-effects, unless topical glucocorticoids such as hydrocortisone aceponate (Cortavance®, Virbac) are adequate.

Acute pinnal urticaria

Peracute pinnal swelling may occasionally be encountered. Most commonly these are small papules, often multiple (Figs 4.49A, B). Insect bites and, perhaps inoculations from vegetation, are the usual cause.

Fig. 4.48 Widespread alopecia, scale and crust in this Cocker Spaniel with drug eruption.

Figs 4.49A, B Two examples of acute pinnal urticaria.

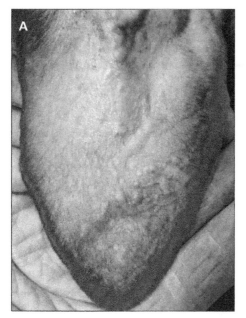

4.7 DERMATOSES PRINCIPALLY CAUSING SYMMETRICAL PINNAL ALOPECIA

Endocrine disorders: hypothyroidism and sex hormone aberrations

Defects in keratinisation associated with hypothyroidism and sex hormone aberrations are often associated with excessive sebaceous gland activity, and a greasy, hyperkeratotic dermatosis results on the concave aspect of the pinna (Figs 4.50A–C). Diagnosis, in the absence of demonstrable gonadal neoplasia, is difficult and

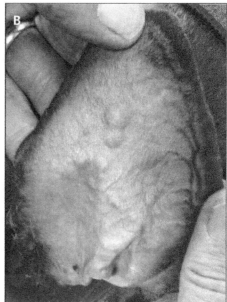

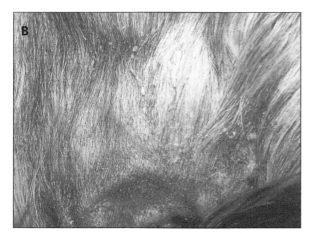

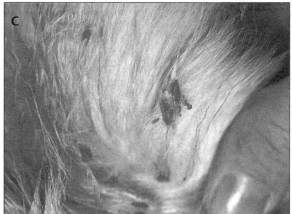

Fig. 4.50 Greasy scale on the concave aspect of the pinna in a Cocker Spaniel with a defect in keratinisation, even after total ear canal ablation has been performed (A); in a Cocker Spaniel with hypothyroidism (B); and in a German Shepherd Dog with Sertoli cell tumour (C).

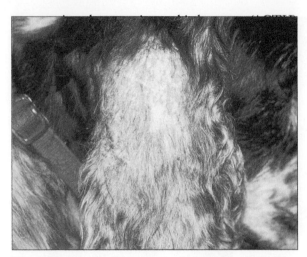

Fig. 4.51 Alopecia of the pinna of a hypothyroid Cocker Spaniel.

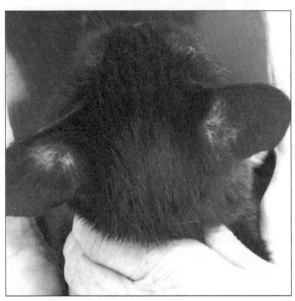

Fig. 4.52 Symmetrical alopecia at the base of the pinnae in a cat with hyperthyroidism.

likely to be associated with alopecia, although they rarely affect the head and extremities. On rare occasions the pinnae may be affected in hypothyroid dogs before other signs become apparent (Fig. 4.51).

Hypothyroidism in the cat may follow thyroidectomy and alopecia may be noted. However, more commonly, alopecia occurs with feline hyperthyroidism, a consequence of overgrooming (Figs. 4.52).

Non-endocrinological symmetrical pinnal alopecia

Symmetrical pinnal alopecia is found in two conditions:
- Pattern alopecia is seen particularly in short-haired breeds such as Miniature Dachshunds, Staffordshire Bull Terriers, Greyhounds, Italian Greyhounds and Whippets. In these animals the symmetrical alopecia may be noted on the pinnae but usually affects the ventrum and caudal medial thighs. The progression of the alopecia is slow, non-inflammatory and not associated with hyperpigmentation. Diagnosis is based on ruling out the differential diagnoses and on demonstrating the histopathological features of miniaturised hair follicles.
- Pattern baldness is principally seen in Dachshunds and, occasionally, in other breeds such as Doberman Pinschers (Figs 4.53A–C). The alopecia is symmetrical and limited to the pinnae. Hair loss begins in the young adult and gradually becomes complete. Pinnal hyperpigmentation then occurs. The aetiology of the condition is unknown and there is no effective treatment.

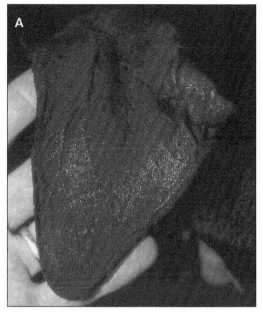

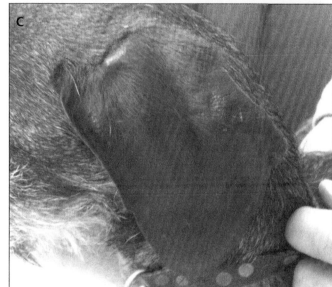

Fig. 4.53 Pattern baldness in a Doberman Pinscher (A), Yorkshire Terrier (B) and a Norfolk Terrier (C).

4.8 REFERENCES

1 Lanz OI, Wood BC. Surgery of the ear and pinna. *Veterinary Clinics of North America Small Animal Practice* 2004;**34**:567–99.

2 Brown C. Surgical management of canine aural haematoma. *Lab Animal (NY)* 2010;**39**:104-5.

3 Joyce JA, Day MJ. Immunopathogenesis of canine aural haematoma. *Journal of Small Animal Practice* 1997;**38**:152–8.

4 August JR. Taking a dermatologic history. *Compendium on Continuing Education* 1986;8:510–18.

5 Matousek JL. Diseases of the ear pinna. *Veterinary Clinics of North America Small Animal Practice* 2004;**34**:511–40.

6 Pin D, Bensignor E, Carlotti DN, Cadiergues MC. Localised sarcoptic mange in dogs: a retrospective study of 10 dogs. *Journal of Small Animal Practice* 2006;**47**:611–14.

7 Bornstein S, Frössling J, Näslund K, Zakrisson G, Mörner T. Evaluation of a serological test (indirect ELISA) for the diagnosis of sarcoptic mange in red foxes (*Vulpes vulpes*). *Veterinary Dermatology* 2006;**17**:411–16.

8 Huang H-P, Lien Y-H. Feline sarcoptic mange in Taiwan: a case series of five cats. *Veterinary Dermatology* 2013;**24**:457–9.

9 Ravera I, Altet L, Francino O, *et al*. Small demodex populations colonize most parts of the skin of healthy dogs. *Veterinary Dermatology* 2013;**24**:168–72.

10 Fondati A, De Lucis M, Furiani N, Monaco M, Ordeix L, Scarampella F. Prevalence of *Demodex canis*-positive healthy dogs at trichoscopic examination. *Veterinary Dermatology* 2010;**21**:146–51.

11 Chesney CJ. Demodicosis in the cat. *Journal of Small Animal Practice* 1989;**30**:689–95.

12 Frank LA, Kania SA, Chung K, Brahmbhatt R. A molecular technique for the detection and differentiation of *Demodex* mites on cats. *Veterinary Dermatology* 2013; **24**:367–9.

13 Moriello KA, Newbury S, Steinberg H. Five observations of a third morphologically distinct feline *Demodex* mite. *Veterinary Dermatology* 2013;**24**:460–2.

14 Morris DO. Contagious demodicosis in three cats residing in a common household. *Journal of the American Animal Hospital Association* 1996;**32**:350–2.

15 Mueller RS. Treatment protocols for demodicosis: an evidence-based review. *Veterinary Dermatology* 2004;**15**:75–89.

16 Mueller RS. An update on the therapy of canine demodicosis. *Compendium on Continuing Education* 2012;**34**:1–4.

17 Mueller RS, Bensignor E, Ferrer L, *et al.* Treatment of demodicosis in dogs: 2011 clinical practice guidelines. *Veterinary Dermatology* 2012;**23**:86–96.

18 Greene RT, Scheidt VJ, Moncol DJ. Trombiculiasis in a cat. *Journal of the American Veterinary Medical Association* 1986;**188**:1054–5.

19 Studdert VP, Arundel JH. Dermatitis of the pinnae of cats in Australia associated with the European rabbit flea (*Spilopsyllus cuniculi*). *Veterinary Record* 1988;**123**:624–5.

20 Harvey RG. Dermatitis in a cat associated with *Spilopsyllus cuniculi*. *Veterinary Record* 1990;**126**:89–90.

21 Angarano DW. Diseases of the pinnae. *Veterinary Clinics of North America Small Animal Practice* 1988;**18**:869–84.

22 Mason KV, Evans AG. Mosquito bite-caused eosinophilic dermatitis in cats. *Journal of the American Veterinary Medical Association* 1991;**198**:2086–8.

23 Nagata M, Takuo I. Cutaneous reactivity to mosquito bites and its antigens in cats. *Veterinary Dermatology* 1997;**8**:19–26.

24 Ciaramella P, Corona M. Canine leishmaniasis: clinical and diagnostic aspects. *Compendium on Continuing Education* 2003;**25**:358–69.

25 Otranto D, Dantas-Torres. The prevention of canine leishmaniasis and its impact on public health. *Trends in Parasitology* 2013;**29**:339–45.

26 Koutinas AF, Polizopoulou ZS, Saridomichelakis MN, *et al.* Clinical considerations on canine visceral leishmaniasis (CVL) in Greece: a retrospective study of 158 spontaneous cases. *Journal of the American Animal Hospital Association* 1999;**35**:376–83.

27 Ferrer L, Rabanal R, Fondevila D, Ramos JA, Domingo M. Skin lesions in canine leishmaniasis. *Journal of Small Animal Practice* 1988;**29**:381–8.

28 Saridomichelakis MN, Koutinas AF. Cutaneous involvement in canine leishmaniasis due to *Leishmania infantum* (syn *L. chagasi*). *Veterinary Dermatology* 2014;**25**:61–71.

29 Dorogi J. Pathological and clinical aspects of the diseases caused by *Malassezia pachydermatitis*. *Acta Microbiolica et Immunologica Hungarica* 2002;**49**:363–9.

30 Bond R. Superficial veterinary mycoses. *Clinical Dermatology* 2010;**28**:226–36.

31 Bond R, Collin NS, Lloyd DH. Use of contact plates for the quantitative culture of *Malassezia pachydermatis* from canine skin. *Journal of Small Animal Practice* 1994;**35**:68–72.

32 Negre A, Bensignor E, Guillot J. Evidence-based veterinary dermatology: a systemic review of intervention for *Malassezia* dermatitis in dogs. *Veterinary Dermatology* 2009;**20**:1–12.

33 Olivry T. A review of autoimmune skin diseases in domestic animals: 1. Superficial pemphigus. *Veterinary Dermatology* 2006;**17**:291–305.

34 Bizikova P, Dean GA, Hashimoto T, Olivry T. Cloning and establishment of canine desmocollin-1 as a major autoantigen in canine pemphigus foliaceus. *Veterinary Immunology and Immunopathology* 2012;**15**:197–207.

35 Mueller RS, Krebs I, Power HT, Fieseler KV. Pemphigus foliaceus in 91 dogs. *Journal of the American Animal Hospital Association* 2006;**42**:189–96.

36 Manning TO, Scott DW, Smith CA, Lewis RM. Pemphigus diseases in the feline: seven case reports and discussion. *Journal of the American Animal Hospital Association* 1982;**18**:433–43.

37 Olivry T, Jackson HA. Diagnosing new autoimmune blistering skin diseases in dogs and cats. *Clinical Techniques in Small Animal Practice* 2001;**16**:225–9.

38 Griffies JD, Mendelsohn CL, Rosenkrantz WS, Muse R, Boord MJ, Griffin CE. Topical 0.1% tacrolumis for the treatment of discoid lupus erythematosus and pemphigus foliaceus in dogs. *Journal of the American Animal Hospital Association* 2004;**40**:29–41.

39 Bhang D-H, Choi U-S, Jung Y-C, *et al.* Topical 0.03% tacrolimus for treatment of pemphigus erythematosus in a Korean Jindo dog. *Journal of Veterinary Medical Science* 2008;**70**:415–17.

40 Parker WM, Foster RA. Cutaneous vasculitis in five Jack Russell Terriers. *Veterinary Dermatology* 1989;**7**:109–15.

41 Innerǎ M. Cutaneous vasculitis in small animals. *Veterinary Clinics of North America Small Animal Practice* 2013;**43**:113–34.

42 Morris DO. Ischemic dermatopathies. *Veterinary Clinics of North America Small Animal Practice* 2013;**43**:99–111.

43 Nuttal TJ, Burrow R, Fraser I, Kipar A. Thrombo-ischaemic pinnal necrosis associated with fenbendazole treatment in a dog. *Journal of Small Animal Practice* 2005;**46**:243–6.

44 Albanese F, Abramo F, Caporali C, Vichi G, Millanta F. Clinical outcome and cyclo-oxygense-2 expression in five dogs with solar/actinic keratosis treated with firocoxib. *Veterinary Dermatology* 2013;**24**:606–12.

45 Lana Se, Ogilvie GK, Withrow SJ, Straw RC, Rogers KS. Feline cutaneous squamous cell carcinoma of the nasal planum and the pinnae: 61 cases. *Journal of the American Animal Hospital Association* 1997;**33**:329–32.

46 Nilula KJ, Benjamin SA, Angleton GM, Saunders WJ, Lee AC. Ultraviolet radiation, solar dermatitis and cutaneous neoplasia in beagle dogs. *Radiation Research* 1992;**129**:11–18.

47 Atwater SW, Powers BE, Straw RC, *et al.* Squamous cell carcinoma of the pinna and nasal planum: 54 cats (1980–1991). *Proceedings of the Veterinary Cancer Society* 1991;**11**:35–6.

48 Bardagi M, Fondevila D, Ferrer L. Immunohistochemical detection of COX-2 in feline and canine actinic keratosis and cutaneous cell carcinoma. *Journal of Comparative Pathology* 2012;**146**:11–17.

49 Swiecicki PL, Hegerova LT, Gertz MA. Cold agglutinin disease. *Blood* 2013;**122**:1114–21.

50 Retamozo S, Brito-Zerö P, Bosch X, Stone JH, Ramos-Casals M. Cryoglobulinemic disease. *Oncology* (Williston Park) 2013;**27**:1110–16.

51 Nagata M, Nanko H, Hashimoto K, Ogawa M, Sakashita E. Cryoglobulinemia and cryofibronogenemia: a comparison of canine and human cases. *Veterinary Dermatology* 1998;**9**:277–81.

52 Kwochka KW, Rademakers AM. Cell proliferation kinetics of epidermis, hair follicles, and sebaceous glands of Cocker Spaniels with idiopathic seborrhea. *American Journal of Veterinary Research* 1989;**50**:1918–22.

53 Kwochka KW. Primary keratinization disorders of dogs. In: *Current Veterinary Dermatology*. Griffin CE, Kwochka KW, MacDonald JM (eds). WB Saunders, Philadelphia 1993, pp. 176–90.

54 Barnhart KF, Credille KM, Ambrus A, Dunstan RW. A heritable keratinization defect of the superficial epidermis of Norfolk Terriers. *Journal of Comparative Pathology* 2004;**130**:246–54.

55 Paradis M, Scott DW. Hereditary primary seborrhea oleosa in Persian cats. *Feline Practice* 1990;**18**:17–20.

56 Kwochka KW. Overview of normal keratinization and cutaneous scaling disorders. In: *Current Veterinary Dermatology*. Griffin CE, Kwochka KW, MacDonald JM (eds). WB Saunders, Philadelphia 1993, pp. 167–75.

57 Miller WH, Griffin CE, Scott DW. *Muller & Kirk's Small Animal Dermatology*, 7th edn. Elsevier, St Louis 2013, pp. 630–46.

58 Miller WH, Griffin CE, Scott DW. *Muller & Kirk's Small Animal Dermatology*, 7th edn. Elsevier, St Louis 2013, p. 642.

59 Sousa CA. Sebaceous adenitis. *Veterinary Clinics of North America Small Animal Practice* 2006;**36**:243–9.

60 Zur G, Botero-Anug AM. Severe ulcerative and granulomatous pinnal lesions with granulomatous sebaceous adentitis in unrelated vizslas. *Journal of the American Animal Hospital Association* 2011;**47**:455–60.

61 Reichler IM, Hauser B, Schiller I, *et al*. Sebaceous adenitis in the Akita: clinical observations, histopathology and heredity. *Veterinary Dermatology* 2001;**12**:243–53.

62 Frazer MM, Schick AE, Lewis TP, Jazic E. Sebaceous adenitis in Havanese dogs: a retrospective study of the clinical presentation and incidence. *Veterinary Dermatology* 2011;**22**:267–74.

63 Rosser EJ, Dunstan RW, Breen PT, Johnson GR. Sebaceous adenitis with hyperkeratosis in the standard poodle: a discussion of 10 cases. *Journal of the American Animal Hospital Association* 1987;**23**:341–5.

64 Lortz J, Favrot C, Mecklenburg L, *et al*. A multicentre placebo-controlled clinical trial on the efficacy of oral ciclosporin A in the treatment of canine idiopathic sebaceous adenitis in comparison with conventional topical therapy. *Veterinary Dermatology* 2010;**21**:593–601.

65 Linek M, Boss C, Haemmerling R, Hewicker-Trautwein M, Mecklenburg L. Effects of cyclosporin A on clinical and histologic abnormalities in dogs with sebaceous adenitis. *Journal of the American Veterinary Medical Association* 2005;**226**:59–64.

66 Palmeiro BS. Cyclosporine in veterinary dermatology. *Veterinary Clinics of North America Small Animal Practice* 2013;**43**:153–71.

67 Anderson WI, Scott DW, Luther PB. Idiopathic benign lichenoid keratosis on the pinna of the ear in four dogs. *Cornell Veterinarian* 1989;**79**:179–84.

68 Mason KV, Halliwell REW, McDougal BJ. Characterization of lichenoid-psoriasiform dermatosis of Springer Spaniels. *Journal of the American Veterinary Medical Association* 1986;**189**:897–901.

69 Werner AH. Psoriasiform-lichenoid-like dermatitis in three dogs treated with microemulsified cyclosporin A. *Journal of the American Veterinary Medical Association* 2003;**223**:1013–19.

70 Theaker AJ, Rest JR. Lupoid dermatosis in a German Short Haired Pointer. *Veterinary Record* 1992;**131**:495.

71 Hensel P. Nutrition and skin diseases in veterinary medicine. *Clinics in Dermatology* 2010;**28**:686–93.

72 White SD, Bourdeau P, Rosychuck RAW, *et al*. Zinc-responsive dermatosis in dogs: 41 cases and literature review. *Veterinary Dermatology* 2001;**12**:101–9.

73 Campbell GA, Crow D. Severe zinc responsive dermatitis in a litter of pharaoh hounds. *Journal of Veterinary Diagnostic Investigation* 2010;**22**:663–6.

74 Gaschen FP, Merchant SR. Adverse food reactions in dogs and cats. *Veterinary Clinics of North America Small Animal Practice* 2011;**41**:361–79.

75 Proverbio D, Perego R, Spada E, Ferro E. Prevalence of adverse food reaction in 130 dogs in Italy with dermatological signs. *Journal of Small Animal Practice* 2010;**51**:370–4.

76 Miller WH, Griffin CE, Scott DW. *Muller & Kirk's Small Animal Dermatology*, 7th edn. Elsevier, St Louis 2013, p. 400.

77 Rosser EJ Jnr. Diagnosis of food allergy in dogs. *Journal of the American Veterinary Medical Association* 1993;**203**:259–62.

78 Harvey RG. Food allergy and dietary intolerance in dogs: a report of 25 cases. *Journal of Small Animal Practice* 1993;**34**:175–9.

79 Jackson HA, Jackson MW, Coblenz L, Hammerberg B. Evaluation of the clinical and allergen specific serum immunoglobulin E responses to oral challenge with cornstarch, corn, soy and soy hydrolysate in dogs with spontaneous food allergy. *Veterinary Dermatology* 2003;**14**:181–7.

80 Olivry T, Bizikova P. A systematic review of the evidence of reduced allergenicity and clinical benefit of food hydrolysates in dogs with cutaneous adverse food reactions. *Veterinary Dermatology* 2010;**21**:32–41.

81 Loeffler A, Soares-Magalhaes R, Bond R, Lloyd DH. A retrospective analysis of case series using home-prepared and chicken hydrolysate diets in the diagnosis of adverse food reactions in 181 pruritic dogs. *Veterinary Dermatology* 2006;**17**:273–9.

82 Bond R. Superficial veterinary mycoses. *Clinics in Dermatology* 2010;**28**:226–36.

83 Frymus T, Gruffyd-Jones T, Pennis MG, *et al*. Dermatophytosis in cats: ABCD guidelines on prevention and treatment. *Journal of Feline Medicine and Surgery* 2013;**15**:598–604.

84 Foust AL, Marsalla R, Akucewich LH, *et al*. Evaluation of persistence of terbinafine in the hair of normal cats after 14 days of daily therapy. *Veterinary Dermatology* 2007;**18**:246–51.

85 Moriello K, Coyner K, Trimmer A, Newbury S, Kunder D. Treatment of shelter cats with oral terbinafine and concurrent lime sulphur rinses. *Veterinary Dermatology* 2013;**24**:618–20.

86 White-Weithers N, Medleau L. Evaluation of topical therapies for the treatment of dermatophyte-infected hairs from dogs and cats. *Journal of the American Animal Hospital Association* 1995;**31**:250–3.

87 Bennett M, Gaskell CJ, Gaskell RM, Baxby D, Gruffyd-Jones TJ. Poxvirus infection in the domestic cat: some clinical and epidemiological investigations. *Veterinary Record* 1986;**118**:387–90.

88 Appl C, von Bomhard W, Hanczaruk M, Meyer H, Bettenay S, Mueller R. Feline cowpoxvirus infections in Germany: clinical and epidemiological aspects. *Berliner und Münchener Tierärztliche Wochenschrift* 2013;**126**:55–61.

89 Herder V, Wohlsein P, Grunwald D, *et al*. Poxvirus infection in a cat with presumptive human transmission. *Veterinary Dermatology* 2011;**22**:220–4.

90 Carletti F, Bordi L, Castilletti C, *et al*. Cat-to-human orthopoxvirus transmission, northeastern Italy. *Emerging Infectious Diseases* 2009;**15**:499–500.

91 Mösti K, Addie D, Belák S, *et al*. Cowpox virus infection in cats: ABCD guidelines on prevention and management. *Journal of Feline Medicine and Surgery* 2013;**15**:557–9.

92 Koch HG. Seasonal incidence and attachment sites of ticks (Acari: Ixodidae) on domestic dogs in southeastern Oklahoma and northwestern Arkansas, USA. *Journal of Medical Entomology* 1982;**19**:293–5.

93 Sucharit S, Rongsiyam Y. *Rhipicephalus sanguinus Latrielle*, the causative agent of foreign body in the ear: distribution on the body of the dog. *Journal of the Medical Association of Thailand* 1990;**63**:535–6.

94 Steiner FE, Pinger RR, Vann CN, *et al*. Infection and co-infection rates of *Anaplasma phagocytophilum* variants, *Babesia* spp., *Borrelia burgdorferi*, and the rickettsial endosymbiont in *Ixodes scapularis* (Acari: Ixodidae) from sites in Indiana, Maine, Pennsylvania, and Wisconsin. *Journal of Medical Entomology* 2008;**45**:289–97.

95 Cotté V, Bonnet S, Cote M, Vayssier-Taussat M. Prevalence of five pathologic agents in questing *Ixodes ricinus* ticks from western France. *Vector Borne Zoonotic Diseases* 2010;**10**:723–30.

96 Searle A, Jensen CJ, Atwell RB. Results of a trial of fipronil as an adulticide on ticks (*Ixodes holocyclus*) naturally attached to animals in the Brisbane area. *Australian Veterinary Practice* 1995;**25**:157–8.

97 Folz SD, Ash KA, Conder GA, Rector DL. Amitraz: a tick and flea repellent and tick detachment drug. *Journal of Veterinary Pharmacology and Therapeutics* 1986;**9**:150–6.

98 Hunter JS, Keister DM, Jeannin P. A comparison of the tick control efficacy of Frontline Spray against the American dog tick and brown dog tick. *Proceedings of the 41st Annual Meeting of the American Association of Veterinary Parasitologists*, Louisville, 1996, p. 51.

99 Stanneck D, Rass J, Radeloff I, *et al*. Evaluation of the long-term efficacy and safety of an imidacloprid 10%flumethrin 4.5% polymer matrix collar (Seresto®) in dogs and cats naturally infected with fleas and/or ticks in multicentre clinical field studies in Europe. *Parasites and Vectors* 2012;**5**:66–71.

100 Stanneck D, Ebbinghaus-Kintscher U, Schoenhense ET, *et al*. The synergistic action of imidacloprid and flumethrin and their release kinetics from collars applied for ectoparasite control in dogs and cats. *Parasites and Vectors* 2012;**5**:73–85.

101 Rohdich N, Roepke RK, Zschieche D. A randomized, blinded, controlled and multi-centred field study comparing the efficacy and safety of Bravecto (fluraner) against Frontline (fipronil) in flea- and tick-infested dogs. *Parasites and Vectors* 2014;**7**:83.

102 Beugnet F, Liebenberg J, Halso L. Comparative efficacy of two oral treatments for dogs containing either afoxolaner or fluraner against *Rhipicephalus sanguinus* sensu lato and *Dermacentor reticulatus*. *Veterinary Parasitology* 2015;**209**:142–5.

103 Goldschmidt MH, Shofer FS. Cutaneous papilloma. In: *Skin Tumors of the Dog and Cat*. Goldschmidt MH, Shofer FS (eds). Pergammon Press, New York 1992, pp. 11–15.

104 Bevier DE, Goldschmidt MH. Skin tumors in the dog. Part 1: Epithelial tumors and tumor-like lesions. *Compendium on Continuing Education* 1981;**3**:389–400.

105 Sula, MJ. Tumors and tumorlike lesions of dog and cat ears. *Veterinary Clinics of North America: Small Animal Practice* 2012;**42**:1161–78.

106 Moore PF, Schrenzel MD. Canine cutaneous histiocytoma represents a Langerhans cell proliferative disorder based on an immunophenotypic analysis. *Proceedings of the American College of Veterinary Pathology* 1991;**42**:119.

107 Fulmer AK, Mauldin GE. Canine histiocytic neoplasia: an overview. *Canadian Veterinary Journal* 2007;**48**:1041–50.

108 Goldschmidt MH, Shofer FS. Canine cutaneous histiocytoma. In: *Skin Tumors of the Dog and Cat*. Goldschmidt MH, Shofer FS (eds). Pergammon Press, New York 1992, pp. 222–30.

109 Welle, M, Bley CR, Howard J, Rüfenacht S. Canine mast cell tumours: a review of the pathogenesis, clinical features, pathology and treatment. *Veterinary Dermatology* 2008;**19**:321–39.

110 Macy DW. Canine and feline mast cell tumors: biologic behavior, diagnosis, and therapy. *Seminars in Veterinary Medicine and Surgery* 1986;**1**:72–83.

111 Hahn KA, Ogilvie G, Rusk T, *et al*. Masitinib is safe and effective for the treatment of canine mast cell tumours. *Journal of Veterinary Internal Medicine* 2008;**22**:1301–9.

112 Smrkovski OA, Essick L, Rohrbach BW, Legendre AM. Masitinib mesylate for metastatic and non-resectable canine cutaneous mast cell tumours. *Veterinary and Comparative Oncology* 2013;**13**:314–21.

113 Fadok VA. Granulomatous dermatitis in dogs and cats. *Seminars in Veterinary Medicine and Surgery* 1987;**2**:186–94.

114 Bloom PB. Canine and feline eosinophilic skin diseases. *Veterinary Clinics of North America Small Animal Practice* 2006;**36**:141–60.

115 Conceiçã LG, Acha LMR, Borges AS, *et al*. Epidemiology, clinical signs, histopathology and molecular characterization of canine leproid granuloma: a retrospective study of cases from Brazil. *Veterinary Dermatology* 2011;**22**:249–56.

116 Carpenter JL, Thornton GW, Moore PF, King NW. Idiopathic periadnexal multinodular granulomatous dermatitis in twenty-two dogs. *Veterinary Pathology* 1987;**24**:5–10.

117 Santoro D, Spaterna A, Mechilli L, Ciaramella P. Cutaneous sterile pyogranuloma/granuloma syndrome in a dog. *Canadian Veterinary Journal* 2008;**49**:1204–7.

118 Miller WH, Griffin CE, Scott DW. *Muller & Kirk's Small Animal Dermatology*, 7th edn. Elsevier, St Louis 2013, pp. 704–6.

119 Scott DW. Observations on canine atopy. *Journal of the American Animal Hospital Association* 1981;**17**:91–100.

120 Wilhelm S, Kovalik M, Favrot C. Breed-associated phenotypes in canine atopic dermatitis. *Veterinary Dermatology* 2011;**22**:143–9.

121 Olivry T. The American College of Veterinary Dermatology Task Force in Canine Atopic Dermatitis. *Veterinary Immunology and Immunopathology* 2001;**81**:255–69.

122 Jacobson LS. Diagnosis and medical treatment of otitis externa in the dog and cat. *Journal of the South African Veterinary Association* 2002;**73**:162–70.

123 Colombo S, Hill PB, Shaw DJ, Thoday KL. Requirement for additional treatment for dogs with atopic dermatitis undergoing allergen-specific immunotherapy. *Veterinary Record* 2007;**160**:861–4.

124 Kimura T. Contact hypersensitivity to stainless steel cages (chromium metal) in hairless descendants of Mexican hairless dogs. *Environmental Toxicology* 2007;**22**:176–84.

125 Scott DW, Miller WH, Erb HN. Feline dermatology at Cornell University: 1407 cases (1988–2003). *Journal of Feline Medicine and Surgery* 2013;**5**:307–16.

126 Olivry T, Prélaud P, Héripret D, Atlee BA. Allergic contact dermatitis in the dog. *Veterinary Clinics of North America* 1990;**20**:1443–56.

CHAPTER 5

AETIOLOGY AND PATHOGENESIS OF OTITIS EXTERNA

5.1 INTRODUCTION

Understanding the relationship between the pathological response of the otic tissues and the various primary and secondary causes of otic disease is the key to successfully managing ear disease, as is knowing when to perform surgery. This chapter discusses the causes of ear disease, from allergy to neoplasia, and describes the limited ways in which the ear canal can respond. Making a simple diagnosis of otitis externa is often not enough. Chronic, or chronic relapsing, ear disease mandates a more detailed aetiological description, if a prognosis is to be given.

5.2 THE CONCEPT OF PRIMARY AND SECONDARY FACTORS, PREDISPOSING FACTORS AND PERPETUATING CHANGE

The concept of primary and secondary factors, predisposing factors and perpetuating factors (PSPP) classification system proposed by Griffin[1] builds on August's original work[2] and further classifies the factors involved in the aetiology and pathogenesis of otitis externa. In addition to being a valuable contribution to clinicians, it also facilitates client communication by clearly illustrating the factors involved in their pet's otitis.

Primary causes can initiate otitis *de novo*, or directly cause inflammation within the external ear canal. Foreign bodies (e.g. grass awns) and otic ectoparasites (*Otodectes cynotis* and *Otobius megnini*) are the most commonly recognised primary causes in practice. Atopy and food allergy, endocrinopathies and defects in keratinisation are often overlooked. In some cases a primary cause may not be readily identified. For example, an otic foreign body may have been dislodged prior to presentation.

Secondary causes do not, in general, cause otitis externa. Rather, they contribute to the ongoing otic pathology, only being present in the abnormal ear, or at least in much larger numbers. Thus, *Malassezia pachydermatis* and *Pseudomonas aeruginosa* are examples of secondary causes.

Predisposing factors do not cause otitis externa but their presence increases the likelihood of otitis occurring. Hirsute ear canals and abnormal amounts of glandular tissue in the ear canal are predisposing problems in Cocker Spaniels; Shar Peis are notorious for having very narrow ear canals.

Perpetuating factors occur in response to primary disease and include otic epithelial hyperplasia, glandular hyperplasia, fibrosis, stenosis on the lumen, tympanic rupture and otitis media. These changes are frustrating to deal with, are often overlooked and may take very long periods of treatment to resolve. In many cases, surgical intervention is the only solution.

Occasionally a case presents that falls outside these rules[1]. For instance, a working Labrador Retriever that habitually enters water may get water in the external ear canals. Maceration of the otic epithelium follows, humidity increases and there is secondary malassezial and bacterial bloom. There is no primary cause. Rather, predisposing factors (water, pendulous pinna) and secondary causes (bacteria, yeast) result in otitis externa.

5.3 PRIMARY CAUSES OF OTITIS EXTERNA

Hypersensitivity
Atopy
Atopy is a genetically-predisposed inflammatory and pruritic allergic skin disease with characteristic clinical features. It is associated most commonly with immunoglobulin E (IgE) antibodies to environmental allergens. There is predisposition to develop IgE to environmental allergens resulting in disease[3]. Affected dogs exhibit pruritus and otitis externa[2–8].

Atopy has been stated to be the most common cause of chronic otitis externa in dogs[3].

The prevalence of otitis externa in two large studies (266 cases[7] and 600 cases[8]) was around 50–60%, and in around 3% of atopic dogs otitis externa may be the only clinical sign[4]. Note that atopic Dalmations were predisposed to otitis externa in one study[7]. Importantly, atopic otitis may not respond to immunotherapy or other non-steroidal treatments. Thus, a study of atopic dogs undergoing immunotherapy, for the management of their atopy[9], found that recurrent, or refractory, otitis externa required ongoing attention.

Cats with atopy show a variety of clinical signs, and in a retrospective study of 45 cases reported by Ravens et al.[10] they found that 16% of the cats exhibited otitis externa. Interestingly, four of the seven cats that exhibited atopic otitis had minimal evidence of associated bacterial overgrowth.

Otitis externa in atopic dogs often begins as a patchy erythematous flush, which may come and go, often unaccompanied by any associated pruritus elsewhere (Figs 5.1, 5.2). As the episodes of otitis become more frequent the signs of early otic hyperplasia may be seen (Fig. 5.3). Note that these early changes affect the base of the concave aspects of the pinnae and on the vertical portions of the external ear canals: The horizontal portion of the external ear canal is minimally affected in early cases.

Fig. 5.1 Right ear of a young German Shepherd dog with patchy erythema typical of early atopic otitis.

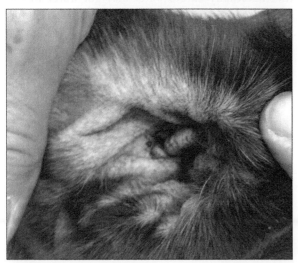

Fig. 5.2 External ear canal showing more confluent erythema.

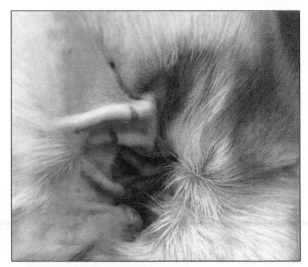

Fig. 5.3 External ear canal of a Labrador Retriever with erythema and early epithelial hyperplasia, the 'cobblestone' changes are clearly visible.

This is a valuable clue that an alert clinician should spot when taking a history.

Atopy is associated with erythema and oedema of the ear canal but there is usually little exudation[4]. Repeated episodes of inflammation result in more widespread erythema and hyperplasia (Fig. 5.4). Animals may be presented with what is described as an acute episode of otitis externa but examination will reveal signs of chronic change. These episodes of repeated inflammation result in progressive stenosis of the external ear canal (Figs 5.5A–C). The epithelial surfaces of the

Fig. 5.4 Acute erythema with much more obvious epithelial hyperplasia.

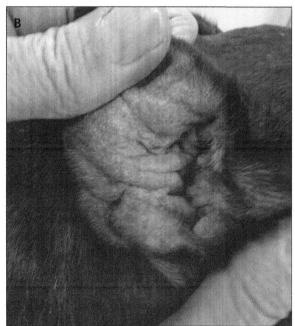

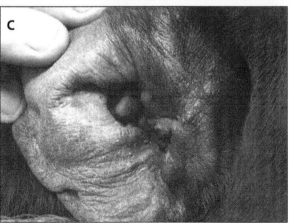

Figs 5.5A–C External ear canals of Staffordshire Bull Terriers with varying degrees of hyperplasia and stenosis. Note the erythema at the base of the pinna and at the top of the vertical ear canal; there is no ulceration and no otic discharge in A and B, atopic ear canals rarely exhibit profuse discharge. The dog in (C) had a ruptured tympanic membrane with chronic purulent discharge.

deeper portions of the vertical ear canal exhibit hyperplasia (Fig. 5.6).

The secondary hyperplastic changes and the increases in humidity associated with stenosis of the external ear canal engenders microbial proliferation. Malassezial proliferation may be particularly troublesome (Fig. 5.7) but gram-negative infection is uncommon, at least in the more temperate climates of Northern Europe. In Central and Southern parts of the USA, for example, gram-negative complication of atopic ears is not uncommon (Prof Lynette Cole, personal communication).

Although chronic otitis externa with staphylococcal/malassezial overgrowth is commonly seen in atopic dogs, note:

- Zur *et al.* found that only 25% of atopic Labrador Retrievers with otitis externa carried rod-shaped bacteria (identified cytologically)[7].
- Saridomichelakis *et al.* found that dogs with atopic ears were significantly less likely to suffer rupture of the tympanic membrane, presumably a reflection of reduced gram-negative burden[5].

The management of recurrent, or chronic recurrent, otitis externa in the atopic dog can be a source of frustration to owner and clinician. The inevitable progression of the pathological changes often results in an otitis that fails to respond to immunotherapy[9] or systemic treatments such as cyclosporine (ciclosporin) (Atopica®). One of the authors (RGH) has, however, noted that oclacitinib (Apoquel®) appears to have a very beneficial effect on otic erythema and otic puritus, although it can do little for the hyperplastic changes.

Lateral wall resection or vertical canal ablation (see Chapter 18 Surgery of the Ear) should not be carried out on an atopic ear, in an attempt to alleviate the otitis externa, without an appreciation of the underlying disease. The inflammation will continue to affect the medial and lateral walls of the residual canal and the proximal aspect of the pinna (Figs 5.8, 5.9).

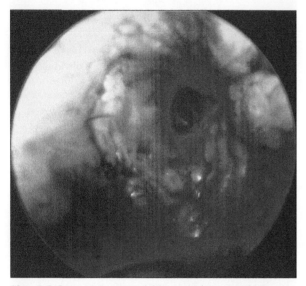

Fig. 5.6 Otoscopic view of the vertical ear canal of an atopic dog. Note the epithelial hyperplasia.

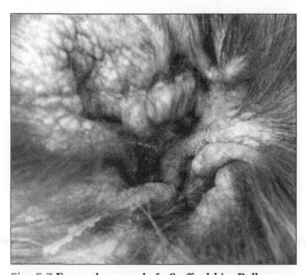

Fig. 5.7 External ear canal of a Staffordshire Bull Terrier exhibiting erythema, hyperplasia and the yellowish, adherent scale typical of malassezial infection, probably with an accompanying staphylococcal infection.

Dietary intolerance

Dietary intolerance is a dermatosis that is much less common than atopy. However, when it occurs there may be concurrent otitis externa. Although otitis externa may, on rare occasions, be the only sign of dietary intolerance[11], it is commonly associated with other signs, typically pruritus[11–13]. The otitis associated with dietary intolerance appears to be more severe that that seen in atopy and it exhibits a rapid progression. Otitis externa is also associated with dietary intolerance in cats[13,14].

Contact dermatitis

Allergic contact dermatitis is a rare dermatosis in dogs and is almost unknown in cats[15]. In dogs with very extensive or generalised allergic contact dermatitis the concave aspects of the pinnae and upper portion of the vertical ear canals may exhibit lesions (Fig. 5.10)[16].

More commonly, allergic contact dermatitis, or irritant contact dermatitis, may occur due to repeated exposure to any of the components of an otic preparation. Irritant contact dermatitis within the confines of the external ear canal may be due to exposure to one of the common vehicles, propylene glycol[17]. However, the most commonly implicated component is neomycin, although documented, published reports are very rare. A chronic, pruritic, erythematous otitis externa would be anticipated (Fig. 5.11). The classic history is a cat or dog with chronic, relapsing otitis externa, which had

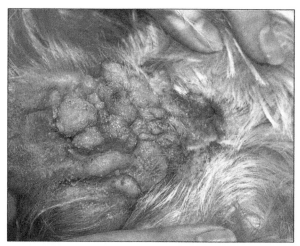

Fig. 5.8 Failed lateral wall resection in a West Highland White Terrier. Dramatic epithelial hyperplasia affecting the residual medial wall of the external ear canal.

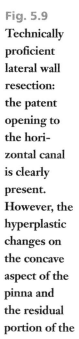
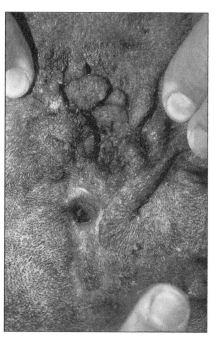

Fig. 5.9 Technically proficient lateral wall resection: the patent opening to the horizontal canal is clearly present. However, the hyperplastic changes on the concave aspect of the pinna and the residual portion of the external ear canal are causing continual problems.

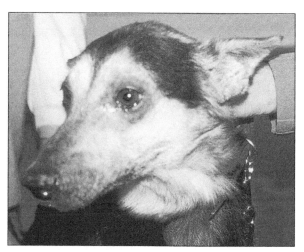

Fig. 5.10 Allergic contact dermatitis. Note the hyperpigmentation affecting the perioral, periocular regions in addition to the concave aspect of the pinna and the upper portion of the vertical ear canal.

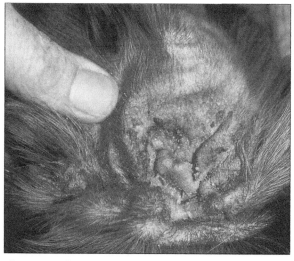

Fig. 5.11 Allergic contact dermatitis to a component in a commercial polypharmaceutical preparation. Oleaginous, erythematous, hyperplastic otitis externa.

previously responded well to a certain medication only to deteriorate when exposed to the medicant again.

Autoimmune and immune-mediated diseases

Pemphigus foliaceus and the uncommon, to rare, diseases such as pemphigus erythematosus and systemic lupus erythematosus, commonly affect the pinna, with vesicles, crusting and patchy alopecia affecting some 50-80% of dogs (Fig. 5.12). Vesicular lesions may also be seen at the entrance to the vertical ear canal though[18-24]. Cytological examination of vesicular contents may reveal acanthocytes and neutrophils (Fig. 5.13), a combination suggestive of pemphigus foliaceous. Note: these immune-mediated diseases very rarely cause otitis externa.

Ectoparasitic diseases

Otodectes cynotis

O. cynotis is a large (0.3–0.4 mm) mite (Fig. 5.14) that lives predominantly in the external ear canal of dogs and cats, and perhaps occasionally on the adjacent skin of the head[25]. The mite does not burrow but lives on the skin surface where it feeds on tissue fluid and debris[26]. It has been suggested that otodectic mites can survive within the household, off the host, for weeks if not months[27].

The physical presence of the mite induces a mechanical irritation that accounts for some of the pruritus experienced by infected animals. However, the saliva is both irritant and immunogenic and in the cat the mite stimulates an IgE-like antibody[26], suggesting that hypersensitivity contributes to the pruritus. The mite produces an antibody that cross-reacts with the house dust mite *Dermatophagoides farinae*[27] and may thus play a part in atopic pruritus. Ear mite antigen may play a part in inducing aural haematoma in both the dog and the cat, for which an autoimmune aetiology has been suggested by some authors [28,29].

Fig. 5.12 Primary lesions of pemphigus foliaceus at the entrance to the external ear canal and on the concave aspect of the pinna.

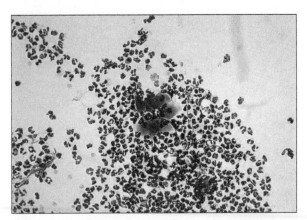

Fig. 5.13 Cytological examination of samples taken from the primary lesions of pemphigus foliaceus reveals beautifully 'clean' neutrophils and 'rounded up' keratinocytes, called acanthocytes.

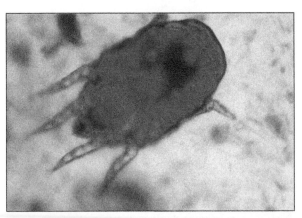

Fig. 5.14 Photomicrograph of an adult *Otodectes cynotis* mite.

Zoonotic lesions may occur on in-contact human members of the household[25]. Vesicles, wheals, erythematous papules and excoriations on the arms and torso have been reported[30].

Lifecycle, transmission, and prevalence

Females lay eggs (Fig. 5.15) and cement them to the epidermal surface. They hatch to yield six-legged larvae that undergo two moults through eight-legged protonymphs and deutonymphs. The emerging deutonymph is approached by, and attached to, an adult male mite (Fig. 5.16) and, if it is female, copulation occurs. Although the life cycle of 3 weeks is confined to the host, it has been suggested that the mite can survive in the environment for long periods[27]. Nevertheless, contact with an infected host is still believed to be the main route of transmission[25].

The prevalence of *O. cynotis* in dogs' ears was assessed as 29.1% in one study of 700 ears, with a significant predisposition in dogs with pendulous and semi-erect pinnae as compared to erect pinnae[31]. This study also reported that there was a highly significant correlation between the presence of mites and otitis externa. Asymptomatic carriage of *Otodectes cynotis* can occur in dogs. For example, in a case series detailing *Otodectes cynotis* in 700 ears, lesions and pruritus were absent in 114. Fewer cases were reported in the summer months. One study suggested a seasonal incidence for the disease[32]; however, a very large study could find no evidence of a seasonal incidence[33]. Note that many of these studies on the prevalence of *O. cynotis* were performed before the widespread use of topical and systemic ectoparasiticides.

A recent study of 187 stray cats in Italy found otitis externa in a rather surprising 55.1% of ears, and otodectic mites accounted for almost all of these[34]. Thus, in an untreated population the mite is still very relevant.

Clinical features

Young dogs appear to be more commonly infected than older animals[33]. This probably reflects the fact that infected dogs are easily diagnosed, effectively treated, and not reinfected. The average number of mites per dogs was only 5.633. *O. cynotis* is typically associated with a pruritic otitis externa[32,33]. However, Scott considered that in the cat, three syndromes (otitis externa, ectopic infection and asymptomatic carriage) might be associated with infection with the mite[32].

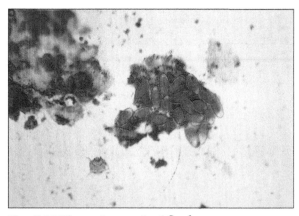

Fig. 5.15 Photomicrograph of *Otodectes* spp. ova collected in cerumen.

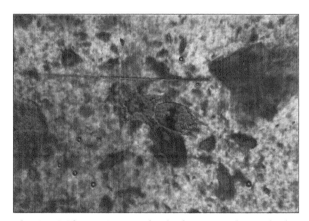

Fig. 5.16 Photomicrograph of a deutonymph attached to an adult otodectic mite.

Very low numbers of mites, even as low as three, may be sufficient to induce clinical signs[31]. This, together with the mite's ability to inhabit the entire external ear canal, can make definitive diagnosis difficult and might make a rule-out of otodectic acariasis, other than by trial therapy, problematic.

Otitis externa

The classic feature of otitis externa due to ear mite infection is moderate to severe otic pruritus. In addition, the external ear canal becomes filled with a crumbly black/brown discharge (Figs 5.17–5.19). Most affected dogs exhibit chronic otic pruritus but Frost reported four dogs out of 200 that had asymptomatic infection[31]. Puppies are most likely to be infected from dams, but in

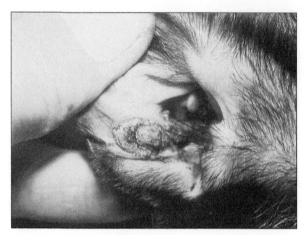

Fig. 5.17 Adult cat with *Otodectes cynotis* infestation. Note the typical dark brown colour and the dry nature of the cerumen. Note also the lack of self-trauma in this asymptomatic case.

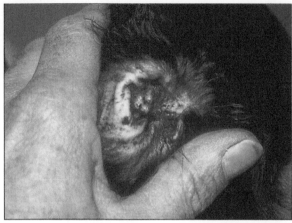

Fig. 5.18 Otodectic mange in a pup. Note the presence of the dryish cerumen and evidence of some self-trauma.

adult dogs the cat is a common cause of contagion[1], particularly since the cat may well be asymptomatic[31,35].

In cats the pruritus associated with infection may be associated with moderate to severe self-trauma to the head (Figs 5.20, 5.21). Scott suggested an age predilection in young cats[32]. In cases of chronic infection there may be hyperplastic changes in the lining of the external ear canal and a predisposition to secondary infection.

Ectopic infection
In some cases, perhaps in cats more than in dogs, the mite causes clinical signs distant from the ear. Scott considered that this might be a consequence of cats sleeping in a curled position so that the ear is apposed to the tail-base[32]. Two syndromes may be associated with ectopic infection:

- Crusted papules, i.e. miliary dermatitis.
- Patchy alopecia.

Asymptomatic infection
This may be a feature of older cats where very high numbers of mites may be found with apparently no associated clinical signs[35]. The presence of asymptomatic carriage in dogs has not been considered a major problem in veterinary dermatological texts, but in light of the discussion above it should be borne in mind, particularly when treatment protocols are discussed.

Fig. 5.19 Cerumen from an ear infested with *Otodectes cynotis*. Note the crumbly nature of the cerumen and the dark brown colour.

Diagnosis
The mite is relatively large and may be easily seen in the external ear canal with the aid of an otoscope (Fig. 5.22). Direct observation may not always result in a diagnosis:

- The degree of discharge may make direct observation difficult.
- There may be so few mites that direct observation is not possible.

In these situations, microscopic examination of discharge may be necessary. Gentle maceration of collected samples in mineral oil will aid diagnosis, and

Fig. 5.20 Small area of crust and self-trauma in the entrance to the external ear canal of a cat with otodectic otitis.

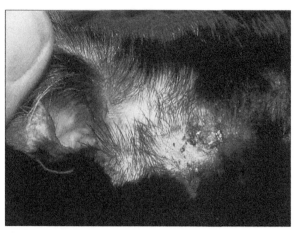

Fig. 5.21 Area of erythema and self-trauma associated with otodectic otitis on the lateral aspect of the head of a cat.

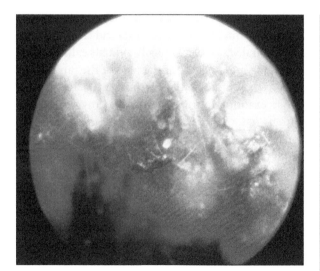

Fig. 5.22 Otoscopic view of ceruminous debris and otodectic mites in a feline external ear canal. The mites appear small in this unmagnified view.

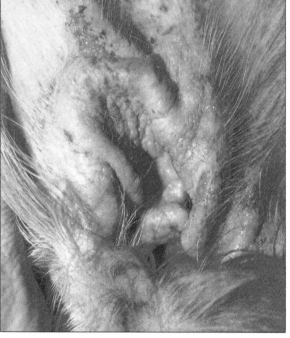

Fig. 5.23 Erythematous, ceruminous otitis externa in a Cavalier King Charles Spaniel with otodemodicosis.

microscopic examination under low power should show evidence of infestation.

Other ectoparasitic causes of otitis externa

Demodex canis, D. felis and *D. gatoi*

D. canis has been reported as a rare cause of otitis externa in dogs. It may occur as part of a generalised condition, in isolation, or as a long-term complication of juvenile-onset generalised demodicosis that has apparently responded to treatment[36]. A history of demodicosis should alert the clinician to the possibility of otodemodicosis but cases arising *de novo* should not be discounted. Typically, otodemodicosis is associated with a ceruminous otitis externa (Fig. 5.23).

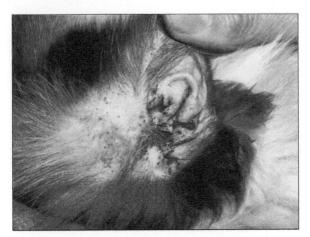

Fig. 5.24 Feline demodicosis causing erythematous dermatitis adjacent to the entrance to the external ear canal.

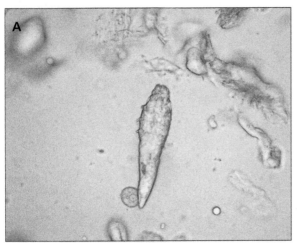

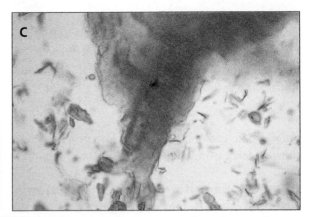

Figs 5.25A–C Photomicrographs of a juvenile demodectic mite (A) and an egg (B) in cerumen from the ear of a dog with otodemodectic mange. Photomicrograph of an adult *Demodex canis* mite in cerumen from the external ear canal of a dog with otodemodectic mange (C) .

In cats, demodicosis is more usually associated with erythema (Fig. 5.24) and crusting on the pinnae and head, rather than otitis externa. However, *D. gatoi* may be associated with a ceruminous otitis externa in cats.

Diagnosis is based on recovery of demodecid mites (Figs 5.25A–C) in skin scrapes and, in this case, on cotton swabs from the external ear canal. Punch biopsy samples would also give appropriate material for a diagnosis.

Harvest mites

Harvest mites such as *Neotrombicula autumnalis* (Fig. 5.26) and *Euotrombicula alfredugesi* are occasional causes of otitis externa in both dogs and cats. The larvae are parasitic and require a mammalian host but they are not species specific. Larvae hatch in rapid succession and usually tens or hundreds are involved in the parasitic attack. Typically, they cause a pruritic crusting dermatitis on the ventrum and face and in the interdigital areas. Occasionally animals exhibit larval clustering and crusting at the base of the pinnae[37] or within the confines of the proximal external ear canal. Close examination usually reveals tiny orange, or orange-red, clusters of larvae. The parasite is a seasonal threat to the hunting or roaming dog and cat and it is more common on ground composed of well-drained, chalky soil.

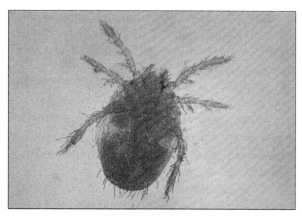

Fig. 5.26 Larva of *Neotrombicula autumnalis*. **Note the red colour and six legs.**

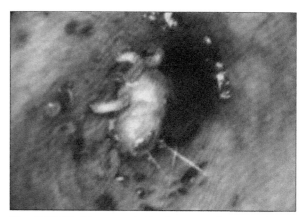

Fig. 5.27 *Otobius megnini* in the external ear canal of a dog. (Courtesy of Dr Louis Gotthelf, Mongomery, USA.)

Ticks

The spinous ear tick, *O. megnini* (Fig. 5.27), is most frequently found in the southern and south-western regions of the USA. However, the increased mobility of owners and their pets means that the tick may be found in almost any region of the USA[38,39]. The larvae (six legs, yellow-pink colour) and adults (eight legs, blue-grey colour) are parasitic and infest the external ear canal of both dogs and cats to the extent that in some cases the external ear canal is entirely filled with the parasites. Acute otitis externa results. Ixodic, hard ticks such as *Demacentor* spp., and the rabbit 'stick tight flea' (*Spillopsyllus cuniculi*) are usually found on the pinnae and head, rather than within the external ear canal.

Endocrinopathy

Endocrinopathies are often cited as underlying causes of chronic ceruminous otitis externa. However, neither hypothyroidism nor hyperadrenocorticism are commonly associated with otitis externa[40–42], allergy is a much more common cause. Diabetes mellitus has been associated with otitis externa[43]. Zur *et al.* found that rod-shaped bacteria (identified cytologically) were correlated with otitis externa in dogs with endocrinopathy[7]. Gonadal hormone changes (e.g. Sertoli cell tumours), in contrast, may have a profound effect on cutaneous glandular function and may therefore be associated with ceruminous otitis (Fig. 5.28) in association with other signs[44].

Fig. 5.28 Otitis externa associated with a Sertoli cell tumour. **Note the ceruminous discharge adhering to the concave aspect of the pinna.**

Disorders of keratinisation

Disorders of keratinsation affect mainly dogs[45], but cats, particularly Persian cats, may be affected[46]. The clinical signs may reflect various aetiologies, such as seborrhoeic diseases, sebaceous adenitis, vitamin A-responsive dermatosis and zinc-responsive dermatosis. Defects in keratinisation may be generalised, localised or multifocal and are classified as either primary or secondary[45]. By far the most common causes of the scaling and crusting dermatoses are secondary causes such as ectoparasites, infectious agents, hypersensitivities and endocrinopathies[45].

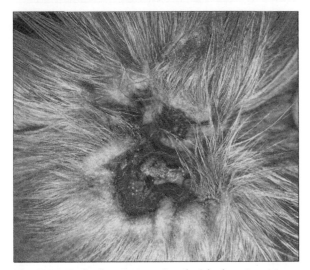

Fig. 5.29 Early changes associated with chronic otitis externa in a Cocker Spaniel. Ceruminous otitis with early hyperplasia.

Otitis externa may be anticipated with these disorders only if the underlying disease is itself a cause of ear disease. For example, superficial pyoderma, dermatophytosis or demodicosis are common causes of crust and scale on the trunk or limbs, but they rarely cause otitis externa. In contrast, the inflammation caused by atopy is generalised, as are the aberrations in cutaneous homeostasis that accompany an endocrinopathy. Thus, these diseases are often associated with otitis externa.

Similarly, some, but not all, of the primary defects in keratinisation (idiopathic seborrhoea) may be associated with otitis externa. Examples include idiopathic seborrhoea in Cocker Spaniels[47] and epidermal dysplasia in West Highland White Terriers[48]. The relationship between epidermal dysplasia and the yeast *M. pachydermatis* in West Highland White Terriers is complex and poorly understood[48,49]. Basset Hounds also suffer from a dermatosis that used to be classified as idiopathic seborrhoea. Many of these dogs suffer from *M. pachydermatis* dermatitis and they show a spectacular response to antimalassezial therapy[50]. Whatever the exact nature of these two disorders, or their relationship, they are both associated with severe otitis externa[48,50].

The otitis externa associated with primary seborrhoea in Cocker Spaniels is initially ceruminous (Fig. 5.29), but epidermal hyperplasia (Figs 5.30A, B) soon follows. The otic discharge is typically thick and oleaginous (Fig. 5.31). Otoscopic examination of early cases reveals hyperplasia (Fig. 5.32). The external ear canal is often of a moister appearance than in the appreci-

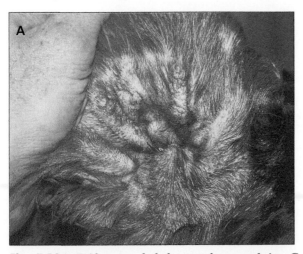

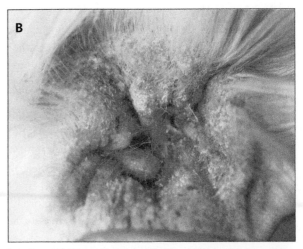

Figs 5.30A, B Almost occluded external ear canals in a Cocker Spaniel (A) and a West Highland White Terrier (B), as a consequence of chronic otitis externa.

ably atopic ear canal (compare with Fig. 5.5A), and has a tendency to bleed easily.

Cytological examination of smears prepared from swabs rolled onto glass slides and stained with, for example, Dif Quik®, will reveal plenty of cerumen and cellular debris but only a few inflammatory cells (Fig. 5.33). Indeed, subsequent bacterial culture from these ears may fail to record any significant bacterial growth at all, illustrating the value of otic cytology. However, in contrast to the otitis associated with atopy, the disease in Cocker Spaniels is often complicated by gram-negative infection. These animals usually exhibit a painful otitis and examination will reveal erosions and ulceration (Fig. 5.34). The accompanying otic discharge is usually thinner, with a more aqueous cerumen, than is found in 'normal' Cocker Spaniels' ears. Furthermore, they are predisposed to rupture of the tympanic membrane, associated with gram-negative infection[5].

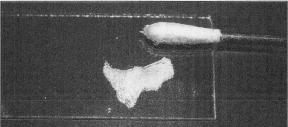

Fig. 5.31 Unstained cytology smear. Note the thick, oleaginous nature of the cerumen.

Fig. 5.32 Otoscopic picture of the external ear canal of a Cocker Spaniel with early changes associated with chronic otitis externa. There is erythema and the ear canal has a moister appearance than the atopic ear (compare with 5.5A). Note that the ear canal has been plucked to facilitate cleaning.

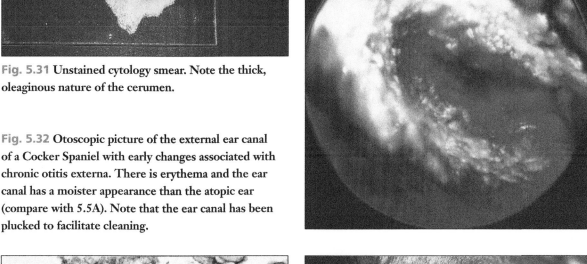

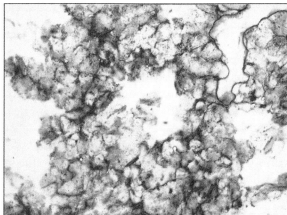

Fig. 5.33 Photomicrograph of a stained cytologic smear from an ear canal of a Cocker Spaniel. Note the increased numbers of squames, the amount of cerumen, the lack of micro-organisms and the absence of inflammatory cells.

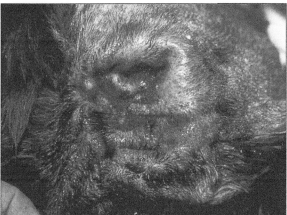

Fig. 5.34 Ulcerated external ear canal due to gram-negative bacterial infection in a Cocker Spaniel.

Hereditary defects in keratinisation have been reported in cats[46]. Persian cats are most commonly affected (Fig. 5.35), although the condition may occur in other breeds. Affected animals show signs from a very early age and either sex may be affected. The ears develop a ceruminous otitis externa and greasy scale

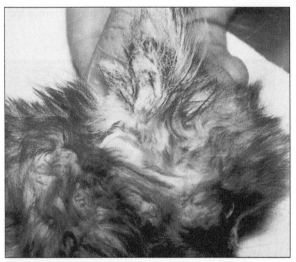

Fig. 5.35 Persian cat with an hereditary defect in keratinisation. There is a greasy otitis externa and greasy tags are apparent on the adjacent skin and hair of the pinna. (Courtesy of Dr Manon Paradis, Faculté de Médicine Vétérinaire, University of Montreal, Canada.)

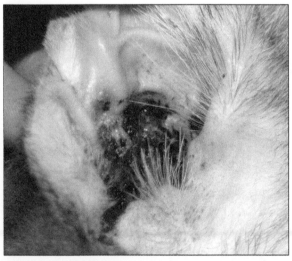

Fig. 5.36 Proliferative otitis in a kitten. (Courtesy of Dr Elizabeth Maudlin, School of Veterinary Medicine, Philadelphia, USA.)

may accumulate on the pinnae. The entire trunk is also affected with scale, grease and malodour. Because of the severity of the disease, many patients are euthanised at an early age as there is no effective treatment.

Proliferative otitis

This is a rare condition, with an almost pathognomonic presentation, that can affect both kittens and adult cats[51]. The condition typically presents as a bilateral, occasionally malodorous, chronic otitis externa. Large, dark red, brownish, almost vegetative, proliferative plaques with necrotic surfaces can be seen at the base of the pinnae and the upper portions of the vertical ear canal (Fig. 5.36). The tissue and exudates may occlude the ear canal. Removal of the rather loosely adherent overlying proliferative tissue results in ulceration and erosions to the underlying tissue. Histopathological examination reveals prominent hyperkeratosis and apoptopic keratinocytes, presumably reflecting an immune-mediated process[52].

Topical tacrolimus ointment twice daily[52] is very effective: some cases appear to go into full time remission, other require maintenance therapy.

Foreign bodies

There is no sex predisposition to otic foreign body penetration but young dogs are predisposed to grass awn penetration[53,54]. In general, all breeds of Spaniels and Golden Retrievers are most commonly affected, while German Shepherd Dogs, Miniature Poodles and Dachshunds are under-represented[54].

The most common foreign body found in the external ear canal of dogs and cats is the grass awn[1]. In the USA the most common species of plant awn is *Hordeum jubatum*, although other members of the genus, such as *H. murinum*, *H. silvestre*, and genera such as Stipa, Setaria, Bromus and Avena, may be involved in other areas of the world[54]. All have a similar shape (Fig. 5.37) with wiry barbs that prevent retrograde movement. Once in the ear canal they can only move forwards (Fig. 5.38).

Hair shafts, particularly if they contact the tympanum, may also act as foreign bodies (Fig. 5.39). In one series of 120 cases of otitis externa, 12.6% were considered to result from matted hair and cerumen in the external ear canal[55]. Other foreign bodies that may enter, or be put into, the external ear canal include other pieces of vegetation and children's toys. Aggregations of otic, usually proprietary, non-veterinary powders and ointments with cerumen may also induce foreign body reactions.

Occasionally a clinician is presented with a young dog exhibiting acute otic discomfort. Examination reveals no evidence of ear disease and no foreign bodies are found. It may be that the dog's violent shaking has dislodged the accumulation of cerumen or debris that was the cause of the irritation.

Foreign body penetration into the ear canal is usually accompanied by acute pain. The dog or cat shakes its head and may attempt to remove the object with a foot. As the object moves down it may induce hyperaemia and ulceration followed by the generation of an otic discharge and secondary bacterial proliferation (Fig. 5.40). If the foreign body penetrates the epithelial lining of

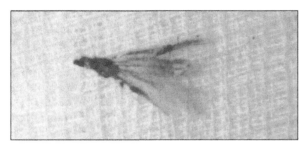

Fig. 5.37 Typical shape of a grass awn; this was removed from the external ear canal of a dog.

Fig. 5.38 Otoscopic picture of grass awn lying adjacent to the tympanic membrane. In this case the grass awn had not punctured the tympanum; however, note the area of erythema and erosion on the tympanic membrane.

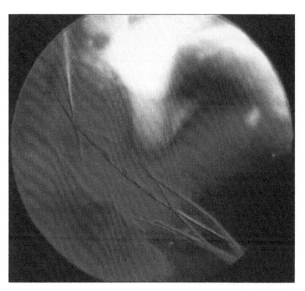

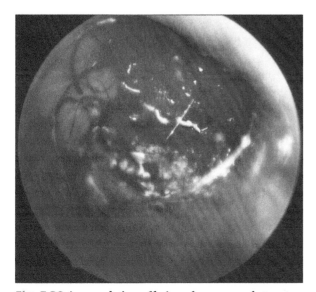

Fig. 5.39 Accumulation of hair and cerumen obstructing the horizontal ear canal at the level of the tympanum.

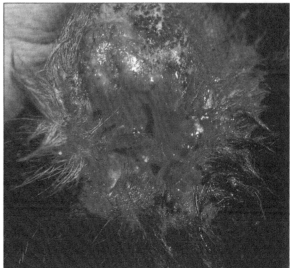

Fig. 5.40 Acute, erythematous, ulcerated otitis externa associated with the penetration of a grass awn into the external ear canal.

the external ear canal, it may become embedded in a pyogranuloma[56]. In one study nearly 20% of cases of otic grass awn penetration were associated with rupture of the tympanum (Figs 5.41A, B), suggesting that otitis media should be considered in long-standing cases, even where the tympanum is intact[54]. Support for this comes from the study by Saridomichelakis *et al.* who found that a grass seed within the external ear canal was positively correlated with rupture of the tympanic membrane[5].

The most common bacteria associated with grass awns are streptococci, although *Staphylococcus* spp., *Pasteurella* spp. and *Actinomycetes* spp. may also be cultured[54].

Obstructive and neoplastic disease

Obstructive disease of the external canal is usually unilateral. It is relatively unusual, in the author's experience, for owners to recognise the presence of a lesion. Most often, the dog or cat is presented for one, or several, associated clinical signs, such as otic discharge and malodour, pruritus, pain or head tilt[57]. Local metastasis may result in vestibular signs.

EAR CANAL NEOPLASIA

Neoplasia of the external ear canal is uncommon[57]. In general, otic neoplasia in cats tends to be malignant[57–60] and they are likely to be found either in the vertical or the horizontal canal with equal frequency[58,59]. Otic discharge, pruritus and pain are common, whereas neurological signs are rare[57,58]. Canine ear canal neoplasia is more likely to be benign than is the case in the cat, but distribution and clinical signs are similar[57–59]. Most benign tumours do not affect the bullae[57–59].

Although the malignant tumours, particularly in the cat, tend to invade locally, it appears that distant metastasis is the exception rather than the rule[57,60]. When neurological signs accompany otic neoplasia, a generally poor prognosis is predicted since this usually indicates middle ear involvement and squamous cell carcinoma, altogether a more malignant tumour than ceruminous gland adenocarcinoma of the external ear canal[58]. Indeed, when squamous cell carcinoma is found in the external ear canal it usually has its origins in the middle ear.

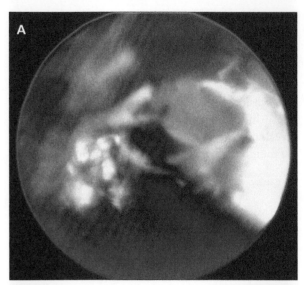

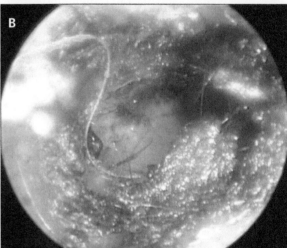

Fig. 5.41A, B A: Grass awn penetration of the tympanum. Note the small piece of vegetation still visible on the left, adjacent to the area of haemorrhage; B: grass awn penetration, picture taken with video-otoscope; note the increase in clarity and depth of field compared to the otoscopic picture in (A). (5.41B courtesy of Dr Susan Paterson.)

Papillomas, basal cell tumours and ceruminous gland adenomas are the most commonly found benign tumours in dogs, while in cats ceruminous gland adenomas are most common[57–59]. Carcinomas, adenocarcinomas and squamous cell carcinomas are the most common malignant tumours in both dogs and cats.

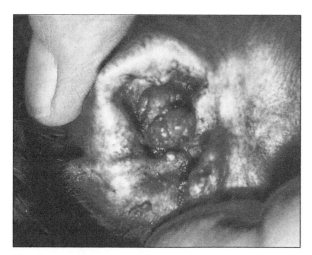

Fig. 5.42 Obstructive otitis secondary to ceruminous gland neoplasia in a cat.

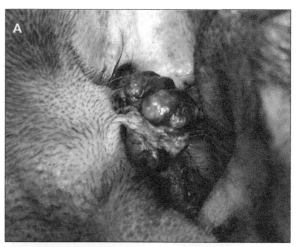

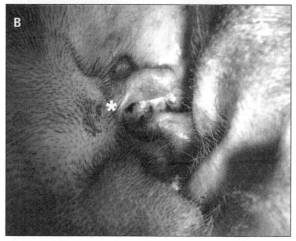

The appearance of these neoplasms is usually that of a raised, frequently ulcerated, mass that may occlude the lumen[59–62].

Ceruminous gland adenoma and ceruminous gland hyperplasia

Ceruminous gland neoplasia is the most common cause of obstructive disease of the external ear canal in dogs and cats[57]. Benign ceruminous gland neoplasia tends to present with signs of obstructive otitis externa (Fig. 5.42): pruritus, head shaking, malodour, otorrhoea, and occasional haemorrhage[60,61]. Ceruminous gland adenomas are most commonly seen in middle-aged to elderly animals[57,61,62]. These benign tumours tend to be raised and occasionally pedunculated (Figs 5.43A, B). They may occlude the external ear canal[62] (Fig. 5.45).

Figs 5.43A, B Pedunculated mass (polypoid fibroma) at the entrance to the vertical ear canal (A), and the post-surgical picture (B), depicting the tiny area of attachment (asterisk).

Note that ceruminous cystomatotis is a non-neoplastic, hyperplastic change characterised by, usually, multiple, often clustered, pigmented papules and nodules within the external ear canal[57,62]. The condition is most often seen in middle aged to older cats[57], and is often, although not always, associated with chronic otitis externa (Fig. 5.44). Ablation of the hyperplastic tissue may also be achieved by surgery, or by tissue ablation

Fig. 5.44 Ceruminous gland hyperplasia in the external ear canal of a cat.

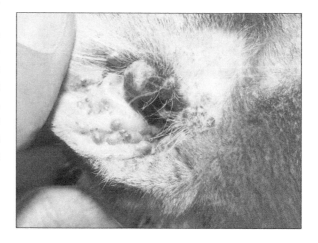

using carbon dioxide lasers[62,63], which produces good cosmetic results.

Ceruminous gland adenocarcinoma

Malignant ceruminous gland tumours tend to be ulcerative and infiltrating rather than occlusive[57,64,65]. Most cases tend to occur in old animals – cats: mean age 12 years; dogs: mean age 9 years[64,65]. Otoscopically they are pinkish in colour (Fig. 5.46), ulcerated and friable in nature. Most dogs and cats exhibit an otic discharge that is commonly malodorous, purulent (Fig. 5.47), and blood stained. Otic pruritus and ipsilateral mandibular lymphadenopathy is also commonly noted. Bulla involvement was demonstrated in nearly half of the cats

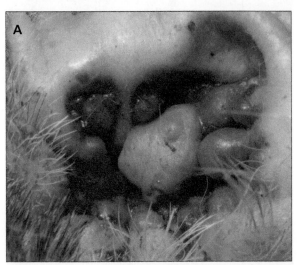
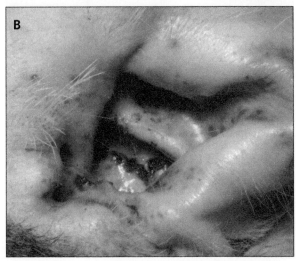

Figs 5.45A, B Ceruminous gland hyperplasia in the external ear canal of a cat, before (A) and after (B) laser ablation. (Courtesy of Dr Lorraine Corriveau, Purdue University Veterinary Teaching Hospital, West Lafayette, USA.)

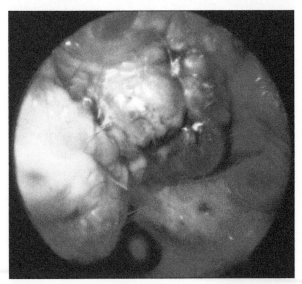
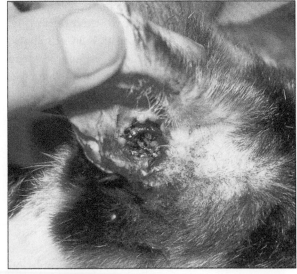

Fig. 5.46 Pinkish, nodular appearance of a ceruminous gland adenocarcinoma in the external ear canal of a cat.

Fig. 5.47 Malodorous, haemorrhagic obstructive otitis secondary to ceruminous gland adenocarcinoma in a cat. The ulcerated mass of tumour may be seen protruding into the lumen at the entrance to the external ear canal.

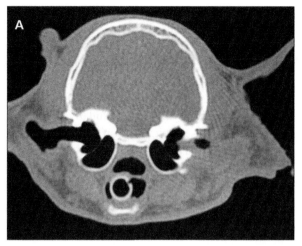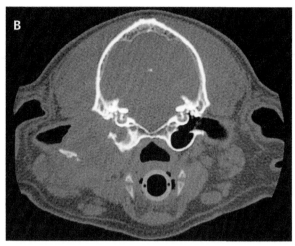

Figs 5.48A, B A: CT scan of a 13-year-old cat. There is increased soft tissue opacity in the right external ear canal extending up to, and perhaps across, the tympanic membrane. This is a ceruminous gland adenocarcinoma. Note that the right bulla appears as normal as the left. The bony septum dividing the feline bulla into lateral and medial compartments is clearly visible with this imaging modality; B: CT scan of a 10-year-old Cocker Spaniel. The left-hand side exhibits an irregular, imprecise outline to the bulla, increased density within the bulla, loss of air within the external ear canal, mineralisation of soft tissue in the external ear canal, and a homogenous soft tissue mass on the ventral aspect of the skull. This is a ceruminous gland adenocarcinoma.

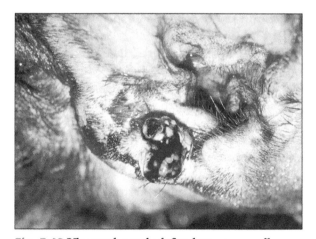

Fig. 5.49 Ulcerated, poorly defined squamous cell carcinoma at the entrance to the external ear canal of a dog.

survival time than simple lateral wall resection. If the tumour has extended through the external ear canal into surrounding soft tissue, adjunctive radiotherapy is indicated[57,64–66].

Squamous cell carcinoma

In cats squamous cell carcinoma appears to be as common as ceruminous gland adenocarcinoma[57,58,67]. The tumours are proliferative and ulcerated (Fig. 5.49) and they have a tendency to grow rapidly. Most ear canal tumours with otoscopically visible evidence of extensive spread and histopathological evidence of local infiltration are squamous cell carcinomas[57,58,61]. Radical resection is necessary and presurgical biopsy may be advantageous.

and dogs in recent studies[64,65]. This tendency to involve the bulla (Figs 5.48A, B) is reflected in the response to surgery; radical, total ear canal ablation (TECA) and bulla osteotomy results in a longer disease-free interval, a lower recurrence rate and longer postoperative

Remember, not all masses within the external ear canal are neoplastic[57,59]. Granulomas[57,59], fungal infections (Fig. 5.50) and proliferative eosinophilic dermatitis have been reported[68]. Biopsy is well worth considering before opting for radical surgery.

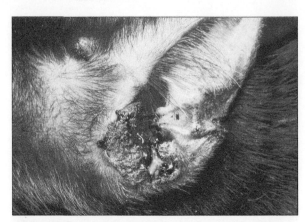

Fig. 5.50 Ulcerated granulomatous lesions due to cryptococcosis should be considered in the differential diagnosis of conchal neoplasia.

Neoplasia within the middle ear

Not all masses within the middle ear are malignant. Polyps, cholesterol granulomas and cholesteatoma[57,62] are examples of benign disease (see Chapter 6 Diseases of the Middle Ear, Sections 6.2, 6.3 for more discussion on these diseases). Malignant neoplasia may arise spontaneously within the middle ear, but more likely it arises as an extension of, or local metastasis from, neoplasia in the external ear canal[57]. Thus, squamous cell carcinoma and adenocarcinoma are most likely to be encountered although fibrosarcoma and lymphoma have been reported[57,64].

5.4 SECONDARY CAUSES OF OTITIS EXTERNA

Bacterial infection
Microbial changes associated with otitis externa
The overall changes in the microbial flora that are associated with otitis externa are qualitative and quantitative:

- The number of bacteria increases and the proportion of the various species changes.
- The incidence of recovery of staphylococci in general, and of coagulase-positive staphylococci in particular, increases[69–71] and more particularly, in some ears the incidence of recovery of gram-negative bacteria such *Pseudomonas* spp. and *Proteus* spp. increases[72–77].
- The frequency of recovery, and the number of *M. pachydermatis* recovered, increases significantly in

otitic external ear canals compared to normal ears, in both dogs and cats[74,78–81].

The detailed pattern of this microbial change depends on the environmental changes within the external ear canal. Thus, in an atopic dog, there is an inflamed, erythematous otitis, which is not exudative, at least at first. Such changes mimic those on the glabrous skin of an atopic dog and might be sufficiently aberrant to favour colonisation by *Staphylococcus pseudintermedius* and *M. pachydermatis*, but not *P. aeruginosa*. Support for this contention comes from a Brazilian study in which the microbial flora of dogs with only bilateral otitis externa was studied[77]. The most frequent organisms isolated were *S. pseudintermedius* and *M. pachydermatis* – exactly what would be expected if the population contained a large number of allergic dogs, who typically manifest bilateral otitis externa. Similarly, Zur *et al.* found that only 25% of Labrador Retrievers with otitis externa had rod-shaped bacteria in their ears[7]. Finally, Saridomichelakis *et al.* found that rupture of the tympanic membrane was statistically under-represented in atopic dogs with otitis externa[5].

In Cocker Spaniels however, there is moister, more ceruminous and rapidly progressive, otitis and gram-negative bacteria rapidly colonise, sometimes with *M. pachydermatis* but less often with staphylococci[82]. Thus, Zur *et al.* found that 9/10 Cocker Spaniels with otitis externa carried rod-shaped bacteria (identified cytologically)[7] and Saridomchelakis *et al.* found that otitis externa in Cocker Spaniels was positively correlated with rod-shaped bacteria and rupture of the tympanic membrane[5].

The bacteria and yeast associated with acute otitis externa, such as grass seed penetration, or recent exposure to water or the accumulation of cerumen and hair are, initially, likely to be those introduced by the foreign bodies or members of the normal otic flora. That is, until the dog tries to alleviate the discomfort with the ipsilateral hind foot and introduces enteric bacteria, at least at first, a mixed flora is likely.

There have been a few recent reports of a novel, gram-positive rod-shaped bacteria, such as *Corynebacterium* spp. and *Arcanobacterium canis*, associated with otitis externa, in five dogs[83–85]. The significance of this isolate is, as yet, uncertain.

The most common bacteria recovered from otitis externa in cats' ears were coagulase-positive staphylococci (54.8%). Gram-negative bacteria such as *Pseudomonas* spp. and *Proteus* spp. were only rarely recovered from feline otitis externa[71–73].

5.5 PERPETUATING FACTORS

Response to insult and injury

The epidermis of the external ear canal reacts to inflammation by increasing its rate of turnover and increasing in thickness, i.e. it becomes hyperplastic[60,84–86]. There may be surface erosions and ulceration, particularly with gram-negative infections. The general pattern for the pathological progression accompanying chronic, or chronic recurrent otitis externa is as follows:

- The dermis becomes infiltrated with inflammatory cells and fibrosis starts to occur[85,87,88].
- However, in Cocker Spaniels with chronic otitis externa, fibrosis within the dermis appears less common, and there appears to be a breed predisposition to hyperplasia of the ceruminous glands, often severe in nature, with marked, almost cystic dilatation being noted[82,88,89]. Morphometric analysis reveals no significant changes in sebaceous gland size or activity[87]. Papillary proliferation of ceruminous glands and ducts may obliterate the lumen of the external ear canal in some cases[89]. In very chronic cases, ossification of the tissues may take place.

Similar changes, hyperplasia and fibrosis, take place in the feline ear canal, although the papillary changes in the ceruminous glands may be sufficiently florid that discrete polyps occur[60].

The consequence of these changes is a reduction in luminal cross-section, a result of increasing soft tissue within the bounds of the containing cartilage[86]. The change in nature of the cerumen, the reduction in luminal diameter and the moisture and warmth that accompany active inflammation contribute to an increase in local humidity[90,91]. These changes in the otic environment result in surface maceration and the creation of a milieu favourable to microbial multiplication, itself a potent inducer of inflammation.

It is not clear at which stage these changes become irreversible, and the progressive changes in glandular architecture correlate with the progression of the otitis externa[89]. Certainly, aggressive medical therapy, initially with antimicrobial agents and then with topical glucocorticoids, can result in significant reduction in soft tissue occlusion of the lumen. However, the structural changes in apocrine ducts and glands are probably irreversible.

Even very early changes in the luminal epithelia have the potential to become permanent and, once these permanent changes occur, simple Zepp resection of the lateral wall of the vertical canal is unlikely to be successful[89,92,93]. Total ablation of the canal is indicated[89,93].

Influence of progressive pathology

The soft tissues surrounding the lumen of the external ear canal react in a predictable sequence to the inflammation associated with chronic otitis externa[87–90]:

- Epidermal hyperplasia (acanthosis and hyperkeratinisation) is an early consequence of otic irritation (Fig. 5.51). The basal cells of the epidermis respond to inflammation by increasing their rate of division and increasing the transit time of cells moving through the epidermis. In addition, keratinisation is affected and a thickened stratum corneum is apparent. This reaction is reversible, provided the initiating cause is alleviated.

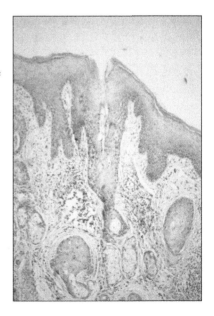

Fig. 5.51 Photomicrograph of a section of canine external ear canal exhibiting marked epidermal hyperplasia.

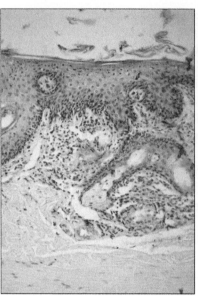

Fig. 5.52 Photomicrograph of a section of canine external ear canal demonstrating epidermal hyperplasia, an inflammatory infiltrate and dermal oedema. Hyperplastic otic epithelium occluding the external ear canal of a West Highland White Terrier.

- Subepithelial infiltration with inflammatory cells, such as lymphocytes, neutrophils, macrophages and plasma cells, occurs in response to inflammation. Chronic cellular infiltration results in local release of inflammatory mediators, cutaneous erythema and oedema (Fig. 5.52). Early cellular infiltration is reversible but the effects of chronic mediator release may engender permanent changes.
- Fibroplasia of the underlying dermis follows chronic inflammatory challenge within the lumen and the epithelium. In long-standing cases the fibrosis may be extensive (Fig. 5.53) and this contributes considerably to the loss of luminal cross-section.
- Early sebaceous gland hyperplasia is followed by massive ceruminous gland hyperplasia, both of the duct and the glandular portion (Fig. 5.54). The changes in the ceruminous glands result in gross thickening of the epidermis, particularly in cats.
- Papillary proliferation of the epithelial lining occurs to such an extent that the lumen becomes occluded (Fig. 5.55). In the external ear canals of cats this papillary proliferation may result in polyp

Fig. 5.53 Photomicrograph of a section of canine external ear canal stained to demonstrate dermal fibrosis, which in this case is extensive.

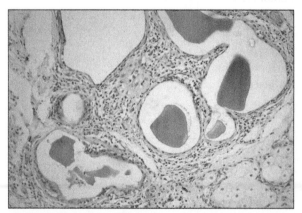

Fig. 5.54 Photomicrograph of a section of canine external ear canal demonstrating massive apocrine gland hyperplasia.

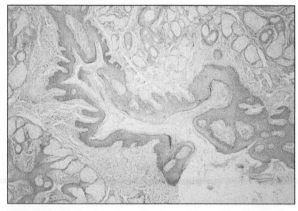

Fig. 5.55 Photomicrograph of a section of hyperplastic external ear canal with papillary fronds almost occluding the lumen.

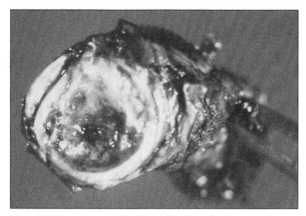

Fig. 5.56 Photograph of a section of a chronically hyperplastic external ear canal with almost no lumen.

formation with trapping of exudate between the polyp and the tympanum.
• Ossification of the dermis, sometimes extending to the auricular cartilage, and complete loss of lumen, occurs as a final stage (Fig. 5.56).

5.6 PRACTICAL MANAGEMENT OF OTITIS – PUTTING THEORY INTO PRACTICE

When faced with a case of otitis externa in a consultation it should be apparent very quickly if this is uni- or bilateral. A glance at the history will show whether this is a repeated or frequent occurrence. However, the clinician must bear in mind that a dog, in particular, which is presented for what the owner thinks is acute otitis externa might be:
• An acute, non-specific otitis externa, perhaps a foreign body or a polyp.
• The first presentation of what may be a long-standing case.
• Or, conversely, a well-known, chronic recurrent case (perhaps a dog with relapsing atopic otitis) with an acute flare.

With these caveats in mind, there are a number of clinical presentations that can act as case models:
• Acute uni- or bilateral otitis externa in a dog.
• Chronic or chronic relapsing otitis in a dog.
• Atopic pattern in the dog.

• Cocker Spaniel pattern.
• Refractory malassezial otitis.
• Refractory *Pseudomonas* spp.-associated otitis externa.
• Refractory methicillin-resistant staphylococcccal otitis externa.
• End-stage hyperplastic otitis externa.

Acute uni- or bilateral otitis externa in a dog

It is not uncommon for first opinion practitioners to be presented with a pup that is showing signs of peracute, unilateral, otic discomfort that is accompanied by signs of intense agitation. However, on examination, there is nothing within the external ear canal. Presumably, the irritation was due to a small accumulation of cerumen, perhaps associated with hair or a small foreign body, which was removed during the shaking process. No treatment is needed.

O. cynotis infestation is also a problem of young dogs, for the most part. It is usually associated with bilateral otitis, although one ear may be more uncomfortable than the other, giving the suggestion of a unilateral problem. The vast majority of affected dogs will exhibit varying amounts of dark brown, dryish, cerumen, with a 'coffee grounds' appearance. It takes only a few minutes to collect some of this discharge, perhaps on a cotton bud or speculum, mix it with liquid paraffin and examine it under a microscope to confirm the diagnosis. Treatment is straightforward.

In regions of the world where the spinous ear tick (*O. megnini*) is found, clinicians will be only too aware of the importance of checking for this ectoparasite in a dog, particularly if it is a young male working dog with acute otitis externa.

Otitis externa associated with secondary infection after exposure to water may be peracute, but the history will identify recent hydrotherapy, swimming or a grooming session, for example.

Almost all other cases of acute otic discomfort are associated with some degree of erythema, inflammation and exudate, making any identification of underlying causes difficult without cleaning the external ear canal and working the case up. Consider the cases illustrated in Figures 5.57A, B. There may be differences in severity and chronicity but cleaning and further examinations are indicated before treatment can be instituted.

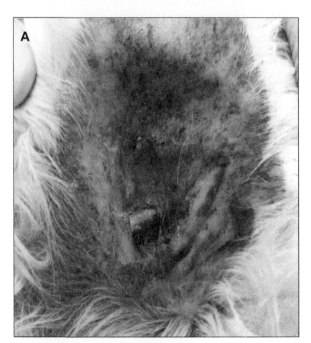

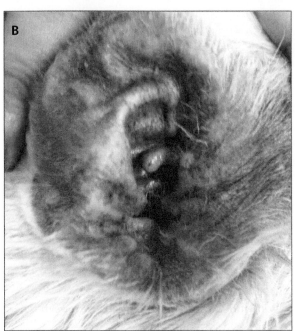

Figs 5.57A, B Non-specific otitis externa in dogs. The clinical picture gives no clue as to the aetiology of the otitis.

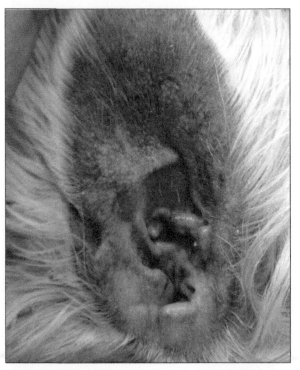

Fig. 5.58 The external ear canal of a West Highland White Terrier that has been cleared of hair, cerumen and debris, flushed and dried and is now ready for examination.

Treatment is based on cleaning and drying the external ear canal (Fig. 5.58), which might entail sedation. Inflammation must be controlled and infection suppressed; typically this will be opportunist secondary infection, which is usually comprised of mixed gram-positive and gram-negative bacteria. A short, 3–5 day, course of proprietary antibacterial/glucocorticoid otic medication (see Table 3.1 , p. 89) is usually sufficient.

Note that in very hot and humid parts of the world, dermatitis in general, and otitis in particular, may be much more aggressive in nature. Furthermore, the otitis has a tendency to progress rapidly and to a predominantly gram-negative and malassezial infection.

Given that grass seeds are the most common foreign body likely to cause acute unilateral otitis, every effort must be made to ensure that there is nothing left in the external ear canal.

Remember that the presence of grass seeds in the external ear canal is correlated positively with rupture of the tympanic membrane and with secondary bacterial infection[5].

General anaesthesia is usually indicated to facilitate removal of the foreign body and to allow thorough examination of the tympanic membrane. Owners may express concern that anaesthesia is suggested. It must be explained that sedation alone rarely gives adequate control, or relief from the discomfort, when attempting to examine the deeper portions of the external ear canal. Failure to find a foreign body or tympanic damage might suggest that the head shaking has dislodged the grass seed. Assuming that there is no damage to the tympanic membrane, the canal may be cleaned, dried and a short course of proprietary antibiotic/glucocorticoid otic medication provided.

If the tympanic membrane has been damaged and perforated then additional measures must be taken:
- Gently insert a shielded swab to obtain a sample for bacteriological culture and sensitivity testing.
- Ensure that there is no remnant of foreign body within the bulla: copious flushing is indicated.
- A 3 week course of systemic antibacterial treatment (adjusted if necessary in the light of the bacteriological report) and systemic prednisolone at an anti-inflammatory dose of 0.5 mg/kg every other day. Pain relief might be indicated (e.g. tramadol 1–5 mg/kg p/o BID to QID
- Book a re-examination date. You decide, not the owner, if the damage has resolved and that the case may be discharged.

If the proximal pinna and the external ear canal is erythematous (Fig. 5.59) or if the otic erythema waxes and wanes, and assuming that there are no ectoparasites or foreign bodies, then allergy may be suspected. Atopy is very common and, as is the case with food allergy, it usually affects both ears, although unilateral otitis has been reported (see below). Breeds that are predisposed to atopy include Terriers in general, German Shepherd Dogs, Labrador Retrievers, Boxers, Shar Pei, Newfoundlands and English Setters. Individuals should be carefully examined for other signs of atopy such as

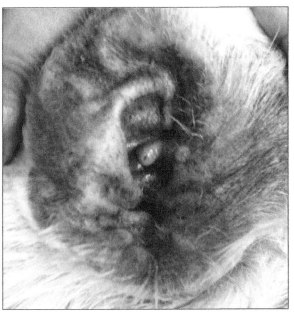

Fig. 5.59 Same ear as in 5.57b. Note that this West Highland White Terrier presented for acute otitis but the ear canals show signs of chronic change (hyperpigmentation and hyperplasia) being apparent in addition to the erythema.

palmer and planter interdigital erythema and erythematous lesions in the axillae and groin[4,6]. While there may be individual and breed variation in the nature and distribution of the clinical signs, otitis is a very common feature in all breeds.

Note that atopic dogs being treated with immunotherapy often have residual ear problems[9], notwithstanding that the pedal and ventral pruritus has resolved, for the most part.

Chronic, or chronic relapsing, otitis
Most cases of acute otitis resolve with appropriate medication. Relapsing otitis implies that something is amiss:
- Unilateral:
  - Poor compliance.
  - Underlying allergy.
  - Chronic, non-specific otitis externa with fibroplasia, glandular hyperplasia, a reduction in luminal volume and microflora overgrowth.
  - Malassezial otitis.
  - Unrecognised otitis media.
  - Neoplasia.

- Bilateral:
  - Poor compliance.
  - Underlying conformational problem, such as in the Shar Pei (Fig. 5.60).
  - Malassezial otitis.
  - Underlying allergy, particularly atopy.
  - Defect in keratinisation.
  - Bilateral otitis media.

When faced with a resigned owner presenting the dog, yet again, for otitis (Fig. 5.61), the temptation to provide another bottle of polypharmaceuticals should be resisted. Before examining the ear it is important to do three things:

1 Get a history:
- How many episodes of otitis? One ear or both? Is it seasonal?
- Does it respond promptly, only to relapse, or does it not quite resolve?
- Are the owners applying the product correctly?
- Are there any obvious predisposing causes?
- Is the individual a breed with anatomical predispositions such as Shar Pei or Bulldog?
- Does the dog have hirsute ear canals and is it regularly groomed?
- Does the dog come into contact with water when exercising or working?

2 Examine the skin:
- Is there any evidence of rhinitis, conjunctivitis, pedal erythema or ventral (axilla, groin, perineum) erythema that might suggest that the otitis is part of the disease spectrum seen with atopy[94]?
- Is the dog a Cocker Spaniel or a Basset Hound, with evidence of widespread (ventral neck folds, ventral trunk) dermatitis with crusted papules or a greasy feel to the skin and coat?
- Is the dog entire and does he have a testicular tumour?
- Are there lesions on the face and pinnae that might suggest pemphigus disease?
- Examine the ears, starting with the unaffected one (if unilateral disease):
- Is there patchy erythema of the pinna and vertical ear canal? Or, more or less symmetrical erythematous 'cobblestone' hyperplasia on the proximal aspect of the concave aspect of the pinna and around the upper portion of the vertical ear canal? These changes are very suspicious of atopy; the deeper portions of the external ear canal are often unaffected in early cases.
- Do the external ear canals show the 'Cocker Spaniel' pattern of early, deep changes associated with a moist, malodorous discharge and stenosis of the vertical ear canal?

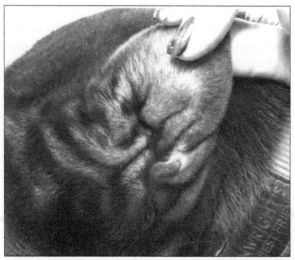

Fig. 5.60 Normal external ear canal of Shar Pei. Note the almost complete absence of a lumen.

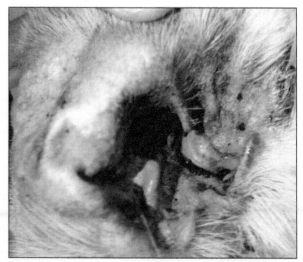

Fig. 5.61 Acute erythema overlying chronic hyperplastic changes in an atopic Labrador Retriever.

- Is ulceration of the epithelial lining of the vertical ear canal visible?
- Is there any discharge?
- Are you able to visualise the deeper portions of the ear canal? Or is the affected ear showing occlusion of the vertical ear canal due to a voluminous discharge, mass, cystic hyperplasia, or chronic inflammatory changes?
- Are the auricular cartilages painful?

3 Triage the patient. The microbial flora of otitis externa is well recognised[75,76,95–99]:

- Probable atopic? Most likely *S. pseudintermedius*, *M. pachydermatis* or both.
- Probable 'Cocker Spaniel pattern' with a tendency to gram-negative overgrowth? There might be some staphylococci and yeast but it will likely be mainly rods.
- An apparently normal conformation but with a pure to predominantly malassezial flora
- *P. aeruginosa* otitis externa.
- Breed-related anatomical predisposition such as Shar Pei or Bulldog, both of which have narrow ear canals that are prone to early stenosis.
- Possible methicillin-resistant *S. pseudintermedius* (MRSP)?
- Sertoli cell tumour with secondary seborrhoeic otitis.
- Pemphigus group disease.
- Vertical ear canal occluded by a mass.
- Impossible to examine the deeper portions of the vertical ear canal, due to stenosis and discharge and patently end-stage proliferative otitis externa.

Poor compliance

In some circumstances owners my find it impossible to apply otic medication, either because they cannot manage single handedly or because the animal, usually a dog, greatly resents it. In such circumstances the clinician might consider admitting the animal and applying product under deep sedation.

- It is important to recognise that the use of these long duration otic medications does not preclude a proper otic work up-they are not short cuts.
- The products, listed below, might be considered, both of which must be applied to a cleaned and dried ear canal, see below.

- Osurnia® which is a gel formulation, contains florfenicaol, terbinafine and betamethasone acetate and is applied at 7 day intervals. Patently, this product is designed to deliver anti-inflammatory and antimicrobial treatment.
- Otic Armor®, in contrast, contains no antibacterial, antifungal or steroid medications but, rather, coats the ear canal with an inert polymer. A single application is reported to last some 3 months. In an atopic dog, for example, it might prove very useful in preventing recurrent allergen induced otitis.

Ear cleaning

Ear cleaning and otic treatment has three phases (see Chapter 2 Diagnostic Procedures, Section 2.4 and Chapter 3 Ear Cleaning for a more detailed discussion on cleaning of the ear canal):

- Ear cleaning to enable initial examination of the external ear canal.
- First-stage management, based on examination and cytology.
- Sampling for bacteriological susceptibility testing.
- Ear cleaning, using ear cleaners, as a means of long-term control of recurrent disease.

Ear cleaning to enable initial examination of the external ear canal

Given that examination of the external ear canal includes assessment of the status of the entire canal and the tympanum, it follows that cleaning of cerumen and purulent debris may be necessary.

Cytological examination of otic discharge should be taken before cleaning is initiated. The presence of erosions or ulceration of the otic epithelium, inflammatory cells and rod-shaped bacteria should prompt suspicion of *P. aeruginosa*. Samples should be submitted for bacteriological assessment and sensitivity testing.

It has been suggested that the application of medication into the external ear canal gives concentrations of antibacterial agents in the mg/ml range[100,101]. This is many times (1,000) more than the µg/ml used in sensitivity testing. Therefore, given that the correlation with outcome (i.e. *in vitro* sensitivity testing and clinical response) is hard to predict, sensitivity testing could be reserved for cases in which there is mixed infection, or a failure to respond to initial therapy.

However, there are two caveats to consider:

- If you suspect MRSP it is mandatory to submit samples for susceptibility testing and to consider the implications *vis-à-vis* staff and owner safety.
- Many dogs with gram-negative infection, in particular, require long courses of antibacterial treatment. Drugs vary in their cost and dosing requirements. Susceptibility testing gives options to allow tailoring the medication for the individual case. Factors to consider are:
  - Cost.
  - Once daily treatment is easier to comply with but carries a cost penalty.
  - Some owners can only manage liquid medication, particularly a problem with cats and small dogs.
  - Injectable only medications.

Sometimes the entire ear canal is so swollen that it is unrealistic to attempt this detailed, initial investigation without instituting some initial treatment, on a symptomatic basis[100]:

- Topical otic glucocorticoids, perhaps within an otic polypharmaceutic, and anti-inflammatory doses of oral prednisolone (0.5 mg/kg daily for 3 days then every other day) for 10–14 days should allow sufficient reduction of epithelial swelling to allow examination of the deep portions of the external ear canal.

Inflamed external ear canals are painful. Animals, not surprisingly, may resent examination, even with the relatively narrow canulae of video-otoscopes. Sedation, and most probably general anaesthesia, will be necessary.

Before examination and staging of the otitis can take place, the external ear canal must be cleaned of matted hair, ceruminous debris and any purulent discharge that have accumulated (see Chapter 3 Ear Cleaners). This may be accomplished in a variety of ways[102]:

- Consider using a depilatory product to minimise the otic trauma associated with plucking an ulcerated and inflamed ear canal.

- Repeated flushing with a soft feeding tube and a 20 ml syringe, or a bulb syringe, using warm water or **very dilute** chlorhexidine.
- As above but using a commercial cleanser, or astringent cleanser, such as Epi-Otic or Panotic.
- Some clinicians prefer the foaming action of 5% dioctyl sodium sulfosuccinate (DSS) for the removal of purulent debris. Note that cleansers such as DSS, and even chlorhexidine, might have some irritant effects on the middle ear. Clinicians should flush the external ear canal copiously after cleaning.
- Using a powered delivery system, such a Otopet Earigator®, Auriflush® or WaterPik® to facilitate removal of otic debris.
- Astringents, such as urea, and mild acids such as boric, lactic, salicylic, are considered to have a drying effect and may be used after flushing and cleaning, provided the tympanic membrane is intact.

First stage management, based on examination and cytology

Based on cytological examinations and visual assessment of the cleaned ear canal and the status of the tympanic membrane, the initial steps in classification of the otitis externa can be put into place:

- Non-specific otitis externa, requiring short-term management with an otic polypharmaceutical (see Table 3.1, p.89) containing glucocorticoids and antimicrobials. This will ensure a prompt return to normal function.
- Atopic type, with a mixed staphylococcal and malassezial flora accompanied by mild erythema and early epithelial hyperplasia. Atopic pattern otitis is rarely complicated by pseudomonal infection in temperate parts of the world:
  - Acetic acid/boric acid solution (Malacetic Ultra Otic®) or Epiotic Advanced Formula®
  - Easotic®, which contains gentamicin, miconazole and hydrocortisone.
- Cocker Spaniel type with a thick, ceruminous discharge and tendency to rapidly develop gram-positive pseudomonal complication: tromethamine ethylenediaminetetraacetic acid (Tris-EDTA) with 0.15% chlorhexidine (Otodine®).

- Possible MRSP: see below for management options.

Management of the atopic type ear

Most cases of atopy in dogs, and in some cats, are associated with otitis and, in some individuals, otitis externa is the only presenting sign[4,6]. These early signs are an occasional, to frequent, relapsing, patchy erythematous flush, sometimes associated with recurrent erythema of the ventral trunk. This early otic erythema is usually confined to the proximal pinna and the upper portions of the vertical ear canal. Examination of the deeper portions of the external ear canal usually reveals no detectable changes at this point.

> This is in stark contrast to the situation in the Cocker Spaniel-type otitis where changes in the deeper portion of the external ear canal occur early[89].

With time, and depending on the individual, more chronic changes set in, and dogs eventually exhibit hyperplastic dermatitis in the affected areas, particularly the proximal pinna and the upper portion of the vertical ear canal. Chronic, hyperplastic, relapsing otitis externa follows. Some individuals exhibit such profound hyperplastic changes that the stenosis becomes sufficient to close the vertical ear canal almost completely, but this is rare if the case management is good.

Cytological and microbiological examination will reveal *S. pseudintermedius* and *M. pachydermatis*, just as is found on the skin of atopic dogs. The balance between staphylococcus and malassezial yeast will depend on the otic environment, and perhaps on factors which we do not, as yet, recognise. Most dogs have a mixed infection whereas others might have a predominantly malassezial microflora. Significant numbers of inflammatory cells are not usually found. Nucleated squames may be seen, reflecting increased epidermal turnover and shedding of immature, nucleated cells, presumably in response to irritation and inflammatory toxins.

In general, submitting samples for bacterial and malassezial culture and sensitivity testing is not indicated[75], unless the animal has received multiple treatments in the past or there is a failure to respond as

anticipated. Whilst resistance is encountered, and occasionally multiple resistance is seen[97–99], the vast majority of gram-positive isolates test sensitive to several of the following antibacterial agents: framycetin, neomycin, gentamicin, enrofloxacin and marbofloxacin[76,95,97]. Failure of an individual to respond as anticipated should prompt complete examination and cleaning under general anaesthesia, in addition to sensitivity testing.

> Remember:
> - Rupture of the tympanum is less likely in atopic dogs[5].
> - Gram-negative, pseudomonas otitis is unusual in atopic dogs in temperate climates. In the southern parts of the USA, where the climate is increasingly humid and warm, the incidence of pseudomonal otitis increases and it becomes more of a problem in atopic dogs.

Treatment of atopic dogs is based on the affected individual's clinical signs. Unfortunately, atopic otitis often fails to respond to immunotherapy[9] or cyclosporine (Atopica). Additional, focused, localised, long-term control may be required to maintain otic health:

- Malacetic Ultra Otic®, Sancerum®, Otodine® or Epiotic Advanced Formula® would be good examples of first-stage treatments. All are good at clearing ceruminous debris and cleaning the ear canal; none contain an antibiotic, making the price reasonable and once weekly flushing is usually adequate as a means of reducing recurrence in dogs prone to repeated episodes of otitis.
- Unfortunately, the continuing inflammation is often associated with increasingly severe otic pruritus and additional control may be required. Combination treatment is quite acceptable – anything to suppress the inflammation and halt the progressive hyperplasia. Case management should be aimed at obviating the requirement for surgery.
  - Cortavance® (Virbac) spray contains hydrocortisone aceponate, which is not significantly absorbed. It may be spritzed onto the pinnal lesions and the upper portions of the external ear canal several times a day. Some

individuals do not tolerate the spray at all but most are not too upset by the process. If the dog finds the direct application uncomfortable then try indirect application using a cotton wool ball.

- Easotic®, which contains gentamicin, miconazole and hydrocortisone, is an example of a product which is the next step up the therapeutic ladder – the addition of antimicrobial agents suppresses secondary infection and this helps to control inflammation.

- Some atopic ears appear to have a malassezial otitis, with little apparent staphylococcal involvement. In this situation the clinician might consider Ketocortic® (Trilogic Pharma). This is a potent antimalassezial preparation with a residual activity. It is applied as a viscous liquid that firms to form an emulsion, which lines the entire ear canal. The medicants (hydrocortisone and ketoconazole) are slowly released over 7 days.

- A liquid bandage (OticArmor®, Allacam Inc.) may be helpful in some cases. The product produces a soft barrier that adheres to the epithelial surface of the external ear canal. In effect, it is a barrier to bacterial and allergen penetration, and it may reduce any increase in humidity. Repeated application will be necessary. There is, as yet, very little published experience with this product but it might prove useful.

- A few individuals require increasingly potent treatment, and on a regular basis, to prevent progressive pathological changes. The authors have had a good response to proprietary otic products that contain more potent glucocorticoids, such as dexamethasone or mometasome, e.g. Aurizon® and Posatex®, pulsed in on 3 days a week. Some individuals in the author's practice have been well-controlled for years with this regime.

- Vertical canal ablation may be required in some individuals if the progressive changes consequent upon chronic inflammation cannot be controlled and are limited to the vertical canal only.

Cocker Spaniel-type otitis

Cocker Spaniels exemplify the type of malodorous, moist otitis (Fig. 5.62), which frequently becomes ulcerated and secondarily infected by gram-negative bacteria, particularly *P. aeruginosa*. The breed exhibits anatomical characteristics thought to favour the onset of otitis externa[89]:

- Compound hair follicles along the entire length of the external ear canal.
- Increased amounts of glandular tissue within the otic epithelium, relative to other breeds.
- Increased humidity within the external ear canal, making a thick, greasy cerumen almost normal in some individuals. Note, however, that this does not imply that infection or inflammation is present.
- A rapid progression to 'end-stage' otitis externa, and a high risk of otitis media.
- Remember, the presence of gram-negative rods correlates positively with a ruptured tympanum[5].

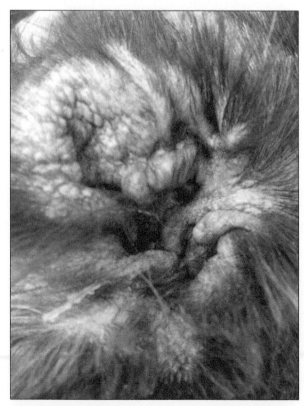

Fig. 5.62 Typical pattern of otitis externa in a Cocker Spaniel with erythema, hyperpigmentation, hyperplasia, luminal stenosis and, in this case, a malassezial otitis.

Not all Cocker Spaniels suffer from otitis externa. However, the development of pseudomonal infection must always be considered a possibility when dealing with otitis externa in this breed. A complete physical examination to assess whether the dog has other dermatological lesions is important, because in many Cocker Spaniels otitis is part of a wider dermatological problem.

Cytological examination should be performed on smears stained, for example with a modified Wright's stain, such as Dif Quik. Most infections are mixed, and malassezial yeast, cocci and rods are to be anticipated. Differentiating between what might be termed otic normal flora, such as low numbers of cocci and yeast, and pathological numbers is difficult; more significant findings are rods, inflammatory cells and blue-staining cellular debris.

Thorough examination is necessary before any treatment is provided, and this may require cleaning to remove any accumulated ceruminous debris. Any suggestion of chronic disease, head tilt or pain, limits severely the therapeutic options and should prompt suspicion of a ruptured tympanic membrane. This mandates thorough cleaning under general anaesthetic, and perhaps radiography.

- Initial cleaning with a ceruminolytic such as Otoclean® or CleanAural® will remove cerumen and debris, as well as bacteria and their toxins. Both of these products scored highly in the studies cited in Chapter 4 Diseases of the Pinna[103].
- After the canal has been cleaned and dried, Tris-EDTA with 0.15% chlorhexidine (e.g. Otodine®) can be applied. This product proved very efficacious in both *in vitro* studies[17] and clinical trials[104–107].
- Other cleaners which might be expected to perform well include the acid-based products such as Sancerum®, Malacetic Otic® and OtiRinse®.
- Occasionally, more potent medication may be necessary. Some authorities have suggested that sampling for antibiotic sensitivity testing cannot be defended in first-diagnosed cases, because:
  - There is demonstrable variability between laboratories with regard to reporting *in vitro* antibiotic susceptibility of *P. aeruginosa*[101,108].
  - The *in vivo* concentration of antibacterial agent within the external ear canal is some thousands of times higher than that on the *in vitro* testing plate. This makes it difficult,

and even unnecessary, to extrapolate *in vitro* minimum inhibitory concentration (MIC) data to the external ear canal, particularly for marbofloxacin[95].

- However, submitting samples for culture and sensitivity testing is advised, if only to obtain a baseline dataset and perhaps to identify a less-expensive treatment option depending on the susceptibility.
- Good empirical choices would be medication containing gentamicin (e.g. Otomax®), polymyxin/miconazole (Surolan®) or a neomycin/nystatin (Panalog®). Some authorities advise against using a fluoroquinolone-based product (e.g. Aurizon®) in cases of acute otitis[12] as the other products mentioned all have activity against both gram-positive and gram-negative bacteria and malazessial yeast[109–114]. They should be reserved for cases that fail to respond to first-line treatment.
  - Note that the effectiveness of fluoroquinolones antibiotics is reduced in an acid environment – important if using an acidified otic flush.

These individuals are not suitable for the long-term pulsed regime that suits many atopic dogs for two reasons:
- The progression of the otitis externa is difficult to predict, and it may deteriorate rapidly.
- The ever present risk of engendering a resistant *P. aeruginosa* infection.

Recurrence of the otitis after successful treatment has been put in place mandates further work-up, and sample submission for bacterial culture sensitivity testing (see below, Chronic relapsing, or multi-resistant, *Pseudomonas aeruginosa* otitis).

Notwithstanding regular and prompt attention to the otic inflammation and infection, some individual dogs, particularly, but not exclusively, Cocker Spaniels, develop chronic hyperplastic changes[15]. Surgical resection (vertical canal ablation or TECA) are options and readers are referred to End-stage hyperplastic otitis externa, below.

Chronic, or relapsing, *Malassezia pachydermatitis* otitis

Malassezial yeast, as has been discussed earlier, can be found in the normal external ear canal, although usually not in high numbers[115]. Chronic or refractory otitis associated with malassezial yeast may reflect a number of different, although not mutually exclusive, aetiologies:

- Localised obstructive disease favouring an increase in humidity within the external ear canal. In these cases the abnormality will be visible once the canal is cleaned and properly examined.
- Underlying atopy or food hypersensitivity. Atopic dogs will usually manifest signs of localised change such as erythema and hyperplasia. In addition, most, although not all, atopic dogs, exhibit facial, pedal and ventral pruritus. Some atopic dogs, however, manifest only otitis externa. These dogs exhibit otitis with the typical erythema and hyperplasia that might be expected with atopy. Dogs with food hypersensitivity may also manifest otitis externa with no other clinical signs.
- Resistance, or at least reduced *in vitro* susceptibility, to the azole antimycotic agents (enilconazole, miconazole and itraconazole) has been demonstrated[116]. *M. pachydermatis* isolates from cases of chronic otitis externa were more likely to demonstrate high MIC than organisms isolated from dogs with acute otitis. However, the authors postulated that, given that the very high concentration of drug delivered by a topical otic therapeutic agent is many orders of magnitude higher (1000×) than even the highest MIC found, resistance is not likely to be a problem. One might add the caveat that the ear canals must be cleaned before topical otic medications are applied.
- Genetic differences between malassezial strains have been demonstrated[117] in isolates from normal dog's ear and ears from dogs with otitis and dermatitis. However, there is, as yet, no evidence that pathogenic strains, virulence factors or the ability to elaborate biofilm or phospholipase[118] play a part either in the aetiopathogenesis or management of chronic malassezial otitis.
- Some dogs, with apparently normal external ear canals may also suffer from recurrent malassezial otitis. Some of these individuals might be, for example, subclinical atopics (i.e. food trial negative but intradermal skin test or enzyme linked immunosorbent assay (ELISA) positive) but in many cases the aetiology is not known.

- The role of IgE-mediated malassezial hypersensitivity is contentious[119]. An IgE-mediated response to malassezial allergens has been demonstrated[120] and equivalence between intradermal testing and ELISA assays has been demonstrated[119]. However, how this IgE response might be related, if at all, to malassezial otitis is not known. Furthermore, there is no evidence that appropriate immunotherapy against malassezia has any clinical benefit.

The approach to recurrent malassezial otitis is, initially, straightforward. Firstly, ensure that there are no local abnormalities or primary factors present. Secondly, establish the integrity of the tympanic membrane and submit samples for malassezial susceptibility testing.

Demonstrating the presence of IgE antibodies to malassezial yeast, either by intradermal skin testing or by serological means, might help to explain some of the pathogenesis. However, as discussed above, the practical relevance is not known.

Treatment options in Europe are subject to the prescribing cascade, unless the licensed product fails to be effective. Veterinarians in the USA have the ability to compound products or choose to use a commercial product. Options include:

- Epi-Otic® and Epi-otic Advanced Formula® (Virbac Animal Health), Otodine® (ICF).
- Miconazole:
  - Aurizon® (Vetoquinol).
  - Surolan® (Vetoquinol, USA; Elanco, Europe).
  - Conofite® (1% miconazole) Schering Plough Animal Health. Bloom advises adding 3 ml dexamethasone SP[100].
- Nystatin:
  - Panalog®, Fort Dodge; Quadritop® and Animax® are generic versions available in the USA.
- Ketoconazole:
  - Ketocortic® Trilogic Pharma.
  - Triz ULTRA™ plus KETO Otic Flush (0.15% ketoconazole (Dechra USA).
  - KC Oto-Pack™ (DermaZoo Pharmaceuticals).

Chronic relapsing, or multi-resistant, *Pseudomonas aeruginosa* otitis

Any individual of any breed, and not just Cocker Spaniels, may be affected by *P. aeruginosa* otitis externa. This includes dogs such as Mastiffs and Newfoundlands with large external ear canals (Fig. 5.63) making it hard to understand the pathogenesis in some cases. Apparently spontaneous peracute onset may be noted, although the majority of cases will have a history of chronic, or chronic relapsing otitis and/or, associated skin disease. Pseudomonas otitis is sufficiently common for most veterinarians to see a case two or three times a year. Many cases resolve with appropriate treatment.

> Note that atopic dogs in Northern Europe rarely suffer from pseudomonal otitis, notwithstanding that many such individuals suffer chronic otitis externa and possess pendulous pinnae. In hotter, more humid environments, pseudomonal complications of atopic otitis are more common.

Dogs with *P. aeruginosa* ear infection typically present with a purulent and malodourous otitis externa. Additionally, there may be extensive ulceration (Fig. 5.64), obvious pain, head shaking or even head tilt, and evidence of otitis media. It is imperative that the external ear canal be cleaned to allow proper examination of the deeper portions of the ear canal. Cleaning also removes cerumen, exudate and organic debris, all of which can deleteriously affect the action of any subsequent antibiotic treatment.

> As many as 83% of cases of pseudomonal otitis externa may have otitis media[121].

The aim of the clinical investigations is to:
- Perform cytological and bacteriological sampling and assess underlying disease.
- Clean and dry the external ear canal.
- Identify if there is otitis media.
- Suppress otic inflammation.
- Eliminate the pseudomonas bacteria.
- Put in place a treatment regime to prevent recurrence.
- Identify and eliminate, or manage, predisposing problems and underlying disease.
- Decide if surgery is indicated as a medical resolution is unlikely.

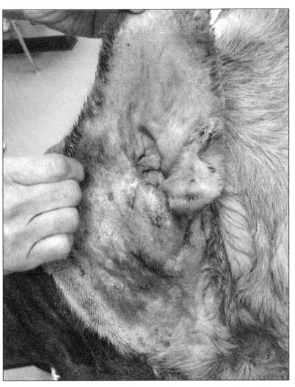

Fig. 5.63 Large external ear canals in a 95 kg Mastiff. Pseudomonal otitis externa of unknown aetiology.

Fig. 5.64 Highland White Terrier with an ulcerated otic epithelium and a gram-negative otitis externa.

Most affected dogs are in such discomfort that they greatly resent anything more than a cursory examination and, perhaps, the gathering of a cytological sample. Given the difficulty in getting adequate access to the ear canals it is appropriate to administer a general anaesthetic to facilitate cleaning, examination and clinical staging of the otitis.

Perform cytological and bacteriological sampling and assess underlying disease

The cytology samples can be assessed and swab samples can be submitted for bacterial culture and sensitivity testing. Notwithstanding the discussion in the previous section on the relevance, or otherwise, of susceptibility testing *vis-à-vis* clinical response, it is useful to characterise the organisms and their *in vivo* susceptibility.

P. aeruginosa otitis is suspected by identifying rods (bacilli) and, usually neutrophils, and often, blue-staining (using a modified Wright's stain such as Dif Quik) whispy strands of degraded nuclear material[122].

Underlying disorders such as otodectic mites, grass seeds and otic neoplasia are easily ruled out. Remember that atopy and food intolerance can cause otitis, either as part of a disease spectrum or as a stand-alone condition. Sertoli cell tumour and hypothyroidism can also affect adversely the otic environment, resulting in otitis externa.

Clean and dry the external ear canal

Once under general anaesthesia the surrounding hairs can be clipped and the canal can be gently flushed with warm saline, perhaps with **very** dilute chlorhexidine added. Given that the discharge in most cases of pseudomonas otitis is rarely oleaginous, an aqueous flush such as this is usually adequate for cleaning, and it is also considered safe should it transpire that the tympanum is ruptured.

If the discharge is difficult to clear with saline then utilise the foaming action of 5% dioctyl sodium sulfosuccinate (DSS) to help removal of the debris. Be ready to flush copiously with saline if rupture of the tympanic membranes is discovered.

Identify if there is otitis media

Otitis media may be inferred on the basis of a ruptured tympanic membrane and suspected if there is an opaque, thickened and bulging tympanum. Radi-ography, or other imaging, may be necessary in some cases. Radiography should be part of the basic work-up, unless middle ear disease can be definitively ruled out.

Suppress otic inflammation

Some dogs manifest such severe otic hyperplasia and inflammatory swelling that the vertical ear canal is almost totally occluded. Trying to insert an otoscope, or even the head of a video-otoscope, would be impossible without causing severe trauma to the inflamed otic tissues. In these cases it may be helpful to take cytology and bacteriological samples and provide systemic prednisolone at an anti-inflammatory dose of 0.5 mg/kg once daily for 3 days and then on alternate days, for 2 weeks[100]. Tris-EDTA and chlorhexidine (Otodine®) can be supplied for otic irrigation *pro tem*.

Eliminate the pseudomonas bacteria

An evidence-based review of the treatment of pseudomonas otitis in dogs[102] concluded that "most, if not all, of therapeutic decisions in this condition are based on inadequate published data, personal experience and anecdote, rather than on evidence-based medicine". Subsequently, some controlled studies have been published that are starting to give scientific support to what has been expert practice and anecdotal reporting.

Whatever otic medication is decided upon, someone, usually the owner, has to instill it. Dogs with painful ears may greatly resent the application of treatment to the external ear canal. Some degree of control is necessary, both to prevent the owner getting bitten and to ensure that any medication is applied correctly.

Consider using ear wicks both to ensure any treatment is, and remains, in place, and to minimise owner input. Consider adjunctive steroids therapy (see below). Repeated anaesthesia will be necessary, but at least the necessary treatment is effected.

Once the discomfort is reduced it may well be the owner has the confidence to apply otic medications again. This is important as several weeks of medication may be necessary to effect cure, perhaps defined as:

- An absence of otic inflammation.
- An absence of cerumen and discharge.
- Negative cytology for rods, at least.

Griffin has highlighted the importance of exploiting synergism between drugs when attempting to eliminate

Pseudomonas spp[1]. Synergistic combinations that have been reported as effective in managing pseudomonas otitis include:

- Tris-EDTA with chlorhexidine (Otodine®). Hosseini *et al.*[107] reported that this was efficacious in treating *P. aeruginosa*, whether or not systemic fluoroquinolones were used. Ghibaudo *et al.*[105] also reported a successful *in vivo* study. Note that Otodine® also scored very highly in the *in vitro* studies reported by Steen and Paterson[103] and Guardabassi *et al.*[104].
- Tris-EDTA and fluoroquinones, both *in vivo*[123] and *in vitro*[124], are synergistic:
  - 1.25 ml 100 mg/ml enrofloxacin in 12.5 ml Tris-EDTA, stable for 28 days[125];
  - Tris-EDTA applied 15 minutes before a fluoroquinolone commercial otic product (such as Aurizon®).
- Tris-EDTA and amikacin (or neomycin) were reported as synergistic when the combinations were used against *S. intermedius*, *P. aeruginosa*, *Proteus mirabilis* and *Escherichia coli*[126].
  - Two × 1.5 ml vials amikacin (250 mg/ml) in 12 ml glycerine. Apply 0.5 ml daily, after Tris-EDTA;
  - Two × 1.5 ml vials amikacin (250 mg/ml) in 97 ml Tris-EDTA.
- Marbofloxacin and gentamicin were demonstrated to act synergistically in an *in vitro* study using 68 pseudomonal strains derived from cases of canine otitis externa: synergism was demonstrated in 33 instances[127].
- Silver sulfadiazine and enrofloxacin demonstrated synergism[128] in an *in vitro* study. This combination is commercially available (Baytril Otic®). There is no reason to think that this synergism would not hold true for other concentration-dependent antibiotics, such as marbofloxacin.
- The combination of polymyxin B and miconazole (as in Surolan®) was shown to be highly effective and synergistic against clinical isolates of *P. aeruginosa*[129]. Two randomised, clinical studies found that the combination was highly effective against otic pathogens, including *P. aeruginosa*[130,131].
- Note that the positive controls, which performed well in the two polymyxin/miconazole studies cited above, were

- Aurizon®: marbofloxacin[130];
- Otomax®: gentamicin[131]. Note that gentamicin in particular, might meet resistance problems in cases of recurrent pseudomonas otitis.

Silver sulfadiazine

Silver sulfadiazine (SSD) is a broad-spectrum antimicrobial agent, with both gram-positive and gram-negative activity. Two recent studies assessed 1% SSD in models of infection in burns. In one study its efficacy was assessed *in vitro* on samples taken from burns[132]. The other study looked at a burn model *in vivo*, infected with multi-drug resistant *P. aeruginosa*[133]. In both studies the SSD scored sufficiently highly to be recommended.

Unfortunately 1% SSD is rather too viscous to be useful in the canine external ear canal. Dilution to 1/10, or even 1/100, with water makes it fluid enough to access the depths of the external ear canal[134] and it has a sufficiently high MIC to kill pseudomonas.

> Make a 0.1% solution by mixing 1% 1.5 ml SSD cream (e.g. Flamazine®) into 13.5 ml water or saline[135].

Synergism of activity against *P. aeruginosa* between SSD and enroflaxacin has been reported using *in vitro* disc diffusion and MIC assessment[128].

Tobramcyin

Tobramycin is used to treat pseudomonal infection in humans with cystic fibrosis, and has been shown to prevent biofilm production *in vitro*[136]. Pseudomonal sensitivity to tobramycin is rarely reported in veterinary studies, but usually it is high[136,137]. Tobramycin is potentially ototoxic and great care is required if the tympanum is ruptured.

Tobramycin should be reserved for intractable cases. The injectable solution is usually presented in 2 ml vials of 40 or 80 mg/ml. Typical concentrations advised are 8 mg/ml, i.e. a five- or tenfold dilution, respectively in saline or Tris-EDTA[138].

Ticarcillin and ceftazadime are two useful drugs, which, like tobramycin, are best reserved for the most refractory cases. Bateman *et al.* reported that when an aqueous solution of ticarcillin was introduced into the external ear canal of dogs it appeared to result in the accumulation of macerated epithelial debris[139]. This was obviated by formulating the ticarcillin into 5% hypromellose solution (e.g. Methopt®, or an equivalent).

Ticarcillin potentiated with clavulanic acid (Timentin)

- 6 g Timentin in 12 ml sterile water. Freeze in 2 ml aliquots and use within 3 months.
- Thaw out one 2 ml aliquot and add to 40 ml saline (or 5% hypromellose).
- Divide into 4×10 ml aliquots for the client to freeze at home.
- Thaw one a week, for twice daily application of 0.25 to 1 ml, depending on the size of the external ear canal.

Ceftazadime (Fortum)

- 1 g injectable solution, with 16 mg dexamethasone into 100 ml Auroclens®.
- 1 g injectable solution, with 24 mg dexamethasone into 100 ml saline.

Another major problem is finding the parenteral preparation of an antibacterial agent to which the multiresistant *P. aeuruginosa* might be sensitive. Those that might be useful are often prodigiously expensive and produced for intravenous use in humans[140]. However, these intravenous agents may be used subcutaneously, allowing for owners to administer them at home, for extended periods[140]. Examples include:

- Ticarcillin, e.g. Ticar 40–80 mg/kg every 6 hours.
- Ceftazidime, e.g. Fortaz 30 mg/kg every 4 hours.
- Meropenem, e.g. Mere, 8 mg/kg every 12 hours.

Bacteriophage

The interest in bacteriophage treatment has surfaced as multiresistant bacteria are increasingly common[141,142], notwithstanding that bacteriophages were discovered almost 100 years ago. Bacteriophage therapy has been used in experimental infections in laboratory animals[143] and there is one report of an open study in 10 dogs[144]. A good response was reported with a significant reduction in the number of viable pseudomonal bacteria. More work is needed before clinical recommendations can be made.

Identify and eliminate, or manage, any predisposing problems

An integral part of the treatment of pseudomonas infection is trying to prevent a relapse occurring soon after you have effected remission.

- Reread the history and try to identify any management issues, such as regular swimming, hydrotherapy sessions, using a water-based ear cleaning regime. Try using, for example, Malacetic Otic® or Orodine® immediately, having dried the ear canal as much as possible after water exposure.
- Have you definitively ruled out otitis media?
- Is there ongoing pathological change, particularly a problem in Cocker Spaniels?
  - Consider alternate day low-dose (0.5 mg/kg) prednisolone in addition to using otic steroid/antibiotic medications in an attempt to reduce glandular and epithelial hyperplasia and accompanying fibrosis.
  - Consider early referral to surgery.
- Reconsider underlying disease:
  - Food allergy and atopy.
  - Hypothyroidism, Sertoli cell tumour.

Put in place a treatment regime to prevent recurrence

Once remission has been achieved and any underlying issues addressed, maintenance of remission is the key objective. A small proportion of cases appear to need no ongoing care. Most require varying degrees of attention and some might best be submitted to surgery.

> Given the high risk for Cocker Spaniels for progression to end-stage otitis[89] it might prove pertinent to consider elective surgery with a reasonable chance of a successful outcome sooner, rather than later, when one is forced to act.

An antibacterial irrigating solution with activity against *P. aeruginosa*, such as Orodine® or EpiOtic Advanced®, Malacetic Otic® would be ideal. DOUXO Micellar®, like the EpiOtic Advanced®, also contains phytosphyngosine. The absence of an antibiotic agent will keep the cost down, as well as greatly reducing the risk of incurring a resistant infection.

Decide if surgery is indicated as effecting a medical resolution is unlikely. Indications for surgery include:

- Pre-emptive surgery before significant pathological changes have taken place.
- Refractory otitis externa complicated by ongoing pathology.
- Refractory otitis externa associated with otitis media.

Methicillin-resistant *Staphylococcus pseudintermedius* otitis externa

Methicillin-resistant *Staphylococcus pseudintermedius* (MRSP) has been recognised as a significant animal health problem since 2006[145]. Resistance is carried by the mecA gene, which codes for a modified cell wall, encompassing cell wall proteins with a low affinity for β-lactam antibiotics[146]. In addition to β-lactam antibiotics, MRSP are frequently resistant to ciprofloxacin, clindamycin, gentamicin, erythromycin, kanamycin, streptomycin and trimethoprim[147,148]. Some strains of MRSP are resistant to oxacillin and cefazolin[147]. Regional differences in resistance and susceptibility likely reflect local antibiotic practice and usage[4].

Although MRSP and methicillin-resistant *Staphylococcus aureus* may be isolated from healthy dogs, and cats[148,149], MRSP is most likely to be recovered from clinical canine and feline samples[150,150a].

Risk factors for acquisition of MRSP include:

- Frequent visits to, or extended stays in, veterinary clinics[6,148,150a–153].
- Regular topical otic medication[151].
- Recent receipt of glucocorticoids[151].
- Prior antimicrobial therapy, particular for dermatitis[143,150a].

Veterinarians and dogs can carry MRSP in their anterior nares[154,155]. The dogs were sampled after MRSP infection, and some cases continued to show nasal carriage, without clinical signs, for over 12 months[154].

Refractory skin infection and otitis externa are the most common clinical findings[155,156].

Treatment options for MRSP otitis externa

- Do not forget to clean the ears thoroughly, but get a sample for susceptibility testing first.
- Ensure that all staff in the vicinity wear a face mask, goggles and gloves to minimise exposure to contaminated aerosols. Try to schedule otic cleaning for the end of the day to minimise cross-contamination to other patients and to facilitate thorough cleaning of the immediate area and kennel accommodation.
- Be prepared to exploit the fact that topical otic medications often reach concentrations 1000 times, or more, higher than the *in vitro* MIC[156]. Thus, Boyen *et al.*[157] ascribed the efficacy of polymyxin B against MRSP, *in vitro*, as likely to be due to the very high concentration, rather than synergism with miconazole, which is thought to come into effect against other otic bacteria.
- Remember that the external ear volume increases as the dog gets bigger[158].
  - A 5.4 kg Shi-Tzu has an external ear canal with approximately 20 cm^2 area and volume 4.5 cm^3, whereas in a 20 kg dog the surface area is 38.5 cm^2 and the volume 9.6 cm^3.
  - Thus, the volume of medication required to deliver the appropriate amount of drug adequately to cover the surface area of the external ear canal will be approximately twice as much in the 20 kg dog compared to that required for the 5 kg dog[158].
- Chlorhexidine and Tris-EDTA (Otodine®). This combination was proved very effective *in vitro*[159,160] and has proven efficacious *in vivo*[105–107], although these studies were not directed at MRSP otitis externa specifically. This product has two main advantages as a first choice:
  - Very useful, pending laboratory susceptibility testing, and as a broad-spectrum antiseptic in its own right.
  - Pretreatment will help render resistant bacteria more susceptible.
  - Recently, the use of medical-grade honey was reported[160a]. The authors instilled 1 ml medical-grade honey per day into the affected external ear canals. Clinical scores were good, although this was an open study. Efficacy against methicillin-resistant *Stapylococcus pseudintermedius* was noted.

End-stage hyperplastic otitis externa

Angus emphasised the problems exemplified by Cocker Spaniels, in particular, of early ceruminous gland hyperplasia and calcification of the otic cartilages[89].

In some circumstances this hyperplastic response is so severe that the entrance to the external ear canal may be obstructed and the deeper portions of the external ear canal are stenosed. Traditionally, TECA has been the only resource available to clinicians[161–163].

Other options to consider include:

- Laser ablation of the hyperplastic tissue[164] can be particularly effective, provided ossification of the auricular cartilages has not taken place. Laser therapy can ablate all exuberant tissue facilitating effective medical treatment (Figs 5.65A, B).
- Hall *et al.*[165] reported a pilot study in which five dogs (four of which were Cocker Spaniels) with end-stage proliferative otitis externa were treated with 5 mg/kg cyclosporine twice daily for at least 12 weeks. All dogs showed a considerable clinical response with an increasing diameter of the previously stenotic external ear canal. All dogs showed significant clinical improvement with a much better life quality and an increasing diameter of the stenotic ear channel. None of the dogs had to go to surgery.
- Intralesional triamcinolone acetonide (2 mg/ml) as cited by Rosychuck[166] in discussions at Vancouver at the World Veterinary Dermatology Congress. This technique is often ineffective, and before considering the procedure the clinician must ensure that calcification of the auricular cartilages has not occurred. If the only alternative is TECA it might be worth trying:

- Use a spinal needle (3.5", 22 gauge) and ensure that the external ear canal is thoroughly cleaned and dried.
- Either 0.1 ml injections into each of the individual proliferative lesions or, if lesions are extensive and encircle the canal use a 'ring block' employing three injection points around the canal. Successive ring blocks are 1–2 cm apart. The maximum triamcinolone dosage recommended in a 15–20 kg dog is 6 mg. Repeat administration may be considered in 3–4 weeks. When intralesional therapy is used, there is usually a lesser need for very aggressive oral glucocorticoid dosages – i.e. instead of starting at 1–2 mg/kg/day, start at 0.5–1 mg/kg/day of prednisone/prednisolone.
- Try a diet enriched with anti-inflammatory and anti-oxidants. A 90 day study investigated the effect of a diet supplemented with extracts of *Melaleuca alternifolia*, *Tilia cordata*, *Allium sativum L.*, *Rosa canina L.*, zinc and an omega 3/6 ratio of 1:0.8 in cases of chronic occlusive otitis externa[166a]. Clinical scores were statistically better in animals fed the supplement diet, notwithstanding both groups receiving the same commercial topical medication, Otomax®.
- Finally, if there is no improvement to any of the above, TECA should be considered.

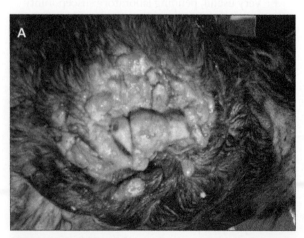 

Figs 5.65A, B Florid, hyperplastic, exuberant, obstructive dermatitis (A) on the concave aspect of pinnal and upper portions of the external ear canal in this Cocker Spaniel. Laser ablation of all hyperplastic tissue resulted in complete resolution (B). (Courtesy of Dr Mona Boord.)

5.7 OTITIS EXTERNA IN CATS

The approach to otitis externa in cats differs to that in dogs[167] in that there appears to be no breed predisposition to otitis, and there is no overt breed-related anatomical or structural predisposition, for the most part. Furthermore, although atopy and food allergy do occur in the cat, they are rarely associated with otitis externa[34].

The most common cause of otitis externa in cats, in both client-owned and stray cats, is *O. cynotis* infestation[34,167]. Older cats, in particular, may carry large numbers of otodectic mites in their ears. The lack of otic pruritus in some of these individuals with large numbers of mites presumably reflects their lack of hypersensitivity to the mite's saliva[168]. Hypersensitivity is thought to be the major cause of otic pruritus in cats with otodectic otitis[169]. The converse is also true, in that cats with pruritic otitis tend to have fewer ear mites in their ear canal[168]. Notwithstanding the widespread use of topical and systemic antiparasiticides, clinicians must make a careful otoscopic examination for ear mites. If in doubt, take a cerumen sample and examine microscopically.

> Take great care when collecting cerumen samples from, or trying to clean the ear canal of cats, as it is very easy to traumatise the otic epithelium[79].

Other causes of otitis externa include:
- Idiopathic ceruminous otitis externa: some Persian cats, particularly, and some older cats, particularly Siamese, appear to produce excessive cerumen[34]. Given the ease with which the feline otic epithelium can be traumatised, it might be best to treat this as conservatively as possible, particularly if there is no pruritus or malodour. Clients should be discouraged from using cotton buds over-enthusiastically[161].
- Otitis associated with a polyp or tumour: tumours and polyps are usually associated with otic malodour and a purulent discharge. The presence of a mass, perhaps an ulcerated mass, is usually easy to detect within the external ear canal. Cytological examination of needle biopsy samples taken from masses in the feline ear canal has been shown to

be sufficiently accurate to distinguish polyps from tumours[170], which will be an aid to management of the case. Surgical removal of polyps is usually straightforward, whereas surgical treatment of feline ear tumours, often malignant, may need to be radical, up to, and including TECA[170].
- Ceruminous cystomatosis: this condition is easy to identify as affected cats exhibit solitary to multiple bluish black, cyst-like papules, often very closely apposed and occasionally sufficiently numerous to occlude the lumen[171,172]. Surgical removal, or laser ablation[85] is indicated (Figs 5.45A, B). Follow-up treatment with topical anti-inflammatory products may be helpful in preventing recurrence[172].

5.8 TRAUMA TO THE EAR CANAL

> - Bleeding from the ear canal after trauma may be caused by separation of ear canal cartilages or fractures of the temporal bone.
> - Trauma patients need to be stabilised first before diagnostic work-up is performed.
> - Traumatic separation of the auricular and annular cartilages is usually amenable to primary surgical repair.

Trauma to the auricle is common in dogs and cats, especially in animals that roam outside (see Chapter 4 Diseases of the Pinna, Section 4.2). Aural hematomas can also result from trauma (see Section 4.2 and Chapter 18 Surgery of the Ear, Section 18.2); the exact cause in most cases, however, remains unknown. Blunt trauma to the external ear canal is most often caused by attempting to remove ear wax[173] or cleaning of the ear canal (see Chapter 2 Diagnostic Procedures, Section 2.4) and the tympanic membrane can be ruptured during these events as well. Trauma to the external ear canal and rupture of the tympanum is usually visible during otoscopy. Most traumatic perforations do not need a specific treatment as they usually close spontaneously[173]. In dogs, the healing of experimentally perforated tympanic membranes, with most of the pars tensa removed, was complete by 21 to 35 days[174]. Persistent infection and inflammation in the middle ear or the external ear canal can lead to a nonhealing membrane[173]. Traumatic injury to the ear, such as road traffic accidents, bite trauma and lifting the animal from the ground by its ears, can lead to trauma of the ear canal and more specifically, avulsion or rupture between the auricular and annular cartilage or between the annular cartilage and external bony meatus. This condition will be discussed here. Trauma to the middle and inner ear is discussed in Chapter 6 Diseases of the Middle Ear, Section 6.6 and Chapter 7 Diseases of the Inner Ear, Section 7.6, respectively.

Ear canal avulsion

Bleeding from the external ear canal after a dog or cat has been hit by a car, or has suffered a similar traumatic event, may be caused by a laceration of the external ear canal or result from fracture of the temporal bone, in which case the blood is escaping through a ruptured tympanic membrane. After stabilisation of the patient, advanced diagnostic imaging and otoscopy are indicated to localise the source of the bleeding and determine whether or not an avulsion of the ear canal is present. At the site of avulsion, an inflammatory reaction leading to the formation of a pseudotympanic membrane, will develop that can lead to obstruction or stenosis of the proximal vertical ear canal or horizontal ear canal, depending on the exact location of the avulsion (Figs 5.66A–C)[175–177]. Ongoing production of ear wax results in accumulation of ceruminous septic

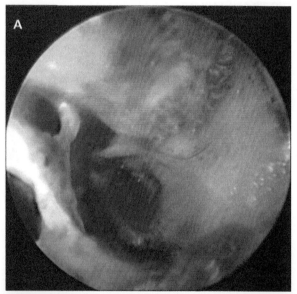

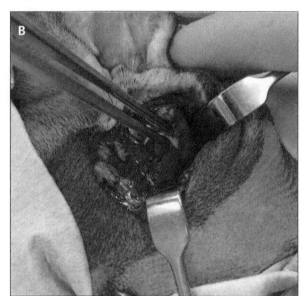

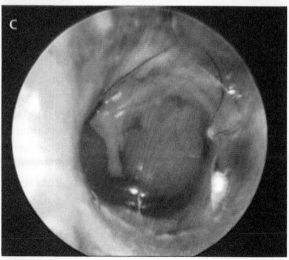

Figs 5.66 A–C European shorthair with otoscopic evidence of traumatic ear canal avulsion at the level of the transition between the vertical and horizontal ear canal. A: Ligamentous avulsion and disruption of the ear canal is visible upon superficial otoscopy; B: view at the level of the avulsion where periauricular tissues can be seen; C: otoscopic view of the normal horizontal ear canal and eardrum proximal to the avulsion.

inflammatory exudate behind the site of obstruction in the deeper part of the intact ear canal[178]. Secondary otitis externa and/or media will develop and can lead to para-aural abscessation, otitis media and otitis interna[175-177].

The diagnosis can be made by palpation of the ear canal and otoscopic examination, but advanced imaging is usually indicated to rule out any concurrent trauma to the middle and inner ear. Interruption of the external ear canal generally occurs at the level of the fibrous connection between the auricular and the annular cartilage and it includes laceration of the skin lining the canal in that area[173]. The continuity of the external ear canal should be restored at an early stage, before spontaneous organisation causes permanent stenosis. Primary repair of the avulsion can be attempted via a caudal approach to the ear[179], or via an incision over the vertical part of the ear canal[173] (Figs 5.67A–D). The two cartilage ends are debrided and then approximated, taking care to oppose them in a natural position. For chronic avulsions with stenotic ear canals, restenosis after primary repair or severe abscessation, horizontal ear canal ablation with lateral bulla osteotomy and preservation of the vertical canal or TECA with lateral bulla osteotomy (LBO) are indicated[175,180–182]. Otoscopic examination is delayed until 4 weeks after surgery at which point complete healing can be expected[173].

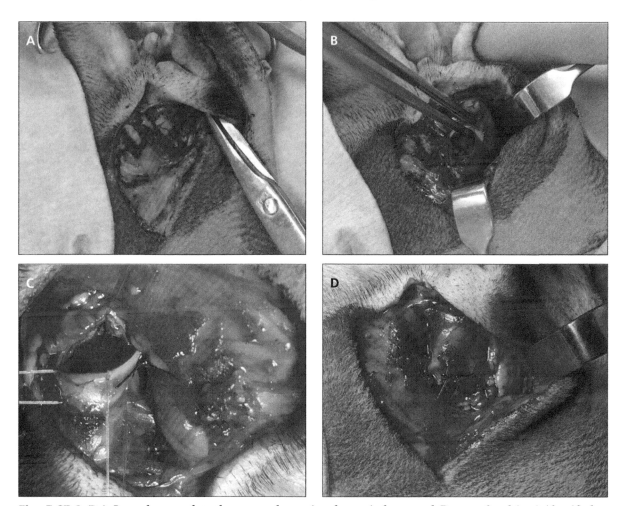

Figs 5.67 A–D A: Lateral approach to the ear canal exposing the vertical ear canal; **B:** area of avulsion is identified and ear canal dissected circumferentially; **C:** using stay sutures the medial aspect of the ear canal parts are reanastomosed using simple interrupted absorbable sutures; **D:** after complete reanastomosis of the avulsion.

5.9 NASOPHARYNGEAL AND MIDDLE EAR POLYPS

Middle ear polyps in cats are relatively common benign, nonneoplastic, pedunculated masses that arise from the mucosal lining of the middle ear, auditory tube or the nasopharynx[183–185]. Polypous mucosal changes can remain confined to the middle ear cavity, but large polyps can grow through the bony external meatus into the ear canal whilst rupturing the tympanic membrane or grow down the auditory tube into the nasopharyngeal area[173,185,186]. Depending on the exact location, these middle ear polyps are also referred to as aural polyps, nasopharyngeal polyps or more general as feline inflammatory or respiratory tract polyps[183–185,187–190].

The exact aetiopathogenesis of middle ear polyps is unknown, but Baker proposed a congenital aetiology in 1982, in which these polyps may arise as aberrant growths from the remnants of the branchial arches[188]. More commonly they are thought to be acquired through, and the result of, chronic inflammation of the middle ear mucosa as part of a primary viral-mediated otitis media or secondary ascending bacterial infection[183–185,191]. Feline calicivirus has been recovered from the polyps as well as from the nasopharynx of some affected cats[183,184,192,193].

Middle ear polyps are most commonly diagnosed in young cats, usually less than 2 years of age, but they have been reported in cats ranging from a few weeks to 15 years of age[183,194,195]. Found in ear canals, they are usually unilateral, but bilateral occurrence has been reported and at times polyps are found both in the ear canal and in the nasopharynx at the same time. No apparent breed or sex predilection has been reported, but it has been noted by many authors that polyp formation is more commonly seen in specific cat breeds like the Norwegian Forest, Sphynx, Maine Coon, Persian, Ragdoll and Abyssinian[183–185,192,194,195].

Many cats with middle ear polyps will not demonstrate any specific clinical signs provided the polypous changes are limited to the tympanic bulla itself and secondary infection does not occur. Cats may exhibit clinical signs consistent with upper respiratory tract infection though, including sneezing, reverse sneezing, nasal and sometimes ocular discharge. Once a polyp perforates the tympanic membrane and continues its growth in the outer ear canal, clinical signs of otitis externa develop. In some cases, a polypous mass may even be visible within the cavum conchae of the auricle (Fig. 5.68). In case of secondary bacterial infection, clinical signs of middle ear disease, including pain, decreased appetite and Horner's syndrome, or inner ear disease, peripheral vestibular ataxia and deafness, may develop[183,185,192,196]. Reports on the incidence of infection have varied widely, ranging from 13% to 83%[196–199]. Polyps exiting the auditory tube and forming a space-occupying mass in the nasopharynx will elicit clinical signs such as gagging, retching, dyspnea and stridor[183,185,192,196].

The actual presence of a middle ear polyp once protruding into the ear canal (Fig. 5.69) or into the nasopharynx can easily be demonstrated using otoscopy and nasopharyngoscopy. With large nasopharyngeal polyps, a ventral deflection of the soft palate can be appreciated in some cases (Fig. 5.70A) or the polyp can be palpated through the soft palate. Polyps of sufficient size in the latter area can also be visualised upon retraction of the soft palate (Fig. 5.70B) with, for instance, a Spay hook. Even though polyps can be visualised on radiographs in many cases[192,200–202], for proper (surgical) treatment planning, computed tomography (CT) of the entire skull is advised (Fig. 5.71). CT offers a higher true-positive diagnostic rate for the detection of otitis media and provides information on concomitant changes in the nasal cavity and frontal sinuses as well[203,204]. For patients with inner ear disease, magnetic resonance imaging (MRI) has distinct advantages over CT in visualising abnormalities of the membranous labyrinth and brainstem[204,205]. Though polyps are usually well recognisable as defined, rounded pale-pink pedunculated masses, fine-needle aspiration biopsy (FNAB) or histological biopsies can be taken if there are atypical features to the history or presentation. Histologically, the epithelial coverage of polyps consists of stratified squamous to pseudostratified ciliated columnar cells (indicating their respiratory tract origin), covering a core of fibrovascular connective tissue containing scattered lymphocytes, plasma cells and macrophages (Fig. 5.72)[61,183,194,195].

The treatment of middle ear polyps that have protruded beyond the middle ear cavity is surgical in all cases. Described techniques include traction-avulsion, ventral bulla osteotomy, lateral ear resection, and TECA with LBO[183–185,190,192,197–199]. However, in the author's opinion, resection of part of or the entire ear

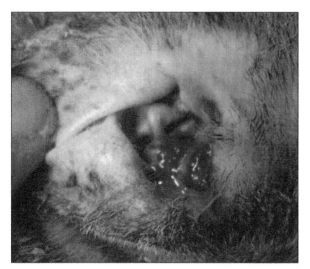

Fig. 5.68 A red polyp is visible in the cavum conchae of the auricle, protruding from the ear canal.

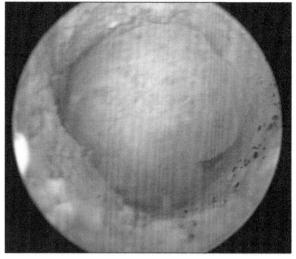

Fig. 5.69 Video-otoscopic image of a pink rounded polypous mass in the horizontal part of the ear canal of a Main Coon.

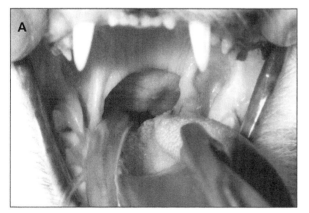

Figs 5.70A, B A: Throat inspection in a cat with stertor, ventral deflection of the caudal soft palate is visible indicating the presence of a large polyp in the nasopharyngeal space; B: the polyp visualised by retracting the soft palate.

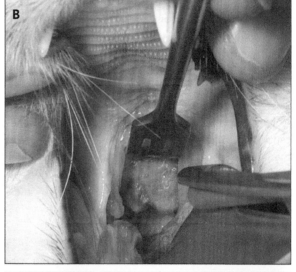

Fig. 5.71 Computed tomographic image of a normal right middle ear cavity and an abnormal left middle ear cavity that is filled with a contrast enhancing soft tissue mass in the dorsolateral compartment protruding through the external bony meatus into and filling the entire horizontal ear canal.

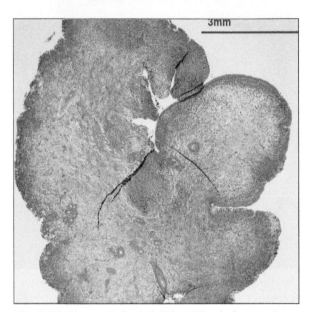

Fig. 5.72 Histopathological slide (HE) of a large poly-poid mass with a central core of mature collagen and numerous small blood vessels and aggregates of plasma cells and lymphocytes covered by non-keratinised squa-mous to ciliated pseudostratified epithelium.

canal is not necessary for complete polyp removal. Polyps in the nasopharynx can usually be visualised with retraction of the soft palate (see Chapter 19 Surgery of the Nose, Section 19.9). The polyp can then be grasped with right-angled forceps as far cranially as possible (close to the auditory tube opening), after which caudal traction is applied until the stalk of the polyp avulses from its origin. Recurrence rates are low after traction-avulsion of nasopharyngeal polyps (11–17%)[184,192,198]. Polyps in the ear canal removed via endoscopy or with the use of polyp-snaring devices have a higher incidence of recurrence, up to 50%[184], although recently better results have been reported using a perendoscopic tran-stympanic traction technique, which resulted in polyp recurrence in only 13.5% of the cases[206]. Using a lateral approach to the ear canal (see Chapter 18 Surgery of the Ear, Section 18.4) and opening the ear canal to allow the polyp to be grasped deep within the bony meatus allows for traction-avulsion of the entire polyp with stalk with a subsequent low recurrence rate, between 13 and 24%[173]. If polyps recur after simple traction-avulsion, ventral bulla osteotomy (VBO) is indicated[184,197,207]. VBO should be the first choice treatment for all patients with

polyps that present clinically with inner ear disease[208]. In addition to surgery, patients with secondary bacterial otitis media and/or interna require a long treatment (4 weeks) with antibiotics based on culture and sensitivity results[205,208,209]. Use of systemic corticosteroids after surgery is controversial. One study found that postoper-ative prednisolone therapy significantly reduced the rate of recurrence of polyps after traction, although the loca-tion of the polyp (pharyngeal versus aural) in patients that received prednisolone was not clearly defined[184]. Horner's syndrome is frequently seen after removal of polyps (40–80%) via either traction-avulsion or VBO (57-81%)[197-199,206]. In most cases, Horner's syndrome resolves in weeks to months.

5.10 PARA-AURAL ABSCESSATION

- Para-aural abscessation is caused by extension of purulent inflammation beyond the boundaries of the external ear canal or middle ear cavity.
- Most commonly, para-aural abscessation follows previous otic surgery, particularly total ear canal ablation and bulla osteotomy.
- Most cases require corrective surgery, which is always challenging.

Para-aural abscessation refers to the extension of puru-lent inflammation and infection outside the deeper parts of the external ear canal or the middle ear cavity, into the surrounding soft tissues. Para-aural absces-sation is more common in dogs, especially in Cocker Spaniels (Figs 5.73, 5.74), than in cats[210-212].

Extension of infection outside the boundaries of the ear canal and/or middle ear cavity can arise from trauma, neoplasia, foreign bodies, chronic otitis externa and previous ear surgery[173,213]. Traumatic separation of the auricular and annular cartilages results in blockage and stenosis of the external ear canal and subsequent absces-sation[175,176,213,214]. Congenital atresia of the external ear canal, though a rare occurrence, should be considered in the differential diagnosis of para-aural abscessation in young animals[210]. Neoplastic obstruction of the external ear canal may result in para-aural abscessation, when sup-purative secretions accumulate in the deeper part of the ear canal and break through the otic cartilage[210,215]. The

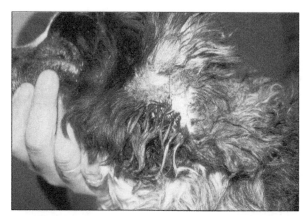

Fig. 5.73 Cocker Spaniel with a para-aural abscess. Note the area of matted hair below the pinna.

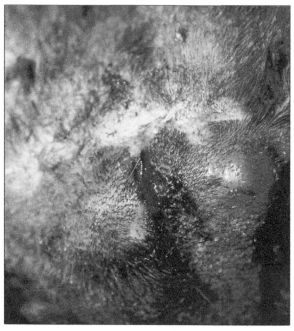

Fig. 5.74 Close up of the area in Fig. 5.73 after clipping. Note the serosanguinous discharge from the sinus and the tense, swollen areas of incipient sinus formation.

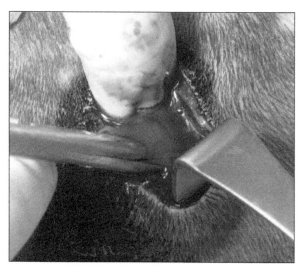

Fig. 5.75 A large para-aural, closed abscess was opened, sampled and drained to reduce local swelling, pain and discomfort prior to definitive treatment.

most common cause of para-aural abscessation however, is incomplete debridement of the epithelial lining of the external bony meatus during previous TECA[213,216]. In a series of 17 dogs with para-aural abscessation reported by Lane and Watkins[210], 50% had undergone previous otic surgery. Even with wide LBO, recurrent deep infection rates are still as high as 2–10%[161,216-220]. When performed for aural cholesteatoma, recurrent deep infection rate approaches 50%[221]. Deep wound infections are much less common in cats[64,222]. Incomplete removal of secretory epithelium lining the tympanic bulla or osseous canal, retained infected ear canal cartilage, osteomyelitis of the ossicles, inadequate drainage of the middle ear through the auditory tube and parotid salivary gland damage are factors implicated in the aetiopathogenesis[211,212,216,219–221]. Performing TECA with VBO instead of LBO does not reduce the risk of deep infection[161].

Clinical signs of pain upon opening the mouth, para-aural swelling and the development of draining tracts in the region of the ear base and parotid gland can appear from 1 month to years after TECA–LBO[211,212,216,221]. In addition there may be signs of otitis interna such as ataxia and head tilt toward the affected side, and of facial nerve paralysis[211,212].

Otoscopy will usually allow recognition of obstruction[210]. In closed abscessation, FNAB can yield purulent material that can be sent for culture and sensitivity testing. Large abscesses can be opened and drained if indicated before definitive treatment (Fig. 5.75). Exploration of any draining tracts with a probe will usually allow confirmation that the discharge is from the middle or external ear[210]. Though contrast radiography (fistulograms), especially using lateral oblique views, may be helpful in making a diagnosis[210–212], CT imaging is recommended as it provides useful information with respect to the cause of the condition and the recommended surgical approach (Fig. 5.76).

> A surgical treatment is recommended for treatment of para-aural abscessation that developed after previous surgery, and can include a lateral approach to the middle ear or VBO[211,212,216,221] (see Chapter 18 Surgery of the Ear, Sections 18.7–18.9).

If diagnostic imaging demonstrates signs of remnant tissue of the horizontal ear canal or tissue suggestive of retained epithelium within the external bony meatus, a lateral approach is preferred. In all other cases it depends on the surgeon's preference[211,216]. VBO in dogs is challenging though, as is a lateral approach in the absence of the external ear canal landmarks and risks of facial nerve neuropraxia or retroglenoid vein haemorrhage are increased[213,218]. Unless avulsions of the ear canal can be repaired or obstructions of the ear canal relieved, TECA with concurrent bulla osteotomy is also probably the treatment of choice in spontaneous cases[210].

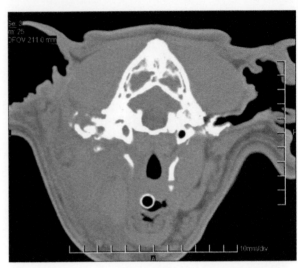

Fig. 5.76 Computed tomographic transverse image of an English Bulldog with a large para-aural abscess as a result of chronic otitis externa and media on the left hand side. Small sclerotic thick-walled (hypoplastic) tympanic bullae are visible on both sides as is calcification of both ear canals (more pronounced on the left hand side). Middle ear effusion is present on the left hand side and severe diffuse swelling of the entire para-aural area, which has lead to loss of visibility of normal facial planes.

5.11 REFERENCES

1 Griffin CE. Otitis: key points and tips. *Proceedings 17th North Carolina Veterinary Conference*, Raleigh 2012.

2 August JR. Otitis externa. A disease of multifactorial etiology. *Veterinary Clinics of North America Small Animal Medicine* 1988;**18**:731–42.

3 Olivry T, DeBoer DJ, Griffin CE. Editorial. American College of Veterinary Dermatology Task Force on Canine Atopic Dermatitis. Olivry T (ed). *Veterinary Immunology and Immunopathology* 2001;**81**:143–6.

4 Scott DW. Observations on canine atopy. *Journal of the American Animal Hospital Association* 1981;**17**:91–100.

5 Saridomichelakis MN, Farmaki R, Leontides LS, Koutinas AF. Aetiology of canine otitis externa: a retrospective study of 100 cases. *Veterinary Dermatology* 2007;**18**:341–7.

6 Wilhelm S, Kovalik M, Favrot C. Breed-associated phenotypes in canine atopic dermatitis. *Veterinary Dermatology* 2011;**22**:143–9.

7 Zur G, Ihrke PJ, White SD, Kass PH. Canine atopic dermatitis: a retrospective study of 266 cases examined at the University of California, Davis, 1992–1998. Part 1. Clinical features and allergy testing results. *Veterinary Dermatology* 2002;**13**:89–102.

8 Tarpataki N, Papa K, Reiczigal J, Vajdovich P, Voros K. Prevalence and features of canine atopic dermatitis in Hungary. *Acta Veterinaria Hungarica* 2006;**54**:353–66.

9 Colombo S, Hill PB, Shaw DJ, Thoday KL. Requirement for additional treatment for dogs with atopic dermatitis undergoing allergen-specific immunotherapy. *Veterinary Record* 2007;**160**:861–4.

10 Ravens PA, Xu BJ, Vogelnest LJ. Feline atopic dermatitis: a retrospective study of 45 cases (2001–2012). *Veterinary Dermatology* 2014;**25**:95–102.

11 Harvey RG. Food allergy and dietary intolerance in dogs: a report of 25 cases. *Journal of Small Animal Practice* 1993;**34**:175–9.

12 Rosser EJ. Diagnosis of food allergy in dogs. *Journal of the American Veterinary Medical Association* 1993;**203**:259–62.

13 Carlotti DN, Remy I, Prost C. Food allergy in dogs and cats: a review and report of 43 cases. *Veterinary Dermatology* 1990;**1**:55–62.

14 White SD, Sequoia D. Food hypersensitivity in cats: 14 cases (1982–1987). *Journal of the American Veterinary Medical Association* 1989;**194**:692–5.

15 Walder EJ, Conroy JD. Contact dermatitis in dogs and cats: pathogenesis, histopathology, experimental induction, and case reports. *Veterinary Dermatology* 1994;**5**:149–62.

16 Nesbitt GH, Schmitz JA. Contact dermatitis in the dog: a review of 35 cases. *Journal of the American Animal Hospital Association* 1997;**13**:155–63.

17 Griffin CE. Otitis externa and otitis media. In: *Current Veterinary Dermatology*. Griffin CE, Kwochka KW, MacDonald JM (eds). Mosby, St Louis 1993, pp. 245–62.

18 Scott DW, Miller WJ Jnr, Lewis RM, Manning TO, Smith CA. Pemphigus erythematosus in the dog and cat. *Journal of the American Animal Hospital Association* 1980;**16**:815–23.

19 Manning TO, Scott DW, Smith CA, Lewis RM. Pemphigus diseases in the feline: seven case reports and discussion. *Journal of the American Animal Hospital Association* 1982;**18**:433–43.

20 Scott DW, Walton DK, Manning TO, Smith CA, Lewis RM. Canine lupus erythematosus. 1: Systemic lupus erythematosus. *Journal of the American Animal Hospital Association* 1983;**19**:561–79.

21 Ihrke PJ, Stannard AA, Ardans AA, Griffin CE. Pemphigus foliaceus in dogs: a review of 37 cases. *Journal of the American Veterinary Medical Association* 1985;**186**:9–66.

22 Mueller RS, Krebs I, Power HT, Fieseler KV. Pemphigus foliaceus in 91 dogs. *Journal of the American Animal Hospital Association* 2006;**42**:189–96.

23 Vaughan DF, Hodgin EC, Hosgood GL, Bernstein JA. Clinical and histopathological features of pemphigus foliaceus with and without eosinophilic infiltrates: a restrospective evaluation of 40 dogs. *Veterinary Dermatology* 2001;**21**:166–74.

24 Peterson A, McKay L. Crusty cats: feline pemphigus foliaceus. *Compendium on Continuing Education* 2010;**32**:E1–4.

25 Scott DW, Horn RT. Zoonotic dermatoses of dogs and cats. *Veterinary Clinics of North America* 1987;**17**:117–44.

26 Powell MB, Weisbroth SH, Roth L, Wilhelmsen C. Reaginic hypersensitivity in *Otodectes cynotis* infestation of cats and mode of mite feeding. *American Journal of Veterinary Research* 1980;**6**:877–81.

27 Larkin AD, Gaillard GE. Mites in cats' ears: a source of cross antigenicity with house dust mites. *Annals of Allergy* 1981;**46**:301–4.

28 Kuwahara J. Canine and feline aural hematoma: clinical, experimental, and clinicopathologic observations. *American Journal of Veterinary Research* 1986;**47**:2300–8.

29 Kuwahara J. Canine and feline aural hematomas: results of treatment with corticosteroids. *Journal of the American Animal Hospital Association* 1986;**22**:641–7.

30 Hewitt M, Walton GS, Waterhouse M. Pet animal infestations and skin lesions. *British Journal of Dermatology* 1971;**85**:215–55.

31 Frost CR. Canine otocariasis. *Journal of Small Animal Practice* 1961;**2**:253–6.

32 Scott DW. Feline dermatology 1900–1978. A monograph. *Journal of the American Animal Hospital Association* 1980;**16**:331–459.

33 Park G-S, Park J-S, Cho B-K, Lee W-K, Cho J-H. Mite infestation rate of pet dogs with ear dermatoses. *Korean Journal of Parasitology* 1996;**34**:143–50.

34 Perego R, Proverbio D, Bagnagatti De Giorgi G, Della Pepa A, Spada E. Prevalence of otitis externa in stray cats in northern Italy. *Journal of Feline Medicine and Surgery* 2013;**16**:483–90.

35 Grono LR. Studies of the ear mite, *Otodectes cynotis*. *Veterinary Record* 1969;**85**:6–8.

36 Knottenbelt MK. Chronic otitis externa due to *Demodex canis* in a Tibetan Spaniel. *Veterinary Record* 1994;**135**:409–10.

37 Greene RT, Scheidt VJ, Moncol DJ. Trombiculiasis in a cat. *Journal of the American Veterinary Medical Association* 1986;**188**:1054–5.

38 Moriello KA. Common ectoparasites of the dog. Part 1: Fleas and ticks. *Canine Practice* 1987;**14**:6–18.

39 White SD, Scott KV, Cheney JM. *Otobius megnini* infestation in three dogs. *Veterinary Dermatology* 1995;**6**:33–5.

40 Pancierra DL. Hypothyroidism in dogs: 66 cases (1987–1992). *Journal of the American Veterinary Medical Association* 1994;**204**:761–7.

41 Ling GV, Stabenfeldt GH, Comer KM, Gribble DH, Schechter RD. Canine hyperadrenocorticism: pretreatment clinical and laboratory evaluation of 117 cases. *Journal of the American Veterinary Medical Association* 1979;**174**:1211–15.

42 White SD, Ceragioli KL, Bullock LP, Mason GD. Cutaneous markers of canine hyperadrenocorticism. *Compendium on Continuing Education* 1989;**11**:446–64.

43 Peikes H, Morris DO, Hess RS. Dermatologic disorders in dogs with diabetes mellitus: 45 cases (1986–2000). *Journal of the American Veterinary Medical Association* 2001;**15**:203–8.

44 Schmeitzel LP, Lothrop CD. Sex hormones and the skin. *Veterinary Medicine Report* 1990;**2**:28-41.

45 Kwochka KW. Overview of normal keratinization and cutaneous scaling disorders of dogs. In: *Current Veterinary Dermatology*. Griffin CE, Kwochka KW, MacDonald JM (eds). Mosby, St Louis 1993, pp. 167–75.

46 Paradis M, Scott DW. Hereditary primary seborrhoea in Persian cats. *Feline Practice* 1990;**18**:17–20

47 Kwochka KW, Rademakers AM. Cell proliferation kinetics of epidermis, hair follicles, and sebaceous glands of Cocker Spaniels with idiopathic seborrhea. *American Journal of Veterinary Research* 1989;**50**:1918–22.

48 Scott DW, Miller WH Jnr. Epidermal dysplasia and *Malassezia pachydermatis* infection in West Highland White Terriers. *Veterinary Dermatology* 1989;**1**:25–36.

49 Maudlin EA, Scott DW, Miller WH Jnr, Smith CA. Malassezia dermatitis in the dog: a retrospective histopathological and immunopathological study of 86 cases (1990–95). *Veterinary Dermatology* 1997;**8**:183–90.

50 Bond R, Rose JF, Ellis JW, Lloyd DH. Comparison of two shampoos for treatment of *Malassezia pachydermatis*-associated seborrhoeic dermatitis in Basset Hounds. *Journal of Small Animal Practice* 1995;**36**:99–104.

51 Maudlin EA, Ness TA, Goldschmidt MH. Proliferative and necrotizing otitis externa in four cats. *Veterinary Dermatology* 2007;**18**:370–6.

52 Vidèmont E, Pin D. Proliferative and necrotizing otitis in a kitten: first demonstration of T cell-mediated apoptosis. *Journal of Small Animal Practice* 2010;**51**:599–603.

53 McKeever PJ, Torres S. Otitis externa. Part 1: The ear and predisposing factors to otitis externa. *Companion Animal Practice* 1988;**2**:7–14.

54 Brennan KE, Ihrke PJ. Grass awn migration in dogs and cats: a retrospective study of 182 cases. *Journal of the American Veterinary Medical Association* 1983;**182**:1201–4.

55 Rycroft AK, Saben HS. A clinical study of otitis externa in the dog. *Canadian Veterinary Journal* 1977;**18**:64–70.

56 Roth L. Pathologic changes in otitis externa. *Veterinary Clinics of North America Small Animal Practice* 1988;**14**:755–64.

57 Sula, MJM. Tumors and tumorlike lesions of dog and cat ears. *Veterinary Clinics of North America Small Animal Practice* 2012;**52**:1161–78.

58 London CA, Dubilzieg RR, Vail DM, *et al*. Evaluation of dogs and cats with tumors of the ear canal: 145 cases (1978–1992). *Journal of the American Veterinary Medical Association* 1966;**208**:1413–18.

59 Legendre AM, Krahwinkel AJ. Feline ear tumours. *Journal of the American Animal Hospital Association* 1981;**17**:1035–7.

60 Goldschmidt MH, Shofer FS. Ceruminous gland tumors. In: *Skin Tumors of the Dog and Cat*. Goldschmidt MH, Shofer FS (eds). Pergammon Press, New York 1992, pp. 96–102.

61 Kirpenstein J. Aural neoplasms. *Seminars in Veterinary Medicine and Surgery (Small Animal)* 1993;**8**:17–23.

62 Corriveal LA. Use of carbon dioxide laser to treat ceruminous gland hyperplasia in a cat. *Journal of Feline Medicine and Surgery* 2012;**14**:413–16.

63 Boord M. Fun with lasers. In: *Advances in Veterinary Dermatology*. Torres SMF, Frank LA, Hargis AM (eds). John Wiley & Sons, New York 2013.

64 Marino DJ, MacDonald JM, Matthiesen DT, Patnaik AK. Results of surgery in cats with ceruminous gland adenocarcinoma. *Journal of the American Animal Hospital Association* 1994;**30**:54–8.

65 Marino DJ, MacDonald JM, Matthiesen DT, Salmeri KR, Patnaik AK. Results of surgery and long-term follow-up in dogs with ceruminous gland adenocarcinoma. *Journal of the American Animal Hospital Association* 1993;**29**:560–3.

66 Theon AP, Barthez PY, Madewell BR, Griffey SM. Radiation therapy of ceruminous gland carcinomas in dogs and cats. *Journal of the American Veterinary Medical Association* 1994;**205**:566–9.

67 Howlett CR, Allan GS. Tumours of the feline ear canal. *Australian Veterinary Practitioner* 1974;**4**:56–7.

68 Poulet FM, Valentine BA, Scott DW. Focal proliferative eosinophilic dermatitis of the external ear canal in four dogs. *Veterinary Pathology* 1991;**28**:171–3.

69 Fraser G. Factors predisposing to canine external otitis. *Veterinary Record* 1961;73:55–8.

70 McCarthy G, Kelly WR. Microbial species associated with the canine ear and their antibacterial sensitivity patterns. *Irish Veterinary Journal* 1982;**36**:53–6.

71 Uchida Y, Tetsuya N, Kitazawa K. Clinicomicrobiological study of the normal and otitis externa ear canals in dogs and cats. *Japanese Journal of Veterinary Science* 1990;**52**:415–17.

72 Baba E, Fukata T. Incidence of otitis externa in dogs and cats in Japan. *Veterinary Record* 1981;**108**:393–5.

73 Harihan H, Coles M, Poole D, Lund L, Page R. Update on antimicrobial susceptibilities of bacterial isolates from canine and feline otitis externa. *Canadian Veterinary Journal* 2006;**47**:253–5.

74 Lyskova L, Vydrzalova M, Mazurova J. Identification and antimicrobial susceptibility of bacteria and yeasts isolated from healthy dogs and dogs with otitis externa. *Journal of Veterinary Medicine A. Physiology, Pathology and Clinical Medicine* 2007;**54**:559–63.

75 Zamamkhan Malayeri H, Jamshidi S, Zahraei Salehi T. Identification and antimicrobial susceptibility patterns of bacteria causing otitis externa in dogs. *Veterinary Research Communications* 2010;**34**:435–44.

76 Bugden D. Identification and antibiotic susceptibility of bacterial isolates from dogs with otitis externa in Australia. *Australian Veterinary Journal* 2013;**91**:43–6.

77 Oliveira LC, Leite CAL, Brihante RSN, Carvalho CBM. Comparative study of the microbial profile from bilateral canine otitis externa. *Canadian Veterinary Journal* 2009;**49**:785–8.

78 Carfachia C, Gallo S, Capelli G, Otranto D. Occurrence and population size of *Malassezia* spp. in the external ear canal of dogs and cats both healthy and with otitis. *Mycopathologia* 2005;**160**:143–9.

79 Girão MD, Prado MR, Brilhante RS, *et al.* *Malassezia pachydermatitis* isolated from normal and diseased external ear canals in dogs: a comparative analysis. *Veterinary Journal* 2006;**172**:544–8.

80 Shokri H, Khosravi A, Rad M, Jamshidi S. Occurrence of *Malassezia* species in Persian and domestic short hair cats with and without otitis externa. *Journal of Veterinary Medicine and Science* 2010;**72**:293–6.

81 Prado MR, Brilhant RS, Cordeiro RA, Monteiro AJ, Sidrim JJ, Rocha MF. Frequency of yeast and dermatophytes from healthy and diseased dogs. *Journal of Veterinary Investigation* 2008;**20**:197–202.

82 Fernando SDA. Certain histopathological features of the external auditory meatus of the cat and dog with otitis externa. *American Journal of Veterinary Research* 1966;**28**:278–82.

83 Henneveld K, Rosychuk RA, Olea-Popelka FJ, Hyatt DR, Zabel S. *Corynebacterium* spp. in dogs and cats with otitis externa and/or media: a retrospective study. *Journal of the American Animal Hospital Association* 2012;**48**:320–6.

84 Hijazin M, Porenger-Berninghoff E, Sammra O, *et al*. *Arcanobacterium canis* sp. Nov., isolated from otitis externa of a dog, and emended description of the genus Arcanobacterium Collins *et al*., 1983 emend. Yassin *et al*. 2011. *International Journal of Systematic and Evolutionary Microbiology* 2012;**62**:2201–5.

85 Sammra O, Balbutskaya A, Zhang S, *et al*. Further characteristics of *Arcanobacterium canis*, a novel species of genus Arcanobacterium. *Veterinary Microbiology* 2013;**167**:619–22.

86 Huang H-P, Little CJL, McNeil PE. Histological changes in the external ear canal of dogs with otitis externa. *Veterinary Dermatology* 2009;**20**:422–8.

87 Fraser G. The histopathology of the external auditory meatus of the dog. *Journal of Comparative Pathology* 1961;**71**:253–8.

88 Stout-Graham M, Kainer RA, Whalen LR, Macy DW. Morphologic measurements of the external ear canal of dogs. *American Journal of Veterinary Research* 1990;**51**:990–4.

89 Angus JC, Lichtensteiger C, Campbell KL, Schaeffer DJ. Breed variations in histopathologic features of chronic severe otitis externa in dogs. *Journal of the American Veterinary Medical Association* 2002;**221**:1000–6.

90 Huang HP, Fixter LM, Little CJL. Lipid content of cerumen from normal dogs and otitic canine ears. *Veterinary Record* 1994;**134**:380–1.

91 Grono LR. Studies of the microclimate of the external auditory canal in the dog. 111: Relative humidity within the external auditory meatus. *Research in Veterinary Science* 1970;**11**:316–19.

92 Fraser G, Gregor WW, Mackenzie CP, Spreull JSA, Withers AR. Canine ear disease. *Journal of Small Animal Practice* 1970;**10**:725–54.

93 Lane JG, Little CJL. Surgery of the external auditory meatus: a review of failures. *Journal of Small Animal Practice* 1986;**27**:247–54.

94 Griffin CE, DeBoer DJ. The ACVD Task Force on Canine Atopic Dermatitis (XIV): Clinical manifestation of canine atopic dermatitis. *Veterinary Immunology and Immunopathology* 2001;**81**:255–69.

95 Blondeau JM, Borsos, Blondeau LD, Blondeau BJ. *In vitro* killing of *Escherichia coli*, *Staphylococcus pseudintermedius* and *Pseudomonas aeruginosa* by enrofloxacin and its active metabolite ciprofloxacin using clinically relevant drug concentrations in the dog and cat. *Veterinary Microbiology* 2012;**155**:284–90.

96 Penna B, Varges R, Madeiros L, *et al*. Species distribution and antimicrobial susceptibility of staphylococci isolated from canine otitis externa. *Veterinary Dermatology* 2009;**21**:292–6.

97 Hariharan H, Coles M, Poole D, Lund L, Page R. Update on antimicrobial susceptibilities of bacterial isolates from canine and feline otitis externa. *Canadian Veterinary Journal* 2006;**47**:253–5.

98 Lilenbaum W, Veras M, Blum E, Souza GN. Antimicrobial susceptibility of staphylococci isolated from otitis externa in dogs. *Letters in Applied Bacteriology* 2000;**31**:42-5.

99 Yamashita K, Shimuzu A, Kawano J, Uchida E, Haruna A, Igimi S. Isolation and characterization of staphylococci from the external auditory meatus of dogs with or without otitis externa with special reference to *Staphylococcus schleiferi* subsp. coagulans isolates. *Japanese Journal of Veterinary Medical Science* 2005;**67**:263–8.

100 Bloom P. Diagnosis and management of otitis in the real world. *Proceedings Ontario Veterinary Medical Association*, Westin Harbour, Toronto 2012, pp. 19–24.

101 Robson DC, Burton GG, Bassett RJ. Correlation between topical antibiotic selection, *in vitro* bacterial antibiotic sensitivity and clinical response in 17 case of canine otitis externa complicated by *Pseudomonas aeriginosa*. In: *College Science Week Proceedings* (Dermatology Chapter), Australian College of Veterinary Scientists, Gold Coast 2010, pp. 101–4.

102 Nuttall T, Cole LK. Evidence-based veterinary dermatology: a systemic review of interventions for treatment of pseudomonas otitis in dogs. *Veterinary Dermatology* 2007;**18**:689–77.

103 Steen SI, Paterson S. The susceptibility of *Pseudomonas* spp. isolated from dogs with otitis to topical ear cleaners. *Journal of Small Animal Practice* 2012;**53**:599–603.

104 Guardabassi L, Ghibaudo G, Damborg P. *In vitro* antimicrobial activity of a commercial ear antiseptic containing chlorhexidine and Tris-EDTA. *Veterinary Dermatology* 2012;**21**:282–6.

105 Ghibaudo G, Cornegliani L, Martino P. Evaluation of the *in vivo* effects of Tris-EDTA and chlorhexidine digluconate 0.15% solution in chronic bacterial otitis externa: 11 cases. *Veterinary Dermatology* 2004;**15**:65.

106 Bouassiba C, Osthold W, Mueller RS. *In vivo* efficacy of a commercial ear antiseptic containing chlorhexidine and Tris-EDTA. A randomised, placebo-controlled, double blinded comparative trial. *Tierärztliche Praxis. Ausgabe K, Kleintiere/ Heimtiere* 2012;**40**:161–70.

107 Hosseini J, Zdovc I, Golob M, *et al.* Effect of treatment with TRIS-EDTA/chlorhexidine topical solution on canine *Pseudomonas aeruginosa* otitis externa with or without concomitant treatment with oral fluoroquinolones. *Slovenian Veterinary Research* 2012;**49**:133–40.

108 Schick AE, August JC, Coyner KS. Variability of laboratory identification and antibiotic susceptibility reporting of *Pseudomonas* spp. isolates from dogs with chronic otitis externa. *Veterinary Dermatology* 2007;**8**:120–6.

109 Seol B, Napliç T, Madiç J, Bedokoviç M. *In vitro* antimicrobial susceptibility of 183 *Pseudomonas aeurginosa* strains isolated from dogs to selected antipseudomonal agents. *Journal of Veterinary Medicine. B, Infectious Diseases and Veterinary Public Health* 2002;**49**:188–92.

110 Maretin Barrassa LJ, Lupiola Gómez P, González Z, Tejedor Junco MT. Antibacterial susceptibility patterns of *Pseudomonas* strains isolated from chronic canine otitis externa. *Journal of Veterinary Medicine. B, Infectious Diseases and Veterinary Public Health* 2000;**47**:191–6.

111 Mekiç S, Matanoviç K, Šeol B. Antimicrobial susceptibility of *Pseudomonas aeruginosa* isolates from dogs with otitis externa. *Veterinary Record* 2011;**169**:125.

112 Rubin J, Walker RD, Blickenstaff K, Bodeis-Jones S, Zhao S. Antimicrobial resistance and genetic characterization of fluoroquinolone resistance of *Pseudomonas aeruginosa* isolated from canine infections. *Veterinary Microbiology* 2008;**131**:164–72.

113 Wildermuth BE, Griffin CE, Rosenkrantz WS, Boord MJ. Susceptibility of *Pseudomonas* isolates from ears and skin of dogs to enrofloxacin, marbofloxacin and ciprofloxacin. *Journal of the American Animal Hospital Association* 2007;**43**:337–41.

114 McKay L, Rose CD, Matousek JL, Schmeitzel LS, Gibson NM, Gaskill JM. Antimicrobial testing of selected fluoroquinolones against *Pseudomonas aeruginosa* isolated from canine otitis. *Journal of the American Animal Hospital Association* 2007;**43**:307–12.

115 Campbell JJ, Coyner KS, Rankin SC, *et al.* Evaluation of fungal flora in normal and diseased canine ears. *Veterinary Dermatology* 2010;**21**:619–25.

116 Chiavassa E, Tizzani P, Peano A. *In vitro* antifungal susceptibility of *Malassezia pachydermatitis* strains isolated from dogs with chronic otitis and acute otitis externa. *Mycopathologia* 2014;**178**:315–19.

117 Han SH, Chung TH, Nam EH, Park SH, Hwang CY. Molecular analysis of *Malassezia pachydermatis* isolated from canine skin and ear in Korea. *Medical Mycology* 2013;**51**:396–404.

118 Figuerodo LA, Carfachia C, Desantis S, Otranto D. Biofilm formation of *Malassezia pachydermatitis* from dogs. *Veterinary Mycology* 2012;**160**:126–31.

119 Oldenhoff WE, Frank GR, DeBoer D. Comparison of the results of intradermal test reactivity and serum allergen-specific IgE measurement for *Malassezia pachydermatitis* in atopic dogs. *Veterinary Dermatology* 2014;**25**:505–11.

120 Morris DO, DrBoer DJ. Evaluation of serum obtained from atopic dogs with dermatitis attributable to *Malassezia pachydermatis* for passive transfer of immediate hypersensitivity to that organism. *American Journal of Veterinary Research* 2003;**64**:262–3.

121 Cole LK, Kwochka KW, Kowalski JJ, *et al.* Microbial flora and antimicrobial susceptibility patterns of isolated pathogens from the horizontal ear canal and middle ear in dogs with otitis media. *Journal of the American Veterinary Medical Association* 1998;**212**:534–8.

122 Griffin CE. Otitis: the basics. *Proceedings 17th North Carolina Veterinary Conference*, Raleigh 2012, pp. 79–83.

123 Farca AM, Piromalli G, Maffei F, Re G. Potentiating effect of EDTA-Tris on the activity of antibiotics against resistant bacteria associated with otitis, dermatitis and cystitis. *Journal of Small Animal Practice* 1997;**38**:243–5.

124 Buckley LM, McEwan NA, Nuttall T. Tris-EDTA significantly enhances antibiotic efficacy against multidrug-resistant *Pseudomonas aeruginosa in vitro*. *Veterinary Dermatology* 2013;**24**:519–26.

125 Metry CA, Maddox CW, Dirikolu L, Johnson YJ, Campbell KL. Determination of enrofloxacin stability and in vitro efficacy against *Staphylococcus pseudintermedius* and *Pseudomonas aeruginosa* in four ear cleaner solutions over a 28 day period. *Veterinary Dermatology* 2011;**23**:23–6.

126 Sparks TA, Kemp DT, Wooley RE, Gibbs PS. Antimicrobial effect of combinations of EDTA-Tris and amikacin or neomycin on the microorganisms associated with otitis externa in dogs. *Veterinary Research Communications* 1994;**18**:241–9.

127 Jerzsele A, Pásztiné-Gere E. Evaluating synergy between marbofloxacin and gentamicin in *Pseudomonas aeruginosa* strains isolated from dogs with otitis externa. *Acta Microbiologica et Immunologica Hungarica* 2015;**62**:45–55.

128 Trott DJ, Moss SM, See AM, Rees R. Evaluation of disc diffusion and MIC testing for determining susceptibility of *Pseudomonas aeruginosa* isolates to topical enrofloxacin/silver sulfadiazine. *Australian Veterinary Journal* 2007;**85**:464–6.

129 Pietschmann S, Meyer M, Voget M, Cieslicki M. The joint *in vitro* action of polymyxin B and miconazole against pathogens associated with canine otitis externa from three different European countries. *Veterinary Dermatology* 2013;**24**:439–45.

130 Rougier S, Borell D, Pheulpin S, Woehrlé F, Boisramé B. A comparative study of two antimicrobial/anti-inflammatory formulations in the treatment of canine otitis externa. *Veterinary Dermatology* 2005;**16**:299–307.

131 Engelen M, De Bock M, Hare J, Goossens L. Effectiveness of an otic product containing miconazole, polymixin B and prednisolone in the treatment of canine otitis externa: Multi field trial in the US and Canada. *International Journal of Applied Research in Veterinary Medicine* 2010;**8**:21–30.

132 Gunjan K, Shobha C, Sheetal C, Nanda H, Vikrant C, Chitnas DS. A comparative study of the effect of different topical agents on burn wound infections. *Indian Journal of Plastic Surgery* 2012;**45**:374–8.

133 Yabanoglu H, Basaran O, Aydogan C, *et al.* Assessment of the effectiveness of silver-coated dressing, chlorhexidine acetate (0.5%), citric acid (3%) and silver sulphadiazine (1%) for topical antibacterial effects against multi-resistant *Pseudomonas aeruginosa* infecting full-skin thickness burn wounds on rats. *International Surgery* 2013;**98**:16–23.

134 Noxon JO, Kinyon JM, Murphy DP. Minimum inhibitory concentration of silver sulfadiazine on *Pseudomonas aeruginosa* and *Staphylococcus intermedius* isolates from the ears of dogs with otitis externa. *Proceedings of the 13th Annual Meeting of the AAVD/ACVD*, Nashville 1997, pp. 12–13.

135 Foster AP, DeBoer DJ. The role of *Pseudomonas aeruginosa* in canine ear disease. *Compendium on Continuing Education* 1998;**20**:909–19.

136 Fernández-Olmos A, Garcia-Castillo M, Maiz L, Lamas A, Baquero F, Canton R. *In vitro* prevention of *Pseudomonas aeruginosa* early biofilm formation with antibiotics used in cystic fibrosis patients. *International Journal of Antimicrobial Agents* 2012;**40**:173–6.

137 Lin D, Foley SL, Qi Y, *et al.* Characterisation of antimicrobial resistance of *Pseudomonas aeuginosa* isolated from canine infections. *Journal of Applied Microbiology* 2012;**113**:16–23.

138 Paterson S, Payne L. Brainstem auditory evoked responses in 37 dogs with otitis media before and after topical therapy. *Veterinary Dermatology* 2008;**19**:S30.

139 Bateman FL, Moss SM, Trott DJ, Shipstone MA. Biological efficacy and stability of diluted ticarcillin-clavulanic acid in the topical treatment of *Pseudomonas aeruginosa* infections. *Veterinary Dermatology* 2011;**23**:97–102.

140 Morris DO. Medical therapy of otitis externa and otitis media. *Veterinary Clinics of North America Small Animal Practice* 2004;**34**:541–55.

141 Knoll BN, Mylonakis E. Antibacterial bio-agents based on principles of bacteriophage biology – an overview. *Clinical Infectious Diseases* 2014;**58**:528–34.

142 Soothill J. Use of bacteriophages in the treatment of *Pseudomonas aeruginosa* infections. *Expert Reviews in Anti-infective Therapy* 2013;**11**:909–13.

143 Saussereau E, Debarbieux L. Bacteriophages in the experimental treatment of *Pseudomonas aeruginosa* infections in mice. *Advances in Virus Research* 2012;**83**:123–41.

144 Hawkins C, Harper D, Burch D, Anggård E, Soothill J. Topical treatment of *Pseudomonas aeruginosa* otitis of dogs with a bacteriophage mixture: a before/after clinical trial. *Veterinary Microbiology* 2010;**146**:309–13.

145 Van Duikeren E, Catry B, Greko C, *et al.* Review on methicillin-resistant *Staphylococcus pseudintermedius*. *Journal of Antimicrobial Chemotherapy* 2011;**66**:2705–14.

146 Frank LA, Loeffler A. Methicillin-resistant *Staphylococcus pseudintermedius*: clinical challenge and treatment options. *Veterinary Dermatology* 2012;**23**:283–92.

147 Ishira K, Shimokubo N, Skagami A, *et al.* Occurrence and molecular characteristics of methicillin-resistant *Staphylococcus aureus* and methicillin-resistant *Staphylococcus pseudintermedius* in an academic teaching hospital. *Applied and Environmental Microbiology* 2010;**76**:5165–74.

148 Morris DO, Rook KA, Shofer FS, *et al.* Screening of *Staphyloccus aureus*, *Staphylococcus intermedius* and *Staphylococcus schleiferi* isolates from small animal for antimicrobial resistance: a retrospective review of 749 isolates (2003–04). *Veterinary Dermatology* 2006;**17**:332–7.

149 Gandolfi-Decristiphoris P, Regula G, Petrini O, Zinsstag J, Schelling E. Prevalence and risk factors for carriage of multi-resistant staphylococci in healthy dogs and cats. *Journal of Veterinary Science* 2013;**14**:449–56.

150 Papich MG. Selection of antibiotics for methicillin-resistant *Staphylococcus pseudintermedius*: time to revisit some old drugs? *Veterinary Dermatology* 2012;**23**:352–61.

150a Grönthal T, Ollilainen M, Eklund M, *et al.* Epidemiology of methicillin resistant *Staphylococcus pseudintermedius* in guide dogs in Finland. *Acta Veterinaria Scandivica* 2015;**57**:37–47.

151 Lehner G, Linek M, Bond R, *et al.* Case-control risk factor study of methicillin-resistant *Staphylococcuus pseudintermedius* (MRSP) infection in dogs and cats in Germany. *Veterinary Microbiology* 2013;**168**:154–60.

152 Hamilton E, Kruger JM, Schall W, Beal M, Manning SD, Kaneene JB. Acquisition and persistence of antimicrobial-resistant bacteria isolated from dogs and cats admitted to a veterinary teaching hospital. *Journal of the American Veterinary Medical Association* 2013;**243**:990–1000.

153 Weese JS, Faires MC, Frank LA, Reynolds LM, Battisti A. Factors associated with methicillin-resistant versus methicillin-susceptible *Staphylococcus pseudintermedius* infection in dogs. *Journal of the American Veterinary Medical Association* 2012;**240**:1450–5.

154 Paul NC, Moodley A, Ghibaudo G, Guardabassi L. Carriage of methicillin-resistant *Staphylococcus pseudintermedius* in small-animal veterinarians: indirect evidence of zoonotic transmission. *Zoonoses and Public Health* 2011;**58**:533–9.

155 Windahl U, Reimegård E, Holst BS, *et al.* Carriage of methicillin-resistant *Staphylococcus pseudintermedius* in dogs – a longitudinal study. *BMC Veterinary Research* 2012;**8**:34–42.

156 Robson DC, Burton GG, Bassett RJ. Correlation between topical antibiotic selection, *in vitro* bacterial antibiotic sensitivity and clinical response in 17 case of canine otitis externa complicated by *Pseudomonas aeriginosa*. In: *College Science Week Proceedings* (Dermatology Chapter), Australian College of Veterinary Scientists, Gold Coast 2010, pp. 101–4.

157 Boyen F, Verstappen KMWH, Dr Bock M, *et al. In vitro* antimicrobial activity of miconazole and polymixin B against canine methicillin-resistant *Staphylococcus aureus* and methicillin-resistant *Staphylococcus pseudintermedius*. *Veterinary Dermatology* 2012;**23**:381–5.

158 Wefstaedt P, Behrens B-A, Nolte I, Bouguecha A. Fine element modelling of the canine and feline outer ear canal: benefits for local drug delivery. *Berliner und Münchener Tierärztliche Wochenschrift* 2011;**124**:78–82.

159 Guardabassi L, Ghibaudo G, Damborg P. *In vitro* antimicrobial activity of a commercial ear antiseptic containing chlorhexidine and Tris-EDTA. *Veterinary Dermatology* 2010;**21**:282–6.

160 Bouassiba C, Osthold W, Mueller RS. *In vivo* efficacy of a commercial ear antiseptic containing chlorhexidine and Tris-EDTA. A randomised, placebo-controlled, double blinded comparative trial. *Tierärztliche Praxis. Ausgabe K, Kleintiere/Heimtiere* 2012;**40**:161–70.

160a Maruhashi E, Braz BS, Nunes T, *et al*. Efficacy of medical grade honey in the management of canine otitis externa – a pilot study. *Veterinary Dermatology* 2016;**27**:93.

161 Doyle RS, Kelly, C, Bellenger CR. Surgical management of 43 cases of chronic otitis externa in the dog. *Irish Veterinary Journal* 2004;**57**:22–30.

162 Sharp NJ. Chronic otitis externa and otitis media treated by total ear canal ablation and ventral bulla osteotomy in thirteen dogs. *Veterinary Surgery* 1990;**19**:162–6.

163 Smeak DD, Kerpsack SJ. Total ear canal ablation and lateral bulla osteotomy for management of end-stage otitis. *Seminars in Veterinary Medicine and Surgery (Small Animal)*1993;**8**:30–41.

164 Boord M. Laser in dermatology. *Practice Clinical Techniques in Small Animal* 2006;**21**:145–9.

165 Hall JA, Waisglass SE, Mathews KA, Tait JL. Oral cyclosporin in the treatment of end-stage ear disease: a pilot study. *Veterinary Dermatology* 2003;**14**:212.

166 Rosychuck RA. Challenges in otitis. *Advances in Veterinary Dermatology*, Volume 7. Torres SMF (ed). Wiley-Blackwell, Ames 2012, pp. 298–304.

166a Di Cerbo A, Centenaro S, Beribè F, *et al*. Clinical evaluation of an antiinflammatory and antioxidant diet effect in 30 dogs affected by chronic otitis externa: preliminary results. *Veterinary Research Communications* 2016;Epublished.

167 Kennis RA. Feline otitis: diagnosis and treatment. *Veterinary Clinics of North America Small Animal Practice* 2013;**43**:51–6.

168 Sotiraki ST, Koutinas AF, Leontides LS, Adamama-Moraitou KK, Himonas CA. Factors affecting the frequency of ear canal and face infestation by *Otodectes cynotis* in the cat. *Veterinary Parasitology* 2001;**96**:309–15.

169 Powell MB, Weisbroth SH, Roth L, Wilhelmsen C. Reaginic hypersensitivity in *Otodectes cynotis* infestation of cats and mode of mite feeding. *American Journal of Veterinary Research* 1980;**41**:877–82.

170 De Lorenzi D, Bonfanti U, Masserdotti C, Tranquillo M. Fine-needle biopsy of external ear canal masses in the cat: cytologic results and histologic correlations. *Veterinary Clinical Pathology* 2005;**34**:100–5.

171 Lanz OI, Wood BC. Surgery of the ear and pinna. *Veterinary Clinics of North America Small Animal Practice* 2004;**34**:567–99.

172 Corriveau LA. Use of a carbon dioxide laser to treat ceruminous gland hyperplasia in a cat. *Journal of Feline Medicine and Surgery* 2012;**14**:413–16.

173 Venker-van Haagen AJ. The Ear. In: *Ear, Nose, Throat, and Tracheobronchial Diseases in Dogs and Cats*. Venker-van Haagen AJ (ed). Schlütersche Verlagsgesellschaft, Hannover 2005, pp. 1–50.

174 Steiss JE, Boosinger TR, Wright JC, Storrs DP, Pillai SR. Healing of experimentally perforated tympanic membranes demonstrated by electrodiagnostic testing and histopathology. *Journal of the American Animal Hospital Association* 1992;**28**:307–10.

175 Boothe HW, Hobson HP, McDonald DE. Treatment of traumatic separation of the auricular and annular cartilages without ablation: results in five dogs. *Veterinary Surgery* 1996;**25**(5):376–9.

176 Connery NA, McAllister H, Hay CW. Para-aural abscessation following traumatic ear canal separation in a dog. *Journal of Small Animal Practice* 2001;**42**(5):253–6.

177 McCarthy PE, Hosgood G, Pechman RD. Traumatic ear canal separations and para-aural abscessation in three dogs. *Journal of the American Animal Hospital Association* 1995;**31**(5):419–24.

178 Bacon NJ. Pinna and External Ear Canal. In: *Veterinary Surgery Small Animal*. Tobias KM, Johnston SA (eds). Elsevier Saunders, St. Louis 2012, pp. 2059–77.

179 Tivers MS, Brockman DJ. Separation of the auricular and annular ear cartilages: surgical repair technique and clinical use in dogs and cats. *Veterinary Surgery* 2009;**38**(3):349–54.

180 Clarke SP. Surgical management of acute ear canal separation in a cat. *Journal of Feline Medicine and Surgery* 2004;**6**(4):283–6.

181 Kyles AE. Traumatic separation of the auricular and annular cartilages in two cats. *Veterinary Record* 2001;**148**(22):696–7.

182 Smeak DD. Traumatic separation of the annular cartilage from the external auditory meatus in a cat. *Journal of the American Veterinary Medical Association* 1997;**211**(4):448–50.

183 Pope ER. Feline inflammatory polyps. *Seminars in Veterinary Medicine and Surgery Small Animal* 1995;**10**(2):87–93.

184 Anderson DM, Robinson RK, White RA. Management of inflammatory polyps in 37 cats. *Veterinary Record* 2000;**147**(24):684–7.

185 Fan TM, de Lorimier L-P. Inflammatory polyps and aural neoplasia. *Veterinary Clinics of North America Small Animal Practice* 2004;**34**(2):489–509.

186 ter Haar G. Inner ear dysfunction in dogs and cats: Conductive and sensorineural hearing loss and peripheral vestibular ataxia. *European Journal of Companion Animal Practice* 2007;**17**:127–35.

187 Bedford PG, Coulson A, Sharp NJ, Longstaffe JA. Nasopharyngeal polyps in the cat. *Veterinary Record* 1981;**109**(25–26): 551–3.

188 Baker G. Nasopharyngeal polyps in cats. *Veterinary Record* 1982;**111**(2):43.

189 Kudnig ST. Nasopharyngeal polyps in cats. *Clinical Techniques in Small Animal Practice* 2002;**17**(4):174–7.

190 Schmidt JF, Kapatkin A. Nasopharyngeal and ear canal polyps in the cat. *Feline Practice* 1990;**18**(4):16–19.

191 Donnelly KE, Tillson DM. Feline inflammatory polyps and ventral bulla osteotomy. *Compendium on the Continuing Education for the Practicing Veterinarian* 2004;446–54.

192 Muilenburg RK, Fry TR. Feline nasopharyngeal polyps. *Veterinary Clinics of North America Small Animal Practice* 2002;**32**(4):839–49.

193 Veir JK, Lappin MR, Foley JE, Getzy DM. Feline inflammatory polyps: historical, clinical, and PCR findings for feline calici virus and feline herpes virus-1 in 28 cases. *Journal of Feline Medicine and Surgery* 2002;**4**(4):195–9.

194 Harvey CE, Goldschmidt MH. Inflammatory polypoid growths in the ear canal of cats. *Journal of Small Animal Practice* 1978;**19**(11):669–77.

195 Davidson JR. Otopharyngeal polyps. In: *Current Techniques in Small Animal Surgery*, 4th edn. Bojrab MJ (ed). Lea & Febiger, Philadelphia 1998, pp. 147–50.

196 Anders BB, Hoelzler MG, Scavelli TD, Fulcher RP, Bastian RP. Analysis of auditory and neurologic effects associated with ventral bulla osteotomy for removal of inflammatory polyps or nasopharyngeal masses in cats. *Journal of the American Veterinary Medical Association* 2008;**233**(4):580–5.

197 Faulkner JE, Budsberg SC. Results of ventral bulla osteotomy for treatment of middle ear polyps in cats. *Journal of the American Animal Hospital Association* 1990;**26**(5):496–9.

198 Kapatkin AS, Matthiesen DT, Noone KE, Church EM, Scavelli TE, Patnaik AK. Results of surgery and long-term follow-up in 31 cats with nasopharyngeal polyps. *Journal of the American Animal Hospital Association* 1990;**26**(4):387–92.

199 Trevor PB, Martin RA. Tympanic bulla osteotomy for treatment of middle-ear disease in cats: 19 cases (1984–1991). *Journal of the American Veterinary Medical Association* 1993;**202**(1):123–8.

200 Lamb CR, Richbell S, Mantis P. Radiographic signs in cats with nasal disease. *Journal of Feline Medicine and Surgery* 2003;**5**(4):227–35.

201 Henderson SM, Bradley K, Day MJ, Tasker S, Caney SM, Hotston Moore A, *et al.* Investigation of nasal disease in the cat – a retrospective study of 77 cases. *Journal of Feline Medicine and Surgery* 2004;**6**(4):245–57.

202 Remedios AM, Fowler JD, Pharr JW. A comparison of radiographic versus surgical diagnosis of otitis media. *Journal of the American Animal Hospital Association* 1991;**27**:183–8.

203 Detweiler DA, Johnson LR, Kass PH, R WE. Computed tomographic evidence of bulla effusion in cats with sinonasal disease: 2001–2004. *Journal of Veterinary Internal Medicine* 2006;**20**(5):1080–4.

204 Bischoff MG, Kneller SK. Diagnostic imaging of the canine and feline ear. *Veterinary Clinics of North America Small Animal Practice* 2004;**34**(2):437–58.

205 Gotthelf LN. Diagnosis and treatment of otitis media in dogs and cats. *Veterinary Clinics of North America Small Animal Practice* 2004;**34**(2):469–87.

206 Greci V, Vernia E, Mortellaro CM. Per-endoscopic trans-tympanic traction for the management of feline aural inflammatory polyps: a case review of 37 cats. *Journal of Feline Medicine and Surgery* 2014;**16**(8):645–50.

207 Booth HW. Ventral bulla osteotomy: dog and cat. In: *Current Techniques in Small Animal Surgery*. Bojrab MJ (ed). Williams & Wilkins, Baltimore 1998, pp. 109–12.

208 Boothe HWJ. Surgical management of otitis media and otitis interna. *Veterinary Clinics of North America Small Animal Practice* 1988;**18**(4):901–11.

209 Bruyette DS, Lorenz MD. Otitis externa and otitis media: diagnostic and medical aspects. *Seminars in Veterinary Medicine and Surgery Small Animals* 1993;**8**(1):3–9.

210 Lane JG, Watkins PE. Para-aural abscess in the dog and cat. *Journal of Small Animal Practice* 1986;**27**(8):521–31.

211 Holt D, Brockman DJ, Sylvestre AM, Sadanaga KK. Lateral exploration of fistulas developing after total ear canal ablations: 10 cases (1989–1993). *Journal of the American Animal Hospital Association* 1996;**32**(6):527–30.

212 Smeak DD, Crocker CB, Birchard SJ. Treatment of recurrent otitis media that developed after total ear canal ablation and lateral bulla osteotomy in dogs: nine cases (1986–1994). *Journal of the American Veterinary Medical Association* 1996;**209**(5):937–42.

213 Bacon NJ. Pinna and External Ear Canal. In: *Veterinary Surgery Small Animal*. Tobias KM, Johnston SA (eds). Elsevier Saunders, St. Louis 2012, pp. 2059–77.

214 McCarthy PE, Hosgood G, Pechman RD. Traumatic ear canal separations and para-aural abscessation in three dogs. *Journal of the American Animal Hospital Association* 1995;**31**(5):419–24.

215 Rogers KS. Tumors of the ear canal. *Veterinary Clinics of North America Small Animal Practice* 1988;**18**(4):859–68.

216 Smeak DD. Management of complications associated with total ear canal ablation and bulla osteotomy in dogs and cats. *Veterinary Clinics of North America Small Animal Practice* 2011;**41**:981–94.

217 Smeak DD, DeHoff WD. Total ear canal ablation clinical results in the dog and cat. *Veterinary Surgery* 1986;**15**(2):161–70.

218 Matthieson DT, Scavelli T. Total ear canal ablation and lateral bulla osteotomy in 38 dogs. *Journal of the American Animal Hospital Association* 1990;**26**:257–67.

219 Beckman SL, Henry WB, Cechner P. Total ear canal ablation combining bulla osteotomy and curettage in dogs with chronic otitis externa and media. *Journal of the American Veterinary Medical Association* 1990;**196**(1):84–90.

220 Mason LK, Harvey CE, Orsher RJ. Total ear canal ablation combined with lateral bulla osteotomy for end-stage otitis in dogs. Results in thirty dogs. *Veterinary Surgery* 1988;**17**(5):263–8.

221 Hardie EM, Linder KE, Pease AP. Aural cholesteatoma in twenty dogs. *Veterinary Surgery* 2008;**37**(8):763–70.

222 Williams JM, White R. Total ear canal ablation combined with lateral bulla osteotomy in the cat. *Journal of Small Animal Practice* 1992;**33**(5):225–7.

6.1 INTRODUCTION

Middle ear effusion as a result of infection, or auditory tube dysfunction, is the most commonly encountered abnormality of the middle ear in dogs and cats. Otitis media results from inflammation within the middle ear. The inflammation may arise *de novo* (primary otitis media) as is most commonly seen in cats, or as a consequence of otitis externa (most commonly seen in dogs), structural and functional abnormalities within the pharynx affecting the auditory tube function, or as a result of brachycephaly-related abnormalities (secondary otitis media). Otitis media is not uncommonly associated with polyp formation from middle ear or auditory tube mucosa in cats and has been discussed in Chapter 5 Aetiology and Pathogenesis of Otitis Externa, Section 5.9. Neoplastic disease originating

from the structures that form the middle ear cavity is rare, and will be addressed here in the sections on otitis media in cats and dogs. Primary secretory otitis media (otitis media with effusion) in the Cavalier King Charles Spaniel (CKCS), cholesteatoma and trauma to the temporal bone are discussed in the last three sections respectively.

The clinical signs of middle ear disease are very similar to those of otitis externa and include local discomfort or even pain, and otic discharge, if the tympanum is ruptured (Fig. 6.1), or if there is associated otitis externa. Pain upon opening the mouth during eating, playing with a ball or stick or during yawning can sometimes be observed. Occasionally, neurological signs such as facial nerve paralysis and/or Horner's syndrome may be seen (Fig. 6.2). In cats, the disease may not become apparent until the inner ear has been involved

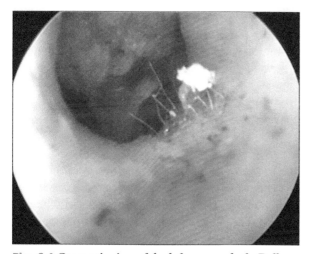

Fig. 6.1 Otoscopic view of the left ear canal of a Bull Mastiff with chronic otitis externa (note the mild hypertrophy and hyperplasia of the ventral skin) and tympanic membrane rupture ventrally with associated inflammation.

Fig. 6.2 Left-sided Horner's syndrome in a shorthaired cat associated with a polyp. The cat was presented for otic smell arising from obstructive otitis externa.

and signs such as head tilt, ataxia and nystagmus can be seen (see Chapter 7 Diseases of the Inner Ear). Note that neurological signs in the dog are uncommon: they occurred in only 5 of 28 cases in one series on otitis media in dogs[1,2].

A histological study of 100 middle ears from 50 normal cats revealed that gross evidence of otitis media was present in 14% cats, and that 48% revealed histological evidence of ongoing or previous inflammation[3]. Support for this comes from a retrospective assessment of 310 cats, which had undergone CT imaging[4]. Some 34% of 101 cats with evidence of middle ear disease did not have a history of ear disease, suggesting subclinical otitis media. This suggests that the incidence of middle ear disease in cats may be higher than either the clinical signs, gross lesions or the literature suggests[3,4].

6.2 OTITIS MEDIA IN THE CAT

Primary otitis media in cats is thought to be of viral origin. Feline upper respiratory viruses (calicivirus and herpesvirus) would be expected to play a major role in the aetiology of otitis media, but, to date, there has been no evidence supporting an infectious cause[5,6]. In most cases of primary otitis media in cats no obvious clinical signs are present until the infection spreads to the inner ear (see Chapter 5 Aetiology and Pathogenesis of Otitis Externa, Section 5.9 and Chapter 7 Diseases of the Inner Ear, Section 7.4). Sometimes Horner's syndrome is seen. Upon otoscopy a red tympanic membrane can be visible (Fig. 6.3A). Computed tomographic (CT) imaging of the skull is recommended (see Chapter 2 Diagnostic Procedures, Section 2.8) to examine the changes in both middle ear cavities (Fig. 6.3B) and nasopharynx and to check for concurrent rhinitis/sinusitis. In the absence of protruding polyps and otitis interna, myringotomy can be performed and the otitis media can be treated as discussed for dogs (see Chapter 6 Diseases of the Middle Ear, Section 6.3).

Polyps are the most commonly reported masses found within the external ear canal of cats[7], and are commonly associated with otitis media in young cats (Fig. 6.4). Inflammatory polyps may arise from within the bulla, from within the auditory tube or from the nasopharynx[7] (see Chapter 5 Aetiology and Pathogenesis of Otitis Externa, Section 5.9). Protruding

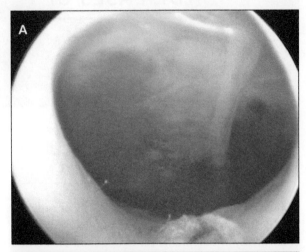

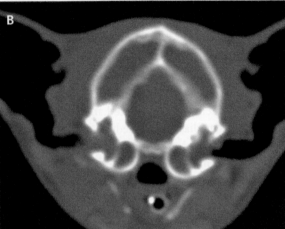

Figs 6.3A, B Otoscopic view of a cat with otitis media and interna (A), note the red discolouration of the pars tensa of the tympanic membrane; B: transverse CT image of the same cat as depicted in (A) with increased soft tissue density within both compartments of the middle ear.

polyps need to be removed surgically, and concurrent otitis media should be treated with a course of systemic antibiotics (see Chapter 6 Diseases of the Middle Ear, Section 6.3). As cats get older the potential for otitis media and neoplasia increases[8]. The mean age of the cats in a study of 19 cats with otitis media subject to bulla osteotomy was as follows:

- Inflammatory polyps, mean age 1.5 years.
- Otitis media, mean age 5.5 years.
- Middle ear neoplasia, mean age 10.25 years.

Fig. 6.4 **Right head tilt in a cat with peripheral vestibular disease associated with otitis media.**

Otitis media associated with soft palate abnormalities was reported in five cats[9]. All five had either marked hypoplasia of the soft palate, such that the free, caudal border of the soft palate did not touch the epiglottis, or they had a cleft palate. Interestingly, all cats presented with neurological signs consistent with otitis media and none exhibited nasal return or oropharyngeal disturbances. All responded to treatment directed to resolving the otitis media, either surgical or medical.

The association of otitis media with oropharyngeal abnormality in the cases described above suggests that ascending infection (oropharynx to middle ear by way of the auditory tube) might occur in cases of severe respiratory viral infection. It is conceivable that severe inflammation within the oropharynx might compromise auditory tube function sufficiently for this to occur.

Cholesterol granulomae are non-neoplastic, granulomatous lesions, containing cholesterol crystals. There are two case reports of cholesterol granuloma within the middle ear in cats, both presenting with neurological signs referable to otitis media[10,11]. Bacteria were recovered from the middle ear in both cases. Removal via ventral bulla osteotomy (VBO) is indicated.

Neoplasia arising from within the middle ear of the cat is rare, typically being squamous cell carcinoma[12,13].

6.3 OTITIS MEDIA IN THE DOG

Otitis media with effusion (OME) appears to be a consequence of the brachycephalic phenotype, particularly in the CKCS[14], but also, for example, in Boxers. This specific condition is discussed in the next section.

Given that the brachycephalic phenotype with severe distortion of the pharynx may be associated with middle ear effusion, it is not surprising that two rare conditions, soft palate hypoplasia[15] and compromised auditory tube function consequent upon trigeminal nerve lesions[16], may also be associated with otitis media.

Inflammatory polyp has been reported in a series of five cases[17], with signs referable to uni- or bilateral otitis externa or media. Surgical excision, VBO or total ear canal ablation with lateral bulla osteotomy (TECA–LBO), was curative.

Cholesterol granuloma, both cases featuring quite aggressive inflammatory masses within the bulla, has been reported twice[18,19]. Surgical removal was curative.

Cholesteatoma has been reported as a serious complication of chronic otitis media in dogs[1,20–22]. It is thought that a pocket of tympanic membrane becomes adherent to inflamed middle ear mucosa, resulting in an inflammatory cystic lesion, lined by stratified keratinising epithelium[22]. Repeated myringotomy may be involved in the aetiopathogenesis of this acquired form of cholesteatoma. This condition and its treatment are discussed in more detail in Chapter 6 Diseases of the Middle Ear, Section 6.5.

Neoplasia of the middle ear is rare in the dog[12]. Most cases reflect transtympanic spread of tumours arising in the external ear canal rather than arising *de novo* within the bulla[12], as demonstrated in a report of 11 dogs with middle ear neoplasia that found[23,24]:
- One papilloma.
- Two basal cell tumours.
- Two papillary adenomas.
- Two adenocarcinomas of sebaceous gland origin.
- Two sebaceous gland adenocarcinomas.
- Three adenocarcinomas of ceruminous gland origin.
- One anaplastic tumour.

Otitis media secondary to otitis externa

Secondary otitis media in association with otitis externa is the most common cause of otitis media in dogs, and is most likely under-recognised[25–27]. In one study of 23 dogs with chronic bilateral otitis externa[28] otitis media was present in 82.6% of the ears and, in 71.1% of these ears, the tympanic membrane was intact. Grass seeds within the external ear canal, in particular, are positively associated with otitis media[29]. A recent history of bulla osteotomy or cleaning of the external ear canal might raise suspicion of otitis media.

> Diagnosis of otitis media is based on:
> - Clinical signs.
> - The appearance and integrity of the tympanic membrane.
> - The demonstration of pathological changes within the bulla using diagnostic modalities.

Inflammation within the middle ear results in the accumulation of exudate, inflammatory cells and, usually, infective organisms. This causes otic pain. This pain might be sufficient to result in discomfort when eating, yawning or playing with toys, or may manifest as a preference to eat on one side of the mouth only. There may be vigorous shaking of the head, sufficient to cause pinnal damage.

Inflammation within the bulla might result in neurological signs such as head tilt, facial nerve paralysis and Horner's syndrome (Figs 6.5, 6.6) depending on which nerves are affected[29,30]:

- Facial nerve paralysis:
  - facial paralysis occurs with ipsilateral drooping and inability to move the ear and lip, a widened palpebral fissure and loss of ability to blink. As long as the parasympathetic nerve supply is unaffected there is no effect on the tear film.
- Vestibulocochlear damage:
  - Vestibular syndrome: head tilt, ataxia, circling, nystagmus.
- Loss of sympathetic innervation to the eye – Horner's syndrome:
  - Miosis, ptosis (drooping of the upper eyelid), protrusion of the third eyelid, enophthalmos (retraction of eyeball as orbital smooth muscle loose tone).

There may be otic discharge and active, ongoing otitis externa.

The changes within the bulla also affect the tympanum: if not already ruptured it may look thickened, dull and discoloured (Fig. 6.7) or it may even appear to bulge outward. Note, however, it is very difficult, if not impossible, to examine the entire tympanic membrane in the conscious dog with chronic painful otitis externa and media[26]. The presence of inflammation, oedema and fibrosis only compounds the difficulty. In one study of 222 ears, some 31% of ear canals were so stenotic that the tympanic membrane could not be visualised, even after cleaning the ear canal[31].

The fact that Little and Lane[26] and Cole *et al.*[28] found that otitis media can be present behind an intact tympanic

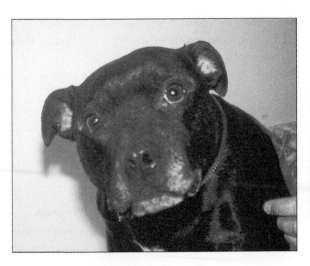
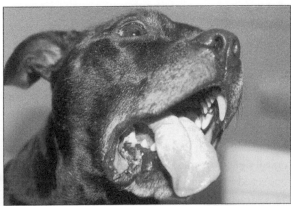

Figs 6.5, 6.6 Bull terrier with Horner's syndrome and facial nerve paralysis on the right hand side.

membrane suggests that tears or punctures can heal, even in the presence of infection. Certainly, experimental perforation and glutaraldehyde soaking (an attempt to induce a model of chronic perforated ear drum in man) was noted to heal within 15 weeks[32]. Many dogs experience chronic recurrent otitis externa for months. In one study, 28 of 42 dogs (66%) had suffered with otitis externa for in excess of 2 years[2]. There is no doubt that the timescale is appropriate for tympanic lesions to repair.

The horizontal ear canal may contain copious, mucous-like exudate from the middle ear if the tympanum is ruptured[27]. However, in other cases this is hard to detect, because of the accumulation of cerumen, inflammatory exudate and cellular debris consequent upon the ongoing otitis externa. Hence, the importance of cleaning the external ear canal in establishing a diagnosis.

Remember:
- The presence of a ruptured tympanic membrane is diagnostic of otitis media.
- The tympanic membrane has the capacity to heal, even in the presence of infection and inflammation.
- Otitis media is often present behind an intact tympanic membrane[2,28].

Establishing the integrity of the tympanic membrane is an important step in both the diagnosis and the management of otitis media. Inflammation and swelling secondary to otitis externa may result in such severe stenosis that it is not possible to insert either an otoscope cone or a video-otoscope head even under general anaesthesia[26,31]. A short course of otic and systemic glucocorticoids (0.5 mg/kg/day for 10–14 days, then tapered over a week) may be necessary to facilitate this important diagnostic stage.

In large dogs with very wide external ear canals, and with the very small working heads of a video-otoscope, it may be possible to visualise adequately the entire tympanum, and to assess it for normal colour, shape and translucency. It may thus be possible to document a rupture of the tympanum.

- Normal tympanic membrane: pearly pink, translucent and gently tense.
- Abnormal tympanic membrane: dull and opaque, thickened rather than translucent.

Given that visual assessment of the tympanum can be difficult, if not impossible, there are some direct and indirect methods that may be used to establish its integrity. Direct methods include, for instance, tympanometry (see Chapter 2 Diagnostic Procedures, Section 2.10). For indirect methods the animal must be under general anaesthesia and with the affected ear uppermost:
- Griffin proposed introducing a thin, soft feeding tube.
- Canalography using iodinated contrast material (e.g. Hypaque) diluted 50% with saline[26]. The

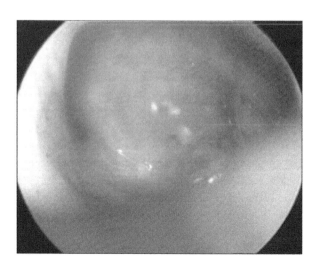

Fig. 6.7 Otoscopic view of the horizontal ear canal and tympanic membrane of a Spaniel with otitis media. Note the red discolouration, lack of transparency of the pars tensa and mild bulging of the tympanic membrane.

diluted contrast is gently introduced, using a catheter if stenosis is present and taking care to avoid spillage. The ear canal is gently agitated to allow air bubbles to clear. Dorsoventral or open-mouth radiographs are then taken. If there is contrast medium within the bulla, the tympanum must be ruptured.

- Dilute fluorescein or very dilute povidone–iodine can be instilled into the external ear canal. The coloured liquid should emerge at the nostril or mouth, after it passes down the auditory canal into the nasopharynx, if it enters the bulla through a ruptured tympanum.

- The affected ear canal is gently filled with warm saline, after the canal is thoroughly cleaned. If the tip of a video-otoscope is gently inserted into the saline and pushed toward the area of the tympanum it may be possible to see small bubbles emerging, as air is pushed up the auditory tube during the respiratory cycle. The presence of bubbles means that the tympanum is ruptured.

Diagnostic imaging

Imaging, be it radiography (Figs 6.8, 6.9), ultrasound or CT (Figs 6.10, 6.11) is utilised in an attempt to document changes within the bulla cavity or within the bony

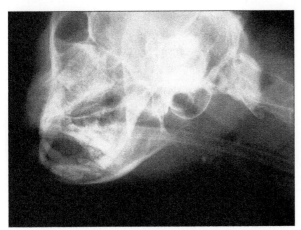
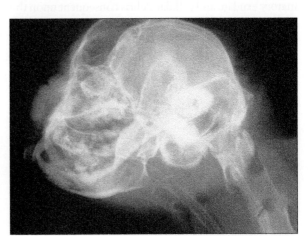

Figs 6.8, 6.9 Lateral oblique radiographs of a 3-year-old Himalayan cat with otitis media associated with a nasopharyngeal polyp. 6.8 shows the normal right bulla; 6.9 shows the increased tissue density within the left bulla.

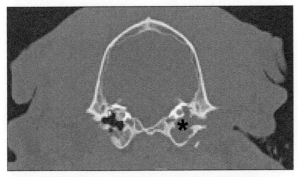
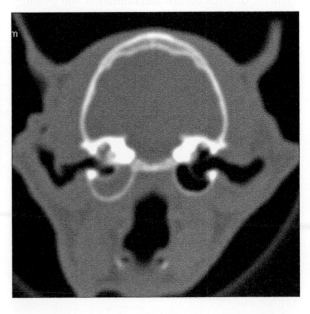

Fig 6.10 Transverse CT image of a dog with mild middle ear effusion after TECA–LBO on the right hand side and severe otitis media on the left hand side (asterisk) completely filling the middle ear cavity.

Fig 6.11 Transverse CT image of a cat with otitis media on the right hand side; note that most of the dorsolateral compartment here is still aerated.

part of the bulla that might support a diagnosis of otitis media:

- Changes within the bulla include increasing opacity due to fluid accumulation or solid tissue.
- Changes within the osseous bulla might include lysis secondary to infection or neoplasia, or proliferative changes resulting in increased thickness of the bulla wall.

Several studies have tried to ascertain whether radiography, ultrasound or CT alone, or in combination, are optimal for visualising disease in the canine and feline bulla[33–36]. CT has become the gold standard in assessing the middle ear for effusion and polyp formation. However, in summary, and bearing in mind that few clinicians have access to CT:

- Radiography tends to underdiagnose otitis media, particularly when the severity is low, or early in the course of the disease, as does CT.
- No technique is absolutely better than any other in detecting both the presence and severity of middle ear disease.
- Ultrasound is a very useful modality, particularly in the case of investigating the feline bulla, but is very operator-dependent.
- Given that many veterinary practices have access to both ultrasound and radiography, the combination of both will give added diagnostic confidence if time is taken to learn the skill.

Myringotomy

Elective puncture of an intact tympanum (myringotomy) is not a step to be taken lightly, but it may be necessary. Indications for myringotomy include:

- To relieve otic pain consequent upon increasing pressure within the middle ear.
- To confirm the diagnosis of otitis media, notwithstanding an intact tympanum.
- To facilitate collection of cytological and bacterial samples from the bulla. Cole *et al.*[28] demonstrated that although *Staphylococcus pseudintermedius*, *Pseudomonas aeruginosa* and *Malassezia pachydermatis* were the most commonly isolated micro-organisms from both the external ear canal and the middle ear, the antimicrobial sensitivity results between samples from the ear canal and those from the middle ear differed in nearly 90% of the cases.

- Remember to take samples for culture and susceptibility testing from both the horizontal portion of the external ear canal and the middle ear cavity.

The procedure is carried out with the anaesthetised dog in lateral recumbency and the affected ear uppermost. The external ear canal is thoroughly cleaned, with e.g. dilute povidone–iodine.

Gotthelf[27] recommends using a sterile, rigid, polypropaline catheter. One end is cut with a blade across such that the angle is around 60°. A 2 mm Buck curette may also be used, although the resulting hole is larger than that produced using the catheter. This may be advantageous if drainage is required. The instruments are used to puncture the tympanum at the '5 o'clock' or '7 o'clock' position, thus avoiding blood vessels, the manubrium of the malleus and the germinal epithelium.

Carbon dioxide lasers may be used to perform the myringotomy[27]. The laser head is passed through a video-otoscope. The laser makes a larger hole than the catheter or Buck curette. This might be advantageous if drainage is anticipated, as the larger defect takes longer to heal.

Once the myringotomy is achieved, samples for bacterial culture and susceptibility testing must be obtained. Most likely the isolates will prove to be one or more of *S. pseudintermedius*, *P. aeruginosa* and *M. pachydermatitis*, but it is the susceptibility spectrum that is most important[27].

If using the catheter, as described above, bulla contents may be aspirated through it. A guarded swab (sterile 3.5 French catheter inserted through a 5 French catheter) may be passed through a laser-produced myringotomy.

> Note, if the bulla contents do not aspirate easily, it may prove necessary to flush with 1–2 ml sterile saline in order to obtain a sufficient sample.

Medical treatment of otitis media

There are several components to the effective medical management of otitis media, and, if performed properly, they can effect a 75% success rate[27]. However, a very high degree of owner commitment is mandatory. Cases can take several months of treatment and repeated episodes of lavage to resolve[37].

Obtain access to the middle ear

Obtain access using a myringotomy if the tympanum is intact. Given that there is otitis media, and with the knowledge that the myringotomy incision will not heal quickly, it is unlikely that repeated incisions into the tympanum will be necessary. This is important because two or even three rechecks may be necessary, at weekly intervals, for cases with severe problems or in which the bulla contains poorly draining debris.

Ensure that there is adequate drainage

Any accumulated discharge must be encouraged to drain and removal of inspissated material must be accomplished, if necessary with repeated flushing. Given that the otitis media has usually been present for a very long time there is little indication for using empirical therapy. However, given that a sample from the bulla must be obtained, it is usually indicated to flush and clean after sampling and to use empirical antibacterial therapy, even if it is anticipated that the bulla will not be effectively cleared at the first attempt and that repeated flushing might be necessary.

Flushing the bulla

Low volume low pressure

Gotthelf recommends a MedRx Earigator connected to a 5 French polypropaline catheter and passed through the 2 mm working channel of a video-otoscope[27].

- If equipment such as this is not available a small rubber feeding tube may be gently inserted into the bulla and used for repeated cycles of flushing and aspirating.
- Use tap water and very dilute povidone–iodine for the flush or consider Tris-EDTA.

Gotthelf anticipates weekly flushing followed by catheter-facilitated direct placement of antibacterial agent into the bulla[27]. Over 2–3 weeks the otitis should resolve and the production of discharge stop.

High volume moderate pressure

Palmeiro *et al.* reported a series of 44 cases in which the bulla was gently but firmly flushed, so as not to damage the inner ear, with a high volume (60 ml) flush liquid comprised of 50/50 warm saline and a mild ceruminolytic (not specified)[37]. Postlavage treatment included systemic and topical antibacterial treatment and an otic cleaner (e.g. EpiOtic or Dermapet ear cleanser). Over half the dogs also received either systemic or locally applied corticosteroids. Otitis media resolved in 36/44 cases (82%), although the mean time to resolution was 118 days (range: 30–360 days) and many experienced transient relapse. Only four dogs required repeated lavage. The statistical analysis performed highlighted some interesting points:

- Neither age at lavage, duration of otitis media, the presence of *P. aeuruginosa* or the use of postlavage corticosteroids had any impact on the time to resolution.

Antibacterial therapy

Although some authors[27] suggest that there is no indication for systemic antibacterial agents[27], others differ[37,38].

- Topical antibacterial agents in aqueous solution, 1–2 ml for a typical bulla, again instilled via the catheter. If necessary select empirically at first, pending susceptibility testing. Good antibacterial choices, with minimal ototoxic potential, include:
  - Fluoroquinolones.
  - Carbenicillin, ticarcillin.
  - Ceftazidime, meropenem.

After the procedure has been accomplished, a topical antibacterial agent is instilled twice a day. Palmeiro *et al.* recommend 1–1.5 ml of antibacterial solution for a medium- to large-breed dog[37]. They also recommend a high volume otic cleanser (e.g. EpiOtic) every 48 hours.

- Remember[39], a 5.4 kg Shi-Tzu has an external ear canal with approximately 20 cm^2 area and volume 4.5 cm^3, whereas in a 20 kg dog the surface area is 38.5 cm^2 and the volume 9.6 cm^3.

- Systemic antibacterial agents are advocated[37,38]. Very high doses of antibacterial agents may be required to ensure that the systemic antibacterial reaches an adequate concentration in the middle ear. For example, Cole *et al.* sampled bulla contents after bulla osteotomy and correlated intravenous dose rate with tissue concentration[40].

- Enrofloxacin is normally administered at 5 mg/kg, once daily. However, to achieve adequate concentrations within the bulla sufficient to overcome a pseudomonal infection with an MIC of 0.5µg/ml, a dose of 20 mg/kg would be necessary.

Another major problem is finding the parenteral preparation of an antibacterial agent to which the multi-resistant *P. aeruginosa* might be sensitive. Those that might be useful are often prodigiously expensive and produced for intravenous use in humans[38]. However, these intravenous agents may be used subcutaneously, allowing for owners to administer them at home, for extended periods[38]. Examples include:
- Ticarcillin, e.g. Ticar, 40–80 mg/kg q6h.
- Ceftazidime, e.g. Fortaz, 30 mg/kg q4h.
- Meropenem, e.g. Mere, 8 mg/kg q12h.

Anti-inflammatory therapy
Prednisolone at 0.5 mg/kg a day for 3–5 days, then reduced to alternate days, helps to reduce inflammation, and helps to reduce oedema and swelling in the external ear canal, which keeps access open and maintains drainage. Anti-inflammatory doses of glucocorticoids do not compromise immunity, but care should be taken in animals with otitis media and interna or brainstem abscessation.

Pain control
Otitis media is painful. Discomfort after myringotomy and lavage may be intense. Control pain, with, for example, tramadol, 2–5 mg/kg 2–3 times day, which is safe in conjunction with prednisolone.

The surgical treatment of otitis media is discussed in Chapter 18 Surgery of the Ear.

6.4 PRIMARY SECRETORY OTITIS MEDIA/ OTITIS MEDIA WITH EFFUSION

Primary secretory otitis media (PSOM), recently renamed otitis media with effusion (OME) is an increasingly recognised disease of unknown aetiopathogenesis that predominantly affects the CKCS[14,41–43]. So far, there is no evidence that infection is part of the pathogenesis, although upper respiratory viral infections have been implicated in humans. Current hypotheses regarding the cause of the disease include:

- Obstruction or dysfunction of the auditory tube with decreased drainage of the mucus from the middle ear.
- Increased production of viscous mucus caused by inflammatory, hypersensitivity or allergic reactions of the middle ear and auditory tube mucosa[14,41–43].

Dogs that present with congenital hypoplasia or malformations of the soft palate are reported to be at risk for middle ear disease and impaired hearing function[15,44,45]. A recent study demonstrated that in the CKCS, greater thickness of the soft palate and reduced nasopharyngeal aperture are significantly associated with OME. The authors suggest that auditory tube dysfunction and OME may represent a previously overlooked consequence of brachycephalic conformation in dogs[46]. A more appropriate name for this disease is therefore probably auditory tube dysfunction, rather than otitis media. Tympanic bulla hypoplasia and middle ear effusion are also commonly seen in brachycephalic dogs, though the effusion is usually not 'glue-like' as in the CKCS, but more thickened and inspissated.

Clinical signs associated with middle ear effusion in CKCS include signs of otitis externa (with or without otoscopic evidence of otitis externa) and impaired hearing[14,41–43]. Middle ear effusion is sometimes a coincidental finding in dogs with clinical signs of cervical disc disease or central nervous system disease including syringomyelia[42,47,48]. In addition, many animals show pharyngeal signs with stridorous breathing (snoring) and sometimes dysphagia.

A thorough and complete clinical examination is mandatory for correct diagnosis and therapy. It should include a general physical, dermatological, neurological (cranial nerves) and otoscopic examination with either a hand-held otoscope or using video-otoscopy[49]. For complete visualisation of the tympanic membrane and prior to therapy, ear flushing is necessary[49]. Video-otoscopy has many advantages over hand-held otoscopy, the most important ones being the higher degree of magnification and a more intense light source that is positioned at the tip of the endoscope. Abnormalities of the tympanic membrane, like increased opacity and hyperemia in cases of otitis media, and small tears can more easily be identified using this technique[50,51]. Findings on otoscopy in dogs with OME range from completely normal ear canals to ear canals with narrowing of the horizontal parts or slight diffuse thickening of the skin lining the ear canals and ceruminous otitis externa[14,41–43]. In most affected ears, a bulging pars flaccida or opaque but intact tympanic membrane can be seen (Figs 6.12A, B).

Diagnostic imaging (CT scan, magnetic resonance imaging [MRI]) is mandatory for proper evaluation of the middle ear, treatment planning and for exclusion of other differential diagnoses (Figs 6.13, 6.14).

The extent and type of hearing impairment should be examined with electrophysiological methods such as brainstem-evoked response audiometry (BERA)[52,53]. Interestingly, BERA findings demonstrate that middle ear effusion in the CKCS is associated with a conductive hearing loss of 10–33 dB in affected dogs, despite the fact that all animals studied were considered to have normal hearing by their owners[53]. A definitive diagnosis can be achieved following myringotomy; typically a

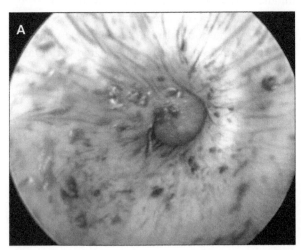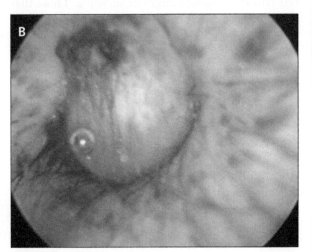

Figs 6.12A, B Otoscopic view of the ear canal of a Cavalier King Charles Spaniel with OME. A: Mild swelling of the skin lining the external ear canal is visible as well as a mild amount of cerumen; at the end of the canal a bulge of the tympanic membrane can be seen; B: a close-up view showing the bulge of the pars flaccida into the ear canal as a result of overpressure within the middle ear cavity, highly suggestive of middle ear effusion.

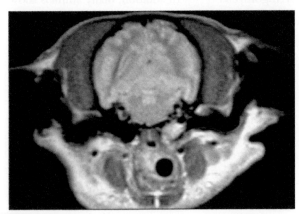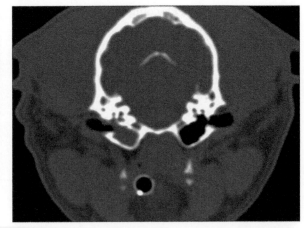

Fig 6.13 MRI transverse image at the level of the middle ear of a Cavalier King Charles Spaniel with left-sided middle ear effusion.

Fig 6.14 CT transverse image at the level of the middle ear of a Cavalier King Charles Spaniel with right-sided middle ear effusion. Note that the pars flaccida can be seen bulging slightly into the ear canal.

highly viscous, opaque mucus effusion is removed from the tympanic bulla (Fig. 6.15). Culture of this mucus is negative in most cases[14,41-43].

Treatment recommendations for OME include:

- Manual removal of the mucoid effusion from the tympanic cavity through a myringotomy incision (with or without assistance of an operating microscope).
- Administration of topical or systemic corticosteroids, systemic antibiotics, mucolytics, topical antibiotics.
- Placement of tympanostomy tubes[14,41-43,54].

Myringotomy (tympanotomy) is a surgical incision of the tympanic membrane used to gain access to the middle ear for draining fluid (for culture and cytology), relieving pressure, and/or instilling medication (see Chapter 2 Diagnostic Procedures, Section 2.9). In most cases of middle ear effusion in dogs, the pars flaccida stretches and bulges into the ear canal and covers most of the pars tensa, making it impossible to visualise the exact location of incision. Paracentesis is therefore preferred by the author and involves the use of a stab incision with special-purpose needles, spear-point knives, swabs, tomcat catheters or small Frazier suction tubes. This can be performed under videoscopic guidance or blindly, depending on the space available in the ear canal. The tomcat catheter and Frazier suction tubes both allow for flushing and aspiration of 1 ml of sterile saline into the middle ear cavity. Retrieved fluid can then be cultured and cytologically evaluated.

Long-term studies are lacking, but it appears that multiple myringotomies usually have to be performed before resolution of clinical signs is obtained[14,41-43]. Tympanostomy tubes provide a more continual tympanic cavity ventilation and drainage and are an acceptable alternative to repeated myringotomy[44,54,55]. Bobbin-reuter type tympanostomy tubes can also be placed in the bulging pars flaccida under endoscopic guidance[55a]. Tubes are pushed out of the tympanic membrane as a result of epithelial migration within 3–6 months, after which decreased hearing can be expected. Current research focuses on evaluation of outcome after tympanostomy tube placement[55a], tuboplasty procedures or stenting of the auditory tube to allow for a more continued drainage and possibly causative treatment of this condition.

6.5 CHOLESTEATOMA

The term aural cholesteatoma implies the formation of a cystic lesion lined with keratinising or exfoliating hyperplastic, hyperkeratotic stratified squamous epithelium filled with a core of keratin debris and expanding within the middle ear cavity (Fig. 6.16)[1,20-22,56]. Though the term 'cholesteatoma' itself is considered a

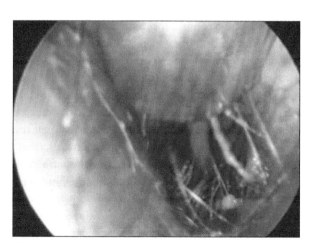

Fig 6.15 Otoscopic view of the ear canal of a Cavalier King Charles Spaniel with OME after myringotomy. A small metal suction tube is used to remove the viscous plug from the middle ear cavity.

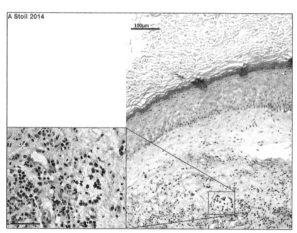

Fig 6.16 Haematoxylin–eosin histopathological slide demonstrating a cholesteatoma expanding within the middle ear cavity, lined with keratinising and exfoliating hyperplastic, hyperkeratotic stratified squamous epithelium with submucosal inflammation.

misnomer because it is neither a granulomatous lesion nor a neoplasm and does not contain cholesterol or fat, it has been used for over 170 years despite efforts to change it[1,57]. Progressive growth of the lesion leads to severe inflammatory changes with enlargement of the tympanic bulla and severe bony changes of its contour such as osteolysis, osteoproliferation and osteosclerosis.

Cholesteatomas in humans are generally classified into acquired or congenital:

- Congenital cholesteatoma is a cyst that forms as a result of misplaced squamous epithelial cells during development of the temporal bone. These have been described in the middle ear, in the petrous apex of the temporal bone, and in the mastoid compartment in humans but have not been reported in dogs[58].
- Acquired cholesteatoma is by far the most commonly encountered form in humans and acquired cholesteatoma has also been reported in dogs.

The exact pathogenesis of cholesteatoma is unclear, but different hypotheses exist. A metaplastic change of the middle ear ciliated epithelium elicited by a severe chronic inflammation of the middle ear has been proposed by some authors (secondary cholesteatoma)[1,57]. Other workers have hypothesised that the squamous epithelium migrated from the external ear through a perforation of the tympanic membrane or as a result of retraction pockets and invaginations of the tympanic membrane due to auditory tube dysfunction (primary cholesteatoma)[57].

In humans myringotomy with tympanostomy tube placement can result in cholesteatoma. Acquired primary and secondary cholesteatoma were found in 1% of nearly 3000 children and were related to longer periods of intubation[59]. Although they are apparently a rare complication of myringotomy alone, they do occur[60]. The risk of cholesteatoma formation in dogs after myringotomy with or without ventilation tube placement is currently unknown.

Cholesteatomas can be experimentally induced by either ligating the external ear canal (in gerbils) or via the insertion of pharmaceuticals, mainly propylene glycol, in the tympanic cavity[61,62]. Concurrent intratympanic glucocorticoid administration inhibits propylene glycol-induced inflammation and sub-

sequent cholesteatoma formation[62]. The role of ear cleaners and otic medications in the aetiopathogenesis of cholesteatoma in dogs is unknown.

Cholesteatoma formation after TECA–LBO has been reported recently in three dogs including two brachycephalic dogs[20,63]. Cholesteatoma formation after TECA may occur because stratified squamous epithelium has not been completely removed during surgery[20]. Complete removal of epithelium can be especially challenging in brachycephalic animals, making them potentially more prone to cholesteatoma development[63]. Another explanation could be the auditory tube dysfunction commonly seen in brachycephalic dogs as a potential cause for otitis media with effusion with subsequent cholesteatoma formation[46]. A recent study on immunohistochemical characteristics of cholesteatoma in dogs, more specifically on the expression of cytokeratins, gave no definitive clues concerning the origin of the epithelial cells[22].

Clinical signs in dogs with cholesteatoma usually reflect outer ear canal disease, such as head shaking, rubbing, otic discharge and otic pain. In addition, facial nerve paralysis and pain on opening or inability to open the mouth can be seen in some animals. In about 15% of the cases symptoms of inner ear disease such as peripheral vestibular ataxia and head tilt were part of the presenting clinical signs[1,20,63,64].

Otoscopic examination under general anaesthesia usually reveals moderate to severe narrowing or stenosis of the horizontal ear canal, making visualisation of the tympanic membrane challenging in most cases. In some animals tympanic membrane perforations can be seen or the membrane can be bulging and/or sclerotic. With video-otoscopy inflammatory tissue, pearly growth or white/yellowish scales can be detected extending within the external bony meatus or in the middle ear cavity. Biopsies can be helpful in differentiating cholesteatoma (keratin flakes, inflammation and metaplastic/hyperkeratotic squamous epithelium) from neoplastic disease.

The diagnosis relies heavily upon advanced diagnostic imaging findings and this is therefore recommended for all patients suspected to have cholesteatoma. On CT imaging cholesteatoma appear as expansile tympanic cavity masses with a mean attenuation value of approximately 55.8 Hounsfield units[65]. There is usually no appreciable contrast enhancement of the tympanic bulla contents (Figs 6.17A–C). Due to the slow progressive growth, the lesion causes severe bone changes at the contour of the tympanic bulla, including osteolysis, osteoproliferation and osteosclerosis, expansion of the tympanic cavity, and sclerosis or osteoproliferation of the ipsilateral temporomandibular joint and paracondylar process. Cholesteatoma can cause lysis of the petrosal part of the temporal bone, leading to intracranial complications[65]. Standard MRI sequences allowed for the identification of a severely expanded bulla containing material that was isointense to brain tissue on T1-weighted images and of mixed intensity on T2-weighted and fluid-attenuated inversion recovery sequences in one dog[66]. No postcontrast enhancement of the content was present, but the lining of the bulla was partially enhanced.

Though endoscopic removal of cholesteatoma lesions and video-assisted surgery have been reported in humans, the only reported treatment options for dogs have been surgical:

- A caudal auricular approach was used successfully in one dog with a relatively small cholesteatoma, with partial reconstruction of the ossicles[64].
- TECALBO is the most frequently performed surgical technique[1,20,21,64],
- whilst VBO has been suggested as the treatment of choice by Venker-van Haagen[56], outcomes have not been reported.

One larger study on the results of surgery for cholesteatoma revealed that surgical approach (VBO versus TECA–LBO) did not influence outcome and suggested that the surgeon can use whichever approach is most appropriate[20]. VBO may be preferable when the ear canal appears normal or mildly changed, but TECA–LBO has the additional advantage that no further migration of ear canal epithelium can occur and no further ear canal treatment is necessary. Recurrence rates after surgery are reportedly high (20–50%[20,21]), particularly if advanced disease is present, as indicated by an inability to open the jaw, neurological disease, or bone lysis on CT imaging[20]. Infection and inflammation are a key part of the disease and, even when all the abnormal epithelium cannot be removed, clinical signs may be controlled for long periods of time with sustained antibiotic and non-steroidal anti-inflammatory therapy.

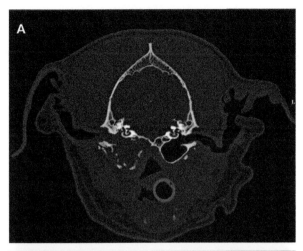

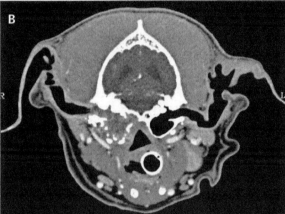

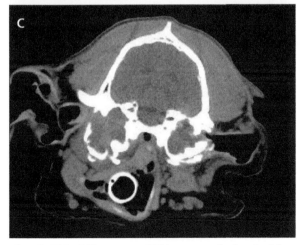

Figs 6.17A–C Transverse CT images at the level of the middle ear of dogs with histopathologically confirmed cholesteatomas. **A:** Precontrast image demonstrating a large cholesteatoma of the right middle ear cavity; **B:** same location as (A) after contrast; **C:** Poodle with bilateral cholesteatoma.

6.6 TRAUMA OF THE TEMPORAL BONE

Trauma is a common cause of morbidity and death in dogs[67-69]. Though outcome of severe blunt trauma in dogs treated with intensive care is very good, head trauma and cranium fractures were significantly associated with a negative outcome in two large studies[70,71]. Head trauma may result from many different events, including motor vehicle accidents, bites, falls, kicks or penetrating injuries such as gunshot wounds[72,73]. The most frequent cause of head trauma noted in small animals is motor vehicle trauma[68,72]. The pathophysiology of trauma to the head, especially involving the nose and nasal sinuses is discussed in Chapter 11 Diseases of the Nasal Cavity and Sinuses, Section 11.12. Less commonly the auricles, ear canals, temporal bones and petrosal bones are involved. However, if fractures of the base of the skull are present, they are often accompanied by petrosal and/or temporal bone damage. Trauma of the auricles and ear canals is discussed in Chapter 4 Diseases of the Pinna, Section 4.2 and Chapter 5 Aetiology and Pathogenesis of Otitis Externa, Section 5.8, respectively. Trauma to the temporal and petrosal bone will be discussed here.

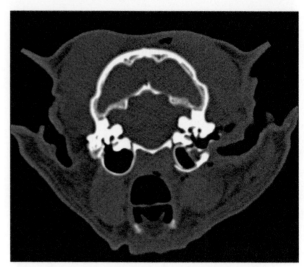

Fig. 6.18 Transverse CT image demonstrating a fracture through the tympanic bulla with subsequent fluid (blood) accumulation within the bulla.

- Bleeding from the external ear canal after a road traffic accident is indicative of either ear canal laceration or avulsion (see Chapter 5 Aetiology and Pathogenesis of Otitis Externa, Section 5.8) or a fracture of the temporal and/or petrosal bone (Fig. 6.18).
- An important rule is that the ear canal should not be flushed in such cases, for fracture of the temporal bone can open the way from the external ear canal to the brain[56].

In addition, animals can demonstrate pain upon opening of the mouth or drool blood-tinged saliva. With inner ear involvement deafness, or more commonly, loss of balance can be seen. Facial nerve paralysis can be an accompanying clinical sign. After a thorough clinical examination for signs of shock, and stabilisation of the patient, further diagnostic imaging can be performed. Temporal bone fractures do not require immediate attention unless brain damage is suspected[56]. If fragments are causing or have the potential to cause additional damage to brain tissue, they will have to be removed surgically. CT and/or MRI of the skull are therefore indicated before any surgery is performed. Diagnostic imaging can be postponed in the absence of signs of brain damage, until other traumatic injuries have been attended to and the patient has been stabilised. Fissures and fractures with minimal dislocation do not need any surgical intervention. With or without surgery, the prognosis for spontaneous recovery of neurological dysfunction is reserved. Other causes of trauma to the middle and inner ear, such as barotrauma and contusio labyrinthi, are discussed in Chapter 7 Diseases of the Inner Ear, Section 7.6.

6.7 REFERENCES

1 Little CJ, Lane JG, Gibbs C, Pearson GR. Inflammatory middle ear disease of the dog: the clinical and pathological features of cholesteatoma, a complication of otitis media. *Veterinary Record* 1991;**128**:319–22.

2 Little CJ, Lane JG, Pearson GR. Inflammatory middle ear disease of the dog: the pathology of otitis media. *Veterinary Record* 1991;**128**:293–6.

3 Sula MM, Njaa BL, Payton ME. Histologic characterization of the cat middle ear. In sickness and in health. *Veterinary Pathology* 2014;**51**:951–67.

4 Shanaman M, Seiler G, Holt DE. Prevalence of clinical and subclinical middle ear disease in cats undergoing computed tomographic scans of the head. *Veterinary Radiology & Ultrasound* 2012;**53**:76–9.

5 Veir JK, Lappin M, Foley JE, Getzy DM. Feline inflammatory polyps: historical, clinical, and PCR findings for feline calici virus and feline herpes virus-1 in 28 cases. *Journal of Feline Medicine and Surgery* 2002;**4**:195–9.

6 Klose TC, MacPhail CM, Schultheiss PC, Rosychuk RA, Hawley JR, Lappin MR. Prevalence of select infectious agents in inflammatory aural and nasopharyngeal polyps from client-owned cats. *Journal of Feline Medicine and Surgery* 2010;**12**:769–74.

7 Sula MJM. Tumors and tumorlike lesions of dog and cat ears. *Veterinary Clinics of North America Small Animal Practice* 2012;**42**:1161–78.

8 Trevor PB, Martin RA. Tympanic bulla osteotomy for treatment of middle-ear disease in cats: 19 cases (1984–1991). *Journal of the American Veterinary Medical Association* 1993;**202**:123–8.

9 Woodbridge NT, Baines EA, Baines SJ. Otitis media in five cats associated with soft palate abnormalities. *Veterinary Record* 2012;**171**:124.

10 Van der Heyden S, Butaye P, Roels S. Cholesterol granuloma associated with otitis media and leptomeningitis in a cat due to a *Streptococcus canis* infection. *Canadian Veterinary Journal* 2013;**54**:72–3.

11 Ilha MRS, Wisell C. Cholesterol granuloma associated with otitis media in a cat. *Journal for Veterinary Diagnostic Investigation* 2013;**25**:515–8.

12 Fan TM, de Lorimier L-P. Inflammatory polyps and aural neoplasia. *Veterinary Clinics of North America Small Animal Practice* 2004;**34**:489–509.

13 Kennis RA. Feline otitis: diagnosis and treatment. *Veterinary Clinics of North America Small Animal Practice* 2013;**43**:51–6.

14 McGuinness SJ, Friend EJ, Knowler SP, Jeffery ND, Rusbridge C. Progression of otitis media with effusion in the Cavalier King Charles spaniel. *Veterinary Record* 2013;**172**:315.

15 White RN, Hawkins HL, Alemi VP, Warner C. Soft palate hypoplasia and concurrent middle ear pathology in six dogs. *Journal of Small Animal Practice* 2009;**50**:364–72.

16 Wessmann A, Hennessey A, Gonçalves R, Benigni L, Hammond G, Volk HA. The association of middle ear effusion with trigeminal nerve mass lesions in dogs. *Veterinary Record* 2013;**173**:449.

17 Pratschke KM. Inflammatory polyps of the middle ear in 5 dogs. *Veterinary Surgery* 2003;**32**:292–6.

18 Fliegner RA, Jubb KVF, Lording PM. Cholesterol granuloma associated with otitis media and destruction of the tympanic bulla in a dog. *Veterinary Pathology* 2007;**44**:547–9.

19 Riedinger B, Albaric O, Gauthier O. Cholesterol granuloma as long-term complication of total ear canal ablation in a dog. *Journal of Small Animal Practice* 2011;**53**:188–91.

20 Hardie EM, Linder KE, Pease AP. Aural cholesteatoma in twenty dogs. *Veterinary Surgery* 2008;**37**:763–70.

21 Greci V, Travetti O, Di Giancamillo M, *et al.* Middle ear cholesteatoma in 11 dogs. *Canadian Veterinary Journal* 2011;**52**:631.

22 Banco B, Grieco V, Di Giancamillo M, *et al.* Canine aural cholesteatoma: a histological and immunohistochemical study. *Veterinary Journal* 2014;**200**:440–5.

23 Lane IF, Hall DG. Adenocarcinoma of the middle ear with osteolysis of the tympanic bulla in a cat. *Journal of the American Veterinary Medical Association* 1992;**201**:463–5.

24 Little C, Pearson GR, Lane JG. Neoplasia involving the middle ear cavity of dogs. *Veterinary Record* 1989;**124**:54–7.

25 Shell LG. Otitis media and otitis interna. Etiology, diagnosis, and medical management. *Veterinary Clinics of North America Small Animal Practice* 1988;**18**:885–99.

26 Little CJ, Lane JG. An evaluation of tympanometry, otoscopy and palpation for assessment of the canine tympanic membrane. *Veterinary Record* 2011;**124**:5–8.

27 Gotthelf LN. Diagnosis and treatment of otitis media in dogs and cats. *Veterinary Clinics of North America Small Animal Practice* 2004;**34**:469–87.

28 Cole LK, Kwochka KW, Kowalski JJ, Hillier A. Microbial flora and antimicrobial susceptibility patterns of isolated pathogens from the horizontal ear canal and middle ear in dogs with otitis media. *Journal of the American Veterinary Medical Association* 1998;**212**:534–8.

29 Saridomichelakis MN, Farmaki R, Leontides LS, Koutinas AF. Aetiology of canine otitis externa: a retrospective study of 100 cases. *Veterinary Dermatology* 2007;**18**:1–7.

30 Garosi LS, Lowrie CT, Swinbourne NF. Neurological manifestations of ear disease in dogs and cats. *Veterinary Clinics of North America Small Animal Practice* 2012;**42**:1143–60.

31 Eom K, Lee H, Yoon J. Canalographic evaluation of the external ear canal in dogs. *Veterinary Radiology & Ultrasound* 2000;**41**:231–4.

32 Truy E, Disant F, Morgon A. Chronic tympanic membrane perforation: an animal model. *American Journal of Otology* 1995;**16**:222–5.

33 Doust R, King A, Hammond G, *et al.* Assessment of middle ear disease in the dog: a comparison of diagnostic imaging modalities. *Journal of Small Animal Practice* 2007;**48**:188–92.

34 Bischoff MG, Kneller SK. Diagnostic imaging of the canine and feline ear. *Veterinary Clinics of North America Small Animal Practice* 2004;**34**:437–58.

35 Rohleder JJ, Jones JC, Duncan RB, Larson MM, Waldron DL, Tromblee T. Comparative performance of radiography and computed tomography in the diagnosis of middle ear disease in 31 dogs. *Veterinary Radiology & Ultrasound* 2006;**47**:45–52.

36 King AM, Weinrauch SA, Doust R, Hammond G. Comparison of ultrasonography, radiography and a single computed tomography slice for fluid identification within the feline tympanic bulla. *Veterinary Journal* 2007;**173**:638–44.

37 Palmeiro BS, Morris DO, Wiemelt SP, Shofer FS. Evaluation of outcome of otitis media after lavage of the tympanic bulla and long-term antimicrobial drug treatment in dogs: 44 cases (1998–2002). *Journal of the American Veterinary Medical Association* 2004;**225**:548–53.

38 Morris DO. Medical therapy of otitis externa and otitis media. *Veterinary Clinics of North America Small Animal Practice* 2004;**34**:541–55.

39 Wefstaedt P, Behrens BA, Nolte I. Finite element modelling of the canine and feline outer ear canal: benefits for local drug delivery? *Berliner and Munchener Tierarztliche Wochenschrift* 2011;**124**:78–82.

40 Cole LK, Papich MG, Kwochka KW, Hillier A, Smeak DD, Lehman AM. Plasma and ear tissue concentrations of enrofloxacin and its metabolite ciprofloxacin in dogs with chronic end-stage otitis externa after intravenous administration of enrofloxacin. *Veterinary Dermatology* 2009;**20**:51–9.

41 Stern-Bertholtz W, Sjostrom L, Hakanson NW. Primary secretory otitis media in the Cavalier King Charles spaniel: a review of 61 cases. *Journal of Small Animal Practice* 2003;**44**:253–6.

42 Rusbridge C. Primary secretory otitis media in Cavalier King Charles spaniels. *Journal of Small Animal Practice* 2004;**45**:222.

43 Cole LK. Primary secretory otitis media in Cavalier King Charles Spaniels. *Veterinary Clinics of North America Small Animal Practice* 2012;**42**:1137–42.

44 White RAS. Middle and inner ear. In: *Veterinary Surgery Small Animal*. Tobias KMJSA (ed). Elsevier Saunders, St. Louis 2012, pp. 2078–89.

45 Gregory SP. Middle ear disease associated with congenital palatine defects in seven dogs and one cat. *Journal of Small Animal Practice* 2000;**41**:398–401.

46 Hayes GM, Friend EJ, Jeffery ND. Relationship between pharyngeal conformation and otitis media with effusion in Cavalier King Charles spaniels. *Veterinary Record* 2010;**167**:55–8.

47 Owen MC, Lamb CR, Lu D, Targett MP. Material in the middle ear of dogs having magnetic resonance imaging for investigation of neurologic signs. *Veterinary Radiology & Ultrasound* 2004;**45**:149–55.

48 Volk HA, Davies ES. Middle ear effusions in dogs: an incidental finding? *Veterinary Journal* 2011;**188**:256–7.

49 ter Haar G. Basic principles of surgery of the external ear (pinna and ear canal). In: *The Cutting Edge: Basic Operating Skills for the Veterinary Surgeon.* Kirpensteijn J, Klein WR (eds). Roman House Publishers, London 2006, pp. 272–83.

50 Rosychuk TAW. Video-otoscopy. In: *Veterinary Endoscopy for the Small Animal Practitioner.* McCarthy TC (ed). Elsevier, St. Louis 2005, pp. 387–411.

51 Cole LK. Otoscopy. In: *Small Animal Endoscopy.* Tams TR, Rawlings CA (eds). Elsevier Health Sciences, St. Louis 2010, pp. 587–605.

52 ter Haar G, Venker-van Haagen AJ, de Groot HN, van den Brom WE. Click and low-, middle-, and high-frequency toneburst stimulation of the canine cochlea. *Journal of Veterinary Internal Medicine* 2002;**16**:274–80.

53 Harcourt-Brown TR, Parker JE, Granger N, Jeffery ND. Effect of middle ear effusion on the brain-stem auditory evoked response of Cavalier King Charles Spaniels. *Veterinary Journal* 2011;**188**:341–5.

54 Corfield GS, Burrows AK, Imani P, Bryden SL. The method of application and short term results of tympanostomy tubes for the treatment of primary secretory otitis media in three Cavalier King Charles Spaniel dogs. *Australian Veterinary Journal* 2008;**86**:88–94.

55 Cox CL, Slack WT, Cox GJ. Insertion of a transtympanic ventilation tube for treatment of otitis media with effusion. *Journal of Small Animal Practice* 1989;**30**:517–9.

55a Guerin V, Hampel R, ter Haar G. Video-otoscopy-guided tympanostomy tube placement for the treatment of middle ear effusion. *Journal of Small Animal Practice* 2015;**56**:606–12.

56 Venker-van Haagen AJ. The ear. In: *Ear, Nose, Throat, and Tracheobronchial Diseases in Dogs and Cats.* Venker-van Haagen AJ (ed). Schlütersche Verlagsgesellschaft, Hannover 2005, pp. 1–50.

57 Olszewska E, Wagner M, Bernal-Sprekelsen M, *et al.* Etiopathogenesis of cholesteatoma. *European Archives of Otorhinolaryngology* 2004;**261**:6–24.

58 Levenson MJ, Michaels L, Parisier SC, Juarbe C. Congenital cholesteatomas in children. *Laryngoscope* 1988;**98**:949–55.

59 Golz A, Goldenberg D, Netzer A, *et al.* Cholesteatomas associated with ventilation tube insertion. *Archives of Otolaryngology and Head & Neck Surgery* 1999;**125**:754–7.

60 Linkov G, Isaacson G. Secondary acquired cholesteatoma after adenoidectomy and myringotomy. *Otolaryngology and Head & Neck Surgery* 2013;**149**:957–8.

61 Kim CS, Chung JW. Morphologic and biologic changes of experimentally induced cholesteatoma in Mongolian gerbils with anticytokeratin and lectin study. *American Journal of Otology* 1999;**20**:13–8.

62 Sennaroglu L, Ozkul A, Gedikoglu G, Turan E. Effect of intratympanic steroid application on the development of experimental cholesteatoma. *Laryngoscope* 1998;**108**:543–7.

63 Schuenemann RM, Oechtering G. Cholesteatoma after lateral bulla osteotomy in two brachycephalic dogs. *Journal of the American Animal Hospital Association* 2012;**48**:261–8.

64 Davidson EB, Brodie HA, Breznock EM. Removal of a cholesteatoma in a dog, using a caudal auricular approach. *Journal of the American Veterinary Medical Association* 1997;**211**:1549–53.

65 Travetti O, Giudice C, Greci V, Lombardo R, Mortellaro CM, Di Giancamillo M. Computed tomography features of middle ear cholesteatoma in dogs. *Veterinary Radiology & Ultrasound* 2010;**51**:374–9.

66 Harran NX, Bradley KJ, Hetzel N, Bowlt KL, Day MJ, Barr F. MRI findings of a middle ear cholesteatoma in a dog. *Journal of the American Animal Hospital Association* 2012;**48**:339–43.

67 Hall KE, Holowaychuk MK, Sharp CR, Reineke E. Multicenter prospective evaluation of dogs with trauma. *Journal of the American Veterinary Medical Association* 2014;**244**:300–8.

68 Kolata RJ, Kraut NH, Johnston DE. Patterns of trauma in urban dogs and cats: a study of 1,000 bases. *Journal of the American Veterinary Medical Association* 1974;**164**:499–502.

69 Hayes G, Mathews K, Doig G, *et al.* The acute patient physiologic and laboratory evaluation (APPLE) score: a severity of illness stratification system for hospitalized dogs. *Journal of Veterinary Internal Medicine* 2010;**24**:1034–47.

70 Simpson SA, Syring R, Otto CM. Severe blunt trauma in dogs: 235 cases (1997–2003). *Journal of Veterinary Emergency and Critical Care* 2009;**19**:588–602.

71 Streeter EM, Rozanski EA, Laforcade-Buress A de, Freeman LM, Rush JE. Evaluation of vehicular trauma in dogs: 239 cases (January–December 2001). *Journal of the American Veterinary Medical Association* 2009;**235**:405–8.

72 Pope ER. Head and facial wounds in dogs and cats. *Veterinary Clinics of North America Small Animal Practice* 2006;**36**:793–817.

73 Platt SR, Radaelli ST, McDonnell JJ. The prognostic value of the Modified Glasgow Coma Scale in head trauma in dogs. *Journal of Veterinary Internal Medicine* 2001;**15**:581–4.

CHAPTER 7

DISEASES OF THE INNER EAR

7.1 INTRODUCTION

The inner ear lies safely protected within the osseous labyrinth of the petrous part of the temporal bone. The membranous labyrinth consists of three parts: the cochlea, responsible for hearing, and the vestibule and the semicircular canals responsible for maintaining balance. Inner ear disease can be either primary with dysfunction of the cochlea or the vestibulum or both as a result of pathology of these end-organs (e.g. ototoxicity and age-related hearing loss) or secondary as a result of extension of disease from surrounding structures (e.g. otitis media, see Chapter 6 Diseases of the Middle Ear, Sections 6.2, 6.3). Inner ear dysfunction is usually demonstrated as loss of balance (vestibular dysfunction) with symptoms of peripheral vestibular ataxia (head tilt, horizontal nystagmus, circling or falling toward the side of the lesion), as hearing loss (cochlear dysfunction) usually goes unnoticed until complete deafness is recognised[1].

Hearing loss is a common disorder in many dog breeds and auditory dysfunction and its clinical consequences can vary from mild to severe. Dogs with unilateral hearing loss can experience difficulty in localising the source of a sound[2]. These animals are not suited as working dogs for blind and deaf people or rescue and police work[3], but are not severely handicapped. In case of bilateral hearing loss however, dogs are unable to anticipate dangers such as motor vehicles and they may consequently fall victim to injury or death[2]. In addition, these animals seem to be easily startled and have an increased tendency to bite[4]. Furthermore, deaf puppies require specialised training and are therefore usually euthanised[4].

Acquired hearing loss in humans is receiving growing attention because of its detrimental effects on the affected individual's psychosocial situation, including social isolation, depression, and loss of self-esteem[5-10]. Hearing impairment has also been implicated as a cofactor in senile dementia[8]. The psychosocial effects of hearing impairment in dogs are not known, but undoubtedly hearing loss contributes to the lethargy, depression and lack of interest in interaction with the environment that is commonly observed in old dogs[11]. Another serious behavioural side-effect of acquired deafness in dogs is exaggerated barking[11]. Similar to acquired hearing loss in humans where the quality of life is not only affected of the person with hearing loss, but also of the person's loved ones, it is expected that hearing disabilities in dogs strongly affects the intimate relationship with the owner, due to the interactive nature of their habitual vocal communication[12]. Timely recognition and correct diagnosis of hearing impairment are therefore mandatory for both audiological rehabilitation of the patient and for owner counselling.

> As is the case in humans, hearing loss in dogs and cats can result from central or peripheral causes.

Central deafness can theoretically result from a variety of retrocochlear lesions but is very rare in veterinary practice[2,13]. Bilateral central deafness requires bilateral lesions of the auditory cortex or lesions of such a significant portion of the brainstem or midbrain, that significant clinical signs beyond deafness are to be expected[2,13]. Peripheral hearing loss in dogs is much more common and has been classified as:

- Inherited or acquired.
- Conductive or sensorineural.
- Congenital or late-onset[2,4,13].

Conductive deafness results from a lack of presentation of sound to the inner ear, usually secondary to otitis externa or media (see Chapter 5 Aetiology and Pathogenesis of Otitis Externa and Chapter 6 Diseases of the Middle Ear), while sensorineural deafness occurs with abnormalities of the cochlear system, cranial nerve (CN) VIII or auditory pathways and higher brain centres[1].

The most frequently observed forms are acquired conductive hearing loss as a result of chronic otitis externa and media, congenital (inherited) sensorineural hearing loss (SNHL), discussed in Section 7.2, and acquired SNHL including age-related hearing loss (ARHL) or presbycusis, noise-induced hearing loss (NIHL) and ototoxicity[1,2]. The latter three will be reviewed in Section 7.7, 7.8 and 7.5 respectively.

With a complete physical exam including otoscopy, the differentiation between conductive and sensorineural hearing loss can usually be made[1]. Advanced imaging with computed tomography (CT) or magnetic resonance imaging (MRI) is necessary however, to rule out definitely conduction deafness and for treatment planning (see Chapter 2 Diagnostic Procedures, Section 2.8)[14]. Though essential for the diagnostic work-up of all patients with hearing disorders, these two techniques can only identify morphological abnormalities of the petrous bone, middle ear and inner ear structures. Functional abnormalities have to be diagnosed with hearing tests (see Chapter 2 Diagnostic Procedures, Section 2.10).

Loss of balance is generally due to disorders of the peripheral vestibular system (the inner ear receptors and/or the vestibular part of the CN VIII), but central vestibular disease (the brainstem vestibular nuclei in the medulla oblongata and neurons in the flocculonodular lobe of the cerebellum) has to be ruled out[15-17]. Though labyrinthitis and inner ear trauma are associated with hearing loss, this is not commonly recognised in dogs and cats as the vestibular abnormalities are usually the most obvious clinical signs. These two conditions will be discussed in Sections 7.4 and 7.6. Congenital peripheral vestibular ataxia will be reviewed in Section 7.3. The cause of the most common form of acquired peripheral vestibular ataxia is, however, unknown; this idiopathic peripheral vestibular ataxia will be the topic of Section 7.9.

7.2 CONGENITAL DEAFNESS

- Congenital deafness in dogs and cats is assumed to be hereditary.
- Deafness is usually associated with genes that cause white or lightened skin and hair.
- Electrophysiological testing is mandatory to identify affected animals so they can be excluded from breeding.

In most dog and cat breeds, inherited congenital sensorineural deafness results from perinatal degeneration of the stria vascularis, which leads to collapse of Reissner's membrane and the cochlear duct and to hair cell degeneration[2,13,18,19]. The strial degeneration appears to result from absence of melanocytes and begins as early as 1 day after birth, but is only clearly evident histologically by 4 weeks[2,4,19,20]. This disease is common in many breeds of dogs and cats, with a predilection for white coat colours and a strong association of deafness with blue irises. Deafness is most closely linked to the recessive alleles of the pigmentation locus S, responsible for Irish spotting, piebald spotting or extreme-white piebald spotting[2,4,19,20].

The most commonly affected breeds include Dalmatians, English Setters, Australian Shepherds, Border Collies and Shetland Sheepdogs, but there are over 80 breeds reported[2,4,19,20]. In cats, deafness in blue-eye animals is widely recognised and is linked to the dominant W gene13. However, not all blue-eyed white cats are deaf[21].

In the Doberman, and probably other dog breeds not carrying the merle or piebald pigment genes, the deafness results from direct loss of cochlear hair cells without any antecedent effects on the stria vascularis[2,22]. The percentage of affected dogs with unilateral deafness ranged from 73% to 90% in one report[23]. In three specific pure cat breeds, the Norwegian forest, Maine Coon and Turkish Angora, the prevalence of deafness was found to be 18%, 17% and 11% respectively[21].

Unilaterally deaf dogs usually exhibit no clinical signs and without electrophysiological testing (as puppies or prior to breeding) using brainstem-evoked response audiometry (BERA) or otoacoustic emissions testing (see Chapter 2 Diagnostic Procedures, Section

2.10) these dogs continue to increase the prevalence of the disorder. Currently, investigations focus on finding genetic markers for the gene or genes responsible for pigment-associated deafness to reduce the prevalence in affected breeds[4,23]. Until these markers are found, electrophysiological testing will be the sole method for identifying affected dogs and cats. Both unilaterally and bilaterally affected animals should be excluded from breeding.

Though cochlear and brainstem implants are used for people with congenital sensorineural deafness, these techniques have not been clinically used in dogs or cats[24].

7.3 CONGENITAL PERIPHERAL VESTIBULAR SYNDROMES

> • Congenital peripheral vestibular ataxia is a relatively uncommon cause of vestibular dysfunction in puppies and kittens and can be associated with deafness.
> • The diagnosis is based on the young age of onset of clinical signs and exclusion of other differentials.
> • There is no known treatment, but most animals make acceptable pets.

Congenital peripheral vestibular syndromes have been reported in the German Shepherd, Doberman Pinscher, Akita, English Cocker Spaniel, Beagle, Smooth Fox Terrier and Tibetan Terrier, as well as in Siamese, Burmese and Tonkinese cats[1,25–29].

The cause is usually unknown, but pathological investigations in affected Doberman puppies revealed a degeneration of the cochlear neuroepithelium and a lack or abnormality of the otoliths in the macula[22]. It is presumed that this disease is inherited as an autosomal recessive disorder[22]. In another study, congenital vestibular dysfunction was described in puppies of related breeding lines, which were affected by a lymphocytic labyrinthitis[29].

Peripheral vestibular signs, usually unilateral, may be present at birth or develop during the first few weeks or months (up to the fourth month) of life. Clinical signs of rolling movements, head tilt, circling and ataxia may initially be severe but are non-progressive. Nystagmus is not seen. Bilateral vestibular dysfunction has only been occasionally reported in some breeds[27]. Deafness may accompany the vestibular signs in Doberman, the Akita and the Siamese cat and can be diagnosed using BERA[16,22]. Signs resolve spontaneously in some animals, whereas others may have residual and permanent head tilts; however, compensation of the ataxia is common and many affected animals make acceptable pets.

The diagnosis is based on the early onset of signs, before 3–4 months of age and exclusion of other differentials. On neurological examination, an abnormal vestibular nystagmus cannot be determined nor can a physiological nystagmus be elicited by passively moving the head[27–29]. Diagnostic imaging findings and cerebrospinal fluid (CSF) analysis are normal, but electrophysiological hearing assessment can reveal unilateral or bilateral hearing loss.

There is no treatment available, but the disease is usually not progressive and improvement is mainly seen at an age of 2–3 months as the animals learn to compensate through visual and proprioceptive clues[25–29]. The affected animals should not be used for breeding.

7.4 LABYRINTHITIS

> • Labyrinthitis or otitis interna in dogs and cats is usually the result of bacterial otitis media, but viral and fungal infections and immune-mediated disease can also cause loss of balance and hearing.
> • Diagnostic imaging is required in all cases not only to assess involvement of the middle and inner ear but to rule out extension towards and into the brainstem.
> • Aggressive surgical treatment with either total ear canal ablation– lateral bulla osteotomy (TECA–LBO) or ventral bulla osteotomy (VBO) is indicated in most patients to avoid bacterial meningoencephalitis.

Labyrinthitis by definition is an acute or chronic serous or purulent inflammatory reaction within the fluid spaces and membranes of the vestibulocochlear labyrinth caused by bacteria, viruses, spirochetes or fungi or with an immune-mediated origin. The route of infection may be otogenic, meningogenic or haematogenic. Whereas both the vestibulum and the cochlea are affected, the clinical signs noticed by the owner are those of loss of balance primarily.

In dogs and cats, labyrinthitis is thought to be mainly otogenic in nature and a continuum of septic otitis media (see Chapter 6 Diseases of the Middle Ear, Sections 6.2, 6.3). Otitis media and interna (OMI) is the most common cause of peripheral vestibular disease seen in dogs and cats, and may account for nearly 50% of all cases of canine peripheral vestibular disease[25–27,30]. It is important to recognise that otitis media alone will not result in vestibular signs. If deficits compatible with peripheral vestibular dysfunction are detected, inner ear involvement is confirmed[28,31].

True infection of the inner ear can cause inflammation of the labyrinth but the inflammatory changes can also be the result of toxin absorption through the oval or round window membranes. In addition, as in humans, viral infections of the upper respiratory tract may be responsible for some cases of labyrinthitis, especially in cats. In humans, picornavirus, influenza virus, parainfluenza virus, respiratory syncytial virus, coronavirus, cytomegalovirus, herpesvirus and adenovirus have been implicated in viral otitis interna[32]. Treatment of patients suspected of having viral labyrinthitis with acyclovir did not improve outcome over treatment with corticosteroids, however[33]. There are no confirmed cases of viral labyrinthitis in dogs and cats. Cryptococcosis has been associated with OMI and peripheral vestibular ataxia in cats[34].

An important, but poorly recognised, cause of sudden SNHL in humans is immune-mediated inner

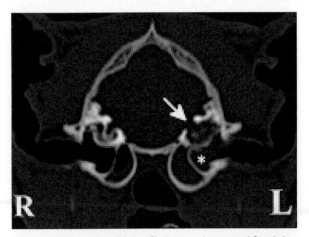

Fig. 7.1 CT septic otitis media/interna in a cat with minimal effusion in the right middle ear cavity and completely filled left middle ear cavity (asterisk), with bony erosion of the petrosal bone (arrow) indicative of otitis interna.

ear disease[35,36]. There are no similar reports available in the veterinary literature.

A thorough otoscopic examination, bulla imaging and myringotomy are the primary tools used to diagnose OMI (see Chapter 6 Diseases of the Middle Ear and Fig. 7.1). Treatment of septic labyrinthitis consists of aggressive medical (parenteral antibiotics, see Chapter 6 Diseases of the Middle Ear, Sections 6.2, 6.3 and Chapter 17 Perioperative Management, Section 17.3) or surgical intervention. If middle ear involvement is apparent, surgical intervention with TECA–LBO in dogs (see Chapter 18, Surgery of the Ear, Sections 18.7, 18.8) or VBO with or without polyp removal in cats (see Chapter 18 Surgery of the Ear, Section 18.9) is indicated to prevent adjacent brainstem involvement and to provide drainage.

Otogenic infections arising from the external or middle ears can extend into the calvarium, causing brain abscessation and bacterial meningoencephalitis (Fig. 7.1)[27,37]. Clinical signs in these cases can indicate central vestibular lesions, but sometimes only peripheral vestibular signs are noted. Aggressive surgical debridement of the affected middle ears and parenteral antibiotic therapy are required in these cases with brainstem involvement.

7.5 OTOTOXICITY

Ototoxicity can be defined as the capacity of certain drugs or chemicals to damage inner ear structures, including the cochlea, the vestibule and the semicircular canals and/or inner ear function[2,13,38–41]. This can result in either hearing impairment and/or vestibular dysfunction (peripheral vestibular ataxia). The clinical signs of ototoxicity are dramatic in acute cases and rapidly bring the owner to the veterinarian[1]. Even with acute treatment, the damage to the inner ear is usually permanent however, rendering a good reason to stress the importance of preventing or recognising early the signs of ototoxicity. To be able to prevent ototoxicity, veterinarians must be aware of the agents with the propensity for ototoxicity, their routes for reaching the inner ear, their toxic effects and clinical signs.

Over 180 compounds and classes of compounds have been identified as ototoxic[2,13,39–41]. Not all of them are equally toxic and some effects are reversible, but in most instances the deficit is permanent. In human medicine, the aminoglycoside antibiotics, the antineoplastic drugs cisplatin and carboplatin, loop diuretics, salicylates, quinine, deferoxamine and various toxic substances are recognised for their propensity to cause ototoxicity[38]. The best recognised, and perhaps most frequent, agents of ototoxicity in veterinary medicine are:

- Aminoglycoside antibiotics, especially gentamicin.
- Polypeptides.
- Chloramphenicol.
- Erythromycin.
- (Oxy)tetracycline[42].

The importance of disinfectant-based (clioquinol, chlorhexidine, cetrimide, iodine, povidone–iodine and 70% ethanol) ototoxicity, for instance used for ear surgery, should not be underestimated, however.

The ototoxicity of drugs and chemicals used in and around the ear has been tested in animals, especially guinea pigs, by introducing the agents into the middle ear after perforating the tympanic membrane or via the tympanic bulla. In addition to the drugs mentioned above, local anaesthetics such as lidocaine, benzocaine, procaine and cocaine, vehicles, detergents and stabilisers including propylene glycol, glycerol and phenol, used in ear drops have also been found to be ototoxic using this method[38–41,43,44].

In order for a drug to exert ototoxicity, it must reach the inner ear. This may be the result of haematogenous spread following oral or parenteral dosage[38–41,43,44]. The severity of ototoxicity depends on:

- The concentration of the drug in the blood.
- The period of time the drug has been used.
- Individual susceptibility (probably determined primarily by heredity).
- Whether other ototoxic drugs are also being used.
- Whether renal function is unimpaired.
- Whether there is concurrent noise exposure[38,39].

Ototoxicity after systemic drug therapies usually follows high dosages, prolonged therapy or concurrent renal failure (affecting drug excretion). In the cat, dihydro-streptomycin and neomycin have been reported to cause ototoxicity during prolonged systemic administration[42,43]. However, more commonly, ototoxicity follows topical application of ototoxic agents into the external ear canal and their subsequent passage into the middle ear via a ruptured tympanum[42,43]. Subsequent

diffusion into the middle ear is enhanced by the presence of otitis media, which induces increased permeability through the round window membrane that is an important portal for the passage of inflammatory mediators, toxins and drugs from the middle ear to the inner ear[38]. The agent passes through the membrane of the round window and enters the perilymph in the tunnel of Corti. It thereby comes in contact with the hair cells of the organ of Corti and causes degeneration of the perceptive cells[42]. This route of entry was demonstrated for gentamicin in the guinea pig[45]. Similar structures in the vestibule make it likely that perilymph also reaches the sensory cells of the vestibular labyrinth.

The mechanism of toxicity is unclear, but the pathology includes hair cell loss with a progression from basal coil outer hair cells to more apical outer hair cells, followed by inner hair cells. The most current evidence points to binding of the drug to glycosaminoglycans of the stria vascularis, causing strial changes and secondary hair cell changes[38,39,41]. Strauss et al. reported degeneration of the spiral ganglion cells and cochlear neurons in patients with documented cisplatin ototoxicity, in addition to outer hair cell degeneration in the basal turn of the cochlea[46]. It is generally assumed that oxygen free radicals are involved in causing injuries to the cochlea by ototoxic substances[38].

Drugs that primarily affect the cochlea, resulting in hearing impairment, are cochleotoxic whereas drugs that affect the vestibular system, resulting in vestibular dysfunction, are vestibulotoxic. Some drugs are both cochleotoxic and vestibulotoxic. The effects may reflect uni- or bilateral toxicity. Clinical signs of vestibular damage may be reflected very early (as soon as 10 minutes!) after the insult and these include horizontal nystagmus, with the fast component to the affected side, strabismus, ataxia, head tilt, circling, nausea and refusal of food[38–41,43,44].

Within 3 days central compensation results in diminishing and eventual disappearance of the nystagmus, gradual attempts to stand, and beginning efforts to eat and drink, but the head tilt is unchanged[1,43]. Within 3 weeks the situation improves, but jumping and walking down stairs often still results in falling. The compensation is optimal after about 3 months. The head tilt, however, may still be obvious and permanent[38–41,43,44].

Clinical signs of cochlear damage usually go unnoticed until complete deafness is recognised. The early signs of cochlear damage in humans include tinnitus and although this would be difficult to document in dogs and cats, it may be that an inappropriate, or unusually strong, response to an auditory stimulus is a reflection of early cochlear damage[38,39]. Bilateral ototoxicity causes bilateral loss of equilibrium and loss of hearing, expressed as a loss of orientation, loss of social contact and sometimes changes in behaviour and general malaise caused by insecurity[38,39].

While the effect on the equilibrium of the animal is evident on clinical examination and can be assessed by simple observation, the effect on hearing, especially with unilateral loss, can only be objectively tested by cochlear microphony, BERA or with the use of otoacoustic emission testing (see Chapter 2 Diagnostic Procedures, Section 2.10)[42,47,48].

In general, ototoxic effects are dose related, therefore avoiding ototoxic chemicals or reducing the dose and frequency of administration are the first principle[38–41,43,44]. Careful observation and regular follow-up examinations of the patient may allow detection of vestibular signs early enough to allow the clinician to suspend therapy.

7.6 INNER EAR TRAUMA

> - Barotrauma and caloric trauma have not been reported in the veterinary literature, but may be more common than currently thought.
> - Barotrauma results from excessive increases in atmospheric pressure in the ear canals beyond the compensating capabilities of the auditory tubes and can lead to temporary or permanent middle and inner ear dysfunction.
> - Concussion of the inner ear can be seen after blunt trauma to the head and is not always associated with fractures of the temporal and petrosal bones.

The pathophysiology of blunt trauma to the head, involving the nose and nasal sinuses is discussed in Chapter 11 Diseases of the Nasal Cavity and Sinuses, Section 11.12. Trauma of the auricles, ear canals and temporal and petrosal bones are discussed in Chapter 4 Diseases of the Pinna, Section 4.2, Chapter 5 Aetiology and Pathogenesis of Otitis Externa, Section 5.8 and Chapter 6 Diseases of the Middle Ear, Section 6.6, respectively. Inner ear trauma

is not necessarily accompanied by fractures of the skull-base and petrosal and temporal bones. Concussion of the inner ear, or contusio labyrinthi, can occur in the absence of fractures. Also, large changes in air pressure (barotrauma) or temperature (caloric trauma) in the ear canal/middle ear can result in dysfunction of the labyrinth. These specific causes of inner trauma will be addressed here. Loud noises can lead to temporary or permanent trauma of the cochlea, but will be discussed in Section 7.8 NIHL.

Barotrauma and caloric trauma

Acute mono- or bilateral loss of ventilation in the middle ear can be caused by a sudden increase of atmospheric pressure beyond the compensating capabilities of the auditory tube[49]. In humans barotrauma may occur in patients with acute rhinitis during descent in an aircraft, but also in healthy people. The external pressure increases rapidly during descent and causes the tympanic membrane to collapse inwards. This leads to traumatic inflammation of the middle ear mucosa with clinical signs as ear fullness, otalgia and deafness[49].

More commonly though, middle ear barotrauma is experienced by divers on deep descent[50–52]. It is important to realise that the auditory tubes irreversibly block with a pressure differential of approximately 90 mmHg, equivalent to a depth of 1.37 m and eardrums may rupture in water as shallow as 1.22 m[50]. Although inner ear barotrauma occurs infrequently, it may lead to persistent hearing loss and vestibular ataxia. Inner ear barotrauma is probably the result of either haemorrhage, a labyrinthine membrane tear or perilymph fistula formation through the round or oval window[50]. No veterinary reports exist on barotrauma, but based on the aforementioned data, flying should probably be discouraged in animals with chronic ear, nose or throat disorders.

In humans and in small animals it is generally recommended to flush the ear canals with warm saline, preferably near body temperature. People have complained about dizziness, nausea and even hearing loss after ear flushes with water below body temperature. Otorhinolaryngologists routinely perform caloric testing to assess the vestibular system and it is well known that infusion of cold water in the ear canals stimulates the semicircular canals[53]. Though no veterinary reports exist, flushing the ear canals with cold water can lead to similar caloric trauma to the middle and inner ear in dogs and cats with (usually temporary) signs of inner ear dysfunction. This is often misinterpreted as an oto-toxic effect of the cleaning solution itself, rather than a caloric insult.

Contusio labyrinthi

Blunt trauma to the skull, even without fractures can lead to bleeding within, and membrane ruptures of the vestibulocochlear organ. SNHL affecting all frequencies above 3 kHz has been reported in humans associated with concussion of the inner ear, as well as vertigo and tinnitus as accompanying symptoms[54]. A work-up as described for trauma to the temporal bone (Chapter 6 Diseases of the Middle Ear, Section 6.6) is recommended after blunt head trauma to rule out any skull-base fractures. There is no specific treatment possible other than the use of anti-inflammatory medication. The prognosis is uncertain in humans. In many cases complete restoration of the cochleovestibular deficit is possible but some patients may exhibit progressive hearing loss and/or long-lasting vertigo[54].

7.7 AGE-RELATED HEARING LOSS

- ARHL is the most common form of acquired hearing loss in dogs.
- The diagnosis is based on BERA characteristics of increased auditory thresholds, mainly in the high-frequency region (8–32 kHz).
- Treatment with middle ear implants is an option if hearing loss is not too advanced; alternatively vibrating collars can be used to improve client–patient communication.

Age-related hearing loss (ARHL, presbyacusis or presbycusis in American texts) is the most common form of hearing loss in people in industrialised countries, affecting approximately 40% by age 65 years[5–8,55]. The disease is more prevalent and severe in men than in women[56–58]. ARHL is also the most common form of acquired hearing loss in dogs[10,59,60], yet in contrast to human medicine, little is known about the prevalence, aetiology, audiometric characteristics and therapeutic options. As in humans, presbycusis in dogs most likely reflects the cumulative effects of heredity, disease, noise and ototoxic agents superimposed upon those of the aging process itself[8]. The physiology of hearing and electrophysiological hearing tests are discussed in Chapter 1 Anatomy and Physiology of the Ear, Section 1.8 and Chapter 2 Diagnostic Procedures, Section 2.10, respectively.

There have been many studies documenting audiometric characteristics of age-related hearing impairment in humans[61–66]. Cross-sectional studies, describing differences between age groups, have shown that pure tone hearing thresholds increase with age, particularly at high frequencies[8,67–69]. From longitudinal studies[63–67] it appears that significant reduction in hearing capacity occurs from the age of 60 years onward and begins at the high frequencies (6–16 kHz), but gradually encompasses the entire frequency range. Longitudinal and cross-sectional studies on ARHL have also been reported for various animal species, including dogs[10,55,70–72].

Most studies have demonstrated elevated hearing thresholds in aged animals. Significantly higher thresholds at all frequencies tested (1–32 kHz) were found in a group with 10 geriatric dogs (mean age 12.7 years) than in the groups with 10 young and 10 middle-aged dogs (mean age of 1.9 and 5.7 years, respectively) (Fig. 7.2)[10].

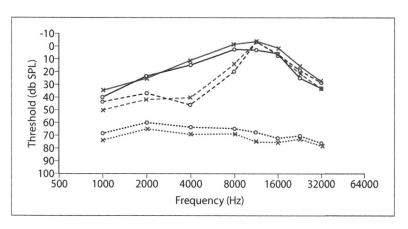

Fig. 7.2 Tone audiogram indicating mean threshold values in 3 groups of 10 dogs: group I (solid lines), mean age 1.9 years; group II (dashed lines), mean age 5.7 years; group III (dotted lines), mean age 12.7 years. (X = left ear, O = right ear.)

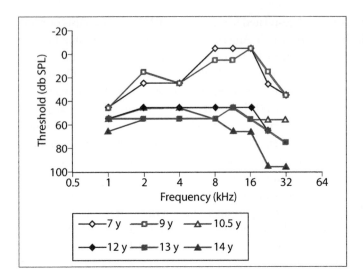

Fig. 7.3 Longitudinal tone audiograms of the left ear of a dog from the longitudinal group at octave frequencies from 1 to 32 kHz. This dog entered the study at 7 years of age.

A longitudinal study revealed a progressive increase in hearing thresholds with aging, starting at 8–10 years of age (Fig. 7.3). The effects in both studies were most pronounced at the middle- to high-frequencies (8–32 kHz)[10].

The most widely referenced framework for describing the histopathological changes associated with aging, is that proposed by Schuknecht[73,74]. He divided ARHL into four types:
- Sensory (predominant loss of outer hair cells [OHCs]).
- Neural (loss of afferent neurons; spiral ganglion cells [SGCs]).
- Metabolic (atrophy of stria vascularis).
- Cochlear conductive.

Later, he added 2 more categories:
- Mixed (sensory, neural and metabolic).
- Indeterminate (no morphological findings).

Although >25% of cases are classified as indeterminate, in most cochleas from aging humans there is a mixture of histopathological changes[8,75,76]. Age-related cochlear changes have been reported in several animal species, including dogs[8,77–84]. Knowles *et al.* found loss of SGCs in a group of deaf dogs in which auditory thresholds were completely absent[77]. Shimada *et al.* reported varying degrees of loss of SGCs, and atrophy of the organ of Corti and the stria vascularis, in all dogs over 12 years of age, predominantly in the basal turn of the cochlea[78]. In a study by ter Haar *et*

al., significant loss of SGCs was found, as well as loss of OHCs and inner hair cells (IHCs) and a reduction in stria vascularis cross-sectional area (SVCA)[84]. The histological abnormalities found were primarily in the basal turn, which was consistent with the occurrence of the largest audiometric threshold shifts and lowest absolute thresholds in the middle- to high-frequency regions (Figs 7.4A, B). It was concluded that the degeneration of the OHCs and SGCs observed in the basal turn was primarily responsible for the elevated hearing thresholds, similar to findings in humans[84].

Remediation of presbycusis is an important contributor to quality of life in geriatric medicine. For hearing impaired people, modern hearing aids, assistive listening devices, middle ear implants, cochlear implants and brainstem implants provide valuable aids to communication[5–8,24,85]. Commonly, in humans with mixed SNHL in which the average hearing thresholds reach 40 dB on the audiogram, amplification is indicated. Amplification is primarily accomplished with conventional hearing aids in people, but the use of conventional hearing aids has only anecdotally been mentioned in dogs[86]. The use of hearing aids has not met with great clinical success and clinical reports on efficacy are not available.

Elderly people with mild to severe SNHL who cannot benefit from conventional external amplification greatly benefit from implantable auditory prostheses[87–92]. Cochlear implants are reserved for patients with severe bilateral hearing loss that is not improved by hearing aids or middle ear implants[8]. Middle ear

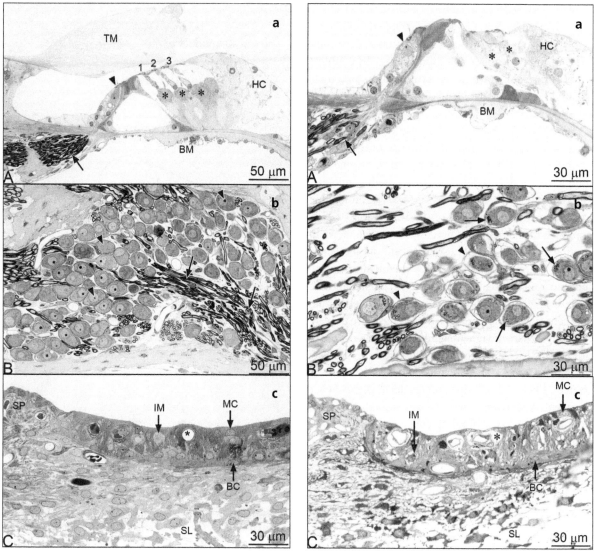

Figs 7.4A–C A: Midmodiolar section through the lower middle turn (M1) of the cochlea of a young dog. (a) In the organ of Corti the OHCs are arranged in three separate rows (1, 2, 3) and the IHCs (arrowhead) are present as a single row. The arrow points to nerve fibres in the osseous spiral lamina. (TM) tectorial membrane, (BM) basilar membrane, (HC) Hensen's cells. Asterisks indicate Deiters' cells. (b) Rosenthal's canal, in its normal appearance, containing the spiral ganglion with SGCs (arrowheads) and nerve fibres (arrows). (c) The lateral wall consists of the spiral ligament (SL), the spiral prominence (SP), and the stria vascularis consisting of marginal cells (MC), intermediate cells (IM), basal cells (BC) and strial capillaries (asterisk). B: Midmodiolar section through the upper basal turn (B2) of the cochlea of a geriatric dog. (a) In the organ of Corti the OHCs have been replaced by supporting cells (asterisks). The IHC (arrowhead) is still present. Note that the number of nerve fibres in the osseous spiral lamina is reduced (arrow). (BM) basilar membrane, (HC) Hensen's cells. (b) In the spiral ganglion there is obvious loss of SGCs and nerve fibres. In the remaining SGCs there is shrinkage of the perikarya (arrowheads), intracellular vacuolation, and lipofuscin inclusions (arrows). (c) In the stria vascularis there is extracellular oedema (asterisks), shrinkage of the intermediate cells (IM), and vacuolation of the marginal cells (MC). (BC) basal cells, (SL) spiral ligament, (SP) spiral prominence. With permission, ter Haar G, de Groot JCMJ, Venker-van Haagen AJ, van sluijs FJ, Smoorenburg GF. Effects of aging on inner ear morphology in dogs in relation to brainstem responses to toneburst auditory stimuli. *Journal of Veterinary Internal Medicine* 2009, 23(3): 536–43.

implants are used for moderate to severe conductive and sensorineural hearing loss. The principle of middle ear implants is to provide acoustic amplification of residual hearing and transmission of sound energy by coupling a vibratory element (implanted transducer) directly to the middle ear ossicular chain or on the round window membrane[24,88,93,94].

Several studies have been reported on the successful short-, medium- and long-term use of the Vibrant Soundbridge (VSB middle ear) implant in people with mild to severe SNHL[87,88,90–92,95–98] or both conductive and sensorineural hearing loss (Fig. 7.5)[99,100]. A feasibility study with implantation of the VSB using a lateral approach to the tympanic bulla was reported, demonstrating that the VSB could also successfully be implanted in dogs[101]. Three dogs with ARHL subsequently successfully received a VSB middle ear implant unilaterally (Fig. 7.6). Recovery from surgery was uneventful, except for transient facial nerve paralysis in two dogs. The implantation procedure did not affect residual hearing as measured with BERA and measurable benefit from the VSB on hearing thresholds in the implanted ear was demonstrated[9]. Owner satisfaction and results in client-owned dogs with ARHL have yet to be established. A cheaper option is the use of a vibrating collar and teaching deaf dogs and cats to respond to vibrations that can be triggered by the client from a remote control. In addition, many dogs can be taught to respond to sign language.

7.8 NOISE-INDUCED HEARING LOSS

- Acoustic trauma is rarely reported in dogs and cats, but flying, hunting associated with firearms, housing in kennels and acoustic noise from MRI are implicated in NIHL.
- The pathophysiological consequences of NIHL in dogs and cats are currently unknown.
- Wearing ear protection is advised for animals subjected to loud environmental noise as there is no treatment available after permanent damage.

Exposure to loud noise causes hearing loss in humans and laboratory animals[55,102–105]. Although reports on acoustic trauma in dogs are scarce, noise-induced hearing loss (NIHL) has been mentioned in the lit-

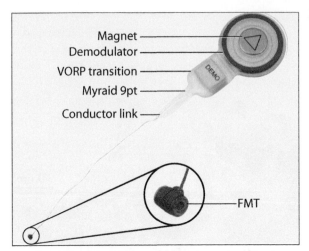

Fig. 7.5 The implanted part of the Vibrant Soundbridge middle ear implant (vibrating ossicular prosthesis or VORP) consists of a magnet, a coil and a demodulator. It is connected to the floating mass transducer (FMT) via a conductor link. The inset shows the FMT in greater detail.

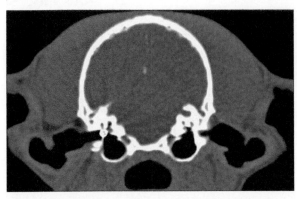

Fig. 7.6 Postoperative CT scan of a dog, 3 months after VSB implant surgery. The scan shows the FMT in the left round window niche and part of the conducting wire.

erature and possibly occurs in dogs that are used for hunting associated with firearms[2,13,60], dogs housed in kennels with high ambient noise levels[106,107], and in dogs shipped by airplane in cargo compartments where there are high ambient noise levels[60]. In addition, noise produced by MRI scanners (that can peak at a sound pressure level of 131 dB) has been shown to cause noise-induced cochlear dysfunction in humans and dogs[108,109]. Protective reflexes exist where middle ear muscles contract in response to loud sounds to reduce sound levels reaching the inner ear (see Chapter 1 Anatomy and Physiology of the Ear, Section 1.8), but the reflexes are too slow for percussive sounds, such as gunfire, and only provide limited protection against very loud noises[60].

Hearing thresholds in humans after NIHL may return to its normal value after minutes, hours or days, depending on the intensity and duration of the noise exposure and the individual's susceptibility. Whether or not NIHL in dogs is transient or permanent is unknown. Genetic variations, age, health status, smoking, drugs and hypertension are factors known to influence the susceptibility to NIHL[105,110,111]. Conductive hearing loss will act as an ear protector and decreases hearing loss from exposure to noise[112]. Hearing loss that resolves leads to temporary threshold shifts, while hearing loss that remains after a recovery period causes a broad set of physical changes in the major cellular systems in the cochlea leading to a permanent threshold shift[104,113,114].

Noise can damage most of the cell populations in the cochlea, but the OHCs are affected primarily, especially at the basal end of the cochlea[113]. OHC stereocilia can be broken, fused or have broken tip links that lead to loss of structural integrity[113,115–117]. With more severe noise exposures, the pathology spreads to include IHCs, loss of auditory nerve fibres and even damage to the stria vascularis[104]. With impulse noise exposure, in addition to the changes described above, the organ of Corti can be ripped from the basilar membrane[118,119]. Damage to the cochlea can be instant, but pathology can also increase over a period of 2–30 days after exposure[105,113,118–120]. Recently, oxygen free radicals have been implicated in causing the injury to hair cells from not only noise exposure, but also aging and ototoxic antibiotics[105,121,122]. Oxygen free radical scavengers can reduce the effect of acoustic trauma, especially when administered before the noise exposure[121]. Adenosine A1 receptor agonists may even be effective postexposure[123].

NIHL affects high frequencies more than low frequencies in humans and usually has a dip at or near 4 kHz[113,124,125]. Hearing thresholds above the affected frequency are usually better, which distinguishes NIHL from ARHL, where hearing loss increases in every frequency above a certain frequency[113]. The amount of hearing loss depends on the intensity of the noise, the duration of the exposure and the frequency spectrum and time pattern of the noise. NIHL in humans is usually associated with continuous loud recreational or occupational noise exposure, and the cumulative effects of repetitive noise exposure exceeding a maximum permissible daily dose are well known[105,113,126].

Unlike continuous environmental and occupational noise, firearms and firecrackers are sources of impulse noise exposure. Most rifles, shotguns, handguns and firecrackers, but also acoustical noise from MRI, produce very high peak sound pressure levels capable of producing permanent NIHL instantly if ear protection is not used[102,126]. In dogs, results of one study demonstrate that exposure to MRI noise results in a significant reduction in frequency-specific cochlear function and ear protection as a routine precautionary measure is therefore recommended[109]. Another study, on the effects of kennel noise on hearing in dogs, suggested that the exposure to high ambient noise had a clear adverse impact on their hearing ability[107]. At the end of a 6-month study, all of the tested dogs had some degree of hearing change and more than half had shifts in their wave V thresholds of ≥20 dB[107]. Not only people working in kennels, but ideally dogs housed in kennels as well, especially if long-term, should wear ear protection. Ear protection in dogs is not commonly used yet, as effective devices for noise protection are not yet commercially available for dogs[60].

7.9 IDIOPATHIC PERIPHERAL VESTIBULAR ATAXIA

Acquired peripheral vestibular ataxia most commonly results from otitis media and interna[17,25–27] (see Chapter 6 Diseases of the Middle Ear, Sections 6.2, 6.3), but can result from ear trauma (Section 7.6), hypothyroidism, ototoxicity (Section 7.5) or neoplastic disease (Chapter 5 Aetiology and Pathogenesis of Otitis Externa 5.3) of ear canal and middle ear as well. Canine idiopathic peripheral vestibular disease is however the second most common cause of peripheral vestibular dysfunction in dogs, and is a common aetiology for unilateral peripheral vestibular dysfunction of peracute onset (head tilt, ataxia, horizontal or rotary nystagmus) in dogs and cats[1,27,127].

Canine vestibular disease may be seen in any aged dog, but geriatric dogs appear to be predisposed. The mean age of onset is 12.5 years and it is very atypical to be seen in dogs younger than 5 years old[25–28]. Clinical signs are unilateral in nature in most animals.

Feline idiopathic peripheral vestibular disease differs slightly in that it can occur in cats of any age, has a higher incidence in outdoor cats in the summer and autumn months in the northeastern and mid-Atlantic regions of the USA, and will occasionally result in bilateral peripheral vestibular signs[16,27,28].

In both dogs and cats, idiopathic peripheral vestibular disease results in clinical signs referable to dysfunction of the peripheral vestibular system only; affected animals do not have concurrent facial nerve paralysis or postganglionic Horner syndrome. In the acute setting the clinical signs can be severe (head tilt, rolling, falling) and some animals may vomit.

The diagnosis is made on exclusion of other causes and on the alleviation of clinical signs with time. Diagnostic imaging studies of the peripheral vestibular apparatus are usually normal in animals with this disease[1,16,26–28].

Therapy is mainly supportive and vestibulosedative drugs are not routinely recommended, as compensation and habituation will be greatly accelerated in animals that are able to and encouraged and assisted to walk. Occasionally however, vomiting is severe, and the following drugs are administered for 2–3 days to alleviate the emesis associated with motion sickness[128]:

- H1 histaminergic receptor antagonists.
- M1 cholinergic receptor antagonists.
- Vestibulosedative drugs.

An empiric course of systemic broad-spectrum antibiotic therapy is advised by some others to treat occult OMI[27]. The prognosis for recovery is excellent, spontaneous improvement is usually seen within 2–3 days, with a complete return to normal within 2–4 weeks. Occasionally, a residual head tilt may persist. The condition may occasionally recur[27].

7.10 REFERENCES

1　ter Haar G. Inner ear dysfunction in dogs and cats: Conductive and sensorineural hearing loss and peripheral vestibular ataxia. *European Journal of Companion Animal Practice* 2007;**17**:127–35.

2　Strain GM. Aetiology, prevalence and diagnosis of deafness in dogs and cats. *British Veterinary Journal* 1996;**152**(1):17–36.

3　Wilson WJ, Mills PC. Brainstem auditory-evoked response in dogs. *American Journal of Veterinary Research* 2005;**66**(12):2177–87.

4　Rak SG, Distl O. Congenital sensorineural deafness in dogs: a molecular genetic approach toward unravelling the responsible genes. *Veterinary Journal* 2005;**169**(2):188–96.

5　Sajjadi H, Paparella MM, Canalis RF. Presbycusis. In: *The Ear; Comprehensive Otology*. Canalis RF, Lambert PR (eds). Lippincott Williams & Wilkins, Philadelphia 2000, pp. 545–57.

6　Jennings CR, Jones NS. Presbyacusis. *Journal of Laryngology and Otology* 2001;**115**(3):171–8.

7　Willott JF, Hnath Chisolm T, Lister JJ. Modulation of presbycusis: current status and future directions. *Audiology and Neuro-otology* 2001;**6**(5):231–49.

8　Gates GA, Mills JH. Presbycusis. *Lancet* 2005;**366**(9491):1111–20.

9　ter Haar G, Mulder JJ, Venker-van Haagen AJ, van sluijs FJ, Snik AF, Smoorenburg GF. Treatment of age-related hearing loss in dogs with the Vibrant Soundbridge middle ear implant: short-term results in 3 dogs. *Journal of Veterinary Internal Medicine* 2010;**24**(3):557–64.

10　ter Haar G, Venker-van Haagen AJ, van den Brom WE, van sluijs FJ, Smoorenburg GF. Effects of aging on brainstem responses to toneburst auditory stimuli: a cross-sectional and longitudinal study in dogs. *Journal of Veterinary Internal Medicine* 2008;**22**(4):937–45.

11　Houpt KA, Beaver B. Behavioral problems of geriatric dogs and cats. *Veterinary Clinics of North America Small Animal Practice* 1981;**11**(4):643–52.

12　Gratton MA, Vázquez AE. Age-related hearing loss: current research. *Current Opinion in Otolaryngology & Head and Neck Surgery* 2003;**1**(5):367–71.

13　Strain GM. *Deafness in Dogs and Cats*. Strain GM (ed). CABI, Cambridge 2011.

14　Bischoff MG, Kneller SK. Diagnostic imaging of the canine and feline ear. *Veterinary Clinics of North America Small Animal Practice* 2004;**34**(2):437–58.

15　Strain GM. Vestibular testing: on balance. *Veterinary Journal* 2010;**185**(3):239–40.

16　LeCouteur RA. Feline vestibular diseases – new developments. *Journal of Feline Medicine and Surgery* 2003;**5**(2):101–8.

17 Garosi LS, Lowrie CT, Swinbourne NF. Neurological manifestations of ear disease in dogs and cats. *Veterinary Clinics of North America Small Animal Practice* 2012;**42**(6):1143–60.

18 Cook LB. Neurologic evaluation of the ear. *Veterinary Clinics of North America Small Animal Practice* 2004;**34**(2):425–35.

19 Strain GM. Congenital deafness and its recognition. *Veterinary Clinics of North America Small Animal Practice* 1999;**29**(4):895–907.

20 Strain GM, Kearney MT, Gignac IJ. Brainstem auditory-evoked potential assessment of congenital deafness in Dalmatians: associations with phenotypic markers. *Journal of Veterinary Internal Medicine* 1992;**6**(3):175–82.

21 Geigy CA, Heid S, Steffen F, Danielson K, Jaggy A, Gaillard C. Does a pleiotropic gene explain deafness and blue irises in white cats? *Veterinary Journal* 2007;**173**(3):548–53.

22 Wilkes MK, Palmer AC. Congenital deafness and vestibular deficit in the Dobermann. *Journal of Small Animal Practice* 1992;**33**(5):218–24.

23 Strain GM. Deafness prevalence and pigmentation and gender associations in dog breeds at risk. *Veterinary Journal* 2004;**167**:23–32.

24 Shinners MJ, Hilton CW, Levine SC. Implantable hearing devices. *Current Opinion in Otolaryngology & Head and Neck Surgery* 2008;**16**(5):416–9.

25 Schunk KL. Disorders of the vestibular system. *Veterinary Clinics of North America Small Animal Practice* 1988;**18**(3):641–65.

26 Schunk KL, Averill DR. Peripheral vestibular syndrome in the dog: a review of 83 cases. *Journal of the American Veterinary Medical Association* 1983;**182**(12):1354–7.

27 Rossmeisl JH. Vestibular disease in dogs and cats. *Veterinary Clinics of North America Small Animal Practice* 2010;**40**(1):81–100.

28 Thomas WB. Vestibular dysfunction. *Veterinary Clinics of North America Small Animal Practice* 2000;**30**:227–46.

29 Forbes S, Cook JR. Congenital peripheral vestibular disease attributed to lymphocytic labyrinthitis in two related litters of Doberman Pinscher pups. *Journal of the American Veterinary Medical Association* 1991;**198**(3):447–9.

30 Gotthelf LN. Diagnosis and treatment of otitis media in dogs and cats. *Veterinary Clinics of North America Small Animal Practice* 2004;**34**(2):469–87.

31 Shell LG. Otitis media and otitis interna. Etiology, diagnosis, and medical management. *Veterinary Clinics of North America Small Animal Practice* 1988;**18**(4):885–99.

32 Beyea JA, Agrawal SK, Parnes LS. Recent advances in viral inner ear disorders. *Current Opinion in Otolaryngology & Head and Neck Surgery* 2012;**20**(5):404–8.

33 Westerlaken BO, Stokroos RJ, Dhooge IJM, Wit HP, Albers FWJ. Treatment of idiopathic sudden sensorineural hearing loss with antiviral therapy: a prospective, randomized, double-blind clinical trial. *Annals of Otology Rhinology and Laryngology* 2003;**112**(11):993–1000.

34 Beatty JA, Barrs VR, Swinney GR, Martin PA, Malik R. Peripheral vestibular disease associated with cryptococcosis in three cats. *Journal of Feline Medicine and Surgery* 2000;**2**(1):29–34.

35 Ryan AF, Harris JP, Keithley EM. Immune-mediated hearing loss: basic mechanisms and options for therapy. *Acta Oto-Laryngologica* 2002;**122**(5):38–43.

36 Schreiber BE, Agrup C, Haskard DO, Luxon LM. Sudden sensorineural hearing loss. *Lancet* 2010;**375**(9721):1203–11.

37 Sturges BK, Dickinson PJ, Kortz GD, *et al.* Clinical signs, magnetic resonance imaging features, and outcome after surgical and medical treatment of otogenic intracranial infection in 11 cats and 4 dogs. *Journal of Veterinary Internal Medicine* 2006;**20**(3):648–56.

38 Rybak LP, Kanno H. Ototoxicity. In: *Otorhinolaryngology: Head and Neck Surgery*, 15th edn. Ballenger JJ, Snow JJB (eds). Williams & Wilkins, Baltimore 1996, pp. 1102–8.

39 Oishi N, Talaska AE, Schacht J. Ototoxicity in dogs and cats. *Veterinary Clinics of North America Small Animal Practice* 2012;**42**(6):1259–71.

40 Pickrell JA, Oehme FW, Cash WC. Ototoxicity in dogs and cats. *Seminars in Veterinary Medicine and Surgery (Small Animal)* 1993;**8**(1):42–9.

41 Merchant SR. Ototoxicity. *Veterinary Clinics of North America Small Animal Practice* 1994;**24**:971–80.

42 Gallé HG, Venker-van Haagen AJ. Ototoxicity of the antiseptic combination chlorhexidine/cetrimide (Savlon): effects on equilibrium and hearing. *Veterinary Quarterly* 1986;**8**(1):56–60.

43 Venker-van Haagen AJ. The ear. In: *Ear, Nose, Throat and Tracheobronchial Diseases in Dogs and Cats*. Venker-van Haagen AJ (ed). Schlütersche Verlagsgesellschaft, Hannover 2005, pp. 1–50.

44 Mills PC, Ahlstrom L, Wilson WJ. Ototoxicity and tolerance assessment of a TrisEDTA and polyhexamethylene biguanide ear flush formulation in dogs. *Journal of Veterinary Pharmacology and Therapeutics* 2005;**28**(4):391–7.

45 De Groot J, Meeuwsen F, Ruizendaal WE, Veldman JE. Ultrastructural localization of gentamicin in the cochlea. *Hearing Research* 1990;**35**(1–2):35–42.

46 Strauss M, Towfighi J, Lipton A, Lord S, Harvey HA, Brown B. Cis-platinum ototoxicity: clinical experience and temporal bone histopathology. *Laryngoscope* 1983;**93**(12):1554–9.

47 ter Haar G, Venker-van Haagen AJ, de Groot HN, van den Brom WE. Click and low-, middle-, and high-frequency toneburst stimulation of the canine cochlea. *Journal of Veterinary Internal Medicine* 2002;**16**(3):274–80.

48 Gonçalves R, McBrearty A, Pratola L, Calvo G, Anderson TJ, Penderis J. Clinical evaluation of cochlear hearing status in dogs using evoked otoacoustic emissions. *Journal of Small Animal Practice* 2012;**53**(6):344–51.

49 Mirza S, Richardson H. Otic barotrauma from air travel. *Journal of Laryngology & Otology* 2005;**119**:366–70.

50 Becker GD, Parell GJ. Barotrauma of the ears and sinuses after scuba diving. *European Archives of Otorhinolaryngology* 2001;**258**(4):159–63.

51 Klingmann C, Praetorius M, Baumann I. Barotrauma and decompression illness of the inner ear: 46 cases during treatment and follow-up. *Otology & Neurotology* 2007;**28**(4):447–54.

52 Sheridan MF, Hetherington HH, Hull JJ. Inner ear barotrauma from scuba diving. *Ear, Nose and Throat Journal* 1999;**78**(3):181–7.

53 Davies RA, Luxon LM. Dizziness following head injury: a neuro-otological study. *Journal of Neurology* 1995;**242**(4):222–30.

54 Schuknecht HF, Davison RC. Deafness and vertigo from head injury. *Archives of Otolaryngology and Head & Neck Surgery* 1956;**63**(5):513–28.

55 Ohlemiller KK. Contributions of mouse models to understanding of age- and noise-related hearing loss. *Brain Research* 2006;**1091**(1):89–102.

56 Hinchcliffe R. Aging and sensory thresholds. *Journal of Gerontology* 1962;**17**(1):45–50.

57 Corso JF. Aging and auditory thresholds in men and women. *Archives of Environmental Health* 1963;**6**:350–6.

58 Pearson JD, Morrell CH, Gordon-Salant S, *et al*. Gender differences in a longitudinal study of age-associated hearing loss. *Journal of the Acoustical Society of America* 1995;**97**(2):1196–205.

59 Delauche AJ. Brain-stem evoked responses as a diagnostic tool for deafness; a neurophysiological test with potential? *British Veterinary Journal* 1996;**152**(1):13–15.

60 Strain GM. Canine deafness. *Veterinary Clinics of North America Small Animal Practice* 2012;**42**(6):1209–24.

61 Robinson DW, Sutton GJ. Age effect in hearing – a comparative analysis of published threshold data. *Audiology* 1979;**18**(4):320–34.

62 Mościcki EK, Elkins EF, Baum HM, McNamara PM. Hearing loss in the elderly: an epidemiologic study of the Framingham Heart Study Cohort. *Ear and Hearing* 1985;**6**(4):184–90.

63 Pedersen KE, Rosenhall U, Metier MB. Changes in pure-tone thresholds in individuals aged 70–81: results from a longitudinal study. *International Journal of Audiology* 1989;**28**(4):194–204.

64 Brant LJ, Fozard JL. Age changes in pure-tone hearing thresholds in a longitudinal study of normal human aging. *Journal of the Acoustical Society of America* 1990;**88**(2):813–20.

65 Enrietto JA, Jacobson KM, Baloh RW. Aging effects on auditory and vestibular responses: a longitudinal study. *American Journal of Otolaryngology* 1999;**20**(6):371–8.

66 Gates GA, Schmid P, Kujawa SG, Nam B, D'Agostino R. Longitudinal threshold changes in older men with audiometric notches. *Hearing Research* 2000;**141**(1-2):220–8.

67 Lee F-S, Matthews LJ, Dubno JR, Mills JH. Longitudinal study of pure-tone thresholds in older persons. *Ear and Hearing* 2005;**26**(1):1–11.

68 Nelson EG, Hinojosa R. Presbycusis: a human temporal bone study of individuals with downward sloping audiometric patterns of hearing loss and review of the literature. *Laryngoscope* 2009;**116**(S112):1–12.

69 Blanchet C, Pommie C, Mondain M, Berr C. Pure-tone threshold description of an elderly French screened population. *Otology & Neurotology* 2008;**4**:432–40.

70 Dum N, Schmidt U, Wedel von H. Age-related changes in the auditory evoked brainstem potentials of albino and pigmented guinea pigs. *Archives of Otorhinolaryngology* 1980;**228**(4):249–58.

71 Harrison J, Buchwald J. Auditory brainstem responses in the aged cat. *Neurobiology of Aging* 1982;**3**:163–71.

72 Cooper WA, Coleman JR, Newton EH. Auditory brainstem responses to tonal stimuli in young and aging rats. *Hearing Research* 1990;**43**(2-3):171–9.

73 Schuknecht HF. Further observations on the pathology of presbycusis. *Archives of Otolaryngology* 1964;**80**:369–82.

74 Schuknecht HF, Gacek MR. Cochlear pathology in presbycusis. *Annals of Otology, Rhinology and Laryngology* 1993;**102**:1–16.

75 Boettcher FA. Presbyacusis and the auditory brainstem response. *Journal of Speech, Language, and Hearing Research* 2002;**45**(6):1249.

76 Ohlemiller KK, Gagnon PM. Apical-to-basal gradients in age-related cochlear degeneration and their relationship to primary loss of cochlear neurons. *Journal of Comparative Neurology* 2004;**479**(1):103–16.

77 Knowles K, Blauch B, Leipold H, Cash W, Hewett J. Reduction of spiral ganglion neurons in the aging canine with hearing loss. *Zentralblatt fur Veterinarmedizin* 1989;**36**(3):188–99.

78 Shimada A, Ebisu M, Morita T, Takeuchi T, Umemura T. Age-related changes in the cochlea and cochlear nuclei of dogs. *Journal of Veterinary Medical Science* 1998;**60**(1):41–8.

79 Ingham NJ, Comis SD, Withington DJ. Hair cell loss in the aged guinea pig cochlea. *Acta Oto-Laryngologica* 1999;**119**(1):42–7.

80 Spicer SS, Schulte BA. Spiral ligament pathology in quiet-aged gerbils. *Hearing Research* 2002;**172**(1):172–85.

81 Popelar J, Groh D, Pelánová J, Canlon B, Syka J. Age-related changes in cochlear and brainstem auditory functions in Fischer 344 rats. *Neurobiology of Aging* 2006;**27**(3):490–500.

82 Schuknecht HF, Icarashi M, Gacek RR. The pathological types of cochleo-saccular degeneration. *Acta Oto-Laryngologica* 1965;**59**(2-6):154–70.

83 Johnsson LG, Hawkins JE. Strial atrophy in clinical and experimental deafness. *Laryngoscope* 1972;**82**(7):1105–25.

84 ter Haar G, de Groot JCMJ, Venker-van Haagen AJ, van sluijs FJ, Smoorenburg GF. Effects of aging on inner ear morphology in dogs in relation to brainstem responses to toneburst auditory stimuli. *Journal of Veterinary Internal Medicine* 2009;**23**(3):536–43.

85 Backous DD, Duke W. Implantable middle ear hearing devices: current state of technology and market challenges. *Current Opinion in Otolaryngology & Head and Neck Surgery* 2006;**14**(5):314–8.

86 Marshall AE. Invited commentary on Knowles, K, 1990. Reduction of spiral ganglion neurons in the aging canine with hearing loss. *Advances in Small Animal Medicine and Surgery* 1990;**12**(2):6–7.

87 Fisch U, Cremers CW, Lenarz T, *et al*. Clinical experience with the Vibrant Soundbridge implant device. *Otology & Neurotology* 2001;**22**(6):962–72.

88 Fraysse B, Lavieille JP, Schmerber S, *et al*. A multicenter study of the Vibrant Soundbridge middle ear implant: early clinical results and experience. *Otology & Neurotology* 2001;**22**(6):952–61.

89 Snik AF, Mylanus EA, Cremers CW, *et al*. Multicenter audiometric results with the Vibrant Soundbridge, a semi-implantable hearing device for sensorineural hearing impairment. *Otolaryngologic Clinics of North America* 2001;**34**(2):373–88.

90 Luetje CM, Brackman D, Balkany TJ, *et al*. Phase III clinical trial results with the Vibrant Soundbridge implantable middle ear hearing device: a prospective controlled multicenter study. *Otolaryngology Head and Neck Surgery* 2002;**126**(2):97–107.

91 Sterkers O, Boucarra D, Labassi S, *et al*. A middle ear implant, the Symphonix Vibrant Soundbridge: retrospective study of the first 125 patients implanted in France. *Otology & Neurotology* 2003;**24**(3):427–36.

92 Vincent C, Fraysse B, Lavieille JP, Truy E, Sterkers O, Vaneecloo FM. A longitudinal study on postoperative hearing thresholds with the Vibrant Soundbridge device. *European Archives of Otorhinolaryngology* 2004;**261**(9):493–6.

93 Spindel JH, Lambert PR, Ruth RA. The round window electromagnetic implantable hearing aid approach. *Otolaryngologic Clinics of North America* 1995;**28**(1):189–205.

94 Kiefer J, Arnold W, Staudenmaier R. Round window stimulation with an implantable hearing aid (Soundbridge) combined with autogenous reconstruction of the auricle – a new approach. *Journal for Otorhinolaryngology and its Related Specialties* 2006;**68**(6):378–85.

95 Snik A, Cremers C. Audiometric evaluation of an attempt to optimize the fixation of the transducer of a middle-ear implant to the ossicular chain with bone cement. *Clinical Otolaryngology and Allied Sciences* 2004;**29**(1):5–9.

96 Schmuziger N, Schimmann F, àWengen D, Patscheke J, Probst R. Long-term assessment after implantation of the Vibrant Soundbridge device. *Otology & Neurotology* 2006;**27**(2):183–8.

97 Mosnier I, Sterkers O, Bouccara D, *et al*. Benefit of the Vibrant Soundbridge device in patients implanted for 5 to 8 years. *Ear and Hearing* 2008;**29**(2):281–4.

98 Truy E, Eshraghi AA, Balkany TJ, Telishi FF, Van De Water TR, Lavieille J-P. Vibrant soundbridge surgery: evaluation of transcanal surgical approaches. *Otology & Neurotology* 2006;**27**(6):887–95.

99 Colletti V, Soli SD, Carner M, Colletti L. Treatment of mixed hearing losses via implantation of a vibratory transducer on the round window. *International Journal of Audiology* 2006;**45**(10):600–8.

100 Beltrame AM, Martini A, Prosser S, Giarbini N, Streitberger C. Coupling the Vibrant Soundbridge to cochlea round window: auditory results in patients with mixed hearing loss. *Otology & Neurotology* 2009;**30**(2):194–201.

101 ter Haar G, Mulder JJ, Venker-van Haagen AJ, van Sluijs FJ, Smoorenburg GF. A surgical technique for implantation of the vibrant soundbridge middle ear implant in dogs. *Veterinary Surgery* 2011;**40**(3):340–6.

102 Smoorenburg GF. Risk of noise-induced hearing loss following exposure to Chinese firecrackers. *Audiology* 1993;**32**(6):333–43.

103 Bartels S, Ito S, Trune DR, Nuttall AL. Noise-induced hearing loss: the effect of melanin in the stria vascularis. *Hearing Research* 2001;**154**(1-2):116–23.

104 Henderson D, Bielefeld EC, Harris KC, Hu BH. The role of oxidative stress in noise-induced hearing loss. *Ear and Hearing* 2006;**27**(1):1–19.

105 Kevin K. Ohlemiller. Recent findings and emerging questions in cochlear noise injury. *Hearing Research* 2008;**245**(1-2):5–17.

106 Coppola CL, Enns RM, Grandin T. Noise in the animal shelter environment: building design and the effects of daily noise exposure. *Journal of Applied Animal Welfare Science* 2006;**9**(1):1–7.

107 Scheifele P, Martin D, Clark JG, Kemper D, Wells J. Effect of kennel noise on hearing in dogs. *American Journal of Veterinary Research* 2012;**73**(4):482–9.

108 Lauer AM, El-Sharkawy A-MM, Kraitchman DL, Edelstein WA. MRI acoustic noise can harm experimental and companion animals. *Journal of Magnetic Resonance Imaging* 2012;**36**(3):743–7.

109 Venn RE, McBrearty AR, McKeegan D, Penderis J. The effect of magnetic resonance imaging noise on cochlear function in dogs. *Veterinary Journal* 2014;**202**(1):141–5.

110 Ferrite S. Joint effects of smoking, noise exposure and age on hearing loss. *Occupational Medicine* 2005;**55**(1):48–53.

111 Toppila E, Pyykkö I, Starck J. Age and noise-induced hearing loss. *Scandinavian Audiology* 2001;**30**(4):236–44.

112 Nilsson R, Borg E. Noise-induced hearing loss in shipyard workers with unilateral conductive hearing loss. *Scandinavian Audiology* 1983;**2**(2):135–40.

113 Moller AR. Disorders of the cochlea. In: *Hearing*. Moller AR (ed). Academic Press, San Diego 2000; pp. 395–434.

114 Pickles JO. The cochlea. In: *An Introduction to the Physiology of Hearing*. Pickles JO (ed). Emerald Group Publishing, Bingley, UK 2008, pp. 25–72.

115 Liberman MC, Dodds LW. Acute ultrastructural changes in acoustic trauma: serial-section reconstruction of stereocilia and cuticular plates. *Hearing Research* 1987;**26**(1):45–64.

116 Liberman MC. Chronic ultrastructural changes in acoustic trauma: serial-section reconstruction of stereocilia and cuticular plates. *Hearing Research* 1987;**26**(1):65–88.

117 Tsuprun V, Schachern PA, Cureoglu S, Paparella M. Structure of the stereocilia side links and morphology of auditory hair bundle in relation to noise exposure in the chinchilla. *Journal of Neurocytology* 2003;**32**(9):1117–28.

118 Hamernik RP, Patterson JH, Turrentine GA, Ahroon WA. The quantitative relation between sensory cell loss and hearing thresholds. *Hearing Research* 1989;**38**(3):199–211.

119 Hamernik RP. The effect of impulse intensity and the number of impulses on hearing and cochlear pathology in the chinchilla. *Journal of the Acoustical Society of America* 1987;**81**(4):1118–29.

120 Ahmad M, Bohne BA, Harding GW. An *in vivo* tracer study of noise-induced damage to the reticular lamina. *Hearing Research* 2003;**175**(1–2):82–100.

121 Lautermann J, Crann SA, McLaren J, Schacht J. Glutathione-dependent antioxidant systems in the mammalian inner ear: effects of aging, ototoxic drugs and noise. *Hearing Research* 1997;**114**(1–2):75–82.

122 Krause K-H. Aging: A revisited theory based on free radicals generated by NOX family NADPH oxidases. *Experimental Gerontology* 2007;**42**(4):256–62.

123 Wong A, Guo CX, Gupta R, Housley GD, Thorne PR. Post exposure administration of A1 adenosine receptor agonists attenuates noise-induced hearing loss. *Hearing Research* 2010;**260**(1–2):81–8.

124 Rabinowitz PM. Hearing loss and personal music players. *British Medical Journal* 2010;**340**:1261.

125 Rabinowitz PM, Pierce Wise J Sr., Hur Mobo B, Antonucci PG, Powell C, Slade M. Antioxidant status and hearing function in noise-exposed workers. *Hearing Research* 2002;**173**(1–2):164–71.

126 Katbamna B, Flamme GA. Acquired hearing loss in adolescents. *Pediatric Clinics of North America* 2008;**55**(6):1391–402.

127 Pentlarge VW. Peripheral vestibular disease in a cat with middle and inner ear squamous cell carcinoma. *Compendium on Continuing Education for the Practicing Veterinarian* 1984;**6**:731–5.

128 Taylor SM. Head tilt. In: *Small Animal Internal Medicine*, 4th edn. Nelson RW, Couto CG (eds). Mosby Elsevier, St. Louis 2002, pp. 1047–53.

THE NOSE

8.1 INTRODUCTION

The respiratory system encompasses the nose and nasal cavities, pharynx, larynx, trachea, bronchi and smaller passageways, and functions to deliver air to the pulmonary alveoli where gas exchange occurs[1]. The nasal cavity and the conchae (also called turbinates) warm and moisten the air, remove foreign material and allow for olfaction. The nose consists of an external part with its associated nasal cartilages and ligaments and an internal part or nasal cavity with scrolls or conchae[1]. The gross and microscopic anatomy of the nose will be discussed first in this chapter, followed by a detailed description of its important functions, regulation and conditioning of the airflow and olfaction. The mechanisms involved in keeping the nose healthy and clean, mucosal cleaning,

sneezing and reverse sneezing will be addressed, as well as the normal microbial flora.

8.2 GROSS AND MICROSCOPIC ANATOMY OF THE NOSE

Gross anatomy of the nose

The external nose consists of a fixed bony case consisting of paired nasal bones dorsally and incisive and maxillary bones laterally (Figs 8.1A, B), and a cartilaginous framework, which is movable by virtue of several skeletal muscles including the levator nasolabialis and levator labii superioris muscles[1,2]. The nasal cartilages include the unpaired septal cartilage with the ventrally located paired vomeronasal cartilages enclosing the vomeronasal organ, the paired dorsal lateral and ventral

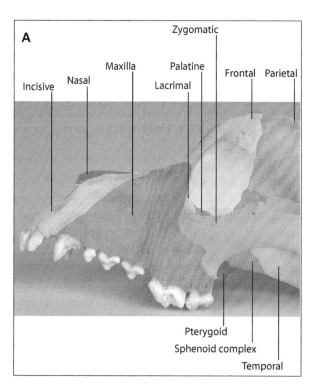

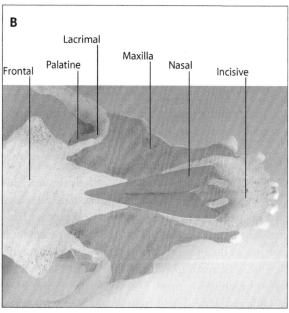

Figs 8.1A, B Bones of the skull that make up the nose: A: lateral aspect; B: dorsal aspect.

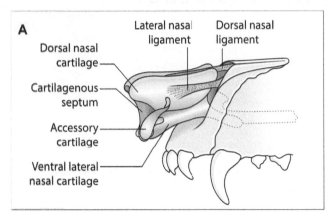

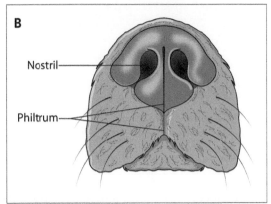

Figs 8.2A, B External nose and nasal cartilages: A: lateral view showing the nasal cartilages; B: rostral view demonstrating the nasal planum of a dog.

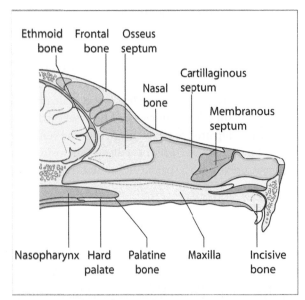

Fig. 8.3 Sagittal section of the nose showing the different components of the nasal septum.

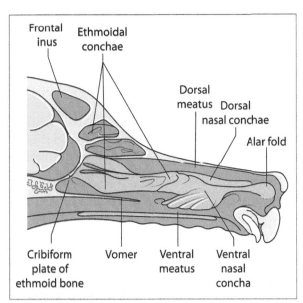

Fig. 8.4 Sagittal section of the nose showing the conchae (turbinates).

lateral nasal cartilages and the paired lateral accessory cartilages (Fig. 8.2A). The paired vomeronasal organs (Jacobson's organ) are tubular pockets of olfactory epithelium that open into an incisive duct that connects the nasal with the oral cavities[3]. The most apical portion of the nose, the nasal plane, is flattened and devoid of hair and glands (Fig. 8.2B). It includes the nares, which are separated from each other by a groove, or philtrum. Three ligaments, the unpaired dorsal nasal ligament and the paired lateral nasal ligaments, attach the mobile part of the nose to the dorsal portion of the osseous nose (Fig. 8.2A).

The nasal cavities extend from the nostrils with the nasal vestibule to the nasopharyngeal meatus and choanae, and are separated by the nasal septum, which consists of a bony, a membranous and a cartilaginous portion (Fig. 8.3). Each nasal cavity is divided into four principal air channels or nasal meatuses by three groups of cartilaginous or ossified nasal scrolls[1]. They include the dorsal, ventral and ethmoidal conchae, which were formerly called the *dorsal nasoturbinates*, the *maxilloturbinates* and the *ethmoturbinates*, respectively (Fig. 8.4). The latter are further subdivided into six small dorsally located ectoturbinates, some of which extend into the frontal sinuses,

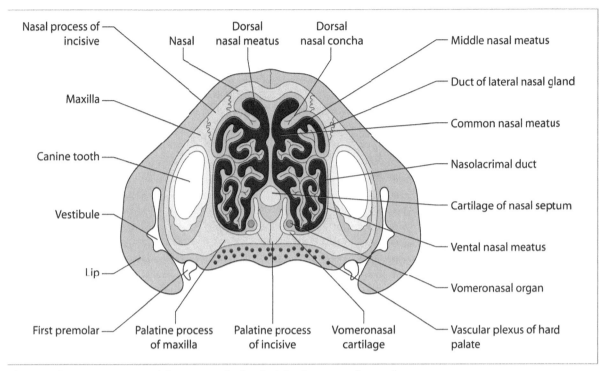

Fig. 8.5 Transverse section of the nose at the level of the first premolar teeth.

and four large endoturbinates that fill the caudal portion of the nasal cavity[1,4]. After the inhaled air leaves the nasal vestibule it traverses the longitudinal nasal meatuses to reach the nasopharynx. The dorsal nasal meatus is the passage between the dorsal nasal concha and the ventral surface of the nasal bone. The middle nasal meatus lies between the dorsal nasal concha dorsally and the dorsal part of the ventral nasal concha, ventrally (Fig. 8.5). At its rostral end it presents a dilatation, the atrium of the middle meatus, that connects the nasal vestibule with the middle nasal meatus and contains the orifice of the duct from the lateral nasal gland[1]. The ventral nasal meatus is located between the ventral nasal concha and the dorsal surface of the hard palate. The common nasal meatus is a longitudinal narrow space on either side of the nasal septum with which all other meatuses communicate. All four nasal meatuses of each nasal cavity converge caudally and continue as the nasopharyngeal meatus to the choanae where the nasopharynx begins[1]. This nasal portion of the pharynx extends to the intrapharyngeal ostium where the digestive and respiratory passageways cross. The rostral part of the nasopharynx is bounded by the hard palate ventrally, the vomer dorsally and the pala-

tine bones bilaterally. The middle and caudal portions of the nasopharynx are bounded dorsally by the base of the skull and the muscles that attach to it, and ventrally by the soft palate. On each lateral wall of the nasopharynx, dorsal to the middle of the soft palate, is an oblique slit-like opening, approximately 5 mm long, that is the pharyngeal opening of the auditory tube (Eustachian tube), leading to the middle ear cavity.

The paranasal sinuses in dogs, which are also connected with the respiratory passageways, include a maxillary recess, a frontal sinus and a sphenoidal sinus. The maxillary recess is not called a sinus because it is not enclosed in the maxilla[1]. The frontal sinus is divided into rostral, medial and lateral compartments. The sphenoidal sinus of the dog is only a potential cavity because it is filled by an endoturbinate scroll; it is however a proper sinus in cats[5,6].

The lateral nasal gland is a serous gland that is located in the mucosa of the maxillary recess near the opening of this recess into the nasal cavity[1]. It has a role in thermoregulation in dogs[7–9]. The lateral nasal gland is thickest at the level of the fourth superior premolar (Fig. 8.6), where its ducts unite and pass rostrally to

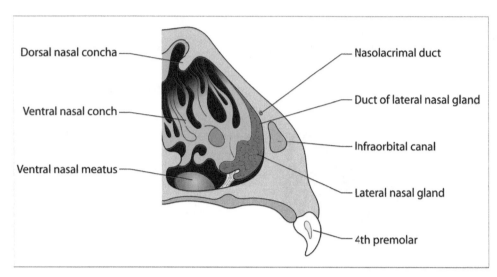

Dorsal nasal concha

Ventral nasal conch

Ventral nasal meatus

Nasolacrimal duct

Duct of lateral nasal gland

Infraorbital canal

Lateral nasal gland

4th premolar

Fig. 8.6 Transverse section of the left nasal cavity at the level of the fourth premolar tooth showing the position of the lateral nasal gland.

form one major duct that opens on the lateral wall of the vestibule, dorsal and caudal to that of the nasolacrimal duct, hidden from view by the alar fold. The nasolacrimal duct carries the serous secretion from the conjunctival sac to the nasal vestibule via an orifice located at the rostral end of the attached margin of the alar fold.

The arterial blood supply to the nose is based on the infraorbital artery that emerges from the infraorbital canal and terminates in lateral and rostral dorsal nasal arteries[1]. The lateral nasal artery is the larger of the two and runs rostrally into the muzzle with many large infraorbital nerve branches and anastomoses with branches of the superior labial and major palatine arteries. The vessel branches profusely and supplies the upper lip and snout. At the philtrum the vessel anastomoses with its fellow and sends a relatively large branch dorsally between the nostrils and another branch caudally in the mucosa of the nasal cartilage. The rostral dorsal nasal artery travels rostrodorsally across the lateral surface of the nose to its dorsal surface and supplies the structures of the dorsal surface of the rostral half of the muzzle. It anastomoses with its fellow of the opposite side, as well as with the nasal and septal branches of the sphenopalatine artery and the septal branches of the major palatine artery[1].

Venous drainage of the nose is via the facial vein, which begins on the dorsolateral surface of the muzzle, covered by the m. levator nasolabialis. It is formed by the confluence of the smaller dorsal nasal vein, which drains the dorsolateral surface of the nose, and the

larger angular vein of the eye. The lateral nasal vein is a satellite of the lateral nasal artery.

There are three lymph centres in the head that drain the nose, nasal cavity and nasopharynx. These are the parotid, mandibular and medial and lateral retropharyngeal lymph centres.

Microscopic anatomy of the nose

Each nasal cavity is divided into a cutaneous region, a respiratory region and an olfactory region[10–12]. Rostrally, the cutaneous region in the nasal vestibule is lined by a relatively thick keratinised stratified squamous epithelium (Fig. 8.7). At midvestibule, the epithelium is thinner and nonkeratinised. The caudal portion of the cutaneous region and the rostral third of the nasal cavity proper are a transitional zone lined by an epithelium that varies from stratified cuboidal to non-ciliated pseudostratified columnar, with microvilli on their free surface[10–12]. The propria–submucosa of the cutaneous region interdigitates via papillae with the epithelium. The papillae contain small vessels, nerves and numerous migratory cells, including mast cells, plasma cells, lymphocytes, macrophages and granulocytes[10–12]. Bundles of collagen fibres, larger blood vessels and nerves and serous glands are located deep in the propria–submucosa.

Epithelium lining the caudal two-thirds of the nasal cavity proper, with the exception of the olfactory region, is classified as respiratory epithelium. This ciliated pseudostratified columnar epithelium of the nasal cavity contains several cell types, including ciliated, secretory, brush and basal cells[10-12]. The morphological and histochemical appearance of these cells is both species and regionally variable[12,13]. Their description as mucous or serous is based on their glycoprotein content. Goblet cells of most species secrete primarily sulfated glycoprotein as a major component of mucus. Granules of serous epithelial cells contain neutral glycoproteins, and are smaller than those of mucous cells. Brush cells have long, thick microvilli and a cytoplasm containing mitochondria and many filaments[10-13]. These cells may be sensory receptors associated with endings of the trigeminal nerve. Basal cells are small polyhedral cells located along the basal lamina, characterised by anchoring attachments (desmosomes) to other cell types and to the basal lamina (hemidesmosomes)[10-13]. Basal cells appear to have some role in replacing other cell types.

The lamina propria–submucosa in the respiratory region is the most vascular of all regions and is called the cavernous stratum. The veins anastomose profusely and are referred to as capacitance vessels because they determine the degree of mucosal congestion and, inversely, nasal patency[10-13]. Constriction of nasal blood vessels is effected by α-adrenergic stimulation via the sympathetic nervous system. Serous or mixed glands are present between the numerous veins of this stratum. Acini of nasal glands also secrete immuno-globulin A, lysozyme and odourant-binding protein in this region[10-13].

The olfactory region comprises the dorsocaudal portion of the nasal cavity, including some of the surfaces of the ethmoid conchae, dorsal nasal meatus and nasal septum. Olfactory mucosa may be discerned from adjacent respiratory mucosa because it has a thicker epithelium, numerous tubular glands, and many bundles of non-myelinated nerve fibres in the lamina propria (Fig. 8.8). The olfactory mucosa is lined by a ciliated pseudostratified columnar epithelium, the olfactory epithelium, consisting of three primary cell types: neurosensory, sustentacular and basal[10-13]. Neurosensory olfactory cells are bipolar neurons with dendrites extending to the lumen, and axons reaching the olfactory bulb of the brain (Fig. 8.9)[14]. A club-shaped apex, the dendritic bulb or olfactory knob, protrudes from each dendrite into the lumen, from which 10–30 cilia emanate. Individual axons converge as they pass into the lamina propria, thereby forming bundles of non-myelinated nerve fibres[10-13]. Sustentacular cells are columnar cells with a narrow base and a wide apical portion covered by microvilli, often branched, at the luminal surface[10-13]. Juxtaluminal junctional complexes occur between sustentacular cells and the adjacent dendrites of neurosensory cells. Neurosensory and sustentacular cells are continuously replaced during life by cells derived from basal cells. Basal cells of the olfactory

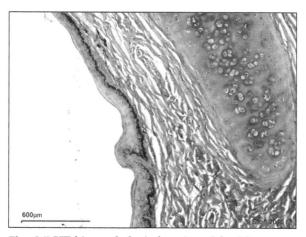

Fig. 8.7 HE histopathological section of the feline nose showing the epithelial lining of the cutaneous region. (Courtesy of Alexander Lewis Stoll, Royal Veterinary College, University of London, UK.)

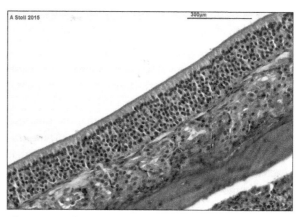

Fig. 8.8 HE histopathological section of the feline nose showing the epithelial lining of the olfactory region. (Courtesy of Alexander Lewis Stoll, Royal Veterinary College, University of London, UK.)

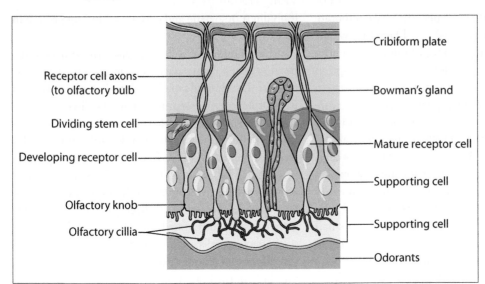

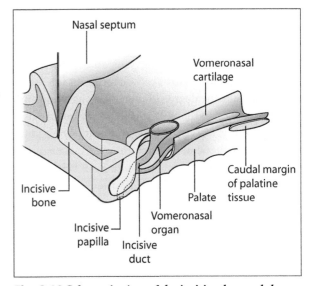

Fig. 8.9 Diagram of the olfactory epithelium, showing the major cell types and projection of the olfactory receptor neurons to the olfactory bulb.

mucosa are similar in structure to those of the nonolfactory epithelium. Olfactory glands, the cells of which contain pigment granules, are located in the propria–submucosa[10–13]. Squamous cells line the intraepithelial portion of their ducts. The glands secrete a watery product, which may serve to enhance the solubility of airborne odourants and cleanse the cilia, facilitating access for new odourants. The olfactory mucosa also has very high levels of cytochrome P-450 mono-oxygenase activity and is the primary site for chemically induced nasal tumours[10–13].

Nerves in the nasal mucosa include sensory fibres arising from the terminal, olfactory, vomeronasal and maxillary division of the trigeminal nerves and efferent fibres of the autonomic nervous system[10–13]. Nerves are distributed throughout all compartments of the nasal mucosa, including within the epithelium. Lymphatic nodules are commonly present in the caudal part of the nasal cavity, adjacent to the choana, the opening between the nasal cavity and nasopharynx. Metabolically active exogenous compounds (xenobiotics) that reach nasal tissues via air or blood pathways may remain firmly bound to tissue elements unless degraded[10–13]. Cytochrome P-450-dependent mono-oxygenase enzymes in surface epithelium and in acinar cells of the lateral nasal gland actively metabolise endogenous compounds (e.g. progesterone and testosterone) and exogenous compounds. These enzymes convert lipid-soluble exogenous compounds, some of which are

Fig. 8.10 Schematic view of the incisive duct and the vomeronasal organ.

highly toxic (e.g. formaldehyde and acetaldehyde), to water-soluble metabolites.

The tubular vomeronasal organ consists of an internal epithelial duct (vomeronasal duct), a middle propria–submucosa and an external cartilaginous support (Figs 8.10, 8.11A). The vomeronasal duct is crescent-shaped in transverse section with a lateral convex and a medial concave mucosa wall. The epithelium transitions from a stratified cuboidal lining ros-

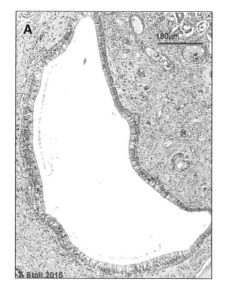

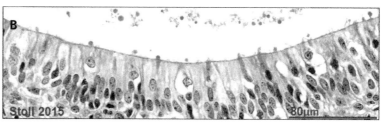

Figs 8.11A, B HE histopathological section of the feline nose showing the vomeronasal organ (A) and its epithelial lining (B). (Courtesy of Alexander Lewis Stoll, Royal Veterinary College, University of London, UK.)

trally near the incisive duct to a ciliated pseudostratified columnar epithelium over much of the caudal portion of the vomeronasal duct[10–13]. The medial pseudostratified columnar epithelium has neurosensory, sustentacular and basal cells (Fig. 8.11B). The dendritic portions of vomeronasal neurosensory cells lack dendritic bulbs. The lateral pseudostratified columnar epithelium has ciliated and non-ciliated columnar, goblet and basal cells[10–13]. Vomeronasal glands, located in the highly vascular propria–submucosa secrete into the vomeronasal duct most commonly through the commissures between lateral and medial mucosal walls. Secretory granules of the acinar cells contain neutral glycoproteins. The vomeronasal organ functions in the chemoreception of liquid-borne compounds of low volatility[10–13]. These substances, dissolved in fluid in the incisive duct, are sucked into the vomeronasal duct by constriction of blood vessels within the propria–submucosa of the vomeronasal organ[10–13]. Upon dilation of these vessels, the dissolved substances are expelled from the vomeronasal lumen.

The mucosae of the paranasal sinuses are thinner than those of the respiratory region of the nasal cavity with which they are continuous. Glands and blood vessels in the propria–submucosa are scant. The epithelium is ciliated pseudostratified columnar, containing a few goblet cells[10–13]. The ciliary beat carries mucus toward openings connecting the sinuses with the nasal cavity. The lateral nasal gland is a relatively large com-

pound gland that secretes neutral glycoproteins via a long duct into the nasal vestibule. The lateral nasal gland is present in the maxillary recess of carnivores, and separate maxillary recess glands can also be found.

The nasopharynx is the portion of the pharynx located dorsal to the soft palate, extending from the nasal cavity to the laryngopharynx. The lining of the nasopharynx consists mostly of respiratory epithelium, but stratified squamous epithelium over the caudodorsal portion of the soft palate makes contact either with the dorsal wall of the nasopharynx during deglutition or with the epiglottis[10–13]. The propria–submucosa is loose connective tissue containing mixed glands. Lymphatic nodules are prominent in the dorsal portion of the nasopharynx, where they aggregate as the pharyngeal tonsil[10–13].

8.3 REGULATION AND CONDITIONING OF THE AIRFLOW

Regulation of the airflow

Dogs and cats breathe through their noses at rest. Upon entering the nasal cavity, inspiratory airflow is well mixed within the nasal vestibule by turbulence, prior to splitting into olfactory and respiratory flow paths, thus ensuring delivery of a representative odour sample to the olfactory region (see Section 8.5). During normal respiration, the respiratory flow path is mainly laminar over the maxilloturbinate mucosa[2,15].

The nose is responsible for 40–70% of the total airway resistance in people and is assumed to be even greater in dogs, especially in brachycephalic breeds[16,17]. The nasal resistance has to be overcome by greater negative pressure in the thorax during inspiration, which leads to better expansion and filling of the alveoli by the inspired air and a greater venous blood flow in the lungs[14,18]. The size and congestion of the turbinate mucosa determine total nasal airway resistance. Several factors can affect this turbinate volume. Sympathetic nervous system stimulation causes vasoconstriction in the nasal mucosa that widens the nasal passageways and decreases respiratory resistance, whereas parasympathetic stimulation causes vasodilation and subsequently increases respiratory resistance[14].

In humans, exercise usually decreases nasal resistance, whereas dust, smoke and alcohol usually increase nasal resistance[16]. Also, pressure on one side of the body induces reflex nasal congestion on that side. In healthy adults, each side of the nose alternatively congests and decongests every 3–7 h, leading to a spontaneous, resistive cycle phenomenon called the nasal cycle[16]. Similar nasal cycles have been documented with the use of advanced imaging modalities in both dogs and cats[19–21]. The reason for the existence of this cycle remains unclear. A prolonged increase in nasal resistance can lead to cor pulmonale, cardiomegaly and pulmonary oedema in humans[18]. In humans and dogs the most common response to increased nasal resistance is, however, breathing through the mouth[14].

Whereas sympathetic and parasympathetic stimulation determines the degree of congestion of the turbinates, sensory receptors of airflow have not been yet identified in nasal mucosa. It is believed however that sensation of airflow could depend on thermal and/or sensitive stimulation of inhaled air[16]. Moreover, several aromatic substances, notably L-menthol, enhance the sensations of chill, airflow and nasal patency, without modification of objective airflow indices. Thus, sensory interpretations of nasal airflow are influenced not only by ambient and pathophysiological conditions, but also by psychological factors[16]. The airflow distribution pattern through the nasal cavities and its characteristics are crucial to effective air conditioning.

Conditioning of the airflow

The nose is responsible for conditioning of inhaled air via thermal and hygrometric regulation, maintaining a constant mean temperature and high relative humidity, regardless of the variation of ambient humidity and temperature. Dilatation of blood vessels that comprise the rich vascular supply to the nasal mucosa enhances

heat exchange. With intense activity or in high ambient temperatures, dogs and cats will start panting (mouth breathing) to increase evaporative cooling by using the larger mucosal surface of the oral cavity[22,23].

Inhaled air is supplemented with nitric oxide (NO) synthesised in the paranasal sinuses in people. NO-induced broncho- and vasodilatation contribute to improve oxygenation as well as blood perfusion in ventilated alveolar areas, resulting in better function in the entire respiratory system[16,18]. Nitric oxide may also have antimicrobial effects. The exact functions of the paranasal sinuses in dogs and cats are unknown.

Evaporation of the seromucous layer covering the mucosa, produced by submucosal glands and lateral nasal glands, humidifies the inhaled air. Both the nasal blood flow and the activity of the nasal glands are regulated by the autonomic nervous system via the vidian nerve[14]. Sympathetic nervous system stimulation decreases nasal secretion whereas parasympathetic system activation increases nasal secretion. Essential in assuring the delicate function of air conditioning are the efficacy of the vascular network in the lamina propria, the contribution of watery secretion, the quantity of seromucous glands, the surface contact between inspired air and mucosa and the beating quality of the cilia[16].

In addition to regulating the airflow, conditioning the air and physical removal of particles by mucociliary clearance (see Section 8.4), the nose actively participates in immunological defence of the airways. The entire respiratory tract is endowed with a range of different immune cell populations, but the highest concentration can be found in the nasal mucosa. Mast cells and lymphocytes T and B are found immediately beneath the epithelial lining[24]. These cells participate in antigenic-particle removal, immunological memory and release of preformed and granule-derived mediators of inflammation. CD1+ dendritic cells are located within the epithelium and are optimally located for capture of inhaled antigen and could act as antigen-presenting cells to infiltrating lymphocytes[24]. In addition, nasal NO and several nasal secretion constituents (peroxidases, interferon, lysozymes, lactoferrin, complement and immunoglobulins [A, G, M and E]) have immunological properties that may act non-specifically to maintain the sterility of the lower airways[16]. Recent studies suggest that qualitatively different immune responses are induced by different types of intranasal carcinomas and between dogs with aspergillosis and those with chronic rhinitis[25,26].

The precise stimulus for secretions of the paired lateral nasal glands is not known. The rate of secretion increases in warm climate conditions and in dogs tempted with food, suggesting physiological and psychological cues and a role in evaporative cooling[7,8,23,27]. The serous fluid also contains IgA, which suggests a defensive function. The lateral nasal glands of cats are smaller, and their secretions are less copious and more mucoid[23].

8.4 MUCOSAL CLEANING, SNEEZING AND REVERSE SNEEZING

Mechanical clearance of mucus (mucosal cleaning) is considered the primary innate airway defence mechanism in mammals[28]. The role of the epithelia lining airway surfaces is to provide the integrated activities required for mucus transport, including ciliary activity and regulation of the proper quantity of salt and water on airway surfaces via transepithelial ion transport[28]. Transport of mucus towards the oropharynx is aided by reverse sneezing. Sneezing is a protective reflex designed to remove both chemical and physical irritants rapidly from the nasal epithelial surface.

Mucosal cleaning

During normal breathing, nasal air filtration is facilitated by the disturbed airflow pattern over the maxilloturbinates (respiratory pathway), allowing particles (micro-organisms and noxious materials) to attach to the mucosa. The number of particles that will attach to the mucosal surface depends on several factors such as physical size, shape, density and hygroscopicity[16]. Mucus film covering nasal epithelium is responsible for trapping particulates greater than 10 μm, which will be transported to the oropharynx and then swallowed and destroyed by gastric enzymes[16,28]. The bilayered mucus blanket, produced mostly by the serous and goblet cells, is sticky, tenacious and adhesive. The outer layer is more viscid than the deeper, periciliary layer. This blanket not only traps insoluble particles and soluble gases, but also functions as a lubricant and protects against desiccation[14,18]. Transport of trapped particles depends on the mucociliary clearance, which is determined by the motion of the blanket of mucus from the front of the nose to the nasopharynx by the coordinated waves of cilia[16,29]. The cilia actively move the overlying blanket of mucus by a to-and-fro movement called the ciliary beat. There is a forceful forward movement and a less forceful recovery. In the recovery stroke the shaft of the cilium curls back on itself so that it does not reach the layer of mucus[14,18]. The energy for this work is derived from adenosine triphosphate found in the dynein arms of the axonemal tubules[18]. Factors affecting nasal mucociliary clearance are either primary abnormality of cilia (Kartagener's syndrome, primary ciliary dyskinesia) and mucus or secondary to viruses, bacteria, fungi, chronic nasal inflammation (allergic and chronic nonspecific rhinitis) and inhaled pollutants.

Sneezing

Sneezing, a sudden, violent, involuntary expulsion of air through the nose, is a radical reflex for clearing the nasal cavity[14]. It is commonly assumed that the sensory elements for this system are free nerve endings of the neurons of the trigeminal ganglion, but in addition an extensive population of trigeminal chemosensory cells exists within the nasal respiratory epithelium[30]. The best defined afferent pathway involves histamine-mediated depolarisation of H1 receptor-bearing type C trigeminal neurons[31]. Other stimuli include, for example, allergens, chemical irritants, sudden exposure to bright lights and cooling of the skin of various parts of the body[32].

These stimuli activate a stereotyped series of actions that are choreographed by activation of a complex array of central pathways and nuclei leading to systemic muscle coordination. Intercostal and accessory respiratory muscle contractions provide a rapid oral inspiration to hyperinflated volumes, followed by closure of the auditory tubes, eyes, glottic and nasopharyngeal structures when at the maximum lung volume[31,33]. Abdominal, neck and other muscles contract in a forceful Valsalva manoeuvre that compresses the thoracic air to pressures of greater than 100 mmHg. Sudden opening of the glottis and anterior flexion of the soft palate opens the nasopharyngeal space so that the pressurised air column can rush through the nose at speeds of over 100 mph (33 m/s)[31]. The shearing force removes mucus and any particulates or other irritants from the epithelial surfaces and blows them out of the nostrils. This process can be rapidly repeated in staccato fashion.

The sneeze reflex may be coordinated by a lateromedullary sneeze centre localised to near the spinal trigeminal tract and nucleus[33]. This centre appears to be bilateral and functionally independent on both sides, based on its unilateral loss in strokes affecting this region in people[33]. In the cat, the sneeze centre is thought to be a strategic continuous strip located in the ventromedial part of the spinal trigeminal nucleus, in close proximity to the bilateral pontino-medullar lateral reticular formation[34,35].

Reverse sneezing

Despite the fact that many nasal disorders in dogs, and to a lesser degree in cats, are accompanied by reverse sneezing, little scientific information is available in the veterinary literature with respect to its physiology, receptors and afferent and efferent pathways. Caudal nasal and nasopharyngeal mucosal irritation leads to 1–2 minute periods of severe inspiratory dyspnoea characterised by repeated short sniffing-snoring sounds, extension of the neck, bulging of the eyes and abduction of the elbows[14]. It is thought that this reflex spasm of the nasopharyngeal and soft palate musculature aids in transportation of mucus from the caudal nasal passages and nasopharynx towards the oropharynx, where it can be swallowed and removed from the airways. Activation of the dog's swallowing reflex leads to culmination of the reverse sneeze, giving further evidence of a trigeminal nerve-mediated mechanism.

Reverse sneezing can be an alarming symptom for owners who often misinterpret it as severe dyspnoea. It is very helpful if the person taking the history recognises the signs and can mimic the sound[14]. Though the clinician should aim to find the underlying cause of reverse sneezing, a symptomatic therapy consists of activating the dog's swallowing reflex. The owner should be given an explanation of the reflex and shown how to induce swallowing by massaging the dog's throat or briefly closing the nares[14].

8.5 OLFACTION AND VOMERONASAL ORGAN FUNCTION

Introduction

Three sensory systems are dedicated to the detection of chemicals in the environment:
- Olfaction.
- Taste.
- The trigeminal chemosensory system[29].

As three quarters of flavour is contributed to by olfaction[36], inappetence and loss of taste in dogs and cats could be related to olfactory loss, but discussion of taste and of the trigeminal chemosensory system is beyond the scope of this book.

Dogs have an extremely sensitive sense of smell, which despite their great hearing capacity, appears to be their primary sense. Dogs use their sense of smell over vision in both light and dark environments[37]. In most vertebrates, including humans, the number of receptor cells exceeds that of any other sensory system except vision. The anatomy of the dog reflects this olfactory acuity: the dog nose has hundreds of millions more olfactory cells lining the epithelium than the human nose. Collectively, the surface area of the cilia is quite large, being estimated as exceeding, for example, 22 cm^2 in the human and 7 m^2 in the German shepherd dog[38].

The canine olfactory bulb and olfactory cortex are highly developed compared to these regions in the human brain. These features lead to a sense of smell that is 10,000–100,000 times more sensitive than humans. The olfactory acuity of the dog can detect odourant concentration levels at 1–2 parts per trillion[39–41]. Moreover, the inhalation of odours in the dog is managed by an adaptive sniffing process. Respiratory and olfactory streams of inhaled air are separated into different flow paths within the nose[42](Fig. 8.12), and odour habituation is prevented through side-nostril exhalation[43]. Olfactory information can influence feeding behaviour, social interaction and reproduction. A dog's sense of smell, together with its personality and intelligence, enables it to function as a nose for humans in many circumstances, for instance with hunting, drug and explosive detection and in detection of cancer[14].

In addition to odourant receptors, a second class of chemosensory receptors was recently discovered in the mouse olfactory epithelium that is apparently associated with the detection of pheromones[44]. Traditionally, pheromone detection in the canine is attributed to receptors in the vomeronasal organ, located in the rostral–ventral part of the nasal cavity[45]. However, given this discovery, the dog may also possess pheromone receptors in the sensory epithelium lining the olfactory recess.

Olfaction

Though olfactory organ size[46], neuronal density[47] and the number of functional genes in the olfactory receptor gene family[48] certainly contribute to the phenomenal sense of smell in dogs, these factors nonetheless fail to take into consideration odourant transport from the external environment, by sniffing, to receptors on the cilia of the olfactory epithelium[42].

Generally, the olfactory mucosa of macrosmats (e.g. dog, cat, rabbit and rat) is relegated to an 'olfactory recess', located in the rear of the nasal cavity and excluded from the main respiratory airflow path by a bony plate, the lamina transversa[2,45]. How odourant molecules reach the olfactory part of the nasal cavity

during sniffing without being filtered by respiratory airways is not completely understood; however, airflow dynamics have been shown to be extremely important[2,15,42](see Section 8.4).

Dogs in particular can be observed to sniff rapidly rather than inhale normally when engaged in hunting or scenting. This staccato inflow of air at high speeds is thought to cause turbulence in the nasal cavity that directs air into the dorsal meatus[49]. The external aerodynamics of canine sniffing was shown to provide bilateral odour samples, which may be used for stereoscopic olfaction[2]. The internal aerodynamics reveal a novel flow pattern in the nasal cavity of the dog during a sniff, where odourant-laden inspired air bypasses the tortuous respiratory airways via the dorsal meatus, along which it is quickly transported to the olfactory recess (Fig. 8.12). Here, the airflow turns 180° and slowly filters through the olfactory airway labyrinth in a forward-lateral direction, permitting efficient odourant deposition. Finally, olfactory airflow either exits the nasal cavity via the nasopharynx or continues to flow forward into the dorsal-most ethmoturbinate extensions of the olfactory recess, where it remains at the conclusion of inspiration. Airflow bypasses the olfactory recess during expiration, which (a) forces unidirectional flow within olfactory airways and (b) prevents premature purging of odourant from the olfactory region[2,42].

In order to interact with the olfactory sensory neurons, hydrophilic odourants are dissolved in the olfactory mucus, and hydrophobic odourant molecules are bound and solubilised by odourant-binding proteins present in the nasal mucus. After interaction between the odourant molecules and receptor proteins, olfactory transduction (transformation of mechanical stimulation into electrical activity) probably involves an olfactory epithelium Golf protein-coupled cascade, with cyclic adenosine monophosphate and/or inositolphosphatidyl-3 as an intracellular second messenger, exciting an ion channel in the cilia, which depolarises the olfactory neuron[16,36,38]. Depolarisation follows the olfactory axons that synapse in the olfactory bulbs, and then projections go to the amygdala, the prepyriform cortex, the olfactory nucleus and entorhinal cortex, as well as the hippocampus, hypothalamus and thalamus.

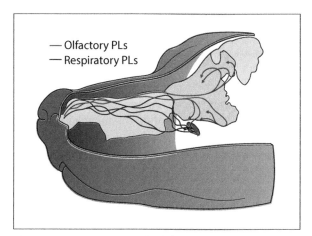

— Olfactory PLs
— Respiratory PLs

Fig. 8.12 The intranasal fluid dynamics of canine olfaction. Unsteady pathlines throughout inspiration reveal distinct respiratory and olfactory flow paths within the nasal cavity[42].

The precise mechanism by which the vast number of smells is recognised and discriminated is unknown, but possible theories include specific odourants exciting specific receptors; differing solubilities of odourants, allowing a temporospatial distribution of the odourant across the olfactory mucosa or a response to the molecules' vibration spectra[36,38]. With functional magnetic resonance imaging (MRI), it was demonstrated that during smelling the olfactory bulb and piriform lobes were commonly activated in both awake and anaesthetised dogs, while the frontal cortex was activated mainly in conscious dogs[50].

There are three major classifications of olfactory disorders:
- Transport (conductive).
- Sensory.
- Neural.

Transport disorders interfere with the access of an odourant molecule to the olfactory receptor, sensory losses result from damage to the olfactory receptor and neural losses result from interruptions in the peripheral or central nervous olfactory pathways[16,36,38]. Olfactory dysfunction can be further classified as follows:
- *Anosmia* is the inability to detect qualitative olfactory sensations (i.e. absence of smell function).
- *Partial anosmia*: ability to perceive some, but not all, odours.
- *Hyposmia* or microsmia: decreased sensitivity to odours.
- *Hyperosmia*: abnormally acute smell function.
- *Dysosmia* (sometimes termed cacosmia or parosmia): distorted or perverted smell perception to odour stimulation.
- *Phantosmia*: a dysosmic sensation perceived in the absence of an odour stimulus (i.e. an olfactory hallucination).
- *Olfactory agnosia*: inability to recognise an odour sensation, even though olfactory processing is essentially intact.

Olfactory dysfunction can be either bilateral (binasal) or, less commonly, unilateral (uninasal)[2,36,38]. There is no simple method to study olfaction in dogs and cats, but the available techniques will briefly be reviewed in Chapter 9 Nose Diagnostic Procedures, Section 9.2. There are only a few reports on normal and abnormal olfactory function in dogs[51–53]. In humans, factors known to alter olfactory function can be nervous, nutritional, local, viral or endocrine[36,38,54]. It is therefore likely that all commonly encountered nasal disorders in dogs and cats will lead to either partial anosmia or hyposmia, but the clinical signs of sneezing, reverse sneezing, nasal discharge and/or epistaxis will dominate the owner-perceived clinical signs. Systemic diseases, especially endocrine disease, can also affect the sense of smell. The olfactory system can be affected by both deficiency[55] and excess of steroids[54], especially glucocorticoids. Dogs that received dexamethasone (iatrogenic Cushing) exhibited a significant elevation in the olfactory detection threshold without any observable structural alteration of the olfactory tissue using light microscopy[54].

Vomeronasal organ

The paired vomeronasal organ (VNO) is located in the rostral base of the nasal septum as a tubular pocket of olfactory epithelium, partially enclosed by a scroll of cartilage[1,56]. It is well established that the mammalian VNO is involved in the control of sexual behaviour and in recognition of kin, via pheromones[14,57–59]; however, the precise functions of this organ remain poorly understood. Basically, the VNO pumps liquids and gases in and out of the vomeronasal duct[60,61], as has been demonstrated experimentally by Eccles[62], Meredith[63] and others[56]. The neural pathways transmitting stimuli from the vomeronasal mucosa to the brain are distinct from those from normal olfactory mucosa[1]. The most intensively studied mammalian olfactory system is that of the mouse, in which olfactory chemical cues of one kind or another are detected in four different nasal areas: the main olfactory epithelium (MOE), the septal organ (SO), Grüneberg's ganglion and the sensory epithelium of the VNO[56]. The extraordinarily sensitive olfactory system of the dog is an important model that is increasingly used, for example in genomic studies of species evolution. In a very recent report no structures equivalent to the Grüneberg ganglion and SO of the mouse were found in dogs[64]. The absence of a vomeronasal component based on VR2 receptors suggests that the VNO may be undergoing a similar involutionary process[64].

Applications

The performance of tracking dogs and drug-, and explosives-detection dogs is a well-known testament to trained dogs' olfactory acuity[37,65–67]. The first evidence of a dog's ability to detect disease came from the well-known report by Williams and Pembroke[68] in the *Lancet*, in which a woman sought medical attention after her dog persistently sniffed a mole on her leg. After clinical examination, the lesion proved to be a malignant melanoma.

Several similar cases have been reported since then. To date, studies have been published on the detection of bladder[69], lung[70,71], breast[71], prostate[72], ovarian[73], melanoma[68,74] and colorectal[75] cancers using a dog's sense of smell. Advances in technology suggest that diseases such as cancer likely have a 'signature scent', characterised by the volatile organic compounds (VOCs) being released that can be detected by dogs[76]. Based on the available literature, it would appear that testing the detection of cancer by dogs is more consistently successful with the use of breath samples[71]; detection of cancer in urine and faecal samples has yielded more conflicting results[69,72,75,77].

In addition to signalling cancers, evidence suggests that dogs (and cats) may be able to signal impending hypoglycaemic events in their owners[78–80]. Researchers hypothesise dogs may be using olfactory cues such as a change in the chemical composition of their owners' sweat (sweating is a common symptom of hypoglycaemia), or that dogs are acutely aware of the behavioural changes accompanying hypoglycaemia[79] or detect physiological stress markers (e.g. increase in adrenalin, cortisol levels, etc.) and not glycaemic VOCs *per se*.

Dogs can also predict and detect seizures and can be trained to do so reliably[81]. As with hypoglycaemia detection dogs, it is not known what signals from the owner alert the dog before a seizure. However, in this case, researchers appear confident that trained dogs are recognising and responding to minute changes in the behaviour of their owner[82], but detection of physiological changes, stress markers as mentioned above, cannot be ruled out without further investigation.

8.6 MICROBIOLOGY OF THE NORMAL NASAL CAVITIES OF DOGS AND CATS

There are only a few studies[83–85] looking at the microflora of the canine nose in general terms, as opposed, for example, to studies looking specifically at the carriage of *Staphylococcus pseudintermedius*, and methicillin-resistant *Staphylococcus aureus* (MRSA).

Clapper and Meade[83] and Bailie *et al*.[84] found broadly similar populations of coagulase-negative staphylococci, variable carriage rates of coagulase-positive staphylococci, *Corynebacterium* spp., *Neisseria* spp., *Pasteurella multocida* and *Bacillus* spp. Both authors sampled only the most rostral areas of the nares. One study looked at the carriage rate of *Staphyloccus intermedius* (now *S. pseudintermedius*) and found carriage rates on the caudal nares (collected by sheathed swabs) of 41%, compared to 34% recovery from the rostral nares[86]. Whether this pertains to other members of the nasal flora is not known.

A few studies have looked at long-term carriage rates of staphylococci in the dog's nose. For example, Cox *et al*.[87] found consistent carriage over 12 months, although they found one dog from which they were consistently unable to recover *S. intermedius*. Another study looked at both pet, owner and veterinarian carriage rates over 6 months and found firm evidence that the canine strain could be recovered from the nares of people in contact with dogs[88]. Recently, two reviews of methicillin-resistant *S. pseudintermedius* were published, highlighting the problems associated with recognition, treatment and zoonotic implications[89,90].

Recently, Hoffman *et al*. used polymerase chain reaction (PCR) technology to look for 16S mRNA of bacteria within the rostral nasal cavities[91]. The 16S mRNA is species specific and has the potential to find evidence of bacteria that cannot be cultured, or are very hard to culture, or are present in very low numbers. Samples from the normal dogs yielded between 25 and 41 species, per nostril sampled. This methodology has the potential to yield much important information, but it is currently at a very early stage.

The microbial population of the normal feline nose has been reported to contain *Streptococcus* spp., *Staphylococcus* spp., *Corynebacterium* spp., *Pasteurella multocida*, *Escherichia coli*, *Pseudomonas* spp., *Enterobacter* spp., *Bordatella bronchiseptica* and *Mycoplasma felis*[92,93]. Feline herpes virus, but not feline calicivirus[95], can also be in the normal feline nose[94]. This is almost identical to the spectrum of bacteria isolated from the upper respiratory tract of 460 samples from cats with upper respiratory infections[96]. Of interest, however, are the studies by Johnson and Kass[97,98], one of which[97] sampled the caudal nasal cavities of cats with upper respiratory disease, by flushing and immediately aspirating. In half the samples, only one species of bacteria was found and the most frequently isolated bacteria were staphylococci, which were recovered three times more frequently than *E. coli*. This might suggest that the enteric bacteria are more commonly carried in the rostral nose with staphylococci occupying niches in the caudal regions.

8.7 REFERENCES

1 Evans HE, De Lahunta A. *Miller's Anatomy of the Dog*, 4th edn. Elsevier, Philadelphia 2013.

2 Craven BA. A Fundamental Study of the Anatomy, Aerodynamics, and Transport Phenomena of Canine Olfaction. Pennsylvania State University dissertation, 2008.

3 Døving KB, Trotier D. Structure and function of the vomeronasal organ. *Journal of Experimental Biology* 1998;**201**:2913–25.

4 Schuenemann R, Oechtering GU. Inside the brachycephalic nose: intranasal mucosal contact points. *Journal of the American Animal Hospital Association* 2014;**50**:149–58.

5 Rycke LMD, Saunders JH, Gielen IM, Bree HJV, Simoens PJ. Magnetic resonance imaging, computed tomography, and cross-sectional views of the anatomy of normal nasal cavities and paranasal sinuses in mesaticephalic dogs. *American Journal of Veterinary Research* 2003;**64**:1093–8.

6 Losonsky JM, Abbott LC, Kuriashkin IV. Computed tomography of the normal feline nasal cavity and paranasal sinuses. *Veterinary Radiology & Ultrasound* 1997;**38**:251–8.

7 Schmidt-Nielsen K, Bretz WL, Taylor CR. Panting in dogs: unidirectional air flow over evaporative surfaces. *Science* 1970;**169**:1102–4.

8 Blatt CM, Taylor CR, Habal MB. Thermal panting in dogs: the lateral nasal gland, a source of water for evaporative cooling. *Science* 1972;**177**:804–5.

9 Habal MB, Blatt CM. The surgical approach to the lateral nasal gland. A new method for studying the canine thermoregulatory response. *Journal of Surgical Research* 1975;**19**:347–9.

10 Banks WK. *Applied Veterinary Histology*, 3rd edn. Williams & Wilkins, Baltimore 1993.

11 Young B, Heath JW. *Wheater's Functional Histology*. Churchill Livingstone, Philadelphia 2000.

12 Plopper CG, Adams DR. Respiratory system. In: *Dellmann's Textbook of Veterinary Histology*. Eurell JA, Frappier BL (eds). Blackwell, Ames, 2006, pp. 153–69.

13 Bacha WJ, Bacha LM. Respiratory system. In: *Color Atlas of Veterinary Histology*. Bacha WJ, Bacha LM (eds). Wiley-Blackwell, Ames, 2012, pp. 195–210.

14 Venker-van Haagen AJ. The nose and nasal sinuses. In: *Ear, Nose, Throat, and Tracheobronchial Diseases in Dogs and Cats*. Venker-van Haagen AJ (ed). Schlütersche Verlagsgesellschaft, Hannover 2005, pp. 51–81.

15 Craven BA, Neuberger T, Paterson EG. Reconstruction and morphometric analysis of the nasal airway of the dog (*Canis familiaris*) and implications regarding olfactory airflow. *Anatomical Record* 2007;**290**:1325–40.

16 Anniko M, Bernal-Sprekelsen M, Bonkowsky V, Bradley P, Iurato S. *European Manual of Medicine; Otorhinolaryngology, Head and Neck Surgery*. Springer, Dordrecht 2010.

17 Oechtering GU. Brachycephalic syndrome – new information on an old congenital disease. *Veterinary Focus* 2010;**20**:2–9.

18 Ballenger JJ. Clinical anatomy and physiology of the nose and paranasal sinuses. In: *Otorhinolaryngology: Head and Neck Surgery*, 15th edn. Snow JJB, Ballenger JJ (eds). BC Decker, Hamilton, Ontario 1996, pp. 3–18.

19 Friling L, Nyman HT, Johnson V. Asymmetric nasal mucosal thickening in healthy dogs consistent with the nasal cycle as demonstrated by MRI and CT. *Veterinary Radiology & Ultrasound* 2014;**55**:159–65.

20 Webber RL, Jeffcoat MK, Harman JT, Ruttimann UE. MR demonstration of the nasal cycle in the beagle dog. *Journal of Computed Assisted Tomography* 1987;**11**:869–71.

21 Eccles R, Lee RL. Nasal vasomotor oscillations in the cat associated with the respiratory rhythm. *Acta Oto-Laryngologica* 1981;**92**:357–61.

22 Goldberg MB, Langman VA, Taylor CR. Panting in dogs: paths of air flow in response to heat and exercise. *Respiratory Physiology* 1981;**43**:327–38.

23 Schmiedt CW, Creevy KE. Nasal planum, nasal cavity, and sinuses. In: *Veterinary Surgery Small Animal*. Tobias KM, Johnston SA (eds). Elsevier Saunders, St. Louis 2012, pp. 1691–706.

24 Peeters D, Day MJ, Farnir F, Moore P, Clercx C. Distribution of leucocyte subsets in the canine respiratory tract. *Journal of Comparative Pathology* 2005;**132**:261–72.

25 Sheahan D, Bell R, Mellanby RJ, *et al*. Acute phase protein concentrations in dogs with nasal disease. *Veterinary Record* 2010;**167**:895–9.

26 Vanherberghen M, Day MJ, Delvaux F, Gabriel A, Clercx C, Peeters D. An immunohistochemical study of the inflammatory infiltrate associated with nasal carcinoma in dogs and cats. *Journal of Comparative Pathology* 2009;**141**:17–26.

27 Adams DR, Deyoung DW, Griffith R. The lateral nasal gland of dog: its structure and secretory content. *Journal of Anatomy* 1981;**132**:29–37.

28 Knowles MR, Boucher RC. Mucus clearance as a primary innate defense mechanism for mammalian airways. *Journal of Clinical Investigation* 2002;**109**:571–7.

29 Purves D. The chemical senses. In: *Neuroscience*, 2nd edn. Purves D, Augustine GJ, Katz LC (eds). Sinaur Associates, Sunderland 2001, p. 317.

30 Finger TE, Bottger B, Hansen A, Anderson KT, Alimohammadi H, Silver WL. Solitary chemoreceptor cells in the nasal cavity serve as sentinels of respiration. *Proceedings of the National Academy of Sciences* 2011;**100**:8981–6.

31 Baraniuk JN, Kim D. Nasonasal reflexes, the nasal cycle, and sneeze. *Current Allergy and Asthma Reports* 2007;**7**:105–11.

32 Garcia-Moreno JM. Sneezing. *Revista de Neurologia* 2005;**41**:615–21.

33 Seijo-Martinez M, Varela-Freijanes A, Grandes J, Vázquez F. Sneeze related area in the medulla: localisation of the human sneezing centre? *Journal of Neurology Neurosurgery and Psychiatry* 2006;**77**:559–61.

34 Batsel HL, Lines AJ. Neural mechanisms of sneeze. *American Journal of Physiology* 1975;229:770–6.

35 Nonaka S, Unno T, Ohta Y, Mori S. Sneeze-evoking region within the brainstem. *Brain Research* 1990;**511**:265–70.

36 Jones N, Rog D. Olfaction: a review. *Journal of Laryngology and Otology* 2007;**29**:112.

37 Gazit I, Terkel J. Domination of olfaction over vision in explosives detection by dogs. *Applied Animal Behaviour Science* 2003;**82**:65–73.

38 Doty RL. Olfaction. *Annual Reviews in Psychology* 2001;**52**:423–52.

39 Walker JC. Human odor detectability: new methodology used to determine threshold and variation. *Chemical Senses* 2003;**28**:817–26.

40 Walker DB, Walker JC, Cavnar PJ, Taylor JL. Naturalistic quantification of canine olfactory sensitivity. *Applied Animal Behaviour Science* 2006;**97**:241–54.

41 Waggoner LP, Johnston JM. Canine olfactory sensitivity to cocaine hydrochloride and methyl benzoate. *Proceedings of SPIE* 1997;**2937**:1–11.

42 Craven BA, Paterson EG, Settles GS. The fluid dynamics of canine olfaction: unique nasal airflow patterns as an explanation of macrosmia. *Journal of The Royal Society Interface* 2010;**7**:933–43.

43 Settles GS. Sniffers: fluid-dynamic sampling for olfactory trace detection in nature and homeland security – the 2004 Freeman scholar lecture. *Journal of Fluids Engineering* 2005;**127**:189–218.

44 Liberles SD, Buck LB. A second class of chemosensory receptors in the olfactory epithelium. *Nature* 2006;**442**:645–50.

45 Evans HE. The respiratory apparatus. In: *Miller's Anatomy of the Dog*, 3rd edn. Evans HE, Christensen GC (eds). WB Saunders, Philadelphia 1993, pp. 507–43.

46 Smith TD, Bhatnagar KP, Tuladhar P, Burrows AM. Distribution of olfactory epithelium in the primate nasal cavity: are microsmia and macrosmia valid morphological concepts? *Anatomical Record A Discoveries in Molecular Cellular and Evolutionary Biology* 2004;**281**:1173–81.

47 Quignon P, Kirkness E, Cadieu E, *et al*. Comparison of the canine and human olfactory receptor gene repertoires. *Genome Biology* 2003;**4**:R80.

48 Rouquier S, Giorgi D. Olfactory receptor gene repertoires in mammals. *Mutation Research* 2007;**616**:95–102.

49 Steen JB, Mohus I, Kvesetberg T, Walløe L. Olfaction in bird dogs during hunting. *Acta Physiologica Scandinavica* 1996;**157**:115–9.

50 Jia H, Pustovyy OM, Waggoner P, *et al*. Functional MRI of the olfactory system in conscious dogs. *PLoS ONE* 2014;**9**:e86362.

51 Hirano Y, Oosawa T, Tonosaki K. Electroencephalographic olfactometry (EEGO) analysis of odour responses in dogs. *Research in Veterinary Science* 2000;**69**:263–5.

52 Myers LJ, Hanrahan LA, Swango LJ. Anosmia associated with canine distemper. *American Journal of Veterinary Research* 1988;**49**:1295.

53 Myers LJ, Nusbaum KE, Swango LJ. Dysfunction of sense of smell caused by canine parainfluenza virus infection in dogs. *American Journal of Veterinary Research* 1988;**49**:188.

54 Ezeh PI, Myers LJ, Hanrahan LA, Kemppainen RJ, Cummins KA. Effects of steroids on the olfactory function of the dog. *Physiology and Behavior* 1992;**51**(6):1183–7.

55 Doty RL, Risser JM, Brosvic GM. Influence of adrenalectomy on the odor detection performance of rats. *Physiology and Behavior* 1991;**49**:1273–7.

56 Salazar I, Quinteiro PS, Cifuentes JM. Comparative anatomy of the vomeronasal cartilage in mammals: mink, cat, dog, pig, cow and horse. *Annals of Anatomy* 1995;**177**:475–81.

57 Estes RD. The role of the vomeronasal organ in mammalian reproduction. *Mammalia* 1972;**36**:315–41.

58 Wysocki CJ. Neurobehavioral evidence for the involvement of the vomeronasal system in mammalian reproduction. *Neuroscience & Biobehavioral Reviews* 1979;**3**:301–41.

59 Meredith M, Fernandez-Fewell G. Vomeronasal system, LHRH, and sex behaviour. *Psychoneuroendocrinology* 1994;**19**:657–72.

60 Hamlin HE. Working mechanisms for the liquid and gaseous intake and output of the Jacobson's organ. *American Journal of Physiology* 1929;**91**:201–5.

61 Meredith M, O'Connell RJ. Efferent control of stimulus access to the hamster vomeronasal organ. *Journal of Physiology* 1979;**286**:301–16.

62 Eccles R. Autonomic innervation of the vomeronasal organ of the cat. *Physiology and Behavior* 1982;**28**:1011–5.

63 Meredith M. Chronic recording of vomeronasal pump activation in awake behaving hamsters. *Physiology and Behavior* 1994;**56**:345–54.

64 Barrios AW, Sánchez-Quinteiro P, Salazar I. Dog and mouse: toward a balanced view of the mammalian olfactory system. *Frontiers in Neuroanatomy* 2014;**8**:1–7.

65 Horowitz A, Hecht J, Dedrick A. Smelling more or less: investigating the olfactory experience of the domestic dog. *Learning and Motivation* 2013;**44**:207–17.

66 Williams M, Waggoner LP. Canine detection odor signatures for mine-related explosives. *Proceedings of SPIE* 1998;**3575**:291–301.

67 Moser E, McCulloch M. Canine scent detection of human cancers: a review of methods and accuracy. *Journal of Veterinary Behavior Clinical Applications and Research* 2010;**5**:145–52.

68 Williams H, Pembroke A. Sniffer dogs in the melanoma clinic? *Lancet* 1989;**1**:734.

69 Willis CM. Olfactory detection of human bladder cancer by dogs: proof of principle study. *British Medical Journal* 2004;**329**:712.

70 Ehmann R, Boedeker E, Friedrich U. Canine scent detection in the diagnosis of lung cancer: revisiting a puzzling phenomenon. *European Respiratory Journal* 2012;**39**:669–76.

71 McCulloch M. Diagnostic Accuracy of canine scent detection in early- and late-stage lung and breast cancers. *Integrative Cancer Therapies* 2006;**5**:30–9.

72 Gordon RT, Schatz CB, Myers LJ. The use of canines in the detection of human cancers. *Journal of Alternative and Complementary Medicine* 2008;**14**:61–7.

73 Horvath G, Jarverud GAK, Jarverud S, Horvath I. Human ovarian carcinomas detected by specific odor. *Integrative Cancer Therapies* 2008;**7**:76–80.

74 Pickel D, Manucy GP, Walker DB, Hall SB. Evidence for canine olfactory detection of melanoma. *Applied Animal Behaviour Science* 2004;**89**:107–16.

75 Sonoda H, Kohnoe S, Yamazato T, Satoh Y. Colorectal cancer screening with odour material by canine scent detection. *Gut* 2011;**60**:814–19.

76 Szulejko JE, McCulloch M, Jackson J, McKee DL, Walker JC, Solouki T. Evidence for cancer biomarkers in exhaled breath. *IEEE Sensors Journal* 2010;**10**:185–210.

77 Cornu J-N, Cancel-Tassin G, Ondet V, Girardet C, Cussenot O. Olfactory detection of prostate cancer by dogs sniffing urine: a step forward in early diagnosis. *European Urology* 2011;**59**:197–201.

78 Chen M, Daly M, Williams N, Williams S, Williams C. Non-invasive detection of hypoglycaemia using a novel, fully biocompatible and patient friendly alarm system. *British Medical Journal* 2000;**321**:1565–6.

79 Wells DL, Lawson SW, Siriwardena AN. Canine responses to hypoglycemia in patients with type 1 diabetes. *Journal of Alternative and Complementary Medicine* 2008;**14**:1235–41.

80 Wells DL, Lawson SW, Siriwardena AN. Feline responses to hypoglycemia in people with type 1 diabetes. *Journal of Alternative and Complementary Medicine* 2011;**17**:99–100.

81 Kirton A, Winter A, Wirrell E, Snead OC. Seizure response dogs: evaluation of a formal training program. *Epilepsy & Behavior* 2008;**13**:499–504.

82 Brown SW, Strong V. The use of seizure-alert dogs. *Seizure* 2001;**10**:39–41.

83 Clapper WE, Meade GH. Normal flora of the nose, throat, and lower intestine of dogs. *Journal of Bacteriology* 1963;**85**:643–8.

84 Bailie WE, Stowe EC, Schmitt AM. Aerobic bacterial flora of oral and nasal fluids of canines with reference to bacteria associated with bites. *Journal of Clinical Microbiology* 1978;**7**:223–31.

85 Abramson AL, Isenberg HD, McDermott LM. Microbiology of the canine nasal cavities. *Rhinology* 1980;**18**:143–50.

86 Harvey RG, Noble WC. Aspects of nasal, oropharyngeal and anal carriage of *Staphylococcus intermedius* in normal dogs and dogs with pyoderma. *Veterinary Dermatology* 1998;**2**:99–104.

87 Cox HU, Hoskins JD, Newman SS, Foil CS, Turnwald GH, Roy AF. Temporal study of staphylococcal species on healthy dogs. *American Journal of Veterinary Research* 1988;**49**:747–51.

88 Goodacre R, Harvey R, Howell SA, Greenham LW, Noble WC. An epidemiological study of *Staphylococcus intermedius* strains isolated from dogs, their owners and veterinary surgeons. *Journal of Analytical and Applied Pyrolysis* 1997;**44**:49–64.

89 Frank LA, Loeffler A. Meticillin-resistant *Staphylococcus pseudintermedius*: clinical challenge and treatment options. *Veterinary Dermatology* 2012;**23**:283–91.

90 Rich M. Staphylococci in animals: prevalence, identification and antimicrobial susceptibility, with an emphasis on methicillin-resistant *Staphyloccus aureus*. *British Journal of Biomedical Science* 2012;**62**:98–105.

91 Hoffmann AR, Patterson AP, Diesel A, *et al*. The skin microbiome in healthy and allergic dogs. *PLoS ONE* 2014;**9**:12.

92 Stein JE. Bacterial diseases. In: *Feline Internal Medicine Questions*. Lappin M (ed). Elsevier Health Sciences, St. Louis 2001, pp. 5–7.

93 Randolph JF, Moise NS, Scarlett JM, Shin SJ, Blue JT, Corbett JR. Prevalence of mycoplasmal and ureaplasmal recovery from tracheobronchial lavages and of mycoplasmal recovery from pharyngeal swab specimens in cats with or without pulmonary disease. *American Journal of Veterinary Research* 1993;**54**:897–900.

94 Lappin M. Viral diseases. In: *Feline Internal Medicine Questions*. Lappin M (ed). Elsevier Health Sciences, St. Louis 2001, pp. 8–11.

95 Adler K, Radeloff I, Stephan B, Greife H, Hellmann K. Bacteriological and virological status in upper respiratory tract infections of cats (cat common cold complex). *Berliner und Munchener Tierarztliche Wochenschrif* 2007;**120**:120–5.

96 Radford AD, Coyne KP, Dawson S, Porter CJ, Gaskell RM. Feline calicivirus. *Veterinary Research* 2007;**38**:319–35.

97 Johnson LR, Kass PH. Effect of sample collection methodology on nasal culture results in cats. *Journal of Feline Medicine and Surgery* 2009;**11**:645–9.

98 Johnson LR, Foley JE, De Cock HEV, Clarke HE, Maggs DJ. Assessment of infectious organisms associated with chronic rhinosinusitis in cats. *Journal of the American Veterinary Medical Association* 2005;**227**:579–85.

9.1 INTRODUCTION AND APPROACH TO THE DIAGNOSIS OF NASAL DISEASE

As is the case with any disease, the history may provide the most important information. While abnormalities of the nasal planum, such as depigmentation, inflammation, crust formation and hyperkeratosis are usually noted early in the course of the disease by the owner, signs of deeper nasal disease may be more insidious.

In dogs, the symptoms most clearly indicating problems in the nose and nasopharynx are unilateral or bilateral nasal discharge, sneezing and reverse sneezing (see Chapter 8 Nose Anatomy and Physiology, Section 8.4). The nature and type of the discharge (serous, mucopurulent, blood = epistaxis) are important in determining the diagnostic approach.

- Unilateral mucopurulent discharge is the clearest indicator of intranasal disease.
- Bilateral nasal discharge in conjunction with systemic illness is usually indicative of lower airway disease.
- Patients presenting with epistaxis without concurrent sneezing or a mucopurulent discharge and in which there is no history of recent trauma, most commonly are suffering from a form of coagulopathy. Clotting profiles need to be examined before imaging and endoscopy of the nose is attempted.

After the history has been taken, a thorough clinical examination of the respiratory tract has to be performed and diagnostic tests may be employed in an attempt to define a diagnosis:
- Examine the nasal planum. See Chapter 10 Diseases of the Nasal Planum for a detailed discussion.
- Assess air passage. In patients with fungal disease air passage is usually normal because of the destruction of conchae. Air passage is usually abnormal or diminished, in patients with tumours or large foreign bodies.
- Assess if there is pain on palpation of the nose. The presence of pain is highly indicative of fungal disease and is almost never encountered in patients with nasal tumours, despite associated bone destruction.
- Check the oral mucosa and teeth by simply lifting the upper lip, and looking for dental disease, which might underlie a nasal discharge. Note that a 'normal' appearance of the teeth does not rule out oronasal fistulae associated with dental disease, dental probing is required for this.

Note: the presence of a bilateral nasal discharge should prompt a check for systemic abnormalities.

Additional examination of the nose and nasal cavities primarily consists of diagnostic imaging (radiography, computed tomography [CT] scan, magnetic resonance imaging [MRI]), endoscopy (rhinoscopy, nasopharyngoscopy) and inspection of the oral cavity with dental probing.

Haematology is invaluable for determining the general condition of an animal and for diagnosing non-respiratory abnormalities influencing the respiratory system such as anaemia and endocrinopathy. Polycythaemia can be seen as a response to chronic hypoxia. Leucocytosis with neutrophilia and a left shift can be seen with acute bacterial bronchopneumonia. Eosinophilia can be associated with pulmonary infiltrates with eosinophils or feline asthma. In case of epistaxis, clotting profiles can rule out or diagnose clotting disorders.

Serology can be used for diagnosing specific viral infections such as feline leukaemia virus (FeLV) and feline immunodeficiency virus (FIV) and for diagnosing fungal infections. Culture, cytology and histology are ideally used in conjunction with nasal endoscopy.

When a diagnosis cannot be made with the diagnostic procedures discussed above, or in case of foreign bodies that cannot be removed via endoscopy, a surgical exploration of the nasal sinuses or nasopharynx might be necessary.

> Note: with the exception of an acute nasal foreign body, all patients with nasal and/or nasopharyngeal disease should be subject to diagnostic imaging under general anaesthesia as a first step. This is because iatrogenic endoscopy-induced bleeding may make it impossible to interpret subsequent imaging findings if performed prior to imaging.

Despite the availability of modern high-quality endoscopic equipment, it is impossible to completely evaluate the entire nasal cavity and all nasal sinuses in dogs and cats. The decision on which endoscopic modality to use, be it rigid or flexible, will depend on the clinical and radiographic findings. More detailed information can be obtained from advanced imaging with CT and MRI.

> In the dog, CT and MRI may be necessary for treatment *planning*, but radiographs are sufficient for *diagnosing* nasal disease. This is not the case in the cat.

Functional tests of the nose to assess nasal patency and, more importantly, tests to examine the sense of smell, are not commonly performed in veterinary practice, but have been limited to referral and research institutions.

The current techniques available with respect to airflow measurements and olfactometry are discussed in the next section. Subsequently, histopathological evaluation of the nasal planum, the radiographic features of the normal and abnormal nose, and advanced imaging of the nasal cavity and nasal sinuses, will be discussed. Finally, the technique of rhinoscopy and nasopharyngoscopy will be reviewed in the last section.

9.2 FUNCTIONAL AND OLFACTORY TESTING

Assessing nasal patency in dogs and cats is relatively simple and straightforward and usually done by holding a piece of cotton wool, a feather or a glass slide in front of the nostrils. Only marked differences in airflow can be detected this way, however. For objective assessment of nasal patency, acoustic rhinometry and rhinomanometry can be used, although except for pharmacological studies, these are not commonly used in veterinary practice. There is also no simple method to study olfaction in dogs and cats. Traditionally, primary studies were based on behavioural observations, later electrophysiological methods were employed. These latter techniques test smell by activating the olfactory receptor neurons and measuring the brain response by electroencephalographic olfactometry analysis[1].

Assessment of nasal patency
Acoustic rhinometry
Acoustic rhinometry (AR) is a measuring technique that is based on the principle of acoustic reflection and can be used to determine the cross-sectional area of the nose on each side, from the nasal vestibule to the nasopharynx[2]. Unlike rhinomanometry, it does not measure dynamic respiratory function but the cross sections of the nasal cavity at various sites, which are averaged together. An increasing number of studies have used AR for study of pharmacological interventions on nasal cavity dimensions in dogs and cats[3–8]. Acoustic rhinometry has been validated and compared with magnetic resonance (MR) imaging and a fluid-displacement method (FDM) using perfluorocarbon[9]. It was found that AR markedly underestimated nasal cavity dimensions in the narrowest parts of the nose, but agreed well with MR, especially in the deeper part of the nasal cavity.

The main advantages of AR over rhinomanometry are that it is faster and easier to perform and does not depend on patient cooperation[2]. While these features are desirable in the examination of veterinary patients, it should always be considered that AR measures static parameters and, unlike rhinomanometry, does not assess the patency of nasal airflow.

Rhinomanometry

Rhinomanometry is the method commonly used to determine nasal resistance (and patency) calculated from simultaneous measurement of nasal airflow and transnasal pressure difference[2,10]. The pressure immediately in front of the nostrils and the pressure in the nasopharynx are measured to determine this transnasal pressure. Airflow is measured by use of a flow meter attached to a breathing mask that has been placed tightly over the nose. The volume of air passing through the nose during active nasal respiration is recorded at the same time as the pressure differential across the nose. Further information can be obtained by performing rhinomanometry before and after decongestion with nose drops (temporary or permanent nasal resistance).

Two methods can be used (posterior and anterior rhinomanometry).

- For posterior rhinomanometry, nasopharyngeal pressure is measured by a pressure-sensing tube placed into the nasopharynx (transorally or through one of the nasal airways). With posterior rhinomanometry, both airways are investigated simultaneously and combined nasal resistance assessed directly[10].

- In anterior rhinomanometry, the nasal passages are investigated unilaterally. Air is fed into one nostril while a pressure probe placed in the contralateral nostril tightly closes that nasal passage. In this manner, pressure measured at the seal of the closed passage equals the pressure at the unification of the two nasal passages in the nasopharynx. Thus, the pressure difference measured between the entrance of the active passage and the closed nostril is the decisive pressure difference of the passage being investigated.

- Active anterior rhinomanometry is widely used in humans and relies on cooperation of the patient[10].

- Passive posterior rhinomanometry performed in anaesthetised animals appears to be the best method for nasal investigations in dogs, even for dogs with brachycephalic syndrome[10].

- Rhinomanometry has often been used in dogs in research and for possible applications of drugs in humans[5,11–13].

Olfactometry

Olfactory detectabilities have been determined in the dog by operant conditioning techniques that are lengthy and impractical in experimental settings[14–20]. Diagnosis of the olfactory function of the dog is extremely important though in field-trial dogs, hunting dogs, and dogs used in the detection of explosives and drugs. Objective assessment of olfactory disorders is only possible with evoked response olfactometry (ERO) where potentials evoked by olfactory stimuli are analysed[21,22]. These electrophysiological tests are not readily obtainable and technically challenging and are therefore primarily used in research. Electro-olfactometry involves application of chemosensory stimuli via a tube inserted into the nasal meatus. Eugenol and benzaldehyde have been most commonly used as scents[16,21]. Electrodes are either placed on the olfactory area directly or on the skull. Good stimulation techniques are mandatory in order to avoid mechanical and thermal artefacts[23]. Though this method is performed under sedation and therefore does not take the whole sniffing process and airway dynamics into consideration, it has been successful in demonstrating disorders of smell in dogs. Both canine distemper and canine influenza virus infections cause dysfunction of smell, though temporarily[24,25]. Also, corticosteroids have been shown to decrease olfactory capability in dogs[26].

9.3 THE HISTOPATHOLOGY OF THE NASAL PLANUM

The nasal planum is heavily pigmented and comprises a very thick epidermis, typically around 1.5 mm thick (Fig. 9.1). This compares strikingly with the thickness of normal haired skin on the immediately adjacent skin of the nose and face, head, trunk and limbs, which averages some 0.1–0.5 mm[27]. The thick epidermis rete ridges project deeply into the underlying dermis, and are almost pseudocarcinomatous in their appearance[28]. These are not found on normal haired skin. The thickness of the epithelium in the nasal planum reflects a very increased spinous layer, comprising the progeny cells of dividing basal epithelial cells, rather than an overall increase in the thickness of all the layers of the epithelium. The overlying stratum corneum is also

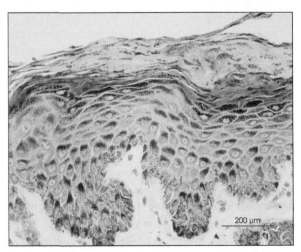

Fig. 9.1 High-power photomicrograph of normal canine nasal planum. Note the thick epidermis, the deep rete pegs extending into the dermis and the pigment content. (Courtesy of Trevor Whitbread, Abbey Vet Services.)

very thick and dense, when compared to normal hairy skin. Pigment-producing melanocytes are found interspersed among the basal cells and account for the dense pigmentation on the nasal planum. The nasal planum also lacks both hair follicles and associated sebaceous and ceruminous glands

Reactions to insult

The thickness of the epidermis and the lack of hair follicles limit the range of response to insult, making the differentiation of the various diseases difficult. Clinically, changes in pigment, hyperkeratosis, erosion, ulceration and crust formation, and fissure can occur. One of the most critical observations is whether the lesions are confined to the nasal planum (i.e. are not affecting the adjacent, haired, skin of the nose), or are part of a more generalised dermatitis – or indeed part of a systemic disease. Specific diseases affecting the nasal planum are discussed in the Chapter 10.

9.4 RADIOGRAPHIC FEATURES OF THE NORMAL AND ABNORMAL NOSE

Radiography is commonly used to evaluate nasal disease in dogs and cats[29,30]. Though the use of radiography permits the diagnosis of all common diseases of the canine nasal cavity, for treatment planning CT

and MRI are far superior techniques for precise localisation and definition of bone and soft tissue pathology, respectively.

Radiography is of less value in the diagnosis of nasal disease in cats, as radiographic abnormalities are not specific for either polyp, tumour or chronic inflammation in many cases, and because of the superiority of CT scan imaging in demonstrating concurrent middle ear abnormalities. For the same reason and because of more superimposition of bony structures, CT is also preferable in brachycephalic dog breeds. In mesaticephalic and dolichocephalic breeds, radiography is more helpful.

When performed, radiographic imaging should always precede endoscopic evaluation, unless the history is indicative of an acute intranasal plant foreign body. Minimal radiographic abnormalities can be expected in this situation. Endoscopy-induced bleeding can render subsequent radiographic imaging useless because it creates an overall increased soft tissue density in the entire nasal cavity. Radiography should always be performed under general anaesthesia to ensure accurate positioning. In addition, during the same anaesthetic procedure, time should be allocated to allow for an endoscopic procedure (with biopsy and culture) and/or oral examination with dental probing to be performed.

In most dogs with nasal disease, the history and clinical examination are so suggestive of a specific abnormality, like foreign body, oronasal fistula or fungal rhinitis, that treatment can also be scheduled during the same session.

Radiographic examination

Standard radiographic examination of the nose and sinuses consisting of a lateral and a dorsoventral, or ventrodorsal, view is usually of limited value because of superimposition of the mandible. To avoid this superimposition, intraoral dorsoventral and open-mouth ventrodorsal projections are commonly used[29,30]. The frontal sinuses can be individually evaluated in the rostrocaudal frontal sinus projection[29,31]. Oblique views enable the maxillary and mandibular dental arcades to be profiled without superimposition of the contralateral side[30].

Nasal diseases have traditionally radiographically be divided in three basic categories:
- Non-destructive nasal disease (rhinitis).
- Destructive rhinitis.
- Neoplasia.

Destruction of surrounding bones (lacrimal, palatine and maxillary) has the greatest predictive value of nasal neoplasia[29-32]. An increasing severity of bone lysis is also associated with a greater likelihood of neoplastic disease. Radiographic signs consistent with a diagnosis of non-destructive rhinitis are a lack of frontal sinus lesions and a lack of lucent foci in the nasal cavity[32]. The location of lesions within the nose, unilateral or bilateral distribution of lesions and involvement of teeth are weak predictors of diagnosis. Radiographic signs in cats are in general less specific and septal deviation and lysis are commonly seen with rhinitis[33,34].

Radiographic views

Many different radiographic views are used to demonstrate individual structures or regions within the skull. The basic views are the lateral, dorsoventral or ventrodorsal, left and right lateral oblique, rostrocaudal, and occlusal or intraoral view (Figs 9.2–9.8).

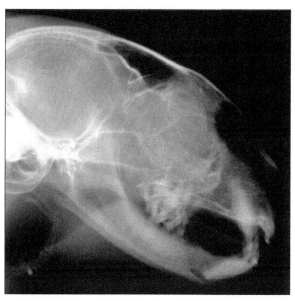

Fig. 9.2 Lateral radiographic view of the skull and nasal cavity, nasopharynx and nasal sinuses in an adult feline.

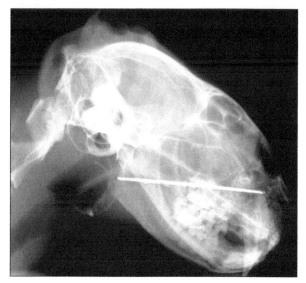

Fig. 9.3 Lateral radiographic view of the skull and nasal cavity, nasopharynx and nasal sinuses in a cat with a metal foreign body (needle).

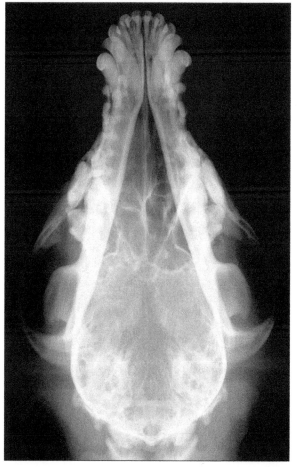

Fig. 9.4 Dorsoventral radiographic view of the skull and nasal cavity of a dog. Overprojection of the mandible and maxilla obscures detailed evaluation of the nasal cavity, but increased soft tissue density within the left nasal cavity is visible.

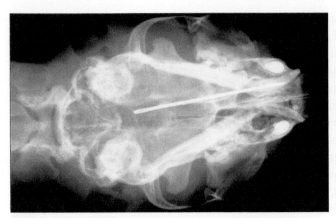

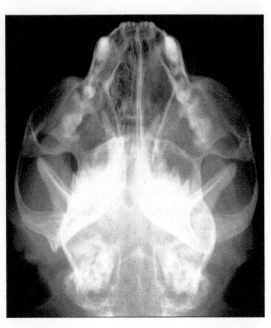

Fig. 9.5 Dorsoventral radiographic view of the skull and nasal cavity, of the same cat as in 9.3 with a metal foreign body (needle).

Fig. 9.6 Intraoral radiographic view of the nasal cavity of a cat with right-sided loss of turbinate detail and increased soft tissue density and mild deviation of the nasal septum.

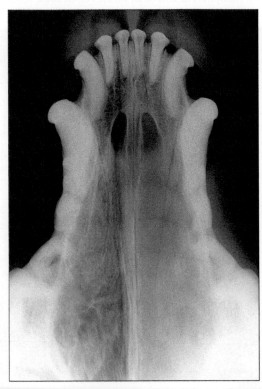

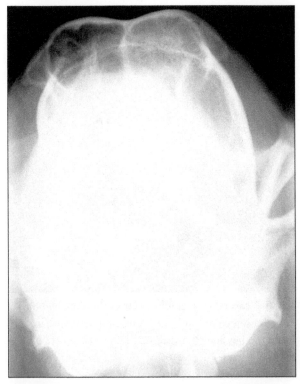

Fig. 9.7 Intraoral radiographic view of the nasal cavity of a dog with right-sided loss of turbinates and increased soft tissue density within the entire nasal cavity without areas of increased radiolucency. This is highly suggestive of nasal neoplasia in the dog.

Fig. 9.8 Caudorostral view and outprojection of the frontal sinuses of a dog with aspergillosis, involvement of the left frontal sinus (loss of radiolucency, filling with soft tissue density) is evident.

- *Lateral view*[29–34]: the patient is placed in lateral recumbency. A foam wedge is placed under the animal's nose and mandible so that the sagittal plane of the skull is parallel to the tabletop. The beam is centred midway between the ear and the eye, dorsal to the zygomatic arch. The jaws should be opened if the temporomandibular joints are the areas of interest.
- *Ventrodorsal view*[29–34]: the animal is placed in dorsal recumbency. A radiolucent block is placed under the neck behind the skull. The occipitoatlantal articulation is extended so that the hard palate lies parallel to the film. The X-ray beam is centred between the eyes and the ears on the midline. The sinuses are more clearly seen on this view than on the dorsoventral view, although it is more difficult to achieve symmetry in the ventrodorsal position. This view is best for demonstrating the cranium because it is closer to the film. Intraoral film or a film–screen combination in a flexible envelope can be used to demonstrate the mandibles. The ventrodorsal position can be used with an intraoral film to demonstrate the mandibular incisor teeth. The X-ray beam is directed rostrocaudally at an angle of 20°. With the animal's mouth opened wide, the maxilla parallel to the tabletop, and the same beam angle centred on the maxilla, the nasal and ethmoid regions can be demonstrated.
- *Dorsoventral view*[29–34]: the animal is placed in sternal recumbency with the head resting on the cassette so that the hard palate lies parallel to the tabletop. This position can be maintained by a bandage passed across the neck, behind the skull, and fixed to the table. On occasion it may be easier to position the head if the cassette is raised off the tabletop on a support. The X-ray beam is centred between the eyes and the ears on the midline. It is easier to achieve symmetry in this position than it is in the ventrodorsal position, but the calvarium is further from the film and is therefore more distorted on this view. However, this is of no practical significance. The maxillary nasal turbinates can also be examined in this position by introducing the intraoral film corner first and placing it as far back in the mouth as possible. Alternatively, a high-detail single-screen film

combination in a plastic lightproof envelope can be used. The X-ray beam is centred over the nasal septum. This position can be used with intraoral film to demonstrate the maxillary incisor teeth. The X-ray beam is directed rostrocaudally at an angle of 20°.

- *Oblique views*[29–34]: oblique views enable some structures to be demonstrated without superimposition of the contralateral side. Oblique views are used to study the temporomandibular joints, the osseous bullae, the frontal sinuses and the dorsal edge of the orbit. In the open-mouth position, the maxillary and mandibular dental arcades are profiled by using oblique views. The structure profiled varies with the oblique study selected.
- *Rostrocaudal (frontal) view*[29–34]: the patient is placed in dorsal recumbency, and the neck is flexed so that the hard palate lies perpendicular to the film. The patient's head is held in position with a bandage or tape around its nose. The beam is directed at right angles to the tabletop, along the line of the hard palate and centred between the eyes. The frontal sinuses, the odontoid process or dens and the foramen magnum can be demonstrated on this view. In cats, a similar technique can be used to demonstrate the tympanic bullae, which lie ventral to the mandibles. The patient is placed in dorsal recumbency, and the head is flexed. Rather than being orientated with the hard palate perpendicular to the tabletop, the head is tilted slightly dorsal so that the hard palate is at an angle of approximately 70–80° to the tabletop and the beam centred just ventral to the mandibular symphysis. This frontal view, with the animal's mouth opened, can be modified to demonstrate the osseous bullae and temporomandibular joints, in which case the beam is directed rostrocaudally at an angle of 20–30° to the hard palate. By varying the angles of the hard palate to the tabletop, the frontal view can also be used to outline the calvarium.
- *Caudorostral view*[29–34]: with the animal in sternal recumbency, the head is supported above the level of the neck with the hard palate parallel

to the tabletop. A horizontal beam, directed caudorostrally, is used parallel to the tabletop and centred on the skull with the cassette placed in front of the animal's nose. The frontal sinuses are profiled. An advantage of the caudorostral view is that it will demonstrate fluid levels in the frontal sinuses.

9.5 ADVANCED IMAGING OF THE NASAL CAVITY AND NASAL SINUSES

Cross-sectional imaging provides more detailed information about the nose's complex architecture and may be better able to differentiate rhinitis from neoplasia[35,36]. CT has been most commonly used for this purpose, but MRI is also increasingly available. Nasal CT and MRI have both proven to be superior to conventional radiography for precise evaluation of diseases within the nasal cavity, paranasal sinuses, periorbital region, tympanic bulla and skull[37,38]. Both modalities excel in determining the extent and severity of the disease process present.

Studies comparing the diagnostic differences between CT and MRI for evaluation of canine nasal neoplasia have not demonstrated superiority of one imaging modality over the other[37,39,40]. If radiation therapy is planned for nasal neoplasia, CT is preferred as it assists with accurate treatment planning[29]. There does not seem to be superiority of MRI over CT for diagnostic evaluation of other chronic nasal disease like aspergillosis in dogs[41]. Since costs associated with MRI are considerably higher when compared with CT, and image acquisition time is much more rapid with CT than with MRI, CT is preferable in most cases.

In addition, up to 28% of cats with inflammatory rhinitis also demonstrate middle ear effusion in one study[42]. Whereas in cats with clinical signs of otitis interna MRI may be preferred to evaluate the inner ear and brainstem, for surgical treatment planning such as ventral bulla osteotomy, CT is more informative and therefore the technique of choice for evaluation of nasal disease in this species as well.

CT is also an excellent tool for evaluation of the brachycephaly-related abnormalities of the nose such as stenotic nares, aberrant nasal turbinates that protrude into the nasopharynx and brachycephaly-related middle ear abnormalities[43–45].

Computed tomography

Cross-sectional imaging findings in chronic nasal disease in dogs are essentially similar to those described for nasal radiography, although more subtle lesions may be detected. The CT appearance of the normal (Figs 9.9, 9.10) and abnormal nasal cavity in dogs has been described in detail[46–50]. Intranasal mass lesions on CT are typically seen with fungal rhinitis or nasal neoplasia in dogs, and masses associated with fungal rhinitis may have a characteristic cavitary appearance[46–50]. Scanning an animal with suspected aspergillosis or neoplastic disease in dorsal recumbency with the nose pointing upward (Fig. 9.11) may be helpful in evaluation of the integrity of the cribriform plate.

In addition to intranasal masses, vomer or paranasal bone lysis are seen in dogs and cats with neoplastic nasal disease[51–53]. Dogs with a final diagnosis of inflammatory rhinitis may also have an apparent mass on cross-sectional imaging, but often this mass fails to enhance in postcontrast imaging sequences[52]. The presence of

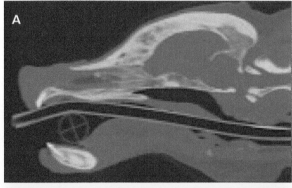

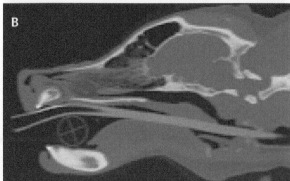

Figs 9.9A, B Computed tomographic sagittal (A) and parasagittal (B) view of the nasal cavity, frontal sinus and nasopharynx of a normal dog.

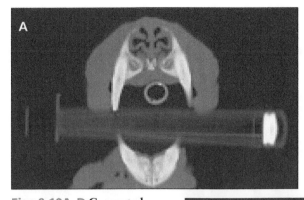

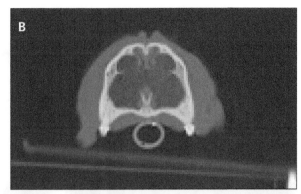

Figs 9.10A–D Computed tomographic transverse images of the nasal cavity, nasopharynx and frontal sinuses of a normal dog. A: At the level of the canines; B: at the level of the lateral nasal gland; C: at the level of the cribriform plate; D: at the level of the frontal sinuses.

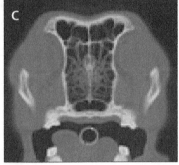

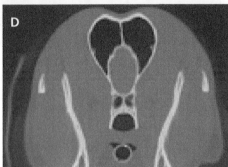

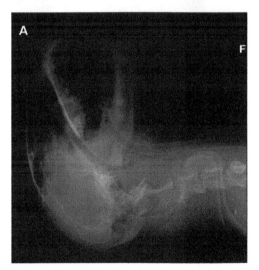

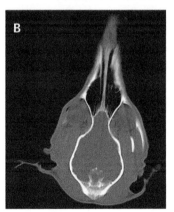

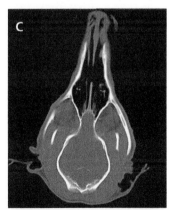

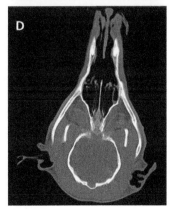

Figs 9.11A–D Computed tomographic transverse images of the nasal cavity in a dog with aspergillosis scanned in dorsal recumbency with the nose pointing upward. A: Position of the animal during scanning; B–D: subsequent transverse sections demonstrating the bilateral atrophy of caudal maxilloturbinates and septal atrophy as a result of aspergillosis. The ethmoturbinates can be seen sprouting from the cribriform plate, which appears to be intact.

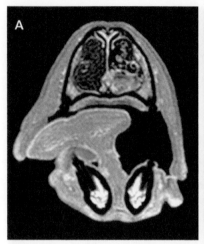
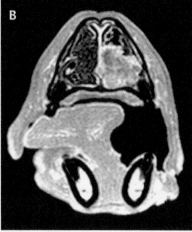
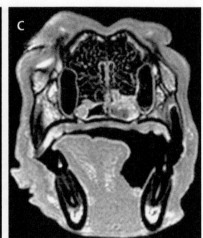

Figs 9.12A–D Magnetic resonance imaging transverse (A–C) and sagittal (D) images of the nasal cavity of a dog with a left-sided nasal carcinoma.

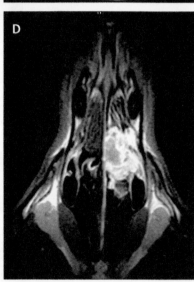

non-destructive, bilateral nasal mucosal thickening and fluid accumulation is associated with a final diagnosis of inflammatory rhinitis[51,52,54,55].

The CT appearance of the normal feline nasal cavity and paranasal sinuses has been described[53]. Studies comparing the CT appearance of nasal neoplasia and inflammatory disease in cats identified that many imaging findings overlapped between these conditions[51,55]. However, lysis of the ventral aspect of the maxilla or vomer bone, unilateral lysis of the ethmoturbinates or dorsal and lateral aspects of the maxilla, bilateral lysis of the orbital lamina and unilateral soft tissue opacification of the frontal sinus, sphenoid sinus, or retrobulbar space are more commonly seen with neoplastic disease[55]. Although septal lysis or cribriform lysis is predictive of a neoplastic diagnosis in dogs, the significance of these findings is debated in cats[34,51,55,56]. Septal deviation and sinus asymmetry can be part of the range of normal for the feline skull[56].

Magnetic resonance imaging

Magnetic resonance imaging findings in chronic nasal disease in dogs are again essentially similar to those described for nasal radiography, though fewer studies report the specific features of the normal and abnormal nasal cavity in dogs. Findings on MRI are similar as those on CT when dealing with neoplastic (Fig. 9.12) and fungal nasal disease[37,40,41,57].

Results of one study suggested that in dogs with nasal disease, the lack of a mass effect on MR images was significantly associated with inflammatory disease, whereas in dogs with a mass effect on MR images, vomer bone lysis, cribriform plate erosion, paranasal bone destruction, sphenoid sinus invasion by a mass and nasopharyngeal invasion by a mass were significantly associated with a diagnosis of neoplasia[52]. Recently common MRI features of inflammatory nasal diseases were reported that could aid in the differentiation of aspergillosis and chronic rhinitis in dogs[58]. When compared with muscle on T1W, the turbinate hypointensity or isointensity was shown to be significantly associated with chronic rhinitis, whereas hyperintensity was associated with a diagnosis of aspergillosis. On T2W, turbinate intensity did not appear to be associated with either pathology[58]. Large case-series or retrospective studies on MRI of nasal diseases in cats are lacking.

9.6 RHINOSCOPY AND NASOPHARYNGOSCOPY

Endoscopy allows for:
- Direct observation of the nasal cavity and nasopharynx.
- Facilitated collection of directed samples.
- Removal of foreign bodies.

Depending on the exact pathology present, the sinuses may also be visualised. A systematic approach to examination of the nasal cavity should be taken to ensure visualisation of all accessible areas. Knowledge of the regional anatomy is therefore very important and should be mastered before engaging in the use of this technique[1]. All rhinoscopic procedures require a deep general anaesthesia and should preferably be preceded by diagnostic imaging.

The nasopharynx and choanae can be observed by use of a spay hook to retract the soft palate and a dental mirror, but retroflexion of a flexible endoscope is preferred as it allows for more complete assessment and biopsy of abnormalities found. The nasal passages themselves are evaluated by rigid or flexible rhinoscopy with or without infusion of saline in the nasal cavities. Technique and equipment for these procedures have been thoroughly described[1,59-61]. Examination of the nasal 'floor' by careful examination of the hard and soft palate and dental probing should be considered part of a standard endoscopic evaluation for nasal diseases in both dogs and cats.

Rhinoscopy

The patient is placed in ventral recumbency with its head supported to lift the nose, which extends slightly over the edge of the examination table. The table height should be adjusted to allow for the investigator to sit down comfortably during the examination.

The simplest method of rhinoscopy is the one using an otoscope[62]. Depending on the size of the animal and otoscopic cones that can be fitted, a fair part of the rostral nasal cavity can be visualised[62]. For a complete

rhinoscopy, a variety of rigid and flexible instruments can be used, but a rigid scope is preferred by the author and provides adequate visualisation of the nasal sinuses. Wolf and Storz endoscopes are the most used types, a 1.9 or 2.7 mm rigid scope can be used for cats and small dogs, while a 3.5 mm or 4.5 mm scope can be used in large to very large dogs[1,59,63].

Outer sheaths and irrigation are not used by the author. A light source has to be attached to the scope in order to provide light and thus sight into the nasal cavity. Fine suction cannulas are used alongside the rigid scope to clear the nasal cavity of mucus[1].

Inspection is started by introducing the endoscope into the nasal vestibule through the dorsomedial part of the orifice (Fig. 9.13A). It is then advanced through the common meatus just medial to the nasal septum, and then first directed dorsally in the dorsal meatus from cranial to caudal to the level of the ethmoturbinates. After inspection of this meatus, the endoscope is placed ventral of the dorsal conchus. While retracting the scope from caudal to cranial, excursions from medial to lateral between the turbinates are made for inspection of all the maxilloturbinates (Figs 9.13B–E). Back in the nasal vestibule the scope is then further advanced ventrally into the ventral meatus under the maxilloturbinates and advanced caudally until the nasopharynx is entered (Figs 9.13F, G), while passing the ventral parts of the ethmoturbinates.

> Note: rhinoscopy should not be an exploratory adventure but should be as short as possible to prevent hypothermia and iatrogenic injury to the nasal mucosa and directed to the abnormalities suggested by imaging studies.

Rhinoscopy enables visual identification of tumours, foreign bodies or fungal granulomas.
- Nasal tumours typically appear as obstructive grey to white soft tissue masses with a reflective surface and bleed easily upon contact with the rhinoscope or with biopsy[64].
- Foreign bodies are usually embedded in exudate and may not be visible until this has been removed by suction or irrigation.

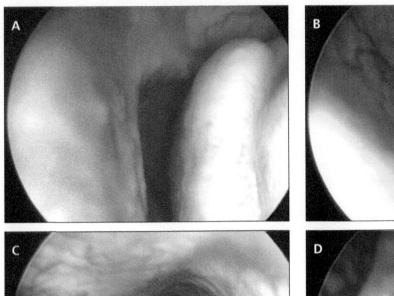

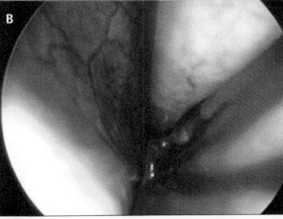

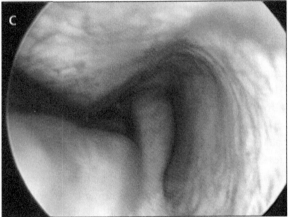

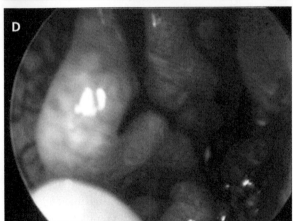

- Fungal infections generally lead to significant atrophy of the turbinates with severe mucosal inflammation and macroscopically visible fungal growth[65].
- In patients with chronic non-specific rhinitis rhinoscopy most commonly reveals nasal discharge and hyperemic mucosa, although in cats neither the presence nor the severity of mucosal abnormalities seen on rhinoscopy predicts the presence or severity of histopathological abnormalities[66].
- Material for diagnostic submission can be collected by swab, flush, or biopsy. Discrete lesions should be sampled directly, but the nasal mucosa should be sampled bilaterally even if no focal lesion is seen[35,67].

Endoscopy is increasingly used for evaluation and scoring of brachycephaly-related abnormalities such as aberrant cranial and caudal nasal turbinates and increased mucosal contact points (see Chapter 11 Diseases of the Nasal Cavity and Sinuses, Section 11.3)[45,68,69].

Nasopharyngoscopy

Though nasopharyngoscopy should be part of the diagnostic work-up for nasal disease in every dog, especially brachycephalic animals, it is of vital importance in cats since they often have primary nasopharyngeal disease. If performed in conjunction with rhinoscopy, retroflexed choanal and nasopharyngeal endoscopic examination is usually performed first to avoid obscuring observation of the choanae and nasopharynx by blood, mucus or flush[70].

Though nasopharyngoscopy can be performed with simple means as retraction of the soft palate and dental mirrors, for complete assessment and for taking biopsies endoscopes should be used. Specific rigid nasopharyngeal endoscopes with retroflexed mirrors are available that allow for visualisation of the nasopharynx, but taking biopsies using this technique is a challenge. The nasopharynx can also be reached in most dogs and large cats antegrade via rhinoscopy, but most clinicians use either small gastrointestinal or bronchoscopic fibreoptic flexible endoscopes[60,61,71].

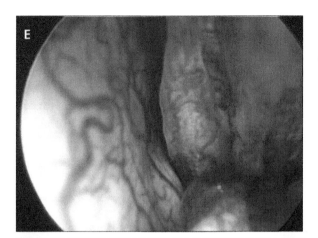

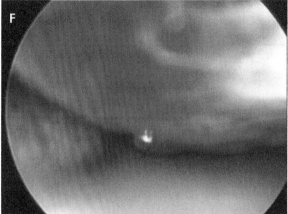

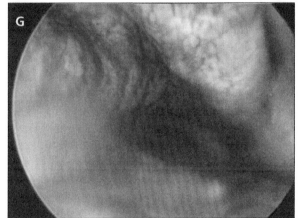

Figs 9.13A–G Rhinoscopic images of a normal left nasal cavity taken using a rigid endoscope. A: Entering the nasal cavity with the septum on the left-hand side and the large ala on the right side; B: nasal vestibule with the large dorsal conchus on the top of the image, the septum on the left and the rostral maxilloturbinates lateroventral; C: at the most lateral level between the maxilloturbinates and nasal wall; D: inbetween the fine-branched midnasal maxilloturbinates; E: at the level of the caudal maxilloturbinates with the rostral ethmoturbinates visible; F: entering the nasopharynx ventral to the ethmoturbinates; G: passed the ethmoturbinates with view of the rostral nasopharyngeal space.

With the patient in sternal or lateral recumbency, the endoscope is introduced straight into the oral cavity and subsequently advanced following the tongue caudally. Once the free edge of the soft palate has been passed, the endoscope is retroflexed 180° into the caudal nasopharynx. With cranial retraction of the endoscope in this position more cranial areas of the nasopharynx and the choanae (Fig. 9.14) can be visualised.

Flushing and biopsies can be taken via the working channels of the scope. Though directed biopsy collection via this approach typically yields diagnostic material when dealing with neoplastic disease[70,72,73], care needs to be taken to take large representative samples. The retroflexed choanal approach also enables observation and retrieval of nasopharyngeal foreign bodies and polyps and allows for assessment of degree of nasopharyngeal stenosis[74–79]. In addition, nasopharyngoscopy allows for identification of aberrant caudal nasal turbinates that can protrude into the nasopharynx[43,80].

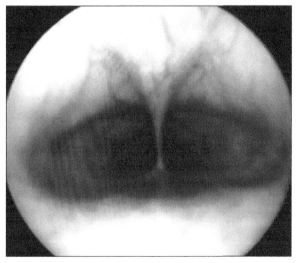

Fig. 9.14 Nasopharyngoscopic image of the nasopharynx of a dog without nasopharyngeal pathology, taken using the flexible endoscope, demonstrating the choanae.

9.7 REFERENCES

1 Venker-van Haagen AJ. The nose and nasal sinuses. In: *Ear, Nose, Throat, and Tracheobronchial Diseases in Dogs and Cats*. Venker-van Haagen AJ (ed). Schlütersche Verlagsgesellschaft, Hannover 2005, pp. 51–81.

2 Probst R, Grevers G, Iro H. Diagnostic evaluation of the nose and nasal sinuses. In: *Basic Otorhinolaryngology: a Step-by-Step Learning Guide*. Probst R, Grevers G, Iro H (eds). Thieme, Stuttgart 2006, pp. 16-25.

3 Erickson CH, McLeod RL, Mingo GG, Egan RW, Pedersen OF, Hey JA. Comparative oral and topical decongestant effects of phenylpropanolamine and d-pseudoephedrine. *American Journal of Rhinology* 2001;**15**:83–90.

4 Koss MC, Yu Y, Hey JA, McLeod RL. Acoustic rhinometry in the dog: a novel large animal model for studies of nasal congestion. *American Journal of Rhinology* 2002;**16**:49–55.

5 Koss MC, Yu Y, Hey JA, McLeod RL. Measurement of nasal patency in anesthetized and conscious dogs. *Journal of Applied Physiology* 2002;**92**:617–21.

6 McLeod RL, Mingo GG, Herczku C, *et al*. Combined histamine H1 and H3 receptor blockade produces nasal decongestion in an experimental model of nasal congestion. *American Journal of Rhinology* 1999;**13**:391–9.

7 McLeod RL, Erickson CH, Mingo GG, Hey JA. Intranasal application of the alpha2-adrenoceptor agonist BHT-920 produces decongestion in the cat. *American Journal of Rhinology* 2001;**15**:407–15.

8 McLeod RL, Mingo GG, Herczku C, *et al*. Changes in nasal resistance and nasal geometry using pressure and acoustic rhinometry in a feline model of nasal congestion. *American Journal of Rhinology* 1999;**13**:375–83.

9 Straszek SP, Taagehøj F, Graff S, Pedersen OF. Acoustic rhinometry in dog and cat compared with a fluid-displacement method and magnetic resonance imaging. *Journal of Applied Physiology* 2003;**95**:635–42.

10 Wiestner TS, Koch DA, Nad N, *et al*. Evaluation of the repeatability of rhinomanometry and its use in assessing transnasal resistance and pressure in dogs. *American Journal of Veterinary Research* 2007;**68**:178–84.

11 Amis TC, O'Neill N, Van der Touw T, Tully A, Brancatisano A. Supraglottic airway pressure-flow relationships during oronasal airflow partitioning in dogs. *Journal of Applied Physiology* 1996;**81**:1958–64.

12 Ohnishi T, Ogura JH. Partitioning of pulmonary resistance in the dog. *Laryngoscope* 1969;**79**:1847–78.

13 Tiniakov RL, Tiniakova OP, McLeod RL, Hey JA, Yeates DB. Canine model of nasal congestion and allergic rhinitis. *Journal of Applied Physiology* 2003;**94**:1821–8.

14 Houpt KA, Davis PP, Hintz HF. Effect of peripheral anosmia in dogs trained as flavor validators. *American Journal of Veterinary Research* 1982;**43**:841–3.

15 Krestel D, Passe D, Smith JC, Jonsson L. Behavioral determination of olfactory thresholds to amyl acetate in dogs. *Neuroscience & Biobehavioral Reviews* 1984;**8**:169–74.

16 Myers LJ, Pugh R. Thresholds of the dog for detection of inhaled eugenol and benzaldehyde determined by electroencephalographic and behavioral olfactometry. *American Journal of Veterinary Research* 1985;**46**:2409–12.

17 Steen JB, Wilsson E. How do dogs determine the direction of tracks? *Acta Physiologica Scandinavica* 1990;**139**:531–4.

18 Gagnon S, Doré FY. Search behavior in various breeds of adult dogs (*Canis familiaris*): object permanence and olfactory cues. *Journal of Comparative Psychology* 1992;**106**:58–68.

19 Schoon G, De Bruin JC. The ability of dogs to recognize and cross-match human odours. *Forensic Science International* 1994;**69**:111–18.

20 Steen JB, Mohus I, Kvesetberg T, Walløe L. Olfaction in bird dogs during hunting. *Acta Physiologica Scandinavica* 1996;**157**:115–19.

21 Myers LJ, Nash R, Elledge HS. Electro-olfactography: a technique with potential for diagnosis of anosmia in the dog. *American Journal of Veterinary Research* 1984;**45**:2296–8.

22 Hirano Y, Oosawa T, Tonosaki K. Electroencephalographic olfactometry (EEGO) analysis of odour responses in dogs. *Research in Veterinary Science* 2000;**69**:263–5.

23 Jones N, Rog D. Olfaction: a review. *Journal of Laryngology and Otology* 2007;**112**:11–24.

24 Myers LJ, Hanrahan LA, Swango LJ. Anosmia associated with canine distemper. *American Journal of Veterinary Research* 1988;**49**:1295–7.

25 Myers LJ, Nusbaum KE, Swango LJ. Dysfunction of sense of smell caused by canine parainfluenza virus infection in dogs. *American Journal of Veterinary Research* 1988;**49**:188–90.

26 Ezeh PI, Myers LJ, Hanrahan LA, Kemppainen RJ, Cummins KA. Effects of steroids on the olfactory function of the dog. *Physiology and Behavior* 1992;**51**:1183–7.

27 Miller WJ, Griffin CE, Campbell KL. Structure and function of the skin. In: *Muller & Kirk's Small Animal Dermatology*, 7th edn. Miller WJ, Griffin CE, Campbell KL (eds). WB Saunders, Philadelphia 2012, pp. 1–55.

28 Bettany S, Mueller R. Histologic features of the normal canine nose. *Proceedings Annual Meeting of the Australian College of Veterinary Scientists, Dermatology Chapter*, Gold Coast, 2003.

29 Thrall DE (ed). *Textbook of Veterinary Diagnostic Radiology*, 6th edn. Elsevier Saunders, St. Louis, 2013.

30 Kealy JK, McAllister H, Graham JP (eds). *Diagnostic Radiology and Ultrasonography of the Dog and Cat*, 5th edn. Elsevier Saunders, St. Louis, 2011.

31 Pownder S, Rose M, Crawford J. Radiographic techniques of the nasal cavity and sinuses. *Clinical Techniques in Small Animal Practice* 2006;**21**:46–54.

32 Russo M, Lamb CR, Jakovljevic S. Distinguishing rhinitis and nasal neoplasia by radiography. *Veterinary Radiology & Ultrasound* 2000;**41**:118–24.

33 Lamb CR, Richbell S, Mantis P. Radiographic signs in cats with nasal disease. *Journal of Feline Medicine and Surgery* 2003;**5**:227–35.

34 O'Brien RT, Evans SM, Wortman JA. Radiographic findings in cats with intranasal neoplasia or chronic rhinitis: 29 cases (1982–1988). *Journal of the American Veterinary Medical Association* 1996;**208**:385–9.

35 Schmiedt CW, Creevy KE. Nasal planum, nasal cavity, and sinuses. In: *Veterinary Surgery Small Animal*. Tobias KM, Johnston SA (eds). Elsevier Saunders, St. Louis 2012, pp. 1691–706.

36 Park RD, Beck ER, LeCouteur RA. Comparison of computed tomography and radiography for detecting changes induced by malignant nasal neoplasia in dogs. *Journal of the American Veterinary Medical Association* 1992;**201**:1720–4.

37 Kuehn NF. Diagnostic imaging for chronic nasal disease in dogs. *Journal of Small Animal Practice* 2014;**55**:341–2.

38 Burk RL. Computed tomographic imaging of nasal disease in 100 dogs. *Veterinary Radiology & Ultrasound* 1992;**33**:177–80.

39 Dhaliwal RS, Kitchell BE, Losonsky JM, Kuriashkin IV, Clarkson RB. Subjective evaluation of computed tomography and magnetic resonance imaging for detecting intracalvarial changes in canine nasal neoplasia. *International Journal of Applied Research in Veterinary Medicine* 2004;**2**:201–8.

40 Drees R, Forrest LJ, Chappell R. Comparison of computed tomography and magnetic resonance imaging for the evaluation of canine intranasal neoplasia. *Journal of Small Animal Practice* 2009;**50**:334–40.

41 Saunders J, Clercx C, Snaps FR. Radiographic, magnetic resonance imaging, computed tomographic, and rhinoscopic features of nasal aspergillosis in dogs. *Journal of the American Veterinary Medical Association* 2004;**225**:1703–12.

42 Detweiler DA, Johnson LR, Kass PH, R WE. Computed tomographic evidence of bulla effusion in cats with sinonasal disease: 2001–2004. *Journal of Veterinary Internal Medicine* 2006;**20**:1080–4.

43 Oechtering GU. Brachycephalic syndrome – new information on an old congenital disease. *Veterinary Focus* 2010;**20**:2–9.

44 Hussein AK, Sullivan M, Penderis J. Effect of brachycephalic, mesaticephalic, and dolichocephalic head conformations on olfactory bulb angle and orientation in dogs as determined by use of *in vivo* magnetic resonance imaging. *American Journal of Veterinary Research* 2012;**73**:946–51.

45 Schuenemann R, Oechtering GU. Inside the brachycephalic nose: intranasal mucosal contact points. *Journal of the American Animal Hospital Association* 2014;**50**:149–58.

46 Saunders J, Van Bree H, Gielen I. Diagnostic value of computed tomography in dogs with chronic nasal disease. *Veterinary Radiology & Ultrasound* 2003;**44**:409–13.

47 Saunders JH, Zonderland J-L, Clercx C, *et al.* Computed tomographic findings in 35 dogs with nasal aspergillosis. *Veterinary Radiology & Ultrasound* 2002;**43**:5–9.

48 Kuehn NF. Nasal computed tomography. *Clinical Techniques in Small Animal Practice* 2006;**21**:55–9.

49 Codner EC, Lurus AG, Miller JB, Gavin PR. Comparison of computed tomography with radiography as a noninvasive diagnostic technique for chronic nasal disease in dogs. *Journal of the American Veterinary Medical Association* 1993;**202**:1106–10.

50 Lefebvre J, Kuehn NF, Wortinger A. Computed tomography as an aid in the diagnosis of chronic nasal disease in dogs. *Journal of Small Animal Practice* 2005;**46**:280–5.

51 Schoenborn WC, R WE, Kass PP, Dale M. Retrospective assessment of computed tomographic imaging of feline sinonasal disease in 62 cats. *Veterinary Radiology & Ultrasound* 2003;**44**:185–95.

52 Miles MS, Dhaliwal RS, Moore MP, Reed AL. Association of magnetic resonance imaging findings and histologic diagnosis in dogs with nasal disease: 78 cases (2001–2004). *Journal of the American Veterinary Medical Association* 2008;**232**:1844–9.

53 Losonsky JM, Abbott LC, Kuriashkin IV. Computed tomography of the normal feline nasal cavity and paranasal sinuses. *Veterinary Radiology & Ultrasound* 1997;**38**:251–8.

54 Michiels L, Day MJ, Snaps F, Hansen P. A retrospective study of non-specific rhinitis in 22 cats and the value of nasal cytology and histopathology. *Journal of Feline Medicine and Surgery* 2003;**5**:279–85.

55 Tromblee TC, Jones JC, Etue AE, Forrester SD. Association between clinical characteristics, computed tomography characteristics, and histologic diagnosis for cats with sinonasal disease. *Veterinary Radiology & Ultrasound* 2006;**47**:241–8.

56 Reetz JA, Maï W, Muravnick KB, Goldschmidt MH, Schwarz T. Computed tomographic evaluation of anatomic and pathologic variations in the feline nasal septum and paranasal sinuses. *Veterinary Radiology & Ultrasound* 2006;**47**:321–7.

57 Avner A, Dobson JM, Sales JI, Herrtage ME. Retrospective review of 50 canine nasal tumours evaluated by low-field magnetic resonance imaging. *Journal of Small Animal Practice* 2008;**49**:233–9.

58 Furtado ARR, Caine A, Herrtage ME. Diagnostic value of MRI in dogs with inflammatory nasal disease. *Journal of Small Animal Practice* 2014;**55**:359–63.

59 Chamness CJ. Introduction to veterinary endoscopy and endoscopic instrumentation. In: *Veterinary Endoscopy for the Small Animal Practitioner*. McCarthy TC (ed). Elsevier, St. Louis, 2005, pp. 1–20.

60 Chamness CJ. Endoscopic instrumentation and documentation for flexible and rigid endoscopy. In: *Small Animal Endoscopy*. Tams TR, Rawlings CA (eds). Elsevier Health Sciences, St. Louis, 2010, pp. 3–26.

61 Noone KE. Rhinoscopy, pharyngoscopy, and laryngoscopy. *Veterinary Clinics of North America Small Animal Practice* 2001;**31**:671–89.

62 Venker-van Haagen AJ. Otoscopy, rhinoscopy, and bronchoscopy in small animal clinics. *Veterinary Quarterly* 1985;**3**:222–4.

63 McCarthy TC. Rhinoscopy: the diagnostic approach to chronic nasal disease. In: *Veterinary Endoscopy for the Small Animal Practitioner*. McCarthy TC (ed). Elsevier, St. Louis, 2005, pp. 137–200.

64 Clercx C, Wallon J, Gilbert S, Snaps F, Coignoul F. Imprint and brush cytology in the diagnosis of canine intranasal tumours. *Journal of Small Animal Practice* 1996;**37**:423–7.

65 Johnson LR, Drazenovich TL, Herrera MA, Wisner ER. Results of rhinoscopy alone or in conjunction with sinuscopy in dogs with aspergillosis: 46 cases (2001–2004). *Journal of the American Veterinary Medical Association* 2006;**228**:738–42.

66 Johnson LR, Clarke HE, Bannasch MJ, De Cock HEV. Correlation of rhinoscopic signs of inflammation with histologic findings in nasal biopsy specimens of cats with or without upper respiratory tract disease. *Journal of the American Veterinary Medical Association* 2004;**225**:395–400.

67 Lent SE, Hawkins EC. Evaluation of rhinoscopy and rhinoscopy-assisted mucosal biopsy in diagnosis of nasal disease in dogs: 119 cases (1985–1989). *Journal of the American Veterinary Medical Association* 1992;**201**:1425–9.

68 Bernaerts F, Talavera J, Leemans J, *et al*. Description of original endoscopic findings and respiratory functional assessment using barometric whole-body plethysmography in dogs suffering from brachycephalic airway obstruction syndrome. *Veterinary Journal* 2010;**183**:95–102.

69 Schuenemann R, Oechtering G. Inside the brachycephalic nose: conchal regrowth and mucosal contact points after laser-assisted turbinectomy. *Journal of the American Animal Hospital Association* 2014;**50**:237–46.

70 Willard MD, Radlinsky MA. Endoscopic examination of the choanae in dogs and cats: 118 cases (1988–1998). *Journal of the American Veterinary Medical Association* 1999;**215**:1301–5.

71 Padrid AP. Laryngoscopy and tracheobronchoscopy of the dog and cat. In: *Small Animal Endoscopy*. Tams TR, Rawlings CA (eds). Elsevier Health Sciences, St. Louis, 2010, pp. 331–59.

72 De Lorenzi D, Bertoncello D. Squash-preparation cytology from nasopharyngeal masses in the cat: cytological results and histological correlations in 30 cases. *Journal of Feline Medicine and Surgery* 2008;**10**:55–60.

73 Little L, Patel R, Goldschmidt M. Nasal and nasopharyngeal lymphoma in cats: 50 cases (1989–2005). *Veterinary Pathology* 2007;**44**:885–92.

74 Tyler JW. Endoscopic retrieval of a large, nasopharyngeal foreign body. *Journal of the American Animal Hospital Association* 1997;**33**:513–6.

75 Henderson SM, Bradley K, Day MJ, *et al*. Investigation of nasal disease in the cat – a retrospective study of 77 cases. *Journal of Feline Medicine and Surgery* 2004;**6**:245–57.

76 Pilton JL, Ley CJ, Voss K, Krockenberger MB, Barrs VR, Beatty JA. Atypical, abscessated nasopharyngeal polyp associated with expansion and lysis of the tympanic bulla. *Journal of Feline Medicine and Surgery* 2014;**16**:699–702.

77 Mitten RW. Nasopharyngeal stenosis in four cats. *Journal of Small Animal Practice* 1988;**29**:341–5.

78 Schafgans KE, Armstrong PJ, Kramek B, Ober CP. Bilateral choanal atresia in a cat. *Journal of Feline Medicine and Surgery* 2012;**14**:759–63.

79 Reed N. Chronic rhinitis in the cat. *Veterinary Clinics of North America Small Animal Practice* 2014;**44**:33–50.

80 Ginn JA, Kumar MS, McKiernan BC, Powers BE. Nasopharyngeal turbinates in brachycephalic dogs and cats. *Journal of the American Animal Hospital Association* 2008;**44**:243–9.

10.1 INTRODUCTION

Many, but not all, of the diseases of the nasal planum are manifestations of internal disease or are part of a wider dermatosis. The limited range of insult that the tissues of the nasal planum can express means that biopsy is often the most cost-effective investigation, and this procedure mandates general anaesthesia. It may prove difficult to persuade an owner that this is indeed the investigation of choice. Management is complicated in that topical treatments are easily removed by licking, necessitating systemic treatment for what may be a focal condition. Hence a definitive diagnosis is important.

The histopathology of the nasal planum has been described in Chapter 9 Nose Diagnostic Procedures, Section 9.3. The key points are that the thickness and the lack of hair follicles[1,2] limit the range of response to insult, making the differentiation of the various diseases difficult. Clinically, one can see a change in pigment, hyperkeratosis, erosion, ulceration and crust formation and fissure. One of the most critical observations is whether the lesions are confined to the nasal planum (i.e. are not affecting the adjacent, haired, skin of the nose), or are part of a more generalised dermatitis – or indeed part of a systemic disease.

In the cat[3] the most common pattern of disease on the nasal planum is ulceroproliferative (Fig. 10.1A) in presentation, a consequence, in part, of the frequency with which squamous cell carcinoma is encountered. In the dog, in contrast, a change in pigment, hyperkeratosis, erosion, ulceration and crust formation and fissure is more usually seen (Fig. 10.1B). Discoid lupus erythematosus is the most common disorder[3].

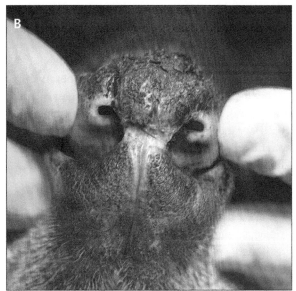

Figs 10.1A, B In the cat, ulcerative lesions of the nasal area are the most common pattern, whereas in the dog a crusting, depigmentary pattern is most common.

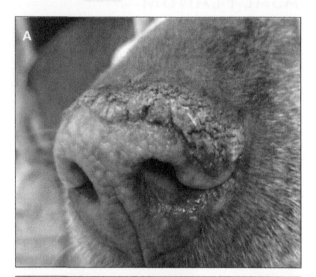

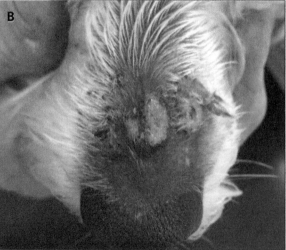

Figs 10.2A, B In (A) the lesions are confined to the nasal planum, making an immune-mediated dermatitis very likely; B: in contrast the lesions in this dog do not affect the nasal planum.

Table 10.1 **Diseases in which lesions affect the adjacent haired skin of the nose, but not usually the nasal planum**
Deep pyoderma
Eosinophilic folliculitis and furunculosis
Dermatophytosis
Feline viral dermatitis
Feline mosquito bite hypersensitivity
Dermoid sinus

Table 10.2 **Conditions in which lesions affect, or are confined to, the nasal planum and alar cartilages**
Idiopathic mucocutaneous pyoderma
Nasal hypopigmentation
Idiopathic nasal hyperkeratosis
Idiopathic nasodigital hyperkeratosis
Hereditary nasal parakeratosis in Labrador Retrievers
Xeromycteria (parasympathetic nose)
Actinic dermatitis
Squamous cell carcinoma
Discoid lupus erythematosus
Pemphigus diseases
Nasal arteritis
Vitiligo
Cryptococcal, Sporotrichial and idiopathic granuloma
Metabolic epidermal necrosis
Uveodermatological syndrome
Carcinoma *in situ*
Epitheliotropic lymphoma
Ulcerative nasal dermatitis of Bengal cats

As a clinician, one of the most critical observations is whether the dermatitis is limited to the nasal planum (Fig. 10.2A), or whether it affects the adjacent nose (Fig. 10.2B), or indeed, the face. Alternatively, is it part of a systemic disease or associated with lesions elsewhere (Tables 10.1, 10.2)?

- Most immune-mediated diseases affect the nasal planum, although the nose and face may also be affected.
- The importance of obtaining a definitive diagnosis, usually by biopsy samples, cannot be stressed enough when presented with a nasal dermatitis.

10.2 DISEASES AFFECTING THE ROSTRAL NOSE BUT NOT THE NASAL PLANUM

Deep pyoderma

Deep pyoderma of the nose presents as a moderately pruritic, occasionally furunculotic, crusting dermatitis on the rostral face. The nasal planum is not affected (Fig. 10.3). Note that superficial pyoderma, which is characterised by follicular and interfollicular papules and pustules, is very uncommonly seen on the face. The major differential diagnoses are the pemphigus diseases, which usually affect the nasal planum and eosinophilic furunculosis, which is peracute in onset, much more pruritic and does not affect the nasal planum. Other considerations are demodicosis and dermatophytosis.

Treatment is directed toward the primary differential with systemic antibacterial therapy for 3 weeks and, perhaps, dilute chlorhexidine soaks, to soften the crust and exert topical antibacterial therapy. No steroids should be given as this will make assessment of the effect of antibacterial treatment impossible. Failure to respond should prompt biopsy for histopathological examination.

Eosinophilic folliculitis and furunculosis

Eosinophilic folliculitis is a peracute nasal dermatitis characterised by early and rapid progression from a papular pattern to a proliferative, highly pruritic, even painful, ulcerated plaque, or plaques (Figs 10.4A, B). Bloom and others[4–6] have pointed out difficulties with this dermatosis being classified as pyoderma, based on a failure to respond to antibacterial treatment and the histopathological pattern. Bloom[4] has also pointed out that if this disease is, indeed, an acute reaction to insect or arthropod bite and stings, as has been proposed, it might not be expected to occur in the winter months and might be anticipated to be more common than it appears to be. Currently, this must be considered an idiopathic dermatopathy.

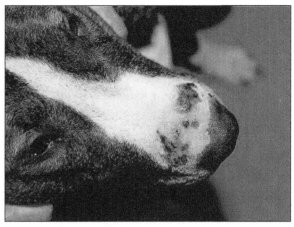

Fig. 10.3 Deep pyoderma on the nose of a Bull Terrier. Crusted papules, furunculosis and mild pruritus.

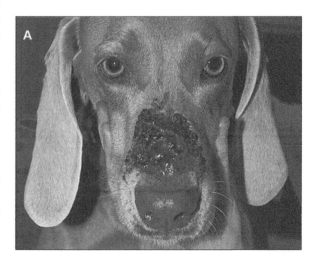

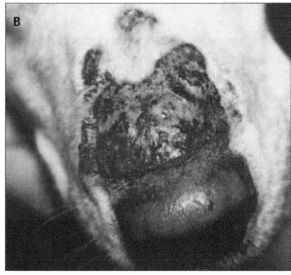

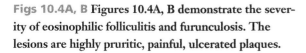

Figs 10.4A, B Figures 10.4A, B demonstrate the severity of eosinophilic folliculitis and furunculosis. The lesions are highly pruritic, painful, ulcerated plaques.

Treatment is directed toward suppressing the acute inflammatory response with supportive therapy:

- Prednisolone 1mg/kg twice daily.
- Covering antibacterial treatment, e.g. cephalosporin 25 mg/kg twice daily.
- Pain control? Tramadol 4 mg/kg twice daily?
- Gentle topical soaks with very dilute chlorhexidine to remove exudate and prevent crust formation.
- Owners should be advised that some scarring is inevitable.

Microsporon canis dermatophytosis

Microsporon canis is a zoophilic dermatophyte[7]. It is the most frequent cause of dermatophytosis in the cat in the UK, Italy, southern United States and Brazil (reviewed by Bond[7]). *M. canis* is classically associated with patches of alopecia associated with fractured hair shafts, a fine scale and variable erythema (Fig. 10.5). It may be mildly pruritic. An inflammatory folliculitis may be seen occasionally. Lesions on the nasal planum appear to be very uncommon[7] but lesions on the rostral face might be encountered occasionally. The management of *M. canis* dermatophytosis is discussed elsewhere.

Trichophyton mentagrophytes, Microsporum gypseum and M. persicolor dermatophytosis

Microsporum gypseum is a geophilic dermatophyte, found in the soil, and it has been most commonly reported as a dermatophytic infection in dogs in the USA[7]. The other two are both zoophilic dermatophytes, which are pathogens of free-living rodents. All three organisms tend to cause a much more inflammatory dermatitis than that seen in *M. canis* infections. Dogs are much more commonly infected than cats.

T. mentagrophytes and *M. gypseum* cause an inflammatory folliculitis and furunculosis[8,9]. The most commonly seen clinical signs are erythema, crusting, alopecia and pruritus. Lesions are usually well-demarcated and commonly found on the face (Fig. 10.6). Partially-healed lesions may have a shiny appearance[7]. *T. mentagrophytes* infection is an uncommon cause of dermatitis in the southern USA (where *M. gypseum* dermatophytosis predominates), but is the second most common cause of dermatophytosis in the UK[7]. Given that these infections are associated with activities such

as hunting and rootling in the ground, it is no surprise that it is seen most frequently in terrier-type dogs.

M. persicolor is a strict epidermophyte that can only grow on keratin. In the dog's case that means the outermost layers of the stratum corneum[10,11]. The disease is mainly seen in adult dogs, and particularly those that

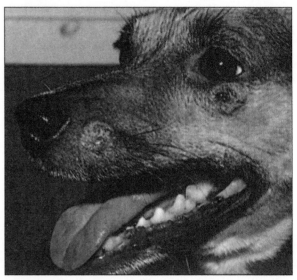

Fig. 10.5 Erythematous patch on the lateral nose of a German Shepherd Dog cross; mildly pruritic and slowly enlarging.

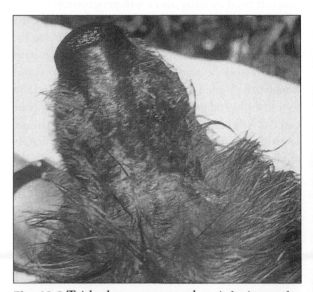

Fig. 10.6 *Trichophyton mentagrophytes* infection on the face of a terrier. Note the shiny appearance of the skin in the partially healed centre of the face.

are hunting or at least very active out of doors, such as terriers, and sporting dogs such as Pointers[10,11].

Clinical signs of *M. persicolor* infection are mainly seen on the face (Figs 10.7A, B) and the rostral aspects of the forelimbs[10]. The dermatological lesion may be an erythematous, papular dermatitis, a 'classic ringworm' lesion of erythema and scale, or a deeper lesion such as an erosion, crust or furuncle[10,11]. The lesions may, rarely, encroach onto the nasal planum, which may then show depigmentation[10]. Pruritus is variable, usually moderate in intensity.

The diagnosis of these sylvatic dermatophyte infections cannot be made on the basis of the clinical signs, as they are not in any way pathognomonic. Histopathological examination of biopsy samples may be able to demonstrate the presence of hyphae in the upper stratum corneum in cases of *M. persicolor* infection, but in one study only in six of the 16 cases was this possible[11]. Similarly, in cases of dermatophytosis due to *T. mentagrophytes*, where infection is follicular rather than epidermal, as in *M. persicolor* infection, the presence of hyphae is diagnostic, but they are not commonly seen[8]. Mycological culture (be careful to include surface scale, not hair shaft material, for *M. persicolor*) onto Sabouraud's medium will allow identification[10]. Molecular methodology using polymerase chain reaction (PCR)

technology might prove extremely useful in the definitive diagnosis of these dermatophytoses.

The treatment of these dermatophytoses is straightforward, once the diagnosis is made, as environmental contamination, cross infection to other pets or zoonotic infection is very uncommon[9]. Owners should be advised that treatment will take 2 or 3 months, and that there may be some scarring and residual alopecia if significant follicular damage has occurred.

With regard to *M. persicolor* infection, one study reported a good response to topical treatment alone, using either miconazole gel or enilconazole cream[10]. The other report used a combination of topical enilconazole and systemic ketoconazole[11]. Topical treatment alone would not be recommended for *T. mentagrophytes* or *M. gypseum* infection as follicular infection is considerable and topical therapeutics would be unable to penetrate sufficiently. Itraconazole (5–10 mg/kg once daily) or terbinafine 30–40 mg/kg once daily) would be good choices, in conjunction with topical enilconazole cream[12].

Feline virus-associated dermatitis

Feline herpesvirus 1 and feline calicivirus are major causes of feline upper respiratory disease[13]. Both have rarely been reported to cause, or at least to be associated

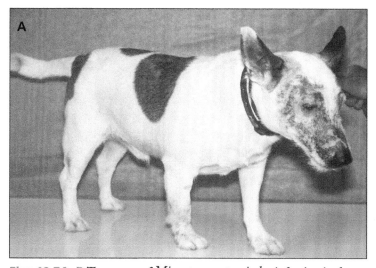

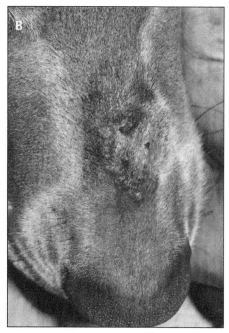

Figs 10.7A, B Two cases of *Microsporum persicorum* infection in dogs illustrating the extent of the disease. (Courtesy of Dr Didier Carlotti, Aquivet, Clinique Vétérinaire, Eysines, 33320 France.)

with nasal dermatitis, particularly erosion and ulceration of the nasal planum and philtrum (Figs 10.8, 10.9). Very rarely they have been associated with dermatitis on the feet and limbs and trunk[14,15]. Neither vaccinal status or a history of recent upper respiratory infection seem to be a precipitating factor.

As the lesions seen in this condition are non-specific in appearance there is a real advantage in biopsy sampling and forwarding for viral antigen detection[16,17]. Administration of non-specific treatment with gluco-

corticoids may not only hinder treatment but might precipitate more serious disease, just as with cowpox infection.

Appropriate non-specific supporting treatment of broad-spectrum antibacterial therapy, and perhaps anabolic steroids, are indicated. Surgical excision of truncal or limb lesion is feasible[13]. There are two reports[18a,b] that describe treatment of cats with FHV-1 dermatitis using antiviral agents, interferon omega or acyclovir, a nucleoside analogue. In the first study[18], a single case report, an injectable preparation was used (Virbac Omega®, VirbacSA, Carros, France) and in the second study[18a], the treatment was oral famciclovir tablets (Famvir®, Novartis). The following dose regime was reported for the injectable preparation::

- Day 1: 1.5 million units (MU), half injected intradermally in a perilesion pattern, and half subcutaneously on the lateral thorax. Sedation or general anaesthesia will be required for the perilesional intradermal injections.
- Day 2 and day 9: 1.5 MU injected subcutaneously on the lateral thorax.
- Days 19, 21, 23: 0.75 MU perilesionally and subcutaneously on the lateral thorax.

Figs 10.8, 10.9 Cutaneous manifestations of feline herpesvirus-1- (10.8) and feline calicivirus-associated (10.9) erosions and ulceration. (Courtesy of Susan Dawson, School of Veterinary Science, University of Liverpool, UK.)

Fig. 10.10 Facial dermatitis due to mosquito bite hypersensitivity. (Courtesy of Dr Ken Mason, Animal Allergy and Dermatology Service, Springwood, Queensland, Australia.)

A slow, but gradual response was noted, although hair regrowth was poor. The oral regime[18a] was used in 10 feline cases, four of which had cutaneous lesions, five cats had refractory ocular FHV-1 ulceration and one cat had chronic upper respiratory disease. All four cats with FHV-1 dermatitis had a favourable response, although one was treated by the owner with 5% acyclovir cream before entering the study. The authors considered that an initial dose of 62.5 mg q8h p/o would be an appropriate starting dose. Patently, if the cat resists oral treatment then the parenteral regime would be necessary, notwithstanding.

Mosquito bite hypersensitivity

This disease has been reported in cats from Australia and California[19,20]. It presents as a summer seasonal, pruritic facial dermatitis (Fig. 10.10). Lesions appear to progressively deteriorate each successive spring and summer, only to regress, or at least partially so, in the winter. Clinically, there is a well-demarcated, erythematous papulocrustous, plaque-like dermatitis affecting the face and nasal planum, with, occasionally, papulocrustous lesions on the pinnae, and feet.

Diagnosis is suggested on the basis of the seasonality, clinical presentation and the histopathological examination of biopsy samples, which show a spongiotic epidermis with a serocellular crust. Most strikingly, there is also a moderate to severe eosinophilic inflammation, with foci of collagen necrosis, often centred around the hair follicles[20]. Diagnosis can be confirmed by restricting the cat indoors, and observing the gradual resolution of the dermatitis over 4–7 days[19].

Treatment, ideally, would involve preventing the cat from accessing areas where the mosquitoes are particularly prevalent. In practice, this is very difficult to achieve. Oral prednisolone, or prednisone, at around 1–2 mg/kg a day can be given until remission is achieved, and then on an alternate day regime. Topical hydrocortisone creams, and even tacrolimus ointment, would be expected to be effective.

Dermoid sinus

A dermoid sinus is a congenital abnormality caused by a failure of the ectodermal neural tube to separate completely from the skin, during embryological development[21]. A tubular connection therefore remains,

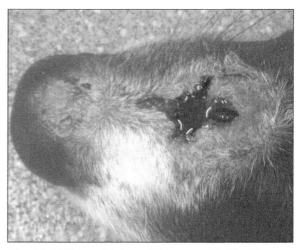

Fig. 10.11 Dermoid sinus. (Courtesy of Dr David Duclos, Animal Skin & Allergy Clinic, Lynnwood, WA, USA.)

usually resulting in a dorsal cervical sinus, as seen most commonly in the Rhodesian Ridgeback[22]. However, small, chronic, sinus formation, with intermittent discharge, immediately caudal to nasal planum (Fig. 10.11) has been reported in seven dogs, three of which were Golden Retrievers[23,24].

Identification is straightforward and surgical dissection and removal is curative, although it might require very skilled surgery.

Mucocutaneous pyoderma

This is an uncommon disorder of dogs characterised by erythema, swelling and subsequent crusting of the mucocutaneous junctions, particularly the lips[25]. The lateral aspects of the nares (not the nasal planum) on one, or both, sides may be affected similarly, but not in a symmetrical manner (Figs 10.12, 10.13). The condition appears to be uncomfortable, perhaps even, painful, and dogs resent examination of the affected areas.

The principal differential diagnoses are lip fold pyoderma and cheilitis, neither of which are associated with nasal lesions, and, perhaps, zinc deficiency although this is usually associated with periorbital lesions and recognised breed predilections.

Treatment with systemic antibacterial therapy and topical cleansing preparations is curative, although relapse appears to be common.

.

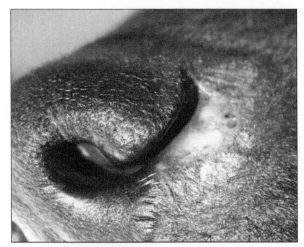

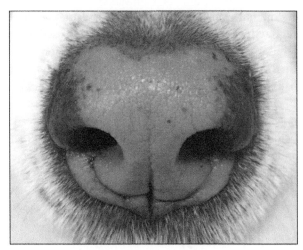

Fig. 10.14 Idiopathic nasal hypopigmentation. Note the lack of crust scale or inflammation.

Figs 10.12, 10.13 Mucocutaneous pyoderma in two dogs, affecting the lateral aspects of the nasal planum. (10.12 courtesy of Dr Jacques Fontaine, Animal Clinic, 1180 Brussels, Belgium.)

10.3 DISEASES AFFECTING THE NASAL PLANUM

Idiopathic nasal hypopigmentation

Melanocytes synthesise melanin pigments and transfer them to basal keratinocytes[26]. The acquired loss of this pigmentation, vitiligo, usually reflects a selective, immune-mediated process[3,27]. When confined to the nasal planum this process results in a non-inflammatory, typically pale brown or pink colour. The loss of pigment is usually symmetrical and often becomes complete (Fig. 10.14). Histopathological examination of biopsy samples from established cases usually fails to demonstrate inflammation. Interestingly, affected areas do not appear to be susceptible to actinic dermatitis, although the reason for this has not been identified.

Idiopathic nasal hypopigmentation (breeder term: Dudley nose) is an acquired defect: affected individuals are normal at birth, but pigment on the nasal planum gradually fades. Certain breeds in particular are affected, including Doberman Pinschers, Golden Retrievers, yellow Labrador Retrievers, German Shepherd Dogs (particularly the white individuals), Samoyeds, Siberian Huskies and Malamutes[28]. Interestingly, in some individuals the nasal hypopigmention appears to be seasonal, pigment is at its maximum in the summer, 'snow nose'. Whether this relates to an actinic radiation-induced transient increase in melanocytes or an increase in melanin deposition, or both, is not known.

There is no treatment.

Idiopathic nasal hyperkeratosis

Idiopathic nasal hyperkeratosis (as opposed to nasodigital hyperkeratosis) appears to be most commonly seen in middle aged and old dogs, and it has been proposed that it is associated with senility[29]. Affected individuals present with variably thickened, often regularly fissured, more or less symmetrical, thick crusting of the nasal planum (Fig. 10.15). The hyperkeratosis does not appear to be painful or pruritic and rarely affects the central philtrum or the extreme lateral aspects of the nasal planum. The fissures are not so deep that bleeding occurs and there is no loss of pigment.

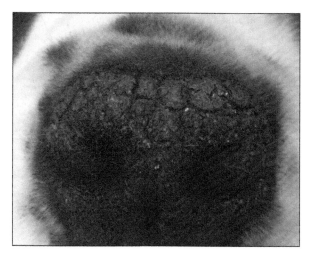

Fig. 10.15 Idiopathic nasal hyperkeratosis. (Courtesy of Dr David Duclos, Animal Skin & Allergy Clinic, Lynnwood, WA, USA.)

Diagnosis is based on clinical signs of non-inflammatory, normally pigmented, nasal hyperkeratosis that is limited to the nasal planum.

Treatment is non-specific and includes topical soaks with propylene glycol, petroleum jelly or a combination of 60% salicylic acid, 5% urea and 5% sodium lactate in a bland propylene glycol vehicle (such as Kerasolv, TEVA Animal Health (DVM)) or topical tretinoin gel (in severe cases)[30,31].

Idiopathic nasodigital hyperkeratosis

In contrast to idiopathic nasal hyperkeratosis where lesions are confined to the nasal planum, in this condition there is concurrent hyperkeratosis of the footpads. Note that foot pad hyperkeratosis, without nasal involvement, has been recognised as familial in two breeds, Irish terrier and Dogues de Bordeaux[32,33]. There is no evidence that nasodigital hyperkeratosis is familial, although the suspicion remains that it might be in some instances.

The clinical signs are straightforward with compacted, often peripherally feathered, and occasionally superficially fissured footpads, in association with nasal hyperkeratosis. The nasal hyperkeratosis is not as regularly fissured as that seen in idiopathic nasal hyperkeratosis of old dogs.

Similar management strategies to those described above for the nasal variant are indicated, although some pain relief might be required if the footpads fissure and become uncomfortable.

Hereditary nasal parakeratosis in Labrador Retrievers

This tardive, autosomal recessive condition was noted to begin around 6–12 months of age[34,35]. The mutation appears to be at a single locus in the *SUV39H2* gene[36]. The condition has been reported in many countries[35]. Lesions were restricted to the nasal planum in all but one of the 25 cases reported in these two publications: one dog was reported to have had parakeratosis of the footpads, but they were normal at the time of specialist examination[35]. Clinical signs consist of hyperkeratosis and loss of pigment of the nasal planum (Figs 10.16A, B). No pruritus was reported, even in those individuals that developed ulceration or fissures of the nasal planum. The

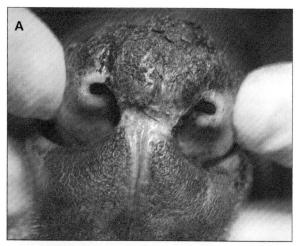

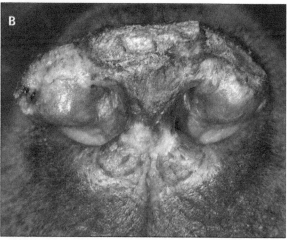

Figs 10.16A, B Two cases of hereditary nasal parakeratosis. Note the ulceration and fissure formation in Fig. 10.16A. (Courtesy of Dr David Duclos, Animal Skin & Allergy Clinic, Lynnwood, WA, USA.)

loss of pigment did not appear to predispose to actinic dermatitis[34]. Diagnosis is based on the clinical features and characteristic histopathological pattern[34,35].

The most effective treatment consists of repeated applications of propylene glycol[35].

Xeromycteria, parasympathetic nose

The paired lateral nasal glands drain through a single duct that opens about 2 cm inside the opening of the nostril[37]. The secretions from these glands, and the lacrimal glands, are essential to avoid desiccation of the nasal tissues. The secretion is under parasympathetic control, mediated by nitrous oxide[38,39]. If the sympathetic nerve supply is compromised (otitis media for example), or the lacrimal gland damaged, then secretion reduces and drying of the nasal tissues may result[40]. Keratoconjunctivitis (KCS) may also be seen.

Clinically, the dogs present with uni- or bilateral nasal drying and nasal hyperkeratosis (Figs 10.17, 10.18). Pain and pruritus do not seem to be a feature. In addition to KCS, there may be xerostomia. Once the condition is suspected the work-up should include thyroid screen and radiographs (or other imaging modality) of the middle ear.

Treatment options over and above resolving any recognised underlying disease:
- Topical emollient therapy designed to rehydrate the nasal tissues may help, although licking may reduce efficacy.

- Pilocarpine administration has been suggested on the basis that it can stimulate secretions from the lateral tarsal gland[39]. Oral pilocarpine has also been recommended as a lacrostimulant in the management of keratoconjunctivitis sicca in the dog[40]. The advised dose is one drop of 2% ophthalmic pilocarpine drops per 10 kg body weight, administered in some food. Owners are then advised to gradually increase the dose, by one drop increments, until signs of pilocarpine toxicity are seen, such as salivation, vomiting and diarrhoea. The dose is then reduced to the previously tolerated amount.

Discoid lupus erythematosus

Discoid lupus erythematosus (DLE), or nasal lupus erythematosus is the most common inflammatory disease of the nasal planum in the dog[3]. Lesions are generally restricted to the nasal planum and the immediately adjacent haired skin of the rostral face (Figs 10.19–10.21). Initially, there is erythema and fine crust, which is followed by loss of pigment, erosion and thicker crust formation. There is loss of the normal 'cobblestone' appearance of the nasal planum[41–43]. More severe cases present with lesions on the face, which may extend caudally as far as the periorbital regions in extreme cases (as in Fig. 10.21). More rarely the lips, prepuce and digits are affected and there is one reported case of generalised disease[44]. Females, generally, and certain breeds

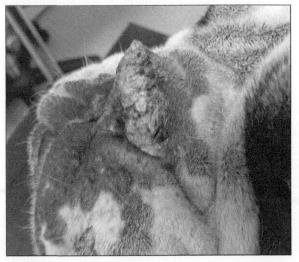
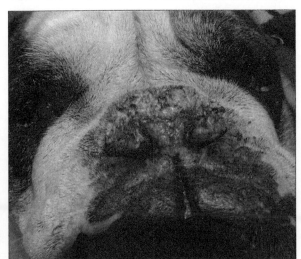

Figs 10.17, 10.18 Xeromycteria, parasympathetic nose. Two cases of nasal crusting. (Courtesy of Dr David C Robson, Animal Skin Ear and Allergy Clinic, Melbourne, Australia.)

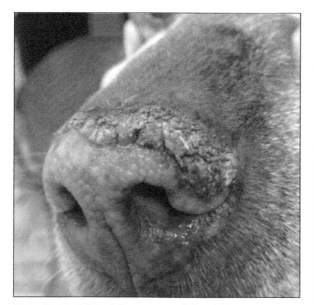

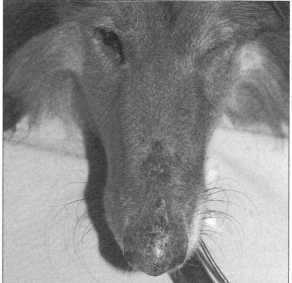

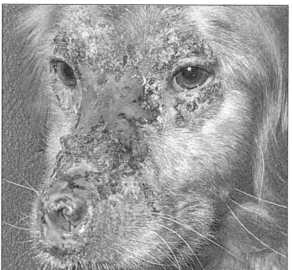

Figs 10.19–10.21 Three cases of discoid lupus erythematosus in the dog, illustrating the spectrum of clinical signs. In 10.19 (top left) lesions are confined to the nasal planum, whereas in 10.20 (top right) the lesions are also affecting the rostral nose. In 10.21 (right) there is extensive facial alopecia, erosion, ulceration and loss of nasal pigmentation.

are predisposed. These include Shetland Sheepdogs, Rough Collies, German Shepherd Dogs and Siberian Huskies[42].

In cats, in which the disease is rare, lesions may affect the nasal planum (Fig. 10.22) the pinnae and the medial canthi of the eyes[45].

Although the condition may be suspected, definitive diagnosis must be confirmed by histopathological examination of biopsy samples[42–44]. The use of a 4 mm biopsy punch, taken with the dog under general anaesthesia, is ideal as the small residual wound is easy to close with a little tissue glue, or even digital pressure – no suture is usually necessary.

The treatment of canine DLE is somewhat problematic because one would like to use a localised, topical, anti-inflammatory medication for what is often a localised, inflammatory condition. However, keeping

Fig. 10.22 Discoid lupus erythematosus in a cat affecting the nasal planum and rostral face.

a topical medication in place in an area that is so easy to lick is difficult. Options therefore are:

- Oral prednisolone, prednisone or methyl prednisolone at low dose, alternate day protocols, i.e. around 0.5 mg/kg every other day[41,42]. This regime is generally very effective and might be appropriate for the induction of remission, particularly in the more severe cases. However, steroid side-effects are likely (Table 10.3) as the condition will require long-term treatment and is subject to actinic radiation-induced exacerbation. Other options are much preferred:
- Topical glucocorticoid preparations, such as hydrocortisone, with, or without, a sunblock cream. The sunblock may help to increase the efficacy of a less potent topical steroid cream, allowing one to avoid a very potent fluorinated preparation. Thinning of the skin, an unwanted side-effect in any case and particularly in this area, is a risk with excessive use of topical steroids, particularly the very powerful preparations.
- Topical tacrolimus (0.1%) twice daily has been reported as efficacious in eight of the ten dogs treated, allowing suspension of all other treatments[46]. Tacrolimus is expensive, but very little is used on a daily basis. After remission is achieved, it may be possible to maintain remission with once daily, or even alternate day, therapy. Owners must be advised to wear gloves when applying this product.
- A combination of oral tetracycline and niacinamide (nicotinamide): this combination was reported to produce a good to an excellent response in 70% of the 20 dogs with discoid lupus in one study[47]. Dogs less than 10 kg in weight received 250 mg of each drug three times daily while dogs over 10 kg in weight received 500 mg of each three times daily. The advantages of this regime are lack of expense, lack of side-effects[48] and good efficacy.
- Hydroxychloroquine at a dose of 5 mg/kg once daily was used successfully to treat a (very rare) case of generalised DLE[44]. No side-effects were noted, in this or another study reporting the use of hydroxychloroquine in dogs[49]. Hydroxychloroquine has a number of effects on the immune system,

Table 10.3 **Steroid-associated side-effects in dogs**
Short term
Polyuria and polydipsia
Panting
Personality changes, 'act like they have a headache'
Medium to long term
Increased weight (around 10%)
Dull coat, thin coat
A fine dry scale
Muscle atrophy
Occult cystitis
Demodicosis
'Steroid' hepatopathy, alterations in glucose and lipid metabolism
Acute complications of high dose (>2.2 mg/kg twice daily)
Gastric ulceration and perforation

such as inhibition of endosomal Toll-like signalling, which downgrades B-cell and dendritic cell activation[50]. The most serious adverse effect in humans is associated with retinopathy. It is thought this might reflect individual idiosyncrasy, although the combination of dose and duration of treatment are also considered major risk factors[51,52]. Notwithstanding, it would be prudent for a veterinary ophthalmologist to make a pre-treatment assessment, and to perform monitoring examinations, at least until we have more experience of this drug in the dog.

- Mix and match of any combination of the above in an attempt to avoid oral glucocorticoids.

Management of DLE in the cat

The treatment of feline DLE is somewhat simplified in that cats appear to tolerate glucocorticoids extremely well[53], and a dose of around 1–2 mg/kg every other day will usually give fair to good control, with minimal side-effects, in the vast majority of cats.

Topical tacrolimus is also a practical possibility as cats appear to tolerate this medication well[54,55]. If the cat will not accept tacrolimus, and either exhibits unacceptable side-effects or the owner objects to 'steroids', then oral cyclosporine (ciclosporin) is a good choice and, again, is well-tolerated by cats[56].

Actinic dermatitis

Dogs with areas of white hair and non-pigmented skin are at risk from actinic (solar) dermatitis, particularly in those areas of the world in which sunlight is intense. If the pale pigmentation is on the rostral nose, where the hair covering is particularly thin, then this area of the body is particularly at risk. The early lesions are erythema associated with a fine scale immediately caudal to the nasal planum, on the haired skin. With time, erosion might occur and the lesions advance rostrally to affect the nasal planum (Fig. 10.23), which loses pigment[57]. Secondary staphylococcal infection is a significant problem in some individuals.

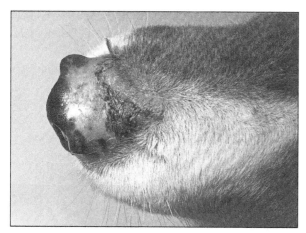

Fig. 10.23 Actinic dermatitis on the rostral face and nose of a Bull Terrier. Note depigmentation of the nasal planum and the ulceration on the adjacent hairy skin.

Patently, the differential diagnosis includes DLE, among others, and definitive diagnosis must be made by histopathological examination of biopsy samples, which exhibit superficial dermal fibrosis and follicular keratosis[57,58].

Management may be difficult if affected animals spend time out of doors. Application of high factor sun block may be helpful, if not immediately licked or rubbed off. Topical indomethacin has been shown to inhibit ultraviolet-induced erythema[59], and if used early in the course of the disease, or indeed early in the season, it might be helpful clinically, although there are, as yet, no published reports to this effect.

Squamous cell carcinoma

There is little doubt that the primary risk for squamous cell carcinoma is cumulative exposure to solar ultraviolet light[60]. In the dog, ulceration, or swelling, of the nasal planum are the most common presenting signs, with nasal bleeding also reported[61,62]. In the cat (Figs 10.24A, B) in contrast, the most common presentation is an erythematous, crusting, erosive lesion[63]. Metastasis is uncommon to rare. Notwithstanding that surgical excision gives the most favourable outcome in both dogs and cats[62,64], obtaining an acceptable cosmetic outcome with radical surgery in this area can be problematic, particularly in cats.

In this light, the report by Bexfield et al.[64] of photodynamic therapy in cats, using 5-aminolaevulinic acid (5-ALA) and a red light source, is most encouraging. Of 55 cases, 53 (96%) responded to therapy, with 47 of the 55 (85%) showing a complete response. Twenty four (51%) of the 47 responders showed recurrence, with

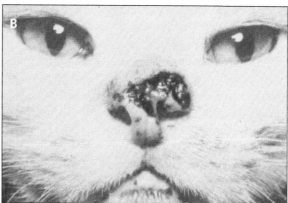

Figs 10.24A, B Examples of nasal squamous cell carcinoma in the cat.

a median representing time of 157 days (Figs 10.25, 10.26). Photodynamic therapy was repeated in 22 of these 24. Twenty three (45%) were disease-free at a median follow-up date of 1,146 days.

Recently, Jarrett et al.[65] described curettage and diathermy in 34 cats with actinic dysplasia and superficial squamous cell carcinoma. The treatment resulted in good cosmetic outcome and a 95% confidence of remaining disease-free at 12 months. Given that this technique utilises equipment found in most small animal practices, it might represent treatment of choice.

Pemphigus diseases

Pemphigus foliaceus (PF) and pemphigus erythematosus (PE) are both superficial pustular dermatoses[66]. The pustules in PF are a result of a loss of cohesion between keratinocytes due to immunoglobulin (Ig) G targeting the protein desmocollin-1, a component of the desmosomes[67]. The detailed pathology of PE has not yet been established, but it is believed to be a disease combining clinical and microscopic lesions of both facial-predominant PF and those of DLE (Olivry and Bizikova, unpublished).

In the case of PF, the pustules are the primary lesion, but they may be very transient and most commonly cases are presented with crusting, erosion and alopecia on the nasal planum, face and pinnae (Figs 10.27–10.29). Mueller et al. reviewed 91 cases of PF in the dog and found pustules in 36 cases, crusts in 79 cases and footpad lesions in 32 cases[68]. Facial lesions were present in 46 cases and were found on the concave aspects of the pinnae in 46 cases. Facial lesions often extend to include the nasal planum and may include crust, erosion and loss of pigment.

In PE, lesions are similar to those of PF, but the ulceration is deeper and the lesions scar, similar to those of DLE (see Fig. 10.31).

Diagnosis, in the large majority of cases, is based on clinical presentation and suspicion based on the cytological examination of pustules. Definitive diagnosis must be based on the histopathological examination of biopsy samples as this will allow the rule-out of differentials as well as allow classification of the depth of the lesions.

Treatment of PF and PE must include immunosuppression.

- Deep nasal lesions can be treated with topical applications of 0.1% and 0.03% tacrolimus ointment, which is reported to give fair to good control with minimal side-effects[69,70].

Figs 10.25, 10.26 Treatment of squamous cell carcinoma in the cat with photodynamic therapy; before (10.25, left) and after (10.26, right) treatment. Note the excellent cosmetic outcome. (Courtesy of Nick Bexfield, University of Cambridge School of Veterinary Science, Cambridge, UK.)

- Prednisolone is the mainstay of control in those cases in which lesions are generalised, or where topical therapy with tacrolimus has failed to be effective[71]. Induction of remission is the first goal and doses of between 2 and 6.6 mg/kg divided twice a day have been variously recommended[66,68,71]. However, 2 mg/kg divided twice a day is a good starting dose, unless the dog is failing to respond, when the higher doses must be used.

- Side-effects are inevitable: polyuria/polydipsia, polyphagia, panting and behavioural changes are common (Table 10.3). Concurrent systemic antibacterial treatment during immunosuppressive induction has been shown to correlate positively with survival. Other factors positively correlated with long-term survival are a low number of steroid-associated adverse effects, and survival for 10 months or more after the start of treatment[72].

- When remission has been achieved, i.e. there are no new lesions (residual crusting is acceptable), a gradual reduction to once daily treatment is

the first target, perhaps over the first month, or 6 weeks. If there is no evidence of relapse, then, again, a gradual reduction to 1 mg/kg per day, and eventually after another 4 weeks, to 1 mg/kg

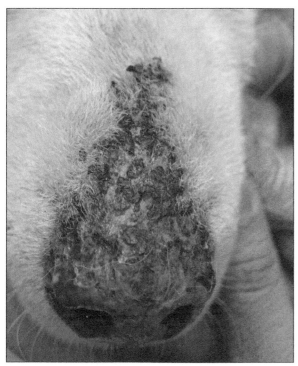

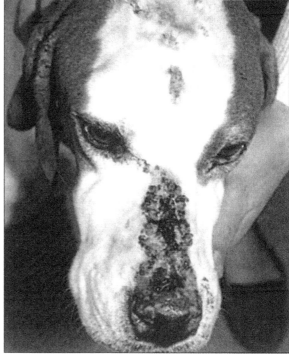

Figs 10.27–10.29 Three examples of pemphigus foliaceus in the dog illustrating the degree of clinical variation that may encountered.

every other day. Even long term, low (ish) doses of prednisolone can result in side-effects. Over and above those described for induction doses, weight gain (perhaps around 10%), reduced rate of hair growth and a dulling of the coat, a fine dry scale, thinning, even atrophy of the skin, muscle weakness, occult cystitis and, perhaps, demodicosis and pancreatitis, can be expected[72].

- Strategies to assist in a gradual reduction in prednisolone to an alternate day regime include:
  - Tetracycline and niacinamide, as described for DLE, above. Some authorities have reported occasional cases that respond to this regime alone, but they are uncommon. However, they are not expensive, have no side-effects and might swing the balance in favour of response.

- Initiate cyclosporine therapy at 5–10 mg/kg once daily. Assuming that the dog does not exhibit unacceptable gastrointestinal side-effects[73], cyclosporine has the potential to reduce or even obviate prednisolone requirement. Anticipate around 4 weeks for cyclosporine to exert its effect. It is expensive, but less toxic then azathioprine.
- Azathioprine at 2–2.5 mg/kg once daily, usually with glucocorticoid therapy, is initiated in anticipation that, over 4–6 weeks, it will facilitate reduction of steroid dose. Azathioprine can result in myelosuppression and one author advises beginning therapy at the lower end of the dose range until the results of blood counts (every two weeks during induction) indicate that it is safe and prudent to increase the dose[69]. However, note that a pilot study investigating the use of azathioprine in atopic dermatitis found that all 12 dogs in their study showed elevations of serum alkaline phosphatase and alanine aminotransferase, and in three of the dogs there was clinical evidence of liver disease[74]. It would be pertinent, therefore, to monitor liver enzymes in addition to haematological parameters.
- Topical tacrolimus or potent steroid creams may be used to 'spot' treat the most severe

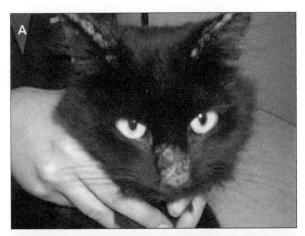

Figs 10.30A, B Crusted lesions of pemphigus foliaceus in the cat. The cat in (A) has nasal and pinnal lesions whereas the cat in (B) has almost generalised facial lesions.

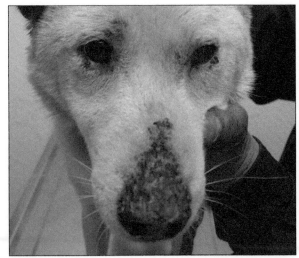

Fig. 10.31 Lesion of pemphigus erythematosus on the nasal planum, face, periorbital regions. (Courtesy of Dr DH Bhang, College of Veterinary Medicine, Seoul National University, Seoul, South Korea.)

lesions in an attempt to obviate increasing the dose. Note, whilst topical tacrolimus has been shown to be useful in cases of PE[70], there is no evidence that it has any benefit in PF.

- Failure to achieve remission and a reduction to alternate day dosing should prompt case review with consideration of referral, or even euthanasia.

Pemphigus foliaceus in cats

In cats, pustules are extremely rare[75]. The most common presenting sign in the cat is symmetrical facial and pinnal crusting (Figs 10.30A, B), although ulcerative lesions on the nasal planum may be seen.

The management of PF in the cat is somewhat simpler than in the dog as prednisolone as a monotherapy (2 mg/kg a day, reducing to 2 mg/kg alternate day in some cases) gave good control[76], with minimal side-effects, as did cyclosporine at around 4–7 mg/kg once daily[77].

Idiopathic nasal arteritis

This disease is pathognomonic: fissure, crust, ulceration of the nasal philtrum, with crusting and occasional fissures extending into the nasal planum (Figs 10.32, 10.33). Most cases (three of four) exhibited episodic, mild, arterial bleeding, although in one case the blood loss was very severe[78]. The case series described four dogs, three St Bernards and one Giant Schnauzer. The condition affected adult dogs and in some animals had been present for several years.

Histopathologically there is proliferative arteritis affecting small and medium sized arterioles in the nasal philtrum. The pathogenesis of the arteritis, the localisation to the nasal philtrum and the chronicity are not understood. However, anti-inflammatory medication appears to be helpful[78]: topical flucinolone in DMSO (one case), alternate-day prednisolone (one case), tetracycline/niacinamide (and fish oil), (one case).

Vitiligo

Vitiligo results from loss of melanin pigmentation in the skin, thought to be a result of selective destruction of melanocytes[79] or at least a proportion of them, in the affected areas. The melanocyte destruction results from immune-mediated cell death[79,80]. The role of zinc-α2-glycoprotein has been highlighted[81] although the pathogenesis of the disease is still unclear.

There is a breed prevalence, for German Shepherd Dog, Newfoundland, German Shorthair Pointer, Rottweiler, Doberman Pinscher, Old English Sheepdog and Dachshund[10,80]. Siamese cats have also been reported to be predisposed, although the condition is rare[82,83].

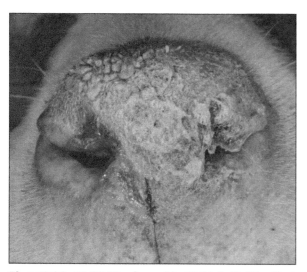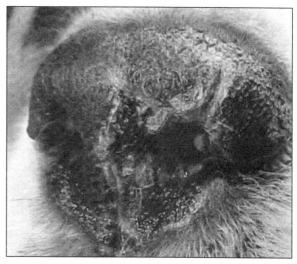

Figs 10.32, 10.33 Nasal arteritis. 10.32 (left): Nasal lesions in a Labrador Retriever (courtesy of Dr David Duclos, Animal Skin & Allergy Clinic, Lynnwood, WA, USA); 10.33 (right): nasal arteritis in a St Bernard. Note the crusting, fissure formation and ulceration.

Non-inflammatory, asymptomatic, symmetrical depigmentation is most commonly noted on the planum nasale, lips, muzzle and buccal mucosa (Fig. 10.34). There may be focal or widespread leucotrichia and/or depigmentation of the nails and foot pads. Initial lesions are usually small, achromic macules[80] that gradually spread and merge (Fig. 10.35). Young dogs appear to be more commonly affected[10]. The natural course of the disease may wax and wane and some repigmentation may occur[80].

There is no recognised treatment.

Cryptococcal, sporotrichial and idiopathic granuloma

These granulomatous diseases present as non-painful nodules, although occasional they may be mildly pruritic, especially if they ulcerate. Fungal nodules on the nasal planum are usually solitary. These nodules are impossible to differentiate on clinical grounds and either biopsy, or excisional biopsy, are necessary. Many of the fungal granulomas occur as a consequence of traumatic inoculation by vegetation, and being an indoor/outdoor cat is a significant risk factor[84].

Cryptococcal granulomas are caused by *Cryptococcus neoformans-gattii* complex, members of a group of closely related fungal organisms that frequently hybridise[85]. There is no age or breed predilection in cats and FeLV/FIV status is not a risk factor, although it may complicate treatment. Trivedi *et al.*[85] reported a case series of 93 dogs and cats and found Cocker Spaniels predisposed. Nasal deformity was noted in 16% of cats (Fig. 10.36) but was not seen in dogs. Diagnosis may be confirmed by histopathological examination of biopsy samples, although it is not possible to define the species without culture[84]. Trivedi *et al.*[85] reported that diagnosis by cryptococcal antigen assay alone was not sufficient to make a diagnosis. Sensitivity testing is recommended, because although ketoconazole, itraconazole, fluconazole and amphotericin B may be effective, treatment may well have to be continued for months. It is therefore appropriate to establish which agent is most likely to be effective. Surgical debulking is recommended, as an adjunct to medical management[85].

Sporotrichosis results from traumatic inoculation of the organism *Sporothrix schenkii*, or from encountering the organism in the draining tracts of infected individuals. In a recent study of 15 affected cats there was a 20% zoonotic infection, raising the question of whether these animals should even be treated. Dermatological presentation, with solitary or multiple nodules (Fig. 10.37), was seen in 10 cats[86]. Fifty percent of the cases responded well to itraconazole, 10 mg/kg, once daily. Recently, the use of crysosurgery, in association with itraconazole (at 10 mg/kg once daily) was report-

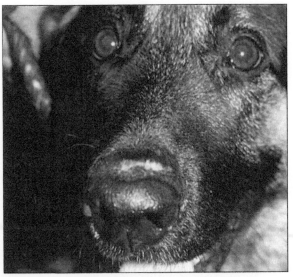

Figs 10.34, 10.35 Vitiligo in a dog (10.34) and a Siamese cat (10.35). (10.35 courtesy of Dr Zeineb Alhaïdari, Clinique Vétérinaire, Roquefort les Pins, France.)

Figs 10.36, 10.37 Solitary fungal granulomas on the face. Both are sufficiently large to cause facial deformity. Figure 10.36 illustrates *Cryptococcus* infection and 10.37 sporotrichosis. It is impossible to differentiate them clinically; biopsy is necessary, and samples should be submitted for culture.

ed[86a]. Three freeze-thaw cycles were performed under general anaesthesia and itraconazole was continued for 4 weeks after lesions has healed. Eleven of the 13 cats treated responded well.

Idiopathic granuloma of the nasal planum of dogs is very uncommon. It usually presents as a solitary nodule

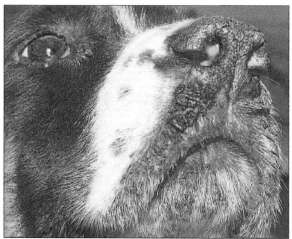

Fig. 10.38 Facial lesions of epidermal metabolic necrosis, demonstrating facial as well as nasal lesions. There is erythema, ulceration and crusting.

on the nasal planum or head[87]. Whether solitary nasal granulomas represent a subset of the idiopathic periadnexal multinodular granulomatous dermatitis is not known. Too few cases have been reported to enable specific therapeutic recommendations but treatment options include topical tacrolimus, systemic cyclosporine or anti-inflammatory doses of glucortiocoids, with or without azathioprine, would be indicated.

Metabolic epidermal necrosis

Metabolic epidermal necrosis (hepatocutaneous syndrome), necrolytic migratory erythema is a rare disease in dogs[88] and is extremely rare in the cat[89]. It is clear that the dermatological lesions reflect internal disease, but the skin lesions can occur either with a demonstrable glucagon-secreting tumour or with hepatic disease without any detectable glucagonoma[90]. All seven cases cited in a case review of dogs with glucagonoma, plus the case cited (i.e. eight cases in total) had skin lesions[90], and there is one case report of the dermatological lesions resolving after successful removal of a glucagon-producing tumour[91].

The clinical signs in dogs include symmetrical erythema, ulceration and crusting on the face and nose (Fig. 10.38), the external genitalia and the feet. There is also marked hyperkeratosis of the foot pads, particularly peripherally. There is no breed, age or sex predisposition, although the disease affects mainly older dogs.

Pruritus can be marked and is often steroid-refractory (Table 10.4). The clinical signs in the single cat in which the disease has been reported were erythema, alopecia and scale in the axillae, extending to the feet. The foot-pads and face were not affected.

Treatment in most cases has been palliative care, diet adjustment (high protein commercial diets, supplemented with eggs, zinc and essential fatty acids [EFAs]), but it is palliative at best and most affected animals are euthanased within a short time of diagnosis. Successful removal of a glucagon-producing tumour can result in cure[91].

Familial canine dermatomyositis

Familial dermatomyositis is an inflammatory disease of the skin and muscles[92]. Certain breeds are predisposed, particularly the Shetland Sheepdog, Rough Collie and Beauceron[93,94,94a]. Although it has been reported occasionally in individuals of other breeds or familial clusters[94a], it is not known if the condition is hereditary in these[94a,95].

The condition occurs at any age, and has been reported, in some cases before 12 weeks of age[92]. There is no sex predisposition. The clinical signs are very variable, as is the course of the disease. In a few individuals the number of lesions is low, they resolve quickly and patchy alopecia is the only consequence. In others, the majority, it is a very severe problem with erosions, crusts, ulcerations and scarring.

The inner surfaces of the pinnae, the face (nose, planum nasale and lips) (Figs 10.39, 10.40) tip of the tail and the distal limbs (including, and distal to, the elbow and stifle) are affected early on in the course of the disease with erythema, alopecia scale, crust and ulceration. Occasionally there may be vesicles, papules and pustules. In most individuals the acute phase slowly resolves over several weeks but there may be patchy scarring and alopecia with patchy depigmentation of the face, lips and nasal planum[92,94]. The myositis usually becomes evident several weeks after the onset of dermatological signs, in some cases after the dermatological lesions have begun to regress. The severity of the myositis correlates with the severity of the dermatological lesions[95], and particularly affects the temporal and limb muscles.

Diagnosis is made on the basis of clinical signs and breed, history, blood work-up and the histopathological examination of biopsy samples. Biopsy of muscle

Table 10.4 **Steroid-refractory dermatoses**
Scabies
Malassezial dermatitis
Calcinosis cutis
Hepatocutaneous syndrome
Syringomelia
Epitheliotropic lymphoma
Some cases of allergic contact dermatitis and food allergy/intolerance?
Psychogenic dermatitis

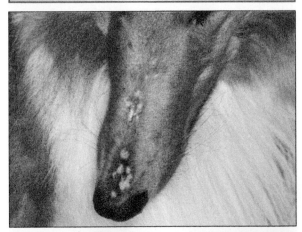

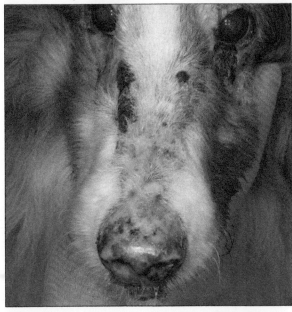

Figs 10.39, 10.40 Typical facial lesions of familial canine dermatomyositis. (Courtesy of Dr David Duclos, Animal Skin & Allergy Clinic, Lynnwood, WA, USA.)

samples is not usually necessary if the definitive diagnosis has already been made.

Initial treatment is aimed at assessing the severity and probable time course as mildly affected animals may need no treatment[95]. Severely affected dogs may require euthanasia. Lesions are aggravated by exposure to strong sunlight and local trauma[95].

Dogs that are judged in need of treatment do show a fair response to oral glucocorticoids but the doses required are high (immunosuppressive doses of 1–2 mg/kg daily). Pentoxifylline (25 mg/kg twice daily) has shown to be of benefit in some of these animals[95,96]. Four of ten dogs in the study[96] showed a complete clinical recovery while the other six showed some response. Most dogs fail to respond completely and many are maintained on pentoxifylline, perhaps with the addition of omega 3 fatty acids and intermittent glucocorticoids.

Uveodermatological syndrome

Uveodermatological syndrome (Vogt–Koyanagi–Harada-like) is a rare autoimmune condition. Numerous studies in the human field support a T-cell-mediated response against melanocytes in genetically predisposed individuals[97]. Certain breeds of dogs are predisposed and these include the Akita, Samoyed, Siberian Husky, Alaskan Malamute and Chow Chow[98]. The condition has also been reported in terriers[99,100]. In the Akita, the increased frequency of uveodermatological syndrome has been shown to be associated with an increased frequency of DQA1*00210, supporting the contention that DLA Class II gene alleles play a part in the pathogenesis[101].

The initial signs are of severe uveitis and panuveitis. Dermatological lesions follow[98,99] and consist of symmetrical depigmentation, ulceration and crusting affecting the nasal planum, lips and the periorbital regions (Fig. 10.41). The external genitalia and footpads may also be affected.

Aggressive treatment with cyclosporine and glucocorticoids, perhaps azathioprine, are required to suppress the ocular and dermatological lesions, in addition to antiglaucoma medications and topical ophthalmic glucocorticoids (or tacrolimus).

Cutaneous reactive and systemic histiocytosis

Cutaneous reactive and systemic histiocytosis are uncommon diseases of the dog[102,103] that are char-

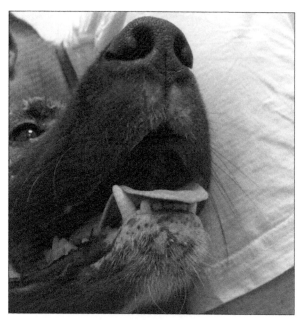

Fig. 10.41 Canine uveodermatological syndrome affecting the face and planum nasale.

acterised by nodular and plaque-like dermatological lesions, accompanied by neoplastic infiltration of the internal organs in the systemic variant. The aetiology is not known.

Cutaneous reactive histiocytosis presents as multiple, soft, non-painful, discrete domed nodules or plaques on the head and trunk[102–104]. The nasal planum may be depigmented, erythematous and swollen[102,103]. There is no breed predilection[103]. Combination treatment with tetracycline and niacinamide (250 mg tid of each for dogs less than 20 kg in weight and 500 mg of each tid for larger dogs) was an effective treatment in several cases[102–104]. Rapid response to systemic prednisolone might be anticipated. Initiating treatment with both prednisolone and the tetracycline/niacinamide regime would allow for gradual withdrawal of the prednisolone as remission is attained, allowing remission to be maintained with the non-steroidal regime.

Systemic histiocytosis is a devastating disease. Breed predilections are found in Bernese Mountain Dogs (Fig. 10.42), Rottweilers and Golden Retrievers[105]. Animals exhibit lethargy, anorexia and weight loss and have malignant infiltration into every organ, but particularly the spleen, lungs and bone marrow[105]. The prognosis is very poor: most dogs are euthanased.

Fig. 10.42 Facial and nasal lesions of systemic histiocytosis in a Bernese Mountain dog. Note the crusted nodules on the skin and the ulceration and crusting around the nares.

Carcinoma *in situ*

Carcinoma *in situ*, Bowen's-like disease, is characterised by multiple, or occasionally solitary, crusting or ulcerated plaques[106,107]. The face (Fig. 10.43), neck and limbs appear to be predisposed[106]. Unlike squamous cell carcinoma, which typically appears in sparsely haired, lightly pigmented, sunlight-exposed sites, these lesions typically appear on haired and pigmented skin[106]. The presence of papillomavirus in many cases of feline viral plaque and in bowenoid carcinoma *in situ*, has led to the proposal that carcinoma *in situ* might evolve from plaque[106,107]. A proportion of carcinoma *in situ* may evolve into true squamous cell carcinoma, but the pathogenesis and likelihood of this happening is unknown[106].

Treatment options include observation[107], photodynamic therapy[108], topical imiquimod cream[109] or strontium Sr90 radiotherapy[110].

Fig. 10.43 Carcinoma *in situ* on the nasal philtrum in a domestic short haired cat. Note the suture at the site of the 4 mm biopsy sample.

Mycosis fungoides

Mycosis fungoides (epitheliotropic cutaneous lymphoma) is cutaneous T-cell lymphoma[111]. This is a very pleomorphic dermatitis, usually associated with truncal lesions of moderate, steroid-refractory pruritus, scale and patchy alopecia or erythroderma. Ulceration and crusting of lesions follows, although there are other manifestations[111–113].

There is one report in the literature of lesions confined to the lips, face and foot pads, planum, and eventually the entire rhinarium (Figs 10.44–10.46)[114]. There was a loss of the normal surface cobblestone texture, a loss of pigment and there were a few patches of fine scale. The lesions crossed onto adjacent haired skin. Histopathological examination of biopsy samples is diagnostic.

There is no effective treatment.

Ulcerative nasal dermatitis of Bengal cats

Bergvall described a non-contagious nasal dermatitis in a series of 48 cats seen over a 4 year period[115]. The clinical signs were confined to the planum nasale and were characterised by fissures, crust, erosion and ulceration (Fig. 10.47). All were young Bengal cats, suggesting an heritable disease. However, the clinical signs responded to anti-inflammatory treatment (systemic prednisolone or topical tacrolimus) suggesting an immune-mediated disease[115]. What relationship, if any, this disease has to the nasal hyperkeratosis in Bengal cats, described by Lee Gross *et al.*[116], is not known.

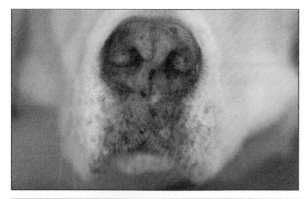

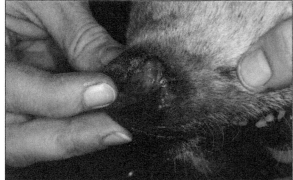

Figs 10.44–10.46 Three cases of epitheliotropic cutaneous lymphoma (mycosis fungoides) affecting the nasal planum. (Courtesy of Dr David Duclos, Animal Skin & Allergy Clinic, Lynnwood, WA, USA.)

10.4 REFERENCES

1 Scott DW, Miller WH, Griffin CE. Structure and function of the skin. In: *Muller & Kirk's Small Animal Dermatology*, 6th edn. Scott DW, Miller WH, Griffin CE (eds). WB Saunders, Philadelphia 2001, pp. 9–24.

2 Bettany S, Mueller R. Histologic features of the normal canine nose. *Proceedings Annual Meeting of the Australian College of Veterinary Scientists, Dermatology Chapter*, Gold Coast, 2003.

3 White SD. Diseases of the nasal planum. *Veterinary Clinics of North America* 1994;**24**:887–95.

4 Bloom PB. Canine and feline eosinophilic skin diseases. *Veterinary Clinics of North America Small Animal Practice* 2006;**36**:141–60.

5 Curtis CF, Bond R, Blunden AS, *et al*. Canine eosinophilic folliculitis and furunculosis in three cases. *Journal of Small Animal Practice* 1995;**36**:119–23.

6 Mauldin EA, Palmeiro BS, Goldschmidt MH, Morris DO. Comparison of clinical history and dermatologic findings in 29 dogs with severe eosinophilic dermatitis: a retrospective analysis. *Veterinary Dermatology* 2006;**17**:338–47.

7 Bond R. Superficial veterinary mycoses. *Clinics in Dermatology* 2010;**28**:226–36.

8 Fairley RA. The histological lesions of *Trichophyton mentagrophytes* var *erinacei* infection in dogs. *Veterinary Dermatology* 2001;**12**:119–22.

9 Scott DW, Miller WH, Griffin CE. Fungal skin diseases. In: *Muller & Kirk's Small Animal Dermatology*, 6th edn. Scott DW, Miller WH, Griffin CE (eds). WB Saunders, Philadelphia 2001, pp. 336–422

10 Carlotti DN, Bensignor E. Dermatophytosis due to *Microsporum gypseum* (20 cases) and *Microsporum persicolor* (13 cases) in dogs: a retrospective study (1988–1996). *Veterinary Dermatology* 1999;**10**:17–27.

Fig. 10.47 Ulcerative nasal dermatitis showing ulceration and crusting, confined to the nasal planum in a young Bengal cat. (Courtesy of Dr Kristen Bergvall, Department of Clinical Sciences, Swedish University of Agricultural Sciences, Uppsala, Sweden.)

11 Muller A, Guaguère E, Degorce-Rubiates F, Bourdoiseau G. Dermatophytosis due to *Microsporum persicolor*: a retrospective study of 16 cases. *Canadian Veterinary Journal* 2011;**52**:385–8.

12 Moriello KA. Treatment of dermatophytosis in dogs and cats: review of published studies. *Veterinary Dermatology* 2004;**15**:99–107.

13 Gaskell RM, Dawson S. Feline respiratory disease. In: *Infectious Diseases of the Dog and Cat*, 2nd edn. Griffin CE (ed). WB Saunders, Philadelphia 1998, pp. 97–106.

14 DeClerke J. Pustular calicivirus dermatitis on the abdomen of two cats following routine ovariohysterectomy. *Veterinary Dermatology* 2005;**16**:395–400.

15 Sánchez MD, Goldschmidt MH, Maudlin EA. Herpesvirus dermatitis in two cats without facial lesions. *Veterinary Dermatology* 2011;**23**:171–3.

16 Persico P, Roccabianca P, Corona A, Vercelli A, Cornegliani L. Detection of feline herpes virus 1 via polymerase chain reaction and immunohistochemistry in cats with ulcerative facial dermatitis, eosinophilicgranulom complex reaction patterns and mosquito bite hypersensitivity. *Veterinary Dermatology* 2011;**22**:521–7.

17 Bosward LM, Norris JM. Immunohistochemical evaluation of feline herpesevirus-1 infection in feline eosinophilic dermatoses or stomatitis. *Journal of Feline Medicine and Surgery* 2010;**12**:72–9.

18 Ricklin-Gutzwiller M, Brachelente C, Taglinger K, Suter MM, Weissenböck H, Roosje PJ. Feline herpes dermatitis treatment with interferon omega. *Veterinary Dermatology* 2007;**18**:50–4.

18a Malik R, Lessels NS, Webb S, *et al*. Treatment of feline herpesvirus-1 associated disease in cats with famciclovir and related drugs. *Journal of Feline Medicine and Surgery* 2009;**11**:40–8.

19 Mason KV, Evans AG. Mosquito bite-caused eosinophilic dermatitis in cats. *Journal of the American Veterinary Medical Association* 1991;**198**:2086–8.

20 Ihrke PJ, Gross TL. Mosquito-bite hypersensitivity in a cat: clinicopathological conference. *Veterinary Dermatology* 1994;**5**:33–6.

21 Fossum TW. Dermoid sinus (pilonidal sinus). In: *Small Animal Surgery*. Fossum TW (ed). Mosby, St Louis 1997, p. 139.

22 Lambrechts N. Dermoid sinus in a crossbred Rhodesian Ridgeback dog involving the second cervical vertebra. *Journal of the South African Veterinary Association* 1996;**67**:155–7.

23 Anderson DM, White RA. Dermoid sinus cysts in the dog. *Veterinary Surgery* 2002;**31**:303–8.

24 Burrow RD. A nasal dermoid sinus in an English Bull terrier. *Journal of Small Animal Practice* 2004;**45**:572–4.

25 Ihrke PJ, Gross TL. Canine mucocutaneous pyoderma. In: *Kirk's Current Veterinary Therapy XI*. Bonague JD (ed). WB Saunders, Philadelphia 1992, pp. 618–19.

26 Sulaimon SS, Kitchell BE. The biology of melanocytes. *Veterinary Dermatology* 2003;**14**:57–65.

27 Lotti TM, Berti SF, Hercogova J, *et al*. Vitiligo: recent insights and new therapeutic approaches. *Giornale Italiano de Dermatologia et Venereologia* 2012;**147**:637–47.

28 Guagure E, Alhaidari Z. Disorders of melanin pigmentation in the skin of dogs and cats. *Kirk's Current Veterinary Therapy X*. Kirk RW (ed). WB Saunders, Philadelphia 1989, pp. 628–32.

29 Scott DW, Miller WH, Griffin CE. Congenital and hereditary defects. In: *Muller & Kirk's Small Animal Dermatology*, 6th edn. Scott DW, Miller WH, Griffin CE (eds). WB Saunders, Philadelphia 2001, pp. 936.

30 Ihrke PJ. Topical therapy – uses, principles and vehicles in dermatological therapy (Part 1). *Compendium on Continuing Education* 1980;**11**:28–35.

31 Kwochka KW. Primary keratinization disorders in dogs. *Current Veterinary Dermatology*. Griffin CE, Kwochka KW, McDonald JM (eds). Mosby Year Book, St Louis 1993, pp. 176–90.

32 Paradis M. Footpad hyperkeratosis in a family of Dogues de Bordeaux. *Veterinary Dermatology* 1992;**3**:75–8.

33 Binder H, Arnold S, Schelling C, Suter M, Wild P. Palmoplantar hyperkeratosis in Irish terriers: evidence of autosomal recessive inheritance. *Journal of Small Animal Practice*, 2000;**41**:52–5.

34 Peters J, Scott DW, Erb HN, Miller WH. Hereditary parakeratosis in Labrador retrievers: 11 new cases and a retrospective study on the presence of accumulations of serum ('serum lakes') in the epidermis of parakeratotic dermatoses and inflamed nasal planum of dogs. *Veterinary Dermatology* 2003;**14**:197–203.

35 Page N, Paradis M, LaPointe, Dunstan RW. Hereditary nasal parakeratosis in Labrador Retrievers. *Veterinary Dermatology* 2003;**14**:103–10.

36 Jagannathan V, Bannoehr J, Plattet P, *et al*. A mutation in the SUV39H gene in Labrador retrievers with hereditary nasal parakeratosis (HNPK) provides insights into the epigenetics of keratinocyte differentiation. *PLOS Genetics* 2013;**9**:1–9.

37 Blatt CM, Taylor CR, Habal MB. Thermal panting in dogs: the lateral nasal gland, a source of water for evaporative cooling. *Science* 1972;**177**:804–5.

38 Lacroix JS, Potter EK, McLachlan E. Nitric oxide and parasympathetic vascular and secretory control of the dog nasal mucosa. *Acta Otolaryngology* 1998;**118**:257–63.

39 Wells U, Widdicombe JG. Lateral nasal gland secretion in the anaesthetized dog. *Journal of Physiology* 1986;**374**:359–74.

40 Mathias FL, Walser-Reinhardt L, Speiss RM. Canine neurogenic keratoconjunctivitis sicca: 11 cases (2006–2010). *Veterinary Ophthalmology* 2012;**15**:288–90.

41 Walton DK, Scott DW, Smith CS, Lewis RM. Canine discoid lupus erythematosus. *Journal of the American Animal Hospital Association* 1982;**17**:851–8.

42 Scott DW, Walton DK, Mannin TK, Smith CA, Lewis RM. Canine discoid lupus erythematosus. *Journal of the American Animal Hospital Association* 1983;**19**:481–6.

43 Wiemelt SP, Goldschmidt JS, Greek JS, Jeffers AP, Wiemelt AP, Maudlin EA. A retrospective study comparing the histopathological features and response to treatment in two canine nasal dermatoses. *Veterinary Dermatology* 2004;**15**:341–8.

44 Oberkirchner U, Linder KE, Olivery T. Successful treatment of a novel generalized variant of canine discoid lupus erythematosus with oral hydroxyquinoline. *Veterinary Dermatology* 2012;**23**:65–70.

45 Willemse T, Koeman JP. Discoid lupus erythematosus in cats. *Veterinary Dermatology* 1989;**1**:19–24.

46 Griffies JD, Mendelsohn CL, Rosenkrantz WS, Muse R, Boord MJ, Griffin CE. Topical 0.1% tacrolimus for the treatment of discoid lupus erythematosus and pemphigus erythematosus in dogs. *Journal of the American Animal Hospital Association* 2004;**40**:29–41.

47 White SD, Rosychuck RA, Reinke SI, Paradis M. Use of tetracycline and niacinamide for treatment of autoimmune skin disease in 31 dogs. *Journal of the American Veterinary Medical Association* 1992;**200**:1497–500.

48 Mueller R, Fieseler KV, Bettany SV, Rosychuck RA. Influence of long-term treatment with tetracycline and niacinamide on antibody production in dogs with discoid lupus erythematosus. *American Journal of Veterinary Research* 2002;**63**:491–4.

49 Mauldin EA, Morris DO, Brown DC, Casal ML. Exfoliative cutaneous lupus erythematosus in German shorthaired pointer dogs: disease development, progression and evaluation of three immunomodulatory drugs (ciclosporin, hydroxychloroquine, and adalimumab) in a controlled environment. *Veterinary Dermatology* 2010;**21**:373–82.

50 Kalia S, Dutz JP. New concepts in antimalarial use and mode of action in dermatology. *Dermatologic Therapy* 2007;**20**:160–7.

51 Tripp JM, Maibach HI. Hydroxychloroquine-induced retinopathy: a dermatologic perspective. *American Journal of Dermatology* 2006;**7**:171–5.

52 Yam JC1, Kwok AK. Ocular toxicity of hydroxychloroquine. *Hong Kong Medical Journal* 2006;**12**: 294–304.

53 Lowe AD, Campbell KL, Barger A, Schaeffer DJ, Borst L. Clinical, clinicopathological and histological changes observed in 14 cats treated with glucocorticoids. *Veterinary Record* 2008;**162**:777–83.

54 Chung TH, Ryu MH Mim, DY, Yoon HY, Hwang CY. Topical tacrolimus (FK506) for the treatment of idiopathic facial dermatitis. *Australian Veterinary Journal* 2009;**87**:417–20.

55 Maudlin EA, Ness TA, Goldschmidt MH. Proliferative and necrotizing otitis externa in four kittens. *Veterinary Dermatology* 2007;**18**:370–7.

56 Wisselink MA, Willemse T. The efficacy of cyclosporine A in cats with presumed atopic dermatitis: a double blind, randomised prednisolone-controlled study. *Veterinary Journal* 2009;**180**:55–9.

57 Frank LA, Calderwood-Mays MB. Solar dermatitis in dogs. *Compendium of Continuing Education* 1994;**16**:465–72.

58 Frank LA, Calderwood-Mays MB, Kunkle GA. Distribution and appearance of elastin fibers in the dermis of clinically normal dogs and dogs with solar dermatitis and other dermatoses. *American Journal of Veterinary Research* 1996;**57**:178–81.

59 Kimura T, Doi K. Effects of indomethacin on sunburn and suntan reactions in hairless descendents of Mexican hairless dogs. *Histology and Histopathology* 1998;**13**:29–36.

60 Ratushny V, Gober MD, Hick R, Seykora JT. From keratinocyte to cancer: the pathogenesis and modelling of cutaneous squamous cell carcinoma. *Journal of Clinical Investigation* 2012;**122**:464–72.

61 Rogers KS, Helman RG, Walker MA. Squamous cell carcinoma of the canine nasal planum: eight cases (1988–1994). *Journal of the American Animal Hospital Association* 1995;**31**:373–8.

62 Lascelles BD, Parry AT, Stidworthy MF, Dobson JM, White RA. Squamous cell carcinoma of the nasal planum in 17 dogs. *Veterinary Record* 2000;**147**:473–6.

63 Lana SE, Ogilvie GK, Withrow SJ, Straw RC, Rogers KS. Feline cutaneous squamous cell carcinoma of the nasal planum and ears: 61 cases. *Journal of the American Animal Hospital Association* 1997;**33**:329–32.

64 Bexfield NH, Stell AJ, Gear RN, Dobson JM. Photodynamic therapy of superficial nasal planum squamous cell carcinomas in cats: 55 cases. *Journal of Veterinary Internal Medicine* 2008;**22**:1385–9.

65 Jarret RH, Norman EJ, Gibson IR, Jarrett P. Curettage and diathermy: a treatment for feline nasal planum actinic dysplasia and superficial squamous cell carcinoma. *Journal of Small Animal Practice* 2013;**54**:92–8.

66 Olivry T. A review of autoimmune skin diseases in domestic animals: 1 – superficial pemphigus. *Veterinary Dermatology* 2006;**17**:291–305.

67 Bizikova P, Dean GA, Hashimoto T, Olivry T. Cloning and establishment of canine desmocollin-1 as a major autoantigenin canine pemphigus foliaceus. *Veterinary Immunology and Immunopathology* 2012;**15**:197–207.

68 Mueller RS, Krebs I, Power HT, Fieseler KV. Pemphigus foliaceus in 91 dogs. *Journal of the American Animal Hospital Association* 2006;**42**:189–96.

69 Griffies JD, Mendelsohn CL, Rosenkrantz WS, Muse R, Boord MJ, Griffin CE. Topical 0.1% tacrolimis for the treatment of discoid lupus erythematosus and pemphigus foliaceus in dogs. *Journal of the American Animal Hospital Association* 2004;**40**:29–41.

70 Bhang D-H, Choi U-S, Jung Y-C, *et al*. Topical 0.03% tacrolimus for treatment of pemphigus erythematosus in a Korean Jindo dog. *Journal of Veterinary Medical Science* 2008;**70**:415–17.

71 Rosenkrantz WS. Pemphigus: current therapy. *Veterinary Dermatology* 2004;**15**:90–8.

72 Gomez SM, Morris DO, Rosenbaum MR, Goldschmidt MH. Outcome and complications associated with treatment of pemphigus foliaceus in dogs: 43 cases (1994–2000). *Journal of the American Veterinary Medical Association* 2004;**224**:1312–16.

73 Forsythe P, Paterson S. Ciclosporin 10 years on: indications and efficacy. *Veterinary Record* 2014;**174**:13–21.

74 Favrot C, Reichmuth P, Olivry T. Treatment of canine atopic dermatitis with azathioprine: a pilot study. *Veterinary Record* 2007;**160**:520–1.

75 Preziosi DE, Goldschmidt MH, Greek JS, *et al*. Feline pemphigus foliaceus: a retrospective analysis of 57 cases. *Veterinary Dermatology* 2003;**14**:313–21.

76 Simpson DL, Burton GG. Use of prednisolone as monotherapy in the treatment of feline pemphigus foliaceus: a retrospective study of 37 cats. *Veterinary Dermatology* 2013;**24**:598–601.

77 Irwin KE, Beale KM, Fadok VA. Us of modified ciclosporin in the management of feline pemphigus foliaceus: a retrospective analysis. *Veterinary Dermatology* 2012;**23**:403–10.

78 Torres SMF, Brien TO, Scott DWM. Dermal arteritis of the nasal philtrum in a Giant Schnauzer and three Saint Bernard dogs. *Veterinary Dermatology* 2002;**13**:275–81.

79 Lotti TM, Hercogová J, Schwartz RA, *et al.* Treatments of vitiligo: what's new at the horizon. *Dermatology Therapeutics* 2012;Suppl 1:S32–40.

80 Alhaidari, Z, Olivry T, Ortonne J-P. Melanocyte genesis and melanogenesis: genetic regulalation and comparative clinical disease. *Veterinary Dermatology* 1999;**10**:3–16.

81 Bagherani N. The newest hypothesis about vitiligo: most of the suggested pathogeneses of vitiligo can be attributed to lack of one factor, zinc-α2-glycoprotein. *ISRN Dermatology* 2012;**2012**:405268.

82 Alhaidari, Z. Vitiligo chez un chat. *Annales de Dermatologieet de Vénéréologie* 2000;**127**:413.

83 López R, Ginel PJ, Molleda JM, Bautista MJ, Pérez J, Moxos E. A clinical, pathological and immunopathological study of vitiligo in a Siamese cat. *Veterinary Dermatology* 1994;**5**:27–32.

84 Lester SJ, Mailk R, Bartlett KH, Duncan CG. Cryptococcus: update and emergence of *Cryptococcus gattii. Veterinary Clinical Pathology* 2011;**40**:4–17.

85 Trivedi SR, Sykes JE, Cannon MS, *et al.* Clinical features and epidemiology of cryptococcosis in cats and dogs in California: 93 cases (1988–2010). *Journal of the American Veterinary Medical Association* 2011;**239**:357–69.

86 Madrid IM, Mattei A, Martins A, Nobre M, Meireles M. Feline sporotrichosis in the southern region of riograndedosul, Brazil: clinical, zoonotic and therapeutic. *Zoonoses and Public Health* 2010;**57**:151–4.

86a Pimental de Souza C, Lucas R, Ramadinha RHR, Pires TBCP. Crysosurgery in association with itraconazole for the treatment of feline sporotrichosis. *Journal of Feline Medicine and Surgery* 2016;**18**:136–143.

87 Santoro D, Prisco M, Ciaramella P. Cutaneous sterile granulomas/pyogranulomas, leishmaniasis and mycobacterial infections. *Journal of Small Animal Practice* 2008;**49**:552–61.

88 Byrne KP. Metabolic epidermal necrosis-hepatocutaneous syndrome. *Veterinary Clinics of North America Small Animal Practice* 1999;**29**:1337–55.

89 Patel A, Whitbread TJ, McNeil PE. A case of metabolic epidermal necrosis in a cat. *Veterinary Dermatology* 1996;**7**:221–5.

90 Langer NB, Jergens AE, Miles KG. Canine glucagonoma. *Compendium on Continuing Education for the Practicing Veterinarian* 2003;**25**:56–63.

91 Torres SM, Caywood DD, O'Brian TD, O'Leary TP, McKeever PJ. Resolution of superficial necrolytic dermatitis following excision of a glucagon-secreting pancreatic neoplasm in a dog. *Journal of the American Animal Hospital Association* 1997;**33**:313–19.

92 Hargis AM, Haupt HK, Hegreberg GA, Prieur DJ, Moore MP. Familial canine dermatomyositis: initial characterization of the cutaneous and muscular lesions. *American Journal of Pathology* 1984;**116**:234–44.

93 Clark LA, Credille KM, Murphy KE, Rees CA. Linkage of dermatomyositis in the Shetland Sheepdog to chromosome 35. *Veterinary Dermatology* 2005;**16**:392–4.

94 Ferguson EA, Cerundolo R, Llyod DH, Rest J, Cappello R. Dermatomyositis in five Shetland Sheepdogs in the United Kingdom. *Veterinary Record* 2000;**146**:214–17.

94a Röthig A, Rüfenacht S, Welle MM, Thom N. Dermatomyositis in a family of Working Kelpies. *Tierarztl Prax Ausg K Kleintiere Heimtiere* 2015;**43**:331–6.

95 Scott DW, Miller WH, Griffin CE. Familial canine dermatomyositis. In: *Muller & Kirk's Small Animal Dermatology*, 6th edn. Scott DW, Miller WH, Griffin CE (eds). WB Saunders, Philadelphia 2001, pp. 940–6.

96 Rees CA, Boothe DM. Therapeutic response to pentoxifylline and its active metabolites in dogs with familial canine dermatomyositis. *Veterinary Therapeutics* 2003;**4**:234–41.

97 Waqar M, Haque WM, Mir MR, Hsu S. Vogt–Koyanagi–Harada syndrome: association with alopecia areata. *Dermatology Online Journal* 2009;**15**:10.

98 Gross TL, Ihrke PJ, Walder EJ. Vogt–Koyanagi–Harada-like syndrome. In: *Veterinary Dermtopathology*. Gross TL, Ihrke, PJ, Walder EJ (eds). Mosby Year Book, St Louis, pp. 148–50.

99 Blackwood SE, Barrie KP, Plummer CE, *et al.* Uveodermatologic syndrome in a rat terrier. *Journal of the American Animal Hospital Association* 2011;**47**:56–63.

100 Balker K, Scurrell, E, Walker D, *et al.* Polymyositis following Vogt–Koyanagi–Harada-like syndrome in a Jack Russell terrier. *Journal of Comparative Pathology* 2011;**144**:317–23.

101 Angles JM, Famula TR, Pederson. Uveodermatologic (VKH-like) syndrome in American Akita dogs is associated with an increased frequency of DQA1*00210. *Tissue Antigens* 2005;**66**:656–65.

102 Schwens Ch, Thom N, Moritz A. Reactive and neoplastic histiocytic diseases in the dog. *Tierärztliche Praxis Ausgabe K Kleintiere/Heimtiere* 2011;**39**:176–90.

103 Affolter VK, More PF. Canine cutaneous and systemic histiocytosis: reactive histiocytosis of dermal dendritic cells. *American Journal of Dermatopathology* 2000;**22**:40–8.

104 Palmeiro BS, Morris DO, Goldschmidt MH, Maudlin EA. Cutaneous reactive histiocytosis in dogs: a retrospective evaluation of 32 cases. *Veterinary Dermatology* 2007;**18**:332–40.

105 Fulmar AK, Maudlin GE. Canine histiocytic neoplasia: an overview. *Canadian Veterinary Journal* 2007;**48**:1041–50.

106 Munday JS. Papillomavirus in felids. *Veterinary Journal* 2014;**199**:340–7.

107 Wilhelm S, Degorce-Rubiales F, Gosdon D, Favrot C. Clinical, histological and immunohistochemical study of feline viral plaques and bowenoid in situ carcinoma. *Veterinary Dermatology* 2006;**17**:424–31.

108 Bucholz J, Walt H. Veterinary photodynamic therapy: a review. *Photodiagnosis and Photodynamic Therapy* 2013;**10**:342–7.

109 Gill VL, Bergamnn PJ, Baer KE, Craft D, Leung C. Use of imiquimod 5% cream (Aldara) in cats with multicentric squamous cell carcinoma in situ: 12 cases (2002–2005). *Veterinary and Comparative Oncology* 2008;**6**:55–64.

110 Hammond GM, Gordon IK, Theon AP, Kent MS. Evaluation of strontium Sr 90 for the treatment of superficial squamous cell carcinoma of the nasal planum in cats: 49 cases (1990–2006). *Journal of the American Veterinary Medical Association* 2007;**231**:736–41.

111 Gross TL, Ihrke PJ, Walder EJ. Lichenoid (interface) diseases of the dermis. In: *Veterinary Dermatopathology*. A macroscopic and microscopic evaluation of canine and feline skin disease. Gross TL, IhrkePJ, Walder EJ (eds). Mosby Year Book, St Louis 2005, pp. 141–62.

112 Scott DW, Miller WH, Griffen CE. Neoplastic and non-neoplastic tumours. In: *Muller & Kirk's Small Animal Dermatology*, 6th edn. Scott DW, Miller WH, Griffin CE (eds). WB Saunders, Philadelphia 2001, pp. 1333–40.

113 Fontaine J, Heimann M, Day MJ. Canine cutaneous epitheliotropic T-cell lymphoma: a review of 30 cases. *Veterinary Dermatology* 2010;**21**:267–75.

114 Duclos DD, Hargis AN. Canine epitheliotropic lymphoma limited to face and footpads. *Veterinary Dermatology* 1996;**7**:243–6.

115 Bergvall K. A novel ulcerative nasal dermatitis of Bengal cats. *Veterinary Dermatology* 2004;**15**:S28.

116 Lee Gross T, Ihrke PJ, Walder EJ, Affolter VK. Diseases with abnormal cornification. In: *Skin Diseases of the Dog and Cat: Clinical and Histopathologic Diagnosis*. Lee Gross T, Ihrke PJ, Walder EJ, Affolter VK (eds). Wiley Blackwell, Hoboken 2005, pp. 169.

11.1 INTRODUCTION

In dogs and cats, diseases of the nasal cavity and nasal sinuses are very common. Sneezing and nasal discharge are commonly encountered in veterinary practice and can be the result of congenital or acquired conditions. Congenital abnormalities such as clefts of the primary palate, which involve the nasal planum, and clefts of the secondary palate that all lead to oronasal communications will be discussed in this chapter. Congenital nasal dermoid sinus cysts have been reviewed in Chapter 10 Diseases of the Nasal Planum, Section 10.2. Choanal atresia as a congenital condition has been reported in both dogs and cats and will be discussed in Chapter 12 Diseases of the Nasopharynx, Section 12.2.

Combined congenital malformations of the nasal planum and nose are increasingly seen in brachycephalic dogs. Brachycephalic obstructive airway syndrome (BOAS) is a complex disease that not only involves the nasal planum and nose, but also the pharynx, larynx and lower airways. All relevant aspects of this disease will be discussed in this chapter, with the exception of surgical treatment of the condition, which will be the topic of Chapter 19 Surgery of the Nose, Section 19.3.

Though epistaxis can result from any nasal pathology, coagulopathies can be the primary cause and this specific form of nasal discharge will therefore be discussed in a separate section. The four most common causes of unilateral nasal discharge in dogs and cats are: nasal foreign body, mycotic rhinitis, oronasal fistulae (including cleft palate) and polyps/neoplastic disease, and these will be discussed in the next four sections respectively. Bilateral nasal discharge is usually the result of either an infectious, allergic or chronic non-specific/idiopathic rhinitis. These conditions will be addressed in the subsequent three sections. Finally, nasal and sinal trauma will be reviewed.

11.2 CONGENITAL DISEASES OF THE NOSE

Congenital cerebrospinal fluid fistula

A congenital cerebrospinal fluid fistula causing bilateral clear nasal discharge, fever and abnormal behaviour with cyst formation on the forehead has been reported in a young cat[1]. With contrast radiography a fistula from the olfactory bulb leading to cerebrospinal rhinorrhoea was diagnosed. Surgical closure of the nasal bone defect resulted in resolution of the fistula and temporary abatement of neurological signs. However, seizures of unknown aetiology but responsive to phenobarbital and antibiotics occurred 26 months after surgical closure of the fistula.

Primary ciliary dyskinesia

Primary ciliary dyskinesia (PCD) is a genetic disorder characterised by abnormally functioning cilia. Cilia are complex structures lining various organs, including the upper and lower respiratory tracts, auditory tubes, ventricles of the brain, spinal canal, oviducts and efferent ducts of the testis[2,3]. Motile cilia are composed of a microtubule backbone, consisting of nine microtubule doublets surrounding a central pair. Inner and outer dynein arms extend from each outer microtubule doublet, and generate the force needed for motility in an ATP-dependent process. Ciliary dysmotility or immotility is often associated with ultrastructural defects of these cilia such as total or partial absence of the outer dynein arms (ODAs), inner dynein arms (IDAs), or both[4,5]. This genetic disease is usually inherited in an autosomal recessive mode[6,7], and has been described in humans as well as in different animals including 19 breeds of dog[3,6–10]. A mutation of the *CCDC39* gene has been identified to be responsible for PCD both in Old English Sheepdogs and in humans[5,11].

The main clinical signs are recurrent or persistent respiratory infections because of the lack of effective ciliary motility, which results in ineffective or abnormal mucociliary clearance[8]. Chronic rhinitis is present in most if not all dogs and begins at the age of a few days to 5 weeks, but some dogs have remained asymptomatic for months[12]. Complications are caused by colonisation of the mucosa and the conchae by *Pasteurella multocida* and *Bordetella bronchiseptica*, which can cause atrophy of conchae via bone resorption[2,3,8]. Disease of the lower airways of dogs with primary dyskinesia varies from mild bronchitis and bronchiolitis, to severe broncho-pneumonia with bronchiectasis and ventral lung lobe consolidation[3]. Male fertility can be impaired because of defects of the spermatozoa flagella[3,8]. Disorders such as hydrocephalus or serous otitis have also occasionally been reported[13]. The dysfunction of the monocilia of the embryonic node might also lead to the randomisation of the left-right body asymmetry, and transposition of the thoracic and abdominal organs[5,10]. The combination of ciliary dysfunction and situs inversus is known as Kartagener's syndrome[10,14].

The diagnosis is based on exclusion of other causes of chronic rhinitis, tracheobronchitis and pneumonia and on functional and ultrastructural analysis of cilia *in vivo* and *in vitro*. This type of analysis requires specialised laboratories. Use of transmission electron microscopy does not always aid in detection of ultrastructural ciliary abnormalities, but mucociliary clearance is always impaired[3,8]. Therefore, functional analysis of cilia should preferably precede ultrastructural examination.

Scintigraphy is used as the diagnostic tool to evaluate mucociliary clearance[8]. Mucociliary clearance in the dog's nasal cavity can be measured by placing a drop of 99mTc macroaggregated albumin deep in the nasal cavity via a catheter, beyond the nonciliated rostral half. The velocity of mucus clearance ranges from 7 to 20 mm/min[3,12]. Though the test is not affected by anaesthesia, not all normal dogs have clearance rates within the described range and inflammation can change the velocity of the ciliary beat. Other inflammatory conditions of the upper airway invariably induce some degree of secondary ciliary dyskinesia. In dogs, reported ranges for normal ciliary beat frequency varied from 9 to 11 Hz in one study[15] and from 4 to 17 Hz in another study[16].

Transmission electron microscopy may reveal various lesions in which dynein arm deficiencies are most prevalent[3,8]. Other lesions reported in dogs with PCD include abnormal microtubular patterns, random orientation of adjacent cilia and electron-dense inclusions in the basal body[3]. With rare exceptions, ultrastructural defects in nasal and bronchial cilia are concordant[8]. This is typically true for spermatozoa flagella as well because spermatozoan tails consist of the same microtubular arrangement as respiratory cilia.

Treatment preferably is based on culture and sensitivity of nasal and bronchoalveolar samples. Continuous treatment with broad-spectrum antibiotics will prolong survival, but the prognosis of PCD is reserved[12]. Most affected dogs that develop severe recurrent broncho-pneumonia eventually die of this.

11.3 BRACHYCEPHALIC OBSTRUCTIVE AIRWAY SYNDROME

Introduction
Brachycephalic dogs, such as English and French Bulldogs, Pugs, Pekingese, Shih Tzus, Shar Peis, Boston Terriers, and Persian and Himalayan cats, frequently present with signs of upper airway obstruction as a result of an anatomical distortion of their faces caused by an exaggerated and incorrect breed selection. Their head shape, a severely compressed face, is the result of an inherited developmental defect of the bones of the base of the skull, which grow to a normal width, but reduced length, without proportionate reduction of the soft tissues of the head (nose, oro- and naso-pharynx). The subsequent increased airway resistance in their hypoplastic airways results in an increased inspiratory effort (obstructive breathing pattern with stertor/stridor), and eventually leads to dyspnoea, heat and exercise intolerance, and secondary gastrointestinal abnormalities. For example, gagging, vomiting and regurgitation as a result of hiatal hernia, oesophagitis, gastritis and pyloric mucosal hyperplasia have been reported[17].

Dogs with these anatomical brachycephalic abnormalities and clinical signs of upper airway obstruction are suffering from 'brachycephalic (obstructive) airway syndrome' (BOAS). The primary components of BOAS are increased nasal resistance as a result of stenotic nares (43–85%) and aberrant or protruding turbinates, pharyngeal hypoplasia (redundant pharyngeal folds) with elongated soft palate (86–96%) and, especially in the English Bulldog, tracheal hypoplasia[18–21]. Secondary components, resulting from the chronic increased negative intra-airway pressure, include everted tonsils, everted laryngeal saccules (55–59%) and laryngeal collapse (8–70%)[20,21].

In addition, shortening of the skull has also led to problems in the back of the skull, of the temporal and occipital bones. Tympanic bulla hypoplasia and middle ear effusion are commonly seen in these breeds, which lead to conductive hearing loss and loss of quality of life.

Even though individual animals can sometimes be effectively managed with medication, surgery or a combination of these, efforts of the veterinary community should be directed at prevention of this condition that strongly affects the welfare of the aforementioned breeds.

Diagnostic work-up

A sniffing or nasal stridor indicates obstruction of airflow through the nasal passages; snoring is typically associated with (naso)pharyngeal disease, whereas a laryngeal stridor (g-sound or sawing sound) is associated with laryngeal disease. Hearing loss often goes unrecognised by the owner. Coughing, gagging, retching, regurgitation and vomiting are frequently present and indicate secondary or concurrent lower airway or gastrointestinal disease.

Physical examination findings are usually unremarkable except for the possible audible stridor and the obvious brachycephalic conformation of the animal. In addition, most patients demonstrate some degree of stenosis of the nares (Fig. 11.1) and increased, referred, respiratory noises upon thoracic auscultation.

Radiographic examination of the head, neck and chest is useful for recognition of obstructing structures in the pharynx and larynx, tracheal hypoplasia (Fig. 11.2), and to detect secondary aspiration pneumonia or pulmonary oedema. The tracheal diameter can be measured at the thoracic inlet and expressed as a percentage of the thoracic inlet diameter. In bulldogs, the tracheal diameter is a mean of 12.7% of the thoracic inlet compared with 20% in non-brachycephalic breeds[22]. In some patients, a sliding hiatal hernia can be observed (Fig. 11.3). Radiography does not, however, provide information on the degree of pharyngeal and laryngeal hypoplasia and associated pharyngitis and laryngitis, nor does it allow for a proper evaluation of the nasal passages (aberrant conchae) and nasopharyngeal diameter.

Computed tomography (CT) imaging is therefore recommended in brachycephalic animals. This allows for a

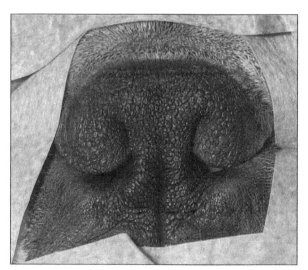

Fig. 11.1 Nasal planum of a French Bulldog demonstrating moderately stenotic nares.

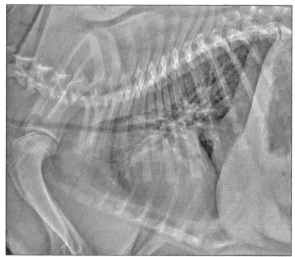

Fig. 11.2 Lateral thoracic radiographs of an English Bulldog demonstrating radiographic evidence of tracheal hypoplasia.

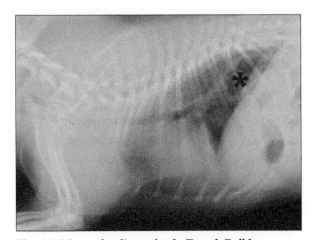

Fig. 11.3 Lateral radiograph of a French Bulldog demonstrating a sliding hiatal hernia (asterisk) upon inspiration.

much more accurate evaluation of the bony abnormalities, measurements of airway diameter, presence of nasopharyngeal turbinates and potential concurrent middle ear disease, also related to brachycephaly (Figs 11.4A–F).

Direct inspection of the pharynx and larynx with a laryngoscope is, however, the most important diagnostic procedure to determine the degree of pharyngeal and laryngeal hypoplasia, length of soft palate (Figs 11.5A, B) and secondary everted tonsils and laryngeal collapse (see Chapter 14 Throat Diagnostic Procedures, Section 14.2). Rigid rhinoscopy allows for assessment of degree of obstruction of air passages by the ventral ala, presence of aberrant rostral turbinates and degree of increased mucosal contact points (Fig. 11.6)[23,24]. With flexible endoscopes the nasopharyngeal

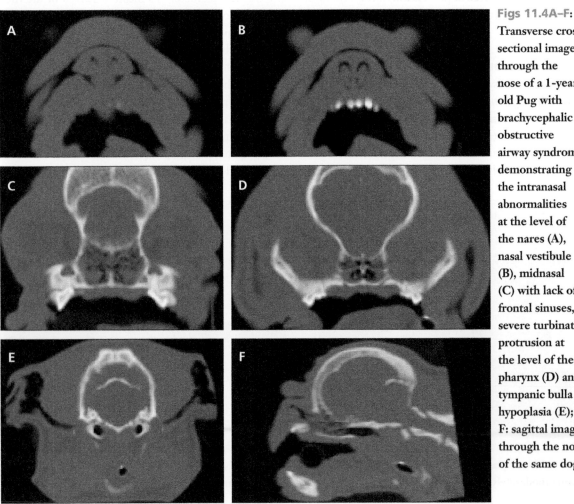

Figs 11.4A–F: Transverse cross-sectional images through the nose of a 1-year-old Pug with brachycephalic obstructive airway syndrome demonstrating the intranasal abnormalities at the level of the nares (A), nasal vestibule (B), midnasal (C) with lack of frontal sinuses, severe turbinate protrusion at the level of the pharynx (D) and tympanic bulla hypoplasia (E); F: sagittal image through the nose of the same dog.

area can be inspected completely, assessed for space and internal dimensions and caudal aberrant nasopharyngeal turbinates can be visualised (Fig. 11.7).

Treatment

Animals in severe respiratory distress need to be evaluated quickly and intubated if respiratory arrest is imminent. Most animals respond to the following treatment to decrease pharyngeal and laryngeal swelling:

- Cold intravenous fluids.

- Sedation with acetylpromazine (0.01 mg/kg IV).
- Oxgygen supplementation.
- Dexamethasone (0.05–0.1 mg/kg IV).

Intravenous access is mandatory in either case but should be obtained with as little restraint and stress to the animal as possible[25].

Long-term treatment of brachycephalic airway syndrome is aimed at reducing airway resistance and alleviating obstruction, either medically and/or surgically:

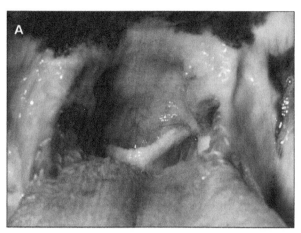
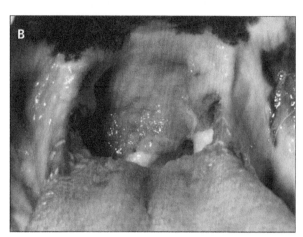

Figs 11.5A, B Pharyngoscopic examination in a 1-year-old French Bulldog showing mild pharyngeal hypoplasia with mildly thickened base of the tongue, mild dorsoventral pharyngeal flattening, mild protrusion of the tonsils, moderate diffuse pharyngeal oedema and moderately elongated soft palate upon inspiration (A) and on expiration (B).

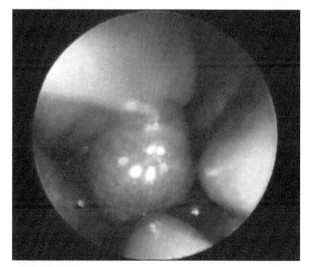
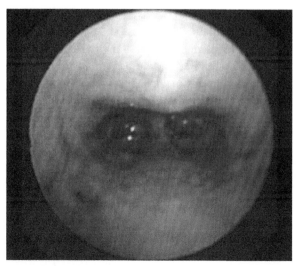

Fig. 11.6 Rhinoscopic assessment in a 1-year-old French Bulldog showing aberrant rostral turbinates and increased mucosal contact points between turbinates.

Fig. 11.7 Nasopharyngoscopic assessment of the same dog as in Fig. 11.6 using a flexible endoscope, demonstrating caudal abberant turbinates protruding into the nasopharynx.

- Maintaining an adequate body weight and condition.
- A clean, fresh and cool environment.
- Regular, controlled exercise is advised.
- Corticosteroids can be used to treat mucosal swelling.
- Broad-spectrum antibiotics are indicated in cases with (aspiration) pneumonia.

Any pre- or postoperative gastrointestinal signs are aggressively treated with:

- A proton pump inhibitor (omeprazole 0.7 mg/kg p/o q24h).
- A prokinetic (cisapride [0.2 mg/kg p/o q8h] or metoclopramide constant rate infusion [1– 2 mg/kg/d i/v]).
- An antacid[17,25].

Cerenia (Maropitant, 1.0 mg/kg SC q24h) is a neurokinin (NK1) receptor antagonist that might be useful in reducing the likelihood of regurgitation during anaesthesia.

The components of the syndrome that are amenable to surgical correction are stenotic nares, aberrant turbinates, elongated soft palate, everted laryngeal saccules and laryngeal collapse. These techniques will be discussed in Chapter 19 Surgery of the Nose, Section 19.3. Currently, there is no long-term successful treatment of tympanic bulla hypoplasia and middle ear effusion.

Management

Brachycephalic animals are gaining in popularity and during the recent decades, the severity of symptoms associated with this malformation has increased[26–28]. As Gerhard Oechtering stated: "Brachycephaly is a manmade disease"[29]. It is therefore up to us to fix it, since we have not only brought respiratory distress and exercise intolerance to these animals by excessively reducing the muzzle, but hearing loss and secondary gastrointestinal abnormalities as well. In addition, since the nose plays a vital role in thermoregulation, it is probable that the heat-related disorders often seen in the affected breeds are caused by restricted temperature regulation as a result of the nasal obstruction[29]. Their sense of smell is also mostly like reduced.

Owners unfortunately are not aware of the severity and implications of these breed-specific problems[26–28].

Veterinarians should play a much more active role in the public discussion and in educating the general public. Social awareness is increasing though, and politicians and the media are increasingly aware of the negative impact that selective breeding for exaggerated features has had on the welfare of pedigree breeds[26–28].

11.4 EPISTAXIS

Epistaxis is defined as haemorrhage originating from the nose and can be acute in onset or follow a more chronic course, and be either unilateral or bilateral. The nose has an extremely abundant blood supply from both the external and internal carotid arteries and blood loss from the nose can therefore appear to be dramatic (Fig. 11.8). Many conditions can cause nasal bleeding and are generally categorised as due to either local or systemic disorders.

Many dog owners and veterinarians consider epistaxis to be an emergency, as evidenced by the fact that 75% (132/176) of the dogs in one large retrospective study were initially examined by an emergency service[30]. Epistaxis accounted for 0.3% of canine emergency visits during the 6-year period of the aforementioned study. Dogs with epistaxis are more likely to be middle-aged to old, male and large (weighing

Fig. 11.8 Spaniel with severe epistaxis.

≥26 kg)[30,31]. Causes of epistaxis are directly related to disease of the nose in about 75% of the cases, except in areas where parasitic and protozoan infections are endemic[12,30–32].

> Remember that nasal arteritis (see Chapter 10 Diseases of the Nasal Planum) might be described as a 'nose bleed' by the client, but it is not epistaxis.

For dogs presenting with epistaxis in the absence of acute trauma or (mucopurulent) nasal discharge, coagulation disorders, hypertension or hyperviscosity syndromes must be ruled out before undertaking nasal imaging or biopsy[33]. If epistaxis is part of the presenting complaint in cats, a coagulation panel should be assessed in addition to a platelet count before work-up of the nose itself. If the cat is older than 8 years old, or if concurrent kidney disease is identified, systolic blood pressure should also be measured[34].

Most nose bleeds stop spontaneously and the short-term outcome of dogs with epistaxis appears to be good, with 90% of dogs discharged in one study[30]. Blood transfusions were indicated in only 10% of the cases in this study[30]. However, if clinical signs of shock due to severe blood loss and/or trauma, such as pale mucous membranes, tachycardia, prolonged capillary refill time, weak pulse, rapid respiration and hypothermia are present, intravenous fluid therapy should be started directly[12,35,36]. Blood transfusion should be given if indicated, but always after collection of a sample for routine haematology and coagulation studies.

Administration of a sedative calms the dog and usually decreases the epistaxis[12]. Local adrenaline drops can be used cautiously to reduce residual bleeding. In life-threatening continuous blood loss, ligation of the ipsilateral external carotid artery can be considered[12]. It has been suggested that in most cases of intermittent epistaxis of unknown origin that do not respond to conservative management, surgical treatment in the form of dorsal rhinotomy with turbinectomy solves the epistaxis[12]. Diagnostic imaging of the nose ideally should be postponed until the bleeding has stopped, as large clots in the nose can hinder interpretation significantly. While rhinoscopy sometimes clearly reveals the site of the bleeding, more often the exact origin is not found[12].

Local causes of epistaxis

Local causes of epistaxis are more likely to cause unilateral signs. However, unilateral epistaxis is not pathognomonic for a local disorder as about 50% of the dogs with systemic disorders in one study had unilateral epistaxis[30]. Also, although dogs with local causes of epistaxis were more likely to have chronic epistaxis in one study, this was not the case in Bissett's study[30]. Chronicity appears to be more associated with the cause of the problem: dogs with nasal neoplasia are more likely to have chronic epistaxis, and dogs with trauma are more likely to have acute epistaxis. Most studies support the assertion that dogs with local causes of epistaxis may have other signs of nasal tract disease in addition to epistaxis, such as sneezing and nasal discharge[30,31]. Forty percent of dogs with a systemic disorder are reported to have other signs of nasal tract disease as well, though this invariably is sneezing and not nasal discharge.

Local disease processes that have previously been reported to cause epistaxis in dogs include nasal and paranasal neoplasia[31,37], mycotic rhinitis[38,39], idiopathic rhinitis[31,33,40], parasitic (*Linguatula, Capillaria, Pneumonyssoides* sp.) rhinitis[31,41–43], nasal foreign body[44], periapical abscesses[45] and arteriovenous malformation[46]. Foreign bodies and oronasal fistula based on periapical dental abscesses seldom cause epistaxis, the predominant signs they cause being instead sneezing and unilateral rhinitis[12]. Nasal neoplasia and mycotic rhinitis have long been regarded as common causes of epistaxis in dogs, and results of most studies support the suggestion that nasal neoplasia is a predominant cause of epistaxis in dogs[30,31,33,47]. Idiopathic rhinitis is commonly associated with, sometimes severe, epistaxis and was the third most common cause of epistaxis in one study[30,40]. Most dogs with traumatic epistaxis also have clear evidence of trauma, such that identifying the cause of epistaxis is not a diagnostic challenge.

Systemic causes of epistaxis

Dogs with systemic disorders are more likely to have clinical signs of systemic disease (lethargy, inappetence, weight loss and bleeding at extranasal sites) than dogs with a local cause for epistaxis[30]. Systemic disorders previously reported to cause epistaxis in dogs include hereditary and acquired bleeding disorders[31,48–50], and various systemic infections[51–54] that usually result in acquired bleeding disorders or affect the integrity of the nasal mucosa and vasculature[55].

Congenital coagulation disorders, such as haemophilia A and B in cats and von Willebrand's disease in dogs, can cause epistaxis. Acquired coagulopathies due to deficiency of clotting factors can occur in liver failure and renal failure, in neoplasia-associated coagulopathy or, for instance, be the result of intoxication with coumarin, a common rat poison[12]. Myelosuppressive drugs may cause epistaxis, as does oestrogen, in both cases due to decreased production of platelets.

Hypertension is not well documented as a cause of epistaxis in dogs, cats or people[31,56,57].

Systemic infections associated with epistaxis include infections with *Leishmania*, *Ehrlichia*, *Rickettsia* and possibly *Bartonella*. Although the protozoan infectious diseases leishmaniasis and ehrlichiosis are the most common systemic causes of epistaxis in dogs in endemic regions[31,32], the most common systemic diseases in non-endemic regions are idiopathic immune-mediated thrombocytopenia, von Willebrand's disease, aspirin-associated thrombocytopathia and neoplasia-associated coagulopathy[30]. Severe thrombocytopenia was the most common systemic disease category associated with epistaxis in one study[30]. Although thrombocytopenic dogs were more likely to have bilateral epistaxis than were dogs with other disorders in this study, the prevalence of unilateral epistaxis among dogs with thrombocytopenia was similar to the prevalence of bilateral epistaxis[30].

11.5 NASAL AND NASOPHARYNGEAL FOREIGN BODY

Nasal foreign bodies are most commonly seen in young dogs with an active outdoor lifestyle, especially Spaniels and Terriers[58]. The history may reveal when the material entered the nasal cavity, as it is sometimes seen by the owner or because of the very acute onset of clinical signs[12].

A nasal foreign body may be associated with extremely variable clinical signs depending on the foreign material, amount, port of entry, location and skull conformation of the patient. Frequent and sometimes ferocious sneezing is the primary clinical sign of an acute nasal foreign body, with rubbing of the front paws at the nose. Gagging, dyspnoea, snoring and reverse sneezing will be noted with foreign bodies that migrate caudally into the nasopharynx. In subacute to chronic cases, unilateral mucopurulent discharge will be the most obvious clinical sign. Some patients may present with bilateral nasal discharge, however, especially when the foreign body is very caudally located or in the rostral nasopharynx. The history can include a response to antibiotic medication and symptom recurrence on treatment cessation.

Grass and other kinds of plant material are the most common offenders, but occasionally unusal foreign bodies are retrieved[12,59]. Foreign material most likely enters the nasal cavity by sniffing, but grass for instance can enter the nasal cavity via the nasopharynx while a cat is chewing on it. Gagging, retching and regurgitation are potential other ports of entry of foreign bodies into the nasopharynx.

When an acute nasal foreign body is expected, imaging is not necessary, but endoscopy can be performed directly. Most nasal foreign bodies are plant material, which has about the same density as soft tissue, thus they are not revealed by radiography[12]. Because of extensive regional soft tissue changes, nasal foreign bodies can be mistaken for tumours, even with advanced imaging[60]. Occasionally, a radiopaque foreign body itself may be visualised[61].

During rhinoscopy, or nasopharyngoscopy under general anaesthesia, the foreign body usually can be seen and be removed with special forceps (Figs 11.9A, B). Flushing and suction of nasal discharge are necessary when foreign bodies are covered in discharge and cannot be visualised immediately. Rigorous flushing of the nasal cavity can dislodge nasal foreign bodies and move them to the nasopharynx or pharynx, where they can be more easily removed[62–65]. Foreign bodies that cannot be removed under endoscopic guidance will have to be removed by surgery[36]. Depending on the degree of damage and secondary bacterial infection a short course of broad-spectrum antibiotics can be prescribed, but in most cases no further treatment is necessary. The prognosis after removal is good.

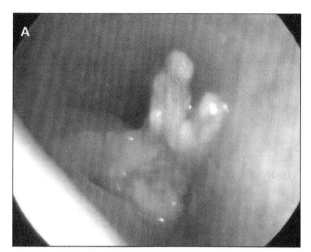
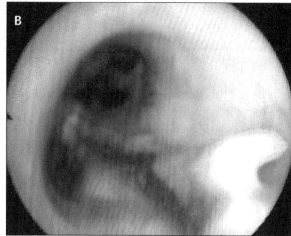

Figs 11.9A, B A: Rhinoscopic view of a foreign body (FB) in the nose; B: FB in the nasopharynx.

11.6 MYCOTIC RHINITIS

Nasal aspergillosis is a condition characterised by destruction of the nasal turbinates associated with the growth of large colonies of fungal hyphae. Masses formed by the fungal hyphae are often referred to as aspergillomas[33,39,66-70]. Although other fungi including *Penicillium, Cryptococcus, Rhinosporidium* and *Trichosporon* species can infect a dog's nose, *Aspergillus* species are the most common ones.

Aspergillus fumigatus is a saprophytic fungus, ubiquitous in the environment, with world-wide distribution[39,69]. *A. fumigatus* is the most common species encountered, although *A. niger, A. clavatus, A. terreus, A. nidulans* and *A. flavus* are occasionally involved. *Aspergillus fumigatus* grows most abundantly in decaying vegetation, sewage sludge compost, decomposing wood chips, mouldy hay and organic compost piles[71]. The *Aspergillus* species produce large numbers of small spores (2–3 μm), which are the source of infection for animals[71]. The capacity of fungal elements to result in infection may depend upon both host immunocompetence and virulence factors associated with the fungal organism[68]. Trapping and removal of inhaled fungal elements by the respiratory tract mucociliary defences usually prevents deeper access. Where this fails, additional innate immune system mechanisms are employed, including the alternative complement system, phagocytic cells, natural killer and T-cells, which work to destroy pathogens intracellularly or by secretion of compounds extracellularly[39,72].

The key factors in the aetiopathogenesis of the infection are:
- Exposure to large fungal inocula.
- Reduced mucociliary clearance.
- Decreased phagocytic cell numbers.
- Impairment in their capacity to destroy organisms.

Though immunologic studies performed before and after treatment have revealed both T- and B-cell dysfunction, canine nasal aspergillosis usually occurs without concomitant malignant or immunosuppressive disease, and affected dogs are otherwise in excellent health[39,66,68,69]. Invasive fungal sinusitis, pulmonary infection and disseminated systemic fungal disease are more likely in immunocompromised patients[68].

There are no documented instances of infection in humans arising from dogs or cats. Infection of all species occurs from common environmental sources. It seems prudent, however, for the clinician to inform an owner that immunosuppressed individuals should not be exposed to affected animals that may be discharging large quantities of fungal hyphae and spores.

Intriguingly cats, known for chronic nasal disease and immunosuppressive viral infections, have been reported to be more rarely infected by *Aspergillus* species[12]. Originally *Aspergillus* in the feline species was reported in immunocompromised animals with concurrent feline parvovirus (panleukopenia), feline leukaemia virus, feline infectious peritonitis virus infections or other debilitating diseases[71]. However, feline upper respiratory tract aspergillosis is now being increasingly reported, often in association with nasal infections[73-84]. Disease occurs over a wide geographic range, including Australia, the USA, the UK, mainland Europe and Japan. Two anatomic forms have been reported, sinonasal and sino-orbital aspergillosis[82,83]. The environmental saprophytic fungi that cause these infections are most commonly from the *A. fumigatus* complex, including *A. fumigatus* and *A. felis*[74,75,82,83,85].

Although nasal, sinus and retrobulbar infection by *Pythium insidiosum*, the aforementioned *Aspergillus* species and various opportunistic fungi have been reported in cats, *Cryptococcus neoformans* is the most common nasal fungal infection, and causes persistent nasal discharge and sneezing and similar destructive rhinitis as aspergillosis in dogs[76,78,86].

The disease is usually seen in dolichocephalic and mesocephalic dogs and is very rare in brachycephalic breeds[66,87]. Specific breed predispositions are not observed. Dogs of any age may be affected, but approximately 40% are 3 years or younger and 80% are 7 years or younger[66]. The disease is very uncommon in dogs younger than 1 year old. Male and female dogs appear to be equally affected a male predisposition is not consistently supported[39,87,88].

The main clinical features are profuse nasal mucopurulent to sanguinopurulent discharge or frank epistaxis, sneezing, reverse sneezing, nasal pain and ulceration of the external nares. Depression as a result of pain and frontal sinus involvement can occur in a later stage of the disease. Additionally, in severe disease signs of systemic illness, facial deformity, epiphora and seizures as a result of erosion through the cribriform plate may be identified[39,66,69].

No single diagnostic test is accurate enough when considered in isolation. False-positive and false-negative results occur with cytological, histopathological, mycological and serological testing. Blind cytological examination or culture of discharge is often unrewarding and can erroneously suggest the disease to be a simple bacterial rhinitis. Cytological examination may also reveal *Aspergillus* or *Penicillium* as contaminants. Therefore caution should also be taken in interpreting

a positive fungal culture result. After culture of nasal swabs taken blindly from normal dogs or from dogs with nasal neoplasia, 30–40% cultured positive for either *Aspergillus* spp. or *Penicillium* spp.[12,66].

Radiographic features of sinonasal aspergillosis are well described and usually demonstrate focal to advanced turbinate destruction within the nasal cavity, evident as wide-spread punctuate lucencies or a generally increased radiolucency (Fig. 11.10). Mixed-density patterns or an overall increase in opacity may be seen with accumulation of fungal plaques, debris or discharge[66,89]. CT improves sensitivity compared to radiography for demonstration of cribriform plate involvement, mucosal thickening and reactive maxillary, vomer and frontal bones (Figs 11.11A–E)[90-92]. Findings on CT were not able to predict the therapeutic success for nasal aspergillosis in dogs treated with a 1-hour infusion of enilconazole[93]. There is no demonstrable difference between CT and magnetic resonance imaging (MRI) for diagnosing nasal cavity mycoses (see Chapter 9 Nose Diagnostic Procedures, Section 9.5).

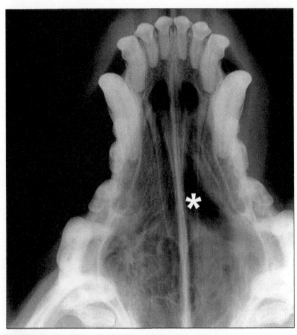

Fig. 11.10 Intraoral radiograph of the nose of a 5-year-old Golden Retriever with left sided aspergillosis. Atrophy of the turbinates (radiolucent areas , asterisk) can be seen midnasally with increased soft tissue density in the caudal nasal cavity.

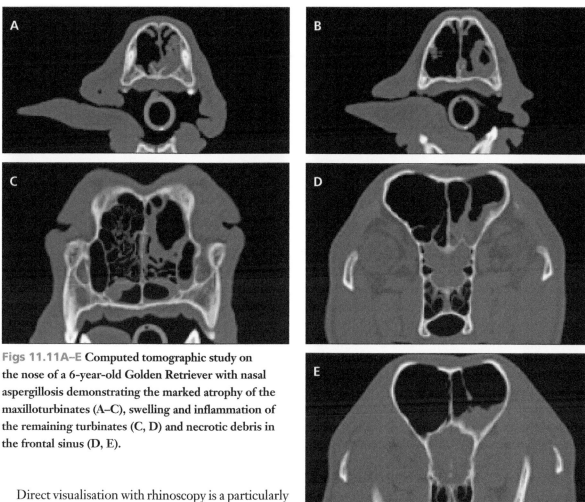

Figs 11.11A–E Computed tomographic study on the nose of a 6-year-old Golden Retriever with nasal aspergillosis demonstrating the marked atrophy of the maxilloturbinates (A–C), swelling and inflammation of the remaining turbinates (C, D) and necrotic debris in the frontal sinus (D, E).

Direct visualisation with rhinoscopy is a particularly valuable diagnostic tool because the fungus destroys turbinate tissue leaving a large airspace within the nasal cavity, which in turn affords good visualisation of fungal plaques. Rhinoscopy may demonstrate marked turbinate destruction, mucoid nasal discharge on chronically inflamed remnant turbinates and fungal plaque that appear as white, yellow or light-green mould lying on the mucosa (Fig. 11.12A, B)[12]. Sometimes masses or granulomas are seen, which should be biopsied in all cases. With marked turbinate destruction, even with rigid endoscopes the frontal sinuses can be evaluated. If the frontosinal ostium is not obviously patent, trephination of the frontal sinus with or without sinuscopy may be necessary to retrieve samples for culture and for treatment[88].

Rhinoscopy-assisted biopsies will allow for the highest detection rates of aspergillosis. Squash preparations of nasal biopsies were 100% sensitive in detection of aspergillosis, whereas mucosal brushings and samples prepared from nasal discharge identified fungal elements in 93% and 13% of the cases, respectively, in one study[94]. Branching, septate fungal hyphae can be seen on direct, 10–20% potassium hydroxide (KOH) or new methylene blue stained wet mounts and with routine haematological stains[71]. Histopathological examination is highly accurate in detecting aspergillosis, and findings are usually consistent with chronic, erosive, non-invasive mycotic rhinosinusitis with a mixed neutrophilic and mononuclear to lymphoplasmacytic infiltrate, though actual fungal hyphae are not seen in all cases[39,68,95,96].

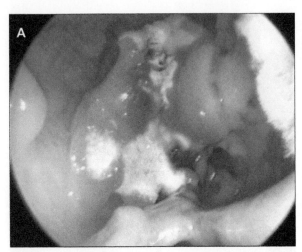
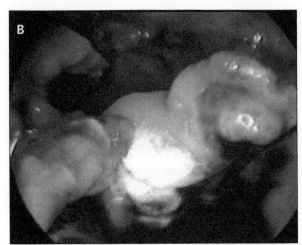

Figs 11.12A, B Rhinoscopic assessment of the nasal cavity in a dog with nasal aspergillosis; severe atrophy of the turbinates can be seen. Remnant turbinates demonstrate a thickened inflamed mucosal lining, and large plaques of fungal material are visible.

Serological diagnosis is possible utilising agar gel immunodiffusion[96,97], counter immunoelectrophoresis or enzyme-linked immunosorbent assay[66,97], but results are variable with respect to sensitivity and specificity and do not affect treatment options or outcome.

Aspergillus spp. grow well on most fungal culture media incubated at room temperature or 35°C, and on blood agar at 37°C[71]. Fungal culture yields from dogs with mycotic rhinosinusitis were found to be greatly enhanced at 37°C compared with incubation at room temperature in a recent study[98].

A variety of conflicting opinions regarding the treatment of sinonasal aspergillosis exist, and often the use of a particular treatment protocol is based upon personal or regional preference. Though the evidence base in support of individual treatment recommendations is weak, topical antifungal administration remains the most widely used method of treatment in dogs.

Poor clinical responses are reported when oral azole antifungal agents alone are prescribed[66,99], but they may be indicated as part of a treatment regimen for infections that have invaded extranasal structures. In addition, these treatments are expensive, prolonged and possibly give rise to hepatotoxicity[33,39,69]. Surgical treatment in the form of rhinotomy is controversial and probably best left for refractory cases only[100,101].

Originally topical therapeutic techniques involved instillation of enilconazole twice daily for 7–14 days via catheters surgically implanted into the nasal cavity. Although very successful (success rates between 80 and 95%)[12,66], prolonged hospitalisation and morbidity led to declining popularity. Administering the antifungal agent not as a flush but instead as a 1-hour soak using general anaesthesia revolutionised topical therapy. The distribution of topical agents after non-invasive infusion via the external nares has been studied in both normal dog skulls and in dogs with fungal rhinitis[102,103]. A Foley catheter (24-French diameter) is retroflexed into the nasopharynx to occlude this area (along with gauze sponges), and the drug is delivered through the external nares into the dorsal nasal meatus with additional catheters (10 French) whilst the external nares are blocked by others (12 French) (Fig. 11.13). A 1% solution of clotrimazole is introduced under pressure (60 ml per side for middle to large breeds, to a pressure of 15 cmH$_2$O) to enhance drug distribution (Fig. 11.14).

Furthermore, by placing the dog on its back to start with and rotating the dog every 15 minutes by 90° to achieve a full 360° after 1 hour, very good distribution is obtained. Care must be taken after the procedure is completed to allow the solution to drain out with the dog in sternal recumbency and its head tilted ventrally, and to remove all gauze sponges and suction the

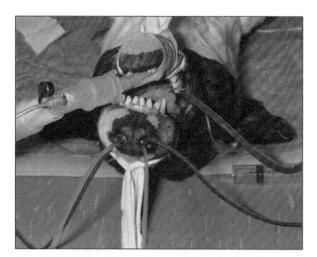

Fig. 11.13 Dog with nasal aspergillosis in dorsal recumbency with all catheters placed for a 1-hour clotrimazole soak.

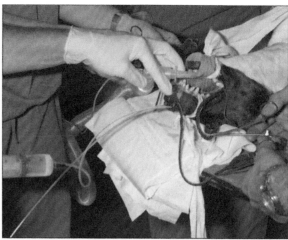

Fig. 11.14 Clotrimazole flush for treatment of nasal aspergillosis.

pharynx. Enilconazole and clotrimazole have both been evaluated at varying concentrations with outcomes varying from 47 to 70% for first treatment outcome, and from 90 to 94% for overall success, albeit in small numbers of dogs for some studies[38,87,93,103–106]. It is the author's opinion that thorough endoscopic debridement before topical treatment is instilled will lead to the highest success rates using this technique.

Further improvements in topical treatment consisted of the use of viscous antifungal creams for application in the frontal sinuses. Combined trephination, short clotrimazole (1%) soak and application of clotrimazole (1%) cream to the frontal sinuses (10–20 g per sinus) has led to a great reduction in procedure duration and hospitalisation and good success rates (50–86% clinical cures)[67,107]. Depot therapy with bifonazole cream via perendoscopic frontal sinus catheters has also been described, and in combination with debridement and enilconazole infusion resulted in nearly 60% clinical cures [108].

11.7 CLEFT PALATE AND ORONASAL FISTULA

A cleft is an abnormal fissure in a body structure resulting from failure of parts to fuse during embryonic development[109]. Cleft palate (or palatoschisis) involves the structures of the secondary palate (hard palate and/or soft palate), whereas cleft lip (cheiloschisis or cheiloalveoloschisis) is a defect involving the structures of the primary palate (lip and premaxilla)[109–112]. Oronasal fistulae are generally acquired communications between the oral cavity and nose, leading to chronic unilateral rhinitis[113–116].

Cleft lip and cleft palate

Defects in the hard and soft palate may result from:
- Congenital abnormalities.
- Resection of neoplasms.
- Traumatic injuries.
- Severe peridontal disease.
- Tooth removals.
- Severe chronic infections.
- Secondary to surgical and radiation therapy[25,110,111,117,118].

Brachycephalic breeds tend to be at higher risk for congenital clefts[25,112]. Intrauterine trauma or stress[110] can also result in incomplete fusion of the palate during fetal development. Clefts can occur if the intrauterine insult (trauma, stress, corticosteroids, antimitotic drugs, nutritional, hormonal, viral and toxic factors) occurs at a very specific time in fetal development (25th to 28th day in dogs)[112]. Congenital defects can result in uni- or bilateral clefts of

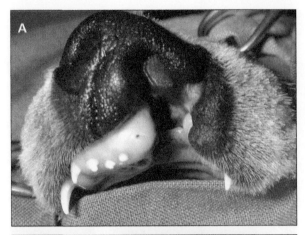

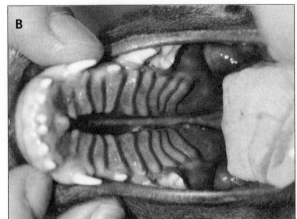

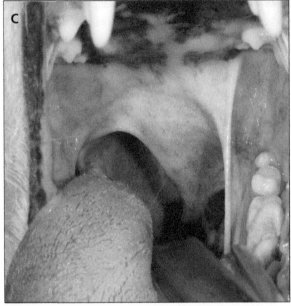

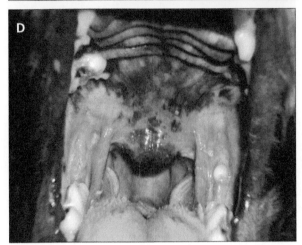

Figs 11.15A–D Cleft palate. A: Cleft of primary palate on the left hand side; B: midline cleft of the secondary palate; C: unilateral soft palate cleft on the right hand side; D: bilateral soft palate cleft.

the upper lip (Fig. 11.15A), lateral area of the most rostral hard palate, midline of the hard and soft palate (Fig. 11.15B), the lateral area of the soft palate (Fig. 11.15C) and rarely in general soft palate hypoplasia with markedly reduced length[25,110,112,119,120].

Cleft lips are externally visible and more frequently found on the left side[25]. Beyond mild local rhinitis they rarely result in clinical signs, and repair may be performed for purely aesthetic reasons, though many puppies and kittens are likely euthanised by the breeder. Cleft lips may be associated with less clearly visible abnormalities of the secondary palate, but obvious clinical signs will be present in those cases. Hard palate clefts are almost always in the midline and usually asso-

ciated with a midline soft palate cleft. The inability to create negative pressure during nursing will result in poor weight gain, and general unthriftiness, whereas the defect itself and chronic irritation of the nasal mucosa by milk or food will result in nasal discharge (drainage of milk from the nares during or after nursing or frank mucopurulent discharge), sneezing and reverse sneezing as a result of rhinitis, gagging and retching as a result of pharyngitis and tonsillitis and coughing as a result of laryngotracheitis or aspiration pneumonia[25,110,111].

Patients with defects of the secondary palate require nursing care by the owner, which includes tube feeding to avoid aspiration pneumonia until the animal can be operated. This is preferably done between 3 and 4 months of age as tissues are less friable at this stage and patients are better anaesthesia candidates. Postponing surgery until after 5 months of age may result in a wider cleft as the animal grows and in compounded management problems, which are not desirable[25]. Traumatic cleft palate is frequently seen in cats with a history of motor vehicle trauma or falling from a height; bilateral epistaxis or dried blood at the nostrils, a visible malalignment along the midline of the maxillary dental arch and a midline hard palate cleft with torn palatal mucoperiosteum are usually evident upon presentation[25].

Unilateral and bilateral congenital defects of the soft palate of the dog without defects of the hard palate have been sporadically reported in the veterinary literature[112,117,120–124]. Whereas small unilateral defects may be incidental findings during throat inspection, large bilateral defects will result in clinical signs, mainly those of bilateral rhinitis and pharyngitis/laryngitis. Bilateral defects result in a laterally unsupported and shortened soft palate that appears as a uvula-like projection (pseudouvula) extending caudally from the hard palate (Fig. 11.15D)[119].

In children, there is a high incidence of middle ear effusions with congenital cleft palates[125]. The effusions are considered to occur as a result of auditory tube dysfunction and/or obstruction leading to the accumulation of mucin-containing fluid and granulation tissue within the affected middle ear[126]. The condition may lead to conductive deafness, cholesteatoma formation, acute infections and sensorineural deafness[119,126]. An association between congenital palatine defects and middle ear disease has been demonstrated both in the dog and in the cat[120]. In one study all six dogs described with soft palate hypoplasia exhibited an abnormal appearance of the middle ear cavities on radiographs (usually small, indistinct bullae with thick walls), though middle ear effusion and hearing loss as measured with brainstem evoked response audiometry was only present in one ear[119]. Techniques for surgical closure of hard and soft palate defects will be discussed in Chapter 19 Surgery of the Nose, Section 19.4.

Oronasal fistulae

Oronasal fistulae are typically caused by loss of incisive or maxillary bone associated with severe periodontal disease or tooth extraction, but palatal defects resulting from electric cord injury, dog bites, foreign body penetration and maloccluding teeth are occasionally seen[113,114,127–129]. An acute oronasal fistula after tooth extraction may be diagnosed by direct visualisation of the nasal cavity and epistaxis at the ipsilateral nostril. Sneezing and ipsilateral nasal discharge are common clinical signs of a chronic oronasal fistula. A defect in the area of a missing tooth (often in the maxillary canine tooth region) that communicates with the nasal cavity may be noted on oral examination[25,110]. However, oronasal fistulae may not be readily apparent upon oral examination and many will go unnoticed unless proper dental probing is performed, even around elements that macroscopically look relatively normal (Figs 11.16A, B). With all traumatic events, it may take several days before the extent of local injury and subsequent necrosis are clearly defined. It is therefore often best to wait and treat trauma patients conservatively until necrotic tissues have sloughed[25]. This approach allows determination of viable tissues available for definitive repair at a later time. Surgical repair of oronasal fistulae will be discussed in Chapter 19 Surgery of the Nose, Section 19.4.

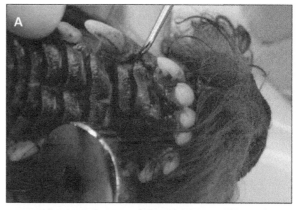

Figs 11.16 A: A dental probe can be seen to effortlessly disappear into an alveolar pocket medial to the canine tooth; B: after tooth removal the inflammation around the tooth is clearly visible.

11.8 POLYPS AND NEOPLASTIC DISEASE

Benign lesions such as polyps, fibromas, intranasally dermoid cysts[130], angiomatous proliferations, nasal angiofibroma[131] and, most recently, an inflammatory myofibroblastic tumour[132] have been reported in dogs, but most tumours of the nasal cavity in dogs are malignant[133]. Clinical signs are similar to those of nasosinal neoplasia (see below). The diagnosis is based on histopathological examination of representative nasal biopsies. The prognosis after surgical resection of benign proliferations via dorsal rhinotomy is usually good.

Inflammatory polyps of the nasal turbinates have been described as a rare, benign disease of the nasal passages of young cats[134]. Cats with nasal polyps are usually younger than 1 year old, and are presented with epistaxis, paroxysmal sneezing and stertorous breathing, but characteristically without mucoid or mucopurulent nasal discharge. Nasal turbinate polyps arise in the nasal passages and should not be confused with nasopharyngeal polyps, which arise from the mucosal lining of the middle ear cavity or auditory tube (see Chapter 12 Diseases of the Nasopharynx, Section 12.2).

The term feline mesenchymal nasal hamartoma has been suggested as a more appropriate description of this disease, based on the histopathological description and comparison with similar lesions in humans[135]. All five cats described in this case-series demonstrated loss of turbinate patterns within the nasal cavity with opalescent radiodensities on radiographic imaging. The changes were unilateral in four cats and bilateral in one. Surgical removal either endoscopically or via open surgery appears to have a good prognosis, though recurrence has been reported[135].

Aetiopathogenesis and symptoms of canine nasosinal neoplasia

Tumours of the nasal cavity and paranasal sinuses (including the nasopharynx) account for approximately 1% of all neoplasms in the dog[136]. Medium-to-large breeds may be more commonly affected[137], with the average age of dogs being approximately 10 years; a slight male predilection has been found[137,138].

Exposure to environmental pollutants and tobacco smoke may be associated with an increased risk[139–141]. Epithelial tumours including adenocarcinoma, squamous cell carcinoma (SCC, Fig. 11.17), and undif-

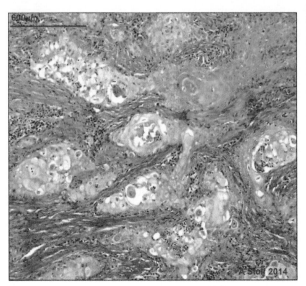

Fig. 11.17 Histopathological cross-sectional view of a squamous cell carcinoma of the rostral nasal septum in an 8-year-old Labrador Retriever.

ferentiated carcinoma represent two-thirds of canine intranasal tumours, with mesenchymal tumours including fibrosarcoma, chondrosarcoma, osteosarcoma and undifferentiated sarcoma making up for most of the remaining part[133,142]. All are characterised by local invasion and in general a low rate of distant metastasis at the time of diagnosis, but up to 40–50% at the time of death[137]. The most common sites of metastasis are the regional lymph nodes and the lungs, but metastatic spread to the bones, kidneys, liver, skin and brain have also been reported[143–145]. Lymphomas are rarely found in dogs, but mast cell tumours, transmissible venereal tumour, haemangiosarcoma, melanoma, neuroendocrine carcinoma, nerve sheath tumour, neuroblastoma, fibrous histiocytoma, multilobular osteochondrosarcoma, hamartoma, rhabdomyosarcoma and leiomyosarcoma have been reported[133]. The importance or relevance of cyclo-oxygenase-2 expression, epidermal growth factor receptor, vascular endothelial growth factor and peroxisome proliferator-activated receptor γ expression in canine nasosinal tumour genesis is unclear[133].

Clinical signs are non-pathognomonic for neoplasia, but reflect general nasosinal disease and therefore most commonly include mucopurulent discharge, epistaxis, sneezing and reverse sneezing, facial deformity, exoph-

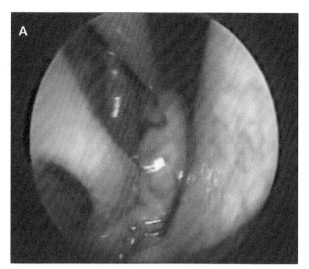

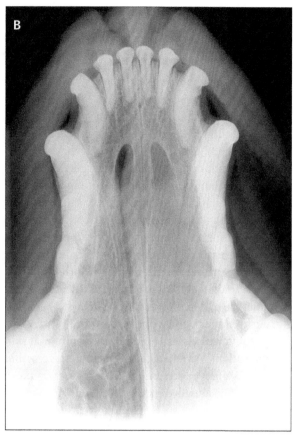

Fig. 11.18A, B **Rhinoscopic assessment of a 9-year-old cross-breed dog with a large pinkish well vascularised tumour extending into the common meatus in the right nasal cavity (A). B: Intra-oral radiograph of the nose of the dog in A, loss of fine turbinate bone is visible on the right hand side, which has been replaced with a mass-type increased soft tissue density filling the right nasal cavity (compare with Fig. 11.10), without radiolucent areas as can be seen with nasal aspergillosis, highly suspicious of neoplastic growth.**

thalmos and ocular discharge as a result of mechanical obstruction of the nasolacrimal duct[37,133,137]. Most dogs with extension of nasal tumours through the cribriform plate do not exhibit neurological signs, but rarely seizures, acute blindness, behaviour change, paresis, circling and obtundation can be found[146].

Diagnosis and staging

The superior imaging value of CT and MRI over conventional radiographs for diagnosing and staging canine and feline nasal neoplasia are well documented[147–153] and has been discussed in Chapter 9 Nose Diagnostic Procedures. A presumptive diagnosis can be made on history, clinical signs and imaging findings, but a definitive diagnosis requires histopathological examination of tissue biopsies, preferably under endoscopic (rhinoscopic, Fig. 11.18) guidance (see Chapter 9 Nose Diagnostic Procedures, Section 9.6).

Of the multiple staging systems for canine nasosinal neoplasia that have been proposed, the Adams staging system[154] is the only one that has a correlation with disease outcome. Dogs with unilateral intranasal involvement without bone destruction beyond the turbinates on CT, had longest median survival (23.4 months), whereas CT evidence of cribriform plate involvement was associated with shortest median survival (6.7 months). Combining CT and histology statistically improved prognostic significance for both survival end-points over the proposed CT staging method alone in this study[154]. Regional lymph node cytology is positive for metastasis in as many as 10–24% of the cases[37,145,155], but thoracic radiographs usually are negative for metastasis at initial presentation[37,133,136,137,145].

Treatment and prognosis

The median survival time without treatment for dogs with nasal carcinoma is approximately 100 days, with epistaxis identified as a negative prognostic indicator in one study[37].

The median survival time after surgery in the form of dorsal rhinotomy with turbinectomy is approximately 3–6 months, thus not much better then for dogs without treatment[136,145,156]. However, surgery can improve quality of life, temporarily, for animals with complete nasal obstruction that are unable to sleep. In addition, two reports report longer median survival times when combining surgery with irradiation compared to other treatment modalities[157,158]. Other studies have failed to demonstrate this positive effect of surgery, and more studies examining the effect and timing of surgery (before or after irradiation) are needed.

High-energy megavoltage cobalt or linear accelerator radiation treatment has become the therapy of choice for canine nasosinal tumours with median survival times ranging from 8 to 19.7 months[157,159–164]. However, radiotherapy is expensive and acute and late toxicities have been reported. Acute toxicities develop during treatment and typically are oral mucositis, keratoconjunctivitis and blepharitis, rhinitis and skin desquamation[159,160,165]. They normally resolve within 2–8 weeks after treatment[166]. Late toxicity is more serious and potentially very detrimental to the patient. It develops months to years after treatment and is usually permanent. Late effects include cataracts, keratoconjunctivitis sicca, uveitis, retinal degeneration, brain necrosis, osteonecrosis and skin fibrosis[165–168]. Although most dogs with nasosinal neoplasia respond favourably to radiation, most dogs die or are euthanised as a result of local disease progression[133].

Chemotherapy is used rarely as a sole treatment and even though it may benefit some dogs, case numbers have been small and further studies are needed[169,170]. Other treatment modalities such as immunotherapy and cryotherapy have not improved survival times[171,172]. Proton-beam therapy does not appear to have any benefits over radiation protocols and is associated with significant acute toxicity[173]. Brachytherapy has so far not shown any improvements in survival times[174,175] and photodynamic therapy needs to be evaluated further before its use can be recommended[176,177].

Feline nasosinal neoplasia

Nasosinal tumours in the cat are malignant in over 90% of cases and occur in cats with a mean age of 9–10 years[62,178,179]. As in dogs the tumours are locally invasive and associated with a low metastatic rate at diagnosis. Clinical signs are non-specific and include nasal discharge, dyspnoea, sneezing, epistaxis, facial swelling, ocular discharge and weight loss. Lymphoma is the most commonly diagnosed tumour type, especially in the nasopharynx, followed by epithelial tumours[180–182]. Fibrosarcoma, osteosarcoma, chondrosarcoma, mast cell tumour, melanoma, plasmacytoma and olfactory neuroblastoma have also been reported[133].

The diagnosis is based on imaging findings and histopathological confirmation of tumour type. Displacement of midline structures, unilateral soft tissue opacity, loss of turbinate detail and evidence of bone invasion are radiographic signs of neoplastic disease[183]. Computed tomographic abnormalities as osteolysis of paranasal bones (vomer and ventral maxilla), extension of disease into the orbit, space-occupying masses and unilateral turbinate destruction favour a diagnosis of tumour over benign disease[78,184,185]. However, nasal biopsies are needed for confirmation. Distinguishing lymphoma from lymphoid inflammatory disease can be difficult and large biopsy samples are recommended[181,186].

Radiotherapy is also the treatment of choice for feline nasosinal neoplasia. Median survival times for nonlymphoproliferative tumours of 12–13 months, with 44–63% of cats alive after 1 year, have been reported[187,188]. The overall response rate for feline nasal and nasopharyngeal lymphoma is high for radiation and/or chemotherapy. The inclusion of radiation appears to enhance overall survival, with median survival for combination therapy ranging from 174 to 955 days[180–182,189,190].

11.9 INFECTIOUS RHINITIS

Infectious upper respiratory disease of dogs and cats is commonly seen in small animal practice. This section will describe the principal organisms involved and will outline practical management strategies. Vaccination is an important component in any control protocol, be it a solitary cat or a rehoming kennel for stray dogs. Although WSAVA has outlined a core vaccination programme for dogs and cats, the availability of various vaccines varies from country to country. Clinicians should be aware of this when reading the chapter.

Feline upper respiratory disease

Multiple studies have shown that perhaps five organisms, acting alone or in various combinations, account for the vast majority of these upper respiratory conditions[76,96,191–200]:

- Feline herpesvirus-1.
- Feline calicivirus.
- *Bordetella bronchiseptica*.
- *Chlamydophilia felis*.
- *Mycoplasma felis*.

Bacteria, other than those mentioned above, are considered secondary, at least at the outset. However, as the upper respiratory tract (URT) disease progresses, bacterial infection may become a relevant component of the problem. Typical isolates include *Pasteurella* spp., *Staphylococcus* spp., *Streptococcus* spp., *Escherichia coli*, *Pseudomonas* spp., *Klebsiella* spp. and *Corynebacterium* spp.[34,192,201]. Johnson *et al.* demonstrated that both *Mycoplasma* spp. and other potentially pathogenic bacteria (such as those detailed above) were isolated significantly more frequently from the noses of affected cats than from the noses of controls[192]. In contrast, the rate of detection of feline herpesvirus-1 (FHV-1) DNA was not significantly different between groups.

Factors such as facial conformation (Persian type), vaccine status, frequent replacement of animals and mean age (especially in a cattery or shelter accommodation, see Gourkow *et al.*[199]), comorbidity (FeLV, FIV, chronic renal disease etc.), stress (social, visitors, renovations, travel etc.), the effect of ongoing pathology and secondary infection and nutritional status can all affect the progression, frequency of relapse and prognosis.

Initiating, primary pathogens

Feline herpesvirus-1

Feline herpesvirus-1 (FHV-1) is a host-specific pathogen of cats, entering and lysing URT epithelial cells[202,203]. Subsequent to the epithelial necrosis an inflammatory reaction ensues and an immune response is mounted[203]. The host's immune response is compromised by two factors:

- Incomplete immunity results in the possibility of, generally less severe, reinfection.
- Integration of viral genome in the nuclei of sensory nerves permits the latent virus to be reactivated, resulting in recrudescence of disease.

Factors predisposing to FHV-1 in catteries were shown to include suboptimal levels of hygiene, stress[199] and the presence of feline calicivirus (FCV) infection[194]. The virus is readily killed by most disinfectants[200].

The clinical signs of acute FHV-1 infection in immunocompetent adult cats reflect rhinitis, with a serous to serosanguinous nasal discharge (Fig. 11.19), sneezing, conjunctivitis with hyperaemia (Fig. 11.20), and the risk of corneal ulceration, pyrexia and, in some

Fig. 11.19 Bilateral serous nasal discharge in a cat with FHV-1 infection.

Fig. 11.20 Bilateral conjunctivitis.

Fig. 11.21 Two small patches of erythematous dermatitis: FHV-1 dermatitis. (Courtesy of Dr Susan Dawson, The University of Liverpool, UK.)

cases, ptyalism and inappetence[204,205]. Herpes viral dermatitis may also occur (Fig. 11.21). Should secondary bacterial infection occur, the nasal discharge changes to purulent, or even a little sanguinous. Kittens, geriatric cats and immunocompromised individuals are predisposed to develop more severe signs. These can include serious ocular complications (such as keratitis, ulceration, corneal sequestration and uveitis), lingual ulceration that is reminiscent of FCV infection, respiratory compromise, anorexia, dehydration and even death. Mono-infections with FHV-1 are uncommon. Co-infection with FCV, *Mycoplasma felis* and *Bordetella bronchiseptica* occur, complicating the acute picture. Mucosal damage, and damage to the underlying turbinates, facilitates secondary bacterial infection and the shift to chronic rhinitis[34] (see Box 1).

Treatment comprises[204,206,207]:

- General nursing support – keeping the nares open and eyes clean; ensuring comfortable, clean, dry bedding.
- Maintaining an adequate nutritional plane. Oral ulceration may make eating difficult and mandate an oesophagotomy.
- Appetite stimulants (see Treatment of chronic rhinosinusitis below).

- Supportive fluid therapy – fluids to correct electrolyte or acid–base imbalances, correct dehydration and maintain hydration. This might be more important in elderly cats.
- Systemic antibacterial therapy if there is any evidence of bacterial infection (for example a purulent nasal discharge).
- Specific antiviral therapy, although this is still in its veterinary infancy:
  - Recombinant feline interferon (Virbagen Omega®) was demonstrated to reduce the severity of the clinical signs, and reduce viral shedding, in naturally infected retrovirus positive cats[207,208]. However, pretreatment of specific pathogen-free (SPF) cats with interferon did not alter the course of FVH-1 infection[206];
  - Famciclovir is one of a number of antiviral drugs which might prove useful, although clinical trials have yet to be performed. Oral famciclovir was administered to 10 cats with various manifestations of FHV-1 disease (rhinosinusitis, dermatitis, ocular disease) with what were deemed promising results[209].

Box 1 Sampling for feline URT pathogens, after Schultz et al[205a].

Samples were taken from the nasal cavity, conjunctiva, the tongue and oropharynx and submitted for PCR (for FHV-1 and *C. felis*) and reverse transcriptase PCR (for FCV). Analysis of samples from 104 cats was made. Statistical evaluation suggested:

- If only one sample can be taken, the oropharynx is the best site
  - but 31% FHV-1-positive, 7.7% FCV-positive and 32.4% *C. felis*-positive cats would be missed
- Sampling the nares in addition improves detection dramatically
  - 94.9% FHV-1 positive, 96.2% FCV-positive and 81.1% of *C. felis*-positive cats would be detected
- Sampling conjunctiva and oropharynx
  - 89.2% of *C. felis*-positive cases identified

Feline calicivirus

Feline calicivirus of cats induces necrosis of epithelial cells, predominantly in the oropharynx, nares and conjunctivae[210]. Unlike FHV-1, infection and subsequent transient viraemia generates an immune response that provides good protection, so that clinical signs associated with subsequent challenge are significantly reduced[211,212].

The clinical signs of FCV infection are of acute oral and URT disease. There may be lingual ulcerations, typically on the lateral or rostral dorsal aspects of the tongue. Upper respiratory signs may be severe enough to resemble FHV-1[210]. Lameness may be noted in some individuals[200]. Rarely, skin ulceration may be seen. Veterinary practitioners were found to overdiagnose FCV infection[212a]. Thus, FCV was isolated in less than half the cases in which it was suspected, based on the clinical signs.

The approach to treatment of FCV infection is the same as for that of FHV-1, although there is no evidence to date that either interferon or RNA-specific antiviral agents are of any value.

Bordetella bronchiseptica

Bordetella bronchiseptica is a potential pathogen of cats, and very rarely, a zoonotic risk[212]. It may be found as a commensal in the normal feline URT[204]. Occasionally *B. bronchiseptica* causes a spectrum of signs, including pyrexia, coughing, sneezing, ocular discharge and lymphadenopathy[212]. More severe cases, usually, seen in kittens rather than adult cats, suffer from pneumonia with cyanosis, dyspnoea and even death. Most commonly *B. bronchiseptica* is found opportunistically associated with other URT pathogens, particularly in association with FCV[193].

Transmission between cats is via oral and nasal secretion. Overcrowding and poor management predispose to infection and disease[212]. The larger the group of cats the more likely it is that *B. bronchiseptica* will be found. It is more likely to be found in shelters than private catteries[194]. The bacterium colonises ciliated epithelium in the respiratory tract: toxins and bacterium-derived proteins result in ciliostasis with destruction of the cilia and modulation of the local immune system[213].

Treatment of *B. bronchiseptica* URT infection is recommended[212] in order to prevent the organism colonising the lower respiratory tract, where serious disease may occur. Doxycycline is the drug of choice[204,206,212]. Potential side-effects in cats[214,215] include oesophageal stricture, vomiting and to a lesser extent diarrhoea and pyrexia. The incidence of vomiting appeared to be greater if cats received concurrent gastric protectants or antiemetics, and the incidence of pyrexia was increased if concurrent corticosteroids were administered[214].

> Note: undoubtedly the most preventable adverse effect of doxycycline treatment is oesophageal stricture. It is due to the severe, local, pH-induced irritation caused by a tablet that has failed to completely clear the oesophagus. Hence it is strongly recommended to use either a liquid formulation or a water bolus 'chaser' following administration of this antibiotic[215,216].

Fluoroquinolones (specifically marbofloxacin, enrofloxacin or ciprofloxacin) have been recommended for the treatment of *B. bronchiseptica* on the basis of proven *in vitro* activity against known feline strains as well as for the postantibiotic effect, which results in extended postdose antibacterial effect[216]. As noted for FHV-1, supportive nutrition as a fluid therapy and correction of any associated electrolyte and acid–base disturbances may be indicated in severe cases.

Chlamydophilia felis

Chlamydophylia felis is gram-negative, cell-associated bacterium[217,218]. *C. felis* affects the feline eye, where it causes acute conjunctivitis. There is hyperaemia (often extreme) of the nictitating membrane, blepharospasm and discomfort[200,217,218]. Corneal ulceration is uncommon[200].

Inoculation of *C. felis* directly into the nose of a SPF cat can result in mild rhinitis, suggesting that it can act as a primary nasal pathogen[218]. However, it is considered[191] that if a cat with URT infection lacks ocular disease, it is unlikely that *C. felis* is involved. In a very large multicentre study involving 573 cats, Bannasch & Foley[193] found *C. felis* in cats with both rhinitis and conjunctivitis, but that in cats lacking conjunctivitis, FHV-1 and/or FCV were identified. Furthermore, the small subset of cats with *C. felis* involvement were reported to have more severe clinical signs[193]. The relative paucity of *C. felis* involvement in feline URT disease is supported by a study of acute URT infection in 52 cats[219] in which the investigators could find no evidence of *C. felis*. Another study into 100 cats with

URT infection only found C felis DNA in the ocular sample, all nasal samples were negative[195].

Although doxycyline and marbofloxacin are good empirical choices, there is evidence that they are unlikely to clear infection[200,220–223]. Amoxycillin-clavulanate and azithromycin might be better agents for treating young animals and for longer-term treatment initiated to clear the infection[204,206,220].

Mycoplasma felis

Although *Mycoplasma felis* is considered a commensal organism of the normal feline URT[199,200,224], it appears that some strains can act as primary pathogens[200,218,224]. Thus, *M. felis* was only recovered from samples taken from cats with chronic rhinosinusitis in one controlled prospective study[192], and the frequency of recovery in samples taken from shelter cats with URT disease was considered significant enough to warrant treatment[193]. In a study of 65 cats conducted by Veir *et al.*[219] *M. felis* was recovered from 47.5% nasal samples and 52.5% of pharyngeal swabs from cats with URT disease.

Doxycycline, marbofloxacin (or pradofloxacin[224] as less risk of ocular toxicity) and clindamycin are thought to be good choices for treating *M. felis* infections[204,224]. Treatment is advised for at least 4 weeks[224].

> Note: Fluoroquinolone-induced retinal degeneration in cats has been reported, following clinical use of marbofloxacin[225,226]. In a review, Wiebe & Hamilton[226] concluded that these drugs were safe in cats provided that the manufacturer's dosing guidelines are followed, and that dose-reduction is appropriate in geriatric cats and those with renal impairment. Rapid intravenous administration should be avoided.

Given the similarity of clinical signs, Table 11.1 may be helpful in suggesting the aetiologic agent(s) involved in a given patient, after Cohn[200]:

Chronic rhinosinusitis in the cat

Chronic rhinosinusitis in the cat usually results in uni- or bilateral nasal discharge[62]. It is generally considered a consequence of infection with one of the primary respiratory pathogens, in particular FHV-1(summarised by Johnson *et al.*[192]), although the virus can be found

Table 11.1 **Likely aetiological agent involved in clinical signs**

Limping	FCV
Oral ulceration	FCV, FHV-1
Keratitis, corneal ulceration	FHV-1
Conjunctivitis without nasal signs	*C. felis*, *Mycoplasma* spp.
Cough	*B. bronchiseptica*

FCV: feline calcivirus; FHV-1: feline herpesvirus-1.

with equal frequency in the noses of normal control cats[192]. Most likely, regardless of the primary aetiology, damage, resulting in inflammation in the delicate tissue of the URT, secondary bacterial infection and even compromised local immune function, are as important as the underlying cause[34,192].

In referral settings, intranasal neoplasia is found in approximately equal frequency to chronic rhinosinusitis as the cause of the chronic nasal discharge[62,183,227]. Whether this holds true for first opinion clinics is unknown, but it is likely that neoplasia would be a less-frequent diagnosis. The differential diagnosis includes[34]:

- Chronic nasosinusitis.
- Allergic rhinitis.
- Nasal foreign body.
- Nasopharyngeal polyp.
- Dental disease (e.g. tooth root infection).
- Intranasal neoplasia.
- Congenital disorders (extreme brachycephalic conformation, cleft palate, nasopharyngeal stenosis).
- Sinonasal and sino-orbital aspergillosis and other fungal disease.

Note: potentially zoonotic diphtheria -toxin producing *Corynebacterium ulcerans* has been isolated from cats with bilateral nasal discharge in the UK[228], on mainland Europe[229] and in Japan[230]. Reports of human disease (diphtheria[228,231,232], pyosialadenitis[228,233], cutaneous ulceration[228,234] and one death[235]) occasionally have been reported associated with domestic cats. The bacterium is sensitive to most systemic antibiotics. Clinicians should be aware of the zoonotic potential.

As previously noted, the clinical signs include nasal discharge, sneezing, respiratory stertor, epistaxis and

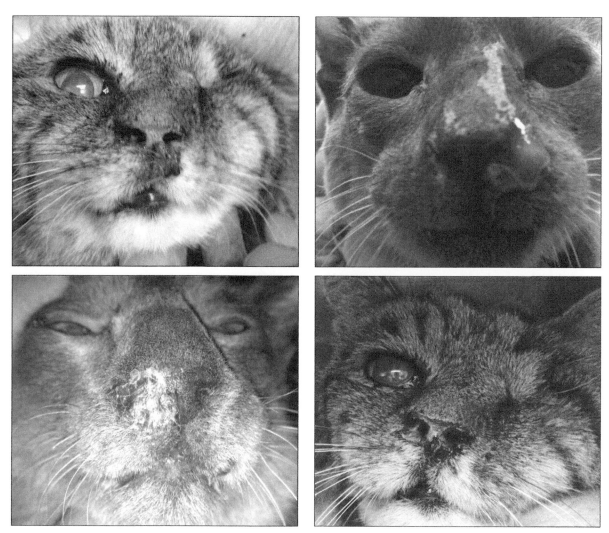

Figs 11.22–11.25 Examples of purulent nasal discharge. Note that the cat in Fig. 11.25 (bottom right) has a sanguinous discharge.

epiphora (Figs 11.22–11.25). Dysphagia and gagging might be seen if the caudal nares and nasopharynx are involved. Appetite may be adversely affected, perhaps because of an inability to smell, or pyrexia or because of difficulty in breathing[34,204,205]. Consideration of the animal's signalment, lifestyle, prior history, including travel and relocation should be established. The duration of the clinical signs, and how these have progressed, is important. For example, was the discharge initially unilateral and is it now bilateral? Was the discharge initially serous and is it now sanguinous?

- For example, a cat that had been sneezing previously, with a chronic bilateral nasal discharge, but has now stopped sneezing. The chronic inflammation results in inhibition of the sneeze response as irritant receptors are damaged and debris accumulates in the nares and sinuses[204].
- Epistaxis or a serosanguinous discharge may reflect hypertension secondary to chronic renal failure or a coagulation defect or a foreign body[34].
- A change from uni to bilateral discharge and a change in the nature of the discharge (e.g. seromucoid to sanguinous or purulent) is suggestive of progressive dental disease, foreign body or neoplasia.

A rational approach to investigation includes the following steps:

- Evaluate the oral cavity for oral and dental disease.
- Rule out a nasal foreign body or an oropharyngeal polyp.
- Perform a complete blood count, serum biochemistries and FeLV/FIV serology.
- Diagnostic imaging to asses for bony changes that might suggest neoplasia.
- Consider submitting serum for *Aspergillus* spp. or *Cryptococcus* spp. titre. This may be prompted by regional awareness or a knowledge of recent travel.
- Submit samples collected using a cytology brush (Cytosoft®) or flushing (see Box 2). Anaerobic culture should be requested in addition to aerobic[192]. Growth of a pure culture (i.e. a single organism) warrants the need for prolonged antimicrobial therapy, perhaps 6–8 weeks in duration. Culture and sensitivity testing (C&S) ensures selection of an appropriate antibacterial agent.
- Consider submitting samples for FHV-1, FCV, *M. felis*, *C. felis*. However, since all four organisms can be found as a commensal in a healthy cat's nose, the value is questionable, particularly since false-negative results confound interpretation[193]. Similarly a 'polymerase chain reaction (PCR) respiratory panel' should be interpreted cautiously due to false-positive findings.
- Perform endoscopic examination of both sides of the nasal cavities followed by the nasopharynx (see Chapter 9 Nose Diagnostic Procedures, Section 9.6). Biopsy samples can be submitted for cytological examination, histopathological examination and C&S.

> Note: authorities agree that there is no one test that will allow for definitive diagnosis of chronic rhinosinusitis.

Treatment of chronic sinusitis

It is important that the owner appreciates that this is a problem that can, at best, be managed, not cured. Palliative treatment can result in an improved quality of

> **Box 2 Flush sampling for investigating the bacterial flora of the deep nose (modified from Scherk[204])**
>
> - Under general anaesthesia with an endotracheal tube in place, pack the pharynx with swabs; count them to ensure retrieval!
> - Hold a tom cat urinary catheter along the cat's face, measure from the tip of the nose to the medial canthus of the eye. Mark this distance to avoid penetrating the cribrifom plate.
> - Gently insert the catheter into the nostril, up to the mark. Inject 2 ml sterile saline through the catheter and aspirate it out. Submit for anaerobe culture.
> - Repeat on the other side.

life and a reduction in the clinical signs. Complete resolution of the problem is unlikely. Secondly, the owner should be informed that it may take time to find the best combination of treatments for any particular cat.

> Ensuring that you maintain the client's confidence is important.

Antibacterial therapy[204,206]

While the results of C&S may indicate specific organisms, many of these are commensal organisms found in healthy cats. Note, and see Box 3:

- If anaerobes are found on C&S they are likely to be significant[192].
- Choosing an antibacterial agent when presented with a mixed growth of varied secondary invaders is problematic[204].
- Select an antibacterial agent that can achieve the required tissue concentration in mucus and must be able to penetrate cartilage.
- Be prepared to treat for 6–8 weeks, if necessary, in order to effect as good as response as possible.
- Choosing an antibacterial preparation that the owner can administer with confidence and that the cat will accept is critical.
- Although an orally-administered liquid preparation might well appear to have an advantage over tablets, there is a concern that a proportion of the administered dose might be spat or drooled out.

- A single injection of covexin sodium (very long acting) was shown to be less effective than oral treatment doxycycline or amoxicillin-clavulanic acid in a study in shelter cats[236].
- There may be no alternative to a tablet.

Box 3 Appropriate first choice antimicrobials

- *M. felis, C. felis, B. bronchiseptica*: tetracycline, particularly doxycycline (10 mg/kg p/o q24h) or pradofloxacin (care with marbofloxacin) (2 mg/kg p/o q24h).
- *C. felis* (not *M. felis*): amoxicillin-clavulanate (12.5–20 mg/kg p/o q12h).
- *M. felis*: clindamycin (5–11 mg/kg p/o q12h with food – risk of oesophagitis) or chloramphenicol (15–25 mg/kg p/o q12h[237]). Azithromycin (5 mg/kg p/o q48h) has been recommended anecdotally.
- Good choices for mixed infection might be:
  - Amoxicillin (10 mg/kg p/o q12h) or amoxicillin-clavulanate 12.5–20 mg/kg p/o q12h;
  - Marbofloxacin 2 mg/kg p/o q24h. See note above re fluroquinolone-associated retinal degeneration when considering these drugs for cats that have impaired renal function[226];
- Note that amoxicillin-clavulanate and azithromycin are not effective against *Pseudomonas* spp.

Improving the air flow through the nares
Nasal therapeutic flushing and nebulisation[34]
- Under general anaesthesia and using a cuffed endotracheal tube, the pharynx is packed with a known number of gauze swabs.
- A 10 ml syringe is filled with sterile saline (at body temperature) and the tip inserted into one of the nostrils.
- The contralateral nares is digitally compressed and the saline is instilled in a steady stream to flush mucous and debris back into the pharynx, where it is captured by the gauze.
- Repeat until the airway is flushed clear and repeat on the contralateral side.
- Nebulisation of saline, and antibacterial agents, may help to loosen debris and deliver antimicrobials topically. This equipment is available at many drug stores. Placing the affected cat in a room with a humidifier may also help[193].

Decongestants and mucolytics
- Xylometazoline hydrochloride 0.05% pediatric nasal drops (Otrivin ®Novartis) may be administered topically, but cats vary greatly in their tolerance to this procedure. Xylometazoline is a long-acting topical decongestant, which outperformed oral pseudoephedrine in reducing airway resistance[238]. In addition, it has no potential to induce any sympathomimetic side-effects, such as might be seen with oral pseudoephedrine decongestants.
- Administer to the cat for 3 days only, to avoid rebound congestion[204]. Rebound congestion is a paradoxical increase in upper airway airflow resistance following several days of topical decongestant treatment. It is most likely a result of receptor down-regulation and uncoupling, facilitating increased blood flow into the mucosa[239].
- Bromhexine (Bisolvon®) 1mg/kg p/o q24h, or 3 mg/kg i/m q24h.
- Saline nasal drops may help keep nasal secretions from building up (Scherk, personal communication).

Antihistamine therapy
- Diphenhydramine is an H-1 antihistamine and its decongestant effect may act via preventing histamine mediated mast cell degranulation within the nasal mucosae. The dose is 2–4 mg/kg p/o q8h. Liquid formulations would be ideal.
- Other antihistamines include:
  - Chlorpheniramine maleate 1–2 mg/kg p/o q8–12h;
  - Clemastine 0.05–0.1 mg/kg p/o q12h;
  - Hydroxyzine 2 mg/kg p/o q8–12h.

Anti-inflammatory treatment
- Prednisolone at standard anti-inflammatory doses (0.5 mg/kg p/o q24h × 7 d then q48h) may be beneficial[204]. By suppressing the cycle of inflammatory cascades, reducing leucocyte migration, and reducing the release and concentration of inflammatory mediators that characterises chronic rhinitis[240].

- The risk of use is recrudescence of FHV-1 in affected cats.
- Glucocorticoids are contraindicated in any cat with keratitis.
- Non-steroidal anti-inflammatory medication such as piroxicam (0.3 mg/kg p/o q48–72h) or meloxicam (0.03 mg/kg p/o q24h) might be of value in some cases. They may help to break the inflammatory cascade and provide analgesia. Patently, one needs to ensure that the cat is eating and, particularly, drinking normally and is optimally hydrated, before considering non-steroidal anti-inflammatory drug (NSAID) therapy.

Antiviral therapy
Conclusive evidence for the benefit from interferon (Virbagen, Virbac: I MegaUnit/kg, q24h × 5 days as a subcutaneous injection has been suggested), famciclovir or lysine therapy is lacking (summarised in[34]).

Nutritional support: appetite stimulation
It is critically important that cats with rhinitis continue to eat to avoid a negative plane of nutrition. Cats with severely compromised nasal function may have a suppressed appetite or be unable to breathe while eating. Hand feeding, off a spoon and offering titbits of highly aromatic foods might help. Appetite stimulants may also be tried, recently reviewed[241].

- Cyproheptadine (1 mg/cat p/o q12h).
- Mirtazapine, 2 mg p/o q24h, was shown to increase the appetite of healthy cats[242]. The authors further advised[243,244] an alternate day dose be used in cats with compromised renal function. By extension, that might also apply to cats with poor hydration status.
- Anecdotally, anabolic steroids (e.g. nandrolone laurate 2–5 mg/kg by subcutaneous injection) may have some value in helping to keep the cat in a positive plane of nutrition. Note, Agnew and Korman[241] regarded nandrolone as palliative only.

Sinonasal and sino-orbital aspergillosis in cats
Barrs *et al.*[83] reported on sinonasal and sino-orbital aspergillosis in 23 cats. Sino-orbital aspergillosis appears to be much more common than sinonasal aspergillosis, and Persian and Himalayan cats appear to be predisposed[82]. Early signs are of chronic serous to mucopurulent nasal discharge, which may occur as much as 6 months before the signs of sino-orbital infection become apparent[82]. These include exophthalmus, and prolapse of the nictitating membrane (Figs 11.26A, B). Barrs *et al.* further reported that the most common cause of this fungal infection was a novel species, *Aspergillus felis*, (an *A. fumigatus*-like fungus)[245]. Molecular identification is required to differentiate *A. felis* from *A. fumigatus*[205,246]. This condition is further discussed in section 11.6 Mycotic rhinitis.

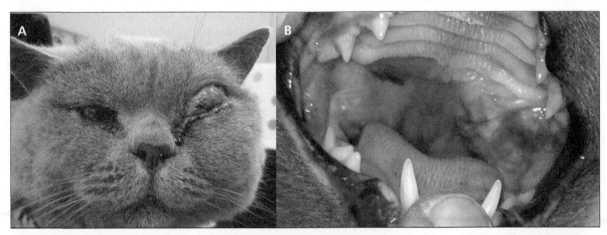

Figs 11.26A, B Feline aspergillosis. In Fig. 11.26A, there is thin serous nasal discharge, facial swelling and exophthalmus. In Fig. 11.26B the ulcerated lesion on the hard palate is plainly visible. (Used with permission, Barrs VR, Halliday C, Martin P, *et al.* Sinonasal and sino-orbital aspergillosis in 23 cats: aetiology, clinicopathological features and treatment outcomes. *Veterinary Journal* 2012;191: 58–64.)

Canine infectious respiratory disease

Canine infectious respiratory disease (CIRD) is commonly known as 'kennel cough'. Most authorities[247–252] agree that two viruses and one bacteria are implicated in most cases of canine upper respiratory disease:

- Canine parainfluenza virus.
- Canine adenovirus type 2.
- *Bordatella bronchiseptica*.

Other viruses, particularly canine influenza virus, canine respiratory coronavirus, canine herpesvirus and canine reovirus, may also prove to be clinically very significant. In addition, other bacteria such as *Streptococcus* spp., *Pasteurella* spp., *Pseudomonas* spp. may also be recovered from clinical cases[250–252], although these are likely to be opportunistic pathogens[253]. The role of *Mycoplasma cynos* is also being reassessed[251–253].

> Note: the presence of one or more of these agents, or the presence of circulating specific antibody, does not imply disease[247–249]. Similarly, although significantly higher recovery rates of canine parainfluenza virus and *B. bronchiseptica*, for example[250], are found in dogs with respiratory disease than in clinically healthy dogs, samples from some healthy dogs do yield positive samples.

The most common signs of canine upper respiratory disease are coughing, nasal discharge, anorexia and, occasionally, dyspnoea[251]. Most cases are self-limiting and recover, although significant secondary infection or immunocompromisation may increase morbidity.

Canine parainfluenzavirus

Canine parainfluenza virus, along with *B. bronchiseptica*, is considered a key component in the canine infectious tracheobronchitis complexes, known as kennel cough. The virus is highly contagious and prevalence of infection is proportional to the density of the dog population[247,253]. The virus replicates in cells of the nares, pharynx, trachea and bronchi. Experimental infection results in rhinitis, pharyngitis and an associated cough. Viral spreading (principally aerosol) persists for 8–10 days postinfection[253]. There may be mild

pyrexia[247]. The cough may have a 'honking' nature, which might reflect vocal fold swelling[253] and tracheal mucosal membrane swelling and there may be a serous nasal discharge. Secondary infection, particularly with *B. bronchiseptica*, is common.

Vaccine: combined intranasal vaccines are available:

- Nobivac® Intra-Trac®3ADT: canine adenovirus type 2, canine parainfluenzavirus and *B. bronchiseptica*.
- Nobivac® In-Trac KC and Nobivac® KC: canine parainfluenzavirus and *B. bronchiseptica*.
- Durvet® Kennel Jec-2: canine parainfluenzavirus and *B. bronchiseptica*.
- Note: Boehringer Ingelheim produce an oral canine parainfluenza vaccine, Bronchi-Shield® ORAL, which is applied into the buccal cavity.

Canine adenovirus type 2

This virus is regarded as one of the contributory components of canine infectious tracheobronchitis[247,250]. The virus replicates in non-ciliated epithelial cells of the nares, pharynx, tonsillar crypts, trachea and bronchi[247]. However, it is only associated with mild, sometimes inapparent, clinical signs in experimental infections[247]. When additional viral or bacterial agents are involved, tracheobronchitis may be observed[247]. A 2 year study of respiratory diseases in dogs in a rehoming kennel found no evidence of adenovirus type 2[254]. The authors theorised that routine vaccination against infectious hepatitis (adenovirus 1) had conferred cross-immunity, citing Fairchild *et al*.[255].

Vaccine: as mentioned above, although there are specific canine adenovirus 2 vaccines, there is cross-protection from canine adenovirus 1, which is part of the WSAVA core-vaccine regime for dogs.

See: http://www.wsava.org/sites/default/files/VaccinationGuidelines2010.pdf[256].

Bordetella bronchiseptica

Bordetella bronchiseptica is a small, gram-negative, flaggelated bacteria[257], which can elaborate several toxins[258], although how these relate to clinical disease is not understood[259]. It is the most common bacteria isolated from clinical cases of infectious tracheobronchitis[250,260,261]. It may be the critical complicating factor in dogs simultaneously infected with a viral pathogen[253]. Clinical signs of *B. bronchiseptica* infection are not specific and would be expected to include a nasal discharge, rhinitis, mild to moderate/severe cough and anorexia[259,262]. Vaccines are available. The intranasal route is preferred, as these have been demonstrated to reduce shedding of bacteria[263] and to protect from experimental infection within 72 hours[264].

Vaccine: a combined canine parainfluenza/*B. bronchiseptica* intranasal vaccine is available:
- Nobivac® Intra-Trac®3ADT: canine adenovirus type 2, canine parainfluenzavirus and *B. bronchiseptica*.
- Nobivac® In-Trac KC and Nobivac® KC: canine parainfluenzavirus and *B. bronchiseptica*.
- Durvet® Kennel Jec-2: canine parainfluenzavirus and *B. bronchiseptica*.
- A subcutaneous vaccine is available (Zoetis® Bronchicine®CAe). Although WSAVA recommends the intranasal route (improved local and systemic response, for example see[263]), there are some circumstances where subcutaneous injection is preferable. For example some dogs become particularly adverse to receiving intranasal vaccination.

Canine respiratory coronavirus

This virus was first described as a canine respiratory pathogen in 2003[248–250,265,266]. It is found worldwide and is considered an important agent of respiratory infection in kenneled dogs[251,252]. Coronavirus is highly contagious and seropositive dogs have been documented wordwide[248–252,260]. For example, in one study 30% of dogs were noted to be seropositive on the day of admission into a rehoming clinic[266]. After 3 weeks, 100% were seropositive. The clinical signs are non-specific and reflect mild upper respiratory disease, with, typically, a nasal discharge, sneezing and a dry cough[251,265]. Highest viral loads are found on the tonsil and in the tracheal epithelium[265]. Histopathological examination of experimentally infected dogs shows epithelial inflammation and damage in the nares and in the trachea, and an associated loss of cilia[265]. This damage to the mucociliary clearance mechanisms may predispose the dog to secondary infections[251].

Vaccines:
- (Nobivac® Canine 1-Cv), recommended for dogs over 6 weeks of age.
- Zoetis® Vanguard CV.

Canine influenzavirus

First described in racing Greyhounds in Florida this H3N8 strain is now recognised in group-housed dogs and rehoming kennels, particularly in the USA, although it is regarded as uncommon[252]. It is considered to be fully adapted to the dog[267a], although the status of the canine H3N2 virus isolated from dogs in China is not known[267b]. Clinical signs occur in around 80% of cases[268] and reflect upper respiratory infection.

Two clinical syndromes have been recognised:
- Severe pneumonia with pyrexia (40–41°C [104–106°F]), moist cough, tachypnoea, dyspnoea, purulent nasal discharge, depression and anorexia. Radiographs may demonstrate severe bronchopneumonia. Although this is a serious disease, mortality is low, 1–8%[268].
- A milder form characterised by a persistent, soft, moist cough and mild pyrexia. A harsh dry hacking cough may persist for 10–30 days[247,267,268] despite treatment, and may be accompanied by a serous to mucopurulent nasal discharge[247].

Suspicion of canine influenzavirus infection should be raised if there is an outbreak of upper respiratory disease in a pool of dogs that have been vaccinated with their core-vaccines and have received intranasal kennel cough vaccine[267]. There is a killed H3N8 vaccine available in the USA, and it is indicated for dogs from 6 weeks of age. It is claimed to significantly reduce the clinical signs, the severity and spread of canine influenza infection and to reduce the incidence and severity of cough. It also offers protection against the formation and severity of lung lesions. The vaccine also reduces the duration and amount of virus shedding.

Vaccine:
- Nobivac® Canine Flu H3N8.

Canine herpesvirus

In dogs older than 3 months of age, canine herpesvirus infection has resulted in mild rhinitis and pharyngitis[269a]. In young puppies infection may be fatal and fatality has rarely been reported in adult dogs[269b]. Although canine herpesvirus has been recovered from dogs with infectious respiratory disease, reproduction of clinical 'kennel cough' has only rarely been reported[269a]. Serological evidence suggests worldwide exposure, and seroprevalence varies from 21.7% (Japan) to 45.8% (Belgium) and 88% (UK), summarised by[269a]. A 2-year study of dogs in a rehoming kennel found a seroprevalence of 12.8% (tracheal sample)[254]. Interestingly, the herpes virus infection was noted 3–4 weeks after entry, after the dogs had acquired, and were recovering from, contagious tracheobronchitis[254]. The dogs were noted to exhibit more severe respiratory signs. A canine herpesvirus vaccine is available in Europe. It is administered to pregnant bitches to ensure good immunity in the newborn pups. It is not known if it would have any impact on respiratory infection in which canine herpesvirus was deemed to play a part.

Vaccine:
- Meriel Eurican Herpes® 205.

Canine reovirus

Reoviruses are common in dogs[247] and they have been isolated from cases of canine contagious respiratory disease. For example, reovirus RNA was demonstrated in 9/12 ocular samples and 10/19 nasal swabs (but not from the oropharynx) in dogs with an oculonasal discharge[247,270]. Currently, reoviruses are not regarded as primary pathogens of the respiratory tract[247].

Mycoplasma cynos
Mycoplasma cynos has been recovered from the respiratory tract of dogs with respiratory disease[251,253,271]. The current status of its pathogenicity is uncertain.

Management of infectious rhinotracheitis

Dogs with infectious upper respiratory disease will typically show a predictable range of signs, with high morbidity and low mortality[253]:
- Serous to purulent nasal discharge.
- Sneezing.
- Cough, dry or productive, sometimes exacerbated by exercise.
- Mild anorexia.
- Mild to moderate/severe pyrexia, particularly with canine influenzavirus infections.
- Dyspnoea, perhaps tachypnoea.

There is no particular correlation between the spectrum of clinical signs an individual may exhibit and aetiological agent(s)[250]. Thus, treatment for affected dogs is, essentially, symptomatic, after[253,268]:
- Antitussives should only be used in cases with dry unproductive cough. They are contraindicated in cases with productive cough as they can inhibit clearance of mucus and inflammatory debris[268].
  - Butorphanol 0.055–0.11 mg/kg p/o q6–12h;
  - Hydrocodone 0.22 mg/kg p/o q6–12h.
- Nebulised delivery of saline can be beneficial, but best performed under veterinary supervision. Nebulised delivery of mucolytics such as N-acetylcysteine is contraindicated as this can result in bronchoconstriction[268].
- Bronchodilators such as the methylxanthine derivatives may be helpful. These agents might help to prevent bronchospasm. There is little clinical data to support their use[253]:
  - Aminophylline 10 mg/kg p/o q12h.
- Antibacterial treatment of secondary bacterial infection, 10–14 days minimum course, first choices[253,272]:
  - Amoxycillin-clavulanate 10–20 mg/kg p/o q12h;
  - Cephalosporin, e.g. cephalexin 20–30 mg/kg p/o q8h;
  - Fluorquinolone, e.g. enrofloxacin 5–20 mg/kg p/o q24h;
  - Trimethoprim sulfonamine 15 mg/kg p/o q12h.
- Prednisolone at an anti-inflammatory dose for 3–5 days can ameliorate cough frequency and intensity, without compromising recovery. CAUTION: do not use bacteriostatic antibacterial agents if using prednisolone.
  - 0.25–0.5 mg/kg p/o q12–24h.
- Advise isolation. Having said that, peak virus shedding occurs 2–3 days after infection, just when clinical signs are often first noticed[267]. The virus is spread by aerosol (cough and sneeze) by both direct (dog to dog) and indirect contact (owner's hands and clothing).

Nasal parasites

Pneumonyssoides caninum (Fig. 11.27) is a mite of around 1.0–1.5 mm in length, and is thus visible to the naked eye. The adult mites are oval in shape and the six legs arise from the rostral half of the body, reviewed[273]. The details of the mite's natural history are not known, but it is widely distributed, having been reported in both Northern Europe, the Middle East and North America[33,274–277]. A feline nasal mite has not been reported. In a Swedish (prospective necropsy) study the prevalence

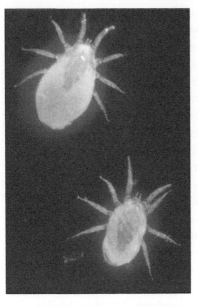

Fig. 11.27 Adult *Pneumonyssoides caninum*. The female is larger than the male. (Courtesy of Professor Lotta Gunnarsson, Swedish National Veterinary Institute; photographer: Bengt Ekberg.)

Fig. 11.28 Adult mites in the nasal cavity of a dog. Postmortem sample. (Courtesy of Professor Lotta Gunnarsson, Swedish National Veterinary Institute; photographer: Bengt Ekberg.)

was reported to be as high as 95% in dogs, with a median number of 13 mites being found on each dog[276].

The mite resides in the nares and sinuses[278] (Figs 11.27, 11.28). Transmission is thought to be via direct contact[279], although indirect transmission is thought to be possible[273] as the mites have been reported to survive for up to 19 days in a cool humid environment[279].

The clinical signs associated with infection are reported to be a serous nasal discharge[278], acute, tending to persistent, sneezing, not necessarily associated with a nasal discharge[273] and reverse sneezing (see Chapter 8 Nose Anatomy and Physiology, Section 8.4). Epistaxis and impaired scenting ability has also been reported[279]. However, the seemingly low morbidity rate (especially when compared with what can be a very high prevalence) suggests that the mite resides in many dogs and is not associated either with pathological or clinical changes.

Treatment with various antiparasitics has been reported as effective:

- Selamectin at 6–24 mg/kg, three times at 2-week intervals[280].
- Milbemycin oxime 0.5–1.0 mg/kg body weight, p/o, once a week for 3 weeks[281].
- Ivermectin at 200µg/kg SC twice at a 2-week interval[282].

Linguatula serrata is an obligate parasite of the upper respiratory tract of terrestrial, carnivorous vertebrates[283,284]. The parasites specifically resides in the nares, frontal sinus and tympanic cavity, although they may also be found sublingually[41]. The parasite is long, flat and annulated in shape (Figs 11.29A, B) and is pale in colour[41]. It measures around 18–20 mm in length and has a maximum width of 10 mm. The intermediate hosts are sheep, cattle or rodents[41,284,285]. Humans may become infected, particularly if they ingest raw, or undercooked, offal from infected cattle and sheep[41,284,285]. The parasite is found worldwide.

The infection is usually subclinical but rhinitis, epistaxis and mucopurulent discharge have been reported[284]. Prevalence has been reported as between 38 and 62%[41,284]. The diagnosis of infection is made by visualising the parasite using endoscopy. There is no proven medical treatment. Manual removal of the parasite is the only proven treatment to date.

Eucoleus boehmi is a parasitic nematode (previously classified as *Capillaria boehmi*) that appears to be highly prevalent in the sinuses and nasal passages of red foxes. It has been suggested that domestic dogs may be infected, presumably via indirect and actual contact[285a]. Canine infection has been reported in North America and Europe[285b–f]. The most common clinical signs are sneezing, rhinitis, impaired scenting ability and a mild to mucopurulent nasal discharge[285c–e].

Diagnosis can be made by direct observation of the white, serpentine worms, some 15–30 mm in length, on the surfaces of the turbinates (Fig. 11.29C), by examination of nasal biopsy samples and by demonstration of the ova in faecal samples (Fig. 11. 29D).

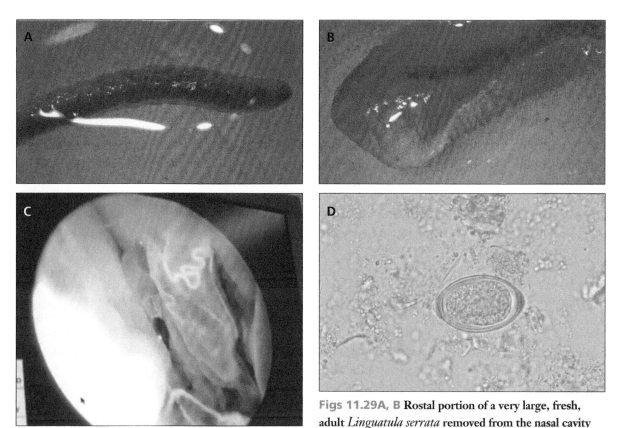

Figs 11.29A, B Rostal portion of a very large, fresh, adult *Linguatula serrata* removed from the nasal cavity of a dog (A). Figure 11.29B shows the extreme rostral portion where the pale colour and spatulate shape is very apparent. (Courtesy of Professor Iraj Sharifi, School of Medicine, Kerman University of Medical Sciences, Kerman, Iran.) Figs 11.29C, D *Eucoleus boehmi* nematode. C: adult *Eucoleus boehmi* nematode on the turbinates of a dog presented with chronic rhinitis. The illustration is taken from an intraoperative rhinoscopic video; D: egg of *Eucoleus boehmi* in a photomicrograph of a faecal preparation from the same dog. (Courtesy Dr Ana Margarida Alho, EVPC Resident, Faculdade de Medicina Veterinária, ULisboa CIISA, Laboratório de Doenças Parasitárias, Lisboa, Portugal.)

Successful treatment with imidacloprid/moxidectin spot on, moxidectin and milbemycin oxime has been reported[285c-e].

11.10 ALLERGIC RHINITIS

Allergic rhinitis is a very well-known and common cause of recurrent sneezing, rhinorrhoea and nasal congestion in humans. Although definite criteria are lacking, loud and frequent sneezing with large amounts of bilateral watery discharge is sometimes seen in dogs. There is therefore speculation that respiratory allergy may manifest with signs of chronic rhinitis as is seen in humans, although evidence of allergic rhinitis as a recognised disease entity in dogs has yet to be established[47].

Experimentally, nasal congestion has been induced using ragweed pollen in Beagle colonies[286–288], and evidence of allergic response to house dustmite antigen characterised by increased interleukin-4 (IL-4) expression and T helper cell 2 (Th2) immune response was identified in the peripheral blood mononuclear cells of three dogs with rhinitis[289].

A presumptive diagnosis is made by performing a thorough history, physical examination, CT, rhinoscopy, with nasal mucosal biopsy to rule out lymphoplasmacytic rhinitis (LPR) and other primary aetiologies of nasal discharge[33,40,47,290]. Ideally histopathological examination of nasal biopsies would identify an eosinophilic inflammation.

Therapy should be aimed at identification and elimination of allergens (mites, dust, pollen, etc). Furthermore, it would appear to be extremely important to improve the air quality and ensure the animal is kept in a fresh, well-ventilated, smoke- and perfume-free environment. Medical therapy consisting of a topical treatment with dexamethasone eyedrops, 3 dd 3 gtt in each nostril can be tried first in patients with presumed allergic rhinitis. With severe congestion, topical xylomethazoline (Otrivin, 0.05%, 3 d 3 gtt in each nostril) can be given additionally. Affected animals may require a systemic dose of corticosteroids, however. Some animals respond favourably to antihistamines; for instance clemastine fumarate can be given in a dosage of 0.1–0.2 mg/kg, twice daily. Careful monitoring and regular re-evaluation of these patients is advised to ensure no other intranasal pathology is present.

11.11 CHRONIC, NONSPECIFIC OR IDIOPATHIC RHINITIS

Inflammatory, idiopathic or lymphoplasmacytic rhinitis or rhinosinusitis is a group of nasal oversensitivity syndromes due to unknown pathomechanisms, frequently diagnosed in dogs and cats. Idiopathic rhinitis appears to be the best term to describe the condition and is a diagnosis of exclusion, when other forms of rhinitis have been thoroughly ruled out. Depending on the report, up to 49% of dogs and 65% of cats may be diagnosed with this type of rhinitis[33,34,40,47,62,63,147,185,291–293].

Three pathomechanisms have been proposed for idiopathic rhinitis in people:
- Neuronal dysfunction (neurogenic).
- Immune inflammatory.
- Mucosal damage.

In neurogenic rhinitis nasal obstruction and hypersecretion are the result of decreased sympathicotonia or increased parasympathicotonia. Occupational rhinitis, drug-induced rhinitis, rhinitis associated with pregnancy, rhinitis due to physical/chemical factors, food-induced or gustatory rhinitis, atrophic rhinitis, non-allergic eosinophilic rhinitis and worsening of rhinitis by emotions, physical stress or sexual arousal have all been documented in humans, but not in dogs or cats. It is likely though that these conditions are responsible for some of the cases of rhinitis in dogs and cats that we currently classify as 'idiopathic'.

There is some evidence that similar pathomechanisms are present in chronic rhinitis in dogs and cats:

severe mucosal damage after surgery, radiation or fungal rhinitis may lead to chronic idiopathic rhinitis[36,133]. Neurogenic rhinitis with dryness of the nasal plane (uni- or bilateral), keratoconjunctivitis (KCS) and otitis media was reported by Venker due to loss of parasympathic innervation to the lacrimal glands[12]. Likely immune-mediated rhinitis has been reported in Irish Wolfhounds, with or without bronchopneumonia[294–296]. In most dogs and cats with chronic rhinosinusitis no cause is identified though, and the condition will be classified as idiopathic. Viral infections are presumed to often be the cause or lead to secondary changes responsible for the syndrome of chronic rhinosinusitis, and this condition in cats is therefore discussed in Section 11.9.

Chronic idiopathic rhinitis in dogs

Chronic inflammatory rhinitis is commonly found in dogs with chronic nasal disease and is characterised by lymphoplasmacytic infiltrates in the nasal mucosa in the absence of an obvious aetiologic process. A recent study investigating the role of infectious organisms in LPR quantified DNA loads of *Chlamydophila*, *Bartonella*, canine adenovirus 2 (CAV-2), and parainfluenza virus 3 (PI-3) in biopsy samples from dogs with LPR by the use of quantitative PCR, and a role for these organisms could not be established[290]. In addition, end-point PCR was used to examine nasal biopsy samples for evidence of *Mycoplasma* spp. and this also failed to identify a specific pathogen in DNA extracted from formalin-fixed nasal biopsy samples[290]. Secondary bacterial infection is, however, common in dogs with chronic nasal disease; nasal cultures from dogs with LPR commonly yield mixed bacterial growth of normal nasal flora including *Staphylococcus*, *Streptococcus*, *Escherichia coli*, *Proteus*, *Pasteurella*, *Corynebacterium*, *Bordetella* and *Pseudomonas* spp.[40].

Candida, *Trichosporum* and *Cladosporidium* have also been cultured from dogs with LPR[40]. Further research is required to determine the significance of increased fungal DNA in dogs with LPR and to assess the immune response of affected dogs to resident fungal organisms[47]. To date, the exact pathogenesis of LPR remains unknown[33,40,47].

Common clinical signs include nasal discharge, sneezing, epistaxis and stertor, though coughing has also been reported, which suggests involvement of larynx, trachea or bronchi.

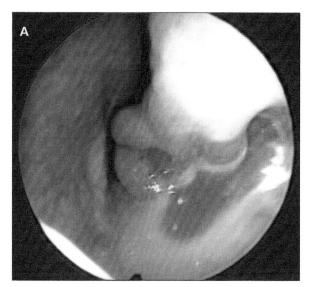

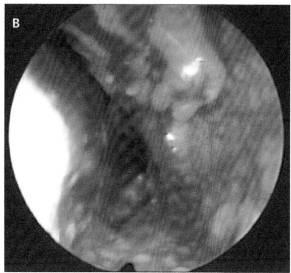

Figs 11.30A, B Rhinoscopic images of a dog with a chronic non-specific lymphoplasmacytic rhinitis. A: Hyperemia and exudate is visible in the nasal vestibule; B: mild atrophy of the turbinates and follicular hyperplasia, hyperaemia and congestion of the overlying mucosa.

The diagnosis is made by performing a thorough history, physical examination, radiography or advanced imaging, rhinoscopy and nasal mucosal biopsy to rule out primary aetiologies of nasal discharge. Imaging studies of affected animals typically reveal increased soft tissue opacity within the nasal cavity, with or without lysis of conchae, but lysis of other bones is not seen[40,147,148,297,298]. Soft tissue opacification consistent with fluid accumulation may also be observed in the sinuses by radiography or CT[150].

The most common rhinoscopic features in dogs with LPR include mucoid or mucopurulent discharge, hyperemic, oedematous and/or friable nasal mucosa and mild turbinate atrophy or destruction[40,47] (Figs 11.30A, B). Nasal biopsy samples of dogs with chronic idiopathic rhinitis are characterised by primarily a LPR, but concurrent neutrophilic or less commonly eosinophilic infiltrate may also be seen[40,47]. The majority of dogs with unilateral clinical signs have bilateral nasal mucosal pathology, and severity of inflammation between the two sides of the nasal cavity often varies.

Treatment strategies have included various antibiotics, antihistamines, oral and inhalant steroids, NSAIDs and antifungal medications. Some dogs may respond partially to doxycycline or azithromycin, although it is unclear whether the response is related to the antimicrobial or anti-inflammatory properties of these drugs[40,47]. Hydration of the nasal cavity through nasal drops or aerosols may limit nasal discharge, and some animals may improve with inhalant (but rarely oral) glucocorticoids[40,47].

11.12 NASAL AND SINAL TRAUMA

Trauma, defined as tissue injury caused by violence or accident that occurs suddenly and includes physical damage to the body[299], is a common cause of morbidity and death in dogs[300]. Results of large-scale epidemiological studies indicate that trauma accounts for approximately 12% of all animals evaluated at urban veterinary teaching hospitals[301,302]. In a large study on causes of death in more then 74,000 dogs, trauma was the second most common cause of death in juvenile dogs (following infectious disease) and adult dogs (following neoplasia)[303]. Though outcome of severe blunt trauma in dogs treated with intensive care is very good, head trauma and cranium fractures were significantly associated with a negative outcome in two large studies[304,305].

Head trauma may result from many different events, including motor vehicle accidents, bites, falls, kicks or penetrating injuries such as gunshot wounds[299,306,307]. The most frequent cause of head trauma noted in small animals is motor vehicle trauma[301,307].

Traumatic events generate pain, stress and fear that individually or together initiate survival-orientated behavioural responses, collectively designed to prevent further tissue injury, and at worst elicits exuberant physiological, immunological and metabolic changes that predispose to organ malfunction, trigger inflammation and coagulopathies, promote infection and trigger an autodestructive inflammatory process[299]. The scale and pattern of the responses to trauma are dependent upon the extent of haemorrhage, tissue injury, pain, and stress and together determine mortality.

Animals with severe head trauma should therefore first be:

- Examined for signs of shock such as tachycardia, hypotension, breathing abnormalities, dilation of pupils, hypothermia, muscle weakness, restlessness, depression and coma.
- Stabilised before further diagnostic work-up[12].

In a retrospective study of dogs with head trauma, modified Glasgow coma scale (MGCS) scores were predictive for non-survival within 48 hours after injury[307]. Once the patient has been stabilised, radiographic examination of the skull may be performed as it provides useful information in head-injured animals, especially if there is the suspicion of fractures or penetrating injuries[12]. Approximately 37% of the dogs in one study had evidence of calvarial fractures on radiographic examination[307]. Computed tomography (CT) or MRI are more valuable if available[307].

The anatomic location and rich blood supply of the nasal passages and maxillary recesses predisposes these structures to several potential complications, such as epistaxis and nasal obstruction[308]. Nasal haemorrhage secondary to trauma to the nasal and frontal bones usually resolves spontaneously, without therapeutic intervention. Nasal obstruction can however, be an immediate or long-term complication of trauma to the nasal passages.

Acute nasal obstruction in maxillofacial trauma patients may be caused by crushing nasal injuries, haemorrhage, contusion and oedema[308]. In some patients supplementation with nasal oxygen or a temporary tracheostomy may be necessary. Exploration of crushing injuries to the nasal passages and frontal sinuses to remove fractured turbinates, bone fragments and devitalised soft tissue has been recommended[12,36]. Early exploration of these injuries may provide an open airway in a shorter time, reduces infection, eliminates prolonged foetid discharge and reduces the chance for airway stenosis due to scarring of malaligned tissue[36]. Special care has to be taken to ensure the nasofrontal ostium is patent to prevent chronic sinusitis as a result of blockage of drainage[12].

Chronic nasal obstruction following maxillofacial trauma is rare. Unilateral chronic nasal obstruction is usually well tolerated and does not require treatment in patients that are minimally compromised. Severe bilateral chronic nasal obstruction or nasopharyngeal stenosis may require surgical intervention[308].

Though fractures of the nasal and frontal bones will merely cause a cosmetic defect that can easily be repaired with miniplate fixation if required, one area of functional importance of the frontal bone is the zygomatic process. This area is the dorsal attachment of the orbital ligament, which serves as the lateral attachment of the lateral palpebral ligament and orbicularis oculi muscle[309]. Loss of support of the zygomatic process of the frontal bone will result in a lack of lateral support of the eye and fractures in this area require surgical stabilisation[309].

Another primary objective for repair of maxillofacial fractures in small animals is restoration of normal dental occlusion in patients with fractures involving the maxilla or mandibula.

11.13 REFERENCES

1 Mason LK, Evans S. Surgical closure of a congenital cerebrospinal fluid fistula causing rhinorrhea in a cat. *Journal of the American Animal Hospital Association* 1990; **26**:153–156.

2 Morrison WB, Wilsman NJ, Fox LE, Farnum CE. Primary ciliary dyskinesia in the dog. *Journal of Veterinary Internal Medicine* 1987;1:67–74.

3 Edwards DF, Patton CS, Kennedy JR. Primary ciliary dyskinesia in the dog. *Problems in Veterinary Medicine* 1992;4:291–319.

4 Zariwala MA, Knowles MR, Omran H. Genetic defects in ciliary structure and function. *Annual Review of Physiology* 2007;69:423–50.

5 Merveille AC, Battaille G, Billen F, *et al*. Clinical findings and prevalence of the mutation associated with primary ciliary dyskinesia in Old English Sheepdogs. *Journal of Veterinary Internal Medicine* 2014;28:771–8.

6 Watson PJ, Herrtage ME, Peacock MA, Sargan DR. Primary ciliary dyskinesia in Newfoundland dogs. *Veterinary Record* 1999;144:718–25.

7 Edwards DF, Kennedy JR, Patton CS. Familial immotile–cilia syndrome in English springer spaniel dogs. *American Journal of Medical Genetics* 1989;33:290–8.

8 Clercx C, Peeters D, Beths T, McEntee K. Use of ciliogenesis in the diagnosis of primary ciliary dyskinesia in a dog. *Journal of the American Veterinary Medical Association* 2000;217:1681–5.

9 De Scally M, Lobetti RG, Van Wilpe E. Primary ciliary dyskinesia in a Staffordshire bull terrier: clinical communication. *Journal of the South African Veterinary Association* 2004;75:150–2.

10 Neil JA, Canapp SO, Cook CR, Lattimer JC. Kartagener's syndrome in a Dachshund dog. *Journal of the American Animal Hospital Association* 2002;38:45–9.

11 Merveille A-C, Davis EE, Becker-Heck A, *et al.* CCDC39 is required for assembly of inner dynein arms and the dynein regulatory complex and for normal ciliary motility in humans and dogs. *Nature Genetics* 2011;43:72–8.

12 Venker-van Haagen AJ. The nose and nasal sinuses. In: *Ear, Nose, Throat, and Tracheobronchial Diseases in Dogs and Cats*. Venker-van Haagen AJ (ed). Schlütersche Verlagsgesellschaft, Hannover 2005, pp. 51–81.

13 Reichler IM, Hoerauf A, Guscetti F. Primary ciliary dyskinesia with situs inversus totalis, hydrocephalus internus and cardiac malformations in a dog. *Journal of Small Animal Practice* 2001;42:345–8.

14 Cavrenne R, De Busscher V, Bolen G, Billen F, Clercx C, Snaps F. Primary ciliary dyskinesia and situs inversus in a young dog. *Veterinary Record* 2008;163:54–5.

15 Edwards DF, Patton CS, Bemis DA. Immotile cilia syndrome in three dogs from a litter. *Journal of the American Veterinary Medical Association* 1983;183:667–72.

16 Edwards DF, Kennedy JR, Toal RL, Maddux JM, Barnhill MA, Daniel GB. Kartagener's Syndrome in a Chow Chow dog with normal ciliary ultrastructure. *Veterinary Pathology* 1989;26:338–40.

17 Poncet CM, Dupre GP, Freiche VG, Estrada MM, Poubanne YA, Bouvy BM. Prevalence of gastrointestinal tract lesions in 73 brachycephalic dogs with upper respiratory syndrome. *Journal of Small Animal Practice* 2005;46:273–9.

18 Lorinson D, Bright RM, White R. Brachycephalic airway obstruction syndrome – a review of 118 cases. *Canine Practice* 1997; 22:18–21.

19 Pink JJ, Doyle RS, Hughes JM, Tobin E, Bellenger CR. Laryngeal collapse in seven brachycephalic puppies. *Journal of Small Animal Practice* 2006;47:131–5.

20 Poncet CM, Dupre GP, Freiche VG, Bouvy BM. Long-term results of upper respiratory syndrome surgery and gastrointestinal tract medical treatment in 51 brachycephalic dogs. *Journal of Small Animal Practice* 2006;47:137–42.

21 Riecks TW, Birchard SJ, Stephens JA. Surgical correction of brachycephalic syndrome in dogs: 62 cases (1991–2004). *Journal of the American Veterinary Medical Association* 2007;230:1324–8.

22 Harvey CEFEA. Tracheal diameter: analysis of radiographic measurements in brachycephalic and nonbrachycephalic dogs. *Journal of the American Animal Hospital Association* 1982;18:570–6.

23 Schuenemann R, Oechtering G. Inside the brachycephalic nose: conchal regrowth and mucosal contact points after laser-assisted turbinectomy. *Journal of the American Animal Hospital Association* 2014;50:237–46.

24 Schuenemann R, Oechtering GU. Inside the brachycephalic nose: intranasal mucosal contact points. *Journal of the American Animal Hospital Association* 2014;50:149–58.

25 Reiter AM, Holt DE. Palate. In: *Veterinary Surgery Small Animal*. Tobias KM, Johnston SA (eds). Elsevier Saunders, St. Louis 2012, pp. 1707–17.

26 Rooney NJ. The welfare of pedigree dogs: cause for concern. *Journal of Veterinary Behaviour and Clinical Applied Research* 2009;4:180–6.

27 Rooney NJ, Sargan DR. Welfare concerns associated with pedigree dog breeding in the UK. *Animal Welfare* 2010;19:133–40.

28 Palmer C. Does breeding a bulldog harm it? Breeding, ethics and harm to animals. *Animal Welfare* 2012;21:157–66.

29 Oechtering GU. Brachycephalic syndrome – new information on an old congenital disease. *Veterinary Focus* 2010;20:2–9.

30 Bissett SA, Drobatz KJ, McKnight A, Degernes LA. Prevalence, clinical features, and causes of epistaxis in dogs: 176 cases (1996–2001). *Journal of the American Veterinary Medical Association* 2007;231:1843–50.

31 Strasser JL, Hawkins EC. Clinical features of epistaxis in dogs: a retrospective study of 35 cases (1999–2002). *Journal of the American Animal Hospital Association* 2005;41:179–84.

32 Mylonakis ME, Saridomichelakis MN. A retrospective study of 61 cases of spontaneous canine epistaxis (1998–2001). *Journal of Small Animal Practice* 2008;49:191–6.

33 Cohn LA. Canine nasal disease. *Veterinary Clinics of North America Small Animal Practice* 2014;44:75–89.

34 Reed N. Chronic rhinitis in the cat. *Veterinary Clinics of North America Small Animal Practice* 2014;44:33–50.

35 Schmiedt C, Danova N, Bjorling DE, Brockman DJ, Holt D. The nose and nasopharynx. In: *BSAVA Manual of Canine and Feline Head, Neck and Thoracic Surgery.* Brockman DJ, Holt DE (eds). BSAVA, Gloucester 2005, pp. 44–55.

36 Schmiedt CW, Creevy KE. Nasal planum, nasal cavity, and sinuses. In: *Veterinary Surgery Small Animal.* Tobias KM, Johnston SA (eds). Elsevier Saunders, St. Louis 2012, pp. 1691–706.

37 Rassnick KM, Goldkamp CE, Erb HN, *et al*. Evaluation of factors associated with survival in dogs with untreated nasal carcinomas: 139 cases (1993–2003). *Journal of the American Veterinary Medical Association* 2006;229:401–6.

38 Schuller S, Clercx C. Long-term outcomes in dogs with sinonasal aspergillosis treated with intranasal infusions of enilconazole. *Journal of the American Animal Hospital Association* 2007;43:33–8.

39 Sharman MJ, Mansfield CS. Sinonasal aspergillosis in dogs: a review. *Journal of Small Animal Practice* 2012;53:434–44.

40 Windsor RC, Johnson LR, Herrgesell EJ, De Cock HEV. Idiopathic lymphoplasmacytic rhinitis in dogs: 37 cases (1997–2002). *Journal of the American Veterinary Medical Association* 2004;224:1952–7.

41 Meshgi B, Asgarian O. Prevalence of *Linguatula serrata* infestation in stray dogs of Shahrekord, Iran. *Journal of Veterinary Medicine Series B* 2003;50:466–7.

42 King RR, Greiner EC, Ackerman N, Woodard JC. Nasal capillariasis in a dog. *Journal of the American Animal Hospital Association* 1990;26:381–5.

43 Conboy G. Helminth parasites of the canine and feline respiratory tract. *Veterinary Clinics of North America Small Animal Practice* 2009;39:1109–26.

44 Brückner GK, Minnaar RJ. Unilateral epistaxis in a dog. *Journal of the South African Veterinary Association* 1979;50:55.

45 Marretta SM. The common and uncommon clinical presentations and treatment of periodontal disease in the dog and cat. *Seminars in Veterinary Medicine and Surgery Small Animal* 1987;2:230–40.

46 Thomas WB, Schueler RO, Kornegay JN. Surgical excision of a cerebral arteriovenous malformation in a dog. *Progress in Veterinary Neurology* 1995; 6:20–3.

47 Windsor RC, Johnson LR. Canine chronic inflammatory rhinitis. *Clinical Techniques in Small Animal Practice* 2006;21:76–81.

48 Brooks MB. A hereditary bleeding disorder of dogs caused by a lack of platelet procoagulant activity. *Blood* 2002;99: 2434–41.

49 Brooks M, Dodds WJ, Raymond SL. Epidemiologic features of von Willebrand's disease in Doberman Pinschers, Scottish Terriers, and Shetland Sheepdogs: 260 cases (1984–1988). *Journal of the American Veterinary Medical Association* 1992;200:1966–71.

50 Brooks MB, Randolph J, Warner K, Center, Sharon. Evaluation of platelet function screening tests to detect platelet procoagulant deficiency in dogs with Scott syndrome. *Veterinary Clinical Pathology* 2009;38:306–15.

51 Keenan KP, Ruhles WC, Huxsoll DL, *et al*. Pathogenesis of infection with *Rickettsia rickettsii* in the dog: a disease model for Rocky Mountain spotted fever. *Journal of Infectious Diseases* 1977;135: 911–17.

52 Breitschwerdt EB, Hegarty BC, Maggi R, Hawkins E, Dyer P. *Bartonella* species as a potential cause of epistaxis in dogs. *Journal of Clinical Microbiology* 2005;43:2529–33.

53 Jüttner C, Rodríguez Sánchez M, Rollán Landeras E, Slappendel RJ, Fragío Arnold C. Evaluation of the potential causes of epistaxis in dogs with natural visceral leishmaniasis. *Veterinary Record* 2001;149:176–9.

54 Troy GC, Vulgamott JC, Turnwald GH. Canine ehrlichiosis: a retrospective study of 30 naturally occurring cases. *Journal of the American Animal Hospital Association* 1980;16:181–7.

55 Maruyama H, Miura T, Sakai M, *et al.* The incidence of disseminated intravascular coagulation in dogs with malignant tumor. *Journal of Veterinary Medical Science* 2004;66:573–5.

56 Ballenger JJ. Epistaxis, septal perforation, and skin of the face. In: *Otorhinolaryngology: Head and Neck Surgery*, 15th edn. Ballenger JJ, Snow JJB (eds). Williams & Wilkins, Philadelphia 1996, pp. 153–7.

57 Littman MP. Spontaneous systemic hypertension in 24 cats. *Journal of Veterinary Internal Medicine* 1994;8:79–86.

58 Brennan KE, Ihrke PJ. Grass awn migration in dogs and cats: a retrospective study of 182 cases. *Journal of the American Veterinary Medical Association* 1983;182:1201–4.

59 Fallon RK, McCaw D, Lattimer J. Unusual nasal foreign body in a dog. *Journal of the American Veterinary Medical Association* 1985;186:710.

60 Saylor DK, Williams JE. Rhinoscopy. In: *Small Animal Endoscopy*. Tams TR, Rawlings CA (eds). Elsevier Health Sciences, St. Louis, 2010, pp. 563–85.

61 Thrall DE. *Textbook of Veterinary Diagnostic Radiology*, 6th edn. Thrall DE (ed). Elsevier Saunders, St. Louis, 2013.

62 Henderson SM, Bradley K, Day MJ, *et al.* Investigation of nasal disease in the cat – a retrospective study of 77 cases. *Journal of Feline Medicine and Surgery* 2004;6:245–57.

63 Hunt GB, Perkins MC, Foster SF. Nasopharyngeal disorders of dogs and cats: a review and retrospective study. *Compendium on Continuing Education for the Practising Veterinarian* 2002;24:184–200.

64 Kirpensteijn J, Withrow SJ, Straw RC. Combined resection of the nasal planum and premaxilla in three dogs. *Veterinary Surgery* 1994;23:341–6.

65 Willard MD, Radlinsky MA. Endoscopic examination of the choanae in dogs and cats: 118 cases (1988–1998). *Journal of the American Veterinary Medical Association* 1999;215:1301–5.

66 Sharp NJH, Harvey CE, Sullivan M. Canine nasal aspergillosis and penicilliosis. *Compendium on Continuing Education for the Practicing Veterinarian* 1991;13:41–6.

67 Sharman M, Paul A, Davies D, *et al.* Multi-centre assessment of mycotic rhinosinusitis in dogs: a retrospective study of initial treatment success (1998–2008). *Journal of Small Animal Practice* 2010;51:423–7.

68 Day MJ. Canine sino-nasal aspergillosis: parallels with human disease. *Medical Mycology* 2009;47:S315–23.

69 Peeters D, Clercx C. Update on canine sinonasal aspergillosis. *Veterinary Clinics of North America Small Animal Practice* 2007;37:901–16.

70 Kuehn NF. Nasal computed tomography. *Clinical Techniques in Small Animal Practice* 2006;21:55–9.

71 Wolf AM, Troy GC. Deep mycotic diseases. In: *Textbook of Veterinary Internal Medicine*. Ettinger SJ, Feldman EC (eds). WB Saunders, Philadelphia 1995, pp. 455–8.

72 Romani L. Immunity to fungal infections. *Nature Reviews Immunology* 2004;4:11–24.

73 Wilkinson GT, Sutton RH, Grono LR. *Aspergillus* spp. infection associated with orbital cellulitis and sinusitis in a cat. *Journal of Small Animal Practice* 1982;23:127–31.

74 Kano R, Itamoto K, Okuda M, Inokuma H, Hasegawa A, Balajee SA. Isolation of *Aspergillus udagawae* from a fatal case of feline orbital aspergillosis. *Mycoses* 2008;51:360–1.

75 Kano R, Shibahashi A, Fujino Y, Sakai H. Two cases of feline orbital aspergillosis due to *Aspergillus udagawae* and *A. viridinutans*. *Journal of Veterinary Medical Science* 2013;75:7–10.

76 Barachetti L, Mortellaro CM, Di Giancamillo M, *et al.* Bilateral orbital and nasal aspergillosis in a cat. *Veterinary Ophthalmology* 2009;12:176–82.

77 Furrow E, Groman RP. Intranasal infusion of clotrimazole for the treatment of nasal aspergillosis in two cats. *Journal of the American Veterinary Medical Association* 2009;235:1188–93.

78 Karnik K, Reichle JK, Fischetti AJ, Goggin JM. Computed tomographic findings of fungal rhinitis and sinusitis in cats. *Veterinary Radiology & Ultrasound* 2009;50:65–8.

79 Giordano C, Gianella P, Bo S, *et al*. Invasive mould infections of the naso-orbital region of cats: a case involving *Aspergillus fumigatus* and an aetiological review. *Journal of Feline Medicine and Surgery* 2010;12:714–23.

80 Smith LN, Hoffman SB. A case series of unilateral orbital aspergillosis in three cats and treatment with voriconazole. *Veterinary Ophthalmology* 2010;13:190–203.

81 Barrs VR, Beatty JA, Dhand NK, *et al*. Computed tomographic features of feline sino-nasal and sino-orbital aspergillosis. *Veterinary Journal* 2014;201:215–22.

82 Barrs VR, Talbot JJ. Feline aspergillosis. *Veterinary Clinics of North America Small Animal Practice* 2014;44:51–73.

83 Barrs VR, Halliday C, Martin P, *et al*. Sinonasal and sino-orbital aspergillosis in 23 cats: aetiology, clinicopathological features and treatment outcomes. *Veterinary Journal* 2012;191:58–64.

84 Declercq J, Declercq L, Fincioen S. Unilaterale sino-orbitale en subcutane aspergillose bij een kat. *Vlaams Diergeneeskundig Tijdschrift* 2012;81:357–62.

85 Whitney J, Beatty JA, Martin P, Dhand NK, Briscoe K, Barrs VR. Evaluation of serum galactomannan detection for diagnosis of feline upper respiratory tract aspergillosis. *Veterinary Microbiology* 2013;162:180–5.

86 Bissonnette KW, Sharp NJ, Dykstra MH, *et al*. Nasal and retrobulbar mass in a cat caused by *Pythium insidiosum*. *Journal of Medical and Veterinary Mycology* 1991;29:39–44.

87 Zonderland J-L, Störk CK, Saunders JH, Hamaide AJ, Balligand MH, Clercx CM. Intranasal infusion of enilconazole for treatment of sinonasal aspergillosis in dogs. *Journal of the American Veterinary Medical Association* 2002;221:1421–5.

88 Johnson LR, Drazenovich TL, Herrera MA, Wisner ER. Results of rhinoscopy alone or in conjunction with sinuscopy in dogs with aspergillosis: 46 cases (2001–2004). *Journal of the American Veterinary Medical Association* 2006;228:738–42.

89 Sullivan M, Lee R, Jakovljevic S, Sharp N. The radiological features of aspergillosis of the nasal cavity and frontal sinuses in the dog. *Journal of Small Animal Practice* 1986;27:167–80.

90 Saunders J, Van Bree H. Comparison of radiography and computed tomography for the diagnosis of canine nasal aspergillosis. *Veterinary Radiology & Ultrasound* 2003;44:414–9.

91 Saunders JH, Zonderland J-L, Clercx C, *et al*. Computed tomographic findings in 35 dogs with nasal aspergillosis. *Veterinary Radiology & Ultrasound* 2002;43:5–9.

92 Saunders J, Clercx C, Snaps FR. Radiographic, magnetic resonance imaging, computed tomographic, and rhinoscopic features of nasal aspergillosis in dogs. *Journal of the American Veterinary Medical Association* 2004;225:1703–12.

93 Saunders JH, Duchateau L, Störk C, van Bree H. Use of computed tomography to predict the outcome of a noninvasive intranasal infusion in dogs with nasal aspergillosis. *Canadian Veterinary Journal* 2003;44:305–11.

94 De Lorenzi D, Bonfanti U, Masserdotti C, Caldin M, Furlanello T. Diagnosis of canine nasal aspergillosis by cytological examination: a comparison of four different collection techniques. *Journal of Small Animal Practice* 2006;47:316–19.

95 Peeters D, Day MJ, Clercx C. An immunohistochemical study of canine nasal aspergillosis. *Journal of Comparative Pathology* 2005;132:283–8.

96 Pomrantz JS, Johnson LR. Repeated rhinoscopic and serologic assessment of the effectiveness of intranasally administered clotrimazole for the treatment of nasal aspergillosis in dogs. *Journal of the American Veterinary Medical Association* 2010;236:757–62.

97 Billen F, Peeters D, Peters IR, Helps CR, Huynen P. Comparison of the value of measurement of serum galactomannan and *Aspergillus*-specific antibodies in the diagnosis of canine sino-nasal aspergillosis. *Veterinary Microbiology* 2009;133:358–65.

98 Billen F, Clercx C, Le Garérrès A, Massart L, Mignon B, Peeters D. Effect of sampling method and incubation temperature on fungal culture in canine sinonasal aspergillosis. *Journal of Small Animal Practice* 2009;50:67–72.

99 Sharp NJH, Harvey CE, O'Brien JA. Treatment of canine nasal aspergillosis/penicilliosis with fluconazole. *Journal of Small Animal Practice* 1991;32:513–16.

100 Claeys S, Lefebvre JB, Schuller S, Hamaide A, Clercx C. Surgical treatment of canine nasal aspergillosis by rhinotomy combined with enilconazole infusion and oral itraconazole. *Journal of Small Animal Practice* 2006;47:320–4.

101 White D. Canine nasal mycosis – light at the end of a long diagnostic and therapeutic tunnel. *Journal of Small Animal Practice* 2006;47:307.

102 Richardson EF, Mathews KG. Distribution of topical agents in the frontal sinuses and nasal cavity of dogs: comparison between current protocols for treatment of nasal aspergillosis and a new noninvasive technique. *Veterinary Surgery* 1995;24:476–83.

103 Mathews KG, Koblik PD, Richardson EF, Davidson AP, Pappagianis D. Computed tomographic assessment of noninvasive intranasal infusions in dogs with fungal rhinitis. *Veterinary Surgery* 1996;25:309–19.

104 Bray JP, White RA, Lascelles BD. Treatment of canine nasal aspergillosis with a new non-invasive technique. Failure with enilconazole. *Journal of Small Animal Practice* 1998;39:223–6.

105 Mathews KG, Davidson AP, Koblik PD, *et al.* Comparison of topical administration of clotrimazole through surgically placed versus nonsurgically placed catheters for treatment of nasal aspergillosis in dogs: 60 cases (1990–1996). *Journal of the American Veterinary Medical Association* 1998;213:501–6.

106 Friend EJ, Williams JM, White RAS. Invasive treatment of canine nasal aspergillosis with topical clotrimazole. *Veterinary Record* 2002;151:298–9.

107 Sissener TR, Bacon NJ, Friend E, Anderson DM, White RA. Combined clotrimazole irrigation and depot therapy for canine nasal aspergillosis. *Journal of Small Animal Practice* 2006;47:312–15.

108 Billen F, Guieu LV, Bernaerts F, *et al.* Efficacy of intrasinusal administration of bifonazole cream alone or in combination with enilconazole irrigation in canine sino-nasal aspergillosis: 17 cases. *Canadian Veterinary Journal* 2010;51:164–8.

109 Kelly KM, Bardach J. Biologic basis of cleft palate and palatal surgery. In: *Oral and Maxillofacial Surgery in Dogs and Cats*. Verstraete FJM, Lommer MJ (eds). Saunders/Elsevier, Toronto 2012, pp. 343–50.

110 Harvey CE. Palate defects in dogs and cats. *Compendium on Continuing Education for the Practicing Veterinarian* 1987;9:404.

111 Sivacolundhu RK. Use of local and axial pattern flaps for reconstruction of the hard and soft palate. *Clinical Techniques in Small Animal Practice* 2007;22:61–9.

112 Warzee CC, Bellah JR, Richards D. Congenital unilateral cleft of the soft palate in six dogs. *Journal of Small Animal Practice* 2001;42:338–40.

113 Smith MM. Oronasal fistula repair. *Clinical Techniques in Small Animal Practice* 2000;15:243–50.

114 Bojrab MJ, Tholen MA, Constantinescu GM. Oronasal fistulae in dogs and cats. *Compendium on Continuing Education for the Practicing Veterinarian* 1986;8:815.

115 Woodward TM. Greater palatine island axial pattern flap for repair of oronasal fistula related to eosinophilic granuloma. *Journal of Veterinary Dentistry* 2006;23:161–6.

116 Cox CL, Hunt GB, Cadier MM. Repair of oronasal fistulae using auricular cartilage grafts in five cats. *Veterinary Surgery* 2007;36:164–9.

117 Sager M, Nefen S. Use of buccal mucosal flaps for the correction of congenital soft palate defects in three dogs. *Veterinary Surgery* 1998;27:358–63.

118 Ferguson MW. Palate development. *Development* 1988;103:41–60.

119 White RN, Hawkins HL, Alemi VP, Warner C. Soft palate hypoplasia and concurrent middle ear pathology in six dogs. *Journal of Small Animal Practice* 2009;50:364–72.

120 Gregory SP. Middle ear disease associated with congenital palatine defects in seven dogs and one cat. *Journal of Small Animal Practice* 2000;41:398–401.

121 Hammer DL, Sacks M. Surgical closure of cleft soft palate in a dog. *Journal of the American Veterinary Medical Association* 1971;158:342–5.

122 Baker GJ. Surgery of the canine pharynx and larynx. *Journal of Small Animal Practice* 1972;13:505–13.

123 Bauer MS, Levitt L, Pharr JW, Fowler JD, Basher AW. Unsuccessful surgical repair of a short soft palate in a dog. *Journal of the American Veterinary Medical Association* 1988;193:1551–52.

124 Sylvestre AM, Sharma A. Management of a congenitally shortened soft palate in a dog. *Journal of the American Veterinary Medical Association* 1997;211:875–7.

125 Muntz HR. An overview of middle ear disease in cleft palate children. *Facial Plastic Surgery* 1993;9:177–80.

126 Bluestone CD, Swarts JD. Human evolutionary history: Consequences for the pathogenesis of otitis media. *Otolaryngology and Head & Neck Surgery* 2010;143:739–44.

127 Tholen MA, Johnson J. Surgical repair of the oronasal fistula. *Veterinary Medicine Small Animal Clinics* 1983;78:1733.

128 Maruo T, Shida T, Fukuyama Y, *et al.* Retrospective study of canine nasal tumor treated with hypofractionated radiotherapy. *Journal of Veterinary Medical Science* 2011;73:193–7.

129 Holmberg DL. Sequelae of ventral rhinotomy in dogs and cats with inflammatory and neoplastic nasal pathology: a retrospective study. *Canadian Veterinary Journal* 1996;37:483–5.

130 Murgia D, Pivetta M, Bowlt K, Volmer C, Holloway A, Dennis R. Intranasal epidermoid cyst causing upper airway obstruction in three brachycephalic dogs. *Journal of Small Animal Practice* 2014;55:431–5.

131 Burgess KE, Green EM, Wood RD, Dubielzig RR. Angiofibroma of the nasal cavity in 13 dogs. *Veterinary and Comparative Oncology* 2011;9:304–9.

132 Swinbourne F, Kulendra E, Smith K, Leo C, ter Haar G. Inflammatory myofibroblastic tumour in the nasal cavity of a dog. *Journal of Small Animal Practice* 2014;55:121–4.

133 Turek MM, Lana SE. Nasosinal tumors. In: *Small Animal Clinical Oncology*, 5th edn. Withrow SJ, Vail DM, Page RL (eds). Elsevier Saunders, St. Louis 2013, pp. 435–51.

134 Galloway PE, Kyles A, Henderson JP. Nasal polyps in a cat. *Journal of Small Animal Practice* 1997;38:78–80.

135 Greci V, Mortellaro CM, Olivero D, Cocci A, Hawkins EC. Inflammatory polyps of the nasal turbinates of cats: an argument for designation as feline mesenchymal nasal hamartoma. *Journal of Feline Medicine and Surgery* 2011;13:213–19.

136 MacEwen EG, Withrow SJ, Patnaik AK. Nasal tumors in the dog: retrospective evaluation of diagnosis, prognosis, and treatment. *Journal of the American Veterinary Medical Association* 1977;170:45–8.

137 Patnaik AK. Canine sinonasal neoplasms: clinicopathological study of 285 cases. *Journal American Animal Hospital Association* 1989;25:103–14.

138 Lefebvre J, Kuehn NF, Wortinger A. Computed tomography as an aid in the diagnosis of chronic nasal disease in dogs. *Journal of Small Animal Practice* 2005;46:280–5.

139 Reif JS, Cohen D. The environmental distribution of canine respiratory tract neoplasms. *Archives in Environmental Health* 1971;22: 136–40.

140 Reif JS, Bruns C, Lower KS. Cancer of the nasal cavity and paranasal sinuses and exposure to environmental tobacco smoke in pet dogs. *American Journal of Epidemiology* 1998;147:488–92.

141 Bukowski JA, Wartenberg D, Goldschmidt M. Environmental causes for sinonasal cancers in pet dogs, and their usefulness as sentinels of indoor cancer risk. *Journal of Toxicology and Environmental Health* 1998;54:579–91.

142 Madewell BR, Priester WA, Gillette EL, Snyder SP. Neoplasms of the nasal passages and paranasal sinuses in domesticated animals as reported by 13 veterinary colleges. *American Journal of Veterinary Research* 1976;37:851–6.

143 Patnaik AK. Canine sinonasal neoplasms: soft tissue tumors. *Journal of the American Animal Hospital Association* 1989;25:491–7.

144 Patnaik AK, Lieberman PH, Erlandson RA, Liu SK. Canine sinonasal skeletal neoplasms: chondrosarcomas and osteosarcomas. *Veterinary Pathology* 1984;21:475–82.

145 Henry CJ, Brewer WG, Tyler JW, *et al.* Survival in dogs with nasal adenocarcinoma: 64 cases (1981–1995). *Journal of Veterinary Internal Medicine* 1998;12:436–9.

146 Smith MO, Turrel JM, Bailey CS, Cain GR. Neurologic abnormalities as the predominant signs of neoplasia of the nasal cavity in dogs and cats: seven cases (1973–1986). *Journal of the American Veterinary Medical Association* 1989;195:242–5.

147 Miles MS, Dhaliwal RS, Moore MP, Reed AL. Association of magnetic resonance imaging findings and histologic diagnosis in dogs with nasal disease: 78 cases (2001–2004). *Journal of the American Veterinary Medical Association* 2008;232:1844–9.

148 Saunders J, Van Bree H, Gielen I. Diagnostic value of computed tomography in dogs with chronic nasal disease. *Veterinary Radiology & Ultrasound* 2003;44:409–13.

149 Thrall DE, Robertson ID, Mcleod DA, Heidner GL, Hoopes PJ, Page RL. A comparison of radiographic and computed tomographic findings in 31 dogs with malignant nasal cavity tumors. *Veterinary Radiology* 2005;30:59–66.

150 Park RD, Beck ER, LeCouteur RA. Comparison of computed tomography and radiography for detecting changes induced by malignant nasal neoplasia in dogs. *Journal of the American Veterinary Medical Association* 1992;201:1720–4.

151 Drees R, Forrest LJ, Chappell R. Comparison of computed tomography and magnetic resonance imaging for the evaluation of canine intranasal neoplasia. *Journal of Small Animal Practice* 2009;50:334–40.

152 Avner A, Dobson JM, Sales JI, Herrtage ME. Retrospective review of 50 canine nasal tumours evaluated by low-field magnetic resonance imaging. *Journal of Small Animal Practice* 2008;49:233–9.

153 Agthe P, Caine AR, Gear RN, Dobson JM, Richardson KJ, Herrtage ME. Prognostic significance of specific magnetic resonance imaging features in canine nasal tumours treated by radiotherapy. *Journal of Small Animal Practice* 2009;50:641–8.

154 Adams WM, Kleiter MM, Thrall DE, *et al.* Prognostic significance of tumor histology and computed tomographic staging for radiation treatment response of canine nasal tumors. *Veterinary Radiology & Ultrasound* 2009;50:330–5.

155 Kondo Y, Matsunaga S, Mochizuki M, *et al.* Prognosis of canine patients with nasal tumors according to modified clinical stages based on computed tomography: a retrospective study. *Journal of Veterinary Medical Science* 2008;70:207–12.

156 Laing EJ, Binnington AG. Surgical therapy of canine nasal tumors: a retrospective study (1982–1986). *Canadian Veterinary Journal* 2006;29:809–13.

157 Adams WM. Outcome of accelerated radiotherapy alone or accelerated radiotherapy followed by exenteration of the nasal cavity in dogs with intranasal neoplasia: 53 cases (1990–2002). *Journal of the American Veterinary Medical Association* 2005;227:936–41.

158 Morris JS, Dunn KJ, Dobson JM, White RAS. Effects of radiotherapy alone and surgery and radiotherapy on survival of dogs with nasal tumours. *Journal of Small Animal Practice* 1994;35:567–73.

159 Theon AP, Madewell BR, Harb MF, Dungworth DL. Megavoltage irradiation of neoplasms of the nasal and paranasal cavities in 77 dogs. *Journal of the American Veterinary Medical Association* 1993; 202:1469–75.

160 Adams WM, Miller PE, Vail DM, Forrest LJ, MacEwen EG. An accelerated technique for irradiation of malignant canine nasal and paranasal sinus tumors. *Veterinary Radiology & Ultrasound* 1998; 39:475–81.

161 Evans SM, Goldschmidt M, McKee LJ, Harvey CE. Prognostic factors and survival after radiotherapy for intranasal neoplasms in dogs: 70 cases (1974–1985). *Journal of the American Veterinary Medical Association* 1989;194:1460–3.

162 McEntee MC, Page RL, Heidner GL. A retrospective study of 27 dogs with intranasal neoplasms treated with cobalt radiation. *Veterinary Radiology & Ultrasound* 1991;32:135–9.

163 Lana SE, Dernell WS, Lafferty MH, Withrow SJ, LaRue SM. Use of radiation and a slow-release cisplatin formulation for treatment of canine nasal tumors. *Veterinary Radiology & Ultrasound* 2004;45:577–81.

164 Hunley DW, Mauldin GN, Shiomitsu K, Mauldin GE. Clinical outcome in dogs with nasal tumors treated with intensity-modulated radiation therapy. *Canadian Veterinary Journal* 2010;51:293–300.

165 Lawrence JA, Forrest LJ, Turek MM, *et al.* Proof of principle of ocular sparing in dogs with sinonasal tumors treated with intensity-modulated radiation therapy. *Veterinary Radiology & Ultrasound* 2010;51:561–70.

166 McEntee MC. Veterinary radiation therapy: review and current state of the art. *Journal of the American Animal Hospital Association* 2006;42:94–109.

167 LaDue TA, Dodge R, Page RL, Price GS, Hauck ML, Thrall DE. Factors influencing survival after radiotherapy of nasal tumors in 130 dogs. *Veterinary Radiology & Ultrasound* 1999; 40:312–17.

168 Gieger T, Rassnick K, Siegel S, *et al.* Palliation of clinical signs in 48 dogs with nasal carcinomas treated with coarse-fraction radiation therapy. *Journal of the American Animal Hospital Association* 2008;44:116–23.

169 Hahn KA, Knapp DW, Richardson RC, Matlock CL. Clinical response of nasal adenocarcinoma to cisplatin chemotherapy in 11 dogs. *Journal of the American Veterinary Medical Association* 1992;200:355–7.

170 Langova V, Mutsaers AJ, Phillips B, Straw R. Treatment of eight dogs with nasal tumours with alternating doses of doxorubicin and carboplatin in conjunction with oral piroxicam. *Australian Veterinary Journal* 2004;82:676–80.

171 Withrow SJ. Cryosurgical therapy for nasal tumors in the dog. *Journal of the American Animal Hospital Association* 1982;18:585–9.

172 Murphy SM, Lawrence JA, Schmiedt C. Image-guided transnasal cryoablation of a recurrent nasal adenocarcinoma in a dog. *Journal of Small Animal Practice* 2011;52:329-33.

173 Mayer-Stankeová S, Fidel J, Wergin MC, *et al.* Proton spot scanning radiotherapy of spontaneous canine tumors. *Veterinary Radiology & Ultrasound* 2009;50:314–18.

174 White R, Walker M, Legendre AM, Hoopes J, Smith J, Horton SB. Development of brachytherapy technique for nasal tumors in dogs. *American Journal of Veterinary Research* 1990;51:1250–6.

175 Thompson JP, Ackerman N, Bellah JR, Beale BS, Ellison GW. 192iridium brachytherapy, using an intracavitary afterload device, for treatment of intranasal neoplasms in dogs. *American Journal of Veterinary Research* 1992;53:617–22.

176 Lucroy MD, Long KR, Blaik MA, Higbee RG, Ridgway TD. Photodynamic therapy for the treatment of intranasal tumors in 3 dogs and 1 cat. *Journal of Veterinary Internal Medicine* 2003;17:727–9.

177 Osaki T, Takagi S, Hoshino Y, Okumura M, Kadosawa T, Fujinaga T. Efficacy of antivascular photodynamic therapy using benzoporphyrin derivative monoacid ring A (BPD-MA) in 14 dogs with oral and nasal tumors. *Journal of Veterinary Medical Science* 2009;71:125–32.

178 Mukaratirwa S, van der Linde-Sipman JS, Gruys E. Feline nasal and paranasal sinus tumours: clinico-pathological study, histomorphological description and diagnostic immunohistochemistry of 123 cases. *Journal of Feline Medicine and Surgery* 2001; 3:235–45.

179 Demko JL, Cohn LA. Chronic nasal discharge in cats: 75 cases (1993–2004). *Journal of the American Veterinary Medical Association* 2007;230:1032–7.

180 Taylor SS, Goodfellow MR, Browne WJ, *et al.* Feline extranodal lymphoma: response to chemotherapy and survival in 110 cats. *Journal of Small Animal Practice* 2009;50:584–92.

181 Little L, Patel R, Goldschmidt M. Nasal and nasopharyngeal lymphoma in cats: 50 cases (1989–2005). *Veterinary Pathology* 2007;44:885–92.

182 Haney SM, Beaver L, Turrel J, *et al.* Survival analysis of 97 cats with nasal lymphoma: a multi-institutional retrospective study (1986–2006). *Journal of Veterinary Internal Medicine* 2009;23:287–94.

183 Lamb CR, Richbell S, Mantis P. Radiographic signs in cats with nasal disease. *Journal of Feline Medicine and Surgery* 2003;5:227–35.

184 Schoenborn WC, R WE, Kass PP, Dale M. Retrospective assessment of computed tomographic imaging of feline sinonasal disease in 62 cats. *Veterinary Radiology & Ultrasound* 2003;44:185–95.

185 Tromblee TC, Jones JC, Etue AE, Forrester SD. Association between clinical characteristics, computed tomography characteristics, and histologic diagnosis for cats with sinonasal disease. *Veterinary Radiology & Ultrasound* 2006;47:241–8.

186 De Lorenzi D, Bertoncello D, Bottero E. Squash-preparation cytology from nasopharyngeal masses in the cat: cytological results and histological correlations in 30 cases. *Journal of Feline Medicine and Surgery* 2008;10:55–60.

187 Theon AP, Peaston AE, Madewell BR, Dungworth DL. Irradiation of nonlymphoproliferative neoplasms of the nasal cavity and paranasal sinuses in 16 cats. *Journal of the American Veterinary Medical Association* 1994;204:78–83.

188 Mellanby RJ, Herrtage ME, Dobson JM. Long-term outcome of eight cats with non-lymphoproliferative nasal tumours treated by megavoltage radiotherapy. *Journal of Feline Medicine and Surgery* 2002;4:77–81.

189 Sfiligoi G, Théon AP, Kent MS. Response of nineteen cats with nasal lymphoma to radiation therapy and chemotherapy. *Veterinary Radiology & Ultrasound* 2007;48:388–93.

190 Elmslie RE, Ogilvie GK, Gillette EL, McChesney-Gillette S. Radiotherapy with and without chemotherapy for localized lymphoma in 10 Cats. *Veterinary Radiology* 2005;32:277–80.

191 Burns RE, Wagner DC, Leutenegger CM, Pesavento PA. Histologic and molecular correlation in shelter cats with acute upper respiratory infection. *Journal of Clinical Microbiology* 2011;49:2454–60.

192 Johnson LR, Foley JE, De Cock HEV, Clarke HE, Maggs DJ. Assessment of infectious organisms associated with chronic rhinosinusitis in cats. *Journal of the American Veterinary Medical Association* 2005;227:579–85.

193 Bannasch MJ, Foley JE. Epidemiologic evaluation of multiple respiratory pathogens in cats in animal shelters. *Journal of Feline Medicine and Surgery* 2005;7:109–19.

194 Helps CR, Lait P, Damhuis A, *et al.* Factors associated with upper respiratory tract disease caused by feline herpesvirus, feline calicivirus, *Chlamydophila felis* and *Bordetella bronchiseptica* in cats: experience from 218 European catteries. *Veterinary Record* 2005;156:669–73.

195 Di Martino B, Di Francesco CE. Etiological investigation of multiple respiratory infections in cats. *New Microbiologica* 2007;30:455–61.

196 Smith TL, Smith JM. Electrosurgery in otolaryngology-head and neck surgery: principles, advances, and complications. *Laryngoscope* 2001;111:769–80.

197 Johnson LR, Clarke HE, Bannasch MJ, De Cock HEV. Correlation of rhinoscopic signs of inflammation with histologic findings in nasal biopsy specimens of cats with or without upper respiratory tract disease. *Journal of the American Veterinary Medical Association* 2004;225:395–400.

198 van der Woerdt A. Adnexal surgery in dogs and cats. *Veterinary Ophthalmology* 2004;7:284–90.

199 Gourkow N, Lawson JH, Hamon SC, Phillips CJC. Descriptive epidemiology of upper respiratory disease and associated risk factors in cats in an animal shelter in coastal western Canada. *Canadian Veterinary Journal* 2013;54:132–8.

200 Cohn LA. Feline respiratory disease complex. *Veterinary Clinics of North America Small Animal Practice* 2011;41:1273–89.

201 Adler K, Radeloff I, Stephan B, Greife H, Hellmann K. Bacteriological and virological status in upper respiratory tract infections of cats (cat common cold complex). *Berliner und Munchener Tierarztliche Wochenschrift* 2007;120:120–5.

202 Thiry E, Addie D, Belák S, *et al.* Feline herpesvirus infection. ABCD guidelines on prevention and management. *Journal of Feline Medicine and Surgery* 2009;11:547–55.

203 Gaskell R, Dawson S, Radford A, Thiry E. Feline herpesvirus. *Veterinary Research* 2007;38:337–54.

204 Scherk M. Snots and snuffles: rational approach to chronic feline upper respiratory syndromes. *Journal of Feline Medicine and Surgery* 2010;12:548–57.

205 Reed N, Gunn-Moore D. Nasopharyngeal disease in cats: 1. Diagnostic investigation. *Journal of Feline Medicine and Surgery* 2012;14:306–15.

205a. Schulz, C, Hartmann K, Mueller RS *et al.* Sampling sites for detection of feline herpesvirus-1, feline calicivirus and Chlamydia felis in cats with feline upper respiratory tract disease. *Feline Medicine and Surgery* 2015, 17: 1012-1019.

206 Reed N, Gunn-Moore D. Nasopharyngeal disease in cats: 2. Specific conditions and their management. *Journal of Feline Medicine and Surgery* 2012;14:317–26.

207 Gil S, Leal RO, Duarte A, *et al.* Relevance of feline interferon omega for clinical improvement and reduction of concurrent viral excretion in retrovirus infected cats from a rescue shelter. *Research in Veterinary Science* 2013;94:753–63.

208 Haid C, Kaps S, Gönczi E, *et al.* Pretreatment with feline interferon omega and the course of subsequent infection with feline herpesvirus in cats. *Veterinary Ophthalmology* 2007;10:278–84.

209 Malik R, Lessels NS, Webb S, *et al.* Treatment of feline herpesvirus-1 associated disease in cats with famciclovir and related drugs. *Journal of Feline Medicine and Surgery* 2009;11:40–8.

210 Radford AD, Addie D, Belák S, *et al.* Feline calicivirus infection. ABCD guidelines on prevention and management. *Journal of Feline Medicine and Surgery* 2009;11:556–64.

211 Radford AD, Coyne KP, Dawson S, Porter CJ, Gaskell RM. Feline calicivirus. *Veterinary Research* 2007;38:319–35.

212 Egberink H, Addie D, Belák S, *et al.* *Bordetella bronchiseptica* infection in cats. ABCD guidelines on prevention and management. *Journal of Feline Medicine and Surgery* 2009;11:610–14.

212a Feline calicivirus and other respiratory pathogens in cats with feline calicivirus-related symptoms and in clinically healthy cats in Switzerland. *BMC Veterinary Research* 2015;11:282–93.

213 Inatsuka CS, Julio SM, Cotter PA. *Bordetella* filamentous hemagglutinin plays a critical role in immunomodulation, suggesting a mechanism for host specificity. *Proceedings of the National Academy of Science USA* 2005;102:18578–83.

214 Schulz BS, Zauscher S, Ammer H, Sauter-Louis C, Hartmann K. Side effects suspected to be related to doxycycline use in cats. *Veterinary Record* 2013;172:184.

215 German AJ, Cannon MJ, Dye C, *et al.* Oesophageal strictures in cats associated with doxycycline therapy. J*ournal of Feline Medicine and Surgery* 2005;7:33–41.

216 Carbone M, Pennisi MG, Masucci M, De Sarro A, Giannone M, Fera MT. Activity and postantibiotic effect of marbofloxacin, enrofloxacin, difloxacin and ciprofloxacin against feline *Bordetella bronchiseptica* isolates. *Veterinary Microbiology* 2001;81:79–84.

217 Gruffydd-Jones T, Addie D, Belák S, *et al.* *Chlamydophila felis* infection. ABCD guidelines on prevention and management. *Journal of Feline Medicine and Surgery* 2009;11:605–9.

218 Masubuchi K, Nosaka H, Iwamoto K, Kokubu T, Yamanaka M, Shimizu Y. Experimental infection of cats with *Chlamydophila felis*. *Journal of Veterinary Medical Science* 2002;64:1165–8.

219 Veir JK, Ruch-Gallie R, Spindel ME, Lappin MR. Prevalence of selected infectious organisms and comparison of two anatomic sampling sites in shelter cats with upper respiratory tract disease. *Journal of Feline Medicine and Surgery* 2008;10:551–7.

220 Gerhardt N, Schulz BS, Werckenthin C, Hartmann K. Pharmacokinetics of enrofloxacin and its efficacy in comparison with doxycycline in the treatment of *Chlamydophila felis* infection in cats with conjunctivitis. *Veterinary Record* 2006;159:591–4.

221 Owen WMA, Sturgess CP, Harbour DA, Egan K, Gruffydd-Jones TJ. Efficacy of azithromycin for the treatment of feline chlamydophilosis. *Journal of Feline Medicine and Surgery* 2003;5:305–11.

222 Hartmann AD, Helps CR, Lappin M, Werckenthin C, Hartmann K. Efficacy of pradofloxacin in cats with feline upper respiratory tract disease due to *Chlamydophila felis* or *Mycoplasma* infections. *Journal of Veterinary Internal Medicine* 2008;22:44–52.

223 Dean R, Harley R, Helps C, Caney S, Gruffydd-Jones T. Use of quantitative real-time PCR to monitor the response of *Chlamydophila felis* infection to doxycycline treatment. *Journal of Clinical Microbiology* 2005;43:1858–64.

224 Lee-Fowler T. Feline respiratory disease: what is the role of *Mycoplasma* species? *Journal of Feline Medicine and Surgery* 2014;16:563–71.

225 Gelatt KN, van der Woerdt A, Ketring KL, *et al.* Enrofloxacin-associated retinal degeneration in cats. *Veterinary Ophthalmology* 2001;4:99–106.

226 Wiebe V, Hamilton P. Fluoroquinolone-induced retinal degeneration in cats. *Journal of the American Veterinary Medical Association* 2002;221:1568–71.

227 Galler A, Shibly S, Bilek A, Hirt R. Chronic diseases of the nose and nasal sinuses in cats: a retrospective study. *Schweizer Archivs fur Tierheilkunde* 2012;154:209–16.

228 De Zoysa A, Hawkey PM, Engler K, *et al.* Characterization of toxigenic *Corynebacterium ulcerans* strains isolated from humans and domestic cats in the United Kingdom. *Journal of Clinical Microbiology* 2005;43:4377–81.

229 Meinel DM, Margos G, Konrad R, Krebs S, Blum H, Sing A. Next generation sequencing analysis of nine *Corynebacterium ulcerans* isolates reveals zoonotic transmission and a novel putative diphtheria toxin-encoding pathogenicity island. *Genome Medicine* 2014;6:113.

230 Saeki J, Katsukawa C, Matsubayashi M, *et al.* The detection of toxigenic *Corynebacterium ulcerans* from cats with nasal inflammation in Japan. *Epidemiology and Infection* 2015;12:1–6.

231 Berger A, Huber I, Merbecks S-S, *et al.* Toxigenic *Corynebacterium ulcerans* in woman and cat. *Emerging Infectious Diseases* 2011;17:1767–9.

232 Kamada T, Hatanaka A, Tasaki A, Honda K, Tsunoda A, Kitamura K. Case of acute pharyngitis caused by *Corynebacterium ulcerans* in Ibaraki Prefecture. *Nippon Jibiinkoka Gakkai Kaiho* 2012;115:682–6.

233 Yoshimura Y, Tachikawa N, Komiya T, Yamamoto A. A case report and epidemiological investigation of axillary lymph node abscess caused by *Corynebacterium ulcerans* in an HIV-1-positive patient. *Epidemiology and Infection* 2014;142:1541–4.

234 Corti MAM, Bloemberg GV, Borelli S, *et al.* Rare human skin infection with *Corynebacterium ulcerans*: transmission by a domestic cat. *Infection* 2012;40:575–8.

235 Vandentorren S, Guiso N, Badell E, *et al.* Toxigenic *Corynebacterium ulcerans* in a fatal human case and her feline contacts. *Euro Surveillance* 2014;19:1–2.

236 Litster AL, Wu CC, Constable PD. Comparison of the efficacy of amoxicillin-clavulanic acid, cefovecin, and doxycycline in the treatment of upper respiratory tract disease in cats housed in an animal shelter. *Journal of the American Veterinary Medical Association* 2012;241:218–26.

237 Taylor-Robinson D, Bebear C. Antibiotic susceptibilities of mycoplasmas and treatment of mycoplasmal infections. *Journal of Antimicrobial Chemotherapy* 1997;40:622–30.

238 Caenen M, Hamels K, Deron P, Clement P. Comparison of decongestive capacity of xylometazoline and pseudoephedrine with rhinomanometry and MRI. *Rhinology* 2005;43:205–9.

239 Vaidyanathan S, Williamson P, Clearie K, Khan F, Lipworth B. Fluticasone reverses oxymetazoline-induced tachyphylaxis of response and rebound congestion. *American Journal of Respiratory and Critical Care Medicine* 2010;182:19–24.

240 Ozturk F, Bakirtas A, Ileri F, Turktas I. Efficacy and tolerability of systemic methylprednisolone in children and adolescents with chronic rhinosinusitis: a double-blind, placebo-controlled randomized trial. *Journal of Allergy and Clinical Immunology* 2011;128:348–52.

241 Agnew W, Korman R. Pharmacological appetite stimulation: rational choices in the inappetent cat. *Journal of Feline Medicine and Surgery* 2014;16:749–56.

242 Quimby JM, Gustafson DL, Samber BJ, Lunn KF. Studies on the pharmacokinetics and pharmacodynamics of mirtazapine in healthy young cats. *Journal of Veterinary Pharmacology and Therapeutics* 2011; 34:388–96.

243 Quimby JM, Gustafson DL, Lunn KF. The pharmacokinetics of mirtazapine in cats with chronic kidney disease and in age-matched control cats. *Journal of Veterinary Internal Medicine* 2011;25:985–9.

244 Quimby JM, Lunn KF. Mirtazapine as an appetite stimulant and anti-emetic in cats with chronic kidney disease: a masked placebo-controlled crossover clinical trial. *Veterinary Journal* 2013;197:651–5.

245 Barrs VR, van Doorn TM, Houbraken J, *et al.* *Aspergillus felis* sp. nov., an emerging agent of invasive aspergillosis in humans, cats, and dogs. *PLoS ONE* 2013;8:e64871.

246 Talbot JJ, Johnson LR, Martin P, *et al.* What causes canine sino-nasal aspergillosis? A molecular approach to species identification. *Veterinary Journal* 2014;200:17–21.

247 Buonavoglia C, Martella V. Canine respiratory viruses. *Veterinary Research* 2007;38:355–73.

248 Schulz BS, Kurz S, Weber K, Balzer H-J, Hartmann K. Detection of respiratory viruses and *Bordetella bronchiseptica* in dogs with acute respiratory tract infections. *Veterinary Journal* 2014;201:365–9.

249 Ellis J, Anseeuw E, Gow S, *et al*. Seroepidemiology of respiratory (group 2) canine coronavirus, canine parainfluenza virus, and *Bordetella bronchiseptica* infections in urban dogs in a humane shelter and in rural dogs in small communities. *Canadian Veterinary Journal* 2011;52:861–8.

250 Mochizuki M, Yachi A, Ohshima T, Ohuchi A, Ishida T. Etiologic study on upper respiratoryinfections of household dogs. *Journal of Veterinary Medical Science* 2008;70:563–9.

251 Priestnall SL, Mitchell JA, Walker CA, Erles K, Brownlie J. New and emerging pathogens in canine infectious respiratory disease. *Veterinary Pathology* 2014;51:492–504.

252 Pesavento PA, Murphy BG. Common and emerging infectious diseases in the animal shelter. *Veterinary Pathology* 2014;51:478–91.

253 Ford RB. Canine infectious respiratory disease. In: *Infectious Diseases of the Dog and Cat*, 4th edn. Green CE (ed). Elsevier, St. Louis 2012, pp. 55–65.

254 Erles K, Dubovi EJ, Brooks HW, Brownlie J. Longitudinal study of viruses associated with canine infectious respiratory disease. *Journal of Clinical Microbiology* 2004;42:4524–9.

255 Fairchild GA, Medway W, Cohen D. A study of the pathogenicity of a canine adenovirus (Toronto A26-61) for dogs. *American Journal of Veterinary Research* 1969;30:1187–93.

256 Day MJ, Horzinek MC, Schultz RD. WSAVA guidelines for the vaccination of dogs and cats. *Journal of Small Animal Practice* 2010;51:1–32, 77–95.

257 Goodnow RA. Biology of *Bordetella bronchiseptica*. *Microbiological Reviews* 1980;44:722–38.

258 Mattoo S, Foreman-Wykert AK, Cotter PA. Mechanisms of *Bordetella* pathogenesis. *Frontiers in Bioscience* 2001;1:168–86.

259 Chalker VJ, Toomey C, Opperman S, *et al*. Respiratory disease in kennelled dogs: serological responses to *Bordetella bronchiseptica* lipopolysaccharide do not correlate with bacterial isolation or clinical respiratory symptoms. *Clinical and Vaccine Immunology* 2003;10:352–6.

260 McCandlish IA, Thompson H, Wright NG. Vaccination against *Bordetella bronchiseptica* infection in dogs using a heat-killed bacterial vaccine. *Research in Veterinary Science* 1978;25:45–50.

261 Thrusfield MV, Aitken C. A field investigation of kennel cough: incubation period and clinical signs. *Journal of Small Animal Practice* 1991;32:215–20.

262 Thompson H, McCandlish IA, Wright NG. Experimental respiratory disease in dogs due to *Bordetella bronchiseptica*. *Research in Veterinary Science* 1976;20:16–23.

263 Davis R, Jayappa H, Abdelmagid OY. Comparison of the mucosal immune response in dogs vaccinated with either an intranasal avirulent live culture or a subcutaneous antigen extract vaccine of *Bordetella bronchiseptica*. *Veterinary Therapeutics* 2007;8:32–40.

264 Gore T, Headley M, Laris R, *et al*. Intranasal kennel cough vaccine protecting dogs from experimental *Bordetella bronchiseptica* challenge within 72 hours. *Veterinary Record* 2005;156:482–3.

265 Erles K, Brownlie J. Canine respiratory coronavirus: an emerging pathogen in the canine infectious respiratory disease complex. *Veterinary Clinics of North America Small Animal Practice* 2008;38:815–25.

266 Erles K, Toomey C, Brooks HW, Brownlie J. Detection of a group 2 coronavirus in dogs with canine infectious respiratory disease. *Virology* 2003;310:216–23

267a Beeler E. Influenza in dogs and cats. *Veterinary Clinics of North America Small Animal Practice* 2009;39:251–64.

267b Zhu H, Hughes J, Murcia PR. Origins and evolutionary dynamics of H3N2 canine influenza virus. *Journal of Virology* 2015;89:5406–18.

268 Hilling K, Hanel R. Canine influenza. *Compendendium on Continuing Education for the Practising Veterinarian* 2010;32:E1–9.

269a Evermann JF, Ledbetter EC, Maes RK. Canine reproductive, respiratory, and ocular diseases due to canine herpesvirus. *Veterinary Clinics of North America Small Animal Practice* 2011; 41:1097–120.

269b Kumar S, DriskellEA, Cooley AJ, *et al*. Fatal canid herpesvirus 1 respiratory infections in 4 clinically healthy adult dogs. *Veterinary Pathology* 2015;52:681–7.

270 Ekia G, Lucente MS, Bellacicci AL. Assessment of reovirus epidemiology in dogs. *Proceedings 16th European Congress of Clinical Microbiology and Infectious Diseases*, Nice 2006, p. 1674.

271 Chalker VJ, Owen WMA, Paterson C, *et al.* Mycoplasmas associated with canine infectious respiratory disease. *Microbiology* 2004;150:3491–7.

272 Rheinwald M, Hartmann K, Hähner M, Wolf G, Straubinger RK, Schulz B. Antibiotic susceptibility of bacterial isolates from 502 dogs with respiratory signs. *Veterinary Record* 2015;176:357.

273 Wills SJ, Arrese M, Torrance A, Lloyd S. *Pneumonyssoides* species infestation in two Pekingese dogs in the UK. *Journal of Small Animal Practice* 2008; 49:107–9.

274 Bredal WP. The prevalence of nasal mite (*Pneumonyssoides caninum*) infection in Norwegian dogs. *Veterinary Parasitology* 1998;76:233–7.

275 Gunnarsson L, Zakrisson G. Demonstration of circulating antibodies to *Pneumonyssoides caninum* in experimentally and naturally infected dogs. *Veterinary Parasitology* 2000;94:107–16.

276 Gunnarsson LK, Zakrisson G, Egenvall A, Christensson DA, Uggla A. Prevalence of *Pneumonyssoides caninum* infection in dogs in Sweden. *Journal of the American Animal Hospital Association* 2001;37:331–7.

277 Movassaghi AR, Mohri M. Nasal mite of dogs *Pneumonyssus (Pneumonyssoides) caninum* in Iran. *Veterinary Record* 1998;142:551–2.

278 Gunnarsson L, Zakrisson G, Lilliehook I, Christensson D, Rehbinder C, Uggla A. Experimental infection of dogs with the nasal mite *Pneumonyssoides caninum*. *Veterinary Parasitology* 1998;77:179–86.

279 Ballweber LR. Respiratory Parasites. Western States Veterinary Conference, Las Vegas, 2004. http://www.vin.com/Members/Proceedings/ Proceedings.plx?CID=wvc2004&PID=pr05466& O=VIN.

280 Gunnarsson L, Zakrisson G, Christensson D, Uggla A. Efficacy of selamectin in the treatment of nasal mite (*Pneumonyssoides caninum*) infection in dogs. *Journal of the American Animal Hospital Association* 2004; 40:400–4.

281 Gunnarsson LK, Möller LC, Einarsson AM, *et al.* Clinical efficacy of milbemycin oxime in the treatment of nasal mite infection in dogs. *Journal of the American Animal Hospital Association* 1999;35:81–4.

282 Mundell AC, Ihrke PJ. Ivermectin in the treatment of *Pneumonyssoides caninum*: a case report. *Journal of the American Animal Hospital Association* 1990; 26:393–6.

283 Drabick JJ. Pentastomiasis. *Clinical Infectious Diseases* 1987;9:1087–94.

284 Oluwasina OS, ThankGod OE, Augustine OO, Gimba FI. *Linguatula serrata* (Porocephalida: Linguatulidae) infection among client-owned dogs in Jalingo, North-eastern Nigeria: prevalence and public health implications. *Journal of Parasitology Research* 2014;2014:1–5.

285 Yazdani R, Sharifi I, Bamorovat M, Mohammadi MA. Human linguatulosis caused by *Linguatula serrata* in the city of Kerman, South-eastern Iran – case report. *Iranian Journal of Parasitology* 2014;9:282–5.

285a Veronesi F, Morgani G, di Cesare A, *et al.* *Eucoleus boehmi* infection in red foxes (*Vulpes vulpes*) from Italy. *Veterinary Parasitology* 2014;**206**:232–9.

285b Magi M, Guardone L, Prati MC, Torracca B, Maccioni F. First report of *Eucoleus boehmi* (syn. *Capillaria boehmi*) in dogs in north-western Italy, with scanning electron microscopy of the eggs. *Parasite* 2012;**19**:433–5.

285c Veronesi F, Leonis E, Morganti G, *et al.* Nasal eucoleosis in a symptomatic dog from Italy. *Veterinary Parasitology* 2013;**195**:187–91.

285d Conboy G, Stewart T, O'Brien S. Treatment of *E. boehmi* infection in a mixed breed dog using milbemycin oxime. *Journal of the American Animal Hopsital Association* 2013;**49**: 204–9.

285e Alho AM, Mouro S, Pissarra H, *et al.* First report of *Eucoleus boehmi* infection in a dog from Portugal. *Parasitology Research* 2016, E-published.

285f Baan M, Kidder AC, Johnson SE, Sherding RG. Rhinoscopic diagnosis of *Eucoleus boehmi* infection in a dog. *Journal of the American Animal Hospital Association* 2011;**47**:60–63.

286 Rudolph K, Bice DE, Hey JA, McLeod RL. A model of allergic nasal congestion in dogs sensitized to ragweed. *American Journal of Rhinology* 2003;**17**:227–32.

287 Cardell LO. Nasal secretion in ragweed-sensitized dogs: effect of leukotriene synthesis inhibition. *Acta Oto-Laryngologica* 2000;120:757–60.

288 Tiniakov RL, Tiniakova OP, McLeod RL, Hey JA, Yeates DB. Canine model of nasal congestion and allergic rhinitis. *Journal of Applied Physiology* 2003;94:1821–8.

289 Kurata K, Maeda S, Yasunaga S, Masuda K. Immunological findings in 3 dogs clinically diagnosed with allergic rhinitis. *Journal of Veterinary Medical Science* 2004;66:25–9.

290 Windsor RC, Johnson LR, Sykes JE, Drazenovich TL, Leutenegger CM, De Cock HEV. Molecular detection of microbes in nasal tissue of dogs with idiopathic lymphoplasmacytic rhinitis. *Journal of Veterinary Internal Medicine* 2006;20:250–6.

291 Meler E, Dunn M, Lecuyer M. A retrospective study of canine persistent nasal disease: 80 cases (1998–2003). *Canadian Veterinary Journal* 2008; 49:71–6.

292 Tasker S, Knottenbelt CM, Munro EA, Stonehewer J, Simpson JW, Mackin AJ. Aetiology and diagnosis of persistent nasal disease in the dog: a retrospective study of 42 cases. *Journal of Small Animal Practice* 1999;40:473–8.

293 Burgener DC, Slocombe RF, Zerbe CA. Lymphoplasmacytic rhinitis in five dogs. *Journal of the American Animal Hospital Association* 1987;23:565–8.

294 Clercx C, Reichler I, Peeters D, *et al*. Rhinitis/bronchopneumonia syndrome in Irish Wolfhounds. *Journal of Veterinary Internal Medicine* 2003;17:843–9.

295 Wilkinson GT. Some observations on the Irish Wolfhound rhinitis syndrome. *Journal of Small Animal Practice* 1969;10:5–8.

296 Leisewitz AL, Spencer JA, Jacobson LS, Schroeder H. Suspected primary immunodeficiency syndrome in three related Irish wolfhounds. *Journal of Small Animal Practice* 1997; 38:209–12.

297 O'Brien RT, Evans SM, Wortman JA. Radiographic findings in cats with intranasal neoplasia or chronic rhinitis: 29 cases (1982–1988). *Journal of the American Veterinary Medical Association* 1996;208:385–9.

298 Russo M, Lamb CR, Jakovljevic S. Distinguishing rhinitis and nasal neoplasia by radiography. *Veterinary Radiology & Ultrasound* 2000;41:118–24.

299 Muir W. Trauma: physiology, pathophysiology, and clinical implications. *Journal of Veterinary Emergency and Critical Care* 2006;16:253–63.

300 Hall KE, Holowaychuk MK, Sharp CR, Reineke E. Multicenter prospective evaluation of dogs with trauma. Journal of the American Veterinary Medical Association 2014;244:300–8.

301 Kolata RJ, Kraut NH, Johnston DE. Patterns of trauma in urban dogs and cats: a study of 1,000 bases. *Journal of the American Veterinary Medical Association* 1974;164:499–502.

302 Hayes G, Mathews K, Doig G, *et al*. The acute patient physiologic and laboratory evaluation (APPLE) score: a severity of illness stratification system for hospitalized dogs. *Journal of Veterinary Internal Medicine* 2010;24:1034–47.

303 Fleming JM, Creevy KE, Promislow DEL. Mortality in North American dogs from 1984 to 2004: an investigation into age-, size-, and breed-related causes of death. *Journal of Veterinary Internal Medicine* 2011;25:187–98.

304 Simpson SA, Syring R, Otto CM. Severe blunt trauma in dogs: 235 cases (1997–2003). *Journal of Veterinary Emergency and Critical Care* 2009;19:588–602.

305 Streeter EM, Rozanski EA, Laforcade-Buress A de, Freeman LM, Rush JE. Evaluation of vehicular trauma in dogs: 239 cases (January–December 2001). Journal of the American Veterinary Medical Association 2009;235:405–8.

306 Pope ER. Head and facial wounds in dogs and cats. *Veterinary Clinics of North America Small Animal Practice* 2006;36:793–817.

307 Platt SR, Radaelli ST, McDonnell JJ. The prognostic value of the modified Glasgow Coma Scale in head trauma in dogs. *Journal of Veterinary Internal Medicine* 2001;15:581–4.

308 Marretta SM. Maxillofacial fracture complications. In: *Oral and Maxillofacial Surgery in Dogs and Cats*. Verstraete FJM, Lommer MJ (eds). Saunders/Elsevier, St. Louis 2012, pp. 333–41.

309 Boudrieau RJ. Maxillofacial fracture repair using miniplates and screws. In: *Oral and Maxillofacial Surgery in Dogs and Cats*. Verstraete FJM, Lommer MJ (eds). Saunders/Elsevier, St. Louis 2012, pp. 292–308.

12.1 INTRODUCTION

Diseases of the nasopharynx have been reported in dogs, but are more common in cats[1-3]. In dogs, diseases of the nasopharynx are often an extension of diseases of the nose (for example rhinitis, brachycephalic obstructive airway syndrome and neoplastic disease), though specific nasopharyngeal abnormalities can be found as well. Cats are often afflicted with primary nasopharyngeal disease. Foreign bodies[4-7] (Figs 12.1A–C), neoplastic disease[1,3,8-11](Fig. 12.2), inflammatory polyps[1,3,12-18], fungal granulomas[19,20], cysts[21-23] and choanal atresia or nasopharyngeal stenosis[24-28] are encountered as causes of nasopharyngeal symptoms in dogs and cats. The middle ears and nasopharynx are commonly inflamed

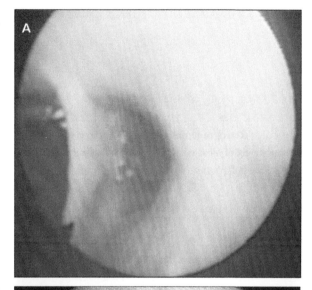

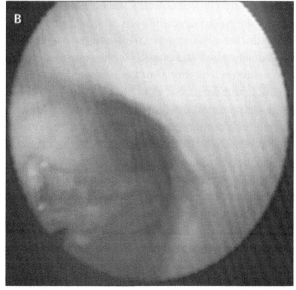

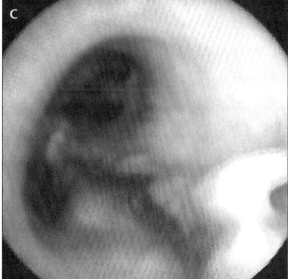

Figs 12.1A–C A, B: Retrograde nasopharyngoscopic view of the feline nasopharynx, demonstrating the caudal end of the nasopharynx and free edge of the soft palate with a piece of grass in the nasopharynx, and close-up view (B); C: foreign body in the nasopharynx with associated inflammation.

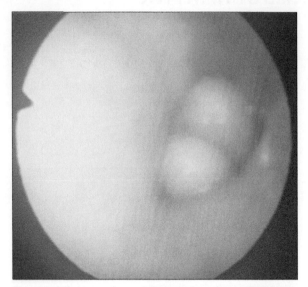

Fig. 12.2 Retrograde nasopharyngoscopic view of the nasopharynx of a cat with malignant lymphoma. In the rostral nasopharynx a bilobular rounded pink mass can be seen.

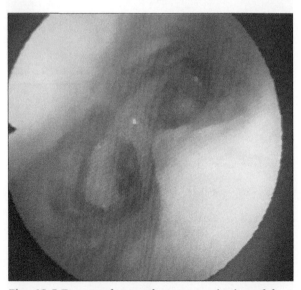

Fig. 12.3 Retrograde nasopharyngoscopic view of the nasopharynx of a cat with chronic rhinitis and sinusitis, involvement of the rostral nasopharyngeal mucosa and accumulation of purulent exudate around the choanae is evident.

in cats with chronic rhinosinusitis (Fig. 12.3), and extension of middle ear disease with or without polyp growth leading to nasopharyngeal compression or abscessation can be found[29,30], though these are rare complications.

Congenital (choanal atresia) and acquired nasopharyngeal stenosis are debilitating diseases that are not uncommonly seen and present a therapeutic challenge. They will be specifically discussed in this chapter. Nasopharyngeal cysts, Rathke's clefts or craniopharyngiomas are rare abnormalities that can cause significant obstruction of the nasopharyngeal airway and will therefore also be discussed here. Nasopharyngeal foreign bodies have been reviewed in Chapter 11 Diseases of the Nasal Cavity and Sinuses, Section 11.5, and for a review of causes of nasopharyngitis the reader is referred to the respective sections on rhinitis and pharyngitis/laryngitis in this book. Neoplastic disease has been discussed in Chapter 11 Diseases of the Nasal Cavity and Sinuses, Section 11.8, nasopharyngeal polyps in Chapter 5 Aetiology and Pathogenesis of Otitis Externa, Section 5.9.

12.2 NASOPHARYNGEAL STENOSIS AND CHOANAL ATRESIA

Nasopharyngeal stenosis

Whereas congenital nasopharyngeal dysgenesis leading to stenosis has been reported[28], most commonly nasopharyngeal stenosis is an acquired condition[18,31]. Congenital stenotic nasopharyngeal dysgenesis has been reported in Dachshunds and is characterised by caudal nasopharyngeal stenosis secondary to abnormally thickened palatopharyngeal musculature[28,32]. In both dogs[2,24,25,27] and cats[33-39] with acquired stenosis, there often is a history of nasopharyngeal surgery, an upper airway infection, or some other inciting cause of nasopharyngeal inflammation, e.g. passage of acid gastric fluids during regurgitation and vomiting. In cats ulcerations caused by herpes- and calicivirus are probably the most common cause of nasopharyngeal inflammation with secondary stenosis.

The clinical signs depend on the degree of obstruction and include chronic nasal discharge, stridorous or open-mouth breathing, lack of nasal airflow, gagging, sneezing and reverse sneezing. Whereas in choanal atresia the obstruction can be osseous or membranous, animals with acquired stenosis have a variably thick, impermeable membrane partially or completely occluding the nasopharynx, either rostrally around the choanae or more caudal in the nasopharynx.

The diagnosis is based on the clinical signs, radiographic or computer tomographic demonstration of an

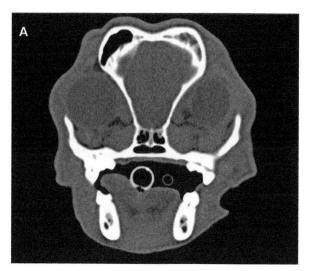

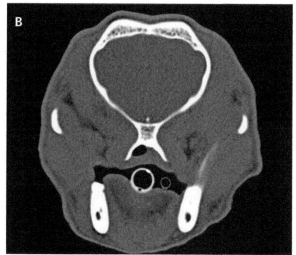

Fig. 12.4A–C CT transverse view of the nasopharynx of a dog with a partial midnasopharyngeal stenosis. A: rostral to area of stenosis; B: at level of stenosis; C: sagittal view.

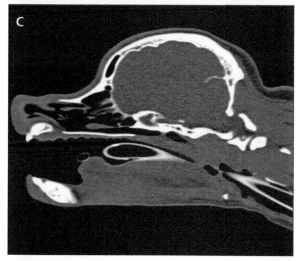

area of soft tissue density in the nasopharynx obliterating the lumen (Fig. 12.4) and nasopharyngoscopy (Fig. 12.5). Upon demonstration of nasopharyngeal stenosis, non-invasive treatment can be employed during the same anaesthetic.

Several treatment strategies for nasopharyngeal stenosis have been described, from open resection, balloon dilatation to nasopharyngeal stenting. Successful surgical resection of the stenotic membrane has been reported[27,36,38]; however, (partial) restenosis after surgical correction is a common complication[27,39]. Restenosis as a result of web formation may be prevented by accurate mucosal apposition using local mucosal flaps after resection of the area of the stenosis[18,36]. Mitomycin may be used topically to reduce the chances of stenosis, as has been advised for the treatment of choanal atresia[40] (see below), but there are no veterinary reports available on topical adjunctive treatment. The use of a stent to augment a surgical revision was successful in one animal[39].

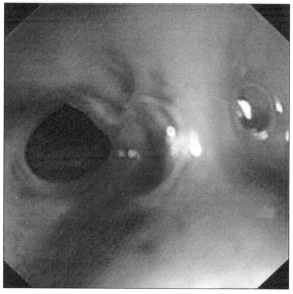

Fig. 12.5 Retrograde nasopharyngoscopic view of the nasopharynx of the dog in 12.4 with a midnasopharyngeal stenosis.

Endoscopic-guided balloon dilatation of the stenotic membrane has been successful in both dogs[24] and cats[33–35], but the treatment may need to be repeated if clinical signs recur. Use of steroids and antibiotics after bougienage has been described[35], but no evidence-based studies exist as to the effectiveness of this adjunctive treatment. Primary treatment with endoscopically and fluoroscopically guided balloon-expandable metallic stenting was reported in six dogs and cats, with all except one breathing normally 12–28 months after the procedure[25].

An elegant and very successful method for treatment of acquired nasopharyngeal stenosis in 15 cats was published very recently[35a]. Using a removable silicone stent sutured in place locally after initial dilatation of the stenosis and removed 3–4 weeks later, resulted in complete resolution of clinical signs in 14 cats and an improvement in 1 cat.

Choanal atresia

Choanal atresia is a congenital condition in which the buccopharyngeal or oronasal membrane fails to resorb, resulting in a lack of communication between the nose and the nasopharynx[31,41]. Choanal atresia has been described in humans where it is the most frequently encountered congenital nasal anomaly[42], but also in dogs[27,41] and cats[43,44]. The abnormality can be unilateral or bilateral and osseous or membranous[41,42].

Puppies and kittens with bilateral choanal atresia probably die young[31], but if they do survive, they will be unable to breathe with their mouths closed or have stridorous breathing when the mouth is closed if some air passage is possible[27]. Affected animals may have difficulty suckling as neonates, develop unilateral or bilateral chronic nasal discharge and experience severe dyspnoea[18].

The diagnosis of choanal atresia is best performed using CT imaging and rhino- and nasopharyngoscopy, but positive-contrast rhinography has also been used[27,41,42]. In addition, failure to pass a catheter through the nose into the nasopharynx, especially when seen poking against a membrane closing the choana as visualised directly via nasopharyngoscopy, is also diagnostic.

Treatment strategies for choanal atresia include balloon dilatation[27] or open excision and choanal reconstruction using a transpalatal approach[43,44]. Development of nasopharyngeal scar tissue and subsequent stenosis is the main concern after any attempt of repair. Establishing adequate mucosal lining of the new choanae after excision of constricting tissue is mandatory to decrease the chances of restenosis[40,42,45]. Placement of soft stents or the use of mitomycin has been advised to reduce the risk of stenosis after treatment in people[40,42,45].

Mitomycin is an antiproliferative and antitumour antibiotic agent that inhibits fibroblast growth and proliferation, and after topical application has been shown to be very efficient in reducing the risk of restenosis in people[40]. The use of topical corticosteroids has only anecdotally been reported, and proper studies of effectiveness are also currently lacking in veterinary literature. Partial restenosis can be managed with repeated balloon dilatations and/or temporary stenting. In a large review on surgical treatment of choanal atresia in children, surgery was effective in establishing a patent airway in all of them[45]. However, multiple procedures (between 3 and 5) were required on average[45].

12.3 NASOPHARYNGEAL CYSTS AND CRANIOPHARYNGIOMA

The pituitary gland is formed in fetal life from two parts: Rathke's pouch or cleft, which is an ectodermal out-pocketing of the stomodeum immediately in front of the buccopharyngeal or oronasal membrane; and the infundibulum, which is a downward extension of the diencephalon[21,31]. Cystic remnants of Rathke's pouch can occur anywhere along the path of migration of the pouch and will result in nasopharyngeal obstruction if large and external to the cranial vault[22,46]. A craniopharyngioma is a tumour derived from rests of the epithelium of the primordial craniopharyngeal canal, or Rathke's pouch[31,46,47] and can also rarely be found to extend into the nasopharynx.

Nasopharyngeal cysts

Four dogs with this condition have been reported[21,22,48]. An 8-year-old Poodle cross dog with an extracranial cyst resulting in nasopharyngeal obstruction was reported by Slatter et al.[48]. A 5-month-old entire male German Shepherd Dog with a history of loud snoring, and a 4-year-old spayed female Kelpie crossbreed with a 3-month history of occasional sneezing, snorting and snoring during sleep were reported by Beck *et al.*[21]. And most recently, a 21-month-old male Boxer dog was reported by Clements *et al.*[22] with a history of depression, dysphagia, pain when opening the mouth and stridorous breathing.

Radiographs taken under general anaesthesia either showed no abnormalities or demonstrated a variable sized soft tissue mass with some calcifications at the margins obstructing the nasopharynx and causing ventral displacement of the soft palate[21,22]. Ultrasonography through the soft palate showed the mass to be cystic with septae dividing the lumen in one dog[21]. Nasopharyngoscopy under general anaesthesia showed a mucosa-covered mass arising from the roof of the nasopharynx in three cases[21,48].

All dogs were treated successfully with surgery. Nasopharyngotomy via a full-thickness longitudinal incision in the soft palate allowed removal of the cyst in three cases[21,48]. The cyst wall was dissected from the roof of the nasopharynx, but no attempts were made to explore or resect the cystic canal towards the cranial cavity and the dorsal nasopharyngeal mucosa was left open to heal by second intention. In the fourth case, the cystic lesion was exposed via a ventral approach to the tympanic bulla as this cyst was located in this area. The spherical cystic mass, enclosed in a thick fibrous and partially mineralised capsule, was easily identified and bluntly dissected from the underlying bulla and basisphenoid.

Histological examination of the cyst wall revealed a dense fibrous capsule with some bone and cartilage formation, lined with an irregular stratified or ciliated columnar epithelium[21,22]. The histological appearance of cystic remnants of Rathke's pouch is similar to the nasopharyngeal epidermal cyst described by Ellison *et al.* in a 12-year-old Miniature Poodle with a history of progressive stridor[23]. Survey radiographs of the skull revealed a large mineralised mass in the nasopharyngeal region, displacing the soft palate ventrally and extending from the right tympanic bulla. This mass was also excised successfully via nasopharyngotomy.

No nasopharyngeal cysts have been described in the literature in cats, but a soft palate cyst was reported recently[49]. A 17-month-old cat presented with a chronic history of frequent swallowing, occasional cough and stridor. Radiographic evaluation revealed a soft tissue mass at the level of the soft palate, which appeared to be within the palate upon oral examination[49]. Histopathology of the surgically excised mass was consistent with a large developmental cyst.

Craniopharyngioma (Rathke's pouch tumour)

Craniopharyngiomas are rare in dogs and cats and only a few case reports exist[47,50–52]. In humans they have been shown to arise at any point along the craniopharyngeal canal and therefore some lie within the sella turcica and others are external to it, in the nasopharynx[46]. In humans craniopharyngiomas are often associated with endocrinopathies[53,54], also described in dogs[50].

The clinical signs of craniopharyngioma in humans depend on the size of the lesion and proximity to the pituitary gland and optic chiasm and include anisocoria, strabismus, progressive depression, somnolence, seizures, extraocular motor paralysis, sudden blindness, and hypopituitarism causing growth hormone deficiency, secondary hypothyroidism and secondary hypoadrenocorticism[53,54]. Both cats reported presented with abnormal breathing, wheezing and nasal discharge and one also had ocular signs; the colour of his eyes changed, mydriasis appeared and he became blind[52]. In dogs, nasopharyngeal signs did not dominate the clinical presentation; one dog presented with lethargy, episodic circling, incoordination and polydypsia[51] and another with persistent anisocoria, progressive depression, somnolence, seizures and anorexia[47].

A presumptive diagnosis can be made on clinical signs and advanced diagnostic imaging, but histopathological examination is required for definitive diagnosis. In humans treatment consists of partial or complete surgical resection with or without radiotherapy[53,54]. In dogs and cats a successful treatment has not been described and all cases reported were diagnosed on necropsy.

The histological pattern is variable and there is a wide spectrum of cell types reported. Some contain cells of the enamel organ of the tooth and are sometimes classified as ademantinomas or ameloblastinomas[31]. Calcification and metaplastic bone formation can occur in the necrotic centres of solid tumours. Often nests of stratified squamous or columnar epithelial cells are found embedded in a loose fibrous stroma[31,47,52,53].

Cramer *et al.* reported a variety of this rare tumour[55] in a 7-year-old, intact male standard Poodle dog who presented with hypothyroidism and atypical hyper-adrenocorticism and developed acute signs of lethargy, weakness, inappetence, vomiting and diarrhoea. Clinical signs progressed to hindlimb proprioceptive deficits, aggressive behaviour with obtundation, and seizures. Necropsy revealed a mass in the sellar region with histological characteristics similar to changes described in the human medical literature as xanthogranuloma of the sellar region and xanthogranulomatous hypophysitis[55].

12.4 REFERENCES

1 Allen HS, Broussard J, Noone K. Nasopharyngeal diseases in cats: a retrospective study of 53 cases (1991–1998). *Journal of the American Animal Hospital Association* 1999;**35**:457–61.

2 Billen F, Day MJ, Clercx C. Diagnosis of pharyngeal disorders in dogs: a retrospective study of 67 cases. *Journal of Small Animal Practice* 2006;**47**:122–9.

3 Hunt GB, Perkins MC, Foster SF, *et al.* Nasopharyngeal disorders of dogs and cats: a review and retrospective study. *Compendium on Continuing Education for the Practising Veterinarian* 2002;**24**:184–200.

4 Papazoglou LG, Patsikas MN. What is your diagnosis? Foreign body in the nasopharynx of a dog. *Journal of Small Animal Practice* 1995;**36**:425–34.

5 Schmidtke HO. Removal of foreign bodies in the nasopharynx. *Journal of Small Animal Practice* 1996;**37**:186.

6 Simpson AM, Harkin KR, Hoskinson JJ. Radiographic diagnosis: nasopharyngeal foreign body in a dog. *Veterinary Radiology & Ultrasound* 2000;**228**:326–8.

7 Tyler JW. Endoscopic retrieval of a large, nasopharyngeal foreign body. *Journal of the American Animal Hospital Association* 1997;**33**:513–6.

8 Carpenter JL, Hamilton TA. Angioleiomyoma of the nasopharynx in a dog. *Veterinary Pathology* 1995;**32**:721–3.

9 Little L, Patel R, Goldschmidt M. Nasal and nasopharyngeal lymphoma in cats: 50 cases (1989–2005). *Veterinary Pathology* 2007;**44**:885–92.

10 Karnik K, Reichle JK, Fischetti AJ, Goggin JM. Computed tomographic findings of fungal rhinitis and sinusitis in cats. *Veterinary Radiology & Ultrasound* 2009;**50**: 65–8.

11 Patnaik AK, Ludwig LL, Erlandson RA. Neuroendocrine carcinoma of the nasopharynx in a dog. *Veterinary Pathology* 2002;**39**:496–500.

12 Anderson DM, Robinson RK, White RA. Management of inflammatory polyps in 37 cats. *Veterinary Record* 2000;**147**:684–7.

13 Bedford P. Origin of the nasopharyngeal polyp in the cat. *Veterinary Record* 1982;**110**:541–2.

14 Bedford PG, Coulson A, Sharp NJ, Longstaffe JA. Nasopharyngeal polyps in the cat. *Veterinary Record* 1981;**109**:551–3.

15 Bradley RL, Noone KE, Saunders GK, Patnaik AK. Nasopharyngeal and middle ear polypoid masses in five cats. *Veterinary Surgery* 1985;**14**:141–4.

16 Brownlie SE, Bedford PG. Nasopharyngeal polyp in a kitten. *Veterinary Record* 1985;**117**:668–9.

17 Venker-van Haagen AJ. The nose and nasal sinuses. In: *Ear, Nose, Throat, and Tracheobronchial Diseases in Dogs and Cats*. Venker-van Haagen AJ (ed). Schlütersche Verlagsgesellschaft, Hannover 2005, pp. 51–81.

18 Schmiedt CW, Creevy KE. Nasal planum, nasal cavity, and sinuses. In: *Veterinary Surgery Small Animal*. Tobias KM, Johnston SA (eds). Elsevier Saunders, St. Louis 2012, pp. 1691–706.

19 Beatty JA, Barrs VR, Swinney GR, Martin PA, Malik R. Peripheral vestibular disease associated with cryptococcosis in three cats. *Journal of Feline Medicine and Surgery* 2000;**2**:29–34.

20 Malik R, Martin P, Wigney DI, *et al.* Nasopharyngeal cryptococcosis. *Australian Veterinary Journal* 1997;**75**:483–8.

21 Beck JA, Hunt GB, Goldsmid SE, Swinney GR. Nasopharyngeal obstruction due to cystic Rathke's clefts in two dogs. *Australian Veterinary Journal* 1999;**77**:94–6.

22 Clements DN, Thompson H, Johnson VS, Clarke SP, Doust RT. Diagnosis and surgical treatment of a nasopharyngeal cyst in a dog. *Journal of Small Animal Practice* 2006;**47**:674–7.

23 Ellison GW, Donnell RL, Daniel GB. Nasopharyngeal epidermal cyst in a dog. *Journal of the American Veterinary Medical Association* 1995;**207**:1590-2.

24 Berent AC, Kinns J, Weisse C. Balloon dilatation of nasopharyngeal stenosis in a dog. *Journal of the American Veterinary Medical Association* 2006;**229**:385–6.

25 Berent AC, Weisse C, Todd K, Rondeau MP, Reiter AM. Use of a balloon-expandable metallic stent for treatment of nasopharyngeal stenosis in dogs and cats: six cases (2005–2007). *Journal of the American Veterinary Medical Association* 2008;**233**:1432–40.

26 Birchard SJ. A simplified method for rhinotomy and temporary rhinostomy in dogs and cats. *Journal of the American Animal Hospital Association* 1988;**24**:69–72.

27 Coolman BR, Marretta SM, McKiernan BC, Zachary JF. Choanal atresia and secondary nasopharyngeal stenosis in a dog. *Journal of the American Animal Hospital Association* 1998;**34**:497–501.

28 Kirberger RM, Steenkamp G, Spotswood TC, Boy SC, Miller DB, van Zyl M. Stenotic nasopharyngeal dysgenesis in the dachshund: seven cases (2002–2004). *Journal of the American Animal Hospital Association* 2006;**42**:290–7.

29 Pilton JL, Ley CJ, Voss K, Krockenberger MB, Barrs VR, Beatty JA. Atypical, abscessated nasopharyngeal polyp associated with expansion and lysis of the tympanic bulla. *Journal of Feline Medicine and Surgery* 2014;**16**:699–702.

30 Cook LB, Bergman RL, Bahr A, Boothe HW. Inflammatory polyp in the middle ear with secondary suppurative meningoencephalitis in a cat. *Veterinary Radiology & Ultrasound* 2003;**44**:648–51.

31 Venker-van Haagen AJ. The Pharynx. In: *Ear, Nose, Throat, and Tracheobronchial Diseases in Dogs and Cats*. Venker-van Haagen AJ (ed). Schlütersche Verlagsgesellschaft, Hannover 2005, pp. 83–120.

32 Dvir E, Spotswood TC, Lambrechts NE, Lobetti RG. Congenital narrowing of the intrapharyngeal opening in a dog with concurrent oesophageal hiatal hernia. *Journal of Small Animal Practice* 2003;**44**:359–62.

33 Boswood A, Lamb CR, Brockman DJ, Mantis P, Witt AL. Balloon dilatation of nasopharyngeal stenosis in a cat. *Veterinary Radiology & Ultrasound* 2003;**44**:53–5.

34 Glaus TM, Tomsa K, Reusch CE. Balloon dilation for the treatment of chronic recurrent nasopharyngeal stenosis in a cat. *Journal of Small Animal Practice* 2002;**43**:88–90.

35 Glaus TM, Gerber B, Tomsa K, Keiser M. Reproducible and long-lasting success of balloon dilation of nasopharyngeal stenosis in cats. *Veterinary Record* 2005;**157**:257–9.

35a De Lorenzi D, Bertoncello D, Comastri S, Bottero E. Treatment of acquired nasopharyngeal stenosis using a removable silicone stent. *Journal of Feline Medicine and Surgery* 2015;**17**:117–24.

36 Griffon DJ, Tasker S. Use of a mucosal advancement flap for the treatment of nasopharyngeal stenosis in a cat. *Journal of Small Animal Practice* 2000;**41**:71–3.

37 Henderson SM, Bradley K, Day MJ, *et al*. Investigation of nasal disease in the cat--a retrospective study of 77 cases. *Journal of Feline Medicine and Surgery* 2004;**6**:245–57.

38 Mitten RW. Nasopharyngeal stenosis in four cats. *Journal of Small Animal Practice* 1988;**29**:341–5.

39 Novo RE, Kramek B. Surgical repair of nasopharyngeal stenosis in a cat using a stent. *Journal of the American Animal Hospital Association* 1999;**35**:251–6.

40 Holland BW, William F McGuirt J. Surgical management of choanal atresia: improved outcome using mitomycin. *Archives of Otolaryngology Head & Neck Surgery* 2001;**127**:1375–80.

41 Jiménez DA, Berry CR, Ferrell EA, Graham JP. Imaging diagnosis – choanal atresia in a dog. *Veterinary Radiology & Ultrasound* 2007;**48**:135–7.

42 Szeremeta W, Parikh TD, Widelitz JS. Congenital nasal malformations. *Otolaryngologic Clinics of North America* 2007;**40**:97–112.

43 Khoo AML, Marchevsky AM, Barrs VR, Beatty JA. Choanal atresia in a Himalayan cat – first reported case and successful treatment. *Journal of Feline Medicine and Surgery* 2007;**9**:346–9.

44 Schafgans KE, Armstrong PJ, Kramek B, Ober CP. Bilateral choanal atresia in a cat. *Journal of Feline Medicine and Surgery* 2012;**14**:759–63.

45 Samadi DS, Shah UK, Handler SD. Choanal atresia: a twenty-year review of medical comorbidities and surgical outcomes. *Laryngoscope* 2003;**113**:254–8.

46 Rottenberg GT, Chong WK, Powell M, Kendall BE. Cyst formation of the craniopharyngeal duct. *Clinical Radiology* 1994;**49**:126–9.

47 Hawkins KL, Diters RW, McGrath JT. Craniopharyngioma in a dog. *Journal of Comparative Pathology* 1985;**95**:469–74.

48 Slatter D, Schirmer RG, Krehbiel JD. Surgical correction of cystic remnants of Rathke's cleft in a dog. *Journal of the American Animal Hospital Association* 1976;**12**:641–3.

49 Peterson NW, Buote NJ. Soft palate cyst in a cat. *Journal of Feline Medicine and Surgery* 2011;**13**:594–6.

50 Neer TM, Reavis DU. Craniopharyngioma and associated central diabetes insipidus and hypothyroidism in a dog. *Journal of the American Veterinary Medical Association* 1983;**182**:519–20.

51 Eckersley GN, Geel JK, Kriek NP. A craniopharyngioma in a seven-year-old dog. *Journal of the South African Veterinary Association* 1991;**62**:65–7.

52 Nagata T, Nakayama H, Uchida K, *et al*. Two cases of feline malignant craniopharyngioma. *Veterinary Pathology* 2005;**42**:663–5.

53 Rémy Van Effenterre, Anne-Laure Boch. Craniopharyngioma in adults and children: a study of 122 surgical cases. *Journal of Neurosurgery* 2002;**97**:3–11.

54 Rivas AM, Sotello D, Lado-Abeal J. Primary and secondary endocrinopathies found in a patient with craniopharyngioma. *American Journal of Medical Science* 2014;**348**:534–5.

55 Cramer SD, Miller AD, Medici EL, Brunker JD, Ritchey JW. Sellar xanthogranuloma in a dog. *Journal for Veterinary Diagnostic Investigation* 2011;**23**:387–90.

CHAPTER 13

THROAT ANATOMY AND PHYSIOLOGY

13.1 INTRODUCTION

The throat comprises the pharynx and larynx and is essential for two vital functions for survival in all species: respiration and digestion. The throat is a relatively unprotected area and prone to trauma that can potentially compromise its important functions and severe trauma may sometimes result in death of the animal. The main function of the pharynx lies in deglutition (swallowing), whereas the main functions of the larynx are regulation of airflow and protection of the lower airways against aspiration. These specific functions of the throat will be discussed in subsequent sections, after the species-specific gross and microscopic anatomy.

13.2 GROSS AND MICROSCOPIC ANATOMY OF THE THROAT

The throat comprises the pharynx and larynx and is an unique crossways of the digestive and respiratory tract[1,2]. The pharynx is further divided into nasal, oral and laryn-

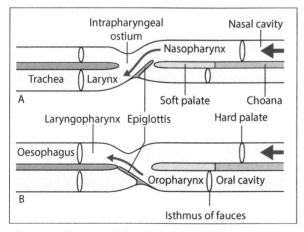

Fig. 13.1 Diagram of the pharyngeal chiasm, during inspiration (A) and expiration (B).

geal parts. The nasopharynx is the respiratory portion of the pharynx dorsal to the soft palate, the anatomy of which has been discussed in Chapter 8 Nose Anatomy and Physiology, Section 8.2. The oropharynx is the space between the oral cavity proper, defined as the space between the lower and upper dental arcades[3], and the laryngopharynx and nasopharynx. It is bound dorsally by the soft palate and ventrally by the root of the tongue. The soft palate separates the nasopharynx from the oropharynx[4]. Caudal to the intrapharyngeal opening the digestive tube is continued dorsal to the larynx as the laryngopharynx, which opens into the oesophagus, and the respiratory tube is continued ventrally as the larynx, which in turn is continuous with the trachea[4].

Gross anatomy of the pharynx and larynx

The pharynx is a fibromuscular tube connecting the oral and nasal cavities rostrally and the oesophagus and larynx caudally. The nasopharynx comprises the respiratory portion of the pharynx and contains the pharyngeal openings of the auditory tubes (see Chapter 8 Nose Anatomy and Physiology, Section 8.2). The intrapharyngeal opening is the opening of the nasopharynx into the laryngopharynx, and is formed by the free caudal border of the soft palate and the right and left palatopharyngeal arches (Figs 13.1, 13.2)[4]. The isthmus of the fauces is the orifice between the oral cavity and the oropharynx. It is bounded on each side by the palatoglossal arch, ventrally by the tongue and dorsally by the soft palate[4].

The oropharynx extends from the isthmus of the fauces to the base of the epiglottis. It is bounded dorsally by the soft palate, ventrally by the root of the tongue and laterally by the tonsillar fossa, with its contained palatine tonsil. The rostral portion of the oropharynx where the palatine tonsil resides in its lateral wall is called the fauces[4]. The palatine tonsil is a long,

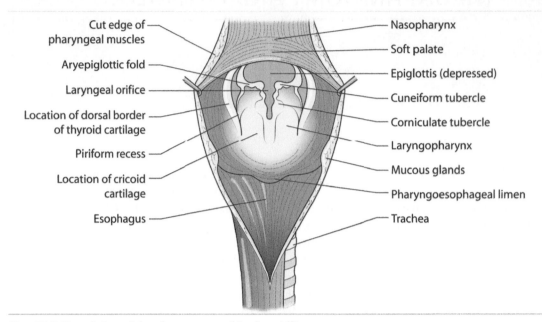

Cut edge of pharyngeal muscles
Aryepiglottic fold
Laryngeal orifice
Location of dorsal border of thyroid cartilage
Piriform recess
Location of cricoid cartilage
Esophagus

Nasopharynx
Soft palate
Epiglottis (depressed)
Cuneiform tubercle
Corniculate tubercle
Laryngopharynx
Mucous glands
Pharyngoesophageal limen
Trachea

Fig. 13.2 Dorsal view of the pharynx opened in the midline.

thin lymphoid organ, located in the tonsillar fossa in the lateral wall of the oropharynx just caudal to the palatoglossal arch. Besides the palatine tonsils, diffuse lymphoid tissue, called the lingual and pharyngeal tonsils, lies in the base of the tongue and in the nasopharynx[4].

The laryngopharynx is that portion of the pharynx that lies dorsal to the larynx. It extends from the intrapharyngeal ostium and nasopharynx rostrally to the beginning of the oesophagus caudally. Although the function of the laryngeal part of the pharynx is both respiratory and alimentary, its main role lies in deglutition[4]. During swallowing the circular closure of the intrapharyngeal ostium by the coordinated action of the pharyngeal muscles separates the nasopharynx from the oropharynx[1,5]. The bolus of food ingested is conveyed to the laryngopharynx by movement of the base of the tongue coordinated by six pairs of extrinsic muscles. These include the hyopharyngeus, thyropharyngeus, cricopharyngeus, stylopharyngeus, palatopharyngeus and pterygopharyngeus muscles, which are all innervated by pharyngeal branches of the glossopharyngeal and vagal nerves[4,5].

- The hyopharyngeus, thyropharyngeus and cricopharyngeus constrict the rostral, middle and caudal parts of the pharynx respectively.

- The stylopharyngeus muscles dilate the pharynx.
- The palatopharyngeus and pterygopharyngeus muscles shorten the pharynx by constricting and drawing it rostrally.

The muscles of the soft palate are closely associated with the muscles of the pharynx and include the tensor veli palatini, levator veli palatini and the palatinus[5,6]. The laryngopharynx receives its blood supply through the paired pharyngeal branches of the cranial thyroid and the ascending pharyngeal arteries. The soft palate and the dorsal wall of the oral part of the pharynx are perfused via the paired minor palatine arteries[4].

The larynx is formed by the following cartilages:
- Epiglottis.
- Thyroid.
- Cricoid.
- Sesamoid.
- Interarytenoid.
- Paired arytenoid[4].

It is attached to the trachea caudally and suspended to the oesophagus by the hyoid apparatus ventrally (Fig. 13.3)[7]. The spade-shaped epiglottis is the most rostral cartilage and rests on the soft palate when in normal

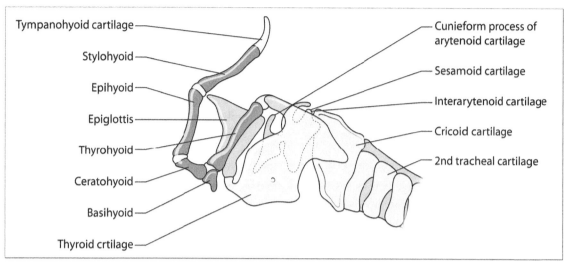

Tympanohyoid cartilage

Stylohyoid

Epihyoid

Epiglottis

Thyrohyoid

Ceratohyoid

Basihyoid

Thyroid crtilage

Cunieform process of arytenoid cartilage

Sesamoid cartilage

Interarytenoid cartilage

Cricoid cartilage

2nd tracheal cartilage

Fig. 13.3 **Left lateral view of the hyoid apparatus, larynx and rostral trachea.**

position. The oral surface of the epiglottis is attached by a short, stout hyoepiglotticus muscle to the middle of the body of the hyoid bone[8,9]. Contraction of this muscle draws the epiglottis downward. A thick stalk of fibrous tissue also unites the midcaudal portion of the epiglottis and dorsal rostral surface of the thyroid cartilage. Laterally, it attaches to the cuneiform processes by aryepiglottic mucosal folds.

The thyroid cartilage is the largest of all cartilages and consists of right and left laminae that cover the sides of the larynx and fuse ventrally to form a short deep trough[4]. Dorsally, the laminae expand to form thin rostral and caudal processes, the horns or cornua, which are separated from the thyroid lamina by notches[4,7]. The cranial laryngeal nerve, responsible for sensory innervation of the larynx, and laryngeal artery pass through the rostral thyroid notch. The cricothyroid ligament unites the thyroid cartilage at its caudal border with the cricoid cartilage that forms a complete ring and subsequently connects caudally to the first tracheal ring[4,10]. The cricoid cartilage articulates via small synovial joints with the thyroid cartilage caudally and the arytenoid cartilage rostrally[4]. The paired arytenoid cartilages consist of cuneiform, corniculate, vocal and muscular processes in dogs[4]; feline arytenoid cartilages lack cuneiform and corniculate processes[7,11].

The ventral aspect of the wedge-shaped cuneiform process lies in the aryepiglottic fold, and its dorsal aspect forms the midportion of the laryngeal inlet

(Fig. 13.4). The ventricular ligament and ventricularis muscle attach to this process. The horn-shaped corniculate process forms the dorsal margin of the laryngeal inlet and lies caudal to the cuneiform process[4,7]. The corniculate processes are interconnected via an interarytenoid ligament, which can contain a sesamoid cartilage. Lateral to the cricoarytenoid articulation is

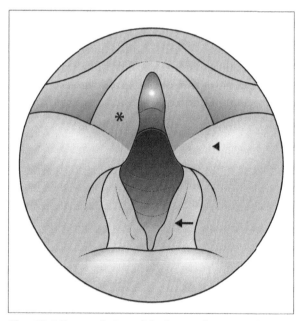

Fig. 13.4 **Intraoral view of the larynx demonstrating the corniculate (asterisk) and cuneiform (arrowhead) processes of the larynx and the vocal folds (arrow).**

the muscular process, which serves as the insertion site for the cricoarytenoideus dorsalis muscle, the abductor of the larynx. The caudal ventral portion of the arytenoid cartilage contains the vocal process, which serves as an attachment of the vocal ligament and, from the thyroid cartilage, the vocalis muscle. A region of everted mucosa, known as laryngeal ventricle or laryngeal saccule, separates mucosal-covered folds over the ventricularis and vocalis muscles[4].

The extrinsic muscles of the larynx include the thyropharyngeus and cricopharyngeus muscles, which cover the outside of the larynx and are involved in swallowing (see above). The intrinsic muscles are responsible for laryngeal function[4].

> Abduction of the arytenoid cartilages to open the glottis is performed by one muscle only, the cricoarytenoideus dorsalis.

This muscle arises from the dorsolateral surface of the cricoid cartilage and inserts on the muscular process of the arytenoid (Fig. 13.5). The cricoarytenoideus lateralis pivots the arytenoid cartilage inward to close the rima glottis[4,7]. The thyroarytenoideus muscle splits into ventricularis and vocalis muscles. The vocalis portion draws the arytenoid cartilage downward, relaxing the vocal cords, and the ventricularis portion constricts the glottis and dilates the laryngeal saccule[4,7]. The arytenoideus transversus assists with vocal fold adduction and glottis constriction. The cricothyroideus muscle lies laterally

between the thyroid lamina and cricoid cartilage, and tenses the vocal cords upon contraction by pivoting the cricoid cartilage on its thyroid articulation[4,7].

Innervation to the larynx is provided by cranial and caudal laryngeal nerves, which originate from the vagus nerve[4,7]. The cranial laryngeal nerve leaves the vagus at the distal ganglion and passes ventral to the larynx. It provides an external branch that supplies the cricothyroideus muscle and an internal branch that receives sensory fibres from laryngeal mucosa cranial to the vocal folds and serves as the afferent limb of the cough reflex[4]. The internal branch of the cranial laryngeal nerve usually anastomoses with the caudal laryngeal nerve.

The caudal laryngeal nerve is the terminal segment of the recurrent laryngeal nerve.

- The left recurrent laryngeal nerve from the left vagus arches around the aorta, ascends along the trachea and gives off the pararecurrent laryngeal nerve before terminating as the left caudal laryngeal nerve.
- The right recurrent laryngeal nerve from the right vagus loops around the right subclavian artery and ascends along the trachea to end as the right caudal laryngeal nerve[4,7].
- The caudal laryngeal nerves provide motor supply to all intrinsic laryngeal muscles except for the cricothyroideus.

Blood supply to the larynx is primarily through branches of the cranial and the caudal thyroid arteries. Lymphatic drainage is through the medial retropharyngeal lymph nodes[4].

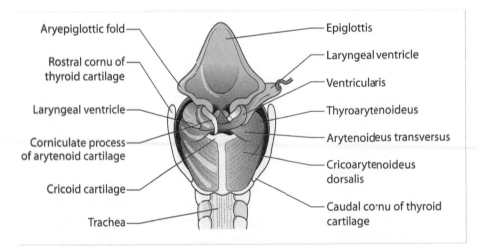

Fig. 13.5 Dorsal aspect of the laryngeal muscles.

Aryepiglottic fold
Rostral cornu of thyroid cartilage
Laryngeal ventricle
Corniculate process of arytenoid cartilage
Cricoid cartilage
Trachea

Epiglottis
Laryngeal ventricle
Ventricularis
Thyroarytenoideus
Arytenoideus transversus
Cricoarytenoideus dorsalis
Caudal cornu of thyroid cartilage

Microscopic anatomy of the throat

The pharyngeal wall is made up of a mucosa, a tunica muscularis of skeletal muscle, and an adventitia (Figs 13.6–13.10)[12,13]. The mucosa is lined by stratified squamous epithelium, except for a portion of the naso-pharynx, which is lined by ciliated, pseudostratified columnar epithelium (see Chapter 8 Nose Anatomy and Physiology, Section 8.2). While there is no lamina muscularis, the lamina propria–submucosa contains collagen and elastic fibres intermingled with lymphatic tissue and mucous glands[12,13]. The underlying tunica muscularis consists entirely of skeletal muscle covered by a dense adventitia consisting of irregular connective tissue that attaches the pharynx to the surrounding tissue. The soft palate consists of a core of skeletal muscle fibres covered by mucosa on both surfaces. The

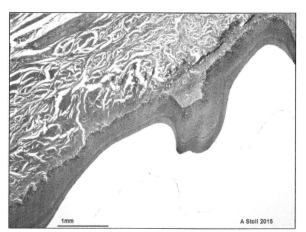

Fig. 13.6 Haeamatoxylin and eosin (HE) histopathological section of the feline hard palate showing the epithelial lining including one ridge. (Courtesy of Alexander Lewis Stoll, Royal Veterinary College, London University, UK.)

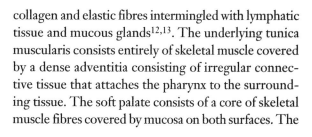

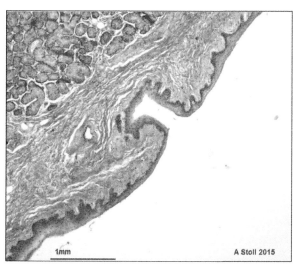

Fig. 13.7 HE histopathological section of the feline soft palate demonstrating its mucosal lining and multiple submucosal glands. (Courtesy of Alexander Lewis Stoll, Royal Veterinary College, London University, UK.)

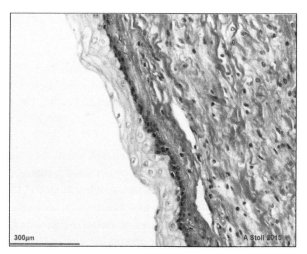

Fig. 13.8 HE histopathological section of the feline pharyngeal wall showing the stratified squamous epithelial mucosal lining. (Courtesy of Alexander Lewis Stoll, Royal Veterinary College University of London, UK.)

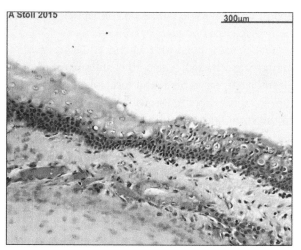

Fig. 13.9 HE histopathological section of the transitional zone in a cat between pharynx and larynx with a change in mucosal architecture. (Courtesy of Alexander Lewis Stoll, Royal Veterinary College University of London, UK.)

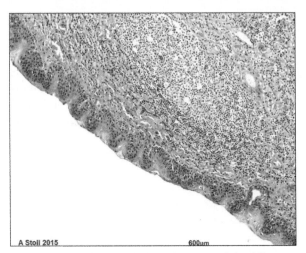

Fig. 13.10 HE histopathological section of the feline tonsil. (Courtesy of Alexander Lewis Stoll, Royal Veterinary College University of London, UK.)

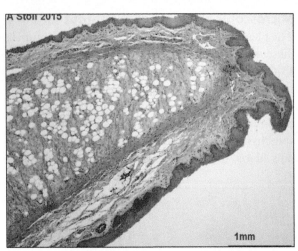

Fig. 13.11 HE histopathological section of the canine epiglottis. (Courtesy of Alexander Lewis Stoll, Royal Veterinary College University of London, UK.)

oropharyngeal side is covered with stratified squamous epithelium, whereas the nasopharyngeal surface is covered caudally by stratified squamous epithelium as well, but ciliated, pseudostratified columnar epithelium rostrally[12,13]. A narrow transition zone consisting of transitional epithelium connects both regions. The lamina propria–submucosa of the soft palate contains branched tubuloacinar mucous and seromucous palatine glands. Lymphatic tissue is present in the mucosa on both surfaces. The longitudinally oriented skeletal palatinus muscle fibres and connective tissue are located between both mucosal surfaces.

The larynx is lined with mucosa and supported by underlying cartilage. The epithelial lining of the epiglottis, laryngeal vestibule and vocal folds is made up of non-keratinised stratified squamous epithelium; caudal to the vocal folds it gradually changes to respiratory epithelium (Figs 13.11, 13.12)[14,15]. The epithelium of the laryngeal surfaces of the epiglottis, aryepiglottic folds and arytenoid cartilages may contain taste buds. Sensory receptors of the cranial laryngeal nerve in the epithelium respond to the presence of fluids: stimulation of these receptors will result in reflex apnoea. Irregular dense connective tissue lies underneath the stratified epithelium, whereas underneath the respiratory epithelium it is loose and rich in elastic fibres, leucocytes, plasma cells, mast cells and diffuse lymphatic tissue or solitary lymphatic nodules[14,15]. Mixed glands

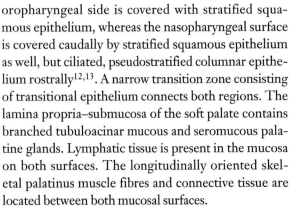

Fig. 13.12 HE histopathological section of the feline laryngeal mucosal lining. (Courtesy of Alexander Lewis Stoll, Royal Veterinary College University of London, UK.)

are found in the submucosa as well, but not in the vestibular and vocal folds. Numerous elastic fibres are present in the vocal ligament and, to a lesser, extent, in the vestibular ligament. Most of the laryngeal cartilages are of the hyaline type and are connected to each other, the trachea and the hyoid apparatus by ligaments. The epiglottis and arytenoid cartilages also contain elastic cartilage[14,15]. The peripheral cartilaginous wall of the epiglottis (Fig. 13.11) encloses white adipose tissue, strands of elastic fibres and small areas of elastic cartilage. A loose connective tissue forms the tunica adventitia and surrounds the laryngeal cartilages and muscles.

13.3 SWALLOWING

Swallowing, or deglutition, is a complex physiological phenomenon allowing transport of material from the mouth to the stomach, whether it is food or saliva and debris from the pharynx. It may be initiated consciously as a voluntary act during eating, but most swallows occur subconsciously between meals, without apparent cerebral participation[1,16]. Swallowing, irrespective of eating, occurs about once a minute in awake individuals and is driven by salivation, which stimulates the sensory receptors in the mouth and pharynx[17]. Salivation and swallowing virtually cease during sleep in humans[18]. Deglutition begins as a voluntary act but during its execution becomes a reflex. It consists of four phases.

- The first two phases of swallowing, the oral preparatory and the oral phase are voluntary
- The last two phases, the pharyngeal and oesophageal phase, are involuntary[1].

Oral preparatory and oral phase

The oral preparatory phase consists of:
- Apprehension of food.
- Chewing it.
- Preparing it for the oral phase.

In this phase, which begins when mastication is complete, the food bolus is moved from the front of the oral cavity to the oropharynx, where swallowing is initiated[1]. The tongue plays a crucial role in this phase as it moves the food bolus caudally and organises it at the base of the tongue in a position that is on midline between the tongue and hard palate. When the food bolus is pushed towards the pharynx, sensory receptors in the oropharynx, dorsal surface of the tongue and laryngopharynx are stimulated that initiate the reflex component of swallowing[17,19].

Stimulation of the receptors in the oropharynx are conveyed via the glossopharyngeal nerve, of those in the laryngopharynx via the cranial laryngeal nerve and of those on the dorsal surface of the tongue via the lingual nerve, a branch of the trigeminal nerve[1]. Sensory feedback also assists tongue movement in the oral and pharyngeal phases in the formation of the food bolus. Motor fibres to the tongue are under control of the hypoglossal nerve[4]. These sensory nerve fibres terminate in the solitary tract and nucleus[20]. Stimulation of the most caudal part of the solitary system, which receives the afferents of the cranial laryngeal nerve, appears to be the most potent in triggering the swallowing reflex[1].

Pharyngeal phase

During the pharyngeal phase the goal is:
- To pass food from the oropharynx into the oesophagus.
- To prevent food from being aspirated into the trachea or moved into the nasopharynx.

This requires well-regulated and coordinated activity of several stages:
- Palatopharyngeal closure.
- Peristalsis of the pharyngeal constrictor muscles.
- Airway protection by elevating and closing the larynx.
- Cricopharyngeal relaxation[1].

These reflex movements occur in sequence and overlap one another. The efferent arm of the reflex involves the motor nuclei of cranial nerves V, VII, IX, X and XII[17,19,21]. These nerves supply the muscles of mastication, tongue, palate, pharynx, larynx and oesophagus. However, the fibres running within the solitary tract are not connected directly to the cranial motor nuclei involved in swallowing, but reflexes are mediated first by the reticular formation in the brain stem[1]. They follow a fixed pattern within an interneuronal network known as the central pattern generator (CPG) for swallowing, which also interacts with the respiratory centre.

There are two CPGs, one in each side of the brainstem, for bilateral organisation of motor activity in the bilateral swallowing muscles[21]. The interneurons are located in two distinct regions of the medulla: a dorsal region that includes the solitary nucleus tract and the adjacent reticular formation, and a ventral region corresponding to the lateral reticular formation around the nucleus ambiguus[1]. Motor neurons of the trigeminal nerve innervate the mylohyoid, rostral digastric and pterygoid muscles and the tensor muscle of the soft palate; motor neurons of the facial nerves innervate the caudal digastric and stylohyoid muscles; and motor neurons of the nucleus ambiguus innervate the muscles of the pharynx and larynx and the striated muscles of the oesophagus[1,19].

The nucleus ambiguus may be considered to be the main motor nucleus for swallowing, together with the hypoglossal nucleus for the tongue[19,21]. The motor supply to the pharyngeal muscles is distributed by the pharyngeal plexus, which receives contributions from the glossopharyngeal nerve and the pharyngeal branch of the vagus nerve[4]. Branches of the glossopharyngeal nerve (IX) were found to join the branches of the pharyngeal branch of the vagus nerve (Xph) in various ways in Beagle dogs, resulting in the combined innervation of the pharyngeal muscles by both nerves[16]. Of the two motor nerves supplying the innervation via this plexus, the pharyngeal branch of the vagus nerve is functionally more important than the glossopharyngeal nerve[16].

The final act during the pharyngeal stage of swallowing occurs when the cricopharyngeal muscle relaxes, the upper oesophageal sphincter opens, the bolus moves into the oesophagus, the sphincter closes and the pharyngeal muscles relax. The cricopharyngeal muscle is innervated by the pharyngoesophageal nerve, which is formed by the glossopharyngeal and vagus nerves[4].

The amplitude and duration of electromyographic (EMG) activity in the mylohyoid, geniohyoid and genioglossus muscles in humans have been shown to depend in part on the consistency of the bolus[21,22]. This strongly suggests that peripheral feedback can control the swallowing motor sequence[1]. In dogs the regulation of swallowing was shown to be markedly disturbed by transection of peripheral components of the pharyngeal plexus[23], while stimulation of the peripheral pathways influenced the contraction timing during swallowing[24].

Oesophageal phase

In the oesophageal stage, the bolus is moved through the oesophagus by the peristaltic contractions of the oesophageal musculature under control of the vagus nerve. In this final stage of deglutition, the bolus is transported from the oesophagus, through the gastroesophageal sphincter and into the stomach[1,19,21].

13.4 COUGHING AND PROTECTION OF LOWER AIRWAYS

The three functions of the larynx, in order of priority: are protection of the lower airways via a glottic closure reflex and coughing, regulation of respiratory airflow and vocalisation[2,7]. Flawless performance of these functions requires an intact neuromuscular system (see Section 13.2) to respond to both volitional and reflex signals presented to the larynx.

Laryngeal reflexes

The glottic closure reflex is a polysynaptic reflex that allows the larynx to protect the lower airway from penetration and aspiration[2,25]. When exaggerated, however, this reflex accounts for the production of laryngospasm. This response is typically seen in cats during endotracheal intubation or after manipulation of the airways. Laryngospasm occurs when stimulation of the cranial laryngeal nerve leads to a prolonged adductor response that is maintained well after the initiating stimulus is removed[25–27].

Initiation of the reflex is triggered by:
- Activation of mechanoreceptors in the superficial mucosa of the glottis and in the muscles and laryngeal joints.
- Chemical and thermal receptors in the supraglottic laryngeal mucosa.
- Many taste buds in the mucosa of the epiglottis and the aryepiglottic folds.
- Touch receptors in the vocal folds[2,25,26,28].

The generated impulses are carried by the sensory afferent internal branch of the ipsilateral cranial laryngeal nerve, through the distal vagal ganglion to the nucleus solitarius in the brainstem[2,29]. The nucleus ambiguus contains the motor neurons of the efferent innervation to the ipsilateral laryngeal intrinsic muscles, the axons of which form the ipsilateral recurrent laryngeal nerve. Even unilateral stimulation of the cranial laryngeal nerve was shown to cause complete closure of the glottis together with evoked EMG responses in both thyroarytenoid muscles[30,31]. Water in the pharynx has been shown to activate the glottis closure reflex[32,33]. Progressively deeper loss of consciousness under anaesthesia abolishes this lower brainstem-mediated reflex by altering central facilitation of the reflex, predisposing the subject to a weakened glottic closure response[34].

Although not classically considered part of the protective reflex, reflex swallowing from stimulation of the cranial laryngeal nerve may protect the airways as well[29,35].

Coughing

Coughing is a reflex triggered by stimulation of cough receptors located in the mucosa of the larynx, trachea and bronchi. It is an important neuronal reflex, which serves to protect the airways from inhaled exogenous micro-organisms and thermal and chemical irritants. Moreover, it prevents the airways from mucus retention.

In guinea pigs, the species most commonly used in cough research, airway sensory nerves can be broadly functionally classified as either:

- Primarily mechanically sensitive (low threshold mechanosensors).
- Primarily chemically sensitive (chemosensors or alternatively, nociceptors)[36–38].

Low threshold mechanoreceptors are readily activated by one or more mechanical stimuli, including lung inflation, bronchospasm or light touch, but generally do not respond directly to chemical stimuli. Conversely, chemosensors are typically activated directly or sensitised by a wide range of chemicals, but are relatively insensible to touch[36–39].

Two classic types of low threshold mechanosensors have been described in the intrapulmonary airways of a number of mammalian species:

- Rapidly adapting receptors (RARs).
- Slowly adapting receptors (SARs)[37].

Chemically-sensitive airway afferent fibres are found throughout the airways and lungs and are generally quiescent in the normal airways, becoming recruited during airway inflammation or irritation. Most intrapulmonary airway chemosensors are C-fibre-type nociceptors. There are few Aδ-fibre nociceptors in the airways, but approximately half of the receptors outside the airways are of this type[36,37].

After stimulation of a cough receptor, the signal is conducted to the cerebral cough centre via vagal sensory neurons[37–39]. The cough itself is mediated by efferent motor neurons. Hence the cough reflex consists of five functionally sequential parts:

- The cough receptors.
- The primary afferent fibres of the nervus vagus (N. vagus), N. trigeminus and N. glossopharyngeus.
- The cough centre in the medulla oblongata (N. tractus solitarius).
- The afferent fibres of the N. phrenicus, spinal nerve and N. laryngeus recurrens.
- The diaphragm and the abdominal, intercostal and laryngeal muscles[37].

Activation of the cough reflex pathway triggers a fixed-action motor pattern[2]:

- First a rapid deep inspiration takes place.
- After which the glottis closes.
- Followed by expiratory muscle activity that raises the pressure against the closed glottis.
- And finally an abrupt opening of the glottis to forcefully expel the air and whatever substance stimulated the cough[2].

As for the glottic closure reflex, coughing is suppressed under a deep level of anaesthesia, but aging and all types of medications, including atropine, can also decrease the cough reflex[37,40].

13.5 REGULATION OF RESPIRATORY AIRFLOW AND VOCALISATION

As stated before, one of the roles of the larynx is regulation of respiratory airflow. The larynx is responsible for controlling airway resistance, primarily by decreasing airway resistance via abduction of the arytenoid cartilages during inspiration[2,7]. In addition, by changing tension on the vocal cords, the larynx is responsible for voice production as well.

Regulation of respiratory airflow

The respiratory centre in the brainstem, with the help of higher central nervous systems and peripheral input, maintains eupnoeic respiration[2]. It drives the synchronous opening of the glottis and relaxation of the diaphragm during inspiration and the closure of the glottis during expiration[25,29,34]. Peripheral input is derived from chemoreceptors for blood O_2 and CO_2 that are located in the glomus caroticum and the glomus aorticum[41–43].

Hypercapnia generally leads to a reduction in airway resistance, but many drugs can affect susceptibility to changes in O_2 and CO_2[41–43]. Afferent fibres of the glossopharyngeal and vagal nerves carry impulses to the respiratory centres in the brainstem from which motor neurons in the left and right nucleus ambiguus are activated, as has been discussed in Section 13.2.

Vocalisation

In humans phonation is of extreme importance and phonatory control again involves a complex mechanism and requires intricate coordination of central and peripheral components. The pathways are well described for humans and are similar in dogs and cats. The dog's and cat's larynx have been used as animal models for more than 2 centuries of phonatory research[44,45], but speech is much more complex in humans than vocalisation in dogs and cats[2].

In dog communication, vocalisation is not very important, except for the advertisement of body size and strength to other dogs[46]. In dominant dogs, large size is exaggerated by the stiff upright threat posture, while the crouched posture of submission decreases apparent size. Low-frequency, broad-band barking or growling often accompanies threats[46]. In one study, significant correlations were found between vocal tract length, body mass and spectral peaks in the vocal signal between different sizes of dog breeds, and it was suggested that the dog's voice can deliver information about the body size of the vocaliser[46].

Cats can express friendly, fearful and aggressive emotions using vocalisation and body posture. Cats' vocalisations largely occur in the context of four types of interactions, specifically, aggressive, sexual, mother–kitten behaviour and human–cat interactions[47]. In comparison to most dogs, domestic cats are unusually vocal.

At least three factors have been proposed to explain the evolution of rich, complex vocal repertoires in domestic cats, which include their social structure, nocturnal activity and the long period of association between mothers and offspring[47]. The vocal repertoire of a cat is considered to be quite large and spectrogram analysis has revealed 23 patterns that could be divided into two major types of feline vocalisations, specifically one-type calls like growls and hisses, and mixed calls like meowing[45]. The vocalisation that is the most common in cat–human interactions is the meow, which is rarely heard during cat–cat interaction[48]. There are considerable variations in frequency, duration and forms of the meow[48,49].

Purring

Purring is commonplace feline behaviour, easily audible and palpable in most domestic cats[50]. Purring has been described in other species of cats and some civets and genets[51,52].

Purring results from the intermittent activation of intrinsic laryngeal muscles as manifested by a very regular, stereotyped pattern of EMG bursts occurring 20–30 times (20–30 Hz) per second[49,50]. Each burst of muscle discharge causes glottal closure and the development of a transglottal pressure, which, when dissipated by glottal opening, generates sound. During inspiration, the diaphragmatic discharge is also chopped, and the diaphragmatic and laryngeal bursts occur asynchronously[7,11].

In non-purring cats, no action potentials were recorded from the EMG electrodes in the cricothyroid muscles[50]. The onset of purring was signalled by the abrupt activation of this recording. This event was usually provoked by some human action; petting was always sufficient, and often a verbal stimulus or simply the approach of a human adequate to provoke purring[50]. Once initiated, the laryngeal EMG usually remained active, displaying a characteristic pattern for prolonged periods without further provocation.

13.6 REFERENCES

1 Venker-van Haagen AJ. The pharynx. In: *Ear, Nose, Throat, and Tracheobronchial Diseases in Dogs and Cats*. Venker-van Haagen AJ (ed). Schlütersche Verlagsgesellschaft, Hannover 2005, pp. 83–120.
2 Venker-van Haagen AJ. The larynx. In: *Ear, Nose, Throat, and Tracheobronchial Diseases in Dogs and Cats*. Venker-van Haagen AJ (ed). Schlütersche Verlagsgesellschaft, Hannover 2005, pp. 121–65.
3 Anderson GM. Soft tissue of the oral cavity. In: *Veterinary Surgery Small Animal*. Tobias KM, Johnston SA (eds). Elsevier Saunders, St. Louis 2012, pp. 1425–38.

4 Evans HE, De Lahunta A. The digestive system. In: *Miller's Anatomy of the Dog*. Evans HE, De Lahunta (eds). Elsevier, St. Louis 2013, pp. 281–337.

5 Dyce KM. The muscles of the pharynx and palate of the dog. *Anatomical Record* 1957;**127**:497–508.

6 Evans HE, De Lahunta A. The respiratory system. In: *Miller's Anatomy of the Dog*. Evans HE, De Lahunta (eds). Elsevier, St. Louis 2013, pp. 338–60.

7 Monnet E, Tobias KM. Larynx. In: *Veterinary Surgery Small Animal*. Tobias KM, Johnston SA (eds). Elsevier Saunders, St. Louis 2012, pp. 1718–33.

8 Amis TC, O'Neill N, Brancatisano A. Influence of hyoepiglotticus muscle contraction on canine upper airway geometry. *Respiratory Physiology* 1996;**104**:179–185.

9 Amis TC, O'Neill N, VanderTouw T, Brancatisano A. Electromyographic activity of the hyoepiglotticus muscle in dogs. *Respiratory Physiology* 1996;**104**:159–67.

10 Burbidge HM. A review of laryngeal paralysis in dogs. *British Veterinary Journal* 1995;**151**:71–82.

11 Dyce KM, Sack WO, Wensing CJG. *Textbook of Veterinary Anatomy*. WB Saunders, Philadelphia 1996.

12 Bacha WJ, Bacha LM. Digestive system. In: *Color Atlas of Veterinary Histology*. Bacha WJ, Bacha LM (eds). Wiley-Blackwell, Ames 2012, pp. 139–82.

13 Frappier BL. Digestive system. In: *Dellmann's Textbook of Veterinary Histology*. Eurell JA, Frappier BL (eds). Blackwell, Ames 2006, pp. 170–211.

14 Bacha WJ, Bacha LM. Respiratory system. In: *Color Atlas of Veterinary Histology*. Bacha WJ, Bacha LM (eds). Wiley-Blackwell, Ames 2012, pp. 195–210.

15 Plopper CG, Adams DR. Respiratory system. In: *Dellmann's Textbook of Veterinary Histology*. Eurell JA, Frappier BL (eds). Blackwell, Ames 2006, pp. 153–69.

16 Venker-van Haagen AJ, Hartman W, Wolvekamp WT. Contributions of the glossopharyngeal nerve and the pharyngeal branch of the vagus nerve to the swallowing process in dogs. *American Journal of Veterinary Research* 1986;**47**:1300–7.

17 Dodds WJ. The physiology of swallowing. *Dysphagia* 1989;**3**:171–8.

18 Lichter I, Muir RC. The pattern of swallowing during sleep. *Electroencephalography and Clinical Neurophysiology* 1975;**38**:427–32.

19 Miller AJ. Neurophysiologic basis of swallowing. *Dysphagia* 1986;**8**:185.

20 Torvik A. Afferent connections to the sensory trigeminal nuclei, the nucleus of the solitary tract and adjacent structures. An experimental study in the rat. *Journal of Comparative Neurology* 1956;**106**:51–141.

21 Jean A. Brain stem control of swallowing: neuronal network and cellular mechanisms. *Physiological Reviews* 2001;**81**:929–69.

22 Hrycyshyn AW, Basmajian JV. Electromyography of the oral stage of swallowing in man. *American Journal of Anatomy* 1972;**133**:333–40.

23 Venker-van Haagen AJ, Hartman W, van den Brom WE, Wolvekamp WT. Continuous electromyographic recordings of pharyngeal muscle activity in normal and previously denervated muscles in dogs. *American Journal of Veterinary Research* 1989;**50**:1725–8.

24 Venker-van Haagen AJ, van den Brom WE, Hellebrekers LJ. Effect of stimulating peripheral and central neural pathways on pharyngeal muscle contraction timing during swallowing in dogs. *Brain Research Bulletin* 1998;**45**:131–6.

25 Sasaki CT, Suzuki M. Laryngeal reflexes in cat, dog, and man. *Archives of Otolaryngology Head and Neck Surgery* 1976;**102**:400–2.

26 Suzuki M, Sasaki CT. Effect of various sensory stimuli on reflex laryngeal adduction. *Annals of Otology Rhinology and Laryngology* 1977;**86**:30–6.

27 Ikari T, Sasaki CT. Glottic closure reflex: control mechanisms. *Annals of Otology Rhinology and Laryngology* 1980;**89**:220–4.

28 Suzuki M, Sasaki CT. Initiation of reflex glottic closure. *Annals of Otology Rhinology and Laryngology* 1976;**85**:382–6.

29 Wadie M, Adam SI, Sasaki CT. Development, anatomy, and physiology of the larynx. Principles of deglutition. In: *Principles of Deglutition: a Multidisciplinary Text for Swallowing and its Disorders*. Shaker R (ed). Springer New York 2012, pp. 175–97.

30 Sasaki CT, Jassin B, Kim Y-H, Hundal J, Rosenblatt W, Ross DA. Central facilitation of the glottic closure reflex in humans. *Annals of Otology Rhinology and Laryngology* 2003;**112**:293–7.

31 Kim YH, Sasaki CT. Glottic closing force in an anesthetized, awake pig model: biomechanical effects on the laryngeal closure reflex resulting from altered central facilitation. *Acta Oto-Laryngologica* 2001;**121**:310–14.

32 Shaker R, Hogan WJ. Reflex-mediated enhancement of airway protective mechanisms. *American Journal of Medicine* 2000;**108**:8S–14S.

33 Shaker R, Medda BK, Ren J, Jaradeh S, Xie P, Lang IM. Pharyngoglottal closure reflex: identification and characterization in a feline model. *American Journal of Physiology – Gastrointestinal and Liver Physiology* 1998;**275**:G521–G525.

34 Sasaki CT, Yu Z, Xu J, Hundal J, Rosenblatt W. Effects of altered consciousness on the protective glottic closure reflex. *Annals of Otology Rhinology and Laryngology* 2006;**115**:759–63.

35 Venker-van Haagen AJ, Barbas-Henry HA, van den Brom WE. CMAPs in pharyngeal and hyoid muscles evoked by nucleus solitarius stimulation in dogs. *Brain Research Bulletin* 1995;**37**:555–9.

36 Canning BJ, Mori N, Mazzone SB. Vagal afferent nerves regulating the cough reflex. *Respiratory Physiology & Neurobiology* 2006;**152**:223–42.

37 Mazzone SB. An overview of the sensory receptors regulating cough. *Cough* 2005;**1**:E1–E9.

38 Widdicombe JG. Neurophysiology of the cough reflex. *European Respiratory Journal* 1995;**8**:1193–202.

39 Kondo T, Hayama N. Cough reflex is additively potentiated by inputs from the laryngeal and tracheobronchial [corrected] receptors and enhanced by stimulation of the central respiratory neurons. *Journal of Physiological Sciences* 2009;**59**:347–53.

40 Tsubouchi T, Tsujimoto S, Sugimoto S, Katsura Y, Mino T, Seki T. Swallowing disorder and inhibition of cough reflex induced by atropine sulfate in conscious dogs. *Journal of Pharmaceutical Sciences* 2008;**106**:452–9.

41 Bisgard GE, Mitchell RA, Herbert DA. Effects of dopamine, norepinephrine and 5-hydroxytryptamine on the carotid body of the dog. *Respiratory Physiology* 1979;**37**:61–80.

42 Bowes G, Townsend ER, Kozar LF, Bromley SM, Phillipson EA. Effect of carotid body denervation on arousal response to hypoxia in sleeping dogs. *Journal of Applied Physiology* 1981;**51**:40–5.

43 Sylvester JT, Scharf SM, Gilbert RD, Fitzgerald RS, Traystman RJ. Hypoxic and CO hypoxia in dogs: hemodynamics, carotid reflexes, and catecholamines. *American Journal of Physiology – Heart and Circulatory Physiology* 1979;**236**:H22–H28.

44 Paniello RC, Dahm JD. Long-term model of induced canine phonation. *Otolaryngology and Head and Neck Surgery* 1998;**118**:512–22.

45 Farley G, Barlow S, Netsell R, Chmelka J. Vocalizations in the cat: behavioral methodology and spectrographic analysis. *Experimental Brain Research* 1992;**89**: 333–40.

46 Riede T, Fitch T. Vocal tract length and acoustics of vocalization in the domestic dog (*Canis familiaris*). *Journal of Experimental Biology* 1999;**202**:2859–67.

47 Yeon SC, Kim YK, Park SJ, Lee SS, Lee SY. Differences between vocalization evoked by social stimuli in feral cats and house cats. *Behavioural Processes* 2011;**87**:183–9.

48 Brown KA, Buchwald JS, Johnson JR, Mikolich DJ. Vocalization in the cat and kitten. *Developmental Psychobiology* 1978;**11**:559–70.

49 Moelk M. Vocalizing in the house-cat; a phonetic and functional study. *American Journal of Psychology* 1944;**57**:184–205.

50 Remmers JE, Gautier H. Neural and mechanical mechanisms of feline purring. *Respiration Physiology* 1972;**16**:351–61.

51 Peters G. Purring and similar vocalizations in mammals. *Mammal Review* 2002;**32**:245–71.

52 Sissom DEF, Rice DA, Peters G. How cats purr. *Journal of Zoology* 1991;**223**:67–78.

14.1 INTRODUCTION

The throat is a conduit for both the respiratory and the digestive tract and comprises the pharynx and larynx. The history usually reveals specific problems caused either by the dysfunction of the airway through the oropharynx or nasopharynx (dyspnoea with or without stridor) or by difficulty in swallowing (dysphagia).

Dysphagia is often accompanied by salivation and/or regurgitation of food or water and is usually caused by pharyngeal or oesophageal abnormalities. Gagging by definition is a reflexive contraction of the constrictor muscles of the pharynx resulting from stimulation of the pharyngeal mucosa, and is commonly associated with other clinical signs, such as coughing, but can be the only presenting complaint. Gagging may signify a problem involving the nasal passage, pharynx, respiratory tract and upper gastrointestinal tract.

Dogs in dyspnoea with stridor per definition have obstructive upper airway disease, whereas dogs in dyspnoea without stridor typically have lower airway disease. Snoring is typically associated with (naso)pharyngeal disease, whereas a laryngeal stridor (g-sound or sawing sound) is associated with laryngeal disease. Coughing and dyspnoea with stridor are usually the most prominent signs of laryngeal disease, but in some cases dysphonia (abnormal voice) is mentioned by the owner.

Clinical approach to the diagnosis of diseases of the throat

Physical examination findings are usually unremarkable except for the possible audible stridor and in some cases masses can be palpated compressing the throat. Direct inspection of the pharynx and larynx with a laryngoscope is the most important diagnostic procedure for diagnosing disorders of the pharynx and larynx. Throat inspection (pharyngolaryngoscopy) will

be discussed in the next section (14.2). However, not just structural abnormalities cause symptoms, functional abnormalities of the pharyngolaryngeal area can also lead to abnormalities. Therefore, direct pharyngolaryngoscopy is only part of the work-up required to diagnose a disorder of the pharynx or larynx.

Standard radiographic examination is useful, for example for the recognition of obstructing structures in the pharynx and larynx, for locating radiopaque foreign bodies and for inspection of fractures of the hyoid bones or laryngeal cartilages. However, it provides no information about, for instance, pharyngitis, laryngitis or laryngeal paralysis. Ultrasonography can be performed without anaesthesia and is increasingly used for examination of laryngeal masses and guidance of fine-needle aspiration biopsy (FNAB) and for evaluation of laryngeal function. Indications and technique for radiographic and ultrasonographic evaluation of the throat are reviewed in Section 14.3.

Contrast videofluorography is indispensable for diagnosing abnormalities of swallowing (dysphagia) in dogs. Recordings are made while the animal eats food and/or drinks fluid mixed with barium. Electromyography (EMG) of the pharyngeal and laryngeal muscles is useful in dogs with signs of dysphagia, dysphonia or dyspnoea with stridor to evaluate muscular dystrophy and neuromuscular disease. These techniques will be discussed in Sections 14.4 and 14.6, respectively.

Advanced diagnostic imaging in the form of computed tomography (CT) and magnetic resonance imaging (MRI) are increasingly used to aid in localising and determining the extent of neoplasms in the (naso)pharyngeal and laryngeal area, but also to evaluate all abnormalities associated with brachycephalic obstructive airway syndrome. These techniques, reviewed in Section 14.5, are indispensable in pre-operative planning, especially for oncological pharyngolaryngeal surgery.

14.2 THROAT INSPECTION (PHARYNGOLARYNGOSCOPY)

Gross inspection of the oropharynx, laryngopharynx and larynx is relatively straightforward and is generally performed by direct visualisation in anaesthetised animals[1–4]. By contrast, investigation of the nasopharynx, and more specifically its cranial portion, is more difficult and requires the use of a flexible endoscope for posterior (or retrograde) nasopharyngoscopy, which has been discussed in Chapter 9 Nose Diagnostic Procedures, Section 9.6.

For inspection of the throat, the patient is placed in ventral recumbency with its head supported by an assistant standing at its side. The assistant opens the animal's mouth and extends its neck, using one hand to raise the upper jaw and the other to depress the lower jaw (Figs 14.1, 14.2)[2], preferably with the help of mouth gags wrapped around the jaws behind the respective canine teeth[3]. Laryngeal function should be assessed under a light plane of anaesthesia so as to ensure the pharyngeal and laryngeal reflexes (swallowing and cough) are intact. Once function has been assessed, further assessment can take place under a deeper level of anaesthesia. It is recommended that, in dogs but certainly in cats, the laryngeal mucosa is anaesthetised locally with lidocaine spray[5].

Instruments needed for pharyngoscopy and laryngoscopy include:

- A laryngoscope fitted with a large blade suitable for the size of the animal that allows endoscopy of the proximal oesophagus (Fig. 14.1).
- A second large blade to aid in lifting the soft palate and allow visualisation of the laryngopharynx.
- A long tissue forceps.
- A Senn retractor with blunt prongs or spayhook for retraction of the soft palate and caudal nasopharyngoscopy.
- Appropriate sizes of endotracheal tubes.

Pharyngoscopy

The laryngoscope is introduced over the tongue and the entire oral cavity, oropharynx, palatine tonsils and tonsillar crypts, soft palate, palatopharyngeal arches and base of the tongue are inspected[2,3]. The tonsils should be inspected immediately as prolonged anaesthesia can lead to protrusion of the tonsils, falsely suggesting enlargement. The tongue should be manipulated from left to right in the oral cavity to allow complete visualisation of the palatoglossal arches and fauces. Particularly in brachycephalic animals, attention should be focused on not just the length of the soft palate, but also

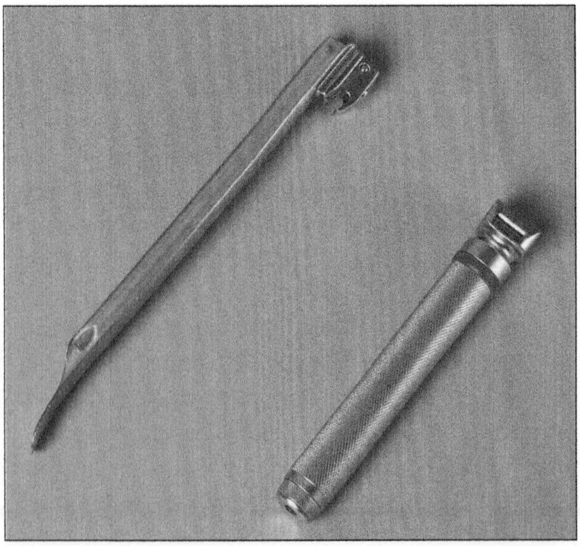

Fig. 14.1 A straight laryngoscope with properly sized blades that allow visualisation of all structures should be used for laryngoscopy in small animals.

Fig. 14.2 With the dog in sternal recumbency, the mouth of the animal should be held open widely, for instance by an assistant, in such a way that the laryngoscope can be introduced without causing harm to the teeth and to allow for full pharyngolaryngoscopy.

on the other concurrent pharyngeal abnormalities such as tonsillar hypertrophy and protrusion, thickening of the base of the tongue, dorsoventral flattening of the pharynx, thickness of the soft palate and presence and degree of diffuse pharyngeal mucosal swelling.

The laryngopharynx can be further evaluated by pushing the soft palate dorsally towards the dorsal nasopharyngeal wall with an additional large laryngoscope blade. After this, the laryngoscope itself can be advanced into the proximal oesophagus to allow inspection of this area. For complete oesophagoscopy rigid or flexible endoscopes are required[1,3].

Finally, the caudal nasopharynx can be evaluated by retracting the soft palate with a spayhook or blunt Senn retractor. Especially in cats, the index finger can be used to feel the ventral surface of the soft palate and evaluate for a nasopharyngeal mass by pushing the soft palate dorsally. If a mass is suspected, blind fine needle aspirates through the soft palate may prove to be diagnostic. However, it is preferable that larger biopsies of abnormalities are taken. For this, the animal should always be intubated and the cuff inflated, to prevent leakage of blood into the trachea and bronchi[2].

Laryngoscopy

As mentioned before, if laryngeal function is to be investigated the anaesthesia should be superficial, for if it is too deep the activity of the intrinsic laryngeal muscles, abduction and adduction, is absent[6]. Either propofol or a short-acting intravenous barbiturate, such as thiopental, is ideal for the dog[3,5]. In cats, ketamine and diazepam in combination is an excellent choice for throat inspection[3].

For diagnostic laryngoscopy, the laryngoscope is introduced over the tongue and advanced to the level of the epiglottis. The ventral side of the epiglottis is inspected first, after which the laryngoscope is allowed to carefully rest on the tip of the epiglottis for inspection of the entire glottis. The size of the laryngeal opening, the rima glottides, is the first concern. In severely dyspnoeic patients with laryngeal dysfunction or obstruction, immediate endotracheal intubation is required. If intubation is not necessary, the glottis can be inspected more carefully.

Under a light plane of anaesthesia, abduction of the laryngeal cartilages upon inspiration should be visible, pulling the vocal folds laterally and opening the glottis. If laryngeal motion seems abnormal, doxapram

hydrochloride (Dopram-V) 0.5–1.0 mg/kg IV can be administered[5]. Within 30 seconds of administration, this medication increases the rate and depth of respiration, and the increased ventilatory effort persists for a few minutes. If laryngeal function remains abnormal after doxapram administration, the endoscopist can reasonably conclude that laryngeal function is abnormal[5]. Laryngeal motion related to vocalisation (e.g. whining) or swallowing should not be confused with laryngeal abduction. If there is still doubt about normal or abnormal function of the larynx, an EMG examination should be performed (see Section 14.6).

After function has been assessed, the level of anaesthesia can be deepened to allow a thorough assessment and subsequent tracheobronchoscopy, if indicated. As described for the pharynx, the overall development of the larynx needs to be assessed first as laryngeal hypoplasia (underdevelopment of the laryngeal cartilages, small and short cartilages that appear less rigid) is commonly seen in brachycephalic animals. A hypoplastic larynx is more likely to develop secondary collapse (see Chapter 11 Diseases of the Nasal Cavity and Sinuses, Section 11.3).

After this the aryepiglottic folds and then the individual parts of the arytenoid cartilages are inspected. From dorsal to ventral, the corniculate processes, the cuneiform processes of the arytenoids and the paired vocal folds are inspected for position, overall morphology, mucosal swelling and presence of focal abnormalities such as foreign bodies, polyps, granulomatous lesions and neoplastic processes. The colour and swelling of the laryngeal mucosa overlying the cartilaginous structures are evaluated, as well as the presence or absence of eversion of the saccules, after which the laryngoscope can gently be advanced through the glottis. The cavity of the larynx caudal to the glottis, can also be evaluated using flexible or rigid endoscopes[3,5].

Though the magnification inherent to the use of these scopes allows more detailed assessment of the laryngeal structures, a properly performed laryngoscopy with the laryngoscope will not miss essential abnormalities. Especially in cats the laryngeal mucosa is prone to develop oedema and hence prior to aspiration biopsies (FNAB or histological biopsy) endotracheal intubation and the intravenous administration of a glucocorticoid should be considered[6]. Laryngeal inspection during the recovery period will reveal when the endotracheal tube can be removed safely[6].

14.3 RADIOGRAPHIC AND ULTRASONOGRAPHIC EVALUATION OF THE THROAT

Cross-sectional imaging modalities such as CT, MRI and diagnostic endoscopy are superior to radiography for the assessment of diseases involving the pharynx and larynx. The limitations of radiography for imaging of the throat are that, while it is sensitive for detecting some diseases it is insensitive in detecting others, and the extent of changes may be underestimated[7,8]. Because the throat is air-filled, ultrasound is of limited use for the evaluation of mural and intraluminal abnormalities, but it has been used for evaluation of laryngeal paralysis and pharyngolaryngeal masses[8,9].

Radiographic examination of the throat

Standard radiographic examination of the pharynx and larynx consists of a lateral and, though of limited value with the exception of demonstration of foreign bodies, a dorsoventral, or ventrodorsal, view (Figs 14.3A, B). For the lateral view, the mouth should be partially opened to reduce superimposition of dental structures and to better assess the air-filled oropharynx[7]. Any degree of obliquity results in underestimation of luminal diameter and overestimation of surrounding soft tissue structures, such as soft palate and laryngeal cartilages, and can even mimic a mass[8]. High-contrast images (relatively low kilovoltage peak and high milliampere second) and measured according to the thickness of the ventral cervical soft tissues are recommended[8]. Lesions of the oropharynx, nasopharynx, larynx, hyoid apparatus and retropharyngeal soft tissues can be best appreciated on a perfect lateral view with these settings.

Radiographic assessment of the throat is focused on recognising soft tissue masses, free gas in fascial planes as a result of penetrating foreign bodies, destruction or fractures of the hyoid bones and displacement of the hyoid or larynx, indicating mass lesions[7]. The soft palate can be seen to divide the oropharynx from the nasopharynx, whereas the palatopharyngeal arches can usually be seen as a soft tissue line extending caudodorsally from the soft palate, dividing the oropharynx and laryngopharynx[10]. The epiglottis is often surrounded by gas and extends to just overlap dorsally or ventrally the tip of the palate[11]. In larger dogs the cuneiform process of the arytenoid can sometimes be seen, but other laryngeal cartilages are usually represented as an indistinct soft tissue opacity superimposed on the laryngeal lumen. Laryngeal cartilages can become identifiable because of mineralisation, the cricoid being the most readily evident[12]. The hyoid bones are clearly identifiable, and the single basihyoid and the paired thyrohyoid, keratohyoid, epihyoid and stylohyoid bones can be distinguished[9]. The normal cricopharyngeus muscle representing the cranial oesophageal sphincter sometimes appears as a distinct soft tissue structure[8].

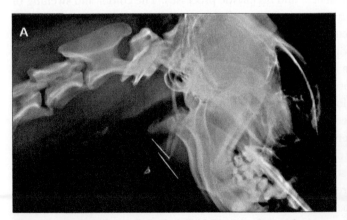

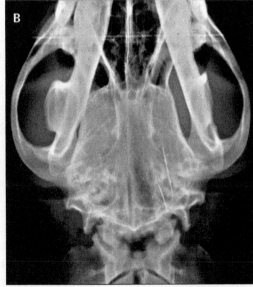

Figs 14.3A, B A: Lateral view of the throat of a dog with two small needle foreign bodies; B: dorsoventral view of the same dog.

Pharyngolaryngeal masses may be extramural, mural, or intramural. Radiographically, extramural masses are usually poorly defined from the surrounding soft tissues and appear as areas of soft tissue thickening, which may cause displacement and/or narrowing of the respiratory tract lumen. Mural masses tend to be smaller when clinical signs appear and can be very difficult to identify[13]. They can appear well defined and/or lobulated if surround by air[14]. Horizontal epiglottic displacement has also been reported[15]. Trauma to the throat can lead to soft tissue thickening with gas opacities in the subcutaneous tissues and between muscle and tissue planes. Linear opacities generally represent emphysema and are commonly reported in dogs with oropharyngeal stick injuries[16], whereas more focal and rounded gas accumulation may indicate abscess[8]. Hyoid bone fractures and laryngeal cartilage fractures can be seen after crushing injury, and foreign bodies can be visible when they are mineral or metallic. Brachycephalic obstructive airway syndrome is evidenced radiographically by extension and thickening of the soft palate beyond the tip of the epiglottis, but elongated soft palate and soft palate oedema cannot easily be distinguished[8]. Epiglottic retroversion is rare and appears radiographically as a more vertically positioned epiglottis that extends dorsally to the dorsal wall of the laryngopharynx[17].

Ultrasonographic examination of the throat

The pharynx and larynx may be imaged with a 7.5–10 MHz high-resolution transducer[9]. The vocal cords can be imaged using the ventral aperture between the cricoid and thyroid cartilages (ligament) or alternatively by using the thyroid cartilage as an acoustic window[9]. Arytenoid cartilage movement can be evaluated noninvasively in real time this way[18,19]. Ultrasonographic diagnosis of laryngeal paralysis is based on the inability to observe lateral motion of the cuneiform processes of the arytenoid cartilages corresponding to normal abduction[19].

Ultrasonographic evaluation of the throat can be beneficial for the evaluation of extramural soft tissues surrounding the upper respiratory tract, determining the origin and nature of mass lesions[20], detecting foreign bodies and guiding needle aspirates and biopsies[8]. It has succesfully been used to detect laryngeal masses in both dogs and cats[18,20]. Ultrasonographic

diagnosis of eversion of the laryngeal saccules has also been reported[21].

14.4 VIDEOFLUOROSCOPY

Dysphagia is a complex problem that can result from many morphological and functional abnormalities including trauma, foreign bodies, masses, regional lymphadenopathy and primary myogenic, neurogenic or neuromuscular disorders. Survey radiographs of the pharynx, cervical region and thorax are always indicated first in order to screen for foreign bodies, masses, soft tissue swelling and gas pockets. However, evaluation of functional disorders of swallowing requires contrast fluoroscopy (videofluoroscopy)[22].

Contrast fluoroscopy

Videofluoroscopic contrast studies allow for recording and dynamic observation of the different phases of swallowing (see Chapter 13 Throat Anatomy and Physiology, Section 13.3). Frame rates of 30–60 per second have been advised for accurate assessment of the rapid events of swallowing[23–25]. With slower acquisition rates, subtle disease may go unnoticed[22]. Positioning is important; lateral body positioning has been shown to significantly increase cervical oesophageal transit time and affected the type of peristaltic wave generated by a swallow compared to sternal recumbency[26]. Swallowing studies have been reported in cats, but oropharyngeal and cricopharyngeal causes of dysphagia have not been reported in the cat. Limitations include the potential need for multiple studies, and the possibility of poor compliance in some cats[27,28].

Fluoroscopic contrast examination assesses:
- Bolus formation.
- Pharyngeal and tongue movement.
- Pharyngeal clearing of barium.
- Cricopharyngeal sphincter function.
- Oesophageal motility[23,24].

Oral dysphagia is usually characterised by the animal dropping food or liquid from its mouth and drooling[29]. In addition, there may be failure of aboral transport of liquid or food and lack of bolus formation. If a swallow is achieved, the pharyngeal and oesophageal phases are usually normal[22]. There may be retention of contrast material in the oropharynx and lack of contrast in the pharynx and oesophagus.

Pharyngeal dysphagia is diagnosed when:
- The oral bolus is inadequately propelled across the pharynx for presentation to the cricopharyngeal sphincter; and/or
- Failure of fully opening of the cricopharyngeal sphincter (cricopharyngeal achalasia); or
- Failure to do so at the appropriate time[9,22,24,30,31].

Videofluoroscopic abnormalities include:
- Slow contraction of the pharynx.
- Incomplete enclosure of the bolus.
- Incomplete rostral and dorsal movement of the larynx.

- Absence of forceful contraction to propel the bolus through the cricopharyngeal sphincter.
- Delayed or nonopening of this sphincter.
- A lack of coordination between pharyngeal contraction and opening of the sphincter.
- Coating of the larynx, trachea, or nasopharynx with contrast material because of misdirection of the bolus or aspiration[24,27,31–33].

Time from onset of swallowing to opening of the cricopharyngeal sphincter is delayed in dogs with cricopharyngeal achalasia compared with normal dogs[34].

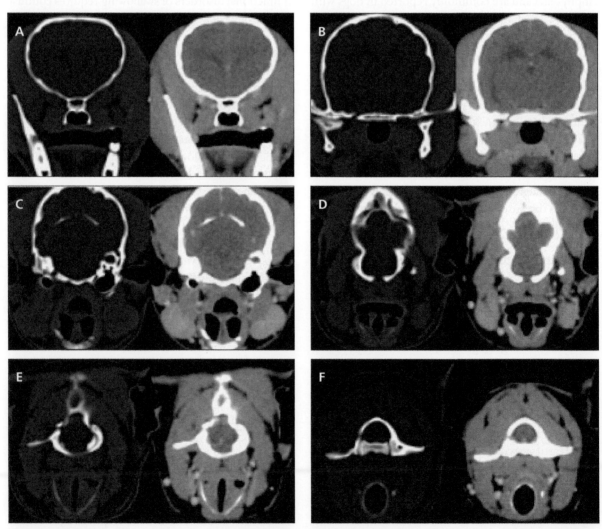

Figs 14.4A–F Transverse computed tomography images of the normal throat of a dog, through midway soft palate (A), end of soft palate (B), hyoid apparatus and epiglottis (C), arytenoids and vocal folds (D), thyroid cartilage (E) and cricoid cartilage (F).

Oesophageal motility disorders may result in abnormal peristalsis and are classified as either primary or secondary dysmotility. Discussion of diseases of the oesophagus is beyond the scope of this book.

14.5 ADVANCED IMAGING OF THE THROAT

CT and MRI are both excellent for evaluating the soft tissues of the throat (Figs 14.4A–F, 14.5A–F)[8,35]. Some

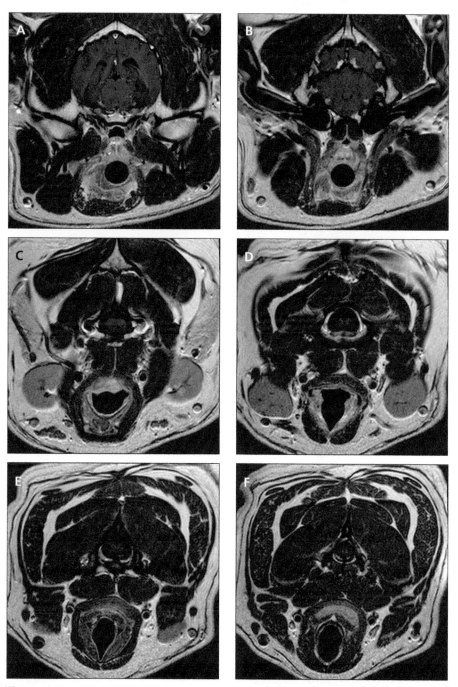

Figs 14.5A–F Transverse magnetic resonance images of the normal throat of a dog, through midway soft palate (A), end of soft palate (B), hyoid apparatus and epiglottis (C), arytenoids (D), thyroid cartilage (E) and cricoid cartilage (F).

CT examinations can be performed under sedation, avoiding the need for general anaesthesia and intubation, which can interfere with detection of lesions. However, brachycephalic dogs should always be intubated. Examination of the throat is best performed with the mouth opened to approximately 30–45° using a mouth gag or syringe to increase the amount of intraluminal air, which improves visualisation[8,36]. Intraluminal air can limit the usefulness of MRI[8].

CT and MRI of the pharynx and larynx

In addition to the previously discussed nasal and middle ear abnormalities that can be found in brachycephalic animals, many abnormalities of the pharynx and larynx can also be found[37-44]. CT findings such as deformed cartilages, reduced pharyngeal and laryngeal diameter, narrow nasopharynx, nasopharyngeal turbinate protrusion, reduced aeration, laryngeal collapse and eversion of the saccules, overlong soft palate, pharyngeal collapse and tracheal hypoplasia have been described in dogs and cats[44a,b]. CT is the diagnostic modality of choice in brachycephalic dogs as it allows for a rapid evaluation of the entire skull, including the middle ears, throat and thorax (see Chapter 11 Diseases of the Nasal Cavity and Sinuses, Section 11.3).

The layered linear appearance of oropharyngeal stick foreign bodies and their contrast-enhancing surrounding inflammatory reaction are seen easily on CT[45]. CT sinography can be helpful in further delineating foreign bodies[46]. Fishhooks and needles can be seen as non-contrast enhancing very high density structures[35], as well as fistulous tracts, granuloma formation, cellulitis or abscess formation. MRI also readily identifies inflammatory reactions in addition to the well-defined linear areas of low-signal intensity representing wooden foreign bodies[47].

Aggressive pharyngolaryngeal neoplastic or infectious processes can also easily be identified using CT[8,35]. It also allows for better assessment of tumour involvement in peripharyngolaryngeal structures, tumour margins and evaluation of the lymph node involvement[35] and pulmonary metastatic disease, making it the diagnostic modality of choice in patients with suspected pharyngolaryngeal neoplasia. CT findings indicative of laryngeal paralysis have been reported to include failure to abduct the arytenoid cartilages, narrowed rima glottis and air-filled laryngeal ventricles[44].

14.6 EMG OF PHARYNX AND LARYNX

Electrophysiological evaluation is the cornerstone of the diagnostic work-up for animals with peripheral nerve, muscle or neuromuscular junction disease[48]. However, results should always be interpreted in light of all findings, including history, physical examination findings, diagnostic imaging, endoscopy and preferably also histopathology.

EMG is the recording and study of insertional, spontaneous and voluntary electric activity of muscle[48]. In contrast to people, most veterinary patients are evaluated under general anaesthesia, where insertional and spontaneous muscle activity are primarily assessed[48]. EMG testing is subject to technical difficulties because of external interference (electric outlets or anaesthesia-related equipment) and ideally should be performed in a Faraday cage or room. Three of the four major types of EMG recording electrodes available, i.e. concentric needle electrodes, monopolar needle electrodes and bipolar concentric needle electrodes, can be used for EMG of the pharynx and larynx[48]. Surface electrodes are not suitable for this purpose.

Normal resting muscle is generally electrically silent, but four types of activity can be seen:
- Insertional activity.
- Miniature end-plate potentials.
- End-plate spikes.
- Motor unit action potentials[49,50].

There are also four patterns of abnormal spontaneous activity in EMG (Fig. 14.6):
- Fibrillation potentials.
- Positive sharp waves.
- Complex repetitive discharges (CRDs).
- Myotonic potentials[2,6,48–50].

The first two can be seen in denervation, polymyositis, muscular dystrophy or other myopathies. CRDs are often seen in chronic denervation and sometimes in some myopathies. Myotonic potentials are characteristic of myotonia congenita[48].

EMG of the pharyngeal muscles is indicated in patients with dysphagia that have no obvious abnormalities on radiographic and when endoscopic examination of the pharynx or as directed by videofluoroscopy. In dogs with signs of laryngeal dysfunction and laryngoscopic examination does not produce a diagnosis, EMG of the intrinsic laryngeal muscles can be helpful, for it can distinguish among normal activity, neurogenic paralysis, ankylotic paralysis and muscular disease[6].

EMG of the pharynx

Under an appropriate level of anaesthesia where a low degree of normal EMG activity in the pharyngeal muscles remains, synchronous with respiration, needle electrodes are inserted in the pharyngeal muscles via throat inspection with the animal in sternal recumbency and mouth held open as described for throat inspection. Unless indicated by videofluoroscopy, all relevant muscles are evaluated. Both halves of the tongue and soft palate and the bilateral thyropharyngeal, hyopharyngeal and cricopharyngeal muscles are tested to record the spontaneous muscle action potentials[2,33]. If necessary, the muscles of the cervical part of the oesophagus can be tested as well[32]. Fibrillation potentials were found to predominate in denervated pharyngeal muscles[51]. In dogs with histological evidence of muscular dystrophy in the pharyngeal muscles there were fibrillation potentials, positive sharp waves and, most characteristic, abundant complex repetitive discharges[32].

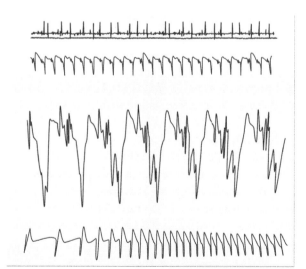

Fig. 14.6 Electromyographic abnormalities such as fibrillation potentials, positive sharp waves, complex repetitive discharges (CRD), and myotonic potentials can be seen from top to bottom.

EMG of the larynx

The dog is anaesthetised and positioned as described for throat inspection and EMG of the pharynx. The epiglottis is depressed with the blade of the laryngoscope, giving access to the main intrinsic laryngeal muscles. Recordings are made via a needle electrode, which is fixed in a long, rigid holder that enables the tip of the electrode to be inserted through the mucosa into the laryngeal muscles under visual guidance[6]. The thyroarytenoid, ventricular, vocal and dorsal cricoarytenoid muscles are accessible beneath the mucosa in this way. Action potentials in normal dogs are observed predominantly during inspiration in the dorsal cricoarytenoid muscles and on expiration in the ventricular muscles. In the thyroarytenoid and vocal muscles, action potentials are observed mainly, but not exclusively, during expiration[6]. No action potentials are observed if the level of anaesthesia is too deep, but abnormal potentials such as fibrillation potentials and CRDs will be observed irrespective of the level of anaesthesia in animals with laryngeal paralysis. Fibrillation potentials are the result of denervation, whereas CRDs are a common finding in neurogenic laryngeal paralysis and in muscular laryngeal disease[6].

14.7 REFERENCES

1 Billen F, Day MJ, Clercx C. Diagnosis of pharyngeal disorders in dogs: a retrospective study of 67 cases. *Journal of Small Animal Practice* 2006;**47**:122–9.

2 Venker-van Haagen AJ. The pharynx. In: *Ear, Nose, Throat, and Tracheobronchial Diseases in Dogs and Cats*. Venker-van Haagen AJ (ed). Schlütersche Verlagsgesellschaft, Hannover 2005, pp. 83–120.

3 Noone KE. Rhinoscopy, pharyngoscopy, and laryngoscopy. *Veterinary Clinics of North America Small Animal Practice* 2001;**31**:671–89.

4 Creevy KE. Airway evaluation and flexible endoscopic procedures in dogs and cats: laryngoscopy, transtracheal wash, tracheobronchoscopy, and bronchoalveolar lavage. *Veterinary Clinics of North America Small Animal Practice* 2009;**39**:869–80.

5 Padrid AP. Laryngoscopy and tracheobronchoscopy of the dog and cat. In: *Small Animal Endoscopy*. Tams TR, Rawlings CA (eds). Elsevier Health Sciences, St. Louis, 2010; pp. 331–59.

6 Venker-van Haagen AJ. The larynx. In: *Ear, Nose, Throat, and Tracheobronchial Diseases in Dogs and Cats*. Venker-van Haagen AJ (ed). Schlütersche Verlagsgesellschaft 2005, pp. 121–65.

7 Holloway A, Avner A. Radiology of the head. In: *BSAVA Manual of Canine and Feline Radiography and Radiology*. Holloway A, McConnell F (eds). British Small Animal Veterinary Association, Gloucester 2013, pp. 302–54.

8 Alexander K. The Pharynx, larynx, and trachea. In: *Textbook of Veterinary Diagnostic Radiology*, 6th edn. Thrall DE (ed). Elsevier Saunders, St. Louis 2013, pp. 489–99.

9 Kealy JK, McAllister H, Graham JP. The thorax. In: *Diagnostic Radiology and Ultrasonography of the Dog and Cat*, 5th edn. Kealy JK, McAllister H, Graham JP (eds). Elsevier Saunders, St. Louis 2011, pp. 199–349.

10 Gaskell CJ. The radiographic anatomy of the pharynx and larynx of the dog. *Journal of Small Animal Practice* 1974;**15**:89–100.

11 Aron DN, Crowe DT. Upper airway obstruction. General principles and selected conditions in the dog and cat. *Veterinary Clinics of North America Small Animal Practice* 1985;**15**:891–917.

12 O'Brien JA, Harvey CE, Tucker JA. The larynx of the dog: its normal radiographic anatomy. *Veterinary Radiology and Ultrasound* 1969;**10**:38–42.

13 Jakubiak MJ, Siedlecki CT, Zenger E, *et al*. Laryngeal, laryngotracheal, and tracheal masses in cats: 27 cases (1998–2003). *Journal of the American Animal Hospital Association* 2005;**41**:310–6.

14 Carlisle CH, Biery DN, Thrall DE. Tracheal and laryngeal tumors in the dog and cat: literature review and 13 additional patients. *Veterinary Radiology & Ultrasound* 1991;**32**:229–35.

15 Wheeldon EB, Suter PF, Jenkins T. Neoplasia of the larynx in the dog. *Journal of the American Veterinary Medical Association* 1982;**180**:642–7.

16 Doran IP, Wright CA, Moore AH. Acute oropharyngeal and esophageal stick injury in forty-one dogs. *Veterinary Surgery* 2008;**37**:781–5.

17 Flanders JA, Thompson MS. Dyspnea caused by epiglottic retroversion in two dogs. *Journal of the American Veterinary Medical Association* 2009;**235**:1330–5.

18 Rudorf H, Barr F. Echolaryngography in cats. *Veterinary Radiology & Ultrasound* 2002;**43**:353–7.

19 Rudorf H, Barr FJ, Lane JG. The role of ultrasound in the assessment of laryngeal paralysis in the dog. *Veterinary Radiology & Ultrasound* 2001;**42**:338–43.

20 Rudorf H, Brown P. Ultrasonography of laryngeal masses in six cats and one dog. *Veterinary Radiology & Ultrasound* 1998;**39**:430–4.

21 Rudorf H, Lane JG, Wotton PR. Everted laryngeal saccules: ultrasonographic findings in a young Lakeland terrier. *Journal of Small Animal Practice* 1999;**40**:338–9.

22 Gaschen L. The canine and feline esophagus. In: *Textbook of Veterinary Diagnostic Radiology*, 6th edn. Thrall DE (ed). Elsevier Saunders, St. Louis 2013, pp. 500–21.

23 Watrous BJ, Suter PF. Normal swallowing in the dog: a cineradiographic study. *Veterinary Radiology & Ultrasound* 1979;**20**:99–109.

24 Watrous BJ, Suter PF. Oropharyngeal dysphagias in the dog: a cinefluorographic analysis of experimentally induced and spontaneously occurring swallowing disorders. *Veterinary Radiology & Ultrasound* 1983;**24**:11–24.

25 Suter PF, Watrous BJ. Oropharyngeal dysphagias in the dog: a cinefluorographic analysis of experimentally induced and spontaneously occurring swallowing disorders *Veterinary Radiology & Ultrasound* 1980;**21**:24-39

26 Bonadio CM, Pollard RE, Dayton PA, Leonard CD, Marks SL. Effects of body positioning on swallowing and esophageal transit in healthy dogs. *Journal of Veterinary Internal Medicine* 2009;**23**:801–5.

27 Pollard RE. Imaging evaluation of dogs and cats with dysphagia. *ISRN Veterinary Science* 2012;**2012**:1–15.

28 Levine JS, Pollard RE, Marks SL. Contrast videofluoroscopic assessment of dysphagic cats. *Veterinary Radiology & Ultrasound* 2014;**55**:465–71.

29 Pollard RE, Marks SL, Leonard R, Belafsky PC. Preliminary evaluation of the pharyngeal constriction ratio (PCR) for fluoroscopic determination of pharyngeal constriction in dysphagic dogs. *Veterinary Radiology & Ultrasound* 2007;**48**:221–6.

30 Pfeifer RM. Cricopharyngeal achalasia in a dog. *Canadian Veterinary Journal* 2003;**44**:993–5.

31 Elliott RC. An anatomical and clinical review of cricopharyngeal achalasia in the dog. *Journal of the South African Veterinary Association* 2010;**81**:75–9.

32 Peeters ME, Haagen AJVV, Goedegebuure SA, Wolvekamp WTC. Dysphagia in Bouviers associated with muscular dystrophy; evaluation of 24 cases. *Veterinary Quarterly* 1991;**13**:65–73.

33 Venker-van Haagen AJ, Hartman W, van den Brom WE, Wolvekamp WT. Continuous electromyographic recordings of pharyngeal muscle activity in normal and previously denervated muscles in dogs. *American Journal of Veterinary Research* 1989;**50**:1725–8.

34 Pollard RE, Marks SL, Davidson A. Quantitative videofluoroscopic evaluation of pharyngeal function in the dog. *Veterinary Radiology & Ultrasound* 2000;**41**:409–12.

35 Taeymans O, Schwarz T. Pharynx, larynx and thyroid gland. In: *Veterinary Computed Tomography*. Schwarz T, Saunders J (eds). Wiley-Blackwell, Chichester 2013, pp. 175–84.

36 Laurenson MP, Zwingenberger AL, Cissell DD, *et al*. Computed tomography of the pharynx in a closed vs. open mouth position. *Veterinary Radiology & Ultrasound* 2011;**52**:357–61.

37 Oechtering GU. Brachycephalic syndrome – new information on an old congenital disease. *Veterinary Focus* 2010;**20**:2–9.

38 Schmidt MJ, Neumann AC, Amort KH, Failing K, Kramer M. Cephalometric measurements and determination of general skull type of Cavalier King Charles Spaniels. *Veterinary Radiology & Ultrasound* 2011;**52**:436–40.

39 Hussein AK, Sullivan M, Penderis J. Effect of brachycephalic, mesaticephalic, and dolichocephalic head conformations on olfactory bulb angle and orientation in dogs as determined by use of *in vivo* magnetic resonance imaging. *American Journal of Veterinary Research* 2012;**73**:946–51.

40 Schuenemann R, Oechtering GU. Inside the brachycephalic nose: intranasal mucosal contact points. *Journal of the American Animal Hospital Association* 2014;**50**:149–58.

41 Oechtering TH, Oechtering GU, Noeller C. Structural characteristics of the nose in brachycephalic dog breeds analysed by computed tomography. *Tieraertzliche Praxis* 2007;**35**:177–87.

42 Schlueter C, Budras KD, Ludewig E, *et al*. Brachycephalic feline noses: CT and anatomical study of the relationship between head conformation and the nasolacrimal drainage system. *Journal of Feline Medicine and Surgery* 2009;**11**:891–900.

43 Grand JG, Bureau S. Structural characteristics of the soft palate and meatus nasopharyngeus in brachycephalic and non-brachycephalic dogs analysed by CT. *Journal of Small Animal Practice* 2011;**52**:232–9.

44 Stadler K, Hartman S, Matheson J, O'Brien R. Computed tomographic imaging of dogs with primary laryngeal or tracheal airway obstruction. *Veterinary Radiology & Ultrasound* 2011;**52**:377–84.

44a Vilaplana Grosso F, ter Haar G, Boroffka SAEB. Prevalence and classification of caudal aberrant nasal turbinates in clinically healthy English Bulldogs: a computed tomograpic study. *Veterinary Radiology and Ultrasound* 2015;**56**:486–93.

44b Kaye BM, Boroffka SAEB, Haagsman AN, ter Haar G. Computed tomographic and radiographic assessment versus endoscopic asssessment of tracheal diameter in 40 non-symptomatic English Bulldogs. *Veterinary Radiology and Ultrasound* 2015;**56**:609–16.

45 Nicholson I, Halfacree Z, Whatmough C, Mantis P, Baines S. Computed tomography as an aid to management of chronic oropharyngeal stick injury in the dog. *Journal of Small Animal Practice* 2008;**49**:451–7.

46 Jones JC, Ober CP. Computed tomographic diagnosis of nongastrointestinal foreign bodies in dogs. *Journal of the American Animal Hospital Association* 2007;**43**:99–111.

47 Dobromylskyj MJ, Dennis R, Ladlow JF, Adams VJ. The use of magnetic resonance imaging in the management of pharyngeal penetration injuries in dogs. *Journal of Small Animal Practice* 2008;**49**:74–9.

48 Cuddon PA. Electrophysiology in neuromuscular disease. *Veterinary Clinics of North America Small Animal Practice* 2002;**32**:31–62.

49 van Nes JJ. An introduction to clinical neuromuscular electrophysiology. *Veterinary Quarterly* 1986;**8**:233–9.

50 van Nes JJ. Clinical application of neuromuscular electrophysiology in the dog: a review. *Veterinary Quarterly* 1986;**8**:240–50.

51 Venker-van Haagen AJ, Hartman W, Wolvekamp WT. Contributions of the glossopharyngeal nerve and the pharyngeal branch of the vagus nerve to the swallowing process in dogs. *American Journal of Veterinary Research* 1986;**47**:1300–7.

15.1 INTRODUCTION

Dogs and cats with diseases of the pharynx can have a variety of historical complaints, mainly relating to swallowing difficulties, but not uncommonly related to the upper respiratory tract including:

- Dysphagia.
- Dyspnoea with stertor or snoring; stridor can be one of the clinical complaints.
- Often gagging, retching and regurgitation are seen.
- Sometimes nasal discharge, reverse sneezing and coughing, as discussed before.

All diseases can be diagnosed after a thorough clinical examination, diagnostic imaging and endoscopy. Cytology and histopathology are essential in the diagnosis of some disorders, but haematology and biochemical profiles will not confirm any diagnosis and are therefore hardly ever indicated, unless for assessment of the general condition of the patient.

A summary of the congenital abnormalities affecting the pharynx is presented first in the next section, after which the most common acquired diseases will be reviewed. These include pharyngeal mucoceles, pharyngitis and tonsillitis, including eosinophilic granulomas, traumatic injuries of the pharynx, pharyngeal and tonsillar neoplasia and finally oropharyngeal dysphagia.

15.2 CONGENITAL ABNORMALITIES OF THE PHARYNX

Cleft lip and palate are common congenital malformations in dogs and cats that affect the pharynx, but most commonly lead to chronic rhinitis. These conditions, including soft palate hypoplasia, have been reviewed in Chapter 11 Diseases of the Nasal Cavity and Sinuses, Section 11.7 and Chapter 19 Surgery of

the Nose, Section 19.4. An overlong or hyperplastic soft palate is associated with brachycephaly and pharyngeal hypoplasia, and has been reviewed in Chapter 11 Diseases of the Nasal Cavity and Sinuses, Section 11.3 and Chapter 19 Surgery of the Nose, Section 19.3. Choanal atresia affects the nasopharynx and is the result of failure or breakdown of the buccopharyngeal membrane between nose and nasopharynx. This condition has been discussed in Chapter 12 Diseases of the Nasopharynx, Section 12.2. Craniopharyngiomas are derived from nests of epithelium of the primordial craniopharyngeal canal, or Rathke's pouch. They are usually benign growths that expand intracranially above the sella turcica. However, expansion into the nasopharynx has also been described. This condition therefore has been reviewed in Chapter 12 Diseases of the Nasopharynx, Section 12.3.

15.3 PHARYNGEAL MUCOCELES

Introduction

Salivary mucocele is the most commonly recognised disease of the salivary glands in dogs but is rarely reported in cats[1-9]. Mucocele refers to a collection of saliva in the subcutaneous tissue near the site of a leaking salivary gland or duct. Lesions are not true cysts as they are lined with nonsecretory granulation tissue that forms secondary to the inflammation caused by the presence of saliva in the tissue[1,3,8,10,11].

Whereas mucoceles can arise in subcutaneous, sublingual, pharyngeal or periorbital locations[3], most commonly they arise from defects in the polystomatic lobules or associated minor ducts of the sublingual salivary gland and/or the major duct of the sublingual gland[11]. The two most common sites of saliva accumulation are therefore the intermandibular region (cervical mucocele, Fig. 15.1), and under the tongue at the base of the lingual frenulum (sublingual mucocele or

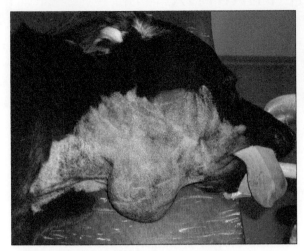

Fig. 15.1 Labrador Retriever with a large submandibular swelling (cervical mucocele).

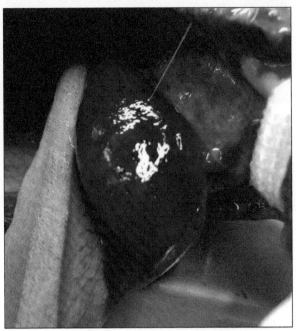

Fig. 15.2 Greyhound with a large sublingual swelling (ranula) after acute oropharyngeal stick injury.

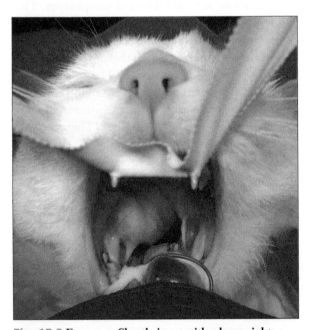

Fig. 15.3 European Shorthair cat with a large right-sided submucosal pharyngeal swelling (pharyngeal mucocele).

Pharyngeal mucoceles

Dogs are more affected than cats, and Miniature Poodles appear to be at increased risk of developing pharyngeal mucoceles[9,14,16,17]. Over-representation of Dachshunds, Australian Silky Terrier, German Shepherd dogs and Greyhounds has also been reported[13,14,16,18]. As is the case for cervical and sublingual mucoceles, a pharyngeal mucocele appears to originate from defects in the sublingual salivary gland[9,13–15], but possibly they can arise from local salivary glands in the pharyngeal wall or soft palate[19].

The clinical signs are inevitably those of pharyngeal obstruction of the airways, and dyspnoea and snoring caused by the voluminous cyst is the most common presenting complaint[9,13–15,19]. However, problems with swallowing can complement the clinical presentation. If the patient is in severe respiratory compromise, a patent airway either via tracheal intubation or temporary tracheotomy is indicated prior to further diagnostic work-up.

Though radiography can be helpful in demonstrating a space-occupying lesion in the pharynx, and sialography can be attempted to localise the source of the mucocele, these procedures are difficult and time-

ranula, Fig. 15.2)[1,3,5,11,12]. Pharyngeal mucoceles are the least common type of mucocele, but are clinically important because most animals present in respiratory distress (Fig. 15.3)[13–15]. While in most cases no inciting cause can be identified, salivary mucoceles have been reported to result from trauma, foreign bodies, sialoliths, mucus retention, neoplasia and oral surgery[10,12,16].

consuming[3,14,20]. Ultrasonographic characteristics of salivary mucoceles[7] have recently been described, and ultrasound is particularly helpful in lateralisation of the lesions. Computed tomography (CT) imaging is especially helpful for evaluation of pharyngeal and zygomatic mucoceles[21]. However, lateralisation of the lesion is usually straightforward for pharyngeal mucoceles, in contrast to sublingual and cervical mucoceles. The definitive diagnosis is made by direct visualisation of a mucocele via oropharyngeal inspection under sedation (Fig. 15.3).

Usually, a soft, non-painful, non-fixed, fluctuant mass is found, that can obscure visualisation of the larynx and complicate intubation. Decompression of the cyst with fine needle aspiration of the contents, or by making a small stab incision in the cyst, will allow intubation in those cases and provides material for cytology and culture to confirm the diagnosis. The fluid is characteristically a thick, tenacious, clear and transparent, yellowish to sanguineous fluid[3,12,16]. Samples are usually hypocellular with small numbers of macrophages, salivary epithelial cells and a varying number of neutrophils in a mucin-blue stained background[3,12,16].

Definitive treatment consists of marsupialisation of the pharyngeal lesion in combination with resection of the mandibular and sublingual salivary gland complex, which leads to excellent results and low recurrence rates[2,3,5,10,13–15].

15.4 PHARYNGITIS AND TONSILLITIS

- Acute pharyngitis is typically caused by viral infections in both dogs and cats.
- In cats, feline herpesvirus and feline calicivirus are commonly implicated in both acute pharyngitis and chronic gingivitis–stomatitis–pharyngitis.
- Severe pharyngitis requires intravenous administration of fluids and antibiotics and sometimes tracheotomy in patients with gross enlargement of the tonsils.
- Chronic gingivitis–stomatitis–pharyngitis in cats has a multifactorial aetiology, in which periodontitis and feline osteoclastic resorptive lesions play an important role.
- Specific forms of pharyngitis include viral papillomatosis and eosinophilic granulomas.

The pharynx and tonsils are rarely primarily inflamed; especially in chronic cases, they are most commonly inflamed secondary to other diseases. The most common cause of acute tonsillitis and pharyngitis is upper respiratory infection in both dogs and cats, in which the disease is never limited to just the pharynx. The tonsils are usually involved in dogs with pharyngitis, but specific solitary tonsillitis has also been reported.

Chronic pharyngitis is commonly seen in cats as part of the gingivitis–stomatitis–pharyngitis complex, but chronic tonsillitis is far more common in dogs than cats. Chronic pharyngitis is very commonly seen in dogs with brachycephalic obstructive airway syndrome, which has been discussed in Chapter 11 Diseases of the Nasal Cavity and Sinuses, Section 11.3. Specific inflammatory conditions such as viral papillomatosis and eosinophilic granulomas in dogs and cats will be discussed below.

Acute pharyngitis and tonsillitis

In dogs and cats, acute pharyngitis is typically caused by viral infections[19]. The most common causes of these viral infections in cats are feline herpesvirus type 1 (FHV-1) and feline calicivirus (FCV, see Chapter 11 Diseases of the Nasal Cavity and Sinuses, Section 11.9), although *Chlamydia psittaci* has also been implicated[22–26]. In dogs, canine distemper virus, canine adenovirus-2 and canine parainfluenza virus may play a role, although they are more commonly incriminated in laryngitis and tracheobronchitis (see Chapter 17 Perioperative Management, Section 17.3). Primary bacterial or fungal infections causing pharyngitis are uncommon, though one case of malassezial pharyngitis has been reported[27]. The characteristic normal tonsillar flora in dogs comprises alpha- and beta-haemolytic streptococci together with *Pasteurella* spp[28,29]. Acute tonsillitis is often associated with bacterial infections, beta-haemolytic streptococcus being the most prominent agent in dogs[19,30].

In cats, the palatine tonsils are also inflamed with FHV-1 and FCV infections, but the tonsillitis is related to the pharyngitis, and specific bacterial tonsillitis has not been described.

Pharyngitis and tonsillitis are characterised by pain, fever, oral pain and extreme discomfort, where the affected animal does not attempt to eat, often salivates, holds the neck extended and makes ineffective swallowing movements[19].

The diagnosis is based on history, physical examination and findings upon inspection of the throat. Fever is common when infectious agents are involved and in severe cases laboratory blood analysis may be indicated. The pharynx will be visibly inflamed and the mucosal lining can show oedema, ulcerations, petechiae or small abscesses. Acute tonsillitis will reveal bright-red, friable, enlarged and protruding tonsils that occasionally have petechiae or small abscesses as well[31]. Bacterial or viral isolation is usually not indicated unless clinical signs are chronic or relapsing.

Symptomatic treatment consists of:

- Parenteral administration of broad-spectrum antibiotics to prevent or treat additional bacterial infections even when a viral disease is suspected[19].
- Intravenous administration of fluids or parenteral/enteral nutrition may be indicated in severely affected animals that are reluctant to eat and drink.
- Analgesics are an important part of the medical treatment and should be given during the first 3–5 days[19].
- Liquid or soft food may be given in small portions, several times a day, until the animal begins to exhibit interest in more usual foods.

Chronic pharyngitis and tonsillitis in dogs

In non-brachycephalic dogs the cause of chronic pharyngitis is usually unclear, but chronic vomiting or regurgitation, food allergies, anal sac diseases and ingestion of caustic or toxic substances need to be excluded[31]. Chronic pharyngitis in dogs is characterised by gagging and retching independent of the intake of food, ptyalism, normal swallowing and sudden periods of pica in dogs[19]. In addition, some animals may snore.

The diagnosis is based on exclusion of other abnormalities and inspection of the pharyngeal mucosa, which reveals thickening and irregular reddening[19]. The tonsils can be enlarged and protrude from the crypts (Fig. 15.4). If the cause of the disease is not clear, symptomatic treatment can be directed at diminishing the irritation of the mucosa by giving moistened food and feeding it in smaller portions[19]. Bouts of pica can be subdued by giving phenobarbital in a moderate dose upon the appearance of the first signs, which are licking and restlessness[19].

Chronic tonsillitis from an unknown cause, not responding to antibiotic treatment, is found more often in dogs than in cats[19]. The enlarged and indurated tonsils

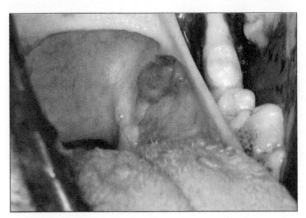

Fig. 15.4 **Chronic pharyngitis and tonsillitis in a 4-year-old Labrador Retriever.**

may cause obstruction and therefore pain during swallowing, and even hinder the passage of air through the oropharynx. In these cases tonsillectomy would be indicated.

Chronic pharyngitis and tonsillitis in cats

In cats, especially in young purebreds, periodontitis and gingivitis is commonly seen in the progressive, inflammatory condition termed gingivitis–stomatitis–pharyngitis complex[23,25,32,33]. Feline chronic gingivostomatitis is considered to have a multifactorial aetiology, with viral[34,35], bacterial[36,37], and immune[34] factors suggested as causal. Dental abnormalities such as periodontitis and feline odontoclastic resorptive lesions have also been suggested to play a role in this complex[38,39].

Clinical signs include oral pain, ptyalism, lethargy, weight loss, unkempt hair and inappetence in affected cats[33,40]. Diagnosis can be made on the basis of clinical findings and ruling out other pharyngeal diseases, but determining the full extent of the dental lesions requires probing and full-mouth dental radiography[33,39,41,42]. Inspection of the oral cavity and pharynx typically reveals inflammation of the oral mucosa, which is disproportionately severe compared with visible dental disease and calculus accumulation. The lesions are commonly concentrated towards the caudal parts of the mouth, involving the palatoglossal folds in particular (Fig. 15.5), with extension rostrally along the buccal and gingival mucosa, crossing the mucogingival junctions. The pharynx and the soft palate may also be involved, and (less commonly) the hard palate and the tongue[43].

Many medications have been reported in treatment, including gold salts (aurothioglucose), azathioprine, chlorambucil, vincristine, 5-fluorouracil, lactoferrin, azithromycin, glucocorticoids, metronidazole, sulodexide, tacrolimus topical, thalomide, zinc sulfate, colchicine, interferon alpha-2A and cyclosporine (ciclosporin)[45–51], with varied success[33,38,44]. Of these, corticosteroids are reported to have the most significant results; however, the chronic inflammation frequently becomes refractory to treatment. Dental scaling can be helpful for a period of time, but in severe cases, extraction of all elements is often the only option[44].

In an evaluation of response to extractions, it was found that 60% of the cats had complete remission of clinical disease, and another 20% had remission with only mild flare-ups not requiring treatment. In the remaining cats, 13% still required medical management and 7% were unresponsive to surgical or medical management[43]. Inflammation and wound healing in cats after extraction of all premolars and molars were not affected by feeding of two diets with different omega-6/omega-3 polyunsaturated fatty acid ratios[45].

Viral papillomatosis

Although the underlying mechanisms are not completely understood, papillomavirus is implicated as an aetiological factor in canine oral papillomatosis[46–48]. Viral papillomatosis is horizontally transmitted by a DNA viral agent (papovavirus) from dog to dog[49]. Affected animals are generally young and present with usually multiple whitish, verrucous, hyperkeratotic lesions in the oral cavity, pharynx, tongue or lips (Figs 15.6A–C).

A biopsy can be performed if necessary, but visual examination is usually diagnostic[49]. Most patients never suffer any significant side-effects of this disease, although occasionally lesions can become so large

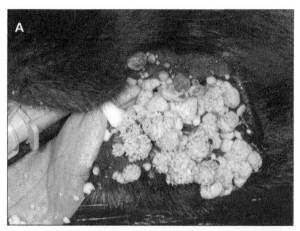

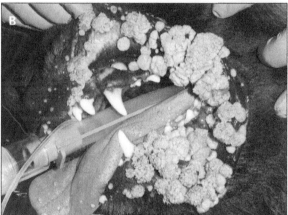

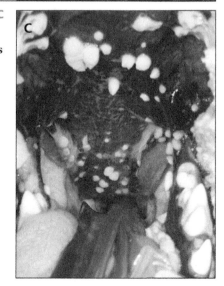

Figs 15.6A–C Viral papillomatosis in a 1-year-old Labrador Retriever with multiple papillomas covering the lips, tongue, oral and pharyngeal mucosa.

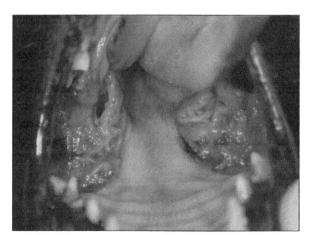

Fig. 15.5 Chronic pharyngitis with severe ulcerative phaucitis, gingivitis and paradontitis in a 6-year-old Domestic Shorthair cat.

that they interfere with swallowing or respiration and require surgical cytoreduction.

Papillomatosis in dogs may be a self-limiting disease requiring no treatment, with the majority regressing spontaneously within 4–8 weeks[46,49]. However, persistent lesions may be associated with defective cell-mediated immunisation. Many therapeutic trials for the treatment of papillomatosis have been reported, including immunomodulators, interleukins, levamisole, thiabendazole, *Propionibacterium acnes*, cimetidine and live papilloma virus vaccine[46,50]. However, these methods are seldom required or effective. Furthermore, autogenous vaccines are not recommended because squamous cell carcinomas have been reported at the site of inoculation[46,50–52]. In a placebo-controlled study, azithromycin in a dose of 10 mg/kg p/o every 24h for 10 days was shown to lead to rapid regressions of lesions in all patients. However, without treatment, the prognosis is usually excellent.

Eosinophilic granulomas in dogs

Canine oral eosinophilic granulomas affect young dogs and may be heritable in the Siberian husky[53,54] and Cavalier King Charles Spaniel[55,56], which is also affected by eosinophilic stomatitis–pharyngitis[57] without granuloma formation (Figs 15.7A, B). On oral examination fiercely red, ulcerated or diffusely swollen mucosa can be seen, often very pronounced in the phauces and pharynx, but the entire oral cavity can be affected. In some patients specific granulomas are formed; these typically occur on the lateral and ventral aspects of the tongue[49]. Granulomas are raised and frequently ulcerated as well, mimicking oropharyngeal cancer. Histologically only eosinophilic, neutrophilic and granulomatous inflammation are seen. Treatment with corticosteroids or surgical excision is generally curative, although spontaneous regression may occur[55,56]. Hypoallergenic diets may be beneficial in some patients.

Eosinophilic granulomas in cats

Eosinophilic granuloma complex, a condition that may manifest as indolent ulcer on the lip, eosinophilic plaque or linear granuloma occurs more commonly in female cats with a mean age of 5 years[49,58,59]. The aetiology is unknown, but many cats have concurrent allergic disease, sometimes associated with food allergy[59,60]. Lesions are most commonly found on the upper lip near the midline, but any oropharyngeal site is at risk. Slowly progressive erosion of the lip is often the only clinical sign, but with severe involvement of the oropharynx inappetence, weight loss, oral pain and ptyalism can be seen. Biopsies are often necessary to differentiate the condition from oropharyngeal cancer[49]. In addition, investigation of an underlying hypersensitivity disorder in affected cats is recommended[59]. If an underlying disorder is found and treated (e.g. hypoallergenic diets), specific treatment may not be necessary. Various treatments for the granulomas have been proposed, including oral prednisone at 1–2 mg/kg twice a day (BID),

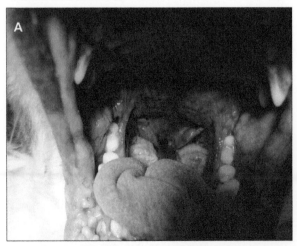
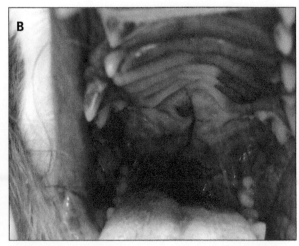

Figs 15.7A, B Cavalier King Charles Spaniel with chronic pharyngitis and tonsillitis with a lymphoplasmacytic to eosinophilic infiltrate on histopathology.

radiotherapy, surgery, immunomodulation or cryo-surgery. The prognosis for complete and permanent recovery is fair[49,59].

15.5 TRAUMATIC INJURIES OF THE PHARYNX

The most common type of pharyngeal trauma in dogs is that caused by penetration by a stick. In cats fissures of the hard and soft palate caused by a fall from a height are well known[19], and both conditions will be reviewed below. However, road traffic accidents, gunshot, knife wounds and fights can also cause trauma to the throat. Blunt trauma as a result of road traffic accidents is especially common. In these cases the entire head is usually involved, not just the pharynx or larynx (see Chapter 11 Diseases of the Nasal Cavity and Sinuses, Section 11.12). If the pharynx is involved, oedema and haematomas can obstruct the pharyngeal airway.

Clinical signs, in addition to the ones described before for nasal and sinal trauma, indicating pharyngeal involvement include:
- Drooling of bloody mucous.
- Increased swallowing.
- Pain in the pharyngeal area, subcutaneous emphysema.
- Dyspnoea with a snoring stridor.

The patient should be evaluated for signs of shock and stabilised before further diagnostic work-up and definitive treatment, as discussed in Chapter 11 Diseases of the Nasal Cavity and Sinuses, Section 11.12. Endotracheal intubation or tracheotomy may be necessary if severe obstruction of the pharyngeal airways is present. CT may be indicated to rule out fractures of the base of the skull, hyoid bones and/or laryngeal cartilages; pharyngeal inspection will reveal any perforating lesions. Surgical repair of minor mucosal lesions and fractures of the hyoid bones is usually not indicated, but deep penetrations, oesophageal wall lacerations and laryngeal cartilage fractures may need neck exploration, debridement, repair and drainage.

Penetrating pharyngeal injuries
Penetrating injuries to the oral cavity and pharyngolaryngeal area from foreign objects is sometimes seen in dogs[61–63] that chew on or carry and retrieve sticks. It is seen rarely in cats[64]. Commonly affected areas include

the sublingual area (Fig. 15.8), lateral pharyngeal wall and tonsillar crypt (Figs 15.9–15.11) and soft palate and dorsal oesophageal wall. In one study the most common site of injury was sublingual[62]. Symptoms, clinical approach and treatment options vary depending on the acute or chronic presentation of the patient, the exact location of the trauma and the severity of it.

Pharyngeal trauma is often witnessed by the owner, though most do not seek medical help immediately[62]. Common presenting complaints in the acute stage are:
- Dysphagia.
- Drooling.
- Depression.
- Oral pain.

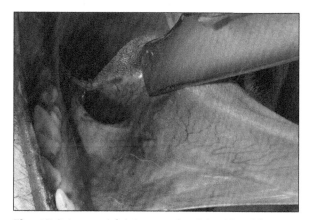

Fig. 15.8 Acute stick injury on the right side of the tongue with a relatively shallow wound that could be explored completely via oral inspection.

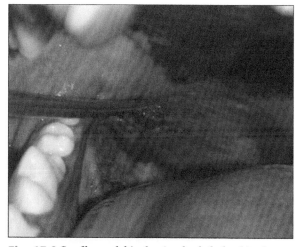

Fig. 15.9 Small woodchip foreign body lodged in the mucosa just medial to the right tonsillar crypt and fold.

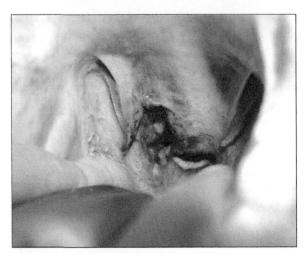

Fig. 15.10 Subacute stick injury to the right tonsillar crypt and fold with partial avulsion of the soft palate and crypt. Inflammation and necrosis of the wound edges is evident.

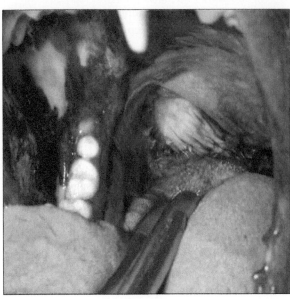

Fig. 15.11 Rottweiler with a chronic stick injury, pain upon flexing the neck and halitosis as presenting signs. Swelling of the right pharyngeal wall with a small fistula can be seen.

- Pain on flexion of the neck.
- Subcutaneous emphysema in the cervical region.
- Blood stained saliva.
- Pain on opening of the mouth[65].

Cases with chronic penetrating stick injuries present with:
- A recurrent swelling or abscess of the head or neck.
- With or without a cutaneous draining sinus in the cervical region[62,63].

The management of acute penetrating pharyngeal stick injury is primarily directed at emergency stabilisation of the patient, and diagnostic work-up should be delayed until the patient has been stabilised. Animals can present in severe dyspnoea as a result of pharyngeal swelling or mediastinal emphysema. Hypoxia, hypovolaemia and shock should be addressed prior to further work-up.

Plain radiographs of the head, neck and thorax are recommended for initial diagnostic work-up in animals with a suspected or confirmed acute penetrating injury to the throat or oesophagus[61–63]. Subcutaneous emphysema, gas between fascial planes in the cervical region, confirms that penetration has occurred and can help

localise the injury (Fig. 15.12A)[65]. However, this gas can dissect away from the original leakage along fascial planes, causing pneumomediastinum or pneumothorax, making localisation difficult[61,62]. Despite being helpful for assessment of the degree of trauma and for localisation, the stick itself, or fragments thereof, do not show on radiographs. Ultrasonography of the throat or neck is not of great use in the acute patient either for this purpose, since subcutaneous emphysema can mask small foreign bodies[66]. Advanced imaging techniques as CT and magnetic resonance imaging (MRI) have not been evaluated for use in pharyngeal stick injury in patients with an acute presentation[65].

Consistent radiographic abnormalities were not seen in patients with chronic presentations[63], but sinography, though only applicable in cases with external draining sinuses, appeared to have a high sensitivity and specificity for identification of foreign bodies[67]. Ultrasonography has also been reported to be a valuable aid in identification of abscesses and wooden foreign bodies in patients with a chronic presentation; it is relatively cheap and can usually be performed without anaesthesia[68,69]. Most foreign bodies are easily recognised when

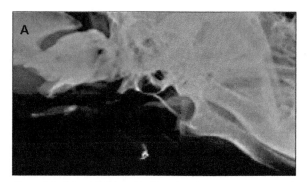

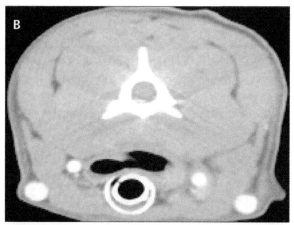

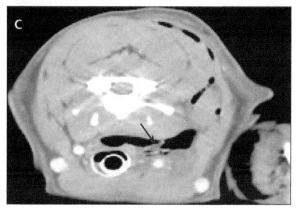

Figs 15.12A–C Diagnostic imaging for penetrating stick injuries. A: A lateral radiograph of a dog demonstrating peripharyngeal subcutaneous emphysema indicative of perforating injury; B: transverse CT image showing disruption of the contours of the dorsal oesophageal wall after acute stick injury; C: transverse CT image of the same dog distal to (B), showing complete dorsal oesophageal rupture and ventrally located foreign body (arrow).

surrounded by exudative fluid. CT with or without IV contrast and MRI are very accurate in recognising wood foreign bodies (Figs 15.12B, C)[70,71]. Depending on the water content of the different layers of the sticks, the foreign bodies show a variable attenuation pattern on CT[71].

General anaesthesia permits thorough examination of the entire oral cavity and inspection of the pharynx, larynx and rostral oesophagus. With the dog in sternal recumbency, and before endotracheal intubation, careful examination of the sublingual areas on the left and right of the frenulum, of the tongue base, the lateral pharyngeal walls and tonsillar crypts and hard and soft palate needs to be performed. After inspection of the larynx, an endotracheal tube can be inserted to secure the airways. After rostral retraction of the soft palate and inspection of the caudal nasopharynx and laryngopharynx, a long laryngoscope blade can be advanced into the rostral oesophagus. When perforations are not found at this time, complete cervical oesophageal endoscopy is advised[66]. The site of penetration is frequently not apparent, particularly in chronic cases.

When a piece of wood is visible in the soft tissues, withdrawal is not recommended because of the risk of fragmentation[66]. Small perforations in the oral cavity to the level of the oropharynx that can be fully inspected

and flushed and do not contain foreign material, do not have to be explored surgically but can be left to heal by second intention. In all cases of perforations of the caudal pharyngeal (laryngopharynx) or oesophageal wall, surgical exploration of the neck via a ventral midline approach is recommended (see Chapter 20 Surgery of the Throat, Section 20.4).

A logical decision as to whether or not surgery is indicated also depends on the point of entry and direction of the stick injury. Trauma in the direction of the dorsal nasopharynx (usually through the soft palate) and brain cannot be explored via a ventral neck approach. Diagnostic imaging findings will help aid in surgical planning in patients with chronic presentations and abscesses. In most of these patients a midline ventral neck approach is indicated.

Acute penetrating injury of the oropharyngeal region, when treated appropriately, has a good prognosis with most or all patients making a complete recovery[61-63]. Acute injuries to the oropharyngeal region have a better prognosis than acute oesophageal penetrations[61]. The most common postoperative complication of the latter is septic mediastinitis and death. Clinical signs resolved in 62% of dogs that presented with chronic signs in one large study[62]. Aggressive surgical debridement of all sinus tracts is essential in obtaining a successful result, but recovery of a foreign body is not necessarily a determinant of success[62].

Pharyngeal trauma in cats

In cats falls from windows or apartment buildings commonly lead to acquired cleft palate[72,73]. The cleft can involve the soft palate as well as the hard palate and results in a fistula between the mouth and the nasal cavity or the oropharynx and the nasopharynx (see Chapter 11 Diseases of the Nasal Cavity and Sinuses, Section 11.7). As for dogs with blunt trauma to the head, these cats need to be examined first for other injuries and brain damage and stabilised prior to further work-up and treatment. Traumatic clefts less than 2 mm wide will usually heal without surgical intervention if soft food is provided for several weeks[19,74]. A cleft wider than 2 mm should be reduced, compressed and fixated using maxillary transfixations pins and circummandibular (upper canine teeth) cerclage wire[74]. If chronic infection develops, leading to bone atrophy or fistula formation, secondary surgical repair involving mucoperiosteal flap reconstruction is indicated[19,75].

15.6 PHARYNGEAL AND TONSILLAR NEOPLASIA

- The most common oropharyngeal cancers in dogs are malignant melanoma, squamous cell carcinoma (SCC) and fibrosarcoma, whereas in cats the most common ones are SCC and fibrosarcoma.
- Lip and tongue melanomas have a better prognosis then hard palate melanoma.
- SCC of the rostral oral cavity has a low metastatic rate and the caudal tongue and tonsil have a high metastatic potential.
- Histologically low-grade but biologically high-grade fibrosarcoma is locally invasive but rate of metastasis is low.

Collectively, oropharyngeal cancer (combining all tumours of the oral cavity and pharynx including those of the gingiva, mandible and maxilla) accounts for 6–7% of canine[76–78] and 3% of feline cancers[78,79]. Male dogs have an increased risk of developing oropharyngeal malignancy in general compared to female dogs[80,81] and most noticeably for malignant melanoma and tonsillar SCC[82,83]. Predisposed breeds appear to be the Cocker Spaniel, German Shepherd dog, German Shorthaired Pointer, Weimaraner, Golden Retriever, Gordon Setter, Miniature Poodle, Chow Chow and Boxer[78,82,84,85].

In dogs, the most common malignant oral tumours are malignant melanoma, SCC and fibrosarcoma[78,86–92]. SCC is the most common oropharyngeal cancer in cats, followed by fibrosarcoma[78,79]. In cats, the risk of developing oral SCC is significantly increased by the use of flea collars, high intake of either canned food in general or canned tuna fish specifically and exposure to household smoke[93,94]. Tonsillar SCC is 10 times more common in dogs living in urban versus rural areas, also implying an aetiological association with environmental pollutants[95]. Other reported malignant oral tumours in dogs include osteosarcoma, chondrosarcoma, anaplastic sarcoma, multilobular osteochondrosarcoma, intraosseous carcinoma, myxosarcoma, haemangiosarcoma, lymphoma, mast cell tumour and transmissible venereal tumour[31,49,96]. Benign tumours of the oral cavity include all forms of epulids, fibromas and extramedullary plasmacytomas (Figs 15.13A, B) and usually have a good prognosis following complete removal.

Most cats and dogs with oral cancer present with a mass in the mouth noticed by the owner (Fig. 15.14). Cancer in the caudal pharynx, however, is rarely seen by the owner, and the animal will present for signs of increased salivation, exophthalmos or facial swelling, epistaxis, weight loss, halitosis, bloody oral discharge, dysphagia or pain on opening the mouth or occasionally cervical lymphadenopathy[1–3,5–10,97–99].

The diagnostic evaluation for oral cancers is critical due to the wide ranges of cancer behaviour and therapeutic options available[1,3,8,10,11,49]. Under general anaesthesia careful palpation of the tumour, inspection of the oral cavity and pharynx, regional and distant imaging and biopsy of the tumour and local lymph nodes can be performed. The most likely cancers to metastasise visibly on thoracic radiographs at the time

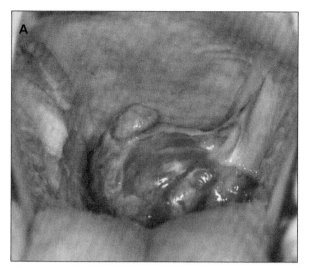

Figs 15.13A, B Tonsillar (A) and tongue (B) extramedullary plasmacytoma.

of diagnosis are melanoma and SCC of the caudal oral, pharyngeal and tonsillar area. CT or MRI can be a very valuable staging tool, especially for evaluation of possible tumour extension in the caudal pharynx, and is preferred to regional radiographs when available[3,100].

Regional lymph nodes, the mandibular, parotid and medial retropharyngeal, should be carefully palpated for enlargement or asymmetry[101–103]. Lymphoscintigraphy or contrast-enhanced ultrasonography can be used to detect the sentinel lymph nodes and guide lymph node aspirates[1,3,5,12,101,104]. Lymph node aspiration should be performed in all animals with oral tumours, regardless of the size or degree of fixation of the lymph nodes[13–15,103,105]. En-bloc resection of the regional lymph nodes provides valuable staging information[10,12,102,103,106]. The last step, under the same anaesthesia, is taking a large incisional biopsy of the actual tumour via the oral cavity[9,14,16,17,49].

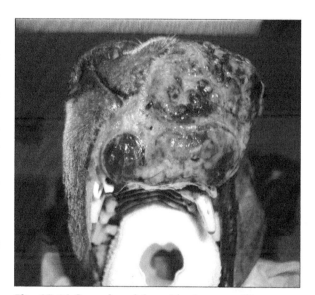

Fig. 15.14 Cross-breed dog with a large maxillary chondroblastic osteosarcoma.

Malignant melanoma

Malignant melanoma tends to occur in smaller dogs and the mean age at presentation is 11.4 years[13,14,16,18,85]. Melanoma of the oral cavity is a highly malignant, highly immunogenic tumour with frequent metastasis to the regional lymph nodes and then the lungs (up to 80% of dogs)[9,13–15,83,105,107–109]. The prognosis is therefore guarded with most dogs dying from metastasis to the lungs (14–67% of dogs)[19,83,86,110]. Surgery or radiation therapy can provide good local control, but chemotherapy or immunotherapy is indicated in the adjunctive management because of the high metastatic risk[9,13–15,19,49]. Tumour control and survival times are best when surgery is included in the treatment plan[3,14,20,111]. The median survival time (MST) for untreated dogs with oral melanoma is 65 days[7,112]. The MST for dogs treated with surgery alone varies from 150 to 318 days with 1-year survival rates less than

35%, and local recurrence rates varying from 22 to 48%[21,83,85,86,110,112,113].

Response rates to hypofractionated radiotherapy are excellent, with a complete response observed in up to 70% of melanomas[3,12,16,114–117]. The MST for dogs treated with radiotherapy is 211–363 days, with a 1-year survival rate of 36–48% and a 2-year survival rate of 21%[3,12,16,114–118]. Risk factors for poor outcome include non-rostral location, bone lysis and macroscopic disease; and in one series of 140 dogs with oral melanoma[2,3,5,10,13–15,118] the MST was 21 months if none of these risk factors were present, compared to a MST of 11 months with one risk factor, 5 months with two risk factors and 3 months with all three risk factors.

The location of malignant melanoma also has some prognostic significance. In one series of 60 dogs with oral melanomas at various sites treated with combinations of surgery, radiotherapy, chemotherapy and immunotherapy, the MST for dogs with lip melanomas (Fig. 15.15) was 580 days and was not reached and greater than 551 days for dogs with tongue melanomas, whereas the MST was 319 days for maxillary melanomas and 330 days for melanomas of the hard palate (Fig. 15.16)[19,83].

Squamous cell carcinoma

The prognosis for dogs with oral SCC is good, particularly for rostrally located tumours (Fig. 15.17). Local tumour control is the main objective, as metastasis to the regional lymph nodes is reported in only 10% of dogs and to the lungs in only 3–36% of dogs[22–26,119]. Surgery[27,87] and radiotherapy[28,29,120] can both be used for local control, though photodynamic therapy has also been reported[19,30,121]. Following mandibulectomy[19,110] or maxillectomy[31,122], the local recurrence rates are 10% and 29% and the MST varies from 19 to 26 months, with a 91% 1-year survival time and from 10 to 19 months with a 57% 1-year survival, respectively. The local tumour recurrence rate after full-course radiotherapy is 31%[19,119,120,123], with the MST for radiotherapy alone being 15–16 months and increasing to 34 months when combined with surgery[19,120,123]. Local tumour control was more successful with smaller lesions, rostral tumour location, maxillary SCC and young age[31,120,123]. Chemotherapy is indicated for dogs

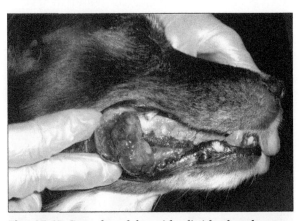

Fig. 15.15 Cross-breed dog with a lip/cheek melanoma.

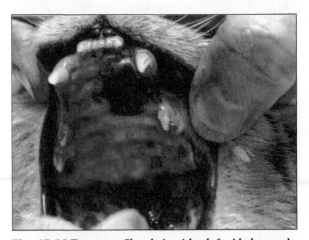

Fig. 15.16 European Shorthair with a left-sided rostral hard palate melanoma.

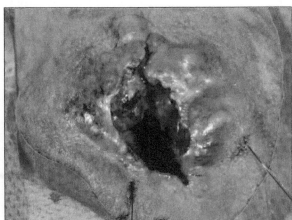

Fig. 15.17 Ulcerative squamous cell carcinoma of the tongue apex in a Flatcoated Retriever.

with metastatic disease, dogs with bulky disease and when owners decline surgery and radiotherapy, with a reported median progression-free interval for dogs of 180 days [19,124].

There is no known effective treatment for cats with oral SCC (Fig. 15.18) that consistently results in durable control or survival [19,88,99,125,126]. Local recurrence is the most challenging problem and the 1-year survival rates are less than 10% [19,99,127]. Radiotherapy alone is generally considered ineffective in the management of cats with oral SCC [19,128]. The combination of radiotherapy with radiation sensitisers or chemotherapy improves response rates and survival times, but results are still poor [19,129,130]. Chemotherapy appears ineffective in the management of cats with oral SCC [23,25,32,33,131], but the administration of a non-steroidal anti-inflammatory drug (NSAID) improved survival times in cats with oral SCC [34,35,127].

Fibrosarcoma

Oral fibrosarcoma tends to occur in large breed dogs (particularly Golden and Labrador Retrievers) at a younger age, with a median of 7.3–8.6 years and with a possible male predisposition [36,37,49,78,86,132]. Oral fibrosarcoma, especially on the hard palate (Fig. 15.19) and maxillary arcade between the canine and carnassial teeth, will often look benign histologically, but behaves biologically very malignantly [34,132]. Treatment should be locally aggressive therefore, as this type of fibrosarcoma is locally invasive but metastasises to the lungs and occasionally regional lymph nodes in less than 30% of dogs [38,39,78,86,97,98,110].

The prognosis for dogs with oral fibrosarcoma is guarded as local recurrence after surgery is common. Multimodality treatment of local disease (surgery with radiotherapy or radiotherapy with hyperthermia) appears to afford the best survival rates [33,40,133,134]. The median disease-free interval (DFI) for five cats treated with mandibulectomy was 859 days [33,39,41,42,133]. Following mandibulectomy or maxillectomy, local recurrence is reported in 40–59% of dogs with a MST of 11–12 months and a 1-year survival rate of 50% [33,38,43,110,122]. Oral fibrosarcomas are considered radiation resistant and reported mean survival times of 17 dogs treated with radiotherapy alone was only 7 months [43,135]. The combination of radiotherapy and surgery resulted in

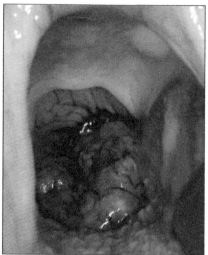

Fig. 15.18
Pharyngeal squamous cell carcinoma obstructing the rima glottis in a 12-year-old Domestic Shorthair cat.

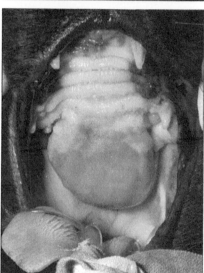

Fig. 15.19
Cat with a large hard palate fibrosarcoma.

local tumour recurrence in 32% of dogs and the MST increased to 18–26 months [43,119,136].

Tonsillar neoplasia

The most common primary tonsillar tumour is SCC (Figs 15.20A, B). However, lymphoma can affect the tonsils as well, but is often bilateral and usually accompanied by generalised lymphadenopathy [19,45,49,137]. Other cancers, especially malignant melanoma, can metastasise to the tonsils. Cervical lymphadenopathy, dysphagia, oropharyngeal pain, cough, lethargy and inappetence are common presenting signs, even with very small primary tonsillar cancers [46–48,137].

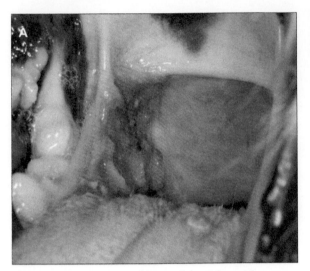

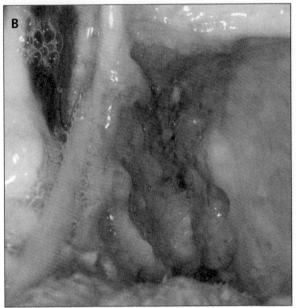

Figs 15.20A, B 9-year-old cross-breed dog with an invasive and metastasised tonsillar squamous cell carcinoma (A); B: close-up view.

Fine-needle aspirates of the regional lymph nodes or excisional biopsy of the tonsil will confirm the diagnosis. Thoracic radiographs are positive for metastasis in 10–20% of cases at presentation, but disease is considered systemic at diagnosis in over 90% of cats and dogs[49]. Simple tonsillectomy is almost never curative[49,86]. Cervical lymphadenectomy should be considered diagnostic only[46,49].

Regional radiotherapy of the pharyngeal region and cervical lymph nodes controls local disease in over 75% of the cases, but survival still remains poor with 1-year survival rates of only 10%[46,50,137,138]. Local tumour control and survival times are better when radiation is combined with a variety of different chemotherapy drugs[46,50–52,138]. Cisplatin, carboplatin, doxorubicin, vinblastine, and bleomycin have been used in different chemotherapy protocols with limited success[53,54,138,139]. In a recent large study of 44 dogs with tonsillar SCC treated with surgery, radiotherapy and/or chemotherapy, the MST was 179 days and dogs presenting with either anorexia or lethargy had a significantly shorter survival time[137].

15.7 OROPHARYNGEAL DYSPHAGIA

Swallowing, or deglutition, is a complex physiological phenomenon divided into four phases, the oral prepar-atory and the oral phase are voluntary, the pharyngeal and oesophageal phase are involuntary (see Chapter 13 Throat Anatomy and Physiology, Section 13.3). Dysphagia refers to abnormal or difficulty in swallow-ing and can be secondary to an isolated lesion or be a manifestation of systemic disease affecting any of these phases. Oropharyngeal dysphagia, indicating an inabil-ity to move a bolus from the oral cavity into the proxi-mal oesophagus, will be discussed in this chapter.

Oropharyngeal dysphagia

Causes of oropharyngeal dysphagia include:
- Anatomic abnormalities.
- Neoplasia (local or brainstem tumours).
- Foreign bodies.
- Pharyngeal pain.
- Systemic disease (rabies, distemper, immune-mediated disorders).
- Neuromuscular disorders[19,31,57].

Central neurogenic lesions causing dysphagia are well known in humans and include cerebral or brain-stem ischaemic or haemorrhagic strokes, Alzheimer's disease, Parkinson's disease, traumatic brain injury and motor neuron disease[19,49,140]. Both brainstem and cer-ebral neoplasms occur in dogs and cats, and dysphagia may be one of the signs[19,55,56]. Local peripharyngeal masses or metastatic disease in the retropharyngeal lymph nodes can cause significant obstruction of the pharyngeal passageway and hence dysphagia[19,49,58,59]. In humans systemic lupus erythematosus, dermatomyositis and Sjögren's syndrome as well as benign mucous membrane pemphigoid and polymyositis have been shown to cause dysphagia[59,60,141–143]; it is likely these diseases also cause dysphagia in dogs.

The most common neuromuscular disorder is myasthenia gravis, a disease of altered neuromuscular transmission secondary to autoantibodies against acetylcholine receptors[49,144]. Manifestations of myasthenia gravis may either be focal (pharyngeal or oesophageal) or generalised (generalised muscular weakness or collapse), and congenital or acquired. Akitas, Scottish Terriers, German Shorthaired Pointers, Chihuahuas and Abyssinians are all breeds identified at increased risk for acquired myasthenia gravis[59,145,146]. In cats megaoesophagus may be a prominent sign of myasthenia gravis[49,59,146] and in dogs myasthenia gravis is recognised as one of the causes of megaoesophagus[19,147].

Neurogenic causes of dysphagia can usually be recognised because the resulting muscular dysfunction also involves muscle groups other than those involved in swallowing. The importance of intact sensory and motor innervation for swallowing has been convincingly demonstrated. Peripheral neurogenic dysphagia caused by loss of sensory innervation (transection of the cranial laryngeal nerve) was studied in 10 dogs[61–63,148,149]. Whereas dysphagia was expected, dogs continued to eat normally after unilateral transection. Bilateral transection appeared to cause a loss of normal triggering of swallowing and possibly allowed leakage of food and saliva into the trachea[64,148].

Peripheral neurogenic dysphagia caused by loss of motor innervation by unilateral or bilateral transection of the glossopharyngeal nerve (IX) or the pharyngeal branch of the vagus nerve (Xph) was studied in 10 dogs[150]. In 9 of the 10 dogs there were no outward signs of dysphagia after unilateral or bilateral transection of either the IX or the Xph nerves. Although dysphagia as demonstrated with contrast videofluorography ranged from absent to severe in these 10 dogs, there was no acute and life-threatening leakage of food and fluid into the trachea[151].

Muscular disorders causing dysphagia include muscular dystrophy, myositis and myogenic disease of unknown origin[19,65,152,153]. Oropharyngeal dysphagia can be one of many clinical signs that are associated with myopathies. The Bouvier des Flandres has been identified with a familial form of muscular dystrophy in which oropharyngeal dysphagia is the primary clinical sign[62,63,152,153]. Cricopharyngeal achalasia can be acquired as a polymyopathy or it can be congenital[61–63,154–156]. Cricopharyngeal achalasia is the failure of the upper oesophageal sphincter to relax and allow a food bolus into the proximal oesophagus. Cocker Spaniels are at increased risk for congenital cricopharyngeal achalasia and usually present at a young age[65,157], but any breed can be affected. Golden Retrievers were over-represented in some studies[61,62,158,159].

Clinical signs of oropharyngeal dysphagia are variable, depending on severity and location of the problem. Signs include:

- Difficulty with prehending food.
- Dropping food from mouth.
- Exaggerated head movements during eating.
- Regurgitation with no correlation with time of eating.
- Repeated attempts to swallow food.
- Occasionally, respiratory signs such as nasal discharge, coughing and aspiration pneumonia will occur secondary to nasopharyngeal or laryngotracheal aspiration[31,66,156].

Accurate historical information is imperative to differentiate regurgitation from vomiting and to help identify the phase of swallowing affected. A standardised questionnaire has been evaluated as a sensitive tool to detect oropharyngeal dysphagia[65,160]. Observing the animal eating and drinking is also extremely helpful.

Diagnosis of oropharyngeal dysphagia requires a thorough evaluation of the entire patient. The physical examination should include a general examination, examination of the respiratory and digestive systems but also a complete neurological examination. Radiographs of the throat and neck can identify radiopaque foreign bodies or masses. Thoracic radiographs can determine if the animal has aspiration pneumonia secondary to the dysphagia and can rule out the presence of a megaoesophagus. Videofluoroscopy is an excellent diagnostic tool to localise the dysphagia[63,152,161]. In a study in 24 Bouviers with muscular dystrophy, in seven dogs the cricopharyngeal muscle did not relax following pharyngeal contractions, whereas in one dog it was permanently relaxed. In five dogs the pharyngeal contractions were very ineffective and uncoordinated, resulting in accumulation of food in the pharyngeal cavity. In 12 dogs there was little or no peristalsis in the cervical part of the oesophagus. In three dogs peristalsis in the thoracic part of the oesophagus was decreased[67,152]. In six dogs more than one stage of the swallowing action was abnormal.

Other possible diagnostics could include antinuclear antibody (ANA) test, evaluation for acetylcholine receptor antibodies, endocrine screening, electromyography (EMG) or biopsy for histopathology[31,68,69,156]. In the study on 24 Bouviers with muscular dystrophy, recordings of spontaneous EMG activity of the tongue, the soft palate, the hyopharyngeal, thyropharyngeal and cricopharyngeal muscles, and the cervical oesophagus revealed a variety of abnormalities, such as fibrillation potentials, positive sharp waves, continuous potentials and complex repetitive discharges (see Chapter 14.6)[70,71,152].

Treatment of oropharyngeal dysphagia should be based on the underlying disease process if present. If no underlying cause can be identified or removed, providing adequate nutrition is necessary. Most patients improve when being fed from an elevation and sitting while eating, whereas others may need gastrotomy tube placement for temporary stabilisation. In severe cases with secondary aspiration pneumonia euthanasia may be indicated.

For congenital cricopharyngeal achalasia the treatment of choice is a cricopharyngeal myotomy or myectomy. It is critical that an accurate diagnosis is made as a myectomy in patients with pharyngeal dysphagia will lead to worsening of the dysphagia and aspiration pneumonia[71,153–155,162,163]. A very poor outcome after surgical treatment in patients with underlying disease has also been reported[66,161,164–166]. The procedure should be performed unilaterally as the muscle forms a band making up the upper oesophageal sphincter. By resecting the band the asynchronous contraction is stopped, thus relieving the dysphagia[66,156]. Some surgeons advocate removal of a 1 cm portion of the cricopharyngeal muscle[61–63,156,167]. This is thought to prevent adhesion formation between the resected ends and reformation of the band by fibrous tissue joining up the ends of the resected muscle. Others simply perform a myotomy[61,158,168]. No differences in outcome between

myotomy or myectomy have been found[62,158]. True achalasia with no underlying disease or concurrent oesophageal hypomotility carries an excellent prognosis after myotomy or myectomy[62,156,167].

15.8 REFERENCES

1 Spangler WL, Culbertson MR. Salivary gland disease in dogs and cats: 245 cases (1985–1988). *Journal of the American Veterinary Medical Association* 1991;**198**:465–9.

2 Brown NO. Salivary gland diseases. Diagnosis, treatment, and associated problems. *Problems in Veterinary Medicine* 1989;**1**:281–94.

3 Waldron DR, Smith MM. Salivary mucoceles. *Problems in Veterinary Medicine* 1991;**3**:270–6.

4 Ritter MJ, Pfeil von DJ, Stanley BJ, Hauptman JG, Walshaw R. Mandibular and sublingual sialocoeles in the dog: a retrospective evaluation of 41 cases, using the ventral approach for treatment. *New Zealand Veterinary Journal* 2006;**54**:333–7.

5 Peeters ME. The treatment of salivary cysts in dogs and cats. *Tijdschrift voor Diergeneeskunde* 1991;**116**:169–72.

6 Harrison JD, Garrett JR. An ultrastructural and histochemical study of a naturally occurring salivary mucocele in a cat. *Journal of Comparative Pathology* 1975;**85**:411–16.

7 Torad FA, Hassan EA. Clinical and ultrasonographic characteristics of salivary mucoceles in 13 dogs. *Veterinary Radiology & Ultrasound* 2013;**54**:293–8.

8 Marsh A, Adin C. Tunneling under the digastricus muscle increases salivary duct exposure and completeness of excision in mandibular and sublingual sialoadenectomy in dogs. *Veterinary Surgery* 2013;**42**:238–42.

9 Feinman JM. Pharyngeal mucocele and respiratory distress in a cat. *Journal of the American Veterinary Medical Association* 1990;**197**:1179–80.

10 Ritter MJ, Stanley BJ. Salivary Glands. In: *Veterinary Surgery Small Animal*. Tobias KM, Johnston SA (eds). Elsevier Saunders, St. Louis 2012, pp. 1439–47.

11 Glen JB. Canine salivary mucocoeles. Results of sialographic examination and surgical treatment of fifty cases. *Journal of Small Animal Practice* 1972;**13**:515–26.

12 Hulland TJ, Archibald J. Salivary mucoceles in dogs. *Canadian Veterinary Journal* 1964;5:109–17.

13 Harvey HJ. Pharyngeal mucoceles in dogs. *Journal of the American Veterinary Medical Association* 1981;**178**:1282–3.

14 Weber WJ, Hobson HP, Wilson SR. Pharyngeal mucoceles in dogs. *Veterinary Surgery* 1986;15:5–8.

15 Dietens A, Spanoghe I, Paepe D, *et al.* Faryngeale sialocele bij een hond. *Vlaams Diergeneeskundig Tijdschrift* 2011;**80**:233–9.

16 Bellenger CR, Simpson DJ. Canine sialocoeles – 60 clinical cases. *Journal of Small Animal Practice* 1992;**33**:376–80.

17 Knecht CD, Phares J. Characterization of dogs with salivary cyst. *Journal of the American Veterinary Medical Association* 1971;**158**:612–3.

18 Glen JB. Salivary cysts in the dog: identification of sub-lingual duct defects by sialography. *Veterinary Record* 1966;**78**:488–92.

19 Venker-van Haagen AJ. The pharynx. In: *Ear, Nose, Throat, and Tracheobronchial Diseases in Dogs and Cats*. Venker-van Haagen AJ (ed). Schlütersche Verlagsgesellschaft, Hannover 2005, pp. 83–120.

20 Harvey CE. Sialography in the dog. *Veterinary Radiology & Ultrasound* 2005;**10**:18–27.

21 McGill S, Lester N, McLachlan A, Mansfield C. Concurrent sialocoele and necrotising sialadenitis in a dog. *Journal of Small Animal Practice* 2009;**50**:151–6.

22 Sykes JE, Anderson GA, Studdert VP, Browning GF. Prevalence of feline *Chlamydia psittaci* and feline herpesvirus 1 in cats with upper respiratory tract disease. *Journal of Veterinary Internal Medicine* 1999;**13**:153–62.

23 Thompson RR, Wilcox GE, Clark WT. Association of calicivirus infection with chronic gingivitis and pharyngitis in cats. *Journal of Small Animal Practice* 1984;**25**:207–10.

24 White SD, Rosychuk RA, Janik TA. Plasma cell stomatitis-pharyngitis in cats: 40 cases (1973–1991). *Journal of the American Medical Veterinary Association* 1992;**200**:1377–80.

25 Reubel GH, Hoffmann DE. Acute and chronic faucitis of domestic cats. A feline calicivirus-induced disease. *Veterinary Clinics of North America Small Animal Practice* 1992;**22**:1347–60.

26 Healey KAE, Dawson S, Burrow R, *et al.* Prevalence of feline chronic gingivo-stomatitis in first opinion veterinary practice. *Journal of Feline Medicine and Surgery* 2007;**9**:373–81.

27 Pinter L, Noble WC. Stomatitis, pharyngitis and tonsillitis caused by *Malassezia pachydermatis* in a dog. *Veterinary Dermatology* 1998;**9**:257–61.

28 Smith JE. The aerobic bacteria of the nose and tonsils of healthy dogs. *Journal of Comparative Pathology and Therapeutics* 1961;**71**:428–33.

29 Baldrias L, Frost AJ, O'Boyle D. The isolation of *Pasteurella*-like organisms from the tonsillar region of dogs and cats. *Journal of Small Animal Practice* 1988;**29**:63–8.

30 Pilot I, Buck C, Davis DJ, Eastman DA. Tonsillitis in dogs due to hemolytic streptococci. *Experimental Biology and Medicine* 1936;**34**:499–502.

31 Parnell N. Diseases of the throat. In: *Textbook of Veterinary Internal Medicine*, 7th edn. Ettinger SJ, Feldman EC (eds). Saunders Elsevier, St. Louis 2010, pp. 1040–6.

32 Baird K. Lymphoplasmacytic gingivitis in a cat. *Canadian Veterinary Journal* 2005;**46**:530–2.

33 Farcas N, Lommer MJ, Kass PH, Verstraete FJM. Dental radiographic findings in cats with chronic gingivostomatitis (2002–2012). *Journal of the American Veterinary Medical Association* 2014;**244**:339–45.

34 Harley R, Gruffydd-Jones TJ, Day MJ. Immu-nohistochemical characterization of oral mucosal lesions in cats with chronic gingivostomatitis. *Journal of Comparative Pathology* 2011;**144**:239–50.

35 Knowles JO, Gaskell RM, Gaskell CJ, Harvey CE, Lutz H. Prevalence of feline calicivirus, feline leukaemia virus and antibodies to FIV in cats with chronic stomatitis. *Veterinary Record* 1989;**124**:336–8.

36 Sims TJ, Moncla BJ, Page RC. Serum antibody response to antigens of oral gram-negative bacteria by cats with plasma cell gingivitis-pharyngitis. *Journal of Dental Research* 1990;**69**:877–82.

37 Dolieslager SMJ, Riggio MP, Lennon A, *et al.* Identification of bacteria associated with feline chronic gingivostomatitis using culture-dependent and culture-independent methods. *Veterinary Microbiology* 2011;**148**:93–8.

38 Hennet PR, Camy GAL, McGahie DM, Albouy MV. Comparative efficacy of a recombinant feline interferon omega in refractory cases of calicivirus-positive cats with caudal stomatitis: a randomised, multi-centre, controlled, double-blind study in 39 cats. *Journal of Feline Medicine and Surgery* 2011;**13**:577–87.

39 Gorrel C, Larsson A. Feline odontoclastic resorptive lesions: unveiling the early lesion. *Journal of Small Animal Practice* 2002;**43**:482–8.

40 Diehl K, Rosychuk RA. Feline gingivitis-stomatitis-pharyngitis. *Veterinary Clinics of North America Small Animal Practice* 1993;**23**:139–53.

41 Lommer MJ, Verstraete FJM. Prevalence of odontoclastic resorption lesions and periapical radiographic lucencies in cats: 265 cases (1995–1998). *Journal of the American Veterinary Medical Association* 2000;**217**:1866–9.

42 Lommer MJ, Verstraete FJM. Radiographic patterns of periodontitis in cats: 147 cases (1998–1999). *Journal of the American Veterinary Medical Association* 2001;**218**:230–4.

43 Hennet P. Chronic gingivo-stomatitis in cats: long-term follow-up of 30 cases treated by dental extractions. *Journal of Veterinary Dentistry* 1997;**14**:15–21.

44 Lyon KF. Gingivostomatitis. *Veterinary Clinics of North America Small Animal Practice* 2005;**35**:891–911.

45 Corbee RJ, Booij-Vrieling HE, van de Lest CHA, *et al*. Inflammation and wound healing in cats with chronic gingivitis/stomatitis after extraction of all premolars and molars were not affected by feeding of two diets with different omega-6/omega-3 polyunsaturated fatty acid ratios. *Journal of Animal Physiology and Animal Nutrition* 2011;**96**:671–80.

46 Yağci BB, Ural K, Öcal N. Azithromycin therapy of papillomatosis in dogs: a prospective, randomized, double–blinded, placebo–controlled clinical trial. *Veterinary Dermatology* 2008;**19**:194–8.

47 Sundberg JP, O'Banion MK, Schmidt-Didier E, Reichmann ME. Cloning and characterization of a canine oral papillomavirus. *American Journal of Veterinary Research* 1986;**47**:1142–4.

48 Sundberg JP, Smith EK, Herron AJ, Jenson AB, Burk RD, Van Ranst M. Involvement of canine oral papillomavirus in generalized oral and cutaneous verrucosis in a Chinese Shar Pei dog. *Veterinary Pathology* 1994;**31**:183–7.

49 Liptak JM, Withrow SJ. Cancer of the gastrointestinal tract. In: *Small Animal Clinical Oncology*, 5th edn. Withrow SJ, Vail DM, Page RL (eds). Elsevier Saunders, St. Louis 2013, pp. 381–98.

50 Derkay CS, Malis DJ, Zalzal G, Wiatrak BJ, Kashima HK, Coltrera MD. A staging system for assessing severity of disease and response to therapy in recurrent respiratory papillomatosis. *Laryngoscope* 1998;**108**:935–7.

51 Dubielzig RR. Proliferative dental and gingival diseases of dogs and cats. *Journal American Animal Hospital Association* 1982;**18**:577–84.

52 Bregman CL, Hirth RS, Sundberg JP, Christensen EF. Cutaneous neoplasms in dogs associated with canine oral papillomavirus vaccine. *Veterinary Pathology* 1987;**24**:477–87.

53 Madewell BR, Stannard AA, Pulley LT, Nelson VG. Oral eosinophilic granuloma in Siberian husky dogs. *Journal of the American Veterinary Medical Association* 1980;**177**:701–3.

54 Potter KA, Tucker RD, Carpenter JL. Oral eosinophilic granuloma of Siberian huskies. *Journal American Animal Hospital Association* 1980;**16**:595–600.

55 German AJ, Holden DJ, Hall EJ, Day MJ. Eosinophilic diseases in two Cavalier King Charles spaniels. *Journal of Small Animal Practice* 2002;**43**:533–8.

56 Bredal WP, Gunnes G, Vollset I, Ulstein TL. Oral eosinophilic granuloma in three Cavalier King Charles spaniels. *Journal of Small Animal Practice* 1996;**37**:499–504.

57 Joffe DJ, Allen AL. Ulcerative eosinophilic stomatitis in three Cavalier King Charles spaniels. *Journal American Animal Hospital Association* 1995;**31**:34–7.

58 Bryan J, Frank LA. Food allergy in the cat: a diagnosis by elimination. *Journal of Feline Medicine and Surgery* 2010;**12**:861–6.

59 Buckley L, Nuttall T. Feline eosinophilic granuloma complex(ities): some clinical clarification. *Journal of Feline Medicine and Surgery* 2012;**14**:471–81.

60 Roosje PJ, Willemse T. Cytophilic antibodies in cats with miliary dermatitis and eosinophilic plaques: passive transfer of immediate-type hypersensitivity. *Veterinary Quarterly* 1995;**17**:66–9.

61 Doran IP, Wright CA, Moore AH. Acute oropharyngeal and esophageal stick injury in forty-one dogs. *Veterinary Surgery* 2008;**37**:781–5.

62 Griffiths LG, Tiruneh R, Sullivan M, Reid SW. Oropharyngeal penetrating injuries in 50 dogs: a retrospective study. *Veterinary Surgery* 2000;**29**:383–8.

63 White R, Lane JG. Pharyngeal stick penetration injuries in the dog. *Journal of Small Animal Practice* 1988;**29**:13–35.

64 Bright SR, Mellanby RJ, Williams JM. Oropharyngeal stick injury in a Bengal cat. *Journal of Feline Medicine and Surgery* 2002;**4**:153–5.

65 Anderson GM. Soft tissue of the oral cavity. In: *Veterinary Surgery Small Animal*. Tobias KM, Johnston SA (eds). Elsevier Saunders, St. Louis 2012, pp. 1425–38.

66 Peeters ME. The management of pharyngeal stick penetrating injuries in dogs. *European Veterinary Conference: The Voorjaarsdagen* 2010, pp. 1–2.

67 Lamb CR, White RN, McEvoy FJ. Sinography in the investigation of draining tracts in small animals: retrospective review of 25 cases. *Veterinary Surgery* 1994;**23**:129–34.

68 Armbrust LJ, Biller DS, Radlinsky MG, Hoskinson JJ. Ultrasonographic diagnosis of foreign bodies associated with chronic draining tracts and abscesses in dogs. *Veterinary Radiology & Ultrasound* 2003;**44**:66–70.

69 Staudte KL, Hopper BJ, Gibson NR, Read RA. Use of ultrasonography to facilitate surgical removal of non-enteric foreign bodies in 17 dogs. *Journal of Small Animal Practice* 2004;**45**:395–400.

70 Dobromylskyj MJ, Dennis R, Ladlow JF, Adams VJ. The use of magnetic resonance imaging in the management of pharyngeal penetration injuries in dogs. *Journal of Small Animal Practice* 2008;**49**:74–9.

71 Nicholson I, Halfacree Z, Whatmough C, Mantis P, Baines S. Computed tomography as an aid to management of chronic oropharyngeal stick injury in the dog. *Journal of Small Animal Practice* 2008;**49**:451–7.

72 Whitney WO, Mehlhaff CJ. High-rise syndrome in cats. *Journal of the American Veterinary Medical Association* 1987;**191**:1399–403.

73 Reiter AM, Holt DE. Palate. In: *Veterinary Surgery Small Animal*. Tobias KM, Johnston SA (eds). Elsevier Saunders, St. Louis 2012, pp. 1707–17.

74 Matis U, Koestlin R. Symphyseal separation and fractures involving the incisive region. In: *Oral and Maxillofacial Surgery in Dogs and Cats*. Verstraete FJM, Lommer MJ (eds). Saunders/Elsevier, St. Louis 2012, pp. 265–73.

75 Marretta SM. Maxillofacial fracture complications. In: *Oral and Maxillofacial Surgery in Dogs and Cats*. Verstraete FJM, Lommer MJ (eds). Saunders/Elsevier, St. Louis 2012, pp.333–41.

76 Hoyt RF, Withrow SJ. Oral malignancy in the dog. *Journal of the American Animal Hospital Association* 1984;**20**:83.

77 Brønden LB, Eriksen T. Oral malignant melanomas and other head and neck neoplasms in Danish dogs – data from the Danish Veterinary Cancer Registry. *Acta Veterinaria Scandinavica* 2009;**51**:54–60.

78 Vos JH, Gaag I. Canine and feline oral-pharyngeal tumours. *Journal of Veterinary Medicine* 1987;**34**:420–7.

79 Stebbins KE, Morse CC, Goldschmidt MH. Feline oral neoplasia: a ten-year survey. *Veterinary Pathology* 1989;**26**:121–8.

80 Dorn CR, Taylor D, Frye FL. Survey of animal neoplasms in Alameda and Contra Costa Counties, California. I. Methodology and description of cases. *Journal of the National Cancer Institute* 1968;**40**:295–306.

81 Dorn CR, Taylor D, Schneider R. Survey of animal neoplasms in Alameda and Contra Costa Counties, California. II. Cancer morbidity in dogs and cats from Alameda County. *Journal of the National Cancer Institute* 1968;**40**:307–18.

82 Cohen D, Brodey RS, Chen SM. Epidemiologic aspects of oral and pharyngeal neoplasms of the dog. *American Journal of Veterinary Research* 1964, **25**: 1776–1779.

83 Kudnig ST, Ehrhart N, Withrow SJ. Survival analysis of oral melanoma in dogs. *Veterinary Cancer Society Proceedings* 2003;**23**:39.

84 Dorn CR, Priester WA. Epidemiologic analysis of oral and pharyngeal cancer in dogs, cats, horses, and cattle. *Journal of the American Veterinary Medical Association* 1976;**169**:1202–6.

85 Ramos-Vara JA, Beissenherz ME, Miller MA, *et al*. Retrospective study of 338 canine oral melanomas with clinical, histologic, and immunohistochemical review of 129 cases. *Veterinary Pathology* 2000;**37**:597–608.

86 Todoroff RJ, Brodey RS. Oral and pharyngeal neoplasia in the dog: a retrospective survey of 361 cases. *Journal of the American Veterinary Medical Association* 1979;**175**:567–71.

87 Withrow SJ, Holmberg DL. Mandibulectomy in the treatment of oral cancer. *Journal of the American Animal Hospital Association* 1983;**19**:273–6.

88 Bradley RL, MacEwen EG, Loar AS. Mandibular resection for removal of oral tumors in 30 dogs and 6 cats. *Journal of the American Veterinary Medical Association* 1984;**184**:460–3.

89 Salisbury SK. Problems and complications associated with maxillectomy, mandibulectomy, and oronasal fistula repair. *Problems in Veterinary Medicine* 1991;**3**:153–69.

90 Lantz GC, Salisbury SK. Partial mandibulectomy for treatment of mandibular fractures in dogs: eight cases (1981–1984). *Journal of the American Veterinary Medical Association* 1987;**191**:243–5.

91 Salisbury SK, Lantz GC. Long-term results of partial mandibulectomy for treatment of oral tumors in 30 dogs. *Journal of the American Animal Hospital Association* 1988;**24**:285–8.

92 White R. Mandibulectomy and maxillectomy in the dog: long term survival in 100 cases. *Journal of Small Animal Practice* 1991;**32**:69–74.

93 Bertone ER, Snyder LA, Moore AS. Environmental and lifestyle risk factors for oral squamous cell carcinoma in domestic cats. *Journal of Veterinary Internal Medicine* 2003;**17**:557–62.

94 Snyder LA, Bertone ER, Jakowski RM, Dooner MS, Jennings-Ritchie J, Moore AS. p53 expression and environmental tobacco smoke exposure in feline oral squamous cell carcinoma. *Veterinary Pathology* 2004;**41**:209–14.

95 Reif JS, Cohen D. The environmental distribution of canine respiratory tract neoplasms. *Archives of Environmental Health* 1971;**22**:136–40.

96 Dernell WS, Straw RC, Cooper MF, Powers BE, LaRue SM, Withrow SJ. Multilobular osteochondrosarcoma in 39 dogs: 1979–1993. *Journal of the American Animal Hospital Association* 1998;**34**:11–18.

97 Schwarz PD, Withrow SJ, Curtis CR, Powers BE, Straw RC. Mandibular resection as a treatment for oral cancer in 81 dogs. *Journal of the American Animal Hospital Association* 1991;**27**:601–6.

98 Schwarz PD, Withrow SJ, Curtis CR, Powers BE, Straw RC. Partial maxillary resection as a treatment for oral cancer in 61 dogs. *Journal of the American Animal Hospital Association* 1991;**27**:617–21.

99 Reeves NP, Turrel JM, Withrow SJ. Oral squamous cell carcinoma in the cat. *Journal of the American Animal Hospital Association* 1993;**29**:438–41.

100 Gendler A, Lewis JR, Reetz JA, Schwarz T. Computed tomographic features of oral squamous cell carcinoma in cats: 18 cases (2002–2008). *Journal of the American Veterinary Medical Association* 2010;**236**:319–25.

101 Glen JB. Canine salivary mucocoeles. Results of sialographic examination and surgical treatment of fifty cases. *Journal of Small Animal Practice* 1972;**13**:515–26.

102 Smith MM. Surgical approach for lymph node staging of oral and maxillofacial neoplasms in dogs. *Journal of the American Animal Hospital Association* 1995;**31**:514–18.

103 Herring ES, Smith MM, Robertson JL. Lymph node staging of oral and maxillofacial neoplasms in 31 dogs and cats. *Journal of Veterinary Dentistry* 2002;**19**:122–6.

104 Lurie DM, Seguin B, Schneider PD, Verstraete FJ, Wisner ER. Contrast-assisted ultrasound for sentinel lymph node detection in spontaneously arising canine head and neck tumors. *Investigative Radiology* 2006;**41**:415–21.

105 Williams LE, Packer RA. Association between lymph node size and metastasis in dogs with oral malignant melanoma: 100 cases (1987–2001). *Journal of the American Veterinary Medical Association* 2003;**222**:1234–6.

106 Bellenger CR, Simpson DJ. Canine sialocoeles – 60 clinical cases. *Journal of Small Animal Practice* 1992;**33**:376–80.

107 Dow SW, Elmslie RE, Willson AP, Roche L, Gorman C, Potter TA. *In vivo* tumor transfection with superantigen plus cytokine genes induces tumor regression and prolongs survival in dogs with malignant melanoma. *Journal of Clinical Investigation* 1998;**101**:2406–14.

108 Bergman PJ. Canine oral melanoma. *Clinical Techniques in Small Animal Practice* 2007;**22**:55–60.

109 Bergman PJ, McKnight J, Novosad A, *et al*. Long-term survival of dogs with advanced malignant melanoma after DNA vaccination with xenogeneic human tyrosinase: a phase I trial. *Clinical Cancer Research* 2003;**9**:1284–90.

110 Kosovsky JK, matthiesen DT, Marretta SM, Patnaik AK. Results of partial mandibulectomy for the treatment of oral tumors in 142 dogs. *Veterinary Surgery* 1991;**20**:397–401.

111 Hahn KA, DeNicola DB, Richardson RC, Hahn EA. Canine oral malignant melanoma: prognostic utility of an alternative staging system. *Journal of Small Animal Practice* 1994;**35**:251–6.

112 Harvey HJ, MacEwen EG, Braun D, Patnaik AK, Withrow SJ, Jongeward S. Prognostic criteria for dogs with oral melanoma. *Journal of the American Veterinary Medical Association* 1981;**178**:580–2.

113 MacEwen EG, Kurzman ID, Vail DM, *et al*. Adjuvant therapy for melanoma in dogs: results of randomized clinical trials using surgery, liposome-encapsulated muramyl tripeptide, and granulocyte macrophage colony-stimulating factor. *Clinical Cancer Research* 1999;**5**:4249–58.

114 Freeman KP, Hahn KA, Harris FD, King GK. Treatment of dogs with oral melanoma by hypofractionated radiation therapy and platinum-based chemotherapy (1987–1997). *Journal of Veterinary Internal Medicine* 2003;**17**:96–101.

115 Bateman KE, Catton PA, Pennock PW, Kruth SA. Radiation therapy for the treatment of canine oral melanoma. *Journal of Veterinary Internal Medicine* 1994;**8**:267–72.

116 Blackwood L, Dobson JM. Radiotherapy of oral malignant melanomas in dogs. *Journal of the American Veterinary Medical Association* 1996;**209**:98–102.

117 Theon AP, Rodriguez C, Madewell BR. Analysis of prognostic factors and patterns of failure in dogs with malignant oral tumors treated with megavoltage irradiation. *Journal of the American Veterinary Medical Association* 1997;**210**:778–84.

118 Proulx DR, Ruslander DM, Dodge RK, *et al*. A retrospective analysis of 140 dogs with oral melanoma treated with external beam radiation. *Veterinary Radiology & Ultrasound* 2003;**44**:352–9.

119 Theon AP, Rodriguez C, Griffey S, Madewell BR. Analysis of prognostic factors and patterns of failure in dogs with periodontal tumors treated with megavoltage irradiation. *Journal of the American Veterinary Medical Association* 1997;**210**:785–8.

120 Evans SM, Shofer F. Canine oral nontonsillar squamous cell carcinoma. *Veterinary Radiology & Ultrasound* 1988;**29**:133–7.

121 McCaw DL, Pope ER, Payne JT, West MK, Tompson RV, Tate D. Treatment of canine oral squamous cell carcinomas with photodynamic therapy. *British Journal of Cancer* 2000;**82**:1297–9.

122 Wallace J, Matthiesen DT, Patnaik AK. Hemimaxillectomy for the treatment of oral tumors in 69 dogs. *Veterinary Surgery* 1992;**21**:337–41.

123 Miller TL, Price GS, Page RL. Radiotherapy of canine non-tonsillar squamous cell carcinoma. *Veterinary Radiology & Ultrasound* 1996;**37**:74–7.

124 Schmidt BR, Glickman NW, DeNicola DB, de Gortari AE, Knapp DW. Evaluation of piroxicam for the treatment of oral squamous cell carcinoma in dogs. *Journal of the American Veterinary Medical Association* 2001;**218**:1783–6.

125 Bostock DE. The prognosis in cats bearing squamous cell carcinoma. *Journal of Small Animal Practice* 1972;**13**:119–25.

126 Cotter SM. Oral pharyngeal neoplasms in the cat. *Journal of the American Animal Hospital Association* 1981;**17**:917–20.

127 Hayes AM, Adams VJ, Scase TJ, Murphy S. Survival of 54 cats with oral squamous cell carcinoma in United Kingdom general practice. *Journal of Small Animal Practice* 2007;**48**:394–9.

128 Fidel JL, Sellon RK, Houston RK, Wheeler BA. A nine-day accelerated radiation protocol for feline squamous cell carcinoma. *Veterinary Radiology & Ultrasound* 2007;**48**:482–5.

129 Fidel J, Lyons J, Tripp C, Houston R, Wheeler B, Ruiz A. Treatment of oral squamous cell carcinoma with accelerated radiation therapy and concomitant carboplatin in cats. *Journal of Veterinary Internal Medicine* 2011;**25**:504–10.

130 Evans SM, LaCreta F, Helfand S, *et al.* Technique, pharmacokinetics, toxicity, and efficacy of intratumoral etanidazole and radiotherapy for treatment of spontaneous feline oral squamous cell carcinoma. *International Journal for Radiatiation Oncology, Biology and Physics* 1991;**20**:703–8.

131 DiBernardi L, Doré M, Davis JA, *et al.* Study of feline oral squamous cell carcinoma: potential target for cyclooxygenase inhibitor treatment. *Prostaglandins, Leukotrienes and Essential Fatty Acids* 2007;**76**:245–50.

132 Ciekot PA, Powers BE, Withrow SJ, Straw RC, Ogilvie GK, LaRue SM. Histologically low-grade, yet biologically high-grade, fibrosarcomas of the mandible and maxilla in dogs: 25 cases (1982–1991). *Journal of the American Veterinary Medical Association* 1994;**204**:610–15.

133 Northrup NC, Selting KA, Rassnick KM, *et al.* Outcomes of cats with oral tumors treated with mandibulectomy: 42 cases. *Journal of the American Animal Hospital Association* 2006;**42**:350–60.

134 Brewer WG, Turrel JM. Radiotherapy and hyperthermia in the treatment of fibrosarcomas in the dog. *Journal of the American Veterinary Medical Association* 1982;**181**:146–50.

135 Thrall DE. Orthovoltage radiotherapy of oral fibrosarcomas in dogs. *Journal of the American Veterinary Medical Association* 1981;**179**:159–62.

136 Forrest LJ, Chun R, Adams WM, Cooley AJ, Vail DM. Postoperative radiotherapy for canine soft tissue sarcoma. *Journal of Veterinary Internal Medicine* 2000;**14**:578–82.

137 Mas A, Blackwood L, Cripps P, *et al.* Canine tonsillar squamous cell carcinoma – a multi-centre retrospective review of 44 clinical cases. *Journal of Small Animal Practice* 2011;**52**:359–64.

138 Brooks MB, Matus RE, Leifer CE, Alfieri AA, Patnaik AK. Chemotherapy versus chemotherapy plus radiotherapy in the treatment of tonsillar squamous cell carcinoma in the dog. *Journal of Veterinary Internal Medicine* 1988;**2**:206–11.

139 Buhles WC, Theilen GH. Preliminary evaluation of bleomycin in feline and canine squamous cell carcinoma. *American Journal of Veterinary Research* 1973;**34**:289–91.

140 Buchholz DW. Neurogenic dysphagia: what is the cause when the cause is not obvious. *Dysphagia* 1994;**9**:245–55.

141 White GN, O'Rourke F, Ong BS, Cordato DJ, Chan DK. Dysphagia: causes, assessment, treatment, and management. *Geriatrics* 2008;**63**:15–20.

142 Foley N, Teasell R, Salter K, Kruger E, Martino R. Dysphagia treatment post stroke: a systematic review of randomised controlled trials. *Age & Ageing* 2008;**37**:258–64.

143 Cook IJ, Kahrilas PJ. AGA technical review on management of oropharyngeal dysphagia. *Gastroenterology* 1999;**116**:455–78.

144 Ertekin C, Yüceyar N, Aydogdu I. Clinical and electrophysiological evaluation of dysphagia in myasthenia gravis. *Journal of Neurology, Neurosurgery and Psychiatry* 1998;**65**:848–56.

145 Shelton GD, Schule A, Kass PH. Risk factors for acquired myasthenia gravis in dogs: 1,154 cases (1991–1995). *Journal of the American Veterinary Medical Association* 1997;**211**:1428–31.

146 Shelton GD, Ho M, Kass PH. Risk factors for acquired myasthenia gravis in cats: 105 cases (1986–1998). *Journal of the American Veterinary Medical Association* 2000;**216**:55–7.

147 Gaynor AR, Shofer FS, Washabau RJ. Risk factors for acquired megaesophagus in dogs. *Journal of the American Veterinary Medical Association* 1997;**211**:1406–12.

148 Venker-van Haagen AJ, van den Brom WE, Hellebrekers LJ. Effect of superior laryngeal nerve transection on pharyngeal muscle contraction timing and sequence of activity during eating and stimulation of the nucleus solitarius in dogs. *Brain Research Bulletin* 1999;**49**:393–400.

149 Venker-van Haagen AJ, van den Brom WE, Hellebrekers LJ. Effect of stimulating peripheral and central neural pathways on pharyngeal muscle contraction timing during swallowing in dogs. *Brain Research Bulletin* 1998;**45**:131–6.

150 Venker-van Haagen AJ, Hartman W, Wolvekamp WT. Contributions of the glossopharyngeal nerve and the pharyngeal branch of the vagus nerve to the swallowing process in dogs. *American Journal of Veterinary Research* 1986;**47**:1300–7.

151 Venker-van Haagen AJ, Hartman W, van den Brom WE, Wolvekamp WT. Continuous electromyographic recordings of pharyngeal muscle activity in normal and previously denervated muscles in dogs. *American Journal of Veterinary Research* 1989;**50**:1725–8.

152 Peeters ME, Venker- van Haagen AJ, Goedegebuure SA, Wolvekamp WTC. Dysphagia in Bouviers associated with muscular dystrophy; evaluation of 24 cases. *Veterinary Quarterly* 1991;**13**:65–73.

153 Peeters ME, Ubbink GJ. Dysphagia-associated muscular dystrophy: a familial trait in the Bouvier des Flandres. *Veterinary Record* 1994;**134**:444–6.

154 Pfeifer RM. Cricopharyngeal achalasia in a dog. *Canadian Veterinary Journal* 2003;**44**:993–5.

155 Ladlow J, Hardie RJ. Cricopharyngeal achalasia in dogs. *Compendium on Continuing Education for the Practicing Veterinarian* 2000;**22**:750–5.

156 Elliott RC. An anatomical and clinical review of cricopharyngeal achalasia in the dog. *Journal of the South African Veterinary Association* 2010;**81**:75–9.

157 Weaver AD. Cricopharyngeal achalasia in Cocker Spaniels. *Journal of Small Animal Practice* 1983;**24**:209–14.

158 Warnock JJ, Marks SL, Pollard R, Kyles AE, Davidson A. Surgical management of cricopharyngeal dysphagia in dogs: 14 cases (1989–2001). *Journal of the American Veterinary Medical Association* 2003;**223**:1462–8.

159 Davidson AP, Pollard RE, Bannasch DL, Marks SL, Hornof WJ, Famula TR. Inheritance of cricopharyngeal dysfunction in Golden Retrievers. *American Journal of Veterinary Research* 2004;**65**:344–9.

160 Peeters ME, Venker-van Haagen AJ, Wolvekamp WT. Evaluation of a standardised questionnaire for the detection of dysphagia in 69 dogs. *Veterinary Record* 1993;**132**:211–13.

161 Pollard RE. Imaging evaluation of dogs and cats with dysphagia. *ISRN Veterinary Science* 2012;**2012**:1–15.

162 Goring RL, Kagan KG. Cricopharyngeal achalasia in the dog: radiographic evaluation and surgical management. *Compendium on Continuing Education for the Practicing Veterinarian* 1982;**4**:438–4.

163 Watrous BJ, Suter PF. Oropharyngeal dysphagias in the dog: a cinefluorographic analysis of experimentally induced and spontaneously occurring swallowing disorders. *Veterinary Radiology & Ultrasound* 1983;**24**:11–24.

164 Allen SW. Surgical management of pharyngeal disorders in the dog and cat. *Problems in Veterinary Medicine* 1991;**3**:290–7.

165 Pollard RE, Marks SL, Davidson A, Hornof WJ. Quantitative videofluoroscopic evaluation of pharyngeal function. *Veterinary Radiology & Ultrasound* 2000;**41**:409–12.

166 Watrous BJ, Suter PF. Normal swallowing in the dog: a cineradiographic study. *Veterinary Radiology & Ultrasound* 1979;**20**:99–109.

167 Niles JD, Williams JM, Sullivan M, Crowsley FE. Resolution of dysphagia following cricopharyngeal myectomy in six young dogs. *Journal of Small Animal Practice* 2001;**42**:32–5.

168 Rosin E, Hanlon GF. Canine cricopharyngeal achalasia. *Journal of the American Veterinary Medical Association* 1972;**160**:1496–9.

16.1 INTRODUCTION

Dogs and cats with diseases of the larynx typically present with dyspnoea. In patients with dyspnoea, the most important question to ask the owner is whether or not they hear obvious abnormal breathing noises (stridor). Dogs and cats in dyspnoea with stridor per definition have obstructive upper airway disease, whereas:

- Dyspnoea without stridor indicates lower airway disease.
- Laryngeal stridor predictably manifests upon inspiration as a g-sound or sawing sound.
- Coughing is the second most prominent sign of laryngeal disease.
- The owner mentions dysphonia in some cases.

The work-up of patients with laryngeal disease consists of a thorough physical examination, which is usually unremarkable except for the possible audible stridor and in some cases masses can be palpated compressing the throat. Direct inspection of the larynx with a laryngoscope is the most important diagnostic procedure for diagnosing disorders of the larynx, but diagnostic imaging also plays an essential role, as has been discussed in Chapter 14 Throat Diagnostic Procedures, Section 14.

Congenital abnormalities affecting the larynx include those associated with brachycephalic obstructive airway syndrome and glottic stenosis, which are discussed in the next section. The most common acquired laryngeal diseases will be reviewed in the subsequent sections. These include acute and chronic laryngitis, laryngeal trauma, epiglottic entrapment and retroversion, laryngeal paralysis and laryngeal neoplasia.

16.2 CONGENITAL ABNORMALITIES OF THE LARYNX

Congenital deformities of the larynx are uncommon in dogs and cats and only those animals with severe obstructions will be presented for evaluation. The most common condition is brachycephalic obstructive airway syndrome. In all brachycephalic dogs the laryngeal structures are hypoplastic to some degree. A hypoplastic larynx is smaller than that of a normal dog of similar weight. On top of this, collapse of the larynx can occur, secondary to the chronic upper airway obstruction. These conditions have been discussed in Chapter 11 Diseases of the Nasal Cavity and Sinuses, Section 11.3 and Chapter 19 Surgery of the Nose, Section 19.3. Congenital glottic and subglottis stenosis will be reviewed in this chapter.

Congenital glottic stenosis

Glottic stenosis is a rare deformity and not recognised to have a breed predisposition[1-3]. It occurs in dogs and cats, and results from webbing of the vocal folds or deformities of the arytenoid cartilages.

Glottic stenosis in one young dog consisted of the joining of the corniculate processes of the left and right arytenoid cartilages[2]. Electromyography revealed no abnormalities. Repeated surgical widening of the glottis by dissecting the cartilaginous connection between the left and right arytenoids was found to lead to a progressive improvement in breathing[2].

Congenital subglottic stenosis

Congenital subglottic stenosis has been reported in humans[5,7], dogs[8,9] and cats[2,4,6,8] as a congenital deformity of the cricoid cartilage. There may also be concurrent deformities in the glottic part of the larynx, including the arytenoid cartilages and vocal folds.

Congenital laryngeal stenosis is a rare malformation in humans and often syndrome associated. It usually leads to neonatal or early respiratory distress, requiring tracheostomy to secure the airway and allow the child to live and then to thrive. Venker reported a 5-month-old male dog of mixed breeding with dyspnoea and aphonia since the age of 6 weeks and cyanosis upon exertion with distinct laryngeal stridor[9]. Laryngoscopy revealed an apparently small larynx with severe subglottic stenosis at the oral margin of the cricoid cartilage. On both arytenoid cartilages the corniculate processes were lacking and the cuneiform process was only half the appropriate size. The vocal folds consisted of no more than mucosa. The abnormalities were bilateral and symmetrical. Pathological examination confirmed the abnormalities and showed that the intrinsic laryngeal muscles were partly absent and the aryepiglottic folds were short[4,9].

In humans, surgical management has been described but is considered to be challenging. Two main approaches have been described, endoscopic and open airway reconstructive surgery[4-7]. Endoscopic laryngoplasty, involving incision of the subglottic laryngeal cartilages with cold steel instruments and subsequent balloon dilation of the stenosis, was shown to be a safe and effective treatment for congenital laryngeal stenosis in humans. The classic treatment however, is open laryngoplasty with either balloon dilation, cricotracheal resection or laryngoplasty with cartilage grafting[4,6,7,11].

16.3 ACUTE AND CHRONIC LARYNGITIS

- Acute laryngitis in dogs and cats is most commonly caused by infectious agents (kennel cough and cat flu).
- Uncomplicated acute laryngitis can be treated conservatively and generally has a good prognosis.
- The cause of chronic laryngitis in dogs and cats is often not found, though most patients respond favourably to tapering courses of corticosteroids.
- Severe obstructive inflammatory laryngeal disease in cats may require a surgical intervention.

Acute inflammatory laryngeal disease is common in both the dog and the cat and is usually caused by infectious agents such as canine infectious tracheobronchitis (CITB), commonly called kennel cough, or the feline upper respiratory agents (i.e. FHV-1, FCV, see Chapter 11 Diseases of the Nasal Cavity and Sinuses, Section 11.9). Other causes of inflammatory laryngeal disease include endotracheal intubation, insect bites or stings (catching bees or wasps), foreign body penetration, trauma from bite wounds, leash and choke-chain injuries, road traffic accidents and vocal fold abuse (constant barking in dogs). In addition, the larynx can also be inflamed as part of a systemic infection or disease. Frequently, however, no cause for acute laryngeal inflammation is found. Chronic laryngitis can result from acute inflammation of the larynx, from brachycephalic obstructive airway syndrome in dogs (see Chapter 11 Diseases of the Nasal Cavity and Sinuses, Section 11.3), or unknown causes in dogs and cats. Chronic inflammation in cats occasionally leads to granuloma formation and severe laryngeal obstruction[10]. Rarely laryngeal abscesses are seen.

Kennel cough

CITB is thought to be the result of coinfection of *Bordetella bronchiseptica* with either canine parainfluenza virus or canine adenovirus-2 (CAV-2)[2,4,6,8,11-13]. However, other viruses such as herpesvirus, respiratory coronavirus and canine reovirus types 1, 2, and 3, mycoplasma infections and *Streptococcus equi* have been incriminated as well[2,14-17]. Most likely CITB is a multifactorial disease, and more than one pathogen needs to be involved in a susceptible host before clinical signs develop.

The clinical signs of CITB are paroxysms of a harsh, dry cough, hoarse voice and sometimes decreased appetite and lethargy. However, if secondary bronchitis or bronchopneumonia develops (Fig. 16.1), more systemic signs as fever and malaise may also be seen. The diagnosis is based on history, clinical examination and laryngoscopy[2,8,16,18]. Contact with a kennel or other coughing dogs in the neighbourhood together with the clinical signs make CITB very likely. Clinical physical examination usually does not reveal any abnormalities except for a hard, dry cough elicited by palpation of the larynx and trachea. If laryngoscopy is performed, the mucosa of the larynx will be found to be red and oedematous. In complicated cases and especially if a productive soft cough develops, radiographic evaluation of neck and thorax is recommended to rule out bronchopneumonia.

CITB is usually self-limiting, and no specific treatment is indicated if the animal has only mild signs.

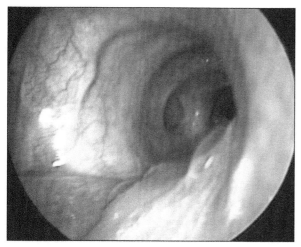

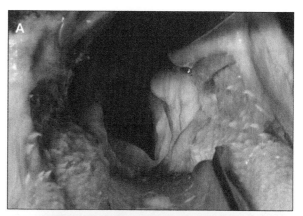

Fig. 16.1 Tracheobronchoscopic image of a 10-month-old Labrador Retriever with septic tracheobronchitis (kennel cough).

Figs 16.2A, B Laryngoscopic abnormalities of the larynx demonstrating chronic laryngitis (diffuse oedematous swelling of the laryngeal mucosal lining) in English Bulldogs. A: Mild hypoplasia with mild eversion of the saccules; B: moderate hypoplasia with mild eversion of the saccules.

- House or kennel rest with only short walks, wearing a harness, and avoidance of excitement is important so long as the dog is coughing.
- Drinking water should be encouraged as it helps to moisten the laryngeal mucosa, diminishing the irritation[2,19].
- Excessive coughing may be treated by sedatives, for example phenobarbital in a dose of 2 mg/kg once or twice daily, especially during the night[2].
- Antitussives, such as butorphanol tartrate or hydrocodone bitartrate, are effective in minimising the severity of the cough in more severe cases but should not be used if pneumonia is suspected.
- If moderate to severe signs of dyspnoea exist, a single dose or short course with an anti-inflammatory dose of glucocorticosteroids can be initiated to decrease laryngeal oedema. Again, the presence of bronchopneumonia needs to be ruled out first before initiating corticosteroid treatment.
- Only in cases of severe secondary bacterial infection or bronchopneumonia should antibiotics be prescribed. Doxycycline 5–10 mg/kg orally once daily is the antimicrobial of choice for

Bordetella bronchiseptica[8,20]. Potentiated amoxicillin or fluoroquinolones should only be prescribed after culture and sensitivity testing of laryngeal swabs or tracheobronchial washes.

Respiratory obstruction secondary to laryngeal inflammation is an uncommon clinical presentation but a tracheostomy is indicated if the patient is dyspnoeic, cyanotic or extremely anxious due to laryngeal inflammation[11,21]. Most dogs recover however without any complications. Prevention relies on vaccination against common pathogens, and vaccines with attenuated canine parainfluenza, CAV-2 and *Bordetella bronchiseptica* are available.

Chronic laryngitis in dogs

Chronic laryngitis in the dog may follow kennel cough, or be associated with chronic tracheobronchitis[22–25]. It is also seen in patients with laryngeal paralysis, or other obstructive laryngeal disease such as laryngeal neoplasia, and in dogs with brachycephalic obstructive airway syndrome (Figs 16.2A, B, 16.3A, B)[26–30]. Chronic irritation, inflammation and contact between both aryt-

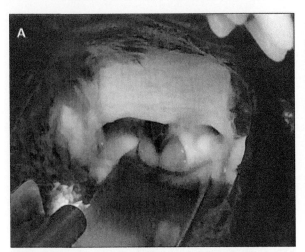

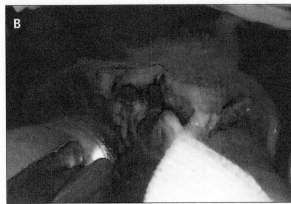

Figs 16.3A, B Laryngoscopic abnormalities of the larynx demonstrating chronic laryngitis in a Pug (A) with significant laryngeal hypoplasia and severe eversion of the saccules and secondary collapse (grade II); B: a French Bulldog with mild to moderate hypoplasia and mild laryngeal collapse, eversion of the saccules (grade I).

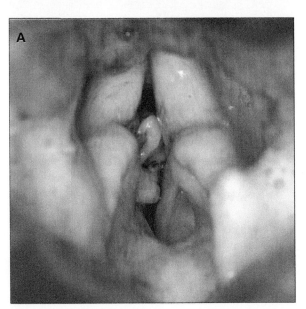

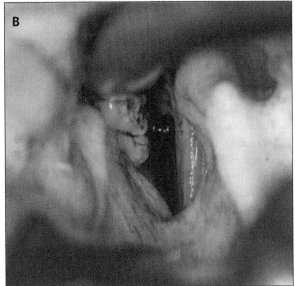

Figs 16.4A, B Laryngoscopic abnormalities of the larynx of an English Bulldog with right sided granulomatous kissing lesion before (A) and (B) after intubation.

enoids can lead to 'kissing lesions', granulopolypous changes of the laryngeal mucosa (Figs 16.4A, B). Excessive straining against the collar when the dog is on a leash and frequent barking or panting can also induce chronic laryngitis[2]. Vocal fold abuse in dogs left alone and barking and squealing all day long is not always readily admitted in the history[18,25].

In humans, gastroesophageal reflux, defined as the entry of the gastric contents into the oesophagus without associated belching or vomiting, has been implicated in the pathogenesis of several laryngological disorders including chronic laryngitis[31-38]. The term laryngopharyngeal reflux was proposed to denote the gastroesophageal reflux that reaches the structures above the upper oesophageal sphincter (UOS)[32,39,40]. It is likely that these conditions can also explain some cases of chronic laryngitis found in the canine population. Chronic regurgitation certainly contributes to

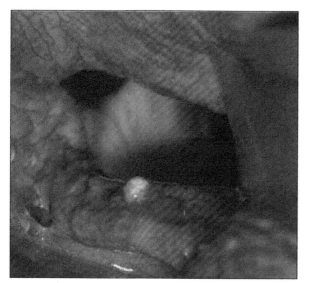

Fig. 16.5 Chronic laryngitis in an old dog with significant redness and swelling of the mucosal lining of the larynx and specifically the aboral side of the epiglottis with a small yellow nodule (calcification) visible near the apex of the epiglottis.

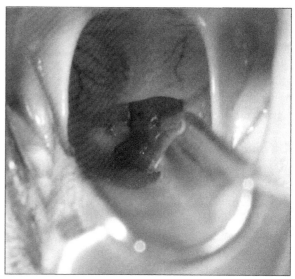

Fig. 16.6 Severe diffuse, but slightly lateralised (left side) laryngeal oedema and swelling associated with acute inflammation in a Domestic Shorthair cat.

pharyngeal and laryngeal oedema in dogs with brachycephalic obstructive airway syndrome[30,41,42].

Clinical signs of chronic laryngitis consist of recurrent coughing and retching, a rasping sound during panting and a roughening of the bark. Palpation of the larynx evokes a cough followed by gagging and swallowing, possibly an indication of pain[2,28,30,43,44].

Further specific examination is required to determine whether there is coexisting tracheitis and bronchitis, and to rule out the presence of neoplastic disease or laryngeal paralysis. Diagnostic imaging of the neck and thorax followed by a complete upper airway endoscopic examination, with sampling for culture and sensitivity and cytology, and oesophagoscopy/gastroscopy are therefore advised.

Calcification of the laryngeal cartilages is occasionally found on laryngeal radiographs in older dogs and is not always associated with clinical signs of laryngitis[2,45–48]. Laryngoscopic abnormalities that can be found in these patients consist of thickening and congestion of the laryngeal mucosa, and in some cases an enlarged and rigid epiglottis is seen with redness of the dorsal mucosal surface that can also contain small yellow nodules (Fig. 16.5). Biopsy of these nodules usually reveals inflammation and calcification, especially in patients concurrently suffering from hyperadrenocorticism.

In all cases of chronic laryngitis in the dog, therapy begins with the use of a harness instead of a collar[2,49,50]. When possible, other contributing habits should be changed and underlying diseases should be treated[2,51]. Non-steroidal anti-inflammatory drugs (NSAIDs) may suppress the clinical signs and should be attempted first. If this fails, a tapering course of corticosteroids can be tried.

Acute laryngitis in cats

Viral upper respiratory disease in cats may cause acute laryngitis, often together with pharyngitis and rhinitis (see Chapter 11 Diseases of the Nasal Cavity and Sinuses, Section 11.9). With significant involvement of the larynx, fever, reluctance to swallow, stridorous breathing and sometimes dyspnoea can be seen. Dyspnoea in cats with laryngeal disease is often more obvious than coughing. Clinical examination and diagnostic imaging is as discussed in Chapter 11 Diseases of the Nasal Cavity and Sinuses, Section 11.9. Inspection of the larynx specifically in these cases will reveal red and oedematous mucosa, sometimes significantly obstructing the glottis (Fig. 16.6). Vesicles and erosions of the mucosa can be seen in some cases with calicivirus infections[1–3,28,30,44,52].

The treatment is symptomatic and aimed at maintaining hydration and supporting the nutritional status. Broad-spectrum antibiotics may be effective for secondary bacterial infections. The use of antivirals, non-specific immunomodulatory agents and the role of vaccinations has been discussed in Chapter 11 Diseases of the Nasal Cavity and Sinuses, Section 11.9.

Chronic laryngitis in cats

As described for dogs, chronic laryngitis in cats is often of unknown origin, though chronic viral infections and chronic bronchitis can be present. The clinical signs consist of a soft laryngeal stridor and occasional swallowing during purring. There may be loss of voice or a change in the voice, but coughing is rare[2,31,33–38]. Underlying systemic disease and feline asthma/chronic bronchitis should be ruled out using diagnostic imaging and/or tracheobronchoscopy with bronchoalveolar lavage (BAL). Laryngoscopic examination reveals thickening of the laryngeal mucosa and sometimes an irregular surface. Treatment with a glucocorticoid is not always satisfactory, but the signs are mild and the cat does not change its habits because of the laryngitis[2,28,30,44,52].

Obstructive chronic inflammatory laryngeal disease has been described in cats[4,6,26,30,44,53]. Although uncommon, it is a disease worth noting because the gross appearance can mimic laryngeal neoplasia (Figs 16.7A, B)[8,39,40]. The underlying cause of inflammation is unknown, but possibly trauma or chronic ulcerative viral infections are involved in the aetiopathogenesis. Feline immunodeficiency virus (FIV) and feline leukaemia virus (FeLV) have not been found to be associated with this disease[4,6,8,30]. Chronic granulomatous laryngitis in humans is associated with micro-organisms, such as leishmania and tuberculosis, that thrive in an immuno-compromised host[2,10]. Though bacteria in low numbers were present in one of the three cases described by Tasker et al., histopathology and specific staining were negative for micro-organisms in the other two cases[2,44,52].

Reported clinical signs in the cat consisted of retching, coughing, dyspnoea and dysphonia. Radiographic examination of the larynx revealed a mass-effect in the larynx in most cases. Direct visualisation of the larynx reveals a laryngeal mass (Fig. 16.7A, B), and histopathology either granulomatous or neutrophilic and lymphoplasmacytic laryngitis[4,6,54].

With severe obstruction, temporary tracheotomy is indicated, but this is associated with a high perioperative morbidity and mortality in this species. Treatment with corticosteroids (dexamethasone, prednisone or prednisolone) has variable success, and occasionally surgical resection of the proliferative tissue is indi-

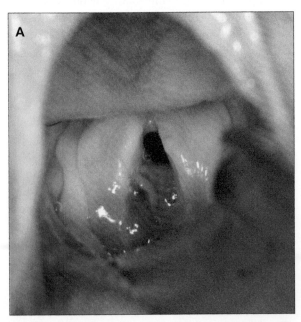 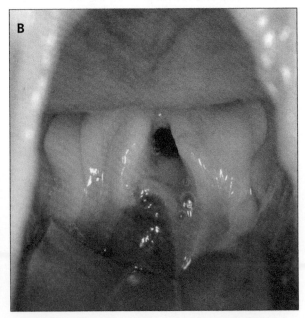

Figs 16.7A, B Granulomatous chronic laryngitis in a Domestic Shorthair cat; the nodular appearance mimics laryngeal neoplasia.

cated[2,4,6,11]. Though based on only a few cases the prognosis appears to be guarded, with a high mortality rate during the initial diagnostic and treatment period[4,6,55].

A laryngeal abscess is sometimes found in cats (Figs 16.8A, B), possibly caused by penetration of the mucosa by a sharp bone, needle or other foreign body, giving entry to oral micro-organisms[2,26]. Clinical signs consist of slowly progressive dyspnoea and laryngeal stridor. Abscesses can usually be recognised on radiographs[2] or ultrasonographic imaging[19,22,31,39,56] of the larynx. Laryngoscopy reveals a round, yellowish mass that partly covers the laryngeal inlet. A biopsy incision is both diagnostic (release of thick yellowish fluid that may yield *Pasteurella multocida* upon culture) and therapeutic[2,52,56–58].

16.4 LARYNGEAL CYSTS, POLYPS AND TUMOURS

- Laryngeal cancer is rare in dogs and cats, and benign masses including polyps and cysts have barely been reported.
- Rhabdomyomas, oncocytomas and extramedullary plasmacytomas carry a good prognosis, but other tumour types usually have a very poor prognosis.
- In cats, malignant lymphoma is not uncommonly found and should be differentiated from squamous cell carcinoma as prognosis is very different for both neoplasms.
- Surgical resection via partial laryngectomy is possible in some cases, whereas while complete laryngectomy is technically feasible, it is not commonly used.

Laryngeal cancer, and particularly squamous cell carcinoma, is common in humans, but is mainly related to smoking and alcohol consumption and therefore risk factors are completely different to those in small animals[20,52]. The disease is treated with partial or complete laryngectomy or chemoradiation, leading to good to excellent local control and cure rates[21,52]. Benign masses such as polyps (Fig. 16.9), cysts[22,24,25,59,60] and neoplastic disease[26,28,30,31,39,56] in the larynx are rare in dogs and cats. Cysts are very rare, but can be recognised by radiographic and ultrasonographic examination of the larynx and localised precisely by laryngoscopy, during which they can also be incised and drained[2,46,51]

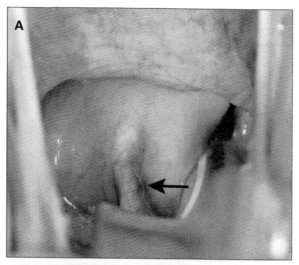

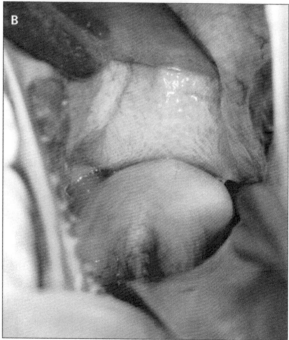

Figs 16.8A, B Laryngeal abscess involving the right arytenoid cartilage (arrow) in a 4-year-old Domestic Shorthair cat.

or surgically excised[25,26]. In cats, obstructive laryngeal disease with granuloma formation can mimic neoplastic disease and the conditions need to be differentiated. This type of disease is reviewed in the previous chapter on laryngitis (see Chapter 16 Diseases of the Larynx, Section 16.3). This chapter will further focus on benign and malignant neoplasms of the larynx in dogs and cats.

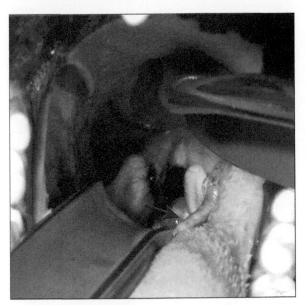

Fig. 16.9 Bilobular polyp (arrow) on the left vocal fold of a 3-year-old male Husky.

Laryngeal neoplasia

Reported canine laryngeal tumours include rhabdomyoma[2,31,33–38,61–64], oncocytoma[2,39,40,61–64], lipoma[12,30,41,65–67], osteosarcoma, chondroma, myxochondroma and chondrosarcoma[2,28,30,43,44,68–75], extramedullary plasmacytoma[45,46,74–76], melanoma[49,50,68,69,75,76], granular cell tumour[26,51,61,77–79], undifferentiated carcinoma, fibropapilloma and fibrosarcoma, mast cell tumour, adenocarcinoma and squamous cell carcinoma[2,28,30,44,52,61,80,81]. Whereas rhabdomyomas in the dog may be large, they are minimally invasive and do not appear to metastasise[31,33–38,82], in contrast with most other laryngeal tumours that are very locally invasive and have a significant metastatic potential[28,30,44,52,82].

Feline laryngeal neoplasms most commonly are lymphomas (Figs 16.10A, B), although squamous cell carcinoma and adenocarcinoma have been reported[26,30,44,53,61,78]. Except for laryngeal oncocytomas, which appear to occur in younger mature dogs[39,40,83,84], most dogs and cats with laryngeal cancer are middle-aged to older[2,30,63,85–88]. Neoplasms such as lymphoma and thyroid adenocarcinoma may secondarily invade the larynx, although the latter usually invades the trachea.

Patients with laryngeal tumours usually present with dyspnoea and a progressive change in voice (dysphonia) or bark (hoarseness), exercise intolerance or dysphagia.

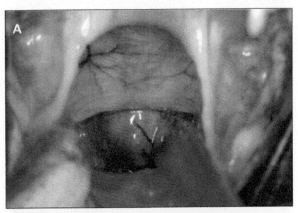
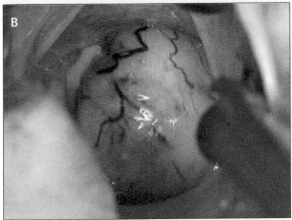

Fig. 16.10A, B A: Laryngeal lymphoma involving the left arytenoid in a 7-year-old Domestic Shorthair cat nearly completely obstructing the rima glottis ; B: close up view after lifting the soft palate dorsally..

What strongly arouses suspicion of tumour is a 'breaking voice', a vocalisation (a bark or meow) that begins normally and then suddenly changes to a breathy sound or is lost completely, while the animal continues the vocalising behaviour[2,64,77,86,89–91].

Whereas the laryngeal stridor is usually only inspiratory at first, when the obstruction is large enough it will be both inspiratory and expiratory.

On physical examination in dogs usually no abnormalities are detected with the exception of the stridor, as most tumours only involve the glottis and do not result in overall enlargement of the larynx. In cats however, tumours can sometimes be palpated as tumours are more likely to involve multiple laryngeal structures, and the larynx is softer and more flaccid than in dogs[2,18,61,63,77].

Ultrasonography, radiography, computed tomography (CT), and magnetic resonance imaging (MRI) can be used to further characterise the process in the larynx. Ultrasonography has been found to provide an accurate indication of the presence and location of laryngeal tumours in cats[54,82]. Regional radiographs of the larynx can reveal the general location of the tumour and give some idea of its extension, but for instance do not clearly distinguish left- and right-sided involvement[2,12,65–67]. Both CT and MRI will show fine anatomical detail and provide useful information about invasion of the cartilage, invasion of the base of the tongue or other extralaryngeal extension and the status of the lymph nodes[55,61,62,86,92].

Laryngoscopy is essential to rule out other abnormalities; it reveals the appearance of the tumour, gives information about exact location and regional ingrowth in surrounding tissues and, more importantly, facilitates biopsy for cytological or histological diagnosis. While small samples or cytology alone may yield false-negative results[26,61,62,70,82,86,92], it can be very helpful in cats where it is important to quickly differentiate malignant lymphoma from carcinoma. Chemotherapy could be started whilst the animal is kept under anaesthesia and with the endotracheal tube in place, which can greatly improve recovery[2,79,87,93,94].

Benign laryngeal cancers such as lipomas, oncocytomas, fibromas and rhabdomyomas can be removed successfully with preservation of function via either an endoscopy-assisted oral approach or via ventral laryngotomy[22,31,39,56,64,87]. Temporary tracheostomy (see Chapter 20 Surgery of the Throat 20.1) is advised for the initial recovery period (2–3 days). Careful mucosal apposition is important to prevent scar tissue formation and secondary webbing. While complete laryngectomy with a permanent tracheostomy is technically feasible in dogs and cats and is the option used in humans, it has had limited use in veterinary medicine[52,56–58,64,87]. The permanent tracheostoma created with this technique will need constant attention and protection. Depending on their suspected radioresponsiveness, invasive cancers can be treated with irradiation to better preserve laryngeal function[52,95,96]. Radiation should control lymphoma or granular cell tumour, mast cell tumour, adenocarcinoma and squamous cell carcinoma[52,55,97], but chemotherapeutic treatment of feline lymphoma also yields excellent results[2,59–61,86,98–100].

Benign lesions of the trachea and larynx have a good prognosis if they can be resected; most animals with rhabdomyomas and oncocytomas for instance can be cured[2,31,39,56,61,86,98–100]. Long-term outcomes may also be expected with treatment of extramedullary plasmacytomas or granular cell tumours[2,46,51]. The prognosis for malignant lesions appears to be poor, but few reports exist on outcome as very few animals have been treated. The median survival time for 27 cats with a variety of laryngeal and tracheal lesions in one study was 5 days and only 7% were alive at 1 year[26,61].

16.5 LARYNGEAL PARALYSIS

Laryngeal paralysis is a complete or partial failure of the arytenoid cartilages and vocal folds to abduct during inspiration, resulting in upper airway obstruction and predisposing the patient to aspiration pneumonia[2,61–64,74,75]. Unilateral paralysis usually does not lead to obvious clinical signs; most animals are diagnosed when bilateral paresis has led to insufficient opening of the rima glottis, which gradually progresses to full blown paralysis. Laryngeal paralysis has been most commonly reported in dogs[2,61–64,101] and rarely in cats[2,12,65–67,102,103] and can be unilateral or bilateral and congenital or acquired.

Laryngeal paralysis

Congenital hereditary laryngeal paralysis is seen in the Bouvier des Flandres, Bull Terriers, Siberian Husky, Rottweiler, Pyrenean Mountain dogs and Dalmatian breeds[2,68–75,102,103]. In Bouviers, the condition is inherited as an autosomal dominant trait[2,74–76,98,104]. In Siberian Huskies, Husky crosses and Bouviers, it starts as a progressive degeneration of neurons within the nucleus ambiguous, with subsequent Wallerian degeneration of the laryngeal nerves[68,69,75,76,88,90,105–116]. Acquired laryngeal paralysis is usually idiopathic but may occur secondary to trauma or disease, such as chronic endocrine, infectious or immune-mediated polyneuropathy[26,61,77–79], or it may be iatrogenic after surgery[61,80,81]. Other causes include organophosphate toxicity, retropharyngeal infection, rabies, polyradiculoneuritis (coonhound paralysis), as part of a paraneoplastic syndrome[82], myasthenia gravis[82] or laryngeal myopathy[61,78]. Laryngeal paralysis has been diagnosed in cats after bilateral thyroidectomy[83,84].

- Clinical signs are more common in large breed dogs than in small breed dogs and males being affected 2–3 times more often than females[2,63,85–88].
- Middle-aged to older (median age of 9 years) Labrador and Golden Retrievers, Afghans, Saint Bernards and Irish Setters are often affected with acquired idiopathic laryngeal paralysis[64,77,86,89–91].
- A progressive inspiratory stridor; and
- Decreased exercise intolerance are noted.
- Usually voice changes (dysphonia).
- Coughing, gagging and restlessness are noted.
- Excitement, stress, obesity and high ambient temperatures exacerbate clinical signs[18,61,63,77].
- Dogs with laryngeal paralysis may present with clinical signs of generalised muscle weakness including difficulty rising, paresis, megaoesophagus and dysphagia, or they may develop these clinical signs several months after the recognition of respiratory disease[82].

In cats tachypnea or dyspnoea, dysphonia, increased upper respiratory noise, inspiratory stridor, exercise intolerance, coughing, dysphagia, weight loss, lethargy and anorexia have been reported[12,65–67].

Physical examination is usually unremarkable with the exception of tachypnoea with a laryngeal stridor, panting and sometimes hyperthermia. Thoracic auscultation may reveal referred upper airway sounds or harsh crackles and wheezes in animals with aspiration pneumonia or pulmonary oedema[61,62,86,92]. Animals with polyneuropathy may have skeletal muscle atrophy or peripheral neurological abnormalities, such as decreased postural reactions, spinal reflex deficits and cranial nerve abnormalities[61,62,70,82,86,92].

Haematological parameters are usually unremarkable; hypothyroidism is sometimes found, but more likely reflects a concurrent rather than a causative disorder[79,87,93,94]. Radiographs of the neck and thorax or CT are made to exclude other causes of the dyspnoea, concurrent pathologies such as aspiration pneumonia (present in 8% of dogs)[64,87], pulmonary oedema or megaoesophagus. Preoperative oesophageal disease was diagnosed in 11% of dogs with laryngeal paralysis and was associated with increased risk of postoperative complications in one study[64,87]. Ultrasonography has been used to evaluate the larynx and laryngeal function, but is not as effective as direct observation of the larynx[95,96]. CT imaging can be used to not only evaluate the laryngeal cartilages and morphological abnormalities in the larynx, but laryngeal function as well[55,97].

The diagnosis can be made on visual evaluation of laryngeal function during a superficial plane of general anaesthesia (see Chapter 14.2)[2,61,86,98–100]. The arytenoids are in a paramedian position in affected animals and show no active abduction during inspiration (Fig. 16.11). Paradoxical motion of the arytenoid cartilages, where they are sucked inwards by the creation of negative intrathoracic pressure during inspiration, should not be confused with normal respiratory motion. Secondary oedema and erythema of the mucosa lining the arytenoid cartilages is also usually present[2,61,86,98–100].

Denervation of the laryngeal muscles can be diagnosed during electromyography (EMG)[2] in dogs as young as 12 weeks of age[61]. EMG of intrinsic laryngeal muscles has been used to demonstrate denervation potentials in suspected cases of laryngeal paralysis (see Chapter 14 Throat Diagnostic Procedures, Section 14.6)[74,75]. Experimentally, denervation potentials are noted within the cricoarytenoideus dorsalis muscles 5 days after iatrogenic trauma to recurrent laryngeal nerves[101]. In dogs with acquired laryngeal paralysis from polyneuropathy, EMG confirms involvement of the cricoarytenoideus dorsalis and other muscles[2,102,103]. Normal laryngeal function can still be observed in some animals, with EMG results indicative of denervation[2,102,103].

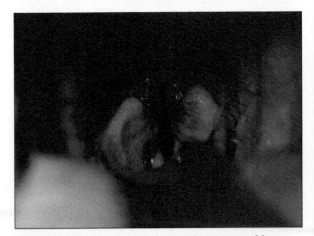

Fig. 16.11 Laryngoscopic image of a 12-year-old Labrador Retriever with laryngeal paralysis. A narrow rima glottis as a result of paramedian position of the arytenoids and secondary laryngitis can be seen.

With acute dyspnoea, emergency medical treatment should be started with:

- Sedation.
- Glucocorticosteroids.
- Supplemental oxygen.
- Cooling when necessary.
- If the obstruction is severe, a temporary tracheostomy should be performed.

Surgical treatment is recommended however, for animals with moderate to severe dyspnoea[2,98,104]. Many surgical techniques have been described, ranging from muscle-nerve pedicle transposition for reinnervation of the larynx to castellated laryngofissures, ventricular cordectomy, partial laryngectomy and permanent tracheostomy[88,90,105–116]. Arytenoid lateralisation procedures are recommended however, because of the consistently good results with minimal complications. Both unilateral cricoid–arytenoid and unilateral thyroid–arytenoid lateralisation procedures give reliable and comparable results and are equally effective from a clinical point of view[2,61,98,116,116a]. Improvement is expected in 90% of animals undergoing unilateral arytenoid lateralisation, and 70% of dogs are still alive 5 years after surgery[2,64,87,116a]. The technique of unilateral thyroid–arytenoid lateralisation will be discussed in Chapter 20 Surgery of the Throat, Section 20.6.

16.6 EPIGLOTTIC ENTRAPMENT AND RETROVERSION

Epiglottic disorders are rare in dogs and cats. Partial obstruction of the rima glottis by displaced folds of the glosso-epiglottic mucosa has been described though as part of the laryngeal abnormalities commonly seen in brachycephalic breeds[2,117]. The diagnosis and importance of the condition remain both difficult and controversial[5,7,118,119]. In the dog, actual entrapment of the epiglottis is not seen as a consequence of this displacement[9,118]. Apparent epiglottic entrapment was reported in five other dogs that were presented for coughing, gagging, dyspnoea and collapse[2,119]. In that report, the rostral third of the epiglottis was amputated, and redundant ventral mucosa was removed when present. Epiglottic retroversion (ER) is a rare cause of inspiratory stridor and dyspnoea and has recently been reported in dogs[9,120,121], but no reports of either condition exist for cats.

Epiglottic retroversion

This condition is characterised by episodic epiglottic retroflexion during inspiration and obstruction of the rima glottidis[120]. The condition has only been described in three dogs and eight horses[9,120–123]. The three canine patients described include an 8-year-old castrated male Boxer, a 10-year-old spayed female Yorkshire Terrier and a 6-year-old male neutered Yorkshire Terrier. All dogs presented with inspiratory dyspnoea and the first two also with a stridor. Two of the three dogs were previously diagnosed with hypothyroidism, a concurrent overlong soft palate was noted in the Boxer and the male Yorkshire terrier had concurrent coughing. It was hypothesised that the condition could be the result of a hypothyroidism-associated peripheral neuropathy and denervation of hypoglossal and/or glossopharyngeal nerves[5,7,120].

Radiographs of the neck and thorax did not reveal any abnormalities, but videofluoroscopy in one patient demonstrated retroversion of the epiglottis into the glottis during inspiration[7,120]. The epiglottis appeared to be of normal size in all patients upon laryngoscopy, but during inspiration the epiglottis would frequently displace caudally (Fig. 16.12) and obstruct the rima glottidis. The epiglottis could be easily returned to the normal horizontal position[2,8,11–13,120,121].

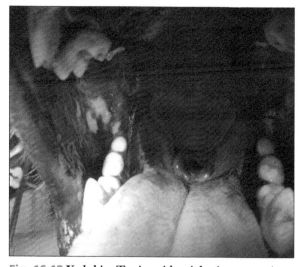

Fig. 16.12 Yorkshire Terrier with epiglottic retroversion. Upon inspiration the tip of the epiglottis is displaced and sucked caudally into the rima glottis. (Courtesy of Victoria Lipscomb, Department of Clinical Science and Services, Royal Veterinary College, University of London, UK.)

Surgical fixation of the epiglottis in a horizontal position (epiglottopexy) resulted in permanent resolution of dyspnoea in two dogs[120]. The most recent reported case reported the need for a subtotal epiglottectomy in order to manage severe ER that did not permanently resolve by epiglottopexy[2,14–17,121]. This dog was reported to be doing well without signs of respiratory tract compromise or dysphagia at 17 months after the procedure.

16.7 TRAUMATIC INJURIES OF THE LARYNX

Road traffic accidents and severe straining on the leash or harsh pulls on the leash are causes of external blunt injury to the larynx. Penetrating wounds to the larynx can be caused by an animal bite, stick, knife, or bullet or foreign bodies (Figs 16.13A, B), and involve the larynx and possibly other cervical structures. Traumatic or incautious insertion of an endotracheal tube can cause blunt laryngeal injury as well and particularly in the cat, laryngeal oedema rapidly occurs after manipulation of the laryngeal mucosa. Using a correct intubation technique should prevent this and laryngospasm can usually be prevented by application of anaesthetic spray.

As described for pharyngeal and nasal and sinal trauma, injuries may be extensive and are generally more extensive than the skin wounds suggests. For general stabilisation, including endotracheal intubation and tracheostomy, and diagnostic work-up of trauma patients, the reader is referred to Chapter 11 Diseases of the Nasal Cavity and Sinuses, Section 11.12 and Chapter 15 Diseases of the Pharynx, Section 15.5. Tracheotomy techniques are reviewed in Chapter 20 Surgery of the Throat, Section 20.1. In this chapter the specific considerations pertaining to blunt and penetrating injuries to the larynx will be discussed.

Blunt laryngeal injuries

Acute injury in a road traffic accident can produce laryngeal contusion and obstruction as a result of haematoma and oedema formation. Obstruction can increase rapidly, and careful observation of breathing rate and pattern and degree of stridor is mandatory for a successful outcome. Diagnostic imaging is always recommended after acute injury. Radiographs will suffice in detecting fractures of the hyoid bones and laryngeal cartilages and will allow assessment of concurrent pulmonary contusions or pulmonary oedema, but with severe and extensive injuries more advanced imaging may be required. Laryngoscopy is always indicated for assessment of intralaryngeal perforations and lacerations, and to estimate the degree of obstruction caused by haematoma and oedema and the need for tracheostomy tube placement. After thorough inspection, patients can be intubated with a small tube before diagnostic imaging commences.

When there are no fractures of the larynx and the laryngeal mucosa is seen to cover the laryngeal carti-

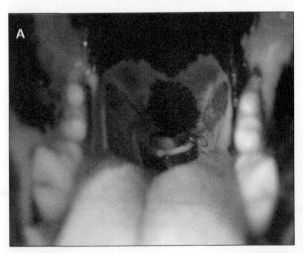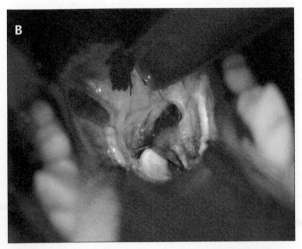

Figs 16.13A, B Laryngoscopic image of a 4-year-old Border Collie with a needle penetrating laryngeal injury. A needle with thread can be seen to have pierced the left arytenoid. Oedema and swelling of the mucosal lining of the left arytenoid is evident.

lages upon laryngoscopy, the dog or cat can be hospitalised for cage rest and observation. In addition, patients should be given analgesics and dogs should be fitted with a harness. In case of severe dyspnoea, if a significant increase in swelling in the postimaging period is expected, or if non-dislocated laryngeal fractures are seen, a tracheotomy should be performed. If there are no complications during hospitalisation, sufficient resolution of the oedema and haematomas can be expected in 7–10 days. Laryngoscopic examination should help in deciding when to remove the cannula[2,8,16,18].

Fractures with discontinuity of the laryngeal cartilages should be repaired immediately or granulation tissue will cause contraction and narrowing of the laryngeal passageway[2]. Depending on their location, most cartilaginous fractures can be stabilised with sutures and fine wire. Apposition of the fragmented parts should be gentle, to avoid tearing the cartilage, but precise, to prevent formation of granulation tissue[2,6,27,61,98]. Associated soft tissue injuries of the skin, subcutaneous tissues and pharyngeal and laryngeal muscles necessitating repair should be attended to in the same surgical session as the repair of the fractured cartilages[2,8].

Sudden or violent straining against a collar or pulling on a choke-chain or collar may cause laryngeal trauma and fractures of the hyoid bones and/or laryngeal cartilages. Management is as described for other blunt injuries to the larynx. These patients specifically need to be monitored for recurrent laryngeal nerve paralysis, presumably caused by trauma at the dorsocaudal border of the thyroid cartilage where the nerve enters the larynx, passing between the thyroid and cricoid cartilages[2,11]. Unilateral laryngeal paralysis rarely results in dyspnoea, but bilateral paralysis will require arytenoid lateralisation as described in Chapter 20 Surgery of the Throat, Section 20.6.

Penetrating laryngeal injuries

Wooden sticks or other foreign bodies that enter the larynx via the oral cavity cause intrinsic penetrating wounds (Figs 16.13A, B)[23,124–126]. Usually the most extensive trauma is found in the pharynx (see Chapter 15 Diseases of the Pharynx, Section 15.5), but occasionally, the stick lacerates the laryngeal mucosa or causes a tear in the epiglottis or cuneiform process of the arytenoid cartilage. Repair of these injuries is as described below for extrinsic penetrating injuries.

Extrinsic penetrating injuries are usually more extensive than suggested by the skin wounds, especially when dealing with dog bites in the neck[27,29,127–129]. Subcutaneous emphysema and dyspnoea are the most important indicators of a penetrating laryngeal wound, but the perforation may include the trachea, pharynx and oesophagus, vessels and nerves and neck muscles.

> Again, in dyspnoeic patients, endotracheal intubation and tracheostomy take precedence over further inventory clinical examination.

Diagnostic imaging and laryngoscopy, tracheobronchoscopy and oesophagoscopy are performed guided by the extent of the trauma. Surgery in patients with penetrating injuries is always recommended to retrieve foreign material, debride necrotic tissue, repair fractures and soft tissue trauma and drain the area as bacterial contamination inevitably will be present[2,127–129].

For all injuries, a ventral midline approach (see Chapter 20.4 ventral neck exploration) is recommended.

- Thrombosis and vessel lacerations that cannot be repaired are treated by permanent ligation of the vessel[18,130].
- Lacerations of the pharynx, trachea and oesophagus are closed after debridement[2,32,61,130].
- Laryngeal mucosal continuity is the single most important factor to protect the cartilage from infection and to reduce formation of granulation tissue and thereby minimise stenosis in the larynx[2,32,61,131]. Advancement flaps of laryngeal mucosa can be created from the piriform area to close rostral cartilage surfaces[42,130].
- Suturing the mucosa on both sides with fine absorbable suture material can usually repair tears in the epiglottis.
- Cartilage fractures that cannot be repaired by simple debridement and reapposition may be repaired using patching techniques (mucoperiosteal) or cartilage or mucosal grafts[2,61,130].
- Caudal cricoid defects can be excised and repaired by advancing the rostral trachea.

- Avulsions of an arytenoid cartilage are best treated by either removal of the arytenoid or by lateralisation. Mucosal flaps and cartilage fragments can be held in position with intraluminal laryngeal stents or Montgomery T-tubes, if indicated[2,47,48,61,130].

Broad-spectrum antibiotics and analgesia are given and the patient is hospitalised in the intensive care unit during the first few days[2,127–129]. Laryngeal webbing is the most serious long-term consequence of laryngeal trauma, and is caused by scar tissue formed between bilateral glottic mucosal lesions. Bilateral vocal fold resection and resection of everted saccules in brachy-cephalic dogs are often the cause of such webs. The treatment of webbing is difficult and repeated surgical procedures, with or without the help of laser, are often necessary to improve this debilitating condition. Despite the potential complications, most patients with penetrating injuries to the throat make a full or near-complete recovery if treated acutely and appropriately[2,127–129].

16.8 REFERENCES

1 Helps CR, Lait P, Damhuis A, *et al*. Factors associated with upper respiratory tract disease caused by feline herpesvirus, feline calicivirus, *Chlamydophila felis* and *Bordetella bronchiseptica* in cats: experience from 218 European catteries. *Veterinary Record* 2005;**156**:669–73.

2 Venker-van Haagen AJ. The larynx. In: *Ear, Nose, Throat, and Tracheobronchial Diseases in Dogs and Cats*. Venker-van Haagen AJ (ed). Schlütersche Verlagsgesellschaft, Hannover 2005, pp. 121–65.

3 Radford AD, Coyne KP, Dawson S, Porter CJ, Gaskell RM. Feline calicivirus. *Veterinary Research* 2007;**38**:319–35.

4 Tasker S, Foster DJ, Corcoran BM, Whitbread TJ, Kirby BM. Obstructive inflammatory laryngeal disease in three cats. *Journal of Feline Medicine and Surgery* 1999;**1**:53–9.

5 Hseu AF, Benninger MS, Haffey TM, Lorenz R. Subglottic stenosis: a ten-year review of treatment outcomes. *Laryngoscope* 2014;**124**:736–41.

6 Costello MF, Keith D, Hendrick M, King L. Acute upper airway obstruction due to inflammatory laryngeal disease in 5 cats. *Journal of Veterinary Emergency and Critical Care* 2001;**11**:205–10.

7 Blanchard M, Leboulanger N, Thierry B, *et al*. Management specificities of congenital laryngeal stenosis: external and endoscopic approaches. *Laryngoscope* 2014;**124**:1013–18.

8 Parnell N. Diseases of the throat. In: *Textbook of Veterinary Internal Medicine*, 7th edn. Ettinger SJ, Feldman EC (eds). Saunders Elsevier, St. Louis 2010, pp. 1040–6.

9 Venker-van Haagen AJ, Engelse EJJ, van den Ingh TSGAM. Congenital subglottic stenosis in a dog. *Journal American Animal Hospital Association* 1981;**17**:223–5.

10 Silva L, Damrose E, Bairão F, Nina MLD, Junior JC, Costa HO. Infectious granulomatous laryngitis: a retrospective study of 24 cases. *European Archives of Otorhinolaryngology* 2008;**265**:675–80.

11 Holt D, Brockman D. Diagnosis and management of laryngeal disease in the dog and cat. *Veterinary Clinics of North America Small Animal Practice* 1994;**24**:855–71.

12 Taylor SS, Harvey AM, Barr FJ, Moore AH, Day MJ. Laryngeal disease in cats: a retrospective study of 35 cases. *Journal of Feline Medicine and Surgery* 2009;**11**:954–62.

13 Keil DJ, Fenwick B. Role of *Bordetella bronchiseptica* in infectious tracheobronchitis in dogs. *Journal of the American Veterinary Medical Association* 1998;**212**:200–7.

14 Thrusfield MV, Aitken CGG, Muirhead RH. A field investigation of kennel cough: Efficacy of different treatments. *Journal of Small Animal Practice* 1991;**32**:455–9.

15 Erles K, Toomey C, Brooks HW, Brownlie J. Detection of a group 2 coronavirus in dogs with canine infectious respiratory disease. *Virology* 2003;**310**:216–23.

16 Buonavoglia C, Martella V. Canine respiratory viruses. *Veterinary Research* 2007;**38**:355–73.

17 Mitchell JA, Brooks HW, Szladovits B, *et al*. Tropism and pathological findings associated with canine respiratory coronavirus (CRCoV). *Veterinary Microbiology* 2013;**162**:582–94.

18 Venker-van Haagen AJ. Diseases of the larynx. *Veterinary Clinics of North America Small Animal Practice* 1992;**22**:1155–72.

19 Rudorf H, Barr F. Echolaryngography in cats. *Veterinary Radiology & Ultrasound* 2002;**43**:353–7.

20 Muscat JE, Wynder EL. Tobacco, alcohol, asbestos, and occupational risk factors for laryngeal cancer. *Cancer* 1992;**69**:2244–51.

21 O'Neill CB, O'Neill JP, Atoria CL, *et al*. Treatment complications and survival in advanced laryngeal cancer: a population-based analysis. *Laryngoscope* 2014;**124**:2707–13.

22 Rudorf H, Lane JG, Brown PJ, Mackay A. Ultrasonographic diagnosis of a laryngeal cyst in a cat. *Journal of Small Animal Practice* 1999;**40**:275–7.

23 Padrid PA, Hornof WJ, Kurpershoek CJ, Cross CE. Canine chronic bronchitis. *Journal of Veterinary Internal Medicine* 1990;**4**:172–80.

24 White RAS. The larynx. In: *Clinical Atlas of Ear, Nose, and Throat Diseases in Small Animals*. Hedlund CS, Tobaoda J (eds). Schlütersche, Hannover 2002, pp. 113–31.

25 Cuddy LC, Bacon NJ, Coomer AR, Jeyapaul CJ, Sheppard BJ, Winter MD. Excision of a congenital laryngeal cyst in a five-month-old dog via a lateral extraluminal approach. *Journal of the American Veterinary Medical Association* 2010;**236**:1328–33.

26 Jakubiak MJ, Siedlecki CT, Zenger E, *et al*. Laryngeal, laryngotracheal, and tracheal masses in cats: 27 cases (1998–2003). *Journal of the American Animal Hospital Association* 2005;**41**:310–16.

27 Fasanella FJ, Shivley JM, Wardlaw JL, Givaruangsawat S. Brachycephalic airway obstructive syndrome in dogs: 90 cases (1991–2008). *Journal of the American Veterinary Medical Association* 2010;**237**:1048–51.

28 Wheeldon EB, Suter PF, Jenkins T. Neoplasia of the larynx in the dog. *Journal of the American Veterinary Medical Association* 1982;**180**:642–7.

29 Pink JJ, Doyle RS, Hughes JM, Tobin E, Bellenger CR. Laryngeal collapse in seven brachycephalic puppies. *Journal of Small Animal Practice* 2006;**47**:131–5.

30 Carlisle CH, Biery DN, Thrall DE. Tracheal and laryngeal tumors in the dog and cat: literature review and 13 additional patients. *Veterinary Radiology & Ultrasound* 1991;**32**:229–35.

31 Meuten DJ, Calderwood-Mays MB, Dillman RC, *et al*. Canine laryngeal rhabdomyoma. *Veterinary Pathology* 1985;**22**:533–9.

32 Ulualp SO, Toohill RJ. Laryngopharyngeal reflux: state of the art diagnosis and treatment. *Otolaryngologic Clinics of North America* 2000;**33**:785–802.

33 Yamate J, Murai F, Izawa T, *et al*. A rhabdomyosarcoma arising in the larynx of a dog. *Journal of Toxicologic Pathology* 2011;**24**:179–82.

34 Barnhart K, Lewis B. Laryngopharyngeal mass in a dog with upper airway obstruction. *Veterinary Clinical Pathology* 2000;**29**:47–50.

35 O'Hara AJ, McConnell M, Wyatt K, Huxtable C. Laryngeal rhabdomyoma in a dog. *Australian Veterinary Journal* 2001;**79**:817–21.

36 Dunbar MD, Ginn P, Winter M, Miller KB, Craft W. Laryngeal rhabdomyoma in a dog. *Veterinary Clinical Pathology* 2012;**41**:590–3.

37 Liggett AD, Weiss R, Thomas KL. Canine laryngopharyngeal rhabdomyoma resembling an oncocytoma: light microscopic, ultrastructural and comparative studies. *Veterinary Pathology* 1985;**22**:526–32.

38 Clercx C, Desmecht D, Michiels L, McEntee K, Hardy N, Henroteaux M. Laryngeal rhabdomyoma in a golden retriever. *Veterinary Record* 1998;**143**:196–8.

39 Calderwood-Mays MB. Laryngeal oncocytoma in two dogs. *Journal of the American Veterinary Medical Association* 1984;**185**:677–9.

40 Pass DA, Huxtable CR, Cooper BJ, Watson AD, Thompson R. Canine laryngeal oncocytomas. *Veterinary Pathology* 1980;**17**:672–7.

41 Brunnberg M, Cinquoncie S, Burger M, Plog S, Nakladal B. Infiltrative laryngeal lipoma in a Yorkshire Terrier as cause of severe dyspnoea. *Tierarztliche Praxis* 2013;**41**:53–6.

42 Poncet CM, Dupre GP, Freiche VG, Estrada MM, Poubanne YA, Bouvy BM. Prevalence of gastrointestinal tract lesions in 73 brachycephalic dogs with upper respiratory syndrome. *Journal of Small Animal Practice* 2005;**46**:273–9.

43 Muraro L, Aprea F, White RAS. Successful management of an arytenoid chondrosarcoma in a dog. *Journal of Small Animal Practice* 2012;**54**:33–5.

44 Saik JE, Toll SL, Diters RW, Goldschmidt MH. Canine and feline laryngeal neoplasia. *Journal of the American Animal Hospital Association* 1986;**22**:359–65.

45 Witham AI, French AF, Hill KE. Extramedullary laryngeal plasmacytoma in a dog. *New Zealand Veterinary Journal* 2012;**60**:61–4.

46 Hayes AM, Gregory SP, Murphy S, McConnell JF, Patterson-Kane JC. Solitary extramedullary plasmacytoma of the canine larynx. *Journal of Small Animal Practice* 2007;**48**:288–91.

47 Péreza B, Gómeza M, Mieresb M, Galeciob JS. Computed tomographic anatomy of the larynx in mesaticephalic dogs. *Archivos de Medicina Veterinaria* 2010;**42**:91–9.

48 Alexander K. The Pharynx, larynx, and trachea. In: *Textbook of Veterinary Diagnostic Radiology*, 6th edn. Thrall DE (ed). Elsevier Saunders, St. Louis 2013, pp. 489–99.

49 McConnell EE, Smit JD, Venter HJ. Melanoma in the larynx of a dog. *Journal of the South African Veterinary Medical Association* 1971;**42**:189–91.

50 Ndikuwera J, Smith DA, Obwolo MJ. Malignant melanoma of the larynx in a dog. *Journal of Small Animal Practice* 1989;**30**:107–9.

51 Rossi G, Tarantino C, Taccini E, Renzoni G, Magi GE, Bottero E. Granular cell tumour affecting the left vocal cord in a dog. *Journal of Comparative Pathology* 2007;**136**:74–8.

52 Withrow SJ. Cancer of the larynx and trachea. In: *Small Animal Clinical Oncology*, 5th edn. Withrow SJ, Vail DM, Page RL (eds). Elsevier Saunders, St. Louis 2013, pp. 451–3.

53 Vasseur PB, Patnaik AK. Laryngeal adenocarcinoma in a cat. *Journal of the American Animal Hospital Association* 1981;**17**:639–41.

54 Rudorf H, Brown P. Ultrasonography of laryngeal masses in six cats and one dog. *Veterinary Radiology & Ultrasound* 1998;**39**:430–4.

55 Stadler K, Hartman S, Matheson J, O'Brien R. Computed tomographic imaging of dogs with primary laryngeal or tracheal airway obstruction. *Veterinary Radiology & Ultrasound* 2011;**52**:377–84.

56 Henderson RA, Powers RD, Perry L. Development of hypoparathyroidism after excision of laryngeal rhabdomyosarcoma in a dog. *Journal of the American Veterinary Medical Association* 1991;**198**:639–43.

57 Crowe DT, Goodwin MA, Greene CE. Total laryngectomy for laryngeal mast cell tumor in a dog. *Journal of the American Animal Hospital Association* 1986;**22**:809–16.

58 Block G, Clarke K, Salisbury SK, DeNicola DB. Total laryngectomy and permanent tracheostomy for treatment of laryngeal rhabdomyosarcoma in a dog. *Journal of the American Animal Hospital Association* 1995;**31**:510–13.

59 Teske E, Straten G, Noort R, Rutteman GR. Chemotherapy with cyclophosphamide, vincristine, and prednisolone (COP) in cats with malignant lymphoma: new results with an old protocol. *Journal of Veterinary Internal Medicine* 2002;**16**:179–86.

60 Simon D, Eberle N, Laacke-Singer L, Nolte I. Combination chemotherapy in feline lymphoma: treatment outcome, tolerability, and duration in 23 cats. *Journal of Veterinary Internal Medicine* 2008;**22**:394–400.

61 Monnet E, Tobias KM. Larynx. In: *Veterinary Surgery Small Animal*. Tobias KM, Johnston SA (eds). Elsevier Saunders, St. Louis 2012, pp. 1718–33.

62 Millard RP, Tobias KM. Laryngeal paralysis in dogs. *Compendium on Continuing Education for the Veterinary Practitioner* 2009;**31**:212–19.

63 Bahr KL, Howe L, Jessen C, Goodrich Z. Outcome of 45 dogs with laryngeal paralysis treated by unilateral arytenoid lateralization or bilateral ventriculocordectomy. *Journal of the American Animal Hospital Association* 2014;**50**:264–72.

64 MacPhail C. Laryngeal disease in dogs and cats. *Veterinary Clinics of North America Small Animal Practice* 2014;**44**:19–31.

65 Schachter S, Norris CR. Laryngeal paralysis in cats: 16 cases (1990–1999). *Journal of the American Veterinary Medical Association* 2000;**216**:1100–3.

66 Hardie EM, Kolata RJ, Stone EA, Steiss JE. Laryngeal paralysis in three cats. *Journal of the American Veterinary Medical Association* 1981;**179**:879–82.

67 Hardie RJ, Gunby J, Bjorling DE. Arytenoid lateralization for treatment of laryngeal paralysis in 10 cats. *Veterinary Surgery* 2009;**38**:445–51.

68 Bennett PF, Clarke RE. Laryngeal paralysis in a rottweiler with neuroaxonal dystrophy. *Australian Veterinary Journal* 1997;**75**:784–6.

69 Braund KG, Shores A, Cochrane S, Forrester D, Kwiecien JM, Steiss JE. Laryngeal paralysis-polyneuropathy complex in young Dalmatians. *American Journal of Veterinary Research* 1994;**55**:534–42.

70 Braund KG, Steinberg HS, Shores A, *et al.* Laryngeal paralysis in immature and mature dogs as one sign of a more diffuse polyneuropathy. *Journal of the American Veterinary Medical Association* 1989;**194**:1735–40.

71 Gabriel A, Poncelet L, Van Ham L, *et al.* Laryngeal paralysis-polyneuropathy complex in young related Pyrenean mountain dogs. *Journal of Small Animal Practice* 2006;**47**:144–9.

72 Mahony OM, Knowles KE, Braund KG, Averill DRJ, Frimberger AE. Laryngeal paralysis-polyneuropathy complex in young Rottweilers. *Journal of Veterinary Internal Medicine* 1998;**12**:330–7.

73 O'Brien JA, Hendriks J. Inherited laryngeal paralysis. Analysis in the husky cross. *Veterinary Quarterly* 2011;**8**:301–2.

74 Venker-van Haagen AJ. Laryngeal paralysis in Bouviers Belge des Flandres and breeding advice to prevent this condition. *Tijdschrift voor Diergeneeskunde* 1982;**107**:21–2.

75 Venker-van Haagen AJ, Hartman W, Goedegebuure SA. Spontaneous laryngeal paralysis in young bouviers [Dogs]. *Journal of the American Animal Hospital Association* 1978;**14**:714–20.

76 Ubbink GJ, Knol BW, Bouw J. The relationship between homozygosity and the occurrence of specific diseases in Bouvier Belge des Flandres dogs in The Netherlands. *Veterinary Quarterly* 1992;**14**:137–40.

77 Burbidge HM. A review of laryngeal paralysis in dogs. *British Veterinary Journal* 1995;**151**:71–82.

78 Gaynor AR, Shofer FS, Washabau RJ. Risk factors for acquired megaesophagus in dogs. *Journal of the American Veterinary Medical Association* 1997;**211**:1406–12.

79 Jaggy A, Oliver JE, Ferguson DC. Neurological manifestations of hypothyroidism: a retrospective study of 29 dogs. *Journal of Veterinary Internal Medicine* 1994;**8**:328–36.

80 Klein MK, Powers BE, Withrow SJ, *et al.* Treatment of thyroid carcinoma in dogs by surgical resection alone: 20 cases (1981–1989). *Journal of the American Veterinary Medical Association* 1995;**206**:1007–9.

81 Salisbury SK, Forbes S, Blevins WE. Peritracheal abscess associated with tracheal collapse and bilateral laryngeal paralysis in a dog. *Journal of the American Veterinary Medical Association* 1990;**196**:1273–5.

82 Thieman KM, Krahwinkel DJ, Sims MH, Shelton GD. Histopathological confirmation of polyneuropathy in 11 dogs with laryngeal paralysis. *Journal of the American Animal Hospital Association* 2010;**46**:161–7.

83 Hardie RJ, Gunby J, Bjorling DE. Arytenoid lateralization for treatment of laryngeal paralysis in 10 cats. *Veterinary Surgery* 2009;**38**:445–51.

84 Mallery KF, Pollard RE, Nelson RW. Percutaneous ultrasound-guided radiofrequency heat ablation for treatment of hyperthyroidism in cats. *Journal of the American Veterinary Medical Association* 2003;**223**:1602–7.

85 Burbridge HM, Goulden BE, Jones BR. An experimental evaluation of castellated laryngofissure and bilateral arytenoid lateralisation for the relief of laryngeal paralysis in dogs. *Australian Veterinary Journal* 1991;**68**:268–72.

86 Gaber CE, Amis TC, LeCouteur RA. Laryngeal paralysis in dogs: a review of 23 cases. *Journal of the American Veterinary Medical Association* 1985;**186**:377–80.

87 MacPhail CM, Monnet E. Outcome of and postoperative complications in dogs undergoing surgical treatment of laryngeal paralysis: 140 cases (1985–1998). *Journal of the American Veterinary Medical Association* 2001;**218**:1949–56.

88 White R. Unilateral arytenoid lateralisation: an assessment of technique and long term results in 62 dogs with laryngeal paralysis. *Journal of Small Animal Practice* 1989;**30**:543–9.

89 LaHue TR. Laryngeal paralysis. *Seminars in Veterinary Medicine and Surgery (Small Animal)* 1995;**10**:94–100.

90 Schofield DM, Norris J, Sadanaga KK. Bilateral thyroarytenoid cartilage lateralization and vocal fold excision with mucosoplasty for treatment of idiopathic laryngeal paralysis: 67 dogs (1998–2005). *Veterinary Surgery* 2007;**36**:519–25.

91 White RAS. Arytenoid lateralization: an assessment of technique, complications and long-term results in 62 dogs with laryngeal paralysis. *Journal of Small Animal Practice* 1989;**30**:543–9.

92 White RA. Laryngeal paralysis: an introduction. *Veterinary Quarterly* 1998;**20**:S2–3.

93 Jaggy A, Oliver JE. Neurologic manifestations of thyroid disease. *Veterinary Clinics of North America Small Animal Practice* 1994;**24**:487–94.

94 Jeffery ND, Talbot CE, Smith PM, Bacon NJ. Acquired idiopathic laryngeal paralysis as a prominent feature of generalised neuromuscular disease in 39 dogs. *Veterinary Record* 2006;**158**:17–21.

95 Radlinsky MG, Williams J, Frank PM, Cooper TC. Comparison of three clinical techniques for the diagnosis of laryngeal paralysis in dogs. *Veterinary Surgery* 2009;**38**:434–8.

96 Rudorf H, Barr FJ, Lane JG. The role of ultrasound in the assessment of laryngeal paralysis in the dog. *Veterinary Radiology & Ultrasound* 2001;**42**:338–43.

97 Wignall JR. *Computed Tomographic Assessment Of Canine Arytenoid Lateralization*. Thesis, Louisiana State University, 2013.

98 White RAS, Monnet E. The larynx. In: *BSAVA Manual of Canine and Feline Head, Neck and Thoracic Surgery*. Brockman DJ, Holt DE (eds). BSAVA, Gloucester 2005, pp. 94–104.

99 Jackson AM, Tobias K, Long C, Bartges J, Harvey R. Effects of various anesthetic agents on laryngeal motion during laryngoscopy in normal dogs. *Veterinary Surgery* 2004;**33**:102–6.

100 Miller CJ, McKiernan BC, Pace J, Fettman MJ. The effects of doxapram hydrochloride (dopram-V) on laryngeal function in healthy dogs. *Journal of Veterinary Internal Medicine* 2002;**16**:524–8.

101 Peterson KL, Graves M, Berke GS, *et al*. Role of motor unit number estimate electromyography in experimental canine laryngeal reinnervation. *Otolaryngology and Head & Neck Surgery* 1999;**121**:180–4.

102 van Nes JJ. An introduction to clinical neuromuscular electrophysiology. *Veterinary Quarterly* 1986;**8**:233–9.

103 van Nes JJ. Clinical application of neuromuscular electrophysiology in the dog: a review. *Veterinary Quarterly* 1986;**8**:240–50.

104 Holden D, Drobatz K. Emergency management of respiratory distress. In: *BSAVA Manual of Canine and Feline Head, Neck and Thoracic Surgery*. Brockman DJ, Holt DE (eds). BSAVA, Gloucester 2005, pp. 73–83.

105 Bureau S, Monnet E. Effects of suture tension and surgical approach during unilateral arytenoid lateralization on the rima glottidis in the canine larynx. *Veterinary Surgery* 2002;**31**:589–95.

106 Gourley IM, Paul H, Gregory C. Castellated laryngofissure and vocal fold resection for the treatment of laryngeal paralysis in the dog. *Journal of the American Veterinary Medical Association* 1983;**182**:1084–6.

107 Greenberg MJ, Bureau S, Monnet E. Effects of suture tension during unilateral cricoarytenoid lateralization on canine laryngeal resistance *in vitro*. *Veterinary Surgery* 2007;**36**:526–32.

108 Greenfield CL, Walshaw R, Kumar K, Lowrie CT, Derksen FJ. Neuromuscular pedicle graft for restoration of arytenoid abductor function in dogs with experimentally induced laryngeal hemiplegia. *American Journal of Veterinary Research* 1987;**49**:1360–6.

109 LaHue TR. Treatment of laryngeal paralysis in dogs by unilateral cricoarytenoid laryngoplasty. *Journal of the American Animal Hospital Association* 1989;**25**:317–25.

110 Lozier S, Pope E. Effects of arytenoid abduction and modified castellated laryngofissure on the rima glottidis in canine cadavers. *Veterinary Surgery* 2011;**21**:195–200.

111 Lussier B, Flanders JA, Erb HN. The effect of unilateral arytenoid lateralization on rima glottidis area in canine cadaver larynges. *Veterinary Surgery* 1996;**25**:121–6.

112 Rosin E, Greenwood K. Bilateral arytenoid cartilage lateralization for laryngeal paralysis in the dog. *Journal of the American Veterinary Medical Association* 1982; **180**:515–18.

113 Sato F, Ogura JH. Functional restoration for recurrent laryngeal paralysis: an experimental study. *Laryngoscope* 1978;**88**:855–71.

114 Smith MM, Gourley IM, Kurpershoek CJ, Amis TC. Evaluation of a modified castellated laryngofissure for alleviation of upper airway obstruction in dogs with laryngeal paralysis. *Journal of the American Veterinary Medical Association* 1986;**188**:1279–83.

115 Smith MM. Treatment of laryngeal paralysis with modified castellated laryngofissure. In: *Current Techniques in Small Animal Surgery*. Bojrab MJ (ed). Williams & Wilkins, Baltimore 1998, pp. 375–6.

116 Griffiths LG, Sullivan M, Reid SWJ. A comparison of the effects of unilateral thyroarytenoid lateralization versus cricoarytenoid laryngoplasty on the area of the rima glottidis and clinical outcome in dogs with laryngeal paralysis. *Veterinary Surgery* 2001;**30**:359–65.

116a ter Haar G. Surgical management of laryngeal paralysis. In: *Complications in Small Animal Surgery*. Griffon D, Hamaide A (eds). Wiley Blackwell, Ames 2016, pp. 178–84.

117 Bedford P. Displacement of the glosso-epiglottic mucosa in canine asphyxiate disease. *Journal of Small Animal Practice* 1983;**24**:199–207.

118 White R. Brachycephalic airway obstructing syndrome: some further controversies. congenital and hereditary diseases of dogs and cats. *Proceedings AVSTS Autumn Scientific Meeting* October 2010, 23–7.

119 Leonard HC. Entrapment of the epiglottis. *Companion Animal Practice* 1989;**19**:16–17.

120 Flanders JA, Thompson MS. Dyspnea caused by epiglottic retroversion in two dogs. *Journal of the American Veterinary Medical Association* 2009;**235**:1330–5.

121 Mullins R, McAlinden AB, Goodfellow M. Subtotal epiglottectomy for the management of epiglottic retroversion in a dog. *Journal of Small Animal Practice* 2014;**55**:383–5.

122 Parente EJ, Martin BB, Tulleners EP. Epiglottic retroversion as a cause of upper airway obstruction in two horses. *Equine Veterinary Journal* 1998;**30**:270–2.

123 Terron-Canedo N, Franklin S. Dynamic epiglottic retroversion as a cause of abnormal inspiratory noise in six adult horses. *Equine Veterinary Education* 2012;**25**: 565–9.

124 Griffiths LG, Tiruneh R, Sullivan M, Reid SW. Oropharyngeal penetrating injuries in 50 dogs: a retrospective study. *Veterinary Surgery* 2000;**29**:383–8.

125 White R, Lane JG. Pharyngeal stick penetration injuries in the dog. *Journal of Small Animal Practice* 1988;**29**:13–35.

126 Peeters ME. The management of pharyngeal stick penetrating injuries in dogs. *European Veterinary Conference: The Voorjaarsdagen* 2010, pp. 1–2.

127 Shamir MH, Leisner S, Klement E, Gonen E, Johnston DE. Dog bite wounds in dogs and cats: a retrospective study of 196 cases. *Journal of Veterinary Medicine* 2002;**49**:107–12.

128 Pavletic MM, Trout NJ. Bullet, bite, and burn wounds in dogs and cats. *Veterinary Clinics of North America Small Animal Practice* 2006;**36**:873–93.

129 Peppler C, Hübler C, Kramer M. Bite injury in the neck in dogs. Retrospective study on extent, diagnostic and therapy options. *Tierärztliche Praxis Kleintiere* 2009;**37**:75–83.

130 Nelson WA. Laryngeal trauma and stenosis. In: *Textbook of Small Animal Surgery*. Slatter D (ed). Saunders, Philadelphia 2003, pp. 845–57.

131 Dohar JE, Stool SE. Respiratory mucosa wound healing and its management. An overview. *Otolaryngologic Clinics of North America* 1995;**28**:897–912.

SURGERY OF THE EAR, NOSE AND THROAT

415

17.1 INTRODUCTION

The immediate postoperative phase is usually defined as the time period from the end of surgery to recovery from anaesthesia, with full return of (protective) reflexes and mentation. This phase is therefore often referred to as 'the recovery period'[1,2]. Following the immediate recovery period comes the postoperative period, which can range from several hours to several days or weeks, in which continued effective pain management and/or supportive care may be required to promote healing and maximal return of function.

The duration of the (immediate) postoperative phase is variable, depending on factors such as the type of surgery performed, health status with underlying specific (co) morbidities of the patient prior to surgery, anaesthetic protocol adopted and the course of the anaesthesia and surgical procedure[1,2]. The Confidential Enquiry into Small Animal Perioperative Fatalities (CEPSAF), conducted in the UK between 2002 and 2004, identified the recovery period to be high risk. Over 60% of deaths attributed to anaesthesia occurred in the first 3 hours after the end of anaesthesia[3], therefore careful monitoring of every patient during the recovery period is imperative, and recovery should be considered an integral part of the anaesthetic procedure.

Pain, nausea, regurgitation, vomiting and ileus, stress-induced catabolism, impaired pulmonary function, increased cardiac demands and risk of thromboembolism can follow surgical procedures. These problems, in addition to possible surgical site infection, can lead to severe complications and need for prolonged hospitalisation at best. Patients that are hyporexic or anorexic are at even greater risk for perioperative morbidity and mortality. The success and outcome of a surgical procedure does not only depend on the knowledge, experience and instrument and tissue handling of the otorhinolaryngological surgeon, but on intensive collaboration with nurses and anaesthetists to provide the best peri- and postoperative care. The following are mandatory for a successful outcome:

- Improved pain relief by early intervention with multimodal analgesia.
- Stress reduction by regional anaesthetic techniques
- The proper use of antibiotics
- Providing nutritional care.
- Protection of the wound.

In this chapter, peri- and postoperative analgesia, peri- and postoperative antibacterial therapy, postoperative nutritional management and postoperative dressings and wound management of patients undergoing ear, nose and throat (ENT) surgery will be discussed in the following sections.

17.2 PERI- AND POSTOPERATIVE ANALGESIA

The ability to detect noxious stimuli is essential to an organism's survival. Detection and interpretation of noxious stimuli (nociception) is done by the nervous system, as a response to intense stimulation of thermal and mechanical stimuli or environmental and endogenous chemical irritants[4]. In addition to specialised nerve endings responding to touch, vibration, movement and proprioception, two major classes of nociceptors have been identified, Aδ afferents and C-fibre afferents[1,4,5]:

- Type I Aδ nociceptors respond to both mechanical and chemical stimuli but have relatively high heat thresholds.
- Type II have a much lower heat threshold, but a very high mechanical threshold.

Unlike other senses, such as vision, receptors for tissue damage do not return to their prestimulus state after activation. Furthermore, the sensitivity of damage-detecting sensory nerve endings may be altered after injury, and their synaptic connections inside the central nervous system (CNS) may be significantly modified and reorganised (plasticity of the CNS)[6,7]. The clinical implications are that once pain is established, analgesic drugs are much less effective and, even with a constant nociceptive stimulation over time, the pain felt by the animal increases. It is the veterinary surgeon's ethical responsibility to try to prevent (pre-emptive analgesia[8,9]) and/or treat pain in our companion animals (using multimodal or balanced analgesia)[10,11] undergoing surgery. Also, the veterinary surgeon needs to be aware of the fact that postoperative pain is influenced by the degree of tissue trauma, which is affected by the choice of surgical technique and the skills of the surgeon[4,12].

Principles of effective pain management

Postoperative analgesia starts preoperatively and is continued based on continuous pain assessment, using objective pain scoring systems. The best-validated and most widely used pain scale for dogs is the Glasgow Composite Pain Scale (GCPS), which is available online, **http://www.gla.ac.uk/media/ media_233876_en.pdf**. The UNESP-Botucatu Multidimensional Composite Pain Scale for assessing postoperative pain in cats[13] is relatively simple to use, and is freely available on the web as well, via **http://www.animalpain.com.br/en-us/**. Alternatively the Colorado State University Feline Acute Pain Scale can be used.

- Use a multimodal analgesia technique. It is recognised that analgesia regimens incorporating different classes of analgesic drugs that act on different receptor and neurotransmitter targets provide better analgesia than using a unimodal technique[10,11]. Multimodal techniques also allow more comprehensive analgesia to be provided, for example, systemic opioids are generally short acting (1–6 hours), but have a quick onset of action. Conversely, nonsteroidal anti-inflammatory drugs (NSAIDs) have a long duration of action (24 hours), but there is a delay of 30–60 minutes after administration before they become effective[2].

- Use loco-regional techniques to supplement systemic analgesia (provided by opioids) if possible. These techniques reduce the magnitude of the surgical stress response, contribute to a balanced anaesthesia technique and allow a reduced concentration of volatile anaesthetic agent during anaesthesia and contribute to postoperative analgesia[2]. In the immediate postoperative period they allow lower doses of opioids to be used, or may allow administration of a partial μ-agonist opioid rather than a full μ-agonist opioid, which will generally be associated with reduced side-effects such as sedation and mild hypothermia.

- Opioids form the backbone of analgesia regimens[2,4]; they have proven efficacy and cause minimal haemodynamic effects. They are very flexible drugs; the dose and dose interval can be adjusted to provide optimal pain relief for the individual patient. Some opioids can be administered by continuous rate infusion (CRI) (e.g. morphine, fentanyl), which can provide better pain relief with a lower total dose of drug compared to bolus dosing.

- NSAIDs provide very effective acute pain relief and form a valuable component of perioperative analgesia regimens[9]. Administer a NSAID to surgical patients unless contraindicated in an individual patient, preferably prior to surgical tissue damage being induced.

- Adjunctive analgesics (e.g. ketamine, lidocaine, dexmedetomidine) can be useful to provide supplementary analgesia to patients in moderate to severe pain that is poorly controlled with opioids and NSAIDs[2] (some patients with very painful ear conditions, such as after bilateral total ear canal ablation with lateral bulla osteotomy [TECA–LBO], or in patients with severe anxiety and stress where panting can exacerbate postoperative swelling of the airways, such as after airway surgery in brachycephalics). There is a poor evidence base on which to justify dose, and studies documenting analgesic efficacy are weak[2]. These drugs generally need to be given by CRI intravenously; careful monitoring of the patient (sedation, pain) is essential. They should not be instigated unless a good background of opioid analgesia is established.

There has been an upsurge in the popularity of oral tramadol for acute pain management in dogs and cats, although the drug is unlicensed in animals. A limited number of studies have shown some benefit of tramadol in dogs[14,15] and cats[16,17], although analgesia is very variable between animals and side-effects (e.g. dysphoria, sedation, nausea) are common. A NSAID should be administered in preference to tramadol if possible; furthermore, most studies fail to show increased analgesia provided by oral tramadol combined with a NSAID compared to a NSAID alone, therefore routine administration of tramadol and a NSAID is unwarranted in most animals. Oral tramadol is very bitter and is therefore not tolerated by most cats, even when given disguised in food or in gelatine capsules.

Endoscopic procedures

Rhinoscopic and video-otoscopic procedures require deep anaesthesia or profound analgesia to prevent patient response to the procedure[18]. Premedication with a μ-opioid is preferable over use of butorphanol[18]. Lidocaine may be instilled through the nares to minimise response upon insertion of the scope in the nasal vestibule[18].

Retroflexion of the scope for nasopharyngoscopy is often accompanied by jerking or twisting of the head or gagging, and can potentially be reduced by spraying the soft palate and endoscope with lidocaine. Bilateral maxillary nerve blocks with lidocaine, approximately 1 mg/kg per side, significantly decrease patient response to this procedure[19].

For otoscopy and ear flushing in very painful ears, the great auricular nerve that innervates the caudolateral part of the ear can be blocked. In addition, the auriculotemporal nerve can be blocked, which innervates the cranial portion of the vertical canal[20]. For most endoscopic procedures postoperative NSAIDs for 2–3 days will provide sufficient analgesia.

Ear, nose and throat surgical procedures

The duration and dose of systemic opioid analgesia in the postoperative period depends largely on pain assessment findings, observed side-effects and how long the patient is hospitalised before discharge to the owner. Most animals that undergo invasive ear or nose surgery (ablation of ear canals, bulla osteotomy, rhinotomy) benefit from local blocks administered preoperatively

as described above for endoscopic procedures, and both opioid and NSAID analgesia postoperatively[21]. Generally, when stopping opioid analgesia, if the animal has been receiving a full μ-opioid agonist (e.g. methadone) postoperatively it is precautionary to transition to a partial μ-agonist (e.g. buprenorphine) before stopping opioid therapy completely[2]. Most animals therefore will receive 1–2 days of methadone, followed by 1 day of buprenorphine before being discharged.

There are currently very limited data describing the optimal duration of NSAID administration after routine surgeries, and there are large differences between veterinary surgeons and practices. It is also important to remember that both inadequate pain control (no NSAID) as well as pain control (NSAIDs) are associated with side-effects[22], therefore unnecessary administration as well as withholding NSAIDs (without an effective alternative) may be detrimental. Decision-making should be informed by the invasiveness of the surgery and expected magnitude of inflammation and the period required for tissue healing. Invasive ear and nose procedures in dogs and cats warrant NSAID therapy during hospitalisation, and ongoing NSAID therapy in the home environment ranging for between 7 and 14 days after surgery. Less invasive procedures without bone resection, such as throat procedures, will require 3–5 days of NSAIDs upon discharge.

17.3 PERI- AND POSTOPERATIVE ANTIBACTERIAL THERAPY

- Most ENT procedures are considered clean-contaminated at best, but many are contaminated or even dirty/infected.
- Prophylactic antibacterial therapy is indicated in most ENT procedures, but can be limited to the first 24 hours after surgery in clean-contaminated procedures.
- Category III or IV wounds require antibacterial therapy beyond prophylaxis, which is best guided by culture and sensitivity results.

Infection of a wound following a surgical intervention is a serious complication and may result in the failure of the procedure. In addition, the healing time may be prolonged, dehiscence of the wound may occur and wound infections may contribute to an exacerbation of the local inflammation (pain, redness and swelling) associated with the primary surgery and may cause systemic illness (fever, anorexia and lethargy)[23]. On occasion, generalised infection can lead to septic shock affecting the function of the vital organs (liver, kidneys, heart and lungs) and may even cause death[24].

Surgical wounds are categorised by the degree of contamination, and this allows a prediction to be made regarding the likelihood of postoperative infection[23,24]:

- Clean wounds (Category I) are reported to have an infection percentage of 0–4.4%.
- In Category II (clean-contaminated wounds) the average chance of wound infection lies between 5 and 10%.
- In Category III (contaminated wounds) there is a 6–29% chance of infection.
- Category IV (dirty wounds) carries the greatest risk of infection at 30%.

Surgical wounds where the aerodigestive tract is entered are considered to be clean-contaminated at best, and most ear procedures in small animals would be classified as contaminated or even dirty (inflammatory process present with purulent discharge). In the human literature there is level I evidence to support the use of prophylaxis in clean-contaminated head and neck procedures (where the aerodigestive tract is opened) and tonsillectomy, while level II evidence fails to support the use of prophylaxis in clean head and neck procedures such as salivary gland resection and thyroidectomy[25–27]. Interestingly, in one report the lowest rate of surgical site infection was observed after otosurgical procedures (6%), and the highest rate (13%) after head and neck (HN) procedures[26], whereas in animals the highest infection rates are found after otological procedures[28]. In a study on HN cancer surgery antibiotic prophylaxis was shown to be effective in reducing the incidence of surgical site infections, but patients undergoing clean-contaminated or contaminated procedures only benefited from perioperative antibiotic prophylaxis restricted to the first 24 postoperative hours. No benefit was demonstrated with its extension beyond these 24 hours[29].

Infection rates of up to 40%[28,30–36] have been reported for dogs undergoing otic surgery, but in cats surgical site infection appears to be less of a problem[37–39]. Although wound drainage after TECA–LBO with active or passive drains has always been recommended to combat incisional complications in early reports, no difference in wound complication rates was seen when TECA–LBO procedures were closed primarily with or without drainage[36]. The use of drains after TECA–LBO therefore depends on the surgeon's preference.

Whereas many complications can arise during and after rhinotomy or pharyngolaryngeal surgery, surgical site infections are rare in small animals and are seldom reported[40].

Antimicrobial prophylaxis

Although antimicrobial substances play an important role in the prevention and treatment of surgical infections, good surgical technique remains the most important objective, with strict adherence to proper aseptic techniques minimising contamination by pathogens. Measures must also be taken to limit other risk factors as much as possible to improve the condition of the patient[23].

- Successful surgery depends on an atraumatic surgical technique and minimising the duration of anaesthesia.
- Mistakes in the basic surgical approach can and should never be corrected by antibiotics!

Antimicrobial prophylaxis is usually defined by the use of antibiotics at the time of surgery in order to decrease the number of micro-organisms to a level that the host defence mechanisms can effectively eradicate to prevent an established surgical site infection. In order to obtain an adequate antimicrobial prophylaxis during an operation, sufficiently high levels of antibiotics must be present at the surgical site during the time of contamination.

For this reason, it is preferable to administer the antibiotic intravenously shortly before the commencement of the operation (30–60 minutes before the incision)[24]. Some antibiotics, however, reach a sufficiently high content quickly after intramuscular administration. Local prophylaxis is not advised because of possible tissue irritation at the site of the wound, and the more reliable effective antibiotic plasma levels that are obtained after intravenous administration[24]. After an operation it is not necessary for tissue levels of antibiotic in the body to remain high; usually a few hours are sufficient, unless there is a good reason to continue (as in the case of infection). In such situations, this is no longer regarded as antimicrobial prophylaxis, but rather as antimicrobial therapy.

In Category II (clean-contaminated wounds), it is advisable to utilise antibiotics as prophylaxis in most circumstances[25-27]. The choice of antibiotic depends on which type of bacteria is likely to be the most significant contaminant. For most patients undergoing ear surgery, cultures will have been part of the diagnostic work-up and a specific prophylactic antibiotic can be selected based on the culture and sensitivity results. In general (see Chapter 1 Anatomy and Physiology of the Ear, Section 1.7) the most likely encountered pathogens are *Staphylococcus pseudintermedius*, *Escherichia coli*, *Proteus mirabilis* and *Pseudomonas aeruginosa*. In the respiratory tract, *Staphylococcus* spp, *Streptococcus* spp and gram-negative bacteria can be found[23]. In Category III and IV wounds antimicrobial prophylaxis should be given, but extra measures to lower the degree of contamination are advised, such as copious lavage of the area, debridement of necrotic or severely inflamed tissue and removal of foreign material and drainage of the wound through the placement of surgical drains. Unless specified by a specific culture result, parenteral prophylaxis with either potentiated amoxicillin (amoxicillin with clavulanic acid) or cephalosporins is indicated in patients undergoing ENT surgery. There is no evidenced-based rationale to continue prophylactic antibiotics for longer then 24 hours after surgery.

Postoperative antibiotic treatment

The antibacterial treatment is continued after surgery in Category III and IV wounds until the wound is healing well (7–28 days after surgery)[23]. For Category II wounds, prophylaxis alone may be sufficient, provided the recovery was uneventful and, for instance, no regurgitation has occurred that could lead to aspiration pneumonia. In patients with pre-existing conditions however, such as low-grade bronchopneumonia, which is commonly seen in brachycephalic dogs, a postoperative course of antibiotics is also indicated.

17.4 POSTOPERATIVE NUTRITIONAL MANAGEMENT AFTER ENT SURGERY

- Nutritional assessment should be performed on any surgical patient because nutrition is critical for maintaining the patient's health and improving response to disease and injury.
- The hypermetabolic state after surgery requires early nutritional support with arginine, taurine, 1-glutamine, omega-6 and omega-3 fatty acids to provide energy and protein to promote recovery.
- Soft canned foods with high moisture, fat and protein content and high palatability are ideal for the postoperative period in patients undergoing pharyngolaryngeal surgery.

In people, nutritional status is an important determinant of outcome after surgery, and a clear association has been noted between malnutrition and poor clinical outcomes[41]. Because malnutrition imparts similar metabolic effects in animals, it is assumed that nutritional support is equally essential for the recovery of critically ill dogs and cats[42]. Veterinary surgical patients receive pre- and postoperative care to optimise the chance of a rapid and uncomplicated recovery, such as oxygen, fluids, anti-inflammatory drugs and other medications to ensure a haemodynamically and metabolically stable patient during and after the surgical procedure. However, nutritional assessment and management of nutrition should be equally considered in the therapeutic plan[43].

Nutritional assessment should be performed on every surgical patient at every consultation/evaluation, because nutrition is critical for maintaining patients' health, as well as for improving their response to disease and injury[43]. Guidelines for assessing the nutritional status of patients can be found on the WSAVA website (**http://www.wsava. org/guidelines/global-nutrition-guidelines**).

Animals requiring ENT surgery may be afflicted with serious conditions such as severe pain (chronic ear problems), swallowing disorders leading to malnutrition, cancer or severe dyspnoea. Metabolic responses to illness or severe injury are complex and place these animals at high risk for malnutrition and its deleterious effects[42]. These effects, which are likely to affect overall survival negatively, include alterations in energy and substrate metabolism, compromised immune function and impaired wound healing[44-46] and are of particular interest in surgical patients, for whom proper wound healing is of paramount importance.

According to the 'ebb/flow' model, animals undergoing surgery demonstrate:

- An initial hypometabolic response ('ebb phase').
- Followed by a more prolonged course of hypermetabolism ('flow phase')[43,45].

The ebb phase is usually a period of haemodynamic instability associated with decreased energy expenditure, hypothermia, mild protein catabolism, decreased cardiac output and poor tissue perfusion. Without intervention, this may progress to a state of refractory or irreversible shock, characterised by severe lactic acidosis, decreased tissue perfusion, multiple organ failure and death[45,47,48]. Careful monitoring of the patient during anaesthesia and appropriate use of intravenous fluids and drugs can address these factors.

After a successful surgery with a stable anaesthesia, patients enter the flow phase, during which profound metabolic alterations occur. Profound protein catabolism, along with increases in cardiac output, energy expenditure, glucose production and insulin and glucagon concentrations are the hallmarks of this response[44,49]. Because of the effects of inflammatory cytokines in this state of hypermetabolism, protein will be catabolised (for energy and gluconeogenesis) and used for the synthesis of acute-phase proteins and immunoglobulins that are needed for wound healing. Given this hypermetabolic state, it is important for early nutritional support to provide energy and protein to promote recovery[43]. Nutritional support (arginine, taurine, l-glutamine, omega-6 fatty acids, omega-3 fatty acids) for at least 3 days is generally needed to change from a catabolic to an anabolic state[43]. This not only necessitates sufficient food intake postoperatively, but also high-quality food intake.

> Most ENT patients suffer from hyporexia instead of anorexia, unless systemic abnormalities are present concurrently or when dealing with metastasised ear, nose or throat cancer.
> Hyporexia refers to a decreased appetite, rather than a total loss of appetite[50].

Especially after painful ear procedures, with painful mastication (otitis media) or painful deglutition (surgery of the throat), patients can be hyporexic if proper analgesia is not provided (see Section 17.2), which should therefore be addressed and corrected first. In anorexic patients, forced feeding will not stimulate spontaneous food intake, and enteral feeding or parenteral feeding should be provided[51]. For hyporectic patients, stimulation of spontaneous food intake may be sufficient and diminishes the need for other assisted feeding techniques, which are more invasive and stressful. However, if stimulation of spontaneous food intake does not result in an adequate amount of food intake, other assisted feeding techniques should be implemented to ensure the administration of sufficient amounts of nutrients during the aforementioned hypermetabolic, postoperative, phase[43].

Enteral provision of nutrients is always preferred to parenteral provision of nutrients and can be provided via nasoesophageal, oesophageal or gastric tubes if necessary. For most ENT patients, oesophageal or gastric tubes are preferred to bypass the diseased area (nose and throat). Parenteral feeding is indicated for a nonfunctional gastrointestinal tract, the inability to place an enteral feeding tube, or excessive vomiting, none of which are the case in uncomplicated ENT cases.

Spontaneous food intake is a sign of recovery, so it is important to offer food and to stimulate spontaneous food intake daily to monitor return of appetite. Hospitalisation is often a stressful event for an animal, so it is therefore important to create a calm and relaxed atmosphere with appropriate and tender care, and a light cycle consistent with the time of year to encourage food intake[43].

Some medications, including opiates, antimicrobials, diuretics, immunosuppressives and chemotherapeutics, can have inhibitory effects on food intake[44,50]. A hospitalised hyporexic animal will typically accept the food that it receives at home, but this can be made more attractive by increasing the moisture content, heating food to body temperature or offering fresh food. Higher moisture content also contributes to rehydration of patients and causes the nutrients to leave the stomach more quickly, making it less likely that the food will be vomited or regurgitated[52]. This is especially important in patients that have undergone throat surgery, such as brachycephalic animals undergoing staphylectomy.

Canned foods have high moisture, high fat and high protein content, which also enhance palatability. These are ideally suited for the first 2 weeks after throat surgery. Soft canned foods have the additional benefit of potentially causing less trauma to intrapharyngeal and intralaryngeal surgical sites. Feeding foods with increased energy density and increased protein concentration will also help prevent additional loss of lean body mass in patients with cancer cachexia and sarcopenia (loss of lean body mass that occurs with aging in absence of disease)[53].

For patients with dysphagia, offering food from a height (see Chapter 15 Diseases of the Pharynx, Section 15.7) and offering several small meals a day will be beneficial in ensuring sufficient energy intake.

17.5 POSTOPERATIVE DRESSINGS AND WOUND MANAGEMENT

The indications for applying a bandage after ENT surgery include[54–56]:
- Protection of wounds from the environment.
- Protection of wounds from the patient (licking, biting etc).
- Absorb exudate.
- Eliminate dead space.
- Apply or relieve pressure.
- Administer topical medications.
- Modulate pain by immobilising soft tissues.

In most cases bandages are applied to deal with a combination of these indications. However, due to the shape of the head, either due to long cranially tapering noses or due to absence of a nose in brachycephalic animals, the inability to eat and drink or see, bandages are not usually placed on patients undergoing nasal surgery. The local area can be protected from automutilation by providing the animal with an Elizabethan collar.

Bandages are also not recommended after throat surgery, as they either usually slip or are applied too tight, which can result in obstruction of venous and lymphatic drainage. With the exception of covering a drain in the neck area, which can usually be protected using tie-over bandages, no bandages are routinely used in this area. Large Elizabethan collars can be used as long as they do not interfere with the surgical wound.

Most commonly bandages are placed after ear surgery, mainly after othaematoma correction or TECA–LBO, the technique of which will be discussed here. Cats do not tolerate bandages very well, but after ventral bulla osteotomy a small Elizabethan collar can be provided if required, though most animals will not scratch at the surgical site.

Ear bandage

After major ear surgery dogs should have their ears dressed to prevent inadvertent removal of drains or sutures and trauma to the wound[32,57–60]. Whereas classically the auricle was reflected back over the head and taped to the local skin[61–63], the author prefers the technique described below[21].

After aural haematoma:
- Cover the auricle on both sides with absorbant bandage material (primary layer) by folding one large piece around it and allow the auricle to hang in a natural position (Fig. 17.1).
- Secure the auricle to the head by a secondary layer that holds the primary dressing in place, wrapping around the head alternating cranial and caudal of the contralateral ear, which is left out of the bandage (Fig. 17.2).

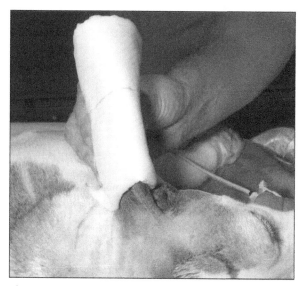

Fig. 17.1 The left auricle is covered on both sides with absorbant bandage material (primary layer) by folding one large piece around it and allowed to hang in a natural position.

- Finish with a tertiary layer of elastic material to stabilise the bandage and provide a comfortable degree of pressure on the ear using the same technique (Fig. 17.3).
- Apply short pieces of tape to secure the bandage (Fig. 17.4).

This bandage should be changed daily to remove any exudate accumulated in the bandage and allow for ventilation of the ear canals and application of topical medication if required. After TECA–LBO, absorbant material can be placed on the surgical wound and exiting drain and folded over the ipsilateral auricle, after which it is secured in the same way as described above.

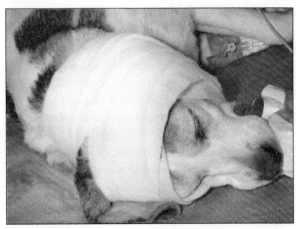

Fig. 17.2 The auricle is secured to the head by a secondary layer that holds the primary dressing in place, leaving the contralateral ear out.

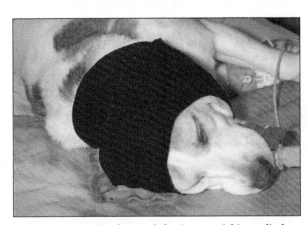

Fig. 17.3 A tertiary layer of elastic material is applied to stabilise the bandage.

Fig. 17.4 Short pieces of tape are applied to the tertiairy layer to secure the bandage, taking care not to apply them circumferentially as this may constrict blood supply. (From ter Haar G. Basic principles of surgery of the external ear [pinna and ear canal]. In: *The Cutting Edge: Basic Operating Skills for the Veterinary Surgeon*. Kirpensteijn J, Klein WR [eds]. Roman House Publishers, London 2006, pp. 272–83.)

17.6 REFERENCES

1 Brainard BM, Hofmeister EH. Anesthesia principles and monitoring. In: *Veterinary Surgery Small Animal*. Tobias KM, Johnston SA (eds). Elsevier Saunders, St. Louis 2012, pp. 248–91.

2 Murell J, Hellebrekers LJ. Post-operative care and pain management. In: *The Cutting Edge: Basic Operating Skills for the Veterinary Surgeon*. Kirpensteijn J, Klein WJ (eds). Roman House Publishers, London 2006, pp. 222–9.

3 Brodbelt DC, Blissitt KJ, Hammond RA, *et al*. The risk of death: the confidential enquiry into perioperative small animal fatalities. V*eterinary Anaesthesia and Analgesia* 2008;**35**:365–73.

4 Lascelles BD. Surgical pain: pathophysiology, assessment, and treatment strategies. In: *Veterinary Surgery Small Animal*. Tobias KM, Johnston SA (eds). Elsevier Saunders, St. Louis 2012, pp. 237–47.

5 Gurney MA. Pharmacological options for intra-operative and early postoperative analgesia: an update. *Journal of Small Animal Practice* 2012;**53**:377–86.

6 Lascelles BD, Cripps PJ, Jones A, Waterman AE. Post-operative central hypersensitivity and pain: the pre-emptive value of pethidine for ovariohysterectomy. *Pain* 1997;**73**:461–71.

7 Lascelles BD, Waterman AE, Cripps PJ, Livingston A, Henderson G. Central sensitization as a result of surgical pain: investigation of the pre-emptive value of pethidine for ovariohysterectomy in the rat. *Pain* 1995;**62**:201–12.

8 Welsh EM, Nolan AM, Reid J. Beneficial effects of administering carprofen before surgery in dogs. *Veterinary Record* 1997;**141**:251–3.

9 Lascelles BDX, Cripps PJ, Jones A, Waterman-Pearson AE. Efficacy and kinetics of carprofen, administered preoperatively or postoperatively, for the prevention of pain in dogs undergoing ovariohysterectomy. *Veterinary Surgery* 1998;**27**:568–82.

10 Williams VM, Lascelles BDX, Robson MC. Current attitudes to, and use of, peri-operative analgesia in dogs and cats by veterinarians in New Zealand. *New Zealand Veterinary Journal* 2005;**53**:193–202.

11 Corletto F. Multimodal and balanced analgesia. *Veterinary Research Communications* 2007;**31**:59–63.

12 Kristiansson M, Saraste L, Soop M, Sundqvist KG, Thörne A. Diminished interleukin-6 and C-reactive protein responses to laparoscopic versus open cholecystectomy. *Acta Anaesthesiologica Scandinavica* 1999;**43**:146–52.

13 Brondani JT, Mama KR, Luna SPL, *et al*. Validation of the English version of the UNESP-Botucatu multidimensional composite pain scale for assessing postoperative pain in cats. *BMC Veterinary Research* 2013;**9**:143.

14 Kongara K, Chambers JP, Johnson CB. Effects of tramadol, morphine or their combination in dogs undergoing ovariohysterectomy on peri-operative electroencephalographic responses and post-operative pain. *New Zealand Veterinary Journal* 2012;**60**:129–35.

15 Mastrocinque S, Fantoni DT. A comparison of preoperative tramadol and morphine for the control of early postoperative pain in canine ovariohysterectomy. *Veterinary Anaesthesia and Analgesia* 2003;**30**:220–8.

16 Cagnardi P, Villa R, Zonca A, *et al*. Pharmacokinetics, intraoperative effect and postoperative analgesia of tramadol in cats. *Research in Veterinary Science* 2011;**90**:503–9.

17 Steagall P, Taylor P, Brondani J, Luna S, Dixon M. Antinociceptive effects of tramadol and acepromazine in cats. *Journal of Feline Medicine and Surgery* 2008;**10**:24–31.

18 Hernandez SM. Anaesthesia of the dog. In: *Veterinary Anaesthesia*, 11th edn. Clark KW, Trim CM, Hall LW (eds). Elsevier, London 2013, pp. 405–98.

19 Cremer J, Sum SO, Braun C, Figueiredo J, Rodriguez-Guarin C. Assessment of maxillary and infraorbital nerve blockade for rhinoscopy in sevoflurane anesthetized dogs. *Veterinary Anaesthesia and Analgesia* 2013;**40**:432–9.

20 Buback JL, Boothe HW, Carroll GL, Green RW. Comparison of three methods for relief of pain after ear canal ablation in dogs. *Veterinary Surgery* 1996;**25**:380–5.

21 ter Haar G. Basic principles of surgery of the external ear (pinna and ear canal). In: *The Cutting Edge: Basic Operating Skills for the Veterinary Surgeon*. Kirpensteijn J, Klein WR (eds). Roman House Publishers, London 2006, pp. 272–83.

22 Gowan RA, Lingard AE, Johnston L, Stansen W, Brown SA, Malik R. Retrospective case–control study of the effects of long-term dosing with meloxicam on renal function in aged cats with degenerative joint disease. *Journal of Feline Medicine and Surgery* 2011;**13**:752–61.

23 Kummeling AK, Klein W. Surgical infections and antimicrobial prophylaxis. In: *The Cutting Edge: Basic Operating Skills for the Veterinary Surgeon*. Kirpensteijn J, Klein WR (eds). Roman House Publishers, London 2006, pp. 89–95.

24 Cimino Brown D. Wound infections and antimicrobial use. In: *Veterinary Surgery Small Animal*. Tobias KM, Johnston SA (eds). Elsevier Saunders, St. Louis 2012, pp. 135–9.

25 Fennessy BG, Harney M, O'Sullivan MJ, Timon C. Antimicrobial prophylaxis in otorhinolaryngology/head and neck surgery. *Clinical Otolaryngology* 2007;**32**:204–7.

26 Rasmussen S, Ovesen T. Insufficient reporting of infections after ear, nose and throat surgery. *Danish Medical Journal* 2014;**61**:A4735.

27 Girod DA, McCulloch TM, Tsue TT. Risk factors for complications in clean-contaminated head and neck surgical procedures. *Head & Neck* 1995;**17**:7–13.

28 Smeak DD. Management of complications associated with total ear canal ablation and bulla osteotomy in dogs and cats. *Veterinary Clinics of North America Small Animal Practice* 2011;**41**:981–94.

29 Garnier M, Blayau C, Fulgencio J-P, *et al*. Rational approach of antibioprophylaxis: systematic review in ENT cancer surgery. *Annales Francaises d'Anesthesie et de Reanimation* 2013;**32**:315–24.

30 Smeak DD, DeHoff WD. Total ear canal ablation clinical results in the dog and cat. *Veterinary Surgery* 1986;**15**:161–70.

31 Sharp NJ. Chronic otitis externa and otitis media treated by total ear canal ablation and ventral bulla osteotomy in thirteen dogs. *Veterinary Surgery* 1990;**19**:162–6.

32 Beckman SL, Henry WB, Cechner P. Total ear canal ablation combining bulla osteotomy and curettage in dogs with chronic otitis externa and media. *Journal of the American Veterinary Medical Association* 1990;**196**:84–90.

33 Mason LK, Harvey CE, Orsher RJ. Total ear canal ablation combined with lateral bulla osteotomy for end-stage otitis in dogs. Results in thirty dogs. *Veterinary Surgery* 1988;**17**:263–8.

34 Matthiesen DT, Scavelli T. Total ear canal ablation and lateral bulla osteotomy in 38 dogs. *Journal of the American Animal Hospital Association* 1990;**26**:257–67.

35 White R, Pomeroy CJ. Total ear canal ablation and lateral bulla osteotomy in the dog. *Journal of Small Animal Practice* 1990;**31**:547–53.

36 Devitt CM, Seim HB, Willer R, McPherron M, Neely M. Passive drainage versus primary closure after total ear canal ablation-lateral bulla osteotomy in dogs: 59 dogs (1985–1995). *Veterinary Surgery* 2008;**26**:210–6.

37 Ader PL, Boothe HW. Ventral bulla osteotomy in the cat. *Journal of the American Animal Hospital Association* 1979;**15**:757–62.

38 Trevor PB, Martin RA. Tympanic bulla osteotomy for treatment of middle-ear disease in cats: 19 cases (1984–1991). *Journal of the American Veterinary Medical Association* 1993;**202**:123–8.

39 Pope ER. Feline inflammatory polyps. *Seminars in Veterinary Medicine and Surgery Small Animal* 1995;**10**:87–93.

40 Mercurio A. Complications of upper airway surgery in companion animals. *Veterinary Clinics of North America Small Animal Practice* 2011;**41**:969–80.

41 Stratton RJ, Elia M. Who benefits from nutritional support: what is the evidence? *European Journal of Gastroenterology and Hepatology* 2007;**19**:353–8.

42 Chan DL. Metabolism and nutritional needs of surgical patients. In: *Veterinary Surgery Small Animal*. Tobias KM, Johnston SA (eds). Elsevier Saunders, St. Louis 2012, pp. 121–4.

43 Corbee RJ, Van Kerkhoven WJS. Nutritional support of dogs and cats after surgery or illness. *Open Journal of Veterinary Medicine* 2014;**4**:44–57.

44 Thatcher CD. Nutritional needs of critically ill patients. *Compendium on Continuing Education for the Practicing Veterinarian* 1996;**18**:1303–9.

45 Chan DL. Nutritional requirements of the critically ill patient. *Clinical Techniques in Small Animal Practice* 2004;**19**:1–5.

46 Lippert AC, Fulton RB, Parr AM. A retrospective study of the use of total parenteral nutrition in dogs and cats. *Journal of Veterinary Internal Medicine* 1993;**7**:52–64.

47 Biffl WL, Moore EE, Haenel JB. Nutrition support of the trauma patient. *Nutrition* 2002;**18**:960–5.

48 Wray CJ, Mammen JMV, Hasselgren P-O. Catabolic response to stress and potential benefits of nutrition support. *Nutrition* 2002;**18**:971–7.

49 Biolo G, Toigo G, Ciocchi B, *et al*. Metabolic response to injury and sepsis: changes in protein metabolism. *Nutrition* 1997;**13**:52S–57S.

50 Seike J, Tangoku A, Yuasa Y, Okitsu H, Kawakami Y, Sumitomo M. The effect of nutritional support on the immune function in the acute postoperative period after esophageal cancer surgery: total parenteral nutrition versus enteral nutrition. *Journal of Medical Investigation* 2011;**58**:75–80.

51 de Aguilar-Nascimento JE, Bicudo-Salomao A, Portari-Filho PE. Optimal timing for the initiation of enteral and parenteral nutrition in critical medical and surgical conditions. *Nutrition* 2012;**28**:840–3.

52 Sachdeva P, Kantor S, Knight LC, Maurer AH, Fisher RS, Parkman HP. Use of a high caloric liquid meal as an alternative to a solid meal for gastric emptying scintigraphy. *Digestive Diseases and Sciences* 2013;**58**:2001–6.

53 Freeman LM. Cachexia and sarcopenia: emerging syndromes of importance in dogs and cats. *Journal of Veterinary Internal Medicine* 2012;**26**:3–17.

54 Swaim SF. Bandages and topical agents. *Veterinary Clinics of North America Small Animal Practice* 1990;**20**:47–65.

55 Grambow Campbell B. Bandages and drains. In: *Veterinary Surgery Small Animal*. Tobias KM, Johnston SA (eds). Elsevier Saunders, St. Louis 2012, pp. 221–30.

56 Campbell BG. Dressings, bandages, and splints for wound management in dogs and cats. *Veterinary Clinics of North America Small Animal Practice* 2006;**36**:759–91.

57 McCarthy RJ. Surgery of the head & neck; total ear canal ablation with lateral bulla osteotomy. In: *Complications in Small Animals Surgery*. Lipowitz AJ, Caywood DD, Cann CC, Newton C, Schwartz A (eds). Lippincott, Williams & Wilkins, Philadelphia 1996, pp. 118–28.

58 Haudequet PH, Gauthier O, Renard E. Total ear canal ablation associated with lateral bulla osteotomy with the help of otoscopy in dogs and cats: retrospective study of 47 cases. *Veterinary Surgery* 2006;**35**:E1–20.

59 Ahirwar V, Chandrapuria VP, Bhargava MK, Madhu S, Apra S, Shobha J. A comparative study on the surgical management of canine aural haematoma. *Indian Journal of Veterinary Surgery* 2007;**28**:98–100.

60 Kolata RJ. A simple method for treating canine aural haematomas. *Canine Practice* 1984;**11**:47–50.

61 Pope ER. Head and facial wounds in dogs and cats. *Veterinary Clinics of North America Small Animal Practice* 2006;**36**:793–817.

62 Swaim SF, Henderson RA. *Small Animal Wound Management*. Williams & Wilkins, Philadelphia 1997, pp. 133–50.

63 Smeak DD, Kerpsack SJ. Total ear canal ablation and lateral bulla osteotomy for management of end-stage otitis. *Seminars in Veterinary Medicine and Surgery Small Animal* 1993;**8**:30–41.

18.1 INTRODUCTION

An otological surgeon must face the particular challenges inherent to performing surgery in an area with complex anatomy and close proximity to major vascular and neurological structures, where small errors can lead to serious complications. The variation in anatomy of the ear within dog and cat breeds and the high number of bacteria inhabiting the ear canal of patients that require surgery, create an additional element of unpredictability that, despite every effort, can lead to undesirable results and outcomes. It is important for otological surgeons to master the complex anatomy and physiology of the outer, middle and inner ear (see Chapter 1 Anatomy and Physiology of the Ear), and to master techniques to avoid surgical pitfalls and methods to manage otological complications prior to performing even 'routine' procedures on the external ear canal. Avoidance of complications is always preferred and is commonly achieved with the use of proper patient selection, precise technique, proper instrument and tissue handling, sound judgment and a realistic goal of what can be achieved in a particular situation.

Surgery of the auricle in both dogs and cats is relatively simple, compared to ear canal and middle ear surgery. Suturing of ear lacerations and amputations of the auricle are discussed in Section 18.3. Probably the most controversial otological surgery is the one for treatment of aural haematoma. Many different opinions exist as to what constitutes the best treatment for othaematoma, and both medical and surgical techniques have been described[1-6]. Aural haematoma, including the surgical technique used by the author, is discussed in Section 18.2. Ear canal surgery in dogs is reserved for those cases in which proper medical treatment for otitis externa has failed, and for ear canal neoplasia. Surgical techniques that have been described for the external auditory canal include lateral ear canal resection (LECR; Zepp and Lacroix modification)[7-10], vertical ear canal resection[1-6,11,12] or ablation (VECA) and total ear canal ablation (TECA) with lateral bulla osteotomy (LBO)[7-10,13-15]. Specific indications and complications are associated with each of these procedures, and will be discussed in Sections 18.5–18.8.

The most commonly performed otological surgeries in cats are those for removal of middle ear polyps. Middle ear polyps may grow into the nasopharynx, ear canal, or both or remain in the tympanic cavity (see Chapter 5 Aetiology and Pathogenesis of Otitis Externa, Section 5.9). Some surgeons perform a ventral bulla osteotomy (VBO) before removing the protruding polyp to facilitate removal of inflammatory tissue and detachment of the pedicle, regardless of whether the polyp is in the external ear canal or nasopharynx. However, recurrence of a nasopharyngeal polyp is uncommon with simple traction-avulsion and outweighs the risks and complications of VBO (see Chapter 11 Diseases of the Nasal Cavity and Sinuses, Section 11.8). Polyps in the ear canal can be removed via endoscopy, with the use of polyp-snaring devices or using a perendoscopic trans-tympanic traction technique (see Chapter 5 Aetiology and Pathogenesis of Otitis Externa, Section 5.9)[16]. Using a lateral approach to the ear canal (discussed here in Section 18.4) and opening the ear canal to allow the polyp to be grasped deep within the bony meatus allows for traction-avulsion of the entire polyp with stalk with a subsequent low recurrence rate of 13.5% in the hands of an experienced surgeon[17, unpublished data]. If polyps recur after simple traction-avulsion, VBO is indicated[18-20]. VBO should be the first choice treatment for all cats with polyps that present clinically with inner ear disease[20,21]. The technique for VBO is discussed in Section 18.9.

Primary secretory otitis media (PSOM), recently renamed otitis media with effusion (OME), is an increasingly recognised disease of unknown aetiopathogenesis that predominantly affects Cavalier King Charles Spaniels (see Chapter 6 Diseases of the Middle Ear, Section 6.4). Tympanostomy tubes provide continual tympanic cavity ventilation and drainage and are now commonly used. Bobbin-reuter type tympanostomy tubes can be placed using an operating microscope[22], or under endoscopic guidance[22a]. The latter technique will be discussed in Section 18.10.

Total ear canal ablation (TECA) and middle ear procedures pose an anaesthetic challenge because of the risk of haemorrhage and the degree of pain the patient experiences postoperatively[23]. Because of the proximity of the caudal auricular and superficial temporal arteries and the retroarticular (retroglenoid) vein to the horizontal ear canal, laceration or transection of these vessels can occur during TECA and LBO, leading to significant haemorrhage. Branches of the carotid artery need to be avoided during VBO. Diligent monitoring of blood loss and blood pressure, preferably via direct arterial blood pressure measurement, is helpful to ensure a successful outcome[23]. Anaesthesia is otherwise relatively straightforward, but as always depending on age and concurrent comorbidities of the patient. Peri- and postoperative analgesia, antibacterial therapy and postoperative nutritional management for patients undergoing otological surgery are discussed in Chapter 17 Perioperative Management.

18.2 AURAL HAEMATOMA

Aural haematoma is a common condition of the canine and feline pinna. Dog breeds with pendulous ears are more commonly affected[24]. Aural haematoma results from bleeding within the cartilage layers of the auricle or cartilage and perichondrium, and is characterised by a fluctuating mass between the concave and convex sides of the auricle[1,17,24,25]. The exact cause is unknown, but it has been thought to occur following trauma secondary to head shaking or otitis externa, which leads to bleeding within the cartilage or cartilage and perichondrium, leading in turn to vessel rupture, haemorrhage and aural haematoma formation[5,24–26]. However, not all cases reported have concurrent otitis externa[5,24–26]. An autoimmune pathogenesis was suggested by Kuwahara *et al.*[1], but proved to be very unlikely by Joyce and Day[27]. The diagnosis is based on the clinical appearance of the heavy auricle, bulging on the concave side (Fig. 18.1), and the absence of clinical signs of an abscess, such as fever and pain[17].

Multiple techniques have been described for treatment of aural haematoma. These include:

- The incision and suture technique[17,28,29].
- A sutureless incisional technique[28,30].
- Closed suction drainage[28].
- Placement of a cannula for intermittent drainage[28].
- Cyanoacrylate adhesives[31].
- Fibrin sealants[2].
- Drainage followed by injection or oral administration of corticosteroids[32,33].

The goals of treatment are to remove the haematoma, prevent recurrence and retain the natural appearance of the ear. Especially for large haematomas, the most predictable outcome is obtained with a surgical treatment, such as that described below[17].

Surgical technique

Surgical treatment is necessary in most cases with large haematomas because without treatment the auricle will be subject to uncontrolled fibrosis and subsequent shrivelling and ossification of the cartilage will cause continuous irritation[34]. Surgery should be postponed until coagulation has taken place, usually within 3 days. The purpose of surgery is to remove the blood clot and to press the layers of the auricle together to eliminate dead space and recurrence of the haematoma.

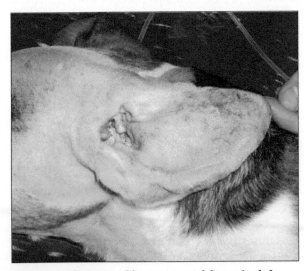

Fig. 18.1 A large aural haematoma of the entire left auricle of a Beagle.

- After surgical preparation of the entire auricle and draping of the patient, the haematoma is opened on the concave side on the most distal aspect of the haematoma with a stab incision through both skin and cartilage using a Bard-Parker scalpel handle with No. 11 blade (Fig. 18.2) until the serohaemorrhagic fluid escapes and can be suctioned[34].
- The incision is enlarged in the shape of an S over the entire haematoma to avoid folding of the auricle during the healing process and blood clots and fibrin are removed[17].
- With gentle use of a curette, fibroangioblastic tissue is removed from the inner surface of the cartilage without causing additional bleeding (Figs 18.3A, B).
- After irrigation of the cavity, interrupted mattress sutures are placed through all layers of the auricle with atraumatic monofilament absorbable suture material on a straight needle (2-0 monofilament absorbable material, such as Monocryl®), leaving the initial incision line open for continued drainage (Fig. 18.4).
- The sutures are placed over the entire surface of the haematoma and should be at least 5–7 mm wide, to avoid excessive tension on the underlying tissue, and no more than 5–7 mm apart[34]. The

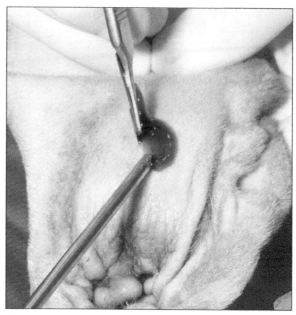

Fig. 18.2 A stab incision is made on the concave surface through skin and cartilage until serohaemorrhagic fluid escapes.

sutures are tied on the convex side, where the skin and subcutis are thicker and thus more resistant to the pressure of the knots[17,34]. Systematic interruption of the blood vessels should be

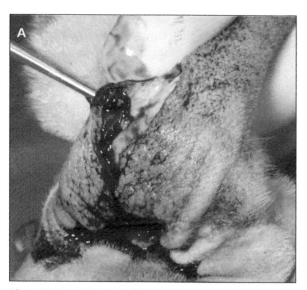

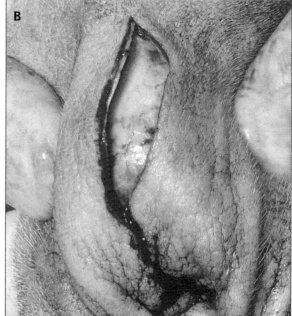

Figs 18.3A, B Existing fibroangioblastic tissue is carefully removed from the cartilage (A) until the surface is clean and slightly oozing with blood (B).

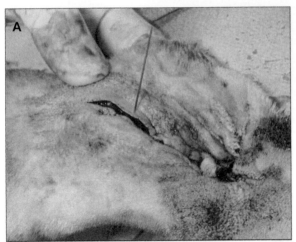

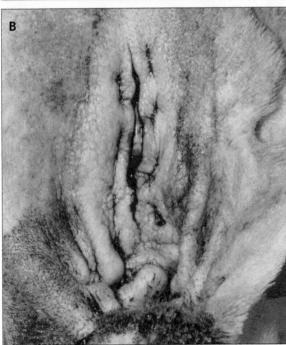

Figs 18.4A, B The incision is left open slightly to heal by second intention and allow for drainage whilst interrupted mattress sutures are placed through all layers of the auricle with atraumatic monofilament absorbable suture material on a straight needle (A), starting adjacent to the incision line (B).

prevented by placing the sutures parallel to the incision line or in a criss-cross manner.
- To prevent suture failure with this monofilament material, a surgical knot should be made with one or two additional throws.

A protective and wound fluid absorbing bandage should be placed over the ear with the ear in a caudoventral position to promote wound drainage and prevent slipping of the bandage (see Chapter 17 Perioperative Management, Section 17.5). Sutures should not be removed before 2 weeks after surgery. Treatment of concurrent otitis externa with ear ointments containing antibiotics and corticosteroids is delayed until the incision has closed to prevent delayed wound healing[34]. The bandage is changed daily until wound fluid production has decreased, usually by 3–5 days postoperatively. After removal of the bandage, the use of an Elizabethan collar is advised to prevent automutilation until all sutures are removed.

18.3 TRAUMATIC INJURY TO THE PINNA AND PINNAL AMPUTATION

Lacerations of the ear are most commonly associated with dog bites or cat scratches and can affect only the skin on one surface of the auricle, skin on one surface and the cartilage or both skin surfaces and cartilage[9]. Because of profuse blood loss, owners usually seek veterinary help in the acute situation and lacerations are therefore usually fresh[34].

- Small lacerations involving only the skin may be left to heal by second intention.
- Primary closure of ear lacerations should be performed when the cartilage is involved, as malalignment or deformity usually develops with second intention healing[6,9,35–37].
- In addition, suturing of the wound will stop acute bleeding.
- Partial amputation of the pinna can be performed for wounds that are large and peripherally located.
- For large wounds near the base of the auricle, single pedicle flaps from the lateral surface of the neck can be raised (two-step procedure)[38].

Technique for closure of traumatic lacerations of the auricle

Because wounds of the auricle are usually contaminated or infected, the wound should be thoroughly clipped and cleaned before suturing. Full-thickness injuries are closed after debridement when necessary and lavage, using fine interrupted sutures with absorbable atraumatic monofilament suture material (4-0 or 5-0)[6,17,34]. The skin on both sides of the defect is closed, avoiding the cartilage (Fig. 18.5). The surgeon should start at the edge of the ear and proceed inwards on both the concave and the convex side. An Elizabethan collar is advised to protect the wound, but with severe trauma an ear bandage is preferred. With severe trauma, abscess formation or necrosis, amputation of the auricle may be necessary.

Pinnectomy

Indications for pinnectomy are:
- Malignant tumours (actinic dermatitis, early neoplastic change and frank squamous cell carcinoma in cats, see Chapter 4 Diseases of the Pinna).
- Severe trauma.
- Pinnal abscess in dogs.

A skin incision is made around the pinna, near to its attachment to the skull in dogs, after which the subcutaneous tissues are dissected to the level of the cartilage[6,17,34]. The cartilage is then cut with scissors[17,34]. In cats, the pinnectomy can be performed in one step with scissors (Figs 18.6A, B), after which branches of the cranial and caudal auricular vein and caudal auricular artery can be coagulated with electrocautery. In dogs these vessels should be ligated. In cats, the dorsal skin can be advanced over the cartilage edge and sutured to the medial skin with interrupted sutures using absorbable material (Figs 18.6C, D). In dogs closure of a subcutaneous layer, if necessary over a Penrose drain, helps in diminishing tension on the skin sutures. The skin is closed with interrupted sutures using absorbable or nonabsorbable material. An Elizabethan collar is advised to protect the wound, or alternatively an ear bandage can be applied.

18.4 LATERAL APPROACH TO THE EAR CANAL FOR POLYP REMOVAL

Middle ear polyps in cats are relatively common masses arising from the mucosal lining of the middle ear, auditory tube or the nasopharynx[18,39,40] (Chapter 5 Aetiology and Pathogenesis of Otitis Externa, Section 5.9. The treatment of middle ear polyps that have protruded beyond the middle ear cavity into the ear canal is surgical in all cases. Described techniques include:
- Traction-avulsion.
- Transtympanic endoscopic traction-avulsion.

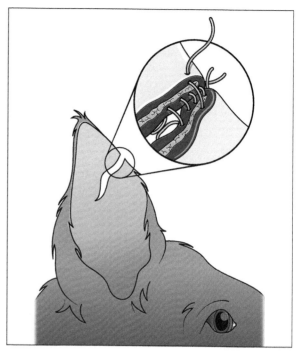

Fig. 18.5 Lacerations of the auricle are closed by suturing skin to skin on both sides, avoiding the cartilage, starting at the edge of the auricle.

- VBO.
- LECR.
- TECA–LBO[16,18,19,39–42].
- The technique described below, however, of a vertical incision in the ear canal after a lateral approach to the ear canal is simple, quick and effective and generally leads to a very low recurrence rate[17].

Surgical technique

After aseptic preparation of the surgical site:
- An incision is made in the skin in a dorsoventral direction over the palpable vertical part of the ear canal, starting just cranioventral to the tragus over approximately 2.5 cm (Fig. 18.7A)[17,34].
- The subcutaneous tissue and parotid gland are dissected with small scissors to free the cartilage of the vertical ear canal to the level of the junction between the auricular and annular cartilages.
- A vertical stab incision is made from ventral to dorsal in the auricular cartilage just above this

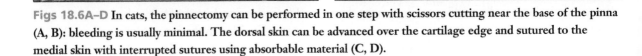

Figs 18.6A–D In cats, the pinnectomy can be performed in one step with scissors cutting near the base of the pinna (A, B): bleeding is usually minimal. The dorsal skin can be advanced over the cartilage edge and sutured to the medial skin with interrupted sutures using absorbable material (C, D).

junction with a Bard Parker scalpel handle with blade no. 11 over 7–10 mm (Fig. 18.7B)[17,34].

- Stay sutures are placed on both sides of the incision in the ear canal cartilage with fine monofilament suture material to increase visualisation and avoiding damage to the cartilage (Fig. 18.7C).
- A small closed haemostatic forceps is then introduced into the ear canal, meticulously following the direction of the horizontal ear canal until the polyp is encountered. This forceps is then opened and advanced deeper over the polyp until it can be grasped as close as possible to the osseous meatus. When a firm

grip has been achieved, the forceps is gently rotated to make sure no other tissue than the polyp itself has been grasped and traction is applied until the polyp breaks loose (Fig. 18.7D).

- With complete removal of a classical middle ear inflammatory polyp, a small stalk at the base of the polyp can usually be identified.
- The middle ear cavity is flushed with warm saline, and with a small curette the osseous meatus and most lateral aspect of the tympanic cavity is 'palpated' to check for additional inflammatory tissue, which is removed with this curette when encountered[34].

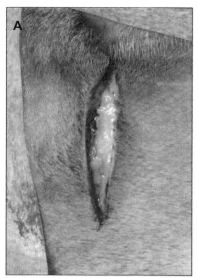

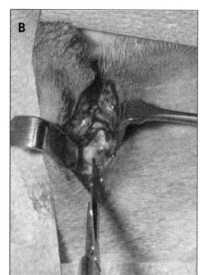

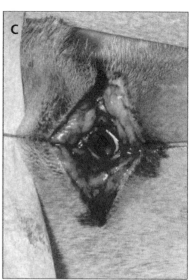

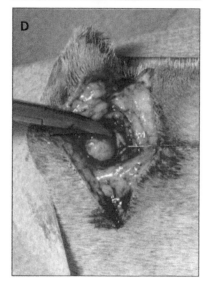

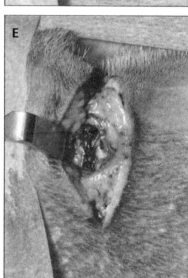

Figs 18.7A–F For ear polyp removal, an incision is made in the skin in a dorsoventral direction over the palpable vertical part of the ear canal (A). After blunt dissection to the ear canal, a vertical stab incision is made in the auricular cartilage (B). Stay sutures are placed on both sides of the incision in the ear canal cartilage (C). The polyp itself has been grasped with forceps and traction is applied until the polyp breaks loose (D). The cartilage of the ear canal is closed in an interrupted pattern (E). The subcutaneous tissues and skin are closed routinely (F).

- The stay sutures are removed and the cartilage of the ear canal is closed with 4-0 monofilament suture material in an interrupted pattern; 3 or 4 knots are usually sufficient (Fig. 18.7E).
- The subcutis is closed in a continuous pattern with 4-0 absorbable monofilament material, and the skin is closed in a subdermal suture pattern using the same material (Fig. 18.7F).
- Postoperative protective collars are usually not necessary when fluids are removed from the ear canal after surgery before recovery of the patient.

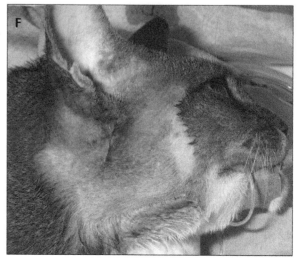

18.5 LATERAL WALL RESECTION

Traditionally, LECR was performed as part of the treatment of medically nonresponsive chronic otitis externa, as opening the lateral ear canal:

- Improves ventilation of the ear canal.
- Improves drainage of exudate.
- Reduces moisture, humidity and temperature[8–10,43–51].

LECR is contraindicated in:

- Patients with ear canal stenosis.
- Horizontal ear canal obstruction.
- Concurrent otitis media.
- Severe proliferative changes of the ear canal with auricular cartilage calcification[51].
- Underlying atopy, wherein ongoing allergic inflammation can be anticipated resulting in continual changes in the now exposed medial and lateral walls of the vertical ear canal.

However, the procedure does not result in a satisfactory outcome in 35–62% of the patients[7,50,52–54], possibly due to improper patient selection. Most animals, especially Cocker Spaniels, did not demonstrate improvement in clinical signs and still required a topical treatment after surgery[7]. Shar Pei however, demonstrated a tendency to have better results compared to other breeds, despite the narrow diameter of their ear canals[7].

For these reasons, in the author's opinion, LECR should be reserved for the removal of small neoplasms restricted to the lateral wall of the vertical ear canal or for *en bloc* resection of skin or subcutaneous neoplasms overlying the lateral ear canal. In these cases, LECR according to Lacroix is indicated, since the lateral ear canal wall is removed during surgery and no ventral drain board is created. LECR according to Zepp, where the lateral wall of the vertical ear canal is reflected to form a ventral drain board, theoretically can be used when dealing with small tumours limited to the very distal part of the lateral ear canal wall, or for the management of early stage otitis externa in Sharpeis with very narrow and swollen entrances to the vertical ear canal.

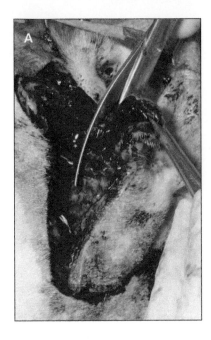

Surgical technique

- A site one half the length of the vertical ear canal is marked below the horizontal ear canal.
- Two parallel skin incisions are then made lateral to the vertical ear canal, extending from the tragus to the marked site and connected ventrally (Fig. 18.8)[6,10,49,53].
- The skinflap is dissected dorsally, exposing the lateral cartilaginous wall. With scissors the vertical canal is cut to the level of the horizontal canal cranially and caudally, from the pretragic incisure and intertragic incisure, respectively.
- The cartilage flap is reflected distally and the distal half with the neoplastic lesion is resected. The proximal half is used as a draining board.
- The skin flap is removed and sutures are now placed from the epithelial tissues to the skin, beginning at the opening of the horizontal canal[6,10,49,53].
- Complications associated with LECR include infection, incisional dehiscence, stenosis of the remaining horizontal ear canal and failure to alleviate clinical signs when performed for otitis externa[8].

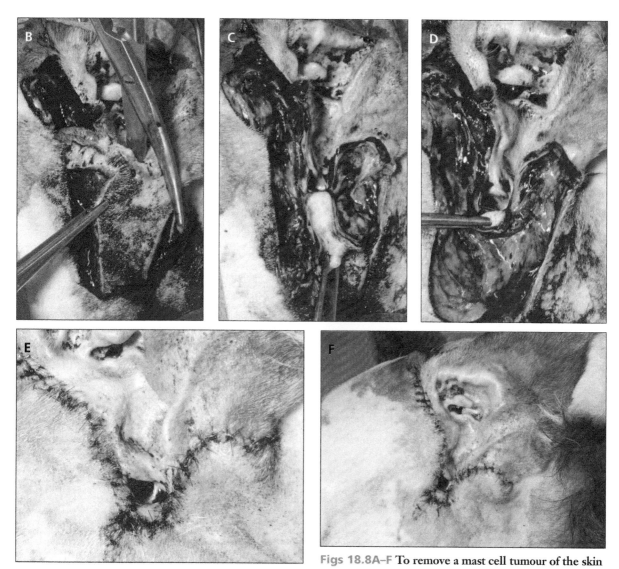

Figs 18.8A–F To remove a mast cell tumour of the skin overlying the ear canal, a Lacroix procedure is performed. After the skin incision around the tumour, two parallel incisions are made in the vertical ear canal (A, B). The cartilage flap is reflected distally (C) and amputated with the neoplastic lesion (D). After approximation of the subcutaneous tissues to reduce tension on the skin (not shown), sutures are placed from the epithelial tissues of the ear canal to the skin (E, F).

18.6 VERTICAL EAR CANAL ABLATION

Vertical ear canal ablation (VECA) can be used as a treatment option for:
- More severe cases of otitis externa.
- Cases with irreversible disease (tumours, polyps or wounds) of the medial wall of the vertical ear canal[8,11,12,55,56].
- However, a patent healthy horizontal ear canal and absence of otitis media is required.

If hearing has already been significantly affected before surgery, TECA is preferred over VECA. Results after VECA are reported to be excellent in 72%, improved in 24% and poor in 4% of animals and complications of the procedure are fewer than those reported for LECR[8,11,12,55,56].

When the 'pull-through' modification of the VECA is used, dehiscence and stenosis may be less likely as the vertical aspect of the incision is eliminated[55,56]. However, this technique does not appear to offer major advantages over the traditional technique and has not gained popularity.

Complications associated with vertical ear canal resections include infection, incisional dehiscence, stenosis of the remaining horizontal ear canal, failure to alleviate clinical signs when performed for otitis externa, facial nerve palsy and failure of erect ears to stand (cosmetic complication)[8].

Surgical technique
- After surgical preparation of the entire auricle and wide surroundings and draping of the patient, a V-shaped incision is made in the skin from the intertragic incisure to the palpable ventral limit of the vertical ear canal and from the tragohelicine incisure to the same ventral point[8,12,17,34,57].
- The triangular skin flap is lifted at this point with Adson forceps, and is dissected free towards the tragus at the level of the dermis and retracted dorsally with Allis forceps (see TECA).
- The cranial, lateral and caudal aspects of the distal vertical ear canal are exposed by blunt and sharp dissection of the subcuteaneous tissue and muscles with scissors. This dissection should be performed carefully and as close to the cartilage as possible to prevent unnecessary haemorrhage, until strong Mayo-Noble dissecting scissors can be advanced from the cranial and caudal side of the vertical ear canal under the cartilage of the auricle on the medial side[17,34].
- The cartilage and the skin of the medial wall of the ear canal is then freed from the remaining auricular cartilage with these strong scissors in an inverted U–V-shape (Fig. 18.9A):
  - One leg starts from the cranial side of the exposed vertical ear canal towards a point just above all diseased tissue, usually close to the anthelix.

- The other leg starts from the caudal side of the vertical ear canal towards and ending in the same point.
- The vertical ear canal is now dissected in a circular fashion with delicate curved Kelly scissors to the level of the horizontal ear canal to free it from all muscular and fascial attachments.
- The vertical canal is then transected ventrally 1–2 cm dorsal to the horizontal canal (Fig. 18.9B)[8–10,57].
- The remnant of the canal is incised cranially and caudally to create dorsal and ventral flaps.
- These flaps are reflected upward and downward respectively and sutured to the skin (Figs 18.9C, D).
- Closure of the remainder of the wound is as described for TECA with remodelling of the pinna.

18.7 TOTAL EAR CANAL ABLATION

Otitis externa is a multifactorial, often systemic disease of dogs and cats, and surgery should be considered to be only a part of the overall treatment plan. Careful consideration should be given to the presence of generalised skin disorders, endocrine dysfunction, allergic diseases and concurrent otitis media. Appropriate treatment of underlying disorders will increase the success rate of the surgical procedure. TECA is reserved for:
- Cases in which proper medical treatment has failed (usually end-stage proliferative otitis externa or chronic ulcerative otitis externa with a *Pseudomonas* spp. superinfection).
- Para-aural abscessation.
- Chronic ear canal avulsions.
- Ear canal neoplasia[6,9,10,13,15,34,43–48,52,58,59].

Overall complication rates of TECA ran as high as 82% with chronic deep wound infection, abscessation and debilitating fistula formation developing in 10% of the cases in early studies[13,60–63]. Since then, it has been shown that retained epithelium within the bony ear canal and/or tympanic bulla was the underlying cause for most persistent infections[15,60,64–68].

Therefore, TECA with wide tympanic cavity exposure has been considered the gold standard treatment for end-stage ear canal disease in the past two decades[66,67]. Over the past few years however, it has

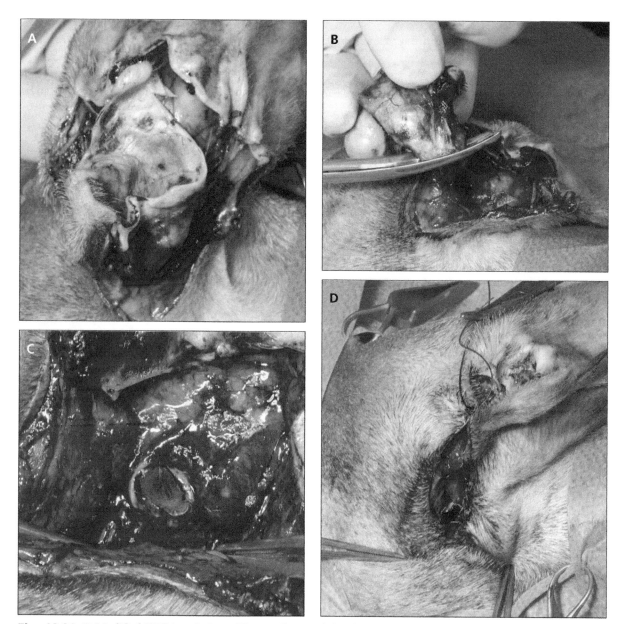

Figs 18.9A–D Modified VECA technique. The cartilage and the skin of the medial wall of the ear canal is freed from the remaining auricular cartilage with strong scissors (A). The vertical ear canal is transected just dorsal to the horizontal canal in this patient with severe stenosis at the level of the transition between horizontal and vertical ear canals (B). The remnant of the canal is sutured to the skin (C, D).

become clear that less invasive bulla osteotomies do not result in increased complication rates, as long as all the epithelium lining the ear canal up to and including the tympanic membrane is removed.

The Venker-van Haagen technique described in this chapter for remodelling of the auricle yields excellent cosmetic results whilst not making concessions to the amount of tissue removed and is therefore the technique of choice for maintaining ear carriage[17,69]. This procedure for TECA will be discussed below, and can be combined with a LBO if required, which is discussed in the next section.

Surgical technique

After surgical preparation of the entire auricle and wide surroundings and draping of the patient:

- A V-shaped incision is made in the skin from the intertragic incisure to the palpable ventral limit of the vertical ear canal and from the tragohelicine incisure to the same ventral point (Fig. 18.10A)[17,34].
- The triangular skin flap is lifted at this point with Adson forceps and is dissected free towards the tragus at the level of the dermis and retracted dorsally with Allis forceps (Fig. 18.10B)[17,34].
- The cranial, lateral and caudal aspects of the distal vertical ear canal are exposed by blunt and sharp

dissection of the subcuteaneous tissue and muscles with scissors (Fig. 18.10C). This dissection should be performed carefully and as close to the cartilage as possible to prevent unnecessary haemorrhage, until strong Mayo-Noble dissecting scissors can be advanced from the cranial and caudal side of the vertical ear canal under the cartilage of the auricle on the medial side.

- The cartilage and the skin of the medial wall of the ear canal is then freed from the remaining auricular cartilage with these strong scissors in an inverted U–V shape (Fig. 18.10D)[17,34]:
  - One leg starts from the cranial side of the exposed vertical ear canal towards a point just

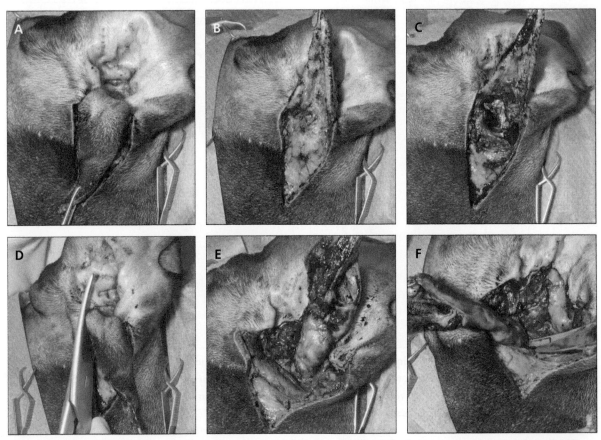

Figs 18.10A–L TECA–LBO. A V-shaped incision is made in the skin from the intertragic incisure to the palpable ventral limit of the vertical ear canal and from the tragohelicine incisure to the same ventral point (A). The triangular skin flap is dissected free towards the tragus at the level of the dermis and retracted dorsally with Allis forceps (B). The cranial, lateral and caudal aspects of the distal vertical ear canal are exposed by blunt and sharp dissection of the subcuteaneous tissue and muscles with scissors (C). The cartilage and the skin of the medial wall of the ear canal is then freed from the remaining auricular cartilage with strong scissors in an inverted U–Vshape (D). The vertical ear canal is now dissected in a circular fashion to the level of the horizontal ear canal to free it from all muscular

above all diseases tissue, usually close to the anthelix.

- The other leg starts from the caudal side of the vertical ear canal towards and ending in the same point.

- The vertical ear canal is now dissected in a circular fashion with delicate curved Kelly scissors to the level of the horizontal ear canal to free it from all muscular and fascial attachments (Fig. 18.10E).

- To control haemorrhage electrocoagulation is mandatory.

- From this level on, appropriate care should be taken to avoid the facial nerve in this area, which can be achieved by using Kelly scissors and staying as close to the cartilage of the ear canal as possible[6,58,70].

- The dissection is continued with freeing the horizontal part of the ear canal from the surrounding tissues to the level of the external acoustic meatus, which can be identified by palpation during manipulation of the ear canal.

- The cartilaginous part is separated from the osseous part with scissors, in a caudal to cranial direction with protection of the facial nerve (Fig. 18.10F).

- The success of the procedure depends on complete removal of all of the skin with ceruminous glands lining the osseous external ear canal and can be accomplished with a small curette, Adson-Brown tissue forceps and Kelly scissors[6,34,58,70].

- Repeated flushing of the external acoustic meatus helps in proper visualisation. The procedure is

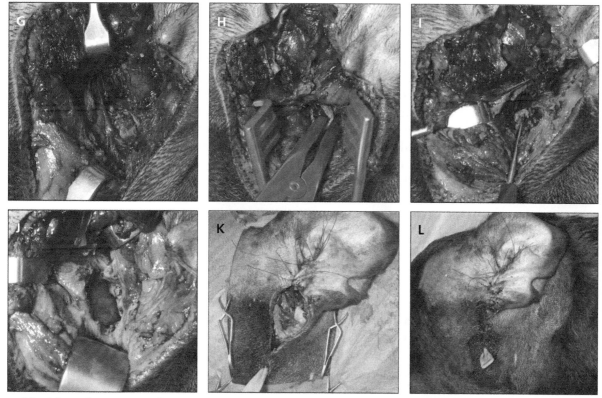

and fascial attachments (E). The cartilaginous part of the ear canal is separated from the osseous part with scissors, in a caudal to cranial direction with protection of the facial nerve (F). After TECA, the external bony meatus with remnant epithelial tissues and cerumen is visible (G). The lateral and ventral aspect of the bulla can now be removed with Kerrison, Böhler and/or Zaufal-Jansen rongeurs (H). A bone curette is used to gently remove any remaining epithelium or debris from within the bulla (I). After curettage the tympanic cavity is copiously lavaged with warm saline (J). Closure of TECA after remodelling of the auricle and placement of Penrose drain (K). After closure of all subcutaneous tissues and skin (L).

completed when after removal of the tympanic membrane, no secretory tissue is left and only bone is visible.

- When this is performed correctly, no LBO (Figs 18.10G–J) is necessary in absence of chronic otitis media[17,34]. With chronic otitis media and accumulation of inflammatory tissue or thick exsudate in the middle ear cavity, a LBO is performed from this point on.
- Closure starts with remodelling of the auricle:
  - The caudal part of the pinna is folded forward towards the cranial part, using the most natural folding point at the base of the pinna as the point of rotation[17,34].
  - These parts are then sutured together, starting at this folding point with monofilament absorbable suture material (2-0), leaving the ends of the sutures long to facilitate removal later on (Fig. 18.10K).
- A Penrose drain is then placed between the (remnants of) the external acoustic meatus and the skin, 1 cm ventral to the incision.
- The subcutaneous tissue is then closed with monofilament absorbable suture material (3-0) in two layers in a continuous fashion.
- The skin under the pinna is closed with nonabsorbable suture material (4-0) in an interrupted pattern or subdermally with monofilament absorbable suture material (4-0) in a continuous pattern (Fig. 18.10L).
- The drain is removed when drainage has decreased to small amounts of serous discharge, which is usually the case within 3–5 days after surgery.
- For the immediate postoperative recovery period an Elisabethan collar is recommended. Further perioperative care is discussed in Chapter 19 Surgery of the Nose.

18.8 LATERAL BULLA OSTEOTOMY

A lateral approach to the bulla can be performed with or without a TECA. Most commonly the procedure is performed in conjunction with TECA though, especially in dogs with chronic otitis externa and media, as the primary problem lies in the ear canal[6,13–15,71,72]. The indications are therefore similar to those for TECA as

discussed in Section 18.7. In cats with primary middle ear disease, a ventral approach to the bulla is recommended (see Section 18.9). A lateral approach to the bulla is also the recommended approach for dogs experiencing recurrent fistulation or abscessation after TECA (see Section 18.7) when residual epithelial tissue in the area of the external bony meatus is suspected[73]. Finally, a lateral approach to the bulla, without TECA, can be employed for surgery of the middle ear ossicles or for implantation of middle ear implants to augment residual hearing in animals with hearing loss[74].

Surgical technique

After TECA, with chronic otitis media and accumulation of inflammatory tissue or thick exudate in the middle ear cavity, a LBO is performed as follows:

- The tissues from the lateral aspect of the bulla are bluntly dissected as close to the bone as possible with small periosteal elevators or raspatories, avoiding damage to the facial nerve and branches of the external carotid artery that travel just ventral to the bulla[6,17,34,72].
- The lateral and ventral aspect of the bulla can now be removed with Kerrison, Böhler and/or Zaufal-Jansen rongeurs, but resection should be limited to the amount that allows adequate visualisation of the middle ear cavity (Figs 18.10G–J).
- There is no need for excessive resections of bone or subtotal bulla ostectomies.
- Samples can be obtained now for culture and susceptibility testing and for cytology or histopathology[6,17,34,72]. A bone curette is used to gently remove any remaining epithelium or debris from within the bulla, taking care to preserve the auditory ossicles and cochlea on the dorsomedial side of the bulla.
- After curettage the tympanic cavity is copiously lavaged with warm saline.
- Closure and postoperative care are as for TECA.

18.9 VENTRAL BULLA OSTEOTOMY

Ventral bulla osteotomy (VBO) is a commonly performed procedure for management of feline middle ear disease such as chronic otitis media and inflammatory polyps (see Chapter 5 Aetiology and Pathogenesis of Otitis Externa, Section 5.9 and Chapter 6 Diseases of the Middle Ear, Section 6.2). Polyps can be removed using traction-avulsion techniques or via nasopharyngotomy (for polyps protruding into the nasopharynx) or a lateral approach to the ear canal (for polyps protruding through the tympanic membrane into the ear canal).

> With severe middle ear involvement demonstrated on diagnostic imaging or recurrence of polyp formation after simple removal, a more invasive surgical treatment is recommended[19,42,75–78].

The ventral approach to the bulla provides superior exposure of the middle ear structures and allows for a more complete removal of inflammatory polyps and abnormal mucosal lining. It does not however, permit access for removal of diseased tissue from the external meatus. Wherever this is indicated, a LBO, with or without a TECA is indicated. In dogs otitis media is usually secondary to otitis externa and therefore surgical treatment of concurrent outer ear canal and middle ear disease in this species consists of TECA–LBO[66,70,76].

Some authors recommend a ventral approach to the middle ear in dogs for the treatment of cholesteatoma and for recurrent fistulation after TECA–LBO[64,70]. The technique in dogs is similar to, yet more difficult than, that in cats. Complications of ventral osteotomy in cats include Horner syndrome and development of vestibular signs[19,70,76,77,79]. Signs of Horner syndrome usually resolve over the course of a few days to a few weeks. Vestibular signs after surgery are usually iatrogenic and the results of overzealous curettage or due to caloric trauma (see Chapter 6 Diseases of the Middle Ear, Section 6.6). These signs resolve in a similar fashion as after any other acute inner ear insult and peripheral vestibular ataxia (see Chapter 7 Diseases of the Inner Ear, Section 7.9).

Surgical technique

- The ventral neck and intermandibular area are clipped and aseptically prepared for surgery[57,76,79]. The patient is placed in dorsal recumbency with a cervical support to extend the neck to maximise exposure of the bullae.
- Care should be taken to avoid overextension of the neck, which can reduce blood flow to the brain and result in severe neurological damage. The tip of the nose should therefore always point slightly upward and not be parallel to the surgery table.
- The ventral bullae can be palpated 1–2 cm caudomedially from the ipsilateral caudal border of the mandible.
- An incision is made parallel with the midline, centred 2–3 cm towards the affected side from halfway along the mandible to the level of the atlas (Fig. 18.11A)[57,77,79].
- The platysma and sphincter colli muscles are incised and linguofacial and maxillary veins are retracted, if required.
- The incision is deepened by blunt dissection between digastricus muscle and hypoglossal and styloglossal muscles until the bulla can be palpated (Fig. 18.11B)[57,77,79]. Small Weitlaner retractors, Gelpi retractors or preferably two small Senn retractors can be used to retain the dissection site.
- After the bulla has been located and the hypoglossal nerve has been identified and protected, a small incision can be made over the bulla periosteum (Fig. 18.11C). Small elevators may be used to remove the periosteal covering over the proposed osteotomy/ostectomy site.
- A Steinmann pin can be used to make a hole on the ventral aspect (Fig. 18.11D), or alternatively a small osteotome can be used. The opening can be enlarged with a small rongeur or Kerrison punch (Fig. 18.11E)[57,77,79].
- The large ventral hypotympanic cavity is opened first.
- Material is collected for culture, sensitivity testing, cytology and histopathology (Figs 18.11F, G).
- After cleaning this compartment, the bony septum separating the ventral cavity from the dorsolateral compartment can be identified and opened with a small osteotome or Kerrison punches (Fig. 18.11H)[57,77,79].
- Again, with a small currette this compartment is emptied, taking care to clean the internal bony meatus and entrance to the auditory tube of polypous tissue.
- Usually polypous tissue can be scooped out and subsequently grabbed with small mosquito

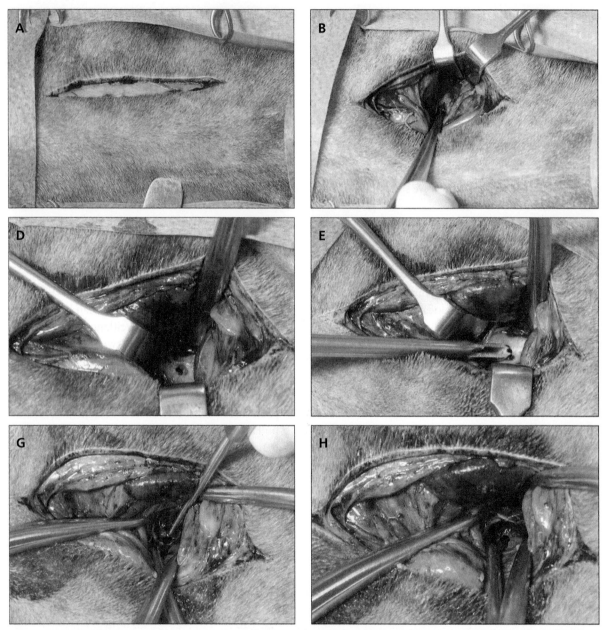

Figs 18.11A–J The head of the cat is on the left side of the images, the procedure is performed on the left bulla. VBO. An incision is made parallel with the midline, centred 2–3 cm toward the affected side from halfway along the mandible to the level of the atlas (A). The incision is deepened by blunt dissection between digastricus muscle and hypoglossal and styloglossal muscles until the bulla can be palpated (B). A small incision can be made over the bulla periosteum (C). A Steinmann pin can be used to make a hole on the ventral aspect (D), and the opening can be enlarged with a small rongeur or Kerrison punch (E). Material is collected from the large ventral compartment for culture, sensitivity testing, cytology and histopathology (F, G). The bony septum separating the ventral cavity from the dorsolateral compartment can be identified and opened with a small osteotome, Steinmann pin or Kerrison punches (H). After cleaning and flushing the dorsolateral compartment, a Penrose drain is placed (I). Subcutaneous tissues and skin are closed routinely (J).

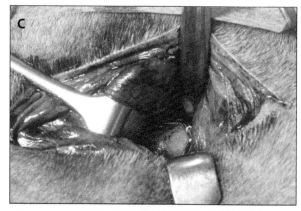

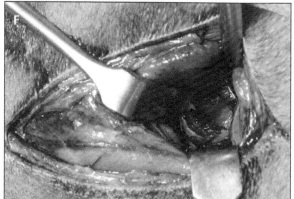

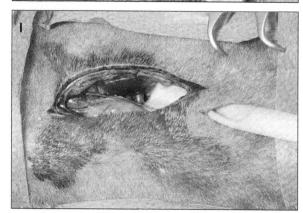

forceps. With a firm but gentle grip on the polyp and subsequent help of the little curette, the entire polyp can usually be removed completely.

- Aggressive curettage should be avoided, especially near the stapes and round window.
- The cavity is flushed and drained with a Penrose drain depending on the surgeon's preference (Fig. 18.11I).
- The muscles do not need to be closed separately, but the subcutaneous tissues are reapposed in a continuous fashion using monofilament absorbable suture material, and the skin is closed intradermally with the same material (Fig. 18.11J).

18.10 VENTILATION TUBES

Primary secretory otitis media (PSOM) or OME is an increasingly recognised disease of unknown aetiopathogenesis that predominantly affects Cavalier King Charles Spaniels[80–83] and can cause hearing loss. auditory tube dysfunction is most likely the primary factor leading to decreased drainage of the mucus from the middle ear (see Chapter 5 Aetiology and Pathogenesis of Otitis Externa, Section 5.4). Described methods of treatment consist of either multiple myringotomies, or placement of tympanostomy tubes for more continual tympanic cavity ventilation and drainage[22,79,84]. Bobbin-reuter type tympanostomy tubes can either be placed with the use of an operating microscope or under endoscopic guidance. Tympanostomy tube insertion so far has only been evaluated in three dogs under direct visualisation of the tympanic membrane with an operating microscope through an otic speculum[22]. Though the authors reported the technique to be easy to perform in a short amount of time, endoscopic placement[22a] is certainly less complicated and less cumbersome. Long-term results have yet to be reported using this technique, which is described below.

Endoscopic technique for grommet placement[22a]

- The patient is positioned in either sternal or lateral recumbency.
- The ear canal is flushed with saline if indicated.
- The surgeon performs otoscopy with a 3.5 mm rigid Storz endoscope without outer sheath. The

small size of the scope allows for the introduction of another relatively large instrument in the ear canal next to the scope, either for myringotomy, suction or manipulation and placement of the grommet.

- Once the bulging pars flaccida has been identified (Fig. 18.12A), an assistant is needed to carefully maintain pulling the auricle lateroventrally, as to allow the surgeon to use both hands for simultaneous handling of the otoscopy (left hand) and myringotomy (right hand) instruments.
- Under endoscopic guidance, using either a Politzer paracentesis needle (Fig. 18.12B) or a 2 mm curved Frazier metal suction cannula, a stab myringotomy is performed in the bulging part of the tympanic membrane.

- At this stage thick viscous mucus usually starts flowing into the ear canal. A thick plug of mucus can be removed by suction with the metal suction cannula alone (Fig. 18.12C) or after flushing the middle ear cavity with saline through the metal tube placed just through the myringotomy site.
- Care should be taken not to enlarge the opening in the tympanic membrane excessively, as it needs to hold the grommet in place.
- Using small crocodile biopsy forceps, the largest size Bobbin-reuter tympanostomy tube possible, usually a 1.5 mm inner diameter grommet, is

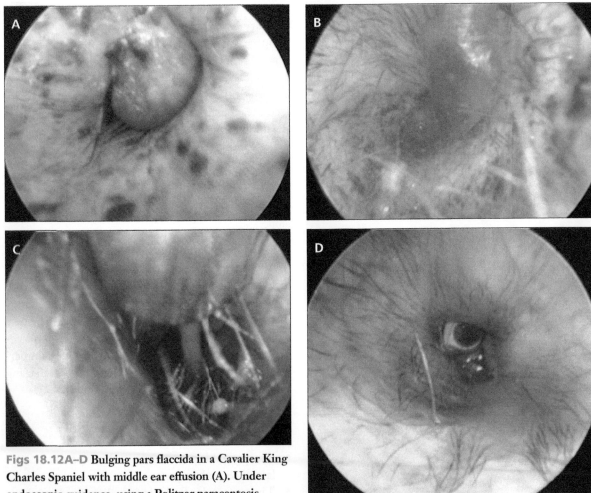

Figs 18.12A–D Bulging pars flaccida in a Cavalier King Charles Spaniel with middle ear effusion (A). Under endoscopic guidance, using a Politzer paracentesis needle (B) a stab myringotomy is performed in the bulging part of the tympanic membrane. A thick plug of mucus is removed by suction with the metal suction cannula (C); a Bobbin-reuter tympanostomy tube of 1.5 mm inner diameter is advanced carefully into the myringotomy site until one flared end is within the middle ear cavity (D).

introduced into the ear canal under endoscopic guidance with the right hand.

- The grommet is advanced carefully into the myringotomy site and further manipulated in place using a small Lucae cerumenhook, the small curved Frazier metal suction cannula (without suction) or the crocodile forceps, until one flared end is within the middle ear cavity (Fig. 18.12D).
- Tubes are pushed out of the tympanic membrane as a result of epithelial migration within 3–9 months after which decreased hearing can be expected.
- Postoperative care consists of a 7-day course of broad-spectrum systemic antibiotics and 3 days of non-steroidal anti-inflammatory drugs.

18.11 REFERENCES

1 Kuwahara J. Canine and feline aural hematoma: clinical, experimental, and clinicopathologic observations. *American Journal of Veterinary Research* 1986;**47**:2300–8.

2 Blattler U, Harlin O, Mattison RG, Rampelberg F. Fibrin sealant as a treatment for canine aural haematoma: a case history. *Veterinary Journal* 2007;**173**:697–700.

3 Ahirwar V, Chandrapuria VP, Bhargava MK, Madhu S, Apra S, Shobha J. A comparative study on the surgical management of canine aural haematoma. *Indian Journal of Veterinary Surgery* 2007;**28**:98–100.

4 Kolata RJ. A simple method for treating canine aural haematomas. *Canine Practice* 1984;**11**:47–50.

5 Joyce JA. Treatment of canine aural haematoma using an indwelling drain and corticosteroids. *Journal of Small Animal Practice* 1994;**35**:341–4.

6 Bacon NJ. Pinna and external ear canal. In: *Veterinary Surgery Small Animal.* Tobias KM, Johnston SA (eds). Elsevier Saunders, St. Louis 2012, pp. 2059–77.

7 Sylvestre AM. Potential factors affecting the outcome of dogs with a resection of the lateral wall of the vertical ear canal. *Canadian Veterinary Journal* 1998;**39**:157–60.

8 McCarthy RJ. Surgery of the head and neck; lateral and vertical ear canal resection. In: *Complications in Small Animals Surgery.* Lipowitz AJ, Caywood DD, Cann CC, Newton C, Schwartz A (eds). Lippincott, Williams & Wilkins, Philadelphia 1996, pp. 114–18.

9 Lanz OI, Wood BC. Surgery of the ear and pinna. *Veterinary Clinics of North America Small Animal Practice* 2004;**34**:567–99.

10 Elkins AD. Surgery of the external ear canal. *Problems in Veterinary Medicine* 1991;**3**:239–53.

11 McCarthy RJ, caywood DD. Vertical ear canal resection for end-stage otitis externa in dogs. *Journal of the American Animal Hospital Association* 1992;**28**:545–52.

12 Siemering GH. Resection of the vertical ear canal for treatment of chronic otitis externa. *Journal of the American Animal Hospital Association* 1980;**16**:753–8.

13 Mason LK, Harvey CE, Orsher RJ. Total ear canal ablation combined with lateral bulla osteotomy for end-stage otitis in dogs. Results in thirty dogs. *Veterinary Surgery* 1988;**17**:263–8.

14 Beckman SL, Henry WB, Cechner P. Total ear canal ablation combining bulla osteotomy and curettage in dogs with chronic otitis externa and media. *Journal of the American Veterinary Medical Association* 1990;**196**:84–90.

15 Smeak DD, Kerpsack SJ. Total ear canal ablation and lateral bulla osteotomy for management of end-stage otitis. *Seminars in Veterinary Medicine and Surgery Small Animal* 1993;**8**:30–41.

16 Greci V, Vernia E, Mortellaro CM. Per-endoscopic trans-tympanic traction for the management of feline aural inflammatory polyps: a case review of 37 cats. *Journal of Feline Medicine and Surgery* 2014;**16**:645–50.

17 Venker-van Haagen AJ. The ear. In: *Ear, Nose, Throat, and Tracheobronchial Diseases in Dogs and Cats.* Venker-van Haagen AJ (ed). Schlütersche Verlagsgesellschaft, Hannover 2005, pp. 1–50.

18 Anderson DM, Robinson RK, White RA. Management of inflammatory polyps in 37 cats. *Veterinary Record* 2000;**147**:684–7.

19 Faulkner JE, Budsberg SC. Results of ventral bulla osteotomy for treatment of middle ear polyps in cats. *Journal of the American Animal Hospital Association* 1990;**26**:496–9.

20 Boothe HWJ. Surgical management of otitis media and otitis interna. *Veterinary Clinics of North America Small Animal Practice* 1988;**18**:901–11.

21 Boothe HW. Surgery of the tympanic bulla (otitis media and nasopharyngeal polyps). *Problems in Veterinary Medicine* 1991;**3**:254–69.

22 Corfield GS, Burrows AK, Imani P, Bryden SL. The method of application and short term results of tympanostomy tubes for the treatment of primary secretory otitis media in three Cavalier King Charles Spaniel dogs. *Australian Veterinary Journal* 2008;**86**:88–94.

22a Guerin V, Hampel R, ter Haar G. Video-otoscopy-guided tympanostomy tube placement for the treatment of middle ear effusion. *Journal of Small Animal Practice* 2015;**56**:606–12.

23 Brainard BM, Hofmeister EH. Anesthesia principles and monitoring. In: *Veterinary Surgery Small Animal*. Tobias KM, Johnston SA (eds). Elsevier Saunders, St. Louis 2012, pp. 248–91.

24 Dye TL, Teague HD, Ostwald DAJ, Ferreira SD. Evaluation of a technique using the carbon dioxide laser for the treatment of aural hematomas. *Journal of the American Animal Hospital Association* 2002;**38**:385–90.

25 Dubielzig RR, Wilson JW, Seireg AA. Pathogenesis of canine aural hematomas. *Journal of the American Veterinary Medical Association* 1984;**185**:873–5.

26 Kagan KG. Treatment of canine aural hematoma with an indwelling drain. *Journal of the American Veterinary Medical Association* 1983;**183**:972–4.

27 Joyce JA, Day MJ. Immunopathogenesis of canine aural haematoma. *Journal of Small Animal Practice* 1997;**38**:152–8.

28 Swaim SF, Bradley DM. Evaluation of closed-suction drainage for treating auricular hematomas. *Journal of the American Animal Hospital Association* 1996;**32**:36–43.

29 Cechner PE. Suture technique for repair of aural hematoma. In: *Current Techniques in Small Animal Surgery*, 4th edn. Bojrab MJ (ed). Williams & Wilkins, Baltimore 1998, pp. 95–7.

30 Bojrab MJCGM. Sutureless technique for repair of aural hematoma. In: *Current Techniques in Small Animal Surgery*, 4th edn. Bojrab MJ (ed). Williams & Wilkins, Baltimore 1998, pp. 97–8.

31 Leftwich MW, Carey DP. Cyanoacrylate adhesive for aural hematoma. *Veterinary Medicine Small Animal Clinics* 1981;**76**:1155.

32 Kuwahara J. Canine and feline aural haematomas: results of treatment with corticosteroids. *Journal of the American Animal Hospital Association* 1986;**22**:641–7.

33 Romatowski J. Nonsurgical treatment of aural hematomas. *Journal of the American Veterinary Medical Association* 1994;**204**:1318.

34 ter Haar G. Basic principles of surgery of the external ear (pinna and ear canal). In: *The Cutting Edge: Basic Operating Skills for the Veterinary Surgeon*. Kirpensteijn J, Klein WR (eds). Roman House Publishers, London 2006, pp. 272–83.

35 Horne RD, Henderson RA. The pinna. In: *Textbook of Small Animal Surgery*, 2nd edn. Slatter D (ed). WB Saunders, Philadelphia 1993, pp. 1545–59.

36 Rice JJ, May BJ, Spirou GA, Young ED. Pinna-based spectral cues for sound localization in cat. *Hearing Research* 1992;**58**:132–52.

37 Henderson RA, Horne R. Pinna. In: *Textbook of Small Animal Surgery*. Slatter D (ed). Saunders, Elsevier Science, Philadelphia 2003, pp. 1737–46.

38 Buiks SC, ter Haar G. Reconstructive techniques of the facial area and head. In: *Reconstructive Surgery & Wound Management of the Dog & Cat*. Kirpensteijn J, ter Haar G (eds). Manson Publishing, London 2012, pp. 95–116.

39 Pope ER. Feline inflammatory polyps. *Seminars in Veterinary Medicine and Surgery Small Animal* 1995;**10**:87–93.

40 Fan TM, de Lorimier L-P. Inflammatory polyps and aural neoplasia. *Veterinary Clinics of North America Small Animal Practice* 2004;**34**:489–509.

41 Muilenburg RK, Fry TR. Feline nasopharyngeal polyps. *Veterinary Clinics of North America Small Animal Practice* 2002;**32**:839–49.

42 Trevor PB, Martin RA. Tympanic bulla osteotomy for treatment of middle-ear disease in cats: 19 cases (1984–1991). *Journal of the American Veterinary Medical Association* 1993;**202**:123–8.

43 Bojrab MJ, Constantinescu GM. Treatment of otitis externa. In: Current *Techniques in Small Animal Surgery*. Bojrab MJ (ed). Williams & Wilkins, Baltimore 1998, pp. 98–101.

44 Bradley RL. Surgical management of otitis externa. *Veterinary Clinics of North America Small Animal Practice* 1988;**18**:813–19.

45 Harvey CE. Ear canal disease in the dog: medical and surgical management. *Journal of the American Veterinary Medical Association* 1980;**177**:136–9.

46 Hobson HP. Surgical management of advanced ear disease. *Veterinary Clinics of North America Small Animal Practice* 1988;**18**:821–44.

47 McCarthy PE, McCarthy RJ. Surgery of the ear. *Veterinary Clinics of North America Small Animal Practice* 1994;**24**:953–69.

48 White RA. The ear: surgery for chronic otitis. *Veterinary Quarterly* 1998;**20**:S7–9.

49 Coffey DJ. Lateral ear drainage for otitis externa. In: *Current Techniques in Small Animal Surgery*. Bojrab MJ (ed). Lea & Febiger, Philadelphia 1975, pp. 64–7.

50 Gregory CR, Vasseur PB. Clinical results of lateral ear resection in dogs. *Journal of the American Veterinary Medical Association* 1983;**182**:1087–90.

51 Layton CE. The role of lateral ear resection in managing chronic otitis externa. *Seminars in Veterinary Medicine and Surgery Small Animal* 1993;**8**:24–9.

52 Doyle RS, Skelly C, Bellenger CR. Surgical management of 43 cases of chronic otitis externa in the dog. *Iranian Veterinary Journal* 2010;**57**:22–30.

53 Tufvesson G. Operation for otitis externa in dogs according to Zepp's method; a statistical analysis of follow-up examinations and a study of possible age, breed, or sex disposition to the disease. *American Journal of Veterinary Research* 1955;**16**:565–70.

54 Lane JG, Little CJL. Surgery of the canine external auditory meatus: a review of failures. *Journal of Small Animal Practice* 1986;**27**:247–54.

55 Tirgari M. Long-term evaluation of the pull-through technique for vertical canal ablation for the treatment of otitis externa in dogs and cats. *Journal of Small Animal Practice* 1988;**29**:165–75.

56 Tirgari M, Pinniger RS. Pull-through technique for vertical canal ablation for the treatment of otitis externa in dogs and cats. *Journal of Small Animal Practice* 1986;**27**:123–31.

57 Fossum TW. Surgery of the ear. In: *Small Animal Surgery*, 3rd edn. Fossum TW, Hedlund CS, Hulse DA, *et al.* (eds). Mosby Elsevier, St. Louis 2007, pp. 289–316.

58 White R, Pomeroy CJ. Total ear canal ablation and lateral bulla osteotomy in the dog. *Journal of Small Animal Practice* 1990;**31**:547–53.

59 Williams JM, White R. Total ear canal ablation combined with lateral bulla osteotomy in the cat. *Journal of Small Animal Practice* 1992;**33**:225–7.

60 Smeak DD, DeHoff WD. Total ear canal ablation clinical results in the dog and cat. *Veterinary Surgery* 1986;**15**:161–70.

61 Anders BB, Hoelzler MG, Scavelli TD, Fulcher RP, Bastian RP. Analysis of auditory and neurologic effects associated with ventral bulla osteotomy for removal of inflammatory polyps or nasopharyngeal masses in cats. *Journal of the American Veterinary Medical Association* 2008;**233**:580–5.

62 Matthieson DT, Scavelli T. Total ear canal ablation and lateral bulla osteotomy in 38 dogs. *Journal of the American Animal Hospital Association* 1990;**26**:257–67.

63 Sharp NJ. Chronic otitis externa and otitis media treated by total ear canal ablation and ventral bulla osteotomy in thirteen dogs. *Veterinary Surgery* 1990;**19**:162–6.

64 Hardie EM, Linder KE, Pease AP. Aural cholesteatoma in twenty dogs. *Veterinary Surgery* 2008;**37**:763–70.

65 Smeak DD, Crocker CB, Birchard SJ. Treatment of recurrent otitis media that developed after total ear canal ablation and lateral bulla osteotomy in dogs: nine cases (1986–1994). *Journal of the American Veterinary Medical Association* 1996;**209**:937–42.

66 Smeak DD, Inpanbutr N. Lateral approach to subtotal bulla osteotomy in dogs. *Compendium on the Continuing Education for the Veterinary Practitioner* 2005;**27**:377–84.

67 McAnulty JF, Hattel A, Harvey CE. Wound healing and brain stem auditory evoked potentials after experimental total ear canal ablation with lateral tympanic bulla osteotomy in dogs. *Veterinary Surgery* 1995;**24**:1–8.

68 Doust R, King A, Hammond G, *et al.* Assessment of middle ear disease in the dog: a comparison of diagnostic imaging modalities. *Journal of Small Animal Practice* 2007;**48**:188–92.

69 Venker-van Haagen AJ. Managing diseases of the ear. In: *Current Veterinary Therapy*. Kirk RW (ed). WB Saunders, Philadelphia 1983, pp. 47–52.

70 Smeak DD. Management of complications associated with total ear canal ablation and bulla osteotomy in dogs and cats. *Veterinary Clinics of North America Small Animal Practice* 2011;**41**:981–94.

71 Mathews KG, Hardie EM, Murphy KM. Subtotal ear canal ablation in 18 dogs and one cat with minimal distal ear canal pathology. *Journal of the American Animal Hospital Association* 2006;**42**:371–80.

72 Smeak DD. Total ear canal ablation and lateral bulla osteotomy. In: *Current Techniques in Small Animal Surgery*. Bojrab MJ (ed). Williams & Wilkins, Baltimore 1998, pp. 102–8.

73 Holt D, Brockman DJ, Sylvestre AM, Sadanaga KK. Lateral exploration of fistulas developing after total ear canal ablations: 10 cases (1989–1993). *Journal of the American Animal Hospital Association* 1996;**32**: 27–30.

74 ter Haar G, Mulder JJ, Venker-van Haagen AJ, van sluijs FJ, Smoorenburg GF. A surgical technique for implantation of the vibrant soundbridge middle ear implant in dogs. *Veterinary Surgery* 2011;**40**:340–6.

75 Ader PL, Boothe HW. Ventral bulla osteotomy in the cat. *Journal of the American Animal Hospital Association* 1979;**15**:757–62.

76 Booth HW. Ventral bulla osteotomy: dog and cat. In: *Current Techniques in Small Animal Surgery*. Bojrab MJ (ed). Williams & Wilkins, Baltimore 1998, pp. 109–12.

77 Donnelly KE, Tillson DM. Feline inflammatory polyps and ventral bulla osteotomy. *Compendium on the Continuing Education for the Veterinary Practitioner* 2004;**26**:446–54.

78 Gotthelf LN. Diagnosis and treatment of otitis media in dogs and cats. *Veterinary Clinics of North America Small Animal Practice* 2004;**34**:469–87.

79 White RAS. Middle and inner ear. In: *Veterinary Surgery Small Animal*. Tobias KM, Johnston SA (eds). Elsevier Saunders, St. Louis 2012, pp. 2078–89.

80 Stern-Bertholtz W, Sjostrom L, Hakanson NW. Primary secretory otitis media in the Cavalier King Charles spaniel: a review of 61 cases. *Journal of Small Animal Practice* 2003;**44**:253–6.

81 Rusbridge C. Primary secretory otitis media in Cavalier King Charles spaniels. *Journal of Small Animal Practice* 2004;**45**:222.

82 McGuinness SJ, Friend EJ, Knowler SP, Jeffery ND, Rusbridge C. Progression of otitis media with effusion in the Cavalier King Charles spaniel. *Veterinary Record* 2013;**172**:315–16.

83 Cole LK. Primary secretory otitis media in Cavalier King Charles Spaniels. *Veterinary Clinics of North America Small Animal Practice* 2012;**42**:1137–42.

84 Cox CL, Slack WT, Cox GJ. Insertion of a transtympanic ventilation tube for treatment of otitis media with effusion. *Journal of Small Animal Practice* 1989;**30**:517–19.

CHAPTER 19

SURGERY OF THE NOSE

19.1 INTRODUCTION

The surgical treatment of rhinological disease via rhinotomy and the management of epistaxis are some of the oldest and most common procedures performed by otorhinolaryngologists. Surgical treatment is only indicated after and based upon (advanced) diagnostic imaging (radiography or computed tomography [CT] scan) and endoscopic examination of the nasal cavity and paranasal sinuses (see Chapter 9 Nose Diagnostic Procedures). Although many diseases of the ears, nose and throat will require specialised treatment, some common disorders of the nose can be managed successfully in private practice with a proper understanding of basic anatomy and physiology. Many aspects of 'general' soft tissue or orthopaedic surgery, for instance concerning preparation of patient and surgeon, delicate tissue handling and basic surgical techniques, are also applicable to otorhinolaryngological surgery. Discussion of the general aspects of surgery is beyond the scope of this book, but special emphasis will be paid to aspects that make rhinological surgery different.

Severe or chronic rhinological disease can lead to weight loss and emaciation, inappetence, dyspnoea, hypoxia and aspiration pneumonia, which can all affect anaesthesia protocols and wound healing[1-3]. In addition, especially if epistaxis has been frequent and/or profuse, anaemia can be present. These factors should be addressed prior to surgery if possible. Brachycephalic dogs are frequently anaesthetised for correction of the brachycephalic obstructive airway syndrome (BOAS) or for other procedures, and must be observed closely following premedication, as apnoea or complete respiratory obstruction can result after relaxation of the pharyngeal muscles[4,5]. Also, dogs with BOAS may have a hypoplastic trachea, and so may require a smaller endotracheal tube diameter than would be predicted

from the animal's size[5]. In general, all patients undergoing rhinological surgery will benefit from:

- Preoxygenation before induction of anaesthesia[5], which can be provided via a mask, nasal prongs, oxygen cage or as flow-by until the airways are secured via endotracheal or per tracheostomy intubation.
- A preoperative anti-inflammatory dose of a corticosteroid (dexamethasone, 0.5–2 mg/kg i/v, i/m or s/c) may reduce upper airway oedema secondary to diagnostic or surgical manipulations, and is routinely given for brachycephaly-related procedures, preferably before intubation.
- Animals in respiratory distress easily become hyperthermic, and need to be cooled actively using cool IV fluids, fans and icepacks.

Sedation is very beneficial in the perianaesthetic period of patients with upper airway obstruction. Opiates and tranquillisers are ideal[5]. Animals with BOAS tend to have elevated vagal tone, and the addition of an anticholinergic to the premedication may be indicated[5]. Propofol allows for a rapid induction, but is associated with apnoea and therefore not ideal for cases in which a difficult intubation is expected. Ketamine with a benzodiazepine would be indicated for those cases[5]. Maintenance, unless an intravenous protocol is indicated, can usually be achieved with inhalant anaesthetics. Monitoring of patients undergoing rhinological surgery should minimally consist of electrocardiogram (ECG), body temperature, oxygenation (SpO_2) and ventilation (end-tidal CO_2).

Biopsy of the nasal planum is indicated in patients with primary nasal planum disease to differentiate between inflammatory, immune-mediated and neoplastic diseases (see Chapter 10 Diseases of the Nasal Planum). Most tumours of the nasal planum can be cured by resection of the nasal planum (see Section 10.3). The techniques for biopsy of the nasal planum and for nasal planum amputation are discussed here in Section 19.2.

The components of BOAS that are amenable to surgical correction are stenotic nares, aberrant turbinates, elongated soft palate, everted laryngeal saccules and laryngeal collapse[2,6,7]. Correction of stenotic nares and staphylectomy are techniques that, provided they are executed meticulously, improve patient welfare significantly and are associated with minimal complications. However, resection of turbinates or laryngoplasty procedures are associated with significantly increased perioperative risks, and therefore should only be employed if conservative treatment and resection of nares and palate do not result in a significant improvement. The former two techniques will be discussed in Section 19.3.

Cleft palate and oronasal fistulae lead to communications between the oral cavity and nose, and subsequent chronic unilateral rhinitis[8,9]. Techniques for closure of these communications are discussed in Section 19.4.

For diagnostic and therapeutic purposes trephination of the frontal sinuses and nasal cavity is sometimes required, especially when dealing with nasal aspergillosis. The technique for trephination and subsequent tube placement for nasal flushes is discussed in Section 19.5.

Rhinotomy is indicated for removal of nasal foreign bodies that cannot be removed using endoscopy, as part of a combination therapy (radiotherapy before or after surgery) for intranasal neoplasia and for the treatment of chronic infectious rhinitis associated with nasal obstruction (chronic rhinitis in cats). Most surgeons prefer a dorsal approach to the nasal cavity and paranasal sinuses because of enhanced accessibility to the cribriform plate and frontal sinuses, but ventral rhinotomy may be indicated in selected patients with focal abnormalities in the ventral meatus and/or nasopharynx[3]. A lateral rhinotomy is indicated in those instances where access is only required in the rostral part of the nasal cavity (nasal vestibule).

Even with proper patient selection and thorough diagnostic work-up and preoperative evaluation of rhinological patients, rhinotomies with unilateral or bilateral turbinectomy are major surgical procedures with many reported complications. Complications reported include:

- Entrance into the cranial vault during surgery.
- Haemorrhage during and after surgery (epistaxis).
- Pneumocephalus and septic meningoencephalitis.
- Subcutaneous emphysema.
- Failure to mouth breath.
- Aspiration pneumonia.
- Persistent anorexia and pain[2,3].
- Persistent nasal discharge or respiratory noise and recurrence of disease are considered late postoperative complications.
- Oronasal fistula formation has been reported after ventral rhinotomy only.
- Chronic fistulation after wound dehiscence and osteomyelitis/osteonecrosis is possible after dorsal rhinotomy in patients that have received radiation therapy prior to surgery.

The different rhinotomy techniques are discussed in Sections 19.6–19.8. Access to the nasopharyngeal area can usually be achieved by nasopharyngoscopy, for instance to retrieve foreign bodies or to take biopsies of masses. Nasopharyngeal polyps can usually be removed via traction-avulsion after retraction of the soft palate, but occasionally, especially for small polyps, nasopharyngotomy is indicated. This approach can also be used to remove foreign bodies that are lodged in the nasopharynx, to surgically treat and excise nasopharyngeal stenosis and to biopsy or remove nasopharyngeal masses. The technique for nasopharyngotomy and subsequent polyp removal is discussed in Section 19.9. Peri- and postoperative analgesia, antibacterial therapy and postoperative nutritional management for patients undergoing rhinological surgery are discussed in Chapter 17 Perioperative Management.

19.2 BIOPSY AND RESECTION OF THE NASAL PLANUM

Biopsies of the nasal planum are performed to determine the presence or extent of disease present in sampled tissue. Samples may be required for histopathological diagnosis, molecular evaluation, immunohistochemical testing, phenotyping and polymerase chain reaction (PCR) testing for DNA analysis[10,11]. Presurgical planning is required to ensure that the appropriate quality and quantity of tissue is harvested to maximise the chance of obtaining a correct diagnosis.

For most inflammatory conditions that are limited to the nasal planum two or three punch samples should be taken from the middle and the margins of the affected area. For neoplastic lesions, samples are taken of the mass and adjacent normal planum or skin, caudal to the lesion as this facilitates cosmetic closure of the biopsy site. Careful sampling is always aimed at avoiding disruption of the architecture of the tissue and altering cosmetic appearance. For sampling of neoplastic lesions, care should be taken not to spread the disease process to adjacent tissues[10,11]. In addition, any incision and tract made for biopsy should always be planned so that they can be removed along with the diseased tissue if surgery is to be done as part of a patient's definitive treatment.

In dogs and cats, squamous cell carcinoma (SCC) is the most common tumour type of the nasal planum. Various methods have been described to treat dogs and cats with SCC of the nasal planum including radiation therapy, hyperthermia, intratumoural administration of carboplatin, cryosurgery, conservative surgery and photodynamic therapy[2,12–14]. With most of these treatment modalities tumour margins cannot be evaluated, special equipment is needed and high rates or tumour recurrence are reported. Most of these techniques may work for early, small lesions or carcinoma *in situ*, but the most cost-effective, reliable treatment for selected patients with invasive SCC is nasal plane resection. This is also the recommended treatment for other tumours of the nasal plane in dogs and cats[15]. For tumours extending into or arising from the nasal septum, invading the nasal floor, maxilla or the upper lip more radical resection techniques are advised, such as lateral rhinotomy or premaxillectomy with nosectomy[16–18].

Nasal planum biopsy techniques

Punch biopsies are the most commonly used biopsy technique for nasal planum abnormalities and provide a cylindrical core of both superficial and deeper layers of the skin[11,19]. The skin should be appropriately prepared for the procedure to avoid altering the skin surface, hence vigorous aseptic preparation should be avoided. The punch, usually 4 mm for the nasal planum, is gently placed into the centre of the affected area.

> Do not attempt to include normal tissue as the histopathologist cannot orientate a punch sample as easily as a wedge.

The punch is then rotated clockwise and counterclockwise while gently pushed and advanced into the tissue to be sampled. Once the desired depth is reached, the punch is removed. The sample is gently lifted with small tissue forceps and the base of the sample is subsequently transected from the underlying tissue. The lesion is closed with skin sutures only, in a caudo-cranial rather than a latero-medial direction for dorsal nasal planum samples and in a latero-medial direction for lateralised samples, in an interrupted pattern with 4-0 absorbable monofilament suture material.

Alternatively, incisional biopsies can be taken, where a wedge of tissue is removed from the target site[11]. Incisional biopsies are preferred to punch or biopsies for superficial, ulcerated and necrotic mass lesions, as larger samples can be obtained, increasing the chance of obtaining a diagnostic sample[10,11]. The wedge resection should be only as large as is necessary to obtain an adequate sample size for diagnosis and should be oriented to allow cosmetic closure, future excisional surgery and adjuvant radiotherapy. A wedge of tissue is sharply excised from the lesion and closure is as described above.

Nasal planum amputation

The original technique for nasal planum amputation was described by Withrow *et al.*, and remains the easiest technique for resection of nasal planum tumours in cats[20]. The cosmetic results of this technique in dogs are less pleasing though. With the animal positioned in sternal recumbency, the surgical area is carefully palpated to estimate tumour extension into adjacent tissue. The nasal planum is completely removed with a 360° skin incision made with a scalpel. The incision is made so it transects the underlying turbinates. The cartilage of the nasal plane and the turbinates should be cut with an incision angled at about 45° to the hard palate[20]. Haemorrhage can be controlled by direct pressure. A pursestring suture of 2-0 or 3-0 absorbable or nonabsorbable suture material is placed through the skin around the incision to cover the exposed nasal conchae partly with skin. The new nasal orifice is closed to approximately 1 cm in diameter[20].

Alternatively, for a more cosmetic result, an advancement flap from the dorsal aspect of the nose can be created to reconstruct the dorsal wound before amputation of the planum (Figs 19.1A–D). Laterally, the exposed turbinate cartilage can be denuded from

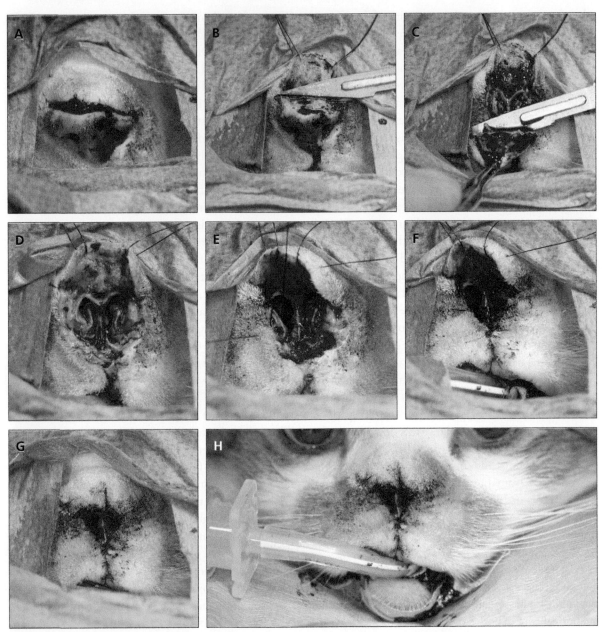

Figs 19.1A–H Nasal planum amputation. An advancement flap from the dorsal aspect of the nose is created (A) and tagged with sutures (B) before removal of the nasal planum with an incision angled at about 45° to the hard palate (C, D). Laterally, the exposed turbinate cartilage has been denuded from the overlying mucosa and sutured to the lateral cartilage (E). The wound is subsequently further closed by suturing the turbinate cartilage on the other side and connecting skin sutures to appose both ventral parts (F, G). Finally, the advancement flap is used to close the dorsal defect (H).

the overlying mucosa and sutured to the lateral cartilage (Fig. 19.1E). The wound is subsequently further closed using interrupted sutures from skin to remaining planum cartilage (Figs 19.1F–H). If wider margins are required, complete nosectomy with premaxillectomy may be required. Postoperative care consists of an Elizabethan collar, broad-spectrum antibiotics for 1 week and analgesia for 3 days.

19.3 SURGICAL CORRECTION OF BRACHYCEPHALIC OBSTRUCTIVE AIRWAY SYNDROME

Brachycephalic dogs such as English and French Bulldogs, Pugs, Pekingese, Shih Tzus, Shar Peis, Boston Terriers and both Persian and Himalayan cats frequently present with signs of upper airway obstruction as a result of an anatomical distortion of their faces caused by an exaggerated and incorrect breed selection (see Chapter 11 Diseases of the Nasal Cavity and Sinuses, Section 11.3). The primary components of BOAS are:

- Increased nasal resistance as a result of stenotic nares (43–85%) and aberrant or protruding turbinates.
- Pharyngeal hypoplasia (redundant pharyngeal folds) with elongated soft palate (86–96%).
- Tracheal hypoplasia, especially in the English Bulldog[21–27].
- Secondary components, resulting from the chronic increased negative intra-airway pressure, include everted tonsils, everted laryngeal saccules (55–59%) and laryngeal collapse (8–70%)[4,23,24].

Correction of stenotic nares and staphylectomy are techniques that, provided they are executed meticulously, improve patient welfare significantly and are associated with minimal complications. Resection of turbinates[28,29] or laryngoplasty procedures are associated with significantly increased perioperative risks, and therefore should only be employed if conservative treatment in combination with resection of nares and palate do not result in a significant improvement.

Technique for correction of stenotic nares and resection of aberrant turbinates

Surgical correction of stenotic nares can be performed at a very early age (3–6 months). It significantly reduces upper airway obstruction and decreases the rate of progression of other components of BOAS[21,30,31]. Several techniques have been described for correction of stenotic nares, using either scalpel blades, laser or electrosurgery.

Patients are placed in sternal recumbency with the nose perpendicular to the table or slightly elevated. Adequate cosmetic and functional results of 'Trader's technique', which involves amputation of the ventral portion of the dorsal lateral nasal cartilage and healing by second intention, have been reported in immature Shih Tzu dogs with stenotic nares[2,30]. In mature dogs, a lateral, vertical or horizontal wedge resection of the dorsal lateral cartilage can be performed[1,2]. The author recommends a modified horizontal wedge resection.

A deep wedge extending into the cartilage is resected following the outer curvature of the ala from its medial dorsal-most aspect ventrally towards the lateral aspect and back over the body of the ala connecting the start and end of the incision (Figs 19.2A, B). A beaver scalpel holder with a 6500 pointed blade allows for accurate incision and determination of adequate depth. The initial suture is then placed from the middle of the remaining ventral part of the nostril to the more dorsolateral aspect of the nasal cartilage to adequately rotate the wing of the nostril laterally and create a maximal opening medially. Closure can be performed with 4-0 or 5-0 monofilament suture in a single layer using a

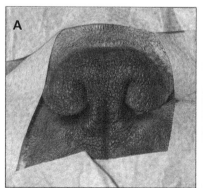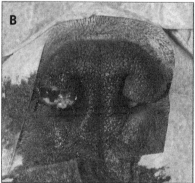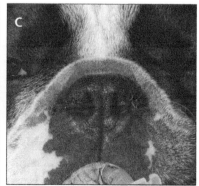

Figs 19.2A–C Modified horizontal wedge resection technique for resection of stenotic nares. Moderately stenotic nares in a French Bulldog (A). A deep wedge is removed from the nare and ala (B) after which the ventral part is sutured laterally to open the medial airpassage (C).

simple interrupted suture pattern with absorbable suture material (Fig. 19.2C).

Guided by preoperative cross-sectional imaging and rhinoscopy, diode laser-assisted turbinectomy can be used to remove aberrant conchae and enlarge the ventral nasal meatus[28,29]. Preliminary results indicate that this technique can reduce intranasal airway resistance by approximately 50% in brachycephalic dogs. Grading of degree of turbinate protrusion has recently been reported in English Bulldogs[32]. At the moment, the degree of protrusion that is clinically tolerated is unknown as is the degree of improvement after resection. Since the procedure is associated with substantial perioperative morbidity and mortality, and regrowth of turbinates after resection has been reported, at this stage standard turbinectomy can therefore not be recommended for every patient.

Staphylectomy for elongated/overlong soft palate

The dog is positioned in sternal recumbency for soft palate resection. A bar is placed over the front of the surgery table to which the upper jaw is suspended by placing gauze, tape or bandage material, around the maxillary canine teeth. The proposed lateral levels of palate resection, the caudal borders of the palatine tonsils when a minimal amount of rostral retraction is applied to the tongue, are tagged with two stay sutures (Fig. 19.3A). An Allis tissue forceps is placed in the caudal edge of the palate on the midline and used to retract the palate rostrally. The palate is then resected in an arch shape, making sure to remove more tissue medially then laterally. Resection is performed with scissors (Figs 19.3B, C)[24,33,34], a carbon dioxide laser[24,35,36] or an electrothermal feedback-controlled bipolar sealing device[37]. The

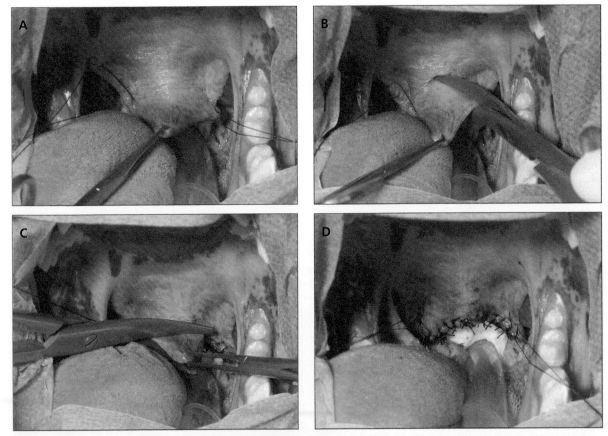

Figs 19.3A–D Staphylectomy. The soft palate at the caudal borders of the palatine tonsils are tagged with two stay sutures and an Allis tissue forceps is placed in the caudal edge of the palate on the midline and used to retract the palate rostrally (A). The palate is then resected in an arch shape with scissors (B, C). The oropharyngeal and nasopharyngeal mucosa are then adapted using 3-0 or 4-0 monofilament rapidly absorbable suture material (D).

oropharyngeal and nasopharyngeal mucosa are then adapted using 3-0 or 4-0 monofilament rapidly absorbable suture material (Fig. 19.3D).

Serious complications of staphlectomy include death as a result of aspiration pneumonia, dyspnoea and cyanosis requiring tracheostomy, or failure to recover from anaesthesia[7,21,34]. Less severe complications include coughing, noisy respiration and gagging and retching[38]. The prognosis of dogs after soft palate resection is good to excellent in 90% of cases, especially in dogs younger than 1 or 2 years of age[24,34,38].

Everted laryngeal saccules and laryngeal collapse

Three different stages of laryngeal collapse are clinically recognised[22,39]:

- Stage I is relatively mild, consisting of laryngeal saccule eversion of varying severity.
- In stage II, medial collapse of the cuneiform process of the arytenoid cartilage as a result of lack of rigidity is seen.
- In stage III, the corniculate processes of the arytenoid cartilages collapse, resulting in significant airway obstruction[22,39].
- Most patients do not follow this specific grading though and present with some collapse of both the cuneiform and corniculate processes.

Stage II and III laryngeal collapse have been reported in dogs as young as 4.5 months of age[22]. Because laryngeal collapse represents a secondary condition, the initial treatment should be focused on surgical correction of the primary disease and should involve the lower risk surgical procedures such as rhinoplasty for stenotic nares and staphylectomy for an elongated soft palate[4].

Resection of everted laryngeal saccules is performed routinely by some surgeons, but can lead to laryngeal webbing[40] and severe postoperative complications and should, in the author's opinion, therefore be reserved for those animals not improving after rhinoplasty and staphylectomy. Medical management, including weight loss, exercise restriction and drugs to reduce airway swelling (e.g. glucocorticoids) or oedema (e.g. furosemide) can be attempted in animals with signs secondary to persistent laryngeal collapse[39]. Patients that do not respond to staged surgical and medical management of airway disorders may require permanent tracheostomy[39]. Alter-

natively, partial laser laryngectomy or laryngeal tie back may be attempted; however, the effectiveness of these procedures for resolution of airway obstruction has not been extensively evaluated[22,39,41].

Postoperative care

In general, after rhinoplasty and staphlectomy, dogs should be observed and kept calm during recovery until at least 1 hour after extubation, which should take place only when they are almost fully awake and consciously aware of the tube[5,39]. The dog's pulse, temperature and respiratory rate and effort are monitored frequently. Food and water are only withheld until complete recovery. Dogs are monitored for any gagging, retching or vomiting, stridor and development of dyspnoea. After recovery dogs are offered water and a small amount of soft food under supervision, and swallowing is carefully observed. After laryngological procedures, dogs have to be observed in an intensive care unit. If postoperative dyspnoea occurs, animals can either be heavily sedated (for 8–12 hours with an additional dose of corticosteroids administered) and reintubated with a small tube or a tracheostomy should be performed.

19.4 CLOSURE OF CLEFT PALATE AND ORONASAL FISTULAE

Defects in the hard and soft palate may result from:
- Congenital abnormalities.
- Resection of neoplasms.
- Traumatic injuries.
- Severe peridontal disease.
- Tooth removals.
- Severe chronic infections.
- Secondarily, to surgical and radiation therapy[38,42–44] (see Chapter 11 Diseases of the Nasal Cavity and Sinuses, Section 11.7).

Animals with clefts of the upper lip and lateral area of the most rostral hard palate (primary palate) are often euthanised at a young age, though techniques have been described to close these defects with acceptable cosmetic outcome[45]. Surgical repair of traumatic clefts in cats (see Chapter 11 Diseases of the Nasal Cavity and Sinuses, Section 11.12) usually only involves reposition and alignment of tissue without the need for reconstructive techniques. Tension on the wound edges is not an issue in these patients as usually no tissue is missing. Some authors recommend surgical closure as described for congenital clefts though, because of a more predictable outcome[38].

Congenital clefts of the secondary palate (hard and soft palate) are commonly seen and closure of these defects, as well as lateral soft palate clefts will be discussed here. For these congenital palate clefts, owners should always be advised that multiple procedures might be required to completely close the defect. Closure of oronasal fistulae is less complex and most can be permanently closed in one procedure[8,46]. Most commonly oronasal fistulae involve the canine teeth; closure of these specific defects will therefore be reviewed here.

Technique for closure of midline secondary palate clefts

Multiple techniques have been described for closure of midline defects[38,43,45,47–49]. Flaps can be:

- Harvested from oral, pharyngeal, and nasal mucosa or skin.
- Can be local or distant.
- Pedicled or free.
- Depend on local blood supply or on direct arteries (axial pattern).
- In addition, flaps can be advanced, rotated, transposed or overlapping[38].
- Though prosthetic devices have been used for closure of palatal defects, the use of autogenous tissue is always preferred[48].

Whichever technique is used, it is essential to close the hard and soft palate in at least two layers. A good surgical technique should prevent suture tension, since this is the most common cause of wound dehiscence and the blood supply of the palate has to remain intact. The minor and major palatine branches of the maxillary artery supply the hard palate and they lie midway between the midline and the lingual side of the teeth. Although there is often considerable haemorrhage during palate surgery, the use of electrosurgical equipment should be avoided and digital pressure is recommended to control bleeding[38,49]. The surgeon should proceed as quickly as possible with flap harvesting to minimise blood loss. Atraumatic handling of the tissues by using stay sutures in the raised flaps are furthermore important to prevent postoperative necrosis.

Closure of large complete clefts of the secondary palate can be accomplished using either an overlapping flap or a medially positioned flap technique. The overlapping flap technique is preferred for midline hard palate clefts that are very wide compared to the available tissue, such as in dolichocephalic dog breeds. It has the advantage that the suture line is not directly located over the defect and the opposing connective tissue area is larger[49].

For the overlapping technique, incisions following the entire defect are made in the palatal mucoperiosteum, to the depth of the bone, along the dental arch about 1–2 mm away from the teeth. At the rostral and the caudal margins of the defect, incisions are made from the primary incision to the midline to allow this overlapping flap to be raised. On the other side an incision is made on the medial margin of the defect; this will form a receiving envelope. Both flaps are carefully undermined with periosteal elevators and handled using stay sutures. Care is taken to avoid damaging the palatine artery when dissecting the overlapping flap. Connective tissue has to be dissected around the vessel carefully to allow rotation of the flap. The overlapped flap is inverted at its base so that its periosteal surface is exposed and secured under the envelope flap with horizontal mattress sutures of absorbable monofilament suture material so that large connective tissue surfaces are in contact. Granulation and epithelialisation of exposed bone with this technique generally are completed in 3–4 weeks[38,49]. Soft palate defects can be closed as described below for the medially positioned flap or with a bilateral overlapping technique[50].

For the medially positioned flap technique[38,45], incisions are made in the mucoperiosteum of the hard palate and the mucosa of the soft palate at a distance of 0.75 × the width of the defect on either side (Figs 19.4A, B). Incisions in the soft palate are made to the level of the caudal end of the tonsils. At the rostral and the caudal margins of the defect, incisions are made from the primary incision to the midline to allow two hinge flaps to be raised. The tissues are carefully dissected from the incisions towards the midline using periosteal elevators and stay sutures in the flap. The raised flaps are dissected towards and along the cleft until they can be hinged at the edge of the flap to provide the new nasal mucosa (Fig. 19.4B). Lateral mucosal relaxing incisions are made along both sides of the hard palate along the dental arch about 1–2 mm away from the teeth, and the soft palate. The hard palate between lateral incision and hinge flap incision lines is

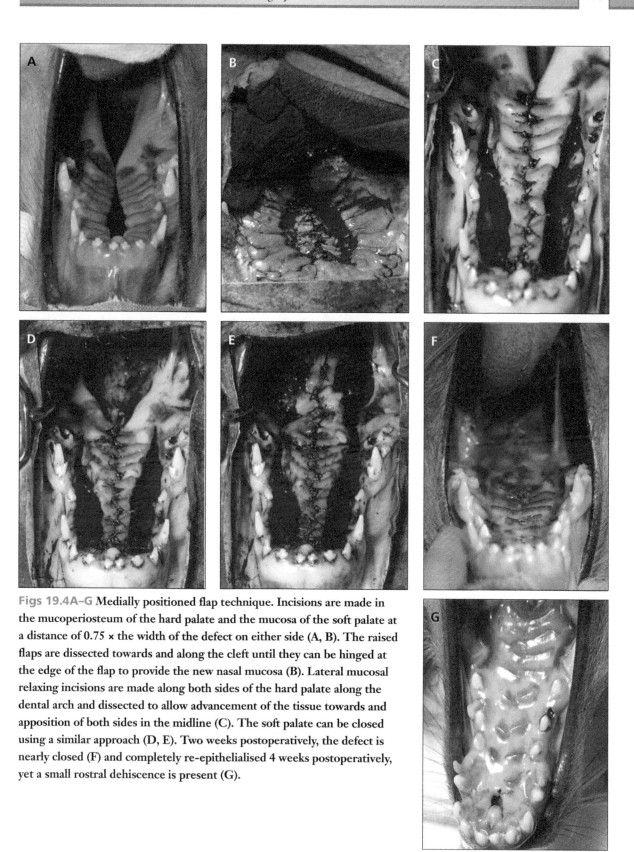

Figs 19.4A–G Medially positioned flap technique. Incisions are made in the mucoperiosteum of the hard palate and the mucosa of the soft palate at a distance of 0.75 × the width of the defect on either side (A, B). The raised flaps are dissected towards and along the cleft until they can be hinged at the edge of the flap to provide the new nasal mucosa (B). Lateral mucosal relaxing incisions are made along both sides of the hard palate along the dental arch and dissected to allow advancement of the tissue towards and apposition of both sides in the midline (C). The soft palate can be closed using a similar approach (D, E). Two weeks postoperatively, the defect is nearly closed (F) and completely re-epithelialised 4 weeks postoperatively, yet a small rostral dehiscence is present (G).

dissected over the palatine bone to allow advancement of the tissue towards and apposition of both sides in the midline (Fig. 19.4C). Oral mucosa at the cleft is then sutured using monofilament absorbable suture material. The soft palate can be closed using a similar approach (Figs 19.4D,E). A buccal mucosal flap may be used to suture into the defect left after undermining and moving the hard palate mucoperiosteal flap. However, if left to

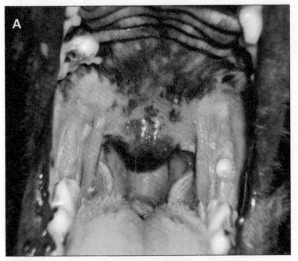

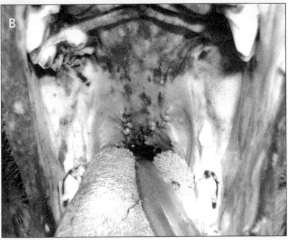

Figs 19.5A, B Soft palate hypoplasia (A). After tonsillectomy, the crypt is dissected to create a ventral and a dorsal flap. On either side of the 'uvula'-shaped soft palate remnant a small rim of mucosa is removed with scissors and incisions on both sides are slightly dissected to create a small ventral and a small dorsal flap in the palate as well. The ventral and dorsal flaps are subsequently sutured separately in a simple interrupted pattern on both sides (B).

heal by second intention, the defect usually fills quickly with granulation tissue and is completely re-epithelialised in 4 weeks (Figs19.4F, G).

Relaxing incisions can also be made at the most lateral aspect of the soft palate, yet undermining of tissue is not necessary as the oral layer usually closes without tension. Alternatively, for closure of the soft palate cleft, incisions can be made along the medial margins of the defect to the level of the caudal end of the tonsils. Palatal tissues are separated with blunt-ended scissors to form a dorsal (nasopharyngeal) and ventral (oropharyngeal) flap on each side[42,49]. The two dorsal and the two ventral flaps are sutured separately in a simple interrupted pattern to the midpoint or caudal end of the palatine tonsils.

Technique for closure of lateral soft palate clefts

Repair of a unilateral soft palate defect can be performed with or without removal of the ipsilateral tonsil[51,52]. The tonsillectomy incisions can be extended rostrally to meet at the most rostral location of the soft palate defect and continued along the medial edge of the soft palate[38]. The pharyngeal and palatal tissues are separated to create two dorsal and two ventral flaps, which are subsequently sutured separately in a simple interrupted pattern to the midpoint or caudal end of the contralateral tonsil.

Congenital hypoplasia of the soft palate can be addressed in a similar fashion with bilateral tonsillectomy and extension and continuation of incisions into the rudimentary, uvula-like soft palate tissue (Figs 19.5A, B)[53,54]. Repair of soft palate hypoplasia in dogs using bilateral buccal mucosa flaps has also been reported[44]. A combination of bilateral pharyngeal advancement flaps and one overlapping hard palate flap has been described for the treatment of a hypoplastic soft palate in a cat[55]. Owners need to be made aware of the fact that even though a mechanical closure of the defect can usually be achieved, restoration of a palatopharyngeal sphincteric ring and normal swallowing function may not be accomplished due to lack of functional soft palate musculature[38,56].

Technique for closure of oronasal fistulae

Periodontal pockets and oronasal fistulae are treated with dental extraction and closure of the defect with mucoperiostal or mucosal flaps and antibiotics. To close a defect after extraction of the canine tooth, an elliptical incision can be made around the fistula opening with removal of the upper and lower triangles. The oral mucoperiosteum can be hinged and brought together and sutured to form the new nasal floor. To create a double-layer closure[8,46,57], a flap can be created from the hard palate mucoperiosteum and rotated over the

defect, or alternatively a buccal mucosal advancement flap can be used (Fig. 19.6A). The oronasal fistula can also be repaired with a single pedicle advancement flap from the buccal mucosa (Fig. 19.6B). A 2–3 mm rim of mucosa is removed around the edge of the fistula. Slightly diverging incisions are made in the adjacent buccal mucosa starting at the rostral and caudal borders

Figs 19.6A, B Closure of oronasal fistula. Closure with mucoperiosteal hinge flap and hard palate mucoperiosteal rotation flap (A). Closure with buccal mucosal advancement flap (B).

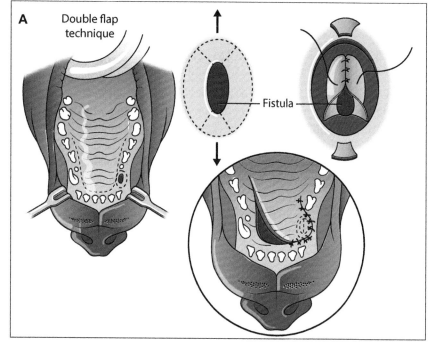

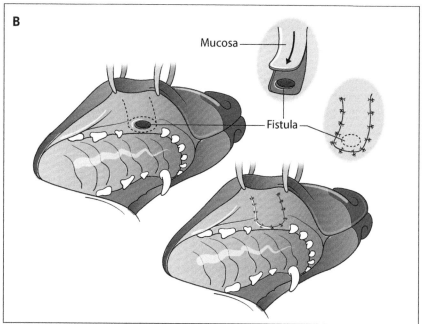

of the defect. The flap is undermined, advanced over the defect and sutured.

Central palate fistulae are best closed with transposition flaps from the hard palate mucoperiosteum, or with axial pattern flaps based on the palatine artery[58]. Caudally located central palate defects can easily be closed with partial thickness advancement flaps from the soft palate. If defects cannot be closed with locally available tissue, cartilage grafts or prosthetic grafts can be used[59–61].

19.5 TREPHINATION OF THE NASAL SINUSES

For several years, the standard treatment for nasal aspergillosis was an enilconazole emulsion delivered via tubes surgically implanted into the nasal chambers and frontal sinuses[1]. Enilconazole is ideal as a topical agent because it is also active in the vapour phase, which enhances its distribution throughout the nasal chamber. Although this approach resulted in the elimination of the fungus in more than 90% of affected dogs, it had several disadvantages. The twice-daily irrigation of drug for 7 days was labour intensive, often very messy, not always well tolerated by the dog and was considerably complicated if the dog removed one or both tubes. The technique has therefore been replaced by 1-hour soaks with clotrimazole (see Chapter 11 Diseases of the Nasal Cavity and Sinuses, Section 11.6)[62–64]. However, trephination of the frontal sinuses and nasal cavity is still indicated in some patients:

- For collection of samples for cytology, histopathology and culture when obvious abnormalities are not seen upon rhinoscopy[65–67].
- If rhinoscopic debridement of the frontal sinus is not possible prior to the soak.
- Another potential contraindication to the soak technique is extension of drug to the brain across a damaged cribriform plate[62–64]. The risks of this appear to be low, but with significant damage to the cribriform plate, trephination and tube placement are indicated.

Technique

Rather than curretting turbinate tissue, the surgeon should view the procedure as diagnostic only; large fungal mats can be removed, but care must be taken to cause as little damage as possible to nasal and turbinate mucosa and to the nasofrontal ostia[2,68]. The animal is placed in sternal recumbency with a roll of bandage material between the upper and lower jaw to avoid injury to the soft tissues of the oral cavity while drilling. A transverse incision is made 1 cm out of the midline in the middle of the triangle formed by the midline, the temporal line caudolaterally and the medial rim of the orbit craniolaterally. Incisions are made through the skin, subcutaneous tissue and the periosteum over the left and right frontal sinuses (Fig. 19.7A). The periosteum is elevated to expose the flat surface of the bone, and a Steinmann pin, trephine or air-driven drill (Fig. 19.7B) is used to create an opening into each frontal sinus. The piece of bone is subsequently discarded and if necessary, the opening can be enlarged with rongeurs[68]. Samples are taken from the frontal sinus with small currettes (Fig. 19.7C) and with flushing and suction and careful curettage the frontal sinus is debrided from obvious fungal material. With a small catheter the patency of the nasofrontal ostium can be verified. An additional opening in the caudal nasal cavity can be made by making a 2 cm incision through skin, subcutis and periosteum 0.5 cm out of the midline on the ridge of the nasal bone (Fig. 19.7D). The incision is started caudally, at the level of the medial canthus of the ipsilateral eye, and extending it cranially over 2 cm. After debridement, long, internally fenestrated tubes can be placed and positioned to lie in the frontal sinus, and caudal nasal passage on each side (Fig. 19.7E). The periosteum is closed on either side of the tube using 3-0 or 4-0 absorbable suture material and the subcutaneous tissues and skin on either side of the tube are closed routinely (Fig. 19.7F). Each tube is attached to the skin with one non-absorbable suture through the skin and tube. The whole construction can be secured using dressings and tape to allow flushing 'from a distance', behind the Elizabethan collar (Figs 19.7G, H).

Treatment involves twice-daily flushing with enilconazole (10 mg/kg total dose for both sides each treatment) for 7–10 days[1,68]. The medication is diluted with an equal volume of warm saline or water, equal volumes are flushed into each tube and the tube is flushed with air at the end of each treatment. Complications are usually limited to premature removal of the tubes. Occasionally, dogs become anorectic and vomit; these signs resolve when the medication is discontinued[1,68].

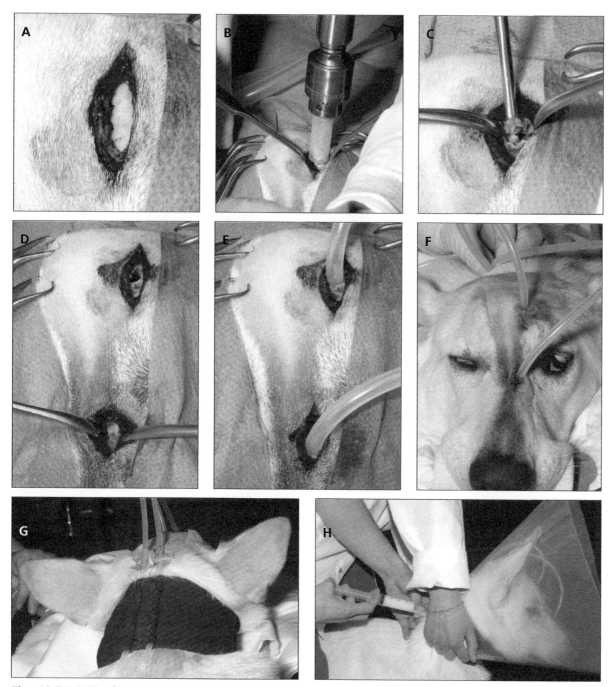

Figs 19.7A–H Trephination of nasal sinuses. Incisions are made through the skin, subcutaneous tissue and the periosteum over the left frontal sinus with the periosteum elevated to expose the flat surface of the bone (A). An air-driven drill (B) is used to create an opening into the frontal sinus. The frontal sinus is carefully debrided using a small currette (C) and with flushing and suction. The ipsilateral caudal nasal cavity is trephined after making an 2 cm incision through skin, subcutis and periosteum 0.5 cm out of the midline on the ridge of the nasal bone (D). After debridement, long tubes are placed that fit exactly through the created holes (E). The periosteum, subcutaneous tissues and skin are closed on either side of the tube (F). The whole construction can be secured using dressings and tape (G) to allow flushing 'from a distance', behind the Elizabethan collar (H).

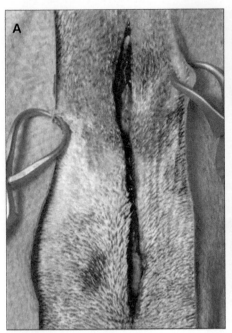
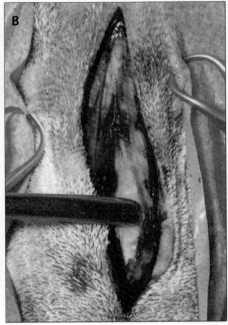
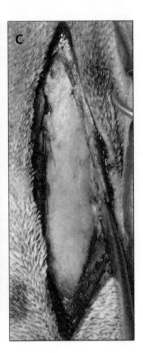

19.6 DORSAL RHINOTOMY WITH TURBINECTOMY

Indications for dorsal rhinotomy are:

- Nasal polyp removal[69].
- Removal of foreign bodies that could not be removed using endoscopy[70].
- Nasal tooth displacement[71,72].
- For diagnosis and/or treatment of neoplastic lesions that are benign or cause severe obstruction of airflow[2,73].
- It can also be used as an adjuvant treatment of malignant intranasal neoplasia[74] (see Chapter 11 Diseases of the Nasal Cavity and Sinuses, Section 11.8).

Significant haemorrhage should be expected during surgery, and coagulation tests and red blood cell cross-match are usually performed preoperatively. Although temporary bilateral carotid artery occlusion has been successfully used to reduce blood loss and enhance visualisation during nasal surgery in dogs[75], it is generally not necessary when surgery is performed rapidly and turbinates are removed as quickly as possible. In one large study no difference in the need for a blood transfusion was noted when comparing animals with or without carotid artery occlusion[76].

In cats the less robust cerebral blood supply and lack of internal carotid artery increase the risk of brain ischaemic damage, so bilateral carotid occlusion should be avoided[77,78]. The airways should be protected with a cuffed endotracheal tube to prevent aspiration of blood during surgery and recovery. In addition, the pharynx and larynx should be packed with several gauze sponges to prevent blood or tissue from entering the airway. The dorsal approach is used most commonly to access the nasal cavity and frontal sinuses as it offers the most complete exposure of the sinuses and nasofrontal openings[3]. For disease limited to the nasal cavity itself, a ventral or lateral approach can be used alternatively (see Sections 19.7 and 19.8).

Technique

The patient is positioned in sternal recumbency, with the dorsum of the head clipped and prepared aseptically for surgery. The tip of the nose should be pointing down slightly to facilitate blood exiting the nose, and a roll of bandage material is positioned between the upper and lower jaw to avoid injury to the soft tissues of the oral cavity by pressure from the teeth while performing the osteotomies.

A dorsal midline skin incision is made from the caudal aspect of the nasal planum to the medial canthus of the eye for both unilateral and bilateral rhinotomy (Fig. 19.8A)[2,3,79]. For sinusotomy, the incision is extended caudal to the zygomatic crests of the frontal bone. The subcutaneous tissue and periosteum are sharply incised

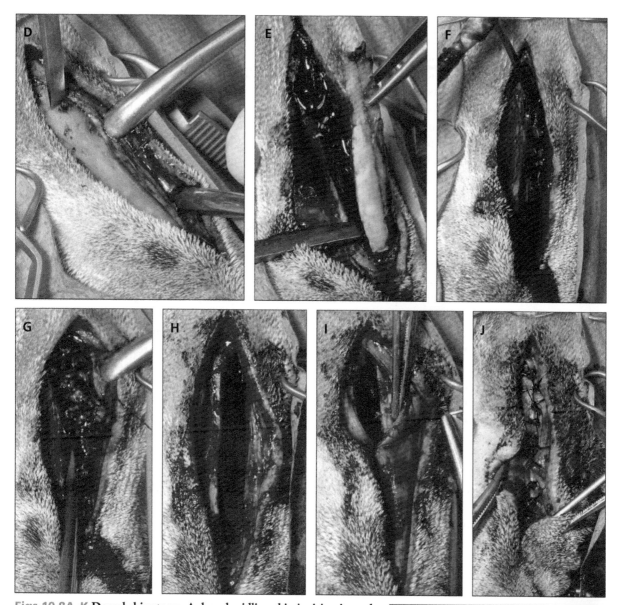

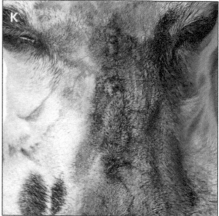

Figs 19.8A–K Dorsal rhinotomy. A dorsal midline skin incision is made from the caudal aspect of the nasal planum to the medial canthus of the eye (A). The subcutaneous tissue and periosteum are sharply incised on the midline and elevated and reflected laterally with small periosteal elevators to expose the entire nasal bone (B, C). With an osteotome and mallet the bone flap is outlined first (D), and subsequently rapidly deepened so the bone can be removed (E). The nasal cavity is then sampled for culture (F) and the turbinates can be removed using a large bone or uterine curette (G). Once the nasal cavity has been completely emptied and haemorrhage has reduced to minimal oozing (H), the rhinotomy side can be closed. The periosteum is closed with 2-0 to 3-0 absorbable monofilament suture material in an interrupted pattern (I, J), followed by the subcutaneous tissues and skin in a routine fashion (K).

on the midline and elevated and reflected laterally on either or both sides of the nasal cavity with small periosteal elevators to expose the entire nasal bone (Figs 19.8B, C). Stay sutures can be placed through skin and subcutis to aid in retraction. A rectangular window of bone should be exposed to allow for removal of the nasal bone and part of the frontal bone. The width of the bone flap should be as narrow as possible, yet allow for introduction of large currettes and rapid turbinectomy. In small dogs, part of the maxillary bone adjacent to the nasal bone can be removed if needed. Whereas the bone can be replaced if desired, it is not recommended if the bone flap is diseased, has been or will be irradiated or is unstable after fixation[2,3]. With an osteotome and mallet the bone flap is outlined first (Fig. 19.8D), by not completely cutting full-thickness. Once this has been done, the incisions can be quickly deepened and the nasal cavity opened to immediately address any bleeding associated with the osteotomy itself. Alternatively, a sagittal saw can be used. As the bone will be discarded (Fig. 19.8E), a window can instead be created with a small burr and subsequently enlarged with rongeurs to the desired length and width[80].

The nasal cavity is then sampled for culture (Fig. 19.8F) and systematically explored gently for foreign body removal or if biopsies of specific areas are to be taken. Usually dorsal rhinotomy is performed in order to remove all diseased turbinates with benign or neoplastic disease. Complete turbinectomy or nasal exenteration refers to complete removal of intranasal structures, but resection is usually limited to the dorsal and maxilloturbinates as excessive haemorrhage and accidental intracranial penetration through the cribriform plate are concerns when removing the ethmoid turbinates[2,3,79]. The turbinates can be removed using a large bone or uterine curette (Fig. 19.8G) that should follow the inside of the maxillary bone ventrally and scoop back a large amount of turbinates. As haemorrhage will occur at this stage, curettage should be as swift and to-the-point as possible. Care should be taken to avoid damaging the cribriform plate caudomedially. Specific attention needs to be paid to removal of the rostral parts of the dorsal and maxilloturbinates within the nasal vestibule and in ensuring a patent nasopharyngeal meatus. Bleeding usually stops once turbinectomy is complete, but any ongoing haemorrhage is best controlled by direct pressure from gauze packing. After removal of the gauzes, active bleeding can usually be controlled with direct electrocautery, using metal Frazier or Adson suction cannulae to use as a conductor while 'fixating' the end of the vessel with the tip of the suction tube and clearing the surgical area of blood at the same time[79a]. Once haemorrhage has reduced to minimal oozing (Fig. 19.8H), the rhinotomy site can be closed. The periosteum is closed with 2-0 to 3-0 absorbable monofilament suture material in an interrupted pattern (Figs 19.8I, J), followed by the subcutaneous tissues and skin in a routine fashion (Fig. 19.8K).

19.7 VENTRAL RHINOTOMY

The indications for ventral rhinotomy are similar to those for dorsal rhinotomy. Whether or not a dorsal or ventral approach is preferred depends on the exact location of the pathology[2,3]. A ventral approach to the nasal cavity is the most cosmetic approach and can be used to more thoroughly explore the region:
- Caudal to the ethmoid turbinates.
- The ventral aspect of the turbinates.
- The rostral nasopharynx[3,78].

Evaluation and evacuation of the frontal sinuses are limited though to the rostral half[2]. The potential advantages of a ventral approach include a more rapid recovery, a more cosmetic closure, a lower risk of subcutaneous emphysema and less postoperative pain[78]. Oronasal fistula formation has been reported as a complication of the ventral approach, but other complications are similar as for dorsal rhinotomy[76,79a]. In young animals, midline ventral rhinotomy may alter muzzle growth[81,82] and reduce transverse, but not sagittal, palatal length[2,83]. Ventral rhinotomy is therefore not recommended in growing animals[2].

Technique
The patient is placed in dorsal recumbency with the nose slightly pointing upward to avoid excessive kinking of the neck and reducing the blood flow to the head. The mandible is taped in a maximally open position to a bar over the table or held open with a mouth gag to facilitate exposure of the hard and soft palate. A mild antiseptic solution (0.05% chlorhexidine or 0.1% povidone-iodine) can be used to cleanse the oral cavity. The pharynx/larynx should be packed with gauze sponges.

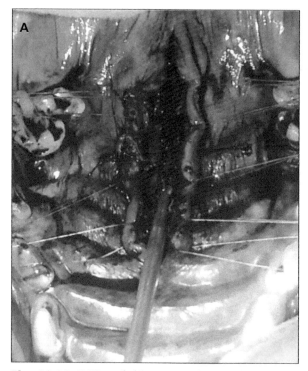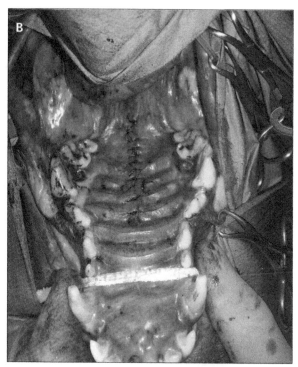

Figs 19.9A, B Ventral rhinotomy. After a midline incision, the mucoperiosteum is undermined bilaterally and hinged open with the help of stay sutures to allow for removal of the hard palate bone (A). Closure is done in two layers, the oral layer with interrupted absorbable suture material (B).

A midline incision in the hard palate is made from 5 mm caudal to the incisive papilla to the transition between hard and soft palate[3,78]. The mucoperiosteum is sharply elevated longitudinally along the incision bilaterally, ensuring the major palatine arteries are not damaged. The edges of the incision can be retracted with stay sutures (Fig. 19.9A). A longitudinally oriented window of hard palate bone is subsequently removed with an osteotome, power-driven burr, oscillating saw or rongeurs[2,3,78]. When using rongeurs, removal of bone is started at the caudal part of the bone from the incision in soft palate. Bone is not replaced at closure. The nasal cavity can now be explored and abnormal tissue or foreign bodies can be removed and samples can be taken for culture and sensitivity. The mucoperiosteum of the hard palate is sutured in two layers (nasal surface and oral surface) in an interrupted or simple continuous pattern with absorbable monofilament material (Fig. 19.9B).

Alternatively, especially if only access to the rostral nasal cavity is required, a U-shaped mucoperiosteal flap can be made[3]. The incision is made parallel to the dental arcade, along the entire lingual edge of the alveolar ridge. After incision, the surgeon elevates and retracts the mucoperiosteum caudoventrally to expose the hard palate while preserving the major palatine arteries. A window of bone can then be removed as described above[2,3].

19.8 LATERAL RHINOTOMY

A lateral approach to the nose is rarely performed[2,3]. Limited lateral rhinotomies have been described in combination with maxillectomy for tumour removal[84] and for recovery of laterally displaced teeth[72]. For the latter indication, an alveolar mucosal approach is described, where an incision is made through the alveolar mucosa from the nasal bone to the rostral end of the interincisive suture[71]. After reflection of the alveolar mucosa the dorsal and ventral lateral nasal cartilages can be reflected medially and the nasal mucosa incised to expose the nasal cavity. A proper lateral rhinotomy

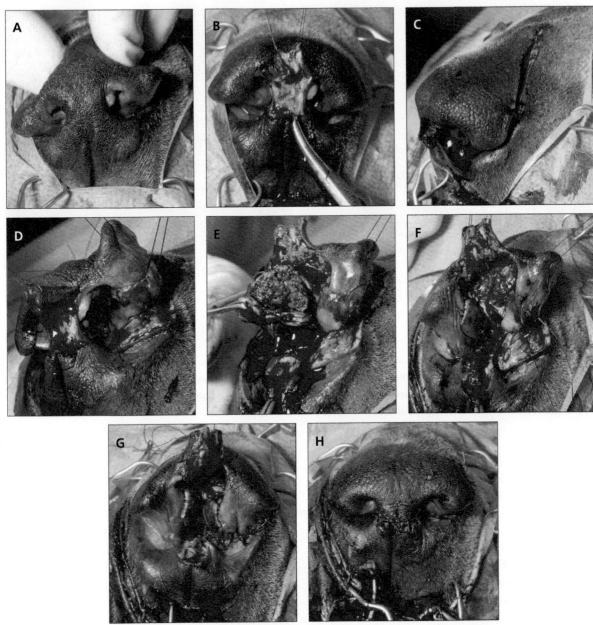

Figs 19.10A–H Lateral rhinotomy. The dog is positioned in sternal recumbency; a small irregular mass can be seen arising from the left rostral nasal septum (A). A horizontal incision is made in the ventral part of the philtrum, from the left to the right sulcus and then extended dorsally towards the point where the alar cartilages meet the central planum (B). A lateral rhinotomy is performed on the ipsilateral side of the nasal septal mass by incising the skin from the angle of the rhinarium to the nasomaxillary notch (C), and subsequently transecting the maxillary cartilage between the dorsal and ventral parietal cartilages (D). The nasal vestibule is thus completely exposed and full-thickness resection of a rostral nasal septal mass can be performed (E). The septal cartilaginous resection sites can be left open to heal by second intention (F). The alar and dorsal lateral nasal cartilage with overlying nasal mucosa are sutured back into place using monofilament absorbable material (G). The nasal planum U-flap is sutured back into its anatomical position (H).

was described by Hedlund[3,79], which gives access to the nasal vestibule and rostral nasal cavity. When combined with elevation of the central planum as described by Pavletic[85], the approach allows access to the entire rostral nasal septum for removal of septal tumours such as SCC[86].

Technique

The dog is positioned in sternal recumbency (Fig. 19.10A). The entire muzzle is clipped and prepared aseptically. The nasal planum is first elevated as a dorsally based U-flap as follows[85]:

- A horizontal incision is made in the ventral part of the philtrum, from the left to the right sulcus and then extended dorsally towards the point where the alar cartilages meet the central planum (Fig. 19.10B). This exposes the very rostral nasal septum and allows for cosmetic closure after resection.
- Then, a lateral rhinotomy is performed on the ipsilateral side of the nasal septal mass as described by Hedlund[3,79]. This requires incising the skin from the angle of the rhinarium to the nasomaxillary notch (Fig. 19.10C), and subsequently transecting the maxillary cartilage between the dorsal and ventral parietal cartilages (Fig. 19.10D). The nasal vestibule is thus completely exposed and full-thickness resection of a rostral nasal septal mass can be performed (Fig. 19.10E)[86].
- The septal cartilaginous resection sites can be left open to heal by second intention (Fig. 19.10F). The lateral rhinotomy incision is closed in three layers[3]. First, the alar and dorsal lateral nasal cartilage with overlying nasal mucosa are sutured back into place using monofilament absorbable material (Fig. 19.10G)[86].
- The subcutaneous tissues of the remainder of the lateral rhinotomy are closed using monofilament absorbable material in a simple continuous pattern, and the skin is closed with simple interrupted nonabsorbable sutures[86].
- The nasal planum U-flap is sutured back into its anatomical position by preplacing sutures laterally and caudally in the submucosal tissues in a simple interrupted pattern, followed by dermal interrupted sutures using absorbable monofilament suture material (Fig. 19.10H)[86].

19.9 NASOPHARYNGOTOMY AND NASOPHARYNGEAL POLYP REMOVAL

Nasopharyngeal polyps, also called otopharyngeal or inflammatory polyps, are benign pedunculated growths that originate from the mucosal lining of the middle ear, auditory tube and nasopharynx[87,88]. The polyp may grow into the nasopharynx, ear canal, or both or remain in the tympanic cavity (see Chapter 5 Aetiology and Pathogenesis of Otitis Externa, Section 5.9). Some surgeons perform a ventral bulla osteotomy (VBO) before removing the protruding polyp to facilitate removal of inflammatory tissue and detachment of the pedicle, regardless of whether the polyp is in the external ear canal or nasopharynx. However, recurrence of a nasopharyngeal polyp is uncommon with simple traction-avulsion and outweighs the risks and complications of VBO. Most nasopharyngeal polyps can be removed by simple traction-avulsion after retraction of the soft palate, but for small polyps nasopharyngotomy may be necessary. Nasopharyngotomy is also indicated for removal of lodged foreign bodies, sampling and removing nasopharyngeal cysts and tumours (see Chapter 12 Diseases of the Nasopharynx, Section 12.3) and for treatment of nasopharyngeal stenosis (see Chapter 12, Section 12.2).

Technique for traction-avulsion of nasopharyngeal polyp

The animal is placed in dorsal, ventral or lateral recumbency with the mouth opened maximally by an assistant using tape or bandage material around the rostral maxilla and mandibula or by using mouth gags (Fig. 19.11A). The caudal edge of the soft palate is gently but firmly retracted with the use of a spay hook or stay suture. In most cases the caudal extension of the polyp can be visualised this way (Fig. 19.11B). A curved Halstedt forceps is introduced over the polyp and advanced rostrally into the nasopharynx towards the opening of the auditory tube (Fig. 19.11C). Here the stalk of the polyp is grasped, after which the forceps are gently rotated. If a firm grip has been achieved and the forceps can be rotated without excessive force, a firm pulling force in a caudal (not ventral) direction can be applied while continuing to rotate the polyp. Once the base of the polyp tears loose, ventral traction can be applied to completely remove the polyp from the nasopharynx

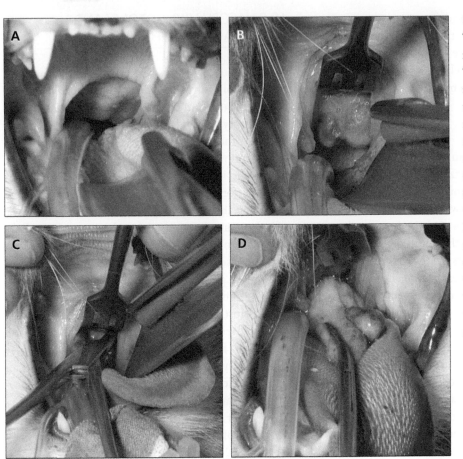

Figs 19.11A–D
Traction-avulsion of nasopharyngeal polyp. The mouth is opened maximally by an assistant using tape or bandage material around the rostral maxilla and mandibula (A). The caudal edge of the soft palate is retracted and the polyp is visualised (B). A curved Halstedt forceps is introduced over the polyp and advanced rostrally into the nasopharynx towards the opening of the auditory tube (C). The polyp is removed with traction-avulsion (D).

(Fig. 19.11D). Any haemorrhage is usually mild and can be stopped by firm digital pressure on the soft palate towards the nasopharyngeal roof. The polyp should be examined to ensure a stalk is present, in which case recurrence is uncommon.

Technique for nasopharyngotomy

The animal can either be placed in sternal recumbency with the maxilla secured to a bar over the table (Fig. 19.12A), or in dorsal recumbency with the mandible taped to the table[2,47]. A longitudinal full-thickness incision is made on the midline of the soft palate leaving the free end of the soft palate intact to facilitate wound closure. The incision is made through the oral mucosa, palatinus muscle and nasopharyngeal mucosa (Fig. 19.12B)[89]. Rostrally the incision extends to the caudal margin of the hard palate or further cranially if required, for instance for treatment of choanal atresia. Stay sutures are used to retract both sides of the palate

laterally for maximal exposure of the nasopharyngeal area (Fig. 19.12C). The polyp is grasped as close to the opening of the auditory tubes as possible, with curved Halstedt forceps (Fig. 19.12D). With traction-avulsion the polyp is removed, foreign bodies are removed or samples are taken for histology and culture and sensitivity. Resecting the membrane or scar tissue responsible for nasopharyngeal stenosis can also be achieved using this approach. However, primary closure of any mucosal defects is mandatory to prevent webbing and secondary restenosis[90]. If primary closure cannot be accomplished, a nasopharyngeal advancement flap can be used[91].

Suturing the nasopharyngeal mucosa, submucosal tissue and oropharyngeal mucosa separately with 4-0 or 5-0 absorbable suture material in a simple continuous pattern performs a three-layer closure. The submucosa can alternatively be incorporated in either the nasopharyngeal or oropharyngeal layer for a two-layer closure.

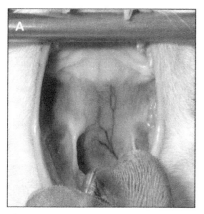
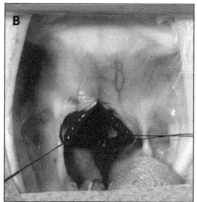
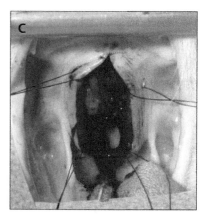

Figs 19.12A–D Nasopharyngotomy. The animal is placed in sternal recumbency with the maxilla secured to a bar over the table (A). A longitudinal full-thickness incision is made on the midline of the soft palate leaving the free end of the soft palate intact (B). Rostrally the incision extends to the caudal margin of the hard palate (C) or further cranially if required, for instance for treatment of choanal atresia. The polyp is grasped as close to the opening of the auditory tubes as possible, with curved Halstedt forceps and removed with traction-avulsion (D).

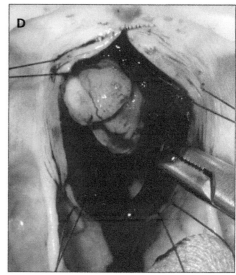

19.10 REFERENCES

1 Venker-van Haagen AJ. The nose and nasal sinuses. In: *Ear, Nose, Throat, and Tracheobronchial Diseases in Dogs and Cats*. Venker-van Haagen AJ (ed). Schlütersche Verlagsgesellschaft, Hannover 2005, pp. 51–81.

2 Schmiedt CW, Creevy KE. Nasal planum, nasal cavity, and sinuses. In: *Veterinary Surgery Small Animal*. Tobias KM, Johnston SA (eds). Elsevier Saunders, St. Louis 2012, pp. 1691–706.

3 Hedlund CS. Rhinotomy techniques. In: *Current Techniques in Small Animal Surgery*. Bojrab MJ (ed). Williams & Wilkins, Baltimore 1998, pp. 346–54.

4 Oechtering GU. Brachycephalic syndrome – new information on an old congenital disease. *Veterinary Focus* 2010;**20**:2–9.

5 Brainard BM, Hofmeister EH. Anesthesia principles and monitoring. In: *Veterinary Surgery Small Animal*. Tobias KM, Johnston SA (eds). Elsevier Saunders, St. Louis 2012, pp. 248–91.

6 Fasanella FJ, Shivley JM, Wardlaw JL, Givaruangsawat S. Brachycephalic airway obstructive syndrome in dogs: 90 cases (1991–2008). *Journal of the American Veterinary Medical Association* 2010;**237**:1048–51.

7 Torrez CV, Hunt GB. Results of surgical correction of abnormalities associated with brachycephalic airway obstruction syndrome in dogs in Australia. *Journal of Small Animal Practice* 2006;**47**:150–4.

8 Smith MM. Oronasal fistula repair. *Clinical Techniques in Small Animal Practice* 2000;**15**:243–50.

9 Kelly KM, Bardach J. Biologic basis of cleft palate and palatal surgery. In: *Oral and Maxillofacial Surgery in Dogs and Cats*. Verstraete FJM, Lommer MJ (eds). Saunders/Elsevier, St. Louis 2012, pp. 343–50.

10 Aiken SW. Principles of surgery for the cancer patient. *Clinical Techniques in Small Animal Practice* 2003;**18**:75–81.

11 Radlinsky MG. Biopsy general principles. I In: *Veterinary Surgery Small Animal*. Tobias KM, Johnston SA (eds). Elsevier Saunders, St. Louis 2012, pp. 231–6.

12 Lascelles BD, Parry AT, Stidworthy MF, Dobson JM, White RA. Squamous cell carcinoma of the nasal planum in 17 dogs. *Veterinary Record* 2000;**147**:473–6.

13 Thomson M. Squamous cell carcinoma of the nasal planum in cats and dogs. *Clinical Techniques in Small Animal Practice* 2007;**22**:42–5.

14 White SD. Diseases of the nasal planum. *Veterinary Clinics of North America Small Animal Practice* 1994;**24**:887–95.

15 Withrow SJ. Cancer of the nasal planum. In: *Small Animal Clinical Oncology*, 5th edn. Withrow SJ, Vail DM, Page RL (eds). Elsevier Saunders, St. Louis 2013, pp. 432–5.

16 Withrow SJ, Nelson AW, Manley PA, Biggs DR. Premaxillectomy in the dog. *Journal of the American Animal Hospital Association* 1985;**21**:49–58.

17 Kirpensteijn J, Withrow SJ, Straw RC. Combined resection of the nasal planum and premaxilla in three dogs. *Veterinary Surgery* 1994;**23**:341–6.

18 Gallegos J, Schmiedt C, McAnulty JF. Cosmetic rostral nasal reconstruction after nasal planum and premaxilla resection: technique and results in two dogs. *Veterinary Surgery* 2007;**36**:669–74.

19 Withrow SJ, Susaneck SJ, Macy DW, Sheetz J. Aspiration and punch biopsy techniques for nasal tumors. *Journal of the American Animal Hospital Association* 1985;**21**:551–4.

20 Withrow SJ, Straw RC. Resection of the nasal planum in nine cats and five dogs. *Journal of the American Animal Hospital Association* 1990;**26**:219–22.

21 Lorinson D, Bright RM, White R. Brachycephalic airway obstruction syndrome – a review of 118 cases. *Canine Practice* 1997;**22**:18–21.

22 Pink JJ, Doyle RS, Hughes JM, Tobin E, Bellenger CR. Laryngeal collapse in seven brachycephalic puppies. *Journal of Small Animal Practice* 2006;**47**:131–5.

23 Poncet CM, Dupre GP, Freiche VG, Estrada MM, Poubanne YA, Bouvy BM. Prevalence of gastrointestinal tract lesions in 73 brachycephalic dogs with upper respiratory syndrome. *Journal of Small Animal Practice* 2005;**46**:273–9.

24 Riecks TW, Birchard SJ, Stephens JA. Surgical correction of brachycephalic syndrome in dogs: 62 cases (1991–2004). *Journal of the American Veterinary Medical Association* 2007;**230**:1324–8.

25 Schuenemann R, Oechtering GU. Inside the brachycephalic nose: intranasal mucosal contact points. *Journal of the American Animal Hospital Association* 2014;**50**:149–58.

26 Bedford PGC. Tracheal hypoplasia in the English Bulldog. *Veterinary Record* 1982;**111**:58–9.

27 Coyne BE, Fingland RB. Hypoplasia of the trachea in dogs: 103 cases (1974–1990). *Journal of the American Veterinary Medical Association* 1992;**201**:768–72.

28 Oechtering GU, Hueber J, Kiefer I, Noeller C. Laser assisted turbinectomy (LATE) – a novel approach to brachycephalic airway syndrome. *Veterinary Surgery* 2007;**36**:E11.

29 Oechtering GU, Hueber J, Oechtering TH, C N. Laser assisted turbinectomy (LATE) – treating brachycephalic airway distress at its intranasal origin. *Veterinary Surgery* 2007;**36**:E18.

30 Huck JL, Stanley, BJ, Hauptman, JG. Technique and outcome of nares amputation (Trader's technique) in immature shih tzus. *Journal of the American Animal Hospital Association* 2008;**44**:82–5.

31 Harvey CE. Upper airway obstruction surgery. 1. Stenotic nares surgery in brachycephalic dogs. *Journal of the American Animal Hospital Association* 1982;**18**:535–7.

32 Vilaplana Grosso F, ter Haar G, Boroffka SAEB. Gender, weight, and age effects on prevalance of caudal aberrant nasal turbinates in clinically healthy English Bulldogs: a computed tomographic study and classification. *Veterinary Radiology & Ultrasound* 2015;**56**:486–93.

33 Bright RM, Wheaton LG. A modified surgical technique for elongated soft palate in dogs. *Journal of the American Animal Hospital Association* 1983;**19**:288.

34 Harvey CE. Upper airway obstruction surgery. 2. Soft palate resection in brachycephalic dogs. *Journal of the American Animal Hospital Association* 1982;**18**:538–44.

35 Clark GN, Sinibaldi KR. Use of a carbon dioxide laser for treatment of elongated soft palate in dogs. *Journal of the American Veterinary Medical Association* 1994;**204**:1779–81.

36 Davidson EB, Davis MS, Campbell GA, *et al.* Evaluation of carbon dioxide laser and conventional incisional techniques for resection of soft palates in brachycephalic dogs. *Journal of the American Veterinary Medical Association* 2001;**219**:776–81.

37 Brdecka DJ, Rawlings CA, Perry AC, Anderson JR. Use of an electrothermal, feedback-controlled, bipolar sealing device for resection of the elongated portion of the soft palate in dogs with obstructive upper airway disease. *Journal of the American Veterinary Medical Association* 2008;**233**:1265–9.

38 Reiter AM, Holt DE. Palate. In: *Veterinary Surgery Small Animal*. Tobias KM, Johnston SA (eds). Elsevier Saunders, St. Louis 2012, pp. 1707–17.

39 Monnet E, Tobias KM. Larynx. In: *Veterinary Surgery Small Animal*. Tobias KM, Johnston SA (eds). Elsevier Saunders, St. Louis 2012, pp. 1718–33.

40 Mehl ML, Kyles AE, Pypendop BH, Filipowicz DE, Gregory CR. Outcome of laryngeal web resection with mucosal apposition for treatment of airway obstruction in dogs: 15 cases (1992–2006). *Journal of the American Veterinary Medical Association* 2008;**233**:738–42.

41 White RN. Surgical management of laryngeal collapse associated with brachycephalic airway obstruction syndrome in dogs. *Journal of Small Animal Practice* 2012;**53**:44–50.

42 Harvey CE. Palate defects in dogs and cats. *Compendium on Continuing Education for the Practicing Veterinarian* 1987;**9**:404–10.

43 Sivacolundhu RK. Use of local and axial pattern flaps for reconstruction of the hard and soft palate. *Clinical Techniques in Small Animal Practice* 2007;**22**:61–9.

44 Sager M, Nefen S. Use of buccal mucosal flaps for the correction of congenital soft palate defects in three dogs. *Veterinary Surgery* 1998;**27**:358–63.

45 Nelson WA. Upper respiratory system. In: *Textbook of Small Animal Surgery*. Slatter D (ed). Saunders, Philadelphia 1993, pp. 733–76.

46 Bojrab MJ, Tholen MA, Constantinescu GM. Oronasal fistulae in dogs and cats. *Compendium on Continuing Education for the Practicing Veterinarian* 1986;**8**:815–20.

47 Nelson AW. Nasal passages, sinus, and palate. I In: *Textbook of Small Animal Surgery*. Slatter D (ed). Saunders, Philadelphia 1993, pp. 824–37.

48 Reiter AM. How to make a palatal obturator. *Proceedings of the Annual Veterinary Dentistry Forum* 2005;**19**:9.

49 Reiter AM, Smith MM. The oral cavity and oropharynx. In: *BSAVA Manual of Canine and Feline Head, Neck and Thoracic Surgery*. Brockman DJ, Holt DE (eds). BSAVA, Gloucester 2005, pp. 24–43.

50 Griffiths LG, Sullivan M. Bilateral overlapping mucosal single-pedicle flaps for correction of soft palate defects. *Journal of the American Animal Hospital Association* 2001;**37**:183–6.

51 Warzee CC, Bellah JR, Richards D. Congenital unilateral cleft of the soft palate in six dogs. *Journal of Small Animal Practice* 2001;**42**:338–40.

52 White RN, Hawkins HL, Alemi VP, Warner C. Soft palate hypoplasia and concurrent middle ear pathology in six dogs. *Journal of Small Animal Practice* 2009;**50**:364–72.

53 Sylvestre AM, Sharma A. Management of a congenitally shortened soft palate in a dog. *Journal of the American Veterinary Medical Association* 1997;**211**:875–7.

54 Bauer MS, Levitt L, Pharr JW, Fowler JD, Basher AW. Unsuccessful surgical repair of a short soft palate in a dog. *Journal of the American Veterinary Medical Association* 1988;**193**:1551–2.

55 Headrick JF, McAnulty JF. Reconstruction of a bilateral hypoplastic soft palate in a cat. *Journal of the American Animal Hospital Association* 2004;**40**:86–90.

56 Venker-van Haagen AJ. The pharynx. In: *Ear, Nose, Throat, and Tracheobronchial Diseases in Dogs and Cats*. Venker-van Haagen AJ (ed). Schlütersche Verlagsgesellschaft, Hannover 2005, pp. 83–120.

57 Tholen MA, Johnson J. Surgical repair of the oronasal fistula. *Veterinary Medicine Small Animal Clinics* 1983;**78**:1733.

58 Woodward TM. Greater palatine island axial pattern flap for repair of oronasal fistula related to eosinophilic granuloma. *Journal of Veterinary Dentistry* 2006;**23**:161–6.

59 Soukup JW, Snyder CJ, Gengler WR. Free auricular cartilage autograft for repair of an oronasal fistula in a dog. *Journal of Veterinary Dentistry* 2009;**26**:86–95.

60 Smith MM, Rockhill AD. Prosthodontic appliance for repair of an oronasal fistula in a cat. *Journal of the American Veterinary Medical Association* 1996;**208**:1410–12.

61 Cox CL, Hunt GB, Cadier MM. Repair of oronasal fistulae using auricular cartilage grafts in five cats. *Veterinary Surgery* 2007;**36**:164–9.

62 Sharman M, Lenard Z, Hosgood G, Mansfield C. Clotrimazole and enilconazole distribution within the frontal sinuses and nasal cavity of nine dogs with sinonasal aspergillosis. *Journal of Small Animal Practice* 2012;**53**:161–7.

63 Sissener TR, Bacon NJ, Friend E, Anderson DM, White RA. Combined clotrimazole irrigation and depot therapy for canine nasal aspergillosis. *Journal of Small Animal Practice* 2006;**47**:312–15.

64 Mathews KG, Davidson AP, Koblik PD, *et al*. Comparison of topical administration of clotrimazole through surgically placed versus nonsurgically placed catheters for treatment of nasal aspergillosis in dogs: 60 cases (1990–1996). *Journal of the American Veterinary Medical Association* 1998;**213**:501–6.

65 White D. Canine nasal mycosis – light at the end of a long diagnostic and therapeutic tunnel. *Journal of Small Animal Practice* 2006;**47**:307.

66 Friend EJ, Williams JM, White RAS. Invasive treatment of canine nasal aspergillosis with topical clotrimazole. *Veterinary Record* 2002;**151**:298–9.

67 Claeys S, Lefebvre JB, Schuller S, Hamaide A, Clercx C. Surgical treatment of canine nasal aspergillosis by rhinotomy combined with enilconazole infusion and oral itraconazole. *Journal of Small Animal Practice* 2006;**47**:320–4.

68 Holt DE. Frontal sinus drainage techniques. In: Current Techniques in Small Animal Surgery, 4th edn. Bojrab MJ (ed). Williams & Wilkins, Philadelphia 1998, pp. 354–5.

69 Greci V, Mortellaro CM, Olivero D, Cocci A, Hawkins EC. Inflammatory polyps of the nasal turbinates of cats: an argument for designation as feline mesenchymal nasal hamartoma. *Journal of Feline Medicine and Surgery* 2011;**13**:213–19.

70 Meler E, Dunn M, Lecuyer M. A retrospective study of canine persistent nasal disease: 80 cases (1998–2003). *Canadian Veterinary Journal* 2008;**49**:71–6.

71 Priddy NH, Pope ER, Cohn LA, Constantinescu GM. Alveolar mucosal approach to the canine nasal cavity. *Journal of the American Animal Hospital Association* 2001;**37**:179–82.

72 Taylor TN, Smith MM, Snyder L. Nasal displacement of a tooth root in a dog. *Journal of Veterinary Dentistry* 2004;**21**:222–5.

73 Turek MM, Lana SE. Nasosinal tumors. In: Small Animal Clinical Oncology, 5th edn. Withrow SJ, Vail DM, Page RL (eds). Elsevier Saunders, St. Louis 2013, pp. 435–51.

74 Adams WM. Outcome of accelerated radiotherapy alone or accelerated radiotherapy followed by exenteration of the nasal cavity in dogs with intranasal neoplasia: 53 cases (1990–2002). *Journal of the American Veterinary Medical Association* 2005;**227**:936–41.

75 Hedlund CS, Tangner CH, Elkins AD, Hobson HP. Temporary bilateral carotid artery occlusion during surgical exploration of the nasal cavity of the dog. *Veterinary Surgery* 1983;**12**:83–5.

76 Holmberg DL. Sequelae of ventral rhinotomy in dogs and cats with inflammatory and neoplastic nasal pathology: a retrospective study. *Canadian Veterinary Journal* 1996;**37**:483–5.

77 Gillilan LA. Extra- and intra-cranial blood supply to brains of dog and cat. *American Journal of Anatomy* 1976;**146**:237–53.

78 Holmberg DL, Fries C, Cockshutt J, Van Pelt D. Ventral rhinotomy in the dog and cat. *Veterinary Surgery* 1989;**18**:446–9.

79 Hedlund CS. Surgery of the upper respiratory system. In: *Small Animal Surgery*, 3rd edn. Fossum TW, Hedlund CS, Hulse DA, *et al*. (eds). Mosby, St. Louis 2007, pp. 817–66.

79a ter Haar G. Dorsal Rhinotomy. In: *Complications in Small Animal Surgery*. Griffon D, Hamaide A (eds). Wiley Blackwell, Ames 2016, pp. 137–43.

80 Birchard SJ. A simplified method for rhinotomy and temporary rhinostomy in dogs and cats. *Journal of the American Animal Hospital Association* 1988;**24**:69–72.

81 Latham RA, Deaton TG, Calabrese CT. A question of the role of the vomer in the growth of the premaxillary segment. *Cleft Palate Journal* 1975;**12**:351–5.

82 Wada T, Kremenak CR, Miyazaki T. Midfacial growth effects of surgical trauma to the area of the vomer in beagles. *Journal of Osaka University Dental School* 1980;**20**:241–76.

83 Freng A. Growth of the middle face in experimental early bony fusion of the vomeropremaxillary, vomeromaxillary and mid-palatal sutural system. A roentgencephalometric study in the domestic cat. *Scandinavian Journal of Plastic and Reconstructive Surgery* 1981;**15**:117–25.

84 Lascelles BD, Thomson MJ, Dernell WS, Straw RC, Lafferty M, Withrow SJ. Combined dorsolateral and intraoral approach for the resection of tumors of the maxilla in the dog. *Journal of the American Animal Hospital Association* 2003;**39**:294–305.

85 Pavletic MM. Nasal Reconstruction techniques. In: *Atlas of Small Animal Wound Management and Reconstructive Surgery*, 3rd edn. Pavletic MM (ed). Wiley-Blackwell, Ames 2010, pp. 573–602.

86 ter Haar G, Hampel R. Combined rostro-lateral rhinotomy for removal of rostral nasal septum squamous cell carcinoma: long-term outcome in 10 dogs. *Veterinary Surgery* 2015;**44**(7):843–51.

87 Galloway PE, Kyles A, Henderson JP. Nasal polyps in a cat. *Journal of Small Animal Practice* 1997;**38**:78–80.

88 Fan TM, de Lorimier L-P. Inflammatory polyps and aural neoplasia. *Veterinary Clinics of North America Small Animal Practice* 2004;**34**:489–509.

89 Hunt GB, Perkins MC, Foster SF. Nasopharyngeal disorders of dogs and cats: A review and retrospective study. *Compendium on Continuing Education for the Practicing Veterinarian* 2002;**24**:184–200.

90 Mitten RW. Nasopharyngeal stenosis in four cats. *Journal of Small Animal Practice* 1988;**29**:341–5.

91 Griffon DJ, Tasker S. Use of a mucosal advancement flap for the treatment of nasopharyngeal stenosis in a cat. *Journal of Small Animal Practice* 2000;**41**:71–3.

20.1 INTRODUCTION AND TEMPORARY TRACHEOTOMY

The throat comprises the pharynx and larynx and is a shared space between the digestive and respiratory tract. Its correct functioning is essential for survival. Deglutition allows transport of food from the mouth to the stomach, which requires well-coordinated activity of the muscles of the oral cavity, pharynx, larynx and oesophagus. The larynx is a complex cartilaginous organ that serves many functions but is primarily used to separate the respiratory tract from the digestive tract. The intrinsic and extrinsic muscles of the cricoid, thyroid and paired arytenoid cartilages and the epiglottis interact under the control of the neuromuscular system of the larynx to move the larynx in swallowing, to protect the lower airways (coughing), to facilitate and regulate the respiratory airflow and to vocalise (see Chapter 13 Throat Diagnostic Procedures and Chapter 14 Throat Anatomy and Physiology). Therefore, the pharyngolaryngological surgeon must be thoroughly aware of the potential effects on all of these functions when performing surgery on the pharynx or larynx. A perturbation of any one of these important tasks may have a marked impact on a patient. A thorough understanding and knowledge of the local anatomy therefore is mandatory before initiating any surgical procedure in this region. In addition, the disease necessitating surgical intervention may have altered normal anatomy and/or have compromised the patient. The exact nature and extent of the disease should be evaluated prior to surgical intervention therefore, using diagnostic imaging, throat inspection and cytology/histopathology and culture as has been described in Chapter 14 Throat Diagnostic Procedures.

Severe or chronic pharyngeal or laryngeal disease can lead to weight loss and emaciation, inappetence, dyspnoea, hypoxia and aspiration pneumonia, which can all affect anaesthesia protocols and wound healing. These factors should be addressed prior to surgery if possible. Surgery of the pharynx or larynx may lead to additional swelling in the airways and to pooling of blood, saliva and lavage fluid that may lead to aspiration. Intubation is therefore always indicated to protect the airways as well as to oxygenate the patient, but it may be complicated by swelling or tumour in the pharynx or larynx. A temporary tracheostomy may be indicated to bypass the throat entirely during surgery if normal endotracheal intubation is problematic or if the tube decreases visibility of and surgical access to the pharynx or larynx[1,2]. Though prolonged intubation is an option after surgery, the animal must be kept sedated or anaesthetised. Furthermore, in a study of prolonged endotracheal intubation in dogs, laryngeal mucosal inflammation was observed after 24 hours, and after 1 week there was loss of mucosal architecture and inflammatory infiltrates were seen in the arytenoid cartilages[3]. In another study, ulceration was found where the tube rubbed and pressed against contact points in the cricoarytenoid region[4]. Both of these reports have shown that tracheostomy is more appropriate than prolonged endotracheal intubation when the pharyngolaryngeal airways are obstructed as result of injury[2]. Therefore, if significant swelling or obstruction of the pharynx or larynx is expected in the immediate postoperative period, a temporary tracheostomy should be performed. The technique for tracheotomy will be discussed at the end of this section.

In general, all patients undergoing pharyngolaryngeal surgery will benefit from preoxygenation before induction of anaesthesia[5]. Preoxygenation is therefore imperative and may provide for an additional 3–5 minutes for intubation of a difficult airway[5]. This can be provided via a mask, nasal prongs, oxygen cage or as flow-by until the airways are secured via endotracheal or per tracheostomy intubation. A preoperative anti-inflammatory dose of a corticosteroid (dexamethasone, 0.5–2 mg/kg i/v, i/m or s/c) may reduce upper

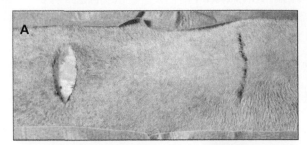

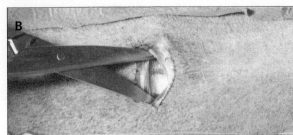

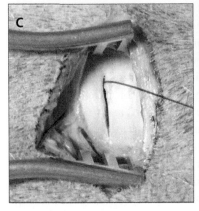

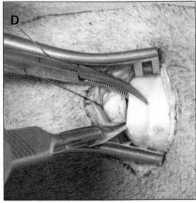

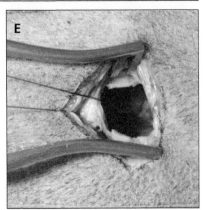

airway oedema secondary to diagnostic or surgical manipulations and is routinely given for intralaryngeal and brachycephaly related procedures, preferably before intubation. Animals in respiratory distress easily become hyperthermic, and need to be cooled actively using cool i/v fluids, fans and icepacks.

Sedation is very beneficial in the perianesthetic period of patients with upper airway obstruction. Opiates and tranquillisers are ideal[5]. Animals with brachycephalic obstructive airway syndrome (BOAS) tend to have elevated vagal tone, and the addition of an anticholinergic to the premedication may be indicated[5]. Propofol allows for a rapid induction, but is associated with apnoea and therefore not ideal for cases in which a difficult intubation is expected. Ketamine with a benzodiazepine would be indicated for those cases[5]. Maintenance, unless an intravenous protocol is indicated, can usually be achieved with inhalant anaesthetics. Monitoring of patients undergoing pharyngolaryngeal surgery should minimally consist of electrocardiography (ECG), body temperature, oxygenation (SpO$_2$) and ventilation (end-tidal CO$_2$).

Peri- and postoperative analgesia, antibacterial therapy and postoperative nutritional management for patients undergoing pharyngolaryngeal surgery are discussed in Chapter 17 Perioperative Management. Pharyngeal surgery is most commonly performed for correction of brachycephaly-related abnormalities such as overlong soft palate, correction of oropharyngeal dysphagia, tonsillectomy, exploration of the throat after stick injury or other traumatic events and for marsupialisation of mucoceles. Surgery for brachycephalic patients has been discussed in Chapter 19 Surgery of the Nose, Section 19.3, the other procedures will be discussed in this chapter. Laryngeal surgery is most commonly performed for acquired laryngeal paralysis in older large and giant breed dogs, or for removal of laryngeal masses (ventral laryngotomy or partial to complete laryngectomy). Partial or complete laryngectomy are rarely performed and are considered technically very demanding. These animals are best served by referral to a specialist surgeon. With proper anatomical knowledge, gentle tissue handling, use of appropriate instruments and technique, cautious use of anaesthetics, and stringent perioperative care and monitoring, succesful thyroid-arytenoid lateralisation and ventral laryngotomy however may be achievable in large small animal practises and will be reviewed in the final two sections of this chapter.

Tracheostomy

A tracheostomy for temporary use is indicated in patients with obstruction of the upper airway in the nose, larynx or cranial part of the trachea. It is also used as an alter-

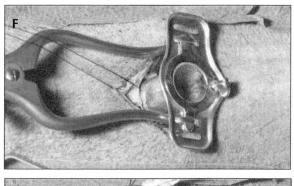

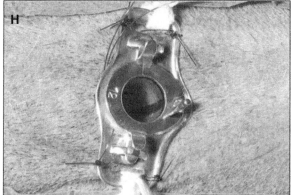
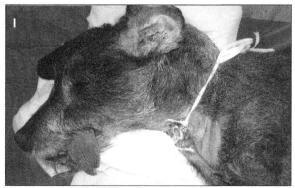

Figs 20.1A–I Temporary tracheostomy. A transverse skin incision of approximately 2 cm is made over the trachea at the midpoint between the larynx (A, blue line) and the thoracic inlet. The subcutaneous fat and the left and right sternothyroid and sternohyoid muscles are divided in the midline by blunt dissection (B). The ligament between two adjacent rings in incised with a no. 11 scalpel and a tag suture can be placed around a tracheal ring to allow for tracheal manipulation (C). A scalpel is used to remove a segment of cartilage that should be grabbed with a small forceps (D). An opening of the same size and shape as the tracheal tube or cannula should be created (E). The outer canula of an endotracheal cannula is inserted through the tracheal window, followed by an inner cannula (F, G). The outer cannula is sutured to the skin with four sutures. In addition, two cotton ribbons are attached to the wings of the tracheal cannula and then tied around the neck (H, I).

native route for endotracheal intubation in laryngeal surgery and as an ancillary procedure to prevent postoperative dyspnoea after extensive surgical procedures on the pharynx or larynx[6]. Temporary tracheostomy includes the introduction of a tracheal cannula that will be left in place until the airways are patent. Silastic tracheal cannulas or stainless steel cannulas with an inner cannula that can be removed and cleaned are preferred[7]. Temporary tracheostomies require inhalation anaesthesia in all patients, including emergency cases.

The patient is placed in dorsal recumbency with a pillow under the neck[8–10]. A transverse skin incision of approximately 2 cm is made over the trachea at the midpoint between the larynx and the thoracic inlet (Fig. 20.1A)[7]. The subcutaneous fat and the left and right ster-

nothyroid and sternohyoid muscles are divided in the midline by blunt dissection (Fig. 20.1B). Care should be taken not to damage the tracheal veins, which lie immediately lateral to the trachea on both sides. A small self-retaining wound retractor is inserted into the wound to expose the trachea. The ligament between two adjacent rings is incised with a no. 11 scalpel and a tag suture can be placed around a tracheal ring to allow for tracheal manipulation (Fig. 20.1C). If intubation via the stoma is only required for the duration of surgery, no tracheal cartilage needs to be removed. If the tracheal tube will stay in place for 12 hours or more, removing part of the cartilage will prevent tracheal necrosis as a result of pressure exerted on the cartilage by the tube. In these cases, a small forceps is placed on one of the tracheal rings beside the incision.

The forceps is locked tight to provide a firm hold on the tracheal ring[7]. This is done to prevent the piece that will be removed from slipping into the trachea. The scalpel is then used to make a circular incision around the forceps (Fig. 20.1D). A round piece of tracheal cartilage and intercartilage ligament is removed to produce an opening of the same size and shape as the tracheal tube or cannula (Fig. 20.1E)[7]. If the window is made too large, air may leak around the cannula and subcutaneous emphysema will form. The oropharyngeal endotracheal tube is removed and replaced by a silastic endotracheal cannula that is inserted through the tracheal window (Figs 20.1F, G). If the tracheostomy is performed prior to a more extensive procedure, an endotracheal tube is used instead. It may be replaced by a cannula at the end of the procedure. The cannula is sutured to the skin with four sutures[7]. In addition, two cotton ribbons are attached to the wings of the tracheal cannula and then tied around the neck (Figs 20.1H, I). The cannula is left in place until the upper airway is patent. Silastic cannulas have an inner cannula that should be cleaned every 2 hours. After removing the cannula, the tracheostomy wound is not sutured, but left open to heal spontaneously. Healing is usually rapid because the incision was made parallel to the natural skin folds, which will result in good apposition of the wound margins.

Whereas complications using this double cannula technique using meticulous tissue and instrument handling are rare, stenosis and infection can occur with any tracheostomy technique[11,12]. In addition, obstruction by mucus plugs and inflammatory debris can occur, especially with single tubes, that leads to a 19% mortality rate in one large study[12], especially in brachycephalic breeds. Any tracheal tube needs intensive nursing to avoid acute blockage of the stoma. Recently, a promising new technique was reported on the use of a novel silicone tracheal stoma. It was successfully used in brachycephalic breeds and offers the potential to reduce complication rates, but it requires further evaluation in a larger case-series before its use can generally be recommended[13].

20.2 SURGICAL CORRECTION OF OROPHARYNGEAL DYSPHAGIA

Introduction

Cricopharyngeal myotomy or myectomy is the treatment of choice for congenital cricopharyngeal acha-lasia. It is critical that an accurate diagnosis is made as a myectomy in patients with pharyngeal dysphagia will lead to worsening of the dysphagia and aspiration pneumonia (see Chapter 15 Diseases of the Pharynx, Section 15.7)[14–18]. The cricopharyngeal muscle is a single, unpaired muscle that originates from the lateral surface of one cricoid cartilage, spreads over the dorsal surface of the oesophagus, and narrows again before attaching on the lateral surface of the contralateral cricoid cartilage[19,20]. The procedure should therefore be performed unilaterally as the muscle forms a band making up the upper oesophageal sphincter. By resecting the band the asynchronous contraction is stopped, thus relieving the dysphagia[21]. Some surgeons advocate removal of a 1 cm portion of the cricopharyngeal muscle[21,22]. This is thought to prevent adhesion formation between the resected ends and reformation of the band by fibrous tissue joining up the ends of the resected muscle. Others simply perform a myotomy[22–25]. No differences in outcome between myotomy or myectomy have been found[19,22].

Cricopharyngeal myotomy/myectomy technique

Cricopharyngeal myotomy can be performed through either a traditional ventral midline or a lateral approach. The lateral approach allows easier access to the dorsal area of the larynx[19,23], but the ventral approach allows for easier identification of regional neurovascular structures[20] and will be discussed here. The patient is placed in dorsal recumbency with a pillow under the neck and the legs pulled caudally. The surgical area to be prepared for aseptic surgery incorporates all the ventral skin from 1 cm cranial to the angle of the mandible to the level of the thoracic inlet[16,20,23]. An orogastric tube passed into the oesophagus will aid identification of the pharyngeal muscles.

The cricopharyngeal muscles are approached via a ventral midline cervical incision from just cranial to the larynx to the midcervical region (Figs 20.2A, B). The paired sternohyoidus muscles are identified and bluntly separated in the midline. The larynx and trachea are gently retracted and rotated laterally 180° using tag sutures in the thyroid cartilage to expose the cricopharyngeal musculature (Fig. 20.2C)[16,20,23]. The cricopharyngeal muscle is subsequently incised on its midline, ensuring that all fibres of the muscle on the oesophagus are elevated and incised, without inadvertently damaging and perforating the oesophageal wall (Fig. 20.2D)[16,20,23]. The thyropharyngeus muscle can be partly or completely incised to ensure complete transection and alleviation of the achalasia[20,25].

A myectomy of the cricopharyngeal muscle may be performed by making a parallel incision and removing a strip of the cricopharyngeal muscle about 1 cm wide. Removing a section of cricopharyngeus muscle on each side of midline, rather than just transecting the muscle, has also been recommended by some authors[17]. The oesophagus should be inspected for damage and any mucosal defect sutured. The larynx and trachea are then returned to their normal positions, and the surgical site is routinely closed. The sternohyoidus muscles are apposed using a simple continuous pattern with 3-0 monofilament absorbable suture material. Closure of the subcutaneous tissues and skin is routine.

Postoperative care and prognosis

Antibiotics are only indicated in the presence of aspiration pneumonia or if gastroesophageal reflux is suspected. In humans a correlation between cricopharyngeal achalasia and gastroesophageal reflux has been found[26]. Following cricopharyngeal myotomy or myectomy gastroesophageal reflux may be increased[26]. Prokinetic agents (metoclopramide) combined with decreasing gastric acidity with H-2 blockers and sucralfate for reflux oesophagitis are recommended[22,26].

With respect to postoperative nutritional support, soft food for 1–2 days after surgery and slow return to normal food over the next 2–3 weeks is recommended. With resolution of the achalasia, dogs should be able to tolerate varying consistencies of food[16,17,25]. Persistent dysphagia after 2–3 days indicates an incomplete myectomy or myotomy was performed, concurrent pharyngeal or oesophageal dysfunction was present or the diagnosis was incorrect[16,20,22].

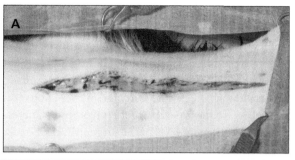

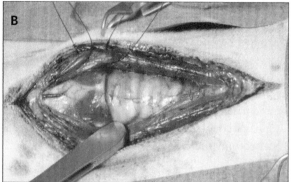

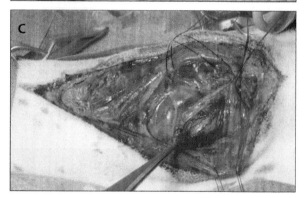

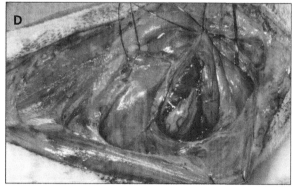

Figs 20.2A–D Cricopharyngeal myotomy. The cricopharyngeal muscles are approached via a ventral midline cervical incision from just cranial to the larynx to the midcervical region (A). The paired cricothyroid muscles are identified (B). The larynx and trachea are gently retracted and rotated laterally 180° using tag sutures in the thyroid cartilage to expose the cricopharyngeal musculature (C). The cricopharyngeal muscle is subsequently incised on its midline, ensuring that all fibres of the muscle on the oesophagus are elevated and incised (D).

The prognosis for surgical treatment is generally considered to be good to excellent for true cricopharyngeal achalasia. Immediate alleviation of clinical signs is reported in most dogs[20]. Fifty percent of dogs had complete resolution of clinical signs in one review article[19]. The outcome in these dogs was not related to the type of surgery (myotomy vs. myectomy) or experience level of the surgeon[19]. In one case series though, the failure rate was high with 57% of dogs euthanised because of dysphagia or aspiration pneumonia[27]. Structural disease, such as fibrosis of the upper oesophageal sphincter or oesophageal stricture, or functional disease, such as laryngeal paralysis, masticatory myositis or myasthenia gravis, concurrent aspiration pneumonia or malnutrition were all suggested to account for this poor outcome[27]. Recurrent dysphagia could result from fibrosis and contracture at the myotomy site[24].

20.3 TONSILLECTOMY

Excision of one or two tonsils is indicated for neoplasia of the palatine tonsils (see Chapter 15 Diseases of the Pharynx, Section 15.6) or when enlarged tonsils contribute to airway obstruction such as in brachycephalic obstructive airway syndrome (see Chapter 11 Diseases of the Nasal Cavity and Sinuses, Section 11.3 and Chapter 19 Surgery of the Nose, Section 19.3) or dysphagia, and in cases of unresponsive chronic tonsillitis (see Chapter 15 Diseases of the Pharynx, Section 15.4)[28,29].

Technique

With the animal in ventral recumbency and positioned as for staphlectomy (see Chapter 19 Surgery of the Nose, Section 19.3), the mouth is opened maximally. The edges of the tonsillar crypt can be retracted caudodorsally to expose the tonsil if neccessary[1,29]. The tonsil itself can be grasped at its base with an Allis tissue forceps and retracted. Alternatively, especially when resecting a friable tonsillar mass, a large stay suture can be placed through the base of the tonsil.

The mucosa at the base of the tonsil can now be transected and the tonsillar artery ligated or coagulated. Transection can be performed with a scalpel blade, sharp scissors, electrosurgery or with laser[1,29]. Haemostasis is subsequently achieved with digital pressure, electrosurgery or by ligation of the tonsillar artery with fine absorbable suture material. Alternatively, especially when dealing with large pedunculated masses, the entire base of the tonsil can be ligated with 2-0 or 3-0 monofilament absorbable suture material (Figs 20.3A, B).

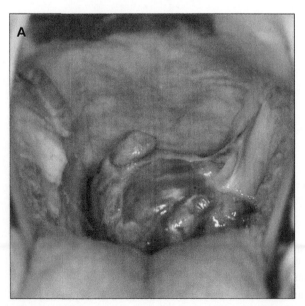

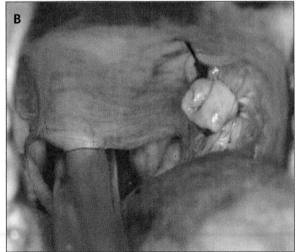

Figs 20.3A, B *En bloc* ligation **(B)** of tonsillar plasmacytoma **(A)**.

After the tonsil has been removed, the edges of the tonsillar crypt can be apposed with a simple continuous suture pattern of 3-0 or 4-0 monofilament absorbable material. Alternatively, after *en-bloc* ligation of the base of the tonsil and transection just above the ligature, the crypt can be left open[1,29]. Tissue is submitted for appropriate diagnostic testing.

Intraoperative and postoperative complications are rare and include haemorrhage, pharyngeal swelling secondary to tissue trauma and postoperative aspiration of blood or fluid[29].

20.4 VENTRAL NECK EXPLORATION

A ventral approach to the pharynx and larynx is the most commonly used approach in pharyngolaryngeal surgery, followed by the intraoral approach. It is the preferred surgical approach for treatment of cricopharyngeal achalasia, resection of pharyngeal and laryngeal masses but especially for exploration of the neck in case of penetrating injuries to the pharyngolaryngeal area from foreign objects such as sticks (see Chapter 15 Diseases of the Pharynx, Section 15.5). A normal ventral approach specifically for the treatment of achalasia or for the removal of laryngeal masses is discussed in Sections 20.2 and 20.7, respectively. In this section exploration of the neck and treatment of stick injuries will be described. A strict ventral approach allows for easy identification of important anatomical structures and minimal tissue dissection.

> A logical decision as to whether or not exploratory surgery of the neck is indicated in trauma patients depends on the point of entry and direction of the stick injury!

Trauma in the direction of the dorsal nasopharynx (usually through the soft palate) and brain cannot be explored via a ventral neck approach. However, for lateralised pharyngeal, tonsillar and sublingual trauma, a midline ventral neck approach is indicated[30–32].

Technique

Surgical exploration of the retropharyngeal space is accomplished through a ventral midline approach from hyoid apparatus to the manubrium of the sternum with the patient in dorsal recumbency and with a support under the neck (Fig. 20.4A). A probe or sterile catheter may be inserted at the oral perforation site to help identification and direction of the penetration tract (Fig. 20.4B)[33]. A gastric tube is very helpful for identification of the oesophagus, yet needs to be carefully inserted so as not to enlarge any possible tears[29]. The goal is to locate and remove foreign material (Figs 20.4C–F), obtain tissue or fluid samples for cytology and culture, debride nonviable tissue and establish drainage if necessary[33]. The exploration is started in the distal neck area and extended rostrally, working from unaffected (clean) towards affected (dirty) tissues[33]. The laryngeal area should be explored with great care to avoid disruption of the recurrent laryngeal nerves or pharyngeal plexus during dissection. Stay sutures in the laryngeal cartilages can help visualise pharyngolaryngeal perforations and rostral oesophageal lesions. Stay sutures in the oesophageal wall can also be used if necessary to rotate the oesophagus and aid in accurate suture placement. After local debridement, the oesophageal tear is sutured in one or two layers with absorbable monofilament suture material[29,32,33]. A local muscular flap can be used as a 'muscular patch' to improve wound healing[29]. After exposure of the penetration tract and after removing the foreign bodies, the tract is flushed with saline, and the surgical wound is sutured over a drain. Drainage can be active or passive; passive drains should exit in a dependent direction through a separate incision. A gastrostomy feeding tube is recommended after oesophageal repair to bypass the area and avoid contamination of the wound with food particles[29,33]. Empirical administration of antibiotics is usually indicated if infection is suspected, see Chapter 17 Perioperative Management, Section 17.3. Therapy can be modified after culture and sensitivity results are available. Acute penetrating injury of the oropharyngeal region, when treated appropriately, has a good prognosis with most or all patients making a complete recovery (see Chapter 15 Diseases of the Pharynx, Section 15.5)[30–32].

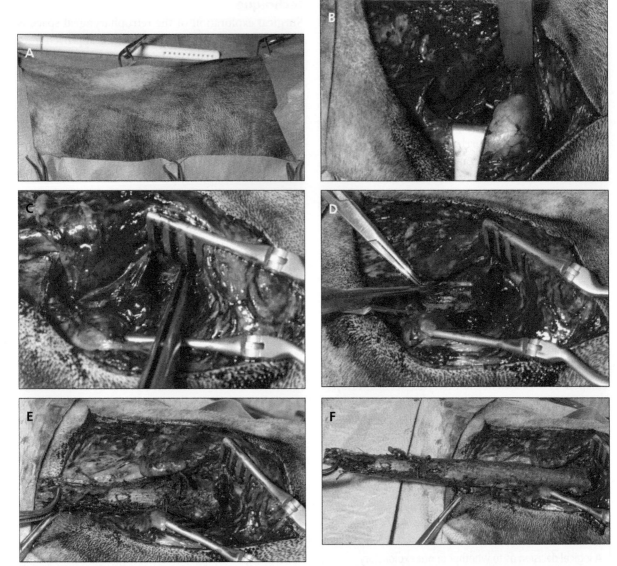

Figs 20.4A–F Ventral neck exploration for stick injury. Surgical exploration of the retropharyngeal space is accomplished through a ventral midline approach from hyoid apparatus to the manubrium of the sternum with the patient in dorsal recumbency and with a support under the neck (A). A probe or sterile catheter may be inserted at the oral perforation site to help identification and direction of the penetration tract; in this case the probe could be palpated and subsequently visualised dorsolateral to the trachea (B). Exploration of the local area revealed a large stick (C–F).

20.5 MARSUPIALISATION OF MUCOCELES

Marsupialisation is the process of incising a mucocele (ranula or pharyngeal mucocele) and suturing the edges to the adjacent mucosa. The interior of the mucocele suppurates and gradually closes by granulation[34]. Definitive treatment of a ranula or pharyngeal mucocele not only consists of marsupialisation of the lesion itself, but needs to be combined with resection of the mandibular and sublingual salivary gland complex responsible for the problem. The combined procedures lead to excellent results and low recurrence rates[35–41] (see Chapter 15 Diseases of the Pharynx, Section 15.3).

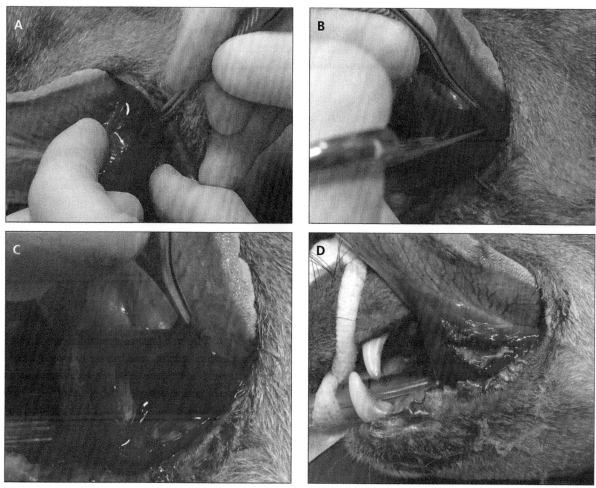

Figs 20.5A–D Marsupialisation of a ranula. The animal is placed in lateral recumbency, with the affected side up, so the tongue falls away medially from the ranula (A). With a stab incision (B) the mucocele is opened, after which the incision can be extended to further open the mucocele (C). The outward folded edge is then sutured to the surrounding mucosa in a simple interrupted pattern with 4-0 monofilament absorbable suture material (D).

Technique

For marsupialisation of pharyngeal mucoceles, the animal is placed in ventral recumbency and positioned as for staphlectomy (see Chapter 19 Surgery of the Nose, Section 19.3) with the mouth opened maximally. For marsupialisation of a ranula, the animal is placed in lateral recumbency, with the affected side up, so the tongue falls away medially from the ranula (Fig. 20.5A)[42]. Swabs are placed in the oropharynx to prevent the contents from entering the larynx.

The dorsolateral surface of the ranula or the craniolateral surface of a pharyngeal mucocele are grasped with Allis forceps or stay sutures are placed. With a stab incision (Fig. 20.5B) the mucocele is opened, after which the incision can be extended to further open the mucocele and subsequently resect a full-thickness portion of the wall to produce a 2–3 cm stoma (Fig. 20.5C)[42–44]. The edges of the remaining wall are then either folded inward upon themselves[42], inverting the oral or pharyngeal mucosal surface into the cystic cavity, or outward. The folded edge is then sutured to itself when folded inward, or to the surrounding mucosa when folded outward (Fig. 20.5D), with a simple continuous or interrupted pattern of 3-0 or 4-0 monofilament absorbable suture material.

20.6 ARYTENOID LATERALISATION FOR LARYNGEAL PARALYSIS

Idiopathic, acquired laryngeal paralysis in large breed dogs is the most common form of laryngeal paralysis encountered in veterinary practice (see Chapter 16 Diseases of the Larynx, Section 16.5)[2,45–47]. With acute dyspnoea, an emergency medical treatment should be started with sedation, corticosteroids, supplemental oxygen and cooling when necessary. If the obstruction and/or mucosal swelling is severe, a temporary tracheostomy should be performed (see Section 20.1). As a definitive treatment, surgery is required. Many surgical techniques have been described, ranging from muscle–nerve pedicle transposition for reinnervation of the larynx to castellated laryngofissures and partial laryngectomy[45,48–50].

- Arytenoid lateralisation procedures are recommended however, because of the consistently good results with minimal complications[2,45,51–53].
- Both unilateral cricoid–arytenoid and unilateral thyroid–arytenoid lateralisation procedures give reliable and comparable results and are equally effective in decreasing symptoms from a clinical point of view[45,51,54–59].

In this section the technique of thyroid–arytenoid lateralisation for laryngeal paralysis will be discussed.

Thyroid–arytenoid lateralisation for laryngeal paralysis

Corticosteroids can be given during the induction of anaesthesia to diminish postoperative mucosal swelling[2,45,60]. The entire neck should be aseptically prepared for surgery. A lateral approach or a ventral paramedian approach can be used, depending on the surgeon's preference. The ventral paramedian approach allows for concurrent tracheotomy, if indicated, without repositioning the patient and will be reviewed here.

The animal is placed in lateral recumbency with a towel under the neck to elevate the laryngeal area. A temporary tracheostomy can be performed as an alternative route for endotracheal intubation and to prevent postoperative dyspnoea, but this prolongs postoperative hospitalisation and should not be routinely performed.

A paramedian skin incision is made on the left side from 2 cm caudal to the larynx extending to the caudal angle of the mandible[2,61,62]. The subcutaneous tissues and platysma are then incised, and blunt dissection is performed lateral to the sternohyoid and sternocephalic muscles until the dorsal edge of the thyroid can be palpated and lifted (Figs 20.6A, B)[2,48]. An incision is made in the thyropharyngeal muscle along the dorsolateral edge of the thyroid (Fig. 20.6C). An atraumatic forceps can be placed on the thyroid cartilage lamina, or stay sutures can be placed, to retract and rotate the larynx laterally. The cricothyroid articulation is disarticulated with scissors after which the dorsal cricoarytenoid muscle and muscular process of the arytenoid can be identified (Fig. 20.6D)[2]. The cricoarytenoid articulation is now disarticulated (Fig. 20.6E), and the arytenoid is dissected free from its attachments without entering the laryngeal lumen, and the interarytenoid ligament is dissected last. Two synthetic nonabsorbable sutures are placed through the muscular process of the arytenoid to the most caudodorsal part of the thyroid to lateralise the arytenoid (Figs 20.6F, G)[2]. The abduction is now verified with the help of the anaesthetist by intraoral visualisation. The thyropharyngeal muscle can be apposed with a simple continuous pattern with absorbable material (Fig. 20.6H). The subcutaneous tissues and the skin are closed routinely. Postoperative care consists of close monitoring of the patient and providing broad-spectrum antibiotics and analgesics[2,45,54].

Complications of lateralisation procedures are haematoma formation, suture avulsion, discomfort during swallowing, temporary glottic dysfunction and coughing after eating or drinking, sometimes resulting in aspiration pneumonia[45,51,54–59a]. The prognosis is good however after lateralisation, with over 90% of the patients having less dyspnoea and increased exercise tolerance[45,51,54–59].

20.7 VENTRAL LARYNGOTOMY FOR REMOVAL OF LARYNGEAL MASSES

Indications for ventral laryngotomy are:
- Polyps and small neoplastic lesions of the vocal folds or arytenoids.
- For benign intraluminal laryngeal tumours that have a base on the mucosa of the cricoid or thyroid cartilage[2,45].

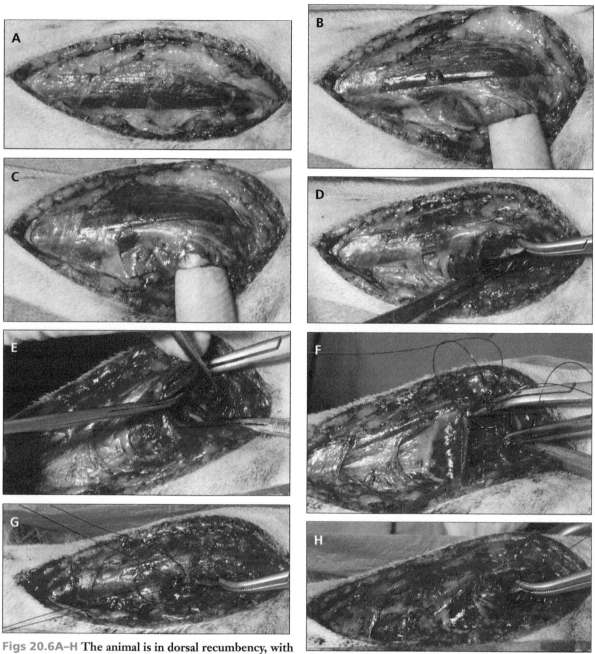

Figs 20.6A–H The animal is in dorsal recumbency, with the head towards the right side of the images. A paramedian skin incision is made on the left side with blunt dissection through subcutis lateral to the sternohyoid and sternocephalic muscles (A) until the dorsal edge of the thyroid can be palpated and lifted (B). An incision is made in the thyropharyngeal muscle along the dorsolateral edge of the thyroid (C). After opening the larynx, the dorsal cricoarytenoid muscle and muscular process of the arytenoid can be identified (D, Allis forceps on dorsal cricoarytenoid muscle). The cricoarytenoid articulation is now disarticulated (E) and the arytenoid is dissected free. Two synthetic nonabsorbable sutures are placed through the muscular process of the arytenoid to the most caudodorsal part of the thyroid to lateralise the arytenoid (F) and tied (G). After the tieback, the thyropharyngeal muscle can be apposed with a simple continuous pattern with absorbable material (H); closure of subcutaneous tissue and skin is routine.

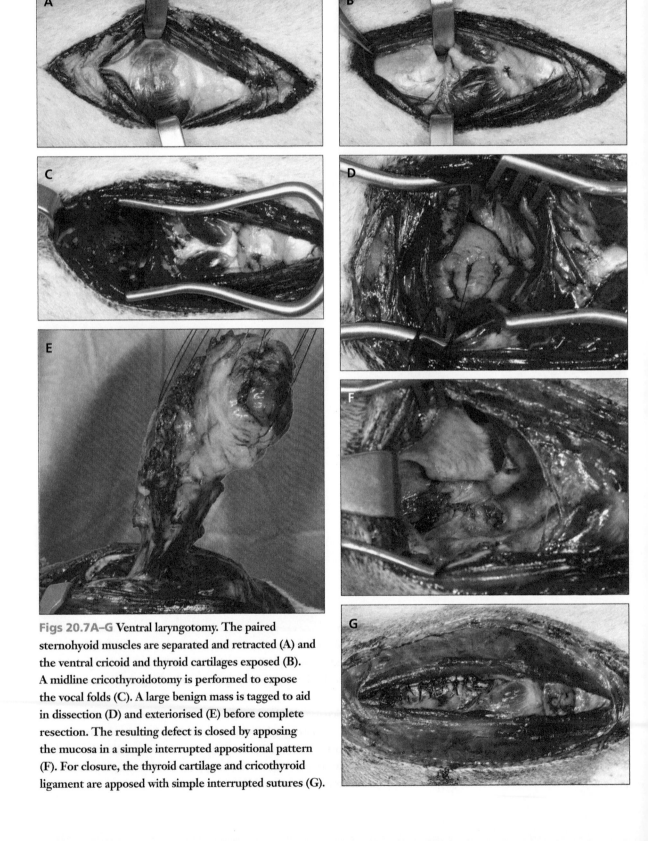

Figs 20.7A–G Ventral laryngotomy. The paired sternohyoid muscles are separated and retracted (A) and the ventral cricoid and thyroid cartilages exposed (B). A midline cricothyroidotomy is performed to expose the vocal folds (C). A large benign mass is tagged to aid in dissection (D) and exteriorised (E) before complete resection. The resulting defect is closed by apposing the mucosa in a simple interrupted appositional pattern (F). For closure, the thyroid cartilage and cricothyroid ligament are apposed with simple interrupted sutures (G).

The ventral approach allows better access for mucosal apposition and reduces the likelihood of scar tissue formation and webbing compared to the transoral approach where generally wounds are left open to heal by second intention[45,60]. Partial or segmental laryngectomy can also be performed using this approach, but will not be discussed here.

Technique

The patient is positioned in dorsal recumbency with the neck extended over a rolled towel or using a vacuum pillow. The larynx is exposed using a ventral midline cervical approach, starting rostral to the basihyoid bone and extending caudally to the proximal trachea[45,60,63]. The paired sternohyoid muscles are separated and retracted (Fig. 20.7A) and the ventral cricoid and thyroid cartilages exposed (Fig. 20.7B). The cricothyroid ligament is incised in the midline, and the incision is extended along the midline of the thyroid cartilage as needed to expose the vocal folds (Fig. 20.7C). In small dogs, the incisions may need to be extended into the cricoid cartilage to improve exposure[45]. Small self-retaining retractors or stay sutures in the thyroid cartilage can be used to improve visualisation. The affected vocal fold can now be excised from the arytenoid cartilage dorsally and the thyroid cartilage ventrally with small scissors or using a small scalpel blade[63]. Benign masses can be tagged and dissected from the surrounding tissues (Figs 20.7D, E). The resulting defect should be closed carefully to prevent webbing, the mucosa from the laryngeal saccules can be used and advanced into the defects created[45]. The mucosa is apposed in a simple continuous or interrupted appositional pattern using 4-0 or 5-0 mono- or multifilament absorbable material (Fig. 20.7F). Tumours extending beyond the vocal cords are resected with removal of the vocal cord and a full-thickness portion of thyroid cartilage[45].

For closure, the thyroid cartilage and cricothyroid ligament are apposed with simple interrupted sutures of a monofilament absorbable suture material of 2-0 to 3-0 and 3-0 to 4-0 respectively (Fig. 20.7G). Sternohyoid muscles are closed in a simple continuous pattern with 3-0 monofilament absorbable material and subcutaneous tissues and skin are closed routinely.

20.8 REFERENCES

1 Venker-van Haagen AJ. The pharynx. In: Ear, Nose, Throat, and Tracheobronchial Diseases in Dogs and Cats. Venker-van Haagen AJ (ed). Schlütersche Verlagsgesellschaft, Hannover 2005, pp. 83–120.

2 Venker-van Haagen AJ. The larynx. In: Ear, Nose, Throat, and Tracheobronchial Diseases in Dogs and Cats. Venker-van Haagen AJ (ed). Schlütersche Verlagsgesellschaft, Hannover 2005, pp.121–65.

3 Bishop MJ, Hibbard AJ, Fink BR, Vogel AM, Weymuller EA. Laryngeal injury in a dog model of prolonged endotracheal intubation. *Anesthesiology* 1985;**62**:770–3.

4 Whited RE. A study of endotracheal tube injury to the subglottis. *Laryngoscope* 1985;**95**:1216–19.

5 Brainard BM, Hofmeister EH. Anesthesia principles and monitoring. In: *Veterinary Surgery Small Animal*. Tobias KM, Johnston SA (eds). Elsevier Saunders, St. Louis 2012, pp. 248–91.

6 Venker-van Haagen AJ. Diseases of the larynx. *Veterinary Clinics of North America: Small Animal Practice* 1992;**22**:1155–72.

7 Venker-van Haagen AJ. Tracheostomy. In: *Atlas of Small Animal Surgery*. van Sluijs FJ (ED). Bunge, Utrecht 1992, pp. 16–17.

8 Harvey CE. Upper airway obstruction surgery. 7. Tracheotomy in the dog and cat: analysis of 89 episodes in 79 animals. *Journal of the American Animal Hospital Association* 1982;**18**:563–6.

9 Aron DN, Crowe DT. Upper airway obstruction. General principles and selected conditions in the dog and cat. *Veterinary Clinics of North America Small Animal Practice* 1985;**15**:891–917.

10 O'Brien JA, Harvey CE, Kelly AM, Tucker JA. Neurogenic atrophy of the laryngeal muscles of the dog. *Journal of Small Animal Practice* 1973;**14**:521–32.

11 Halum SL, Ting JY, Plowman EK, et al. A multi-institutional analysis of tracheotomy complications. *Laryngoscope* 2011;**122**:38–45.

12 Nicholson I, Baines S. Complications associated with temporary tracheostomy tubes in 42 dogs (1998–2007). *Journal of Small Animal Practice* 2012;**53**:108–14.

13 Trinterud T, Nelissen P, White RAS. Use of silicone tracheal stoma stents for temporary tracheostomy in dogs with upper airway obstruction. *Journal of Small Animal Practice* 2014;**55**:551–9.

14 Peeters ME, Ubbink GJ. Dysphagia-associated muscular dystrophy: a familial trait in the Bouvier des Flandres. *Veterinary Record* 1994;**134**:444–6.

15 Pfeifer RM. Cricopharyngeal achalasia in a dog. *Canadian Veterinary Journal* 2003;**44**:993–5.

16 Ladlow J, Hardie RJ. Cricopharyngeal achalasia in dogs. *Compendium on Continuing Education for the Practicing Veterinarian* 2000;**22**:750–5.

17 Goring RL, Kagan KG. Cricopharyngeal achalasia in the dog: radiographic evaluation and surgical management. *Compendium on Continuing Education for the Practicing Veterinarian* 1982;**4**:438–44.

18 Watrous BJ, Suter PF. Oropharyngeal dysphagias in the dog: a cinefluorographic analysis of experimentally induced and spontaneously occurring swallowing disorders. *Veterinary Radiology & Ultrasound* 1983;**24**:11–24.

19 Papazoglou LG, Mann FA, Warnock JJ, Song KJE. Cricopharyngeal dysphagia in dogs: the lateral approach for surgical management. *Compendium on Continuing Education for the Practicing Veterinarian* 2006;**28**:696–705.

20 Kyles AE. Esophagus. In: Veterinary Surgery Small Animal. Tobias KM, Johnston SA (eds). Elsevier Saunders, St. Louis 2012, pp. 1461–83.

21 Suter PF, Watrous BJ. Oropharyngeal dysphagias in the dog: a cinefluorographic analysis of experimentally induced and spontaneously occurring swallowing disorders. *Veterinary Radiology & Ultrasound* 1980;**21**:24–39.

22 Elliott RC. An anatomical and clinical review of cricopharyngeal achalasia in the dog. *Journal of the South African Veterinary Association* 2010;**81**:75–9.

23 Niles JD, Williams JM, Sullivan M, Crowsley FE. Resolution of dysphagia following cricopharyngeal myectomy in six young dogs. *Journal of Small Animal Practice* 2001;**42**:32–5.

24 Rosin E, Hanlon GF. Canine cricopharyngeal achalasia. *Journal of the American Veterinary Medical Association* 1972;**160**:1496–9.

25 Allen SW. Surgical management of pharyngeal disorders in the dog and cat. *Problems in Veterinary Medicine* 1991;**3**:290–7.

26 Jain V, Bhatnagar V. Cricopharyngeal myotomy for the treatment of cricopharyngeal achalasia. *Journal of Pediatric Surgery* 2009;**44**:1656–8.

27 Warnock JJ, Marks SL, Pollard R, Kyles AE, Davidson A. Surgical management of cricopharyngeal dysphagia in dogs: 14 cases (1989–2001). *Journal of the American Veterinary Medical Association* 2003;**223**:1462–8.

28 Dean PW. Surgery of the tonsils. *Problems in Veterinary Medicine* 1991;**3**:298–302.

29 Anderson GM. Soft Tissue of the Oral Cavity. In: Veterinary Surgery Small Animal. Tobias KM, Johnston SA (eds). Elsevier Saunders, St. Louis 2012, pp. 1425–38.

30 Doran IP, Wright CA, Moore AH. Acute oropharyngeal and esophageal stick injury in forty-one dogs. *Veterinary Surgery* 2008;**37**:781–5.

31 Griffiths LG, Tiruneh R, Sullivan M, Reid SW. Oropharyngeal penetrating injuries in 50 dogs: a retrospective study. *Veterinary Surgery* 2000;**29**:383–8.

32 White R, Lane JG. Pharyngeal stick penetration injuries in the dog. *Journal of Small Animal Practice* 1988;**29**:13–35.

33 Peeters ME. The management of pharyngeal stick penetrating injuries in dogs. *European Veterinary Conference: The Voorjaarsdagen* 2010, pp. 1–2.

34 Radlinsky MA. Surgery of the digestive system. In: Small Animal Surgery, 4th edn. Fossum TW, Dewey CW, Horn CV, *et al.* (eds). Elsevier Mosby, St. Louis 2013, pp. 386–460.

35 Brown NO. Salivary gland diseases. Diagnosis, treatment, and associated problems. *Problems in Veterinary Medicine* 1989;**1**:281–94.

36 Waldron DR, Smith MM. Salivary mucoceles. *Problems in Veterinary Medicine* 1991;**3**:270–6.

37 Ritter MJ, Stanley BJ. Salivary Glands. In: Veterinary Surgery Small Animal. Tobias KM, Johnston SA (eds). Elsevier Saunders, St. Louis 2012, pp. 1439–47.

38 Peeters ME. The treatment of salivary cysts in dogs and cats. *Tijdschrift voor Diergeneeskunde* 1991;**116**:169–72.

39 Harvey HJ. Pharyngeal mucoceles in dogs. *Journal of the American Veterinary Medical Association* 1981;**178**:1282–3.

40 Weber WJ, Hobson HP, Wilson SR. Pharyngeal mucoceles in dogs. *Veterinary Surgery* 1986;**15**:5–8.

41 Dietens A, Spanoghe I, Paepe D, *et al.* Faryngeale sialocele bij een hond. *Vlaams Diergeneeskundig Tijdschrift* 2011;**80**:233–9.

42 Tobias KM. Sialoceles. In: *Manual of Small Animal Soft Tissue Surgery*. Tobias KM (ed). Wiley-Blackwell, Ames 2010, pp. 393–400.

43 Ritter MJ, Pfeil von DJ, Stanley BJ, Hauptman JG, Walshaw R. Mandibular and sublingual sialocoeles in the dog: a retrospective evaluation of 41 cases, using the ventral approach for treatment. *New Zealand Veterinary Journal* 2006;**54**:333–7.

44 Spangler WL, Culbertson MR. Salivary gland disease in dogs and cats: 245 cases (1985–1988). *Journal of the American Veterinary Medical Association* 1991;**198**:465–9.

45 Monnet E, Tobias KM. Larynx. In: Veterinary Surgery Small Animal. Tobias KM, Johnston SA (eds). Elsevier Saunders, St. Louis 2012, pp. 1718–33.

46 Millard RP, Tobias KM. Laryngeal paralysis in dogs. *Compendium on Continuing Education for the Practicing Veterinarian* 2009;**31**:212–19.

47 Busch DS, Noxon JO, Miller LD. Laryngeal paralysis and peripheral vestibular disease in a cat. *Journal of the American Animal Hospital Association* 1992;**28**:82–6.

48 Nelson WA. Upper respiratory system. In: *Textbook of Small Animal Surgery*. Slatter D (ed). Saunders, Philadelphia 1993, pp. 733–76.

49 Hedlund CS. Surgery of the upper respiratory system. In: *Small Animal Surgery*, 3rd edn. Fossum TW, Hedlund CS, Hulse DA, *et al.* (eds). Mosby, St. Louis 2007, pp. 817–66.

50 Monnet E. Laryngeal paralysis and devocalization. In: *Textbook of Small Animal Surgery*. Slatter D (ed). Saunders, Philadelphia 2003, pp. 837–45.

51 Griffiths LG, Sullivan M, Reid SWJ. A comparison of the effects of unilateral thyroarytenoid lateralization versus cricoarytenoid laryngoplasty on the area of the rima glottidis and clinical outcome in dogs with laryngeal paralysis. *Veterinary Surgery* 2001;**30**:359–65.

52 Hardie RJ, Gunby J, Bjorling DE. Arytenoid lateralization for treatment of laryngeal paralysis in 10 cats. *Veterinary Surgery* 2009;**38**:445–51.

53 White RAS. Arytenoid lateralization: an assessment of technique, complications and long-term results in 62 dogs with laryngeal paralysis. *Veterinary Surgery* 1989;**18**:72.

54 MacPhail C. Laryngeal disease in dogs and cats. *Veterinary Clinics of North America Small Animal Practice* 2014;**44**:19–31.

55 MacPhail CM, Monnet E. Outcome of and postoperative complications in dogs undergoing surgical treatment of laryngeal paralysis: 140 cases (1985–1998). *Journal of the American Veterinary Medical Association* 2001;**218**:1949–56.

56 White R. Unilateral arytenoid lateralisation: an assessment of technique and long term results in 62 dogs with laryngeal paralysis. *Journal of Small Animal Practice* 1989;**30**:543–9.

57 Hammel SP, Hottinger HA, Novo RE. Postoperative results of unilateral arytenoid lateralization for treatment of idiopathic laryngeal paralysis in dogs: 39 cases (1996–2002). *Journal of the American Veterinary Medical Association* 2006;**228**:1215–20.

58 Bureau S, Monnet E. Effects of suture tension and surgical approach during unilateral arytenoid lateralization on the rima glottidis in the canine larynx. *Veterinary Surgery* 2002;**31**:589–95.

59 Demetriou JL, Kirby BM. The effect of two modifications of unilateral arytenoid lateralization on rima glottidis area in dogs. *Veterinary Surgery* 2002;**32**:62–8.

59a ter Haar G. Surgical management of laryngeal paralysis. In: *Complications in Small Animal Surgery*. Griffon D, Hamaide A (eds). Wiley Blackwell, Ames 2016, pp. 178–84.

60 White RAS, Monnet E. The larynx. In: *BSAVA Manual of Canine and Feline Head, Neck and Thoracic Surgery*. Brockman DJ, Holt DE (eds). BSAVA, Gloucester 2005, pp. 94–104.

61 Harvey CE, Venker-van Haagen AJ. Surgical management of pharyngeal and laryngeal airway obstruction in the dog. *Veterinary Clinics of North America Small Animal Practice* 1975;**5**:515–35.

62 Harvey CE. Upper airway obstruction surgery. 5. Treatment of laryngeal paralysis in dogs by partial laryngectomy. *Journal of the American Animal Hospital Association* 1982;**18**:551–6.

63 Zikes C, McCarthy T. Bilateral ventriculocordectomy via ventral laryngotomy for idiopathic laryngeal paralysis in 88 dogs. *Journal of the American Animal Hospital Association* 2012;**48**:234–44.